Mass units

pounds × 16	= ounces
ounces × 0.0625	= pounds
ounces × 28.35	= grams
grams × 0.0353	= ounces
pounds × 0.454	= kilograms
kilograms × 2.205	= pounds
short tons × 2000	= pounds
pounds × 0.0005	= short tons
long tons (metric tons) × 1000	= kilograms
kilograms × 0.001	= long tons (metric tons)
short tons × 0.907	= metric tons
metric tons × 1.10	= short tons
milligrams × 0.001	= grams
grams × 1000	= milligrams
micrograms × 0.000001	= grams
grams × 1,000,000	= micrograms
nannograms × 0.000000001	= grams
grams × 1,000,000,000	= nannograms

Energy units

horsepower × 745.7	= watts
Btu × 0.293	= watt-hours
kilowatt-hours × 3416	= Btu

One quad is one quadrillion Btu or 10^{15} Btu.
One pound of coal yields 13,000 Btu when burned.
One kilogram of coal yields 28,600 Btu when burned.
One metric ton of coal yields 28.6 million Btu when burned.
One pound of coal burned in a 40%-efficient electric power plant yields 1.52 kilowatt-hours of electrical energy.
One barrel of oil yields 5.5 million Btu when burned.
One billion barrels of oil yields 5.5 Quads when burned.
One ton of oil = approximately 300 gallons
                = approximately 7 barrels
One cubic foot of natural gas yields 1,032 Btu when burned.

Units used in water resources

million acre-feet × 1.23	= cubic kilometers
cubic kilometers × 0.813	= million acre-feet
million acre-feet per year × 0.893	= billion gallons per day
billion gallons per day × 1.12	= million acre-feet per year

Concentration units

milligrams/liter	= parts per million (ppm)
grams/liter × 1000	= parts per million (ppm)
ppm × 0.001	= grams/liter
micrograms/liter (μg/l)	= parts per billion (ppb)
grams/liter × 1,000,000	= parts per billion (ppb)
ppb × 0.000001	= grams/liter
nannograms/liter (ng/l)	= parts per trillion (ppt)
grams/liter × 1,000,000,000	= parts per trillion (ppt)
ppt × 0.000000001	= grams/liter

THE ENVIRONMENT

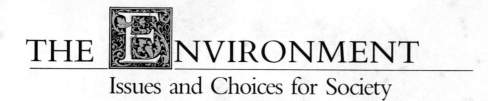

Issues and Choices for Society

SECOND EDITION

Penelope ReVelle
Essex Community College

Charles ReVelle
The Johns Hopkins University

WG WILLARD GRANT PRESS BOSTON

PWS PUBLISHERS

Prindle, Weber & Schmidt • ☘ • Willard Grant Press • **WG** • Duxbury Press • ♠
Statler Office Building • 20 Park Plaza • Boston, Massachusetts 02116

LIBRARY OF CONGRESS CATALOGING IN PUBLICATION DATA

ReVelle, Penelope.
    The environment.

    Bibliography: p.
    Includes index.
    1. Environmental protection.   2. Human ecology.
I. ReVelle, Charles.   II. Title.
TD174.R49   1984   363.7   83-16560

ISBN 0-87150-788-9

Printed in the United States of America.

84 85 86 87 88 — 10 9 8 7 6 5 4 3 2 1

Cover photograph by Richard H. Gross,
    Motlow State Community College

Production editor: David M. Chelton

Text designer: Megan Brook

Text compositor: Bi-Comp, Inc.

Artists: Wayne Clark, Cindie Clark-Huegel

Cover printer: Lehigh Press Lithographers

Text printer and binder: Rand McNally & Company

To Parents and Children,
the two ends of the living arrow of which we are a part

PREFACE

When we wrote the first edition of *The Environment: Issues and Choices for Society,* our goal was to produce a stimulating text for one-term introductory courses in Environmental Studies. The text was designed to accommodate students with various academic backgrounds. In addition, we wanted to write a book that would complement a variety of teaching approaches—for instance, lecture, discussion, or case study. The response from users of the first edition, professors and students alike, was gratifying. In this edition we have kept the major features that contributed to the success of the first edition and have made some changes we think will make the book even more useful.

We have organized most chapters so that there are three types of materials. First, in the basic text portion of the chapter, material is presented at a level suitable for beginning college students. Here students are given a solid foundation in the topic and a background that will enable them to participate in class discussions.

A second type of material consists of controversial issues set off from the general text. The controversies are examples of disagreements on environmental policy and contain capsule quotes from opposing parties. In these debates, sometimes surprising and other times predictable conflicts arise. The identities of the hero and villain are not at all clear; even experts may disagree on environmental matters. Differing viewpoints are argued, popular and unpopular causes are debated, and the difficulties inherent in making fair, effective laws are illustrated. We hope this exposure will encourage students to sift facts and come to their own conclusions rather than take the word of any one expert. In doing so, students will discover a surprising fact: many questions do not have answers. There are only social agreements, sometimes expressed in laws. The agreements represent society's consensus at the moment and are open to change.

The third section contains supplementary material, designed to enrich study of a specific topic. This section is noted by the symbol ⚘. Those desiring a more technical treatment, more of the basic scientific information on a particular topic, or further information on an issue, will find such material useful. Use of this supplementary material will vary, depending on the particular interests of the students and instructors. The optional material in one of the air pollution chapters, for example, allows biologists to read more about the effects of carbon monoxide on oxygen transport to the cells. In another air pollution chapter, chemistry majors can study the chemistry of photochemical air pollution. In the chapter on nuclear power, physics majors will find a description of the breeder reactor.

We have retained and expanded in this edition the separate chapters on ecology. Chapters 1, 2, 5, 8, and 19 introduce the ecological concepts necessary to appreciate the environmental topics that immediately follow them. For example, Chapter 8 on the properties of water and the communities found in an aquatic environment begins Part III on water resource problems. In similar fashion, a chapter on weather and climate precedes the chapter on air pollution. In this way we hope the reader will see how an understanding of environmental issues is enhanced by some knowledge of the ecological concepts involved.

Finally, at the end of each chapter are questions, references, and suggestions for further reading covering the subjects in the chapter. All readings are described briefly to help students choose the ones that will be most helpful on a particular topic.

We made the following changes, suggested by an extensive survey of users, as well as of non-users, of the first edition. All ecological principles are now covered earlier, but they are still introduced where they are immediately applied. The discussion of resources of land and food now follows right after the discussion of human population issues.

New controversy boxes have replaced some older ones, and all material has been updated as necessary. Through judicious tightening and editing we have tried to strike a balance between the need to add new material and the necessity to keep the text at a manageable length.

A major and more obvious change in this edition is the new graphic design. All drawings and graphs have been redone and a second color has been used to enhance their effectiveness. We have made a considerable effort to provide illustrations that enhance the text and illuminate complex relationships. The wide-open page design also ensures that all illustrations appear closer to the concepts they are supporting.

For the instructor, we have prepared an instructional package that contains a variety of materials intended to help organize and enrich environment courses. The Instructor's Manual, available separately from the publisher, contains additional references, more controversies, suggestions for class projects, and a set of questions with answers to test comprehension. The Manual also includes names and addresses of distributors of environmental films. Transparency masters are included at the back of the manual.

ACKNOWLEDGMENTS

Many people influenced this book and made it possible. There is, first of all, our good friend and agent, John Riina, who recognized in our ideas and prose new possibilities. John steered us with a gentle hand toward creation of a new and better shape for our proposed text and was always on hand for advice and counsel.

Julie Allen, originally secretary to "the work," continued as a valued research assistant in the preparation of this second edition. She typed the manuscripts, all of the drafts, proofread galleys, tracked down numerous articles and other references, and penetrated federal agencies in search of information and photographs. All of this was performed graciously and with good cheer. Katharine Lacher, our good neighbor, aided us in proofreading the endless stream of galleys.

Dr. Bruce MacBryde very graciously reviewed the material on endangered plants. Along with Dr. Faith Campbell, he answered a variety of questions and helped us find other wildlife experts.

We would like to acknowledge the following professors, who reviewed the manuscript: Richard Andren, Montgomery Community College; David Ashley, Missouri Western State College; Robert Bertin, Miami University; David Gates, University of Michigan—Ann Arbor; Tom Hellier, University of Texas—Arlington; Kipp Kruse, Eastern Illinois University; John Meyers, Middlesex Community College; John Peck, St. Cloud University; Lee Rockett, Bowling Green State University; Don Scoby, North Dakota State University. Their assistance was invaluable and contributed greatly to the quality of this revision; to them, our grateful appreciation. We also wish to thank, once again, the reviewers of the first edition: Ted L. Hanes, California State University at Fullerton; John Mikulski, Oakton Community College; Clyde W. Hibbs, Ball State University; Jerry Howell, Morehead State University; Arthur C. Borror, University of New Hampshire.

Jean-François Vilain, biology editor at Willard Grant Press, devoted extensive efforts to analysis of the book, seeking to preserve its strengths and extend its areas of excellence. David Chelton, production editor for the text, worked tirelessly with us to improve the work and to meet our deadline.

Our major sources of controversy in environmental issues were articles we found in scientific journals and in the popular press. In this area, two parties contributed particularly. Peter and Linda Rottmann called our attention to the fire at Baxter Park and the sharp debate in the *Bangor Daily News*. It was an archetype of a problem without an answer. And Donald W. Taube, University of California at Santa Barbara, mentioned his own interest in Palau, the Pacific paradise on the verge of becoming a super-tanker port. His bibliography and articles supplemented our own, enabling us to develop the Palau story more deeply.

Finally, Jeff Wright, an environmentally concerned colleague and fellow educator, helped us form the portion of the final chapter that deals with the opportunities we all have to contribute to the environmental movement. His ideas helped to prevent us from preaching, for which we are all grateful.

Penelope and Charles ReVelle

CONTENTS IN BRIEF

Introduction: On Solving Environmental
     Problems   1

PART ONE
**Humans and Other Nations that Inhabit the
Earth**   8
Ch. 1 ～ Lessons from Ecology: Structure in
     Ecosystems   11
Ch. 2 ～ Lessons from Ecology: Dynamics of
     Ecosystems   25
Ch. 3 ～ Human Population Problems   47
Ch. 4 ～ Wildlife Resources   72

PART TWO
Resources of Land and Food   94
Ch. 5 ～ Lessons from Ecology: Land Habitats and
     Communities   97
Ch. 6 ～ Soil Erosion and Pesticides Usage   109
Ch. 7 ～ Feeding the World's People   130

PART THREE
Water Resource Problems   154
Ch. 8 ～ Lessons from Ecology: Water Habitats and
     Communities   157
Ch. 9 ～ Water Resources   172
Ch. 10 ～ Purifying Water   193
Ch. 11 ～ Chemicals in Drinking Water   214
Ch. 12 ～ Organic Wastes and Dissolved
     Oxygen   238
Ch. 13 ～ Eutrophication   247
Ch. 14 ～ Water Pollution Control   257

PART FOUR
**Conventional Sources of Energy: Resources and
Issues**   276
Ch. 15 ～ How We Use Energy Resources   280
Ch. 16 ～ Coal: A Mixed Blessing   294
Ch. 17 ～ Oil and Natural Gas   319
Ch. 18 ～ Nuclear Power   340

PART FIVE
**Air Pollution and Energy-Related Water
Pollution**   366
Ch. 19 ～ Lessons from Ecology: Atmosphere and
     Climate on Earth   369
Ch. 20 ～ Air-Pollution Episodes, Carbon Dioxide,
     and Carbon Monoxide   380
Ch. 21 ～ Sulfur Oxides, Acid Rain, and Particulate
     Air Pollutants   394
Ch. 22 ～ Photochemical Air Pollution   416
Ch. 23 ～ Energy Use and Oil Pollution   431
Ch. 24 ～ Energy as a Pollutant: Thermal Pollution;
     Noise Pollution; High-Voltage Power Lines   446

PART SIX
**Natural Sources of Power and Energy
Conservation**   466
Ch. 25 ～ Power from Falling Water   470
Ch. 26 ～ Power from the Wind and Power from
     the Heat in the Earth   483
Ch. 27 ～ Solar Energy   502
Ch. 28 ～ Energy Conservation   527

PART SEVEN
Human Health and the Environment   546
Ch. 29 ～ Science and Public Protection   550
Ch. 30 ～ Environmental Carcinogens   563
Ch. 31 ～ The Voluntary Factors: Diet, Drugs, and
     Smoking   585
Ch. 32 ～ Toxic Substances Control and Hazardous
     Waste Disposal   614

PART EIGHT
Land Resource Issues   626
Ch. 33 ～ Private Land-Use Decisions   629
Ch. 34 ～ Preserving Public Natural Areas   644
Conclusion   676
Glossary   A1
Index   A14

CONTENTS

Introduction: On Solving Environmental Problems   1

Three Points of View   2
Environmental Decisions   3
How Decisions Are Based and Who Is Making Them   3
Making Sound Environmental Decisions   5

PART ONE

Humans and Other Nations that Inhabit the Earth   8

CHAPTER 1 ∿ Lessons from Ecology: Structure in Ecosystems   11

Peregrine Falcons and the Definition of a Species   12
How Organisms Live Together   15
Abiotic Factors in Ecosystems   17
Habitats and Niches   21
Diversity   21
Photosynthesis   23
Respiration   23

CHAPTER 2 ∿ Lessons from Ecology: Dynamics of Ecosystems   25

Trophic Structure   27
The Flow of Materials in Ecosystems   30
Energy Flow in Ecosystems   37
Population Dynamics   42
r- and K-Selected Populations   46

CHAPTER 3 ∿ Human Population Problems   47

Growing and Changing Populations   48
Effects of Population Growth   54
Limiting Population Growth   60
CONTROVERSY 3.1 *Family Planning in India—A Lesson on the Meaning of Freedom*   62

CONTROVERSY 3.2 *The Ethics of Limiting Family Size*   64
Outlook for Population Growth in the Future   65
CONTROVERSY 3.3 *Population Growth in Africa*   67
Demography—The Study of Populations   68

CHAPTER 4 ∿ Protecting Wildlife Resources   72

Is a Little Fish Worth More than a Big Dam?   73
The Value of Species   74
How Species Become Endangered   76
CONTROVERSY 4.1 *The Most Important Environmental Problem: Extinction of Species*   78
Protecting Wildlife Resources   80
CONTROVERSY 4.2 *Do We Need Flexibility in Laws Protecting Endangered Species?*   83
CONTROVERSY 4.3 *Endangered Species versus Human Health Benefits*   85
Endangered Ocean Mammals versus Endangered Native Cultures   87
CONTROVERSY 4.4 *The Ethics of Hunting Whales*   91

PART TWO

Resources of Land and Food   94

CHAPTER 5 ∿ Lessons from Ecology: Land Habitats and Communities   97

Succession and Climax   98
World Biomes   101
Soils and Ecosystems   105
Influences on Succession   107

CHAPTER 6 ∿ Soil Erosion and Pesticides Usage   109

Soil Erosion   110
Pesticides and Food Production   117

CONTROVERSY 6.1 *Should the U.S. Sell Dangerous Pesticides Overseas?* 120

CONTROVERSY 6.2 *Organic Gardening Economics* 123

🐚 Which Are the Safest Pesticides? 127

🐚 How Much Is a Part per Billion? 127

CHAPTER 7 ∿ Feeding the World's People 130

Is There Really a Food Crisis? 131

How Can We Grow More Food? 135

The Green Revolution 142

Social, Economic, and Political Aspects of Food Production 143

The Pessimist Looks at the Future 144

CONTROVERSY 7.1 *The Lifeboat Ethic* 146

The Optimist Looks at the Future 147

Are Grain Reserves a Solution? 149

CONTROVERSY 7.2 *Not Triage, But Investment in People* 150

PART THREE
Water Resource Problems 154

CHAPTER 8 ∿ Lessons from Ecology: Water Habitats and Communities 157

The Hudson River 158

Water—A Limiting Factor for Life 159

Water Habitats 159

🐚 How Lake and Reservoir Waters Can Become Oxygen-Poor 170

CHAPTER 9 ∿ Water Resources 172

How We Obtain Fresh Water 173

Surface Water and Reservoirs 175

Groundwater Resources 177

Extending Water Supplies without More Reservoirs 178

Water Resources Decisions 179

CONTROVERSY 9.1 *Should Water Be Transferred to the High Plains?* 184

CONTROVERSY 9.2 *Has the West Had Enough Water Projects? And Enough Growth?* 191

CHAPTER 10 ∿ Purifying Water 193

Learning from Past Mistakes 194

CONTROVERSY 10.1 *Should Water-Supply Reservoirs Be Used for Recreation?* 198

How Can We Tell if Water Is Unsafe to Drink? 202

Water Treatment 202

CONTROVERSY 10.2 *Is There Such a Thing as a Zero-Risk Environment?* 206

🐚 Testing for Indicator Organisms 209

🐚 A Simple Test for Free Chlorine 211

🐚 Ozonation to Purify Water 211

CHAPTER 11 ∿ Chemicals in Drinking Water 214

The Problem of Chemicals in Drinking Water 215

Inorganic Chemicals in Drinking Water 216

CONTROVERSY 11.1 *Minimata Revisited: Mercury* 218

Organic Chemicals in Drinking Water 222

CONTROVERSY 11.2 *The Economics of Environmental Improvements: PCBs* 228

How Safe Is U.S. Drinking Water? 230

CONTROVERSY 11.3 *Shaping Up Utility Companies: Safe Drinking-Water Laws* 231

🐚 National Interim Primary Drinking-Water Standards 234

🐚 Why Nitrites Are Poisonous 234

CHAPTER 12 ∿ Organic Wastes and Dissolved Oxygen 236

Why Organic Wastes Are Pollutants 239

Stream Health and Dissolved Oxygen 240

🐚 Organic Wastes: How to State Their Levels and How to Measure Them 243

CHAPTER 13 ∿ Eutrophication 247

Feeding Lakes 248

How Water Becomes Eutrophic 248

How Can Eutrophication Be Controlled? 250

CONTROVERSY 13.1 *The Scientist and Social Responsibility: Optical Brighteners* 252

Lake Erie: A Case History 254

🐚 Other Problems with Detergents: Foaming Waters and Biodegradability 255

CHAPTER 14 ∿ Water Pollution Control 257

The Difference Between Water Treatment and Water Pollution Control 258

Controlling Water Pollution 258

Water Pollution Control at Point Sources 259

CONTROVERSY 14.1 *Can You Save a River without Removing the Pollutants?* 263

Control of Water Pollution from Non-Point Sources 267

CONTROVERSY 14.2 *Economics and Environmental Regulation: Pretreatment Standards* 268

🌊 Other Ways of Treating Wastewater 270

🌊 Tertiary Treatment—How It Works 271

🌊 Problems from Chlorination of Sewage 273

🌊 Control of Storm Waters 273

PART FOUR

Conventional Sources of Energy: Resources and Issues 276

CHAPTER 15 ～ **How We Use Energy Resources** 280

Energy Conversions and the Second Law of Thermodynamics 281

Electric Power 283

CONTROVERSY 15.1 *Should Electric Companies Be Publicly or Privately Owned?* 288

The Uses of Energy and the Fuels that Supply It 289

🌊 New Developments in Conventional Electrical Generation 292

CHAPTER 16 ～ **Coal: A Mixed Blessing** 294

Introduction 295

Coal and Its Uses 295

Environmental and Social Impact of Coal 298

CONTROVERSY 16.1 *Strip Mining: What Are the Implications of Expansion on Jobs? on the Environment? on Human Health?* 309

Health and Safety of the Coal Miner 311

Impact of the Coal-Fired Electric Plant 313

CONTROVERSY 16.2 *What Should Be Done with the Coal in the West?* 314

🌊 Transportation of Coal 315

CHAPTER 17 ～ **Oil and Natural Gas** 319

A Brief History of Oil in the United States 320

Finding and Producing Oil and Natural Gas 320

What Are Oil and Gas? 321

Oil and Gas Resources 322

Synthetic Fuels 327

CONTROVERSY 17.1 *Social Effects of Natural Gas Pricing* 328

CONTROVERSY 17.2 *The Economics of Synthetic Fuels: Will Synthetic Fuel Provide Energy Security or Become a White-Elephant Investment?* 332

🌊 The Power of the Cartel and Its Impact on Oil Companies (Economic and Political Aspects of the Global Oil Situation) 334

🌊 Oil Conservation and the Control of Inflation—More than One Reason to Save 336

CHAPTER 18 ～ **Nuclear Power** 340

Introduction 341

Light-Water Nuclear Reactors 342

CONTROVERSY 18.1 *Is Nuclear Energy an Inevitable Result of Technical Progress?* 347

Problems in the Nuclear Fuel Cycle 348

CONTROVERSY 18.2 *Does Reprocessing Produce Weapons-Grade Plutonium?* 353

CONTROVERSY 18.3 *Does the Public Have the Right to Know (How Easy It Is to Steal Nuclear Material and Build an Atomic Weapon)?* 356

Economic and Social Aspects of Nuclear Power 360

🌊 Alternatives to Reprocessing 361

🌊 Liquid-Metal Fast Breeder Reactors 362

PART FIVE

Air Pollution and Energy-Related Water Pollution 366

CHAPTER 19 ～ **Lessons from Ecology: Atmosphere and Climate on Earth** 369

The Donora Story, October, 1948 370

The Earth's Atmosphere 371

Climate on Earth 372

Changing Climates 375

Inversions 378

CHAPTER 20 ～ **Air-Pollution Episodes, Carbon Dioxide, and Carbon Monoxide** 380

Air-Pollution Episodes: The Awakening 381

Carbon Dioxide 382

Carbon Monoxide 387

CONTROVERSY 20.1 *Should Society Protect the Unhealthy from Air Pollution?* 388

🌊 The Effects of Carbon Monoxide on the Body 391

CHAPTER 21 ∽ **Sulfur Oxides, Acid Rain, and Particulate Air Pollution** 394

Sulfur Oxides 395

CONTROVERSY 21.1 *Costs of Air Pollution Control* 398

Acid Rain 402

Particulate Matter 406

Lead Compounds in the Air 411

🔊 Trends in Air Quality—Challenge in Display 412

🔊 The Health Effects of Sulfur Oxides and Particles 414

🔊 Children and Lead Poisoning 415

CHAPTER 22 ∽ **Photochemical Air Pollution** 418

What Is Photochemical Pollution? 419

Control of Photochemical Air Pollution 421

Progress in Limiting Photochemical Air Pollution 422

CONTROVERSY 22.1 *Air Pollution Control: A "Policy beyond Capability"?* 424

Controlling Pollutants from Automobiles 425

Pollutant Standards Index and Summary of Standards 426

🔊 Effects of Nitrogen Dioxide on Our Health 428

🔊 Chemistry of Photochemical Air Pollution 431

CHAPTER 23 ∽ **Energy Use and Oil Pollution** 433

How Oil Pollution Starts 434

Biological Effects of Oil 436

Economic and Social Effects of Mixing Oil and Water 439

Lessening Oil Pollution and Its Effects 440

CONTROVERSY 23.1 *Save Palau: For What? For Whom? (Effects of Deep-Water Ports)* 441

CONTROVERSY 23.2 *Are National Interests More Important than Local Concerns?* 443

CHAPTER 24 ∽ **Energy as a Pollutant: Thermal Pollution; Noise Pollution; High-Voltage Power Lines** 446

Thermal Pollution 447

CONTROVERSY 24.1 *Economics and Thermal Pollution Control* 455

Noise Pollution 458

High-Voltage Power Lines 462

PART SIX

Natural Sources of Power and Energy Conservation 466

CHAPTER 25 ∽ **Power from Falling Water** 470

Conventional Hydroelectric Generation 471

Small-Scale and Low-Head Hydropower 473

The Redevelopment of Hydropower 474

Power from the Tides 475

🔊 Pumped Storage 479

🔊 Methods to Increase and Steady Tidal Energy 481

CHAPTER 26 ∽ **Power from the Wind and Power from the Heat in the Earth** 483

Power from the Wind 484

Geothermal Energy 492

CONTROVERSY 26.1 *Geothermal Energy Development—Should It Take Precedence over Risk to a National Park?* 498

CHAPTER 27 ∽ **Solar Energy** 502

Introduction and History 503

Applications of Solar Energy 505

CONTROVERSY 27.1 *Should Energy Sources Be Centralized or Decentralized?* 514

Prospects for Solar Energy 518

🔊 The Sun as a Source of Heat for Water and Buildings 521

CHAPTER 28 ∽ **Energy Conservation** 527

Looking at Two Possible Energy Futures 528

How We Heat Homes and Make Hot Water 532

The Manner of Transport of People and Goods 534

CONTROVERSY 28.1 *How Do We Get Fuel-Efficient Cars on the Road?* 538

🔊 Peak-Load Pricing—Decreasing the Need for New Power Plants 542

PART SEVEN

Human Health and the Environment 546

CHAPTER 29 ∽ **Science and Public Protection** 550

Risks and Benefits 551

The Epidemiologist as a Detective 552

Experiments to Determine Potency of Carcinogens 554

Risk–Benefit Analysis: Policy or Panacea? 556

CONTROVERSY 29.1 *The Murder of the Statistical Person: Risk–Benefit Analysis Applied to Carcinogens* 558

Carcinogenesis 559

CONTROVERSY 29.2 *Economics and Cancer Protection: Removing Chloroform from Drinking Water* 560

CHAPTER 30 ∼ **Environmental Carcinogens** 563

Asbestos 564

CONTROVERSY 30.1 *When Do We Have Enough Information to Act?* 566

Toxic Substances in the Workplace 567

CONTROVERSY 30.2 *The Cost of Cancer Control* 571

CONTROVERSY 30.3 *The Economics of Protecting Workers* 572

Radiation: Microwaves, Radiowaves 573

Radiation: x-Rays, Gamma Rays, and Particles 575

CONTROVERSY 30.4 *Women's Rights and the Workplace* 576

Ultraviolet Light and Chlorofluorocarbons 580

CONTROVERSY 30.5 *Is Unilateral Regulation Similar to Unilateral Disarmament?* 581

Skin Cancer 583

CHAPTER 31 ∼ **The Voluntary Factors: Diet, Drugs, and Smoking** 585

Smoking: A Personal Form of Air Pollution 586

Diet and Disease 590

CONTROVERSY 31.1 *Government Protection for an Illegal Activity? Marijuana Smoking* 591

CONTROVERSY 31.2 *What Is the Government's Role in Controlling Smoking?* 592

CONTROVERSY 31.3 *Does Everything Cause Cancer?* 597

CONTROVERSY 31.4 *Protecting People from Themselves: Saccharin* 598

CONTROVERSY 31.5 *Life-Style and Disease: Health Foods* 602

Toxic Substances in Drugs and Cosmetics 602

CONTROVERSY 31.6 *Pitfalls of Regulation: Tris, Pajamas, and Children's Safety* 610

CHAPTER 32 ∼ **Toxic Substances Control and Hazardous Waste Disposal** 614

Toxic Substances Control 615

Disposing of Hazardous Wastes 616

CONTROVERSY 32.1 *Toxic Substance Control in the Soviet Union* 620

PART EIGHT

Land Resource Issues 626

CHAPTER 33 ∼ **Private Land-Use Decisions** 629

A Changing Tradition 630

Enforcing Urban/Suburban Land-Use Plans 630

Preserving Rural Land 634

Other Methods to Secure the Preservation of Rural Land 642

CHAPTER 34 ∼ **Preserving Public Natural Areas** 644

The Public Preservation Movement 645

Federal Natural Areas 648

CONTROVERSY 34.1 *The Baxter Fire—Would You Let It Burn?* 656

CONTROVERSY 34.2 *Wilderness* 662

Problems of Accessibility 667

Reviving an Old Tradition of Land Preservation 668

CONTROVERSY 34.3 *Should Wilderness Be Accessible?* 669

CONTROVERSY 34.4 *The ORV Fight* 670

Expanding the System of Natural Areas 672

Conclusion 676

Two Voices 677

Restoring the Environment—Personal Choices 679

Glossary A1

Index A14

Introduction:
On Solving Environmental Problems

Three Points of View

Environmental Decisions

How Decisions Are Based and Who Is Making Them
Inputs to the Decision Process/Who Makes the Decisions?

Making Sound Environmental Decisions
Basic Considerations/Examining Specific Solutions

Three Points of View

People who are concerned about the environment today sometimes believe that we must break out of the traditional mold, abandon old methods of problem-solving, and not rely on technological solutions. We must invent a new life-style based on a new environmental ethic. Indeed, it is true that technological ways of solving problems may create new problems while solving the old ones. For instance, a farmer can increase production by using chemical pesticides, but this use may also contaminate foods and lead to the poisoning of harmless creatures.

Another feeling, held by parts of the general public, is that problems are best left to the experts, to the investigators who study environmental science. For example, suppose it were necessary to set limits for an acceptable level of a certain air pollutant. Some people think it best to leave this sort of job to a group of scientists. In this case, the group might include specialists in the chemistry of the atmosphere, specialists in how winds transport and mix pollutants, and still others who study the frequency of illnesses related to air pollution.

However, such experts cannot actually decide the appropriate standards for the air pollutant. They can set out the choices by predicting within fairly wide limits the results of various control actions. But only society as a whole can decide whether particular levels of disease are acceptably small and whether the cost of a certain control measure is a reasonable burden to taxpayers. Only society, and you as a member of society, can make decisions on what to achieve, what risks to bear, and what money to spend.

Here is the way one decision maker viewed the responsibility of his office:

> I am convinced that if a decision regarding the use of a particular chemical is to have credibility with the public, and with the media who may strongly influence that public judgment, then the decision must be made in the full glare of the public limelight. It no longer suffices for me to call a group of scientists to my office and, when we have finished, to announce that based on their advice I have arrived at a certain decision. Rather, it is necessary for me to lay my scientific evidence and advice on the table where it may be examined and, indeed, cross-examined by other scientists and the public alike before I make a final decision.[1]

Some people believe a new life-style is necessary; others say leave the decisions to the experts; still others have given up hope. This last group sees the problems as so overwhelming and the effort to change life-styles so enormous, that they see little future for the world. If we felt that way, we would not have written this book. We personally believe there is hope for improving the quality of the environment. Humans are a creative lot, as you will see in this book; we only need

---

1 William D. Ruckelshaus, then Administrator of the Environmental Protection Agency, in a speech to the American Chemical Society, 13 September 1971.

more people who care. And that number is increasing all the time.

Environmental Decisions

Whether or not you feel that changes in life-style are essential to cure and to prevent environmental ills, you are needed, your ideas and energy are important, in an on-going decision process. Your involvement is needed as soon as possible because actions to resolve environmental problems are being taken right now. In some cases, immediate action is demanded by problems so severe that no thinking person would expect to wait for a change in life-style or for an improved philosophical framework. As an example, when it was discovered that vinyl chloride was almost certainly responsible for cancer deaths among industrial workers, government officials moved swiftly to reduce the amount of this chemical in workplaces. They also prohibited its use in products sold to the general public. In the case of vinyl chloride, this included hair spray and food wraps.

In a similar case of a present danger, California health officials have tried to reduce the level of smog in Los Angeles by requiring special air-pollution-control devices on automobiles sold in California. The smog, which is caused mainly by automobile emissions, is the worst in the nation because of the low level of mixing of the air over Los Angeles and the unbelievable crush of traffic on the city's freeways. Almost anyone who has measured or experienced smog in Los Angeles feels the level is too high for good health. It is possible that if people could be convinced to walk more, use carpools, and use public transit, smog could be reduced. But no one wants to wait a long time before taking action to see if these life-style changes can be accomplished. Instead, control actions, mostly technical and offering quick results, are tried. Public opinion and public health demand this.

These actions, which are taken to deal with clear-cut environmental problems, must be sought and achieved within a political and bureaucratic framework that is unlikely to change in the near future. For a solution to work, it must be acceptable to, and seen as reasonable by, the people who make up this framework.

An example of a solution that has not worked, because it has not been politically acceptable, is one of the air-pollution-control policies of the Environmental Protection Agency (EPA). In the early 1970s, traffic control was seen as a bright hope for reducing air pollution in our major cities. By the end of the decade, however, there had been little if any progress in limiting traffic in those cities. *The New York Times,* in reviewing the national situation, headlined:

Cities' Pollution and Traffic Snarls Are Worse Despite Decade's Effort.

Efforts to Encourage More Use of Mass Transportation Founder on Resistance of Motorists and Officials[2]

One reason for this failure is that public transportation has been inadequate to serve the scattered pattern of housing that has developed since the end of World War II. Plans for limiting city traffic without adequate public transportation have been politically unacceptable.

The need for immediate responses to some problems and the limited range of politically acceptable actions does not mean that we should not search for new ways to solve environmental problems. Looking for ways to live so that the environmental ills that beset us do not occur in the first place is of utmost importance. Of course, we should continue to look to the long-term, but we must also look to the short-term. We ought to know how environmental decisions are made and who is making them. The important question is: How can we best influence and become part of this decision-making process *while* we look for the long-term solutions and for environmentally sound ways of living?

How Decisions Are Based and Who Is Making Them

Inputs to the Decision Process

Until they examine the situation closely, many people have the comfortable feeling that decisions being made on environmental matters by elected or appointed government officials are taken on the basis of definite information provided by scientists and other experts. Unfortunately, even a brief examination of some envi-

2 *The New York Times,* 21 May 1979, p. 1.

ronmental issues reveals that in many cases we have conflicting or inadequate scientific information. We are usually not able to predict with any precision the long-term effects (or sometimes even the short-term effects) on environment and health of many actions. In short, many decisions have to be made on less than complete evidence. The experts often disagree, even violently.

The scientific aspects of the problem are simply one component of decisions. The social, moral, and economic aspects also carry a great deal of weight. For instance, what if an otherwise reasonable energy conservation measure put a heavier burden on the poor than on the rich? One such measure is a large new federal tax on gasoline, over and above that being collected for the Highway Trust Fund. The poor who own cars will be hurt more than the middle-income groups by a gasoline tax since more of their money is tied up in purchasing the essentials of living. Is the economist or the scientist alone qualified to judge what amount of tax should be levied? Their concerns may be with tax revenues or the decrease in oil imports resulting from a lower demand. Yet someone must speak for the urban and rural poor.

Who Makes the Decisions?

Decisions on matters affecting the environment can be divided into two kinds: public decisions, such as those made by Congress or federal and state agencies, and private decisions, such as those made by business executives or by consumers. All of these decision makers are important. Public decisions are contained primarily in laws passed by elected officials in federal, state, or local governments. These laws are influenced by, among other things, public opinion and industry pressure groups. Once passed, the laws are then interpreted by appointed officials in the Environmental Protection Agency or in the state conservation departments and by elected officials such as state governors. As an example, Congress, by law, directed the Environmental Protection Agency to set air-quality standards. After extensive study and public hearings, EPA set these standards.

The courts also are involved in the process of environmental decision making. When two groups have different interpretations of the law, a suit forces the courts to resolve the situation. For instance, the Environmental Defense Fund, a citizens conservation group, sued EPA in court because they felt the agency was not acting as quickly as it should to investigate and ban dangerous chemicals, as it was directed to do by the Toxic Substances Control Act. With the help of the court, a compromise schedule of action was agreed on.

Private decisions are no less important than public ones, but here it is less easy to see the decision-making process. As an example, during the early 1950s, Hooker Chemical and Plastics Corporation used several pieces of land in the Niagara Falls area as dumping sites for waste chemicals. One area, the Love Canal site, was later sold to the local board of education, and a school was built there. Hooker was informed in 1958 when several children were burned by chemical wastes at the site. In addition, the company apparently knew that chemicals from this and the other dump sites might be draining into the surrounding areas where homes (and also the city's water supply) were located. Company officials notified the school board of the problem, but decided not to notify local residents. In 1979, the Love Canal area was declared a disaster area because chemicals were seeping into basements and yards. Hundreds of families were moved out of the area by the state. In other cases, of course, corporate officials have made environmentally sound decisions. Sometimes these are made on moral grounds, and sometimes they are made to avoid the sort of publicity that Hooker received after the Love Canal incident.

Consumers, too, make environmental decisions all the time. Should we buy an energy-efficient air conditioner, even if it costs more? Should we buy a small car that uses less gas, even though we like the comfort and handling of a larger one? Should we bother to read ingredient listings on labels in order to avoid synthetic chemicals in foods? Should we support and encourage those legislators working toward a "bottle bill" requiring deposits on soft drinks and beer?

As a citizen, you may be asked to take part in solid wastes recycling programs that require the separation of glass and metals in household garbage. You may wonder if you should go to the trouble of carrying the used oil to a collection center after you change the oil in your car, when it is simpler to let it run down the sewer. You might be considering a boycott on Japanese products to help save whales. We could go on with the list. At this point, however, you may see that as a voter and a consumer, the individual citizen is a vital link in environmental decision making.

Making Sound Environmental Decisions

Whether a citizen is being asked to vote on an environmentally related bond issue or to cast a more subtle but equally important vote by buying (or not buying) a particular product, the citizen–consumer needs some background to help make an educated decision. Whether you are working for an environmentally conscientious political candidate, a conservation group, or a responsible corporation, it helps to know what you are talking about.

In part, this text is meant to help you gain the necessary background for understanding environmental problems. But this is only a portion of what is needed. Just as important is the ability to evaluate proposed approaches to environmental problems. Does the approach, in fact, solve the problem? Can it cause new problems? Will some portions of society have to accept more burdens than others? Is there a better way to approach the whole issue?

Whatever the final answers to such questions, one fact is clear: No one has more of a right to exert influence on decisions in environmental matters than you do.

> Crucial value decisions have to be made, and they should not be made only by involved scientists closeted with financially interested industrialists and governmental authorities. They should be made by unbiased and informed members of the general public after hearing all sides of the questions, with balanced input from scientists, humanists, historians, philosophers, theologians and most of all, from ordinary citizens.[3]

Basic Considerations

Before describing how to go about evaluating solutions in a specific way, some basic philosophical issues ought to be considered. Your agreement or disagreement on these issues is very likely to influence how you feel environmental problems should be approached.

Trusting people. The first issue is one of trust— whether people, as individuals, can be trusted to act in an environmentally sound way without some economic push. There is some vogue, especially in the area of population control, for believing that people act only in their own short-term best interests. The thought is that plans involving long-term benefits for society as a whole, such as limiting population growth, must be accomplished by law.

> Conscience is self eliminating. . . . It is a mistake to think that we can control the breeding of mankind in the long run by an appeal to conscience. . . .
>
> People vary. Confronted with appeals to limit breeding, some people will undoubtedly respond to the plea more than others. Those who have more children will produce a larger fraction of the next generation than those with more susceptible consciences. The difference will be accentuated, generation by generation.[4]

The opposing opinion, stated perhaps most eloquently by Roger Revelle (no relation), is that an educated people can be trusted to see that their long-term best interests lie in smaller families and slower population growth for their country.

> Some scientists and publicists have seriously advocated a "lifeboat ethic," saying that nations which do not *compel* human fertility control (by what means is never stated) are endangering the survival of our species—hence they should be starved out of the human race by denying them food aid. This obscene doctrine assumes that men and women will not voluntarily limit their own fertility when they have good reasons and the knowledge and means to do so.[5]

Note that Roger Revelle is talking about people who are not only educated but who also can afford the environmentally sound choice.

The issue of trust comes up often, not only in the area of population control. One can even ask whether corporations, which are, after all, run by individuals, can be trusted to act in environmentally responsible ways. Or, since environmental protection generally costs money, will corporation executives always choose profits at the expense of the environment?

Trusting technology. The second issue also involves trust; in this case, trust in technology. We must

---

3 John C. Cobb, Letter to the Editor, *Science* **194** (12 November 1976), 674.

4 Garrett Hardin, *Science* **162** (13 December 1968), 1246.
5 Roger Revelle, *Science* **186** (15 November 1974), 589.

ask whether technology can be relied on to solve specific environmental problems and get us out of environmental jams. It is not difficult to find examples of cases in which solutions, especially technological ones, solve one set of problems, but in doing so, create new problems. This was stated more cynically by H. L. Mencken, who said "Every problem has a solution, simple, neat and wrong."

As an example of a technical solution that can cause problems, electric cars are thought by many to be the ideal way to get rid of air pollution in the city. However, cars consuming vast quantities of electricity may not be such a good idea. The number of new power plants needed would be enormous—doubling our current power-generating capacity might be necessary if a complete switch to electric autos were made. And the plants would be fired by either coal, a fuel with undesirable effects on air, water, and land, or by nuclear power, with all its accompanying problems. A better solution might be excellent public transportation for citizens in order to eliminate some of the need for heavy auto use. Such transport is far more energy-efficient than individual automobiles of any sort.

Weighing risks and benefits. Recognizing that actions involving the environment can cause harm as well as good brings us to the final issue: our willingness to weigh the benefits of environmental actions against the risks involved. For instance, we may be willing to risk some of the long-term effects of a pesticide if the pesticide is used to save lives threatened by insect-carried diseases such as malaria. We may also justify using pesticides to gain the benefits of increased food production. At the same time, we need to ask whether such uses may significantly increase our chances of dying from a particular form of cancer. We must also ask whether harmless species are threatened by our use of pesticides and whether the stability of the agricultural ecosystem may be upset. The ability to weigh the benefits of an action against the risks it brings is essential to achieving and preserving a satisfactory environment. Nuclear power can provide electricity to the nation and reduce oil imports, but are the benefits of nuclear power worth the risks, both known and unknown, that accompany it? (See Chapter 18).

This is not to say that such weighing of risks and benefits is easy, or free of conflict. Some individuals see modern society as too concerned about risk:

> No child, no idea is born into this world without the possibility of causing harm as it grows older. Shall we then abort all birth and all innovation for fear of possible environmental damage?[6]

Yet some environmental risks can be reduced and are well worth reducing:

> Such spectacular technical failures [as the wreck of the oil tanker *Torrey Canyon,* or the blowout of an oil well in the Santa Barbara Channel] also brought home to the general public something else: . . . that environmentalists— those impractical people with their feet in a swamp and their head in a cloud—are not necessarily wrong.[7]

These, then, are the three issues: whether you trust people; whether you trust technology; and your willingness to tolerate risk, known and unknown. These basic issues will arise again and again as you evaluate solutions to environmental problems, and you will find that your stance on these issues influences to a very large degree how you approach environmental problems.

In this book you will find environmental controversies set out for you to study. The controversies, however, are not simply environmental. You will be making political, economic, and social decisions as you come to grips with the issues in these controversies. You will be expressing trust or distrust in people, institutions, and technology, and you will be grappling with risks and benefits, asserting economic, social, and political opinions.

Examining Specific Solutions

The following questions are designed to help you examine specific proposals for solving environmental problems. They are intended to help you determine the worth of the solution itself. But they are also meant to help us all move toward the definition of new ways of living, ways that do not ignore natural laws and limits (see, for example, Chapter 2 on energy laws). These ways of living may make it possible for us to do

---

6 Cyrus Adler of the Electric Whale Company, Letter to the Editor, *Science* **178** (3 November 1972), 450.
7 John B. Oakes, Editor, *The New York Times* Editorial page. Speech given to National Audubon Society, November 1976.

more than react to a problem after it occurs. They may make it possible for us to prevent some environmental problems from occurring in the first place.

It is helpful first to classify a given solution as to type, then to go on to ask questions depending on the type of solution.

1. *Is it a "technical fix"?* Is this a technical solution *added on* to a current technology? (For instance, adding scrubbers to coal-fired power plants to remove sulfur oxides from stack gases.)
   Will it solve the problem with reasonable certainty?
   Will the solution cause new problems: technical, economic, social?
   Could we, instead, substitute another technology that does not cause this new problem?
   Could we do without the technology that causes the problem? Or could we do with less of it?

2. *Is the solution an alternate form of technology?* (For instance, substituting nuclear power plants for coal-fired power plants to eliminate sulfur oxide emissions.)
   Will it solve the problem with reasonable certainty?
   Will it cause new problems?
   Could we do without the technology that caused the problem in the first place? Or could we do with less?

3. *Is the solution a nontechnological one?* (For instance, a gasoline tax could be levied to increase the cost of gasoline so that people will use their automobiles less, thus saving fuel resources and decreasing air pollution.)

Does this represent a real solution? That is, will it work? Is it politically acceptable?
Is the idea economically justifiable? Is the cost reasonable compared to benefits gained?
Is the idea socially justifiable? Who gains and who loses? Who bears the hardship?

In some cases nontechnological solutions are the result of an attempt to go directly to the root of a problem. For instance, we could ask if, rather than attempting to decrease automobile use, there are ways to make the use of automobiles unnecessary. New towns are based partly on the philosophy that when shops, recreational areas, and workplaces are located within easy walking distance, people will automatically use their cars less often. For these sorts of ideas, we must ask, in addition to the questions above: Will people accept this new idea?

All of the solutions to environmental problems presented in this book can and should be examined in the light of these questions. You may decide that some are good solutions. Some will fail your examination. However, once you've examined a solution and come to a decision, what do you do then? How can you put what you've learned into practice? Some brief thoughts on this are found in the summary material at the end of the book.

We invite you now to read and study the material that follows. We hope it will help you become an environmentally informed citizen, ready to take a needed position on behalf of the environment, ready to enter the debate and influence decisions.

(Courtesy of F. C. Sunquist)

ART ONE

Humans and Other Nations that Inhabit the Earth

It is undeniable that there is too much of many things in the world today: too much mercury in the water, too much sulfur in the air, too many wrecked automobiles. But surely the most curious excess is people. How has it happened that human populations have grown so rapidly that food and job shortages have gripped some countries, that both wildlife and wilderness are threatened, and that cities have grown to unmanageable sizes?

There are many reasons. Throughout the history of humankind, children have been valued. They have provided hands to work on the farm, to help in the blacksmith and carpentry shops, and in the kitchen. Children represent security for parents in old age because they provide for parents too aged and infirm to work. And, for different reasons to different people, children are desirable in and of themselves. Children are a form of immortality, both of genes and of values. Through their children, a people's physical characteristics and also their beliefs can survive into the future. In traditional African religion, people are immortal as long as their descendents remember them. All societies, from the most primitive to the most cultured, consider the rearing and teaching of children one of their most important tasks.

Children are not only a precious resource; they have also been a most fragile one. In the United States and other countries, before the development of modern medicine and sanitation, five or six children, on the average, were born to each family. Of these, perhaps only two or three lived to adulthood. Because the survival of a child was so uncertain, people customarily had large families. Among other considerations, this helped ensure that at least two children would live to care for the parents when they grew old.

In many societies, especially where most of the population is engaged in farming, the tradition of large families has remained. At the same time, however, modern medicine has made child survival much more certain. Within the past 50 years, the **death rate** (or the number of people who die, each year, per 1000 people in the population) has been reduced in most societies by one-half or more. This reduction in the death rate is due only partly to the medical advances that help adults to live longer; it is due in greater part to the survival of infants and young children who now live to become adults. This means that, in many countries, the number of people who die each year is much smaller than the number of people who are born. And so the populations of these countries grow rapidly.

As the population of a country grows, many facets of life undergo stress. Food supplies must grow or shortages, malnutrition, and finally starvation will result. Job opportunities must increase or unemployment spreads. New housing must be constructed, new parks opened, more roads, hospitals, and schools built or unbearable crowding will occur. Even if construction needs are met, the stresses of city living are forced on many people.

How, and how well, a country meets the stresses of rapid population growth depends to a large extent on how technically advanced that country is. For instance, technically underdeveloped countries may find it difficult or impossible to feed, house, and educate a rapidly growing population. Technically developed countries, on the other hand, can often expand production of food, goods, and services, but cannot cope with increased pollution, vanishing wilderness, or the mushrooming of social problems in growing cities.

Population growth is thus an underlying cause of, or contributes to, many of the environmental problems we know by other names: food crisis; disappearance of wilderness; energy crisis; air and water pollution; urban sprawl.

We begin this part with two chapters that discuss the ecological terms and principles related to populations and their interactions in ecosystems. Next the problem of rapid growth in human populations is explored. How has the population problem come about? What are the effects of rapid population growth in developed and undeveloped nations? What means do we have of controlling population growth? And finally, what is the outlook for the future?

The last chapter in this part considers wildlife resource problems because the human population is, of course, not alone on this planet. Human populations cannot grow without inevitably affecting other populations, both plant and animal, living on the earth. For instance, the Indian tiger, once the most coveted prize of the trophy hunter, is now gravely endangered. Although hunters used to be the most serious danger the tiger faced, now India's rapidly growing population is taking over the little habitat left to the tiger. Before World War II, India literally teemed with great herds of wildlife. Although hunting by wealthy Indian royalty and foreigners made some impact on the numbers of game, it was nothing compared to what happened after independence in 1947. According to Hari Dang, an ecologist: "Postwar exploitation—open season on all resources! Shoot everything, burn what's left, destroy the rest. It was our disaster period. You had the same thing in the American West—in the 1880s everybody started shooting, and the great herds were destroyed." But even worse was to come.

As the human population grew, competition with wildlife increased for food, water, living space, and forest products. John Putman of *National Geographic* tells of a bird-watching walk he took in India's Borivli National Park with Dr. Salim Ali, in 1976: "There was the rustle of palms in a light breeze. We spotted palm swifts, then a rufous-backed shrike, black-headed orioles, racket-tailed drongos. Then out of the forest came a column of tribesmen, each carrying branches. They filed by, looking at us only briefly. 'What is this! What is happening! An army coming out of the forest!' Dr. Salim Ali cried. Part of Borivli's habitat was disappearing before our eyes. 'What can you do? What should you do?' Dr. Salim Ali added. 'They must eat, and need wood for their cooking fires.' He shook his head: 'Population—it is at the root of every problem.' " (Quotes from *National Geographic,* September 1976).

 HAPTER ONE

Lessons from Ecology: Structure in Ecosystems

Peregrine Falcons and the Definition of a Species
What Is a Species?/Populations and Subspecies

How Organisms Live Together
Communities/Ecosystems

Abiotic Factors in Ecosystems
Light/Moisture/Salinity/Temperature/Oxygen Supply/Fire/Soil Type/Abiotic Factors Working Together

Habitats and Niches

Diversity
Ecological Meaning/Evolution and Diversity/Peregrines and Diversity

 Photosynthesis

 Respiration

A student of the environment uses knowledge from many other fields. Biology, law, sociology, mathematics, anthropology, and physics are only some of the subjects useful in understanding environmental problems. The discipline most closely tied to environmental studies, however, is probably ecology.

The word **ecology** is made up of the Greek words "oikos," meaning house, and "logos," to study. Thus ecology means, roughly, the study of where organisms live, or their environment. Perhaps a better way to put it is that ecology is concerned with the relationship between organisms and their environment.

Many ecological terms are used in environmental studies. The purpose of the ecology chapters, which appear in several main sections of this book, is to explain ecological principles, which you will need to understand environmental problems.

In this first chapter, we examine the parts of natural ecosystems: both the living organisms and the nonliving environment that make up these systems.

Peregrine Falcons and the Definition of a Species

Birds belonging to the species called peregrine falcons are spectacular hunters (Figure 1.1). Sailing with wings outspread, peregrines search for the smaller birds that are their prey. When it spots something that looks as if it would make a meal, the falcon plunges downward at speeds over 200 mi/hr (320 km/hr) and snatches its unsuspecting dinner in mid-air.

Figure 1.1 A mature peregrine falcon grows to 15–20 in. (37–50 cm) in length and has a wingspread of up to 43 in. (108 cm). The birds are noted for their beautiful markings and for the spectacular dives they make in mid-air. (The Peregrine Fund)

Since the early 1960s there has been no natural population of peregrine falcons east of the Mississippi. The birds fell victim to poisons in their ecosystem, most likely DDT (see Figure 1.7). But DDT, which was first used in this country in 1945, is now banned and levels in the environment seem to be decreasing. For this reason, Dr. Tom Cade, a Cornell University ornithologist, thinks it is a good time to try to restore the Eastern peregrine falcon population. He breeds the birds in captivity and then trains them to live successfully in the wild areas in which they were once found (Figure 1.2).

During this project, Dr. Cade and the U.S. Fish and Wildlife Service came into conflict over the definition of a species. Dr. Cade breeds his falcons from European peregrine falcons as well as from native North American birds. Because the European birds live in fairly populous areas, Dr. Cade feels that European falcons have characteristics that would be valuable to falcons attempting to settle along the populous eastern coast of the United States. The U.S. Fish and Wildlife Service was uneasy about allowing the result of this cross-breeding, which might be called a foreign species, to be established in this country because an executive order prohibits the introduction of "exotic species" into the United States. Many scientists came to Dr. Cade's defense by pointing out that all of the different birds Dr. Cade used were simply subspecies of peregrine falcons. Who was right?

What Is a Species?

An animal **species** is most often defined in terms of reproduction. That is, a species consists of a group of individuals who can successfully breed with each other; who share ties of common parentage; and who therefore possess a common pool of genes, or hereditary material. In most cases it is possible to tell species apart on the basis of their different appearance or behavior or physiological makeup. However, these differences in themselves do not define species. If two apparently similar groups of organisms cannot breed successfully when given the opportunity, the two groups make up two separate species. Similarly, if two different-seeming groups of organisms are capable of interbreeding, there can be a flow of genes between them. They are thus members of the same species, no matter how different they are in appearance.

Figure 1.2 Young falcons compared to a more mature bird held by Tom Cade. Many peregrine falcons flew in North America until about the early 1960s, when they fell victim to pesticides such as DDT in the environment. Now that DDT is no longer used in the U.S., Tom Cade, a Cornell University ornithologist, is breeding the birds in captivity and then reintroducing them to the wild. Over 200 birds have been successfully released. This is equal to as many birds as the whole Eastern population of peregrines once raised in a year, before DDT came into use. (The Peregrine Fund)

Nevertheless, we should keep in mind that by defining species, scientists are attempting to describe part of the natural world in an orderly way. The natural world does not always fit neatly into such categories. Thus there are exceptions to these rules for defining species. Remmert (1980) points out that two species that have developed in different parts of the world but with similar requirements for food, climate, living space, etc. can sometimes breed if they eventually

Figure 1.3 The grizzly bear is a prized trophy for hunters, but few are left in the 48 states. Some 1000 are believed to live in small populations in wilderness areas of Idaho, Wyoming, and Montana. The Interior Department has listed grizzlies as "threatened," a category that indicates an animal needs less protection than an endangered animal. Larger populations still exist in Canada (11,000–18,000) and Alaska (8,000–10,000), where hunting is still allowed. Proposed Katmai National Park in Alaska would provide increased protection for the Alaskan population. These four are salmon fishing in the McNeil River just outside the proposed park. (National Park Service photograph by Keith Trexler)

meet in natural surroundings. This has been noted in several species of birds, fish, and insects.

In a similar way, plants cannot always be classified into species on the basis of whether or not they interbreed. Plants as a group have more diverse reproductive methods and genetic systems.

What, then, is a species? If we remember that the definition of a species is not based on hard and fast rules, we can say that as a general rule, a species is a group of organisms who can breed successfully, who share ties of parentage, and who therefore possess a common pool of hereditary material.

An important part of the species concept is that, since species are not able to interbreed, they will follow different evolutionary paths as they adapt to their environment. In this way the adaptive zone of a species, which is made up of the resources available to that species and the parasites and predators[1] it encounters, becomes different from that of any other species.

Populations and Subspecies

Populations, such as the 50 or so birds that make up the entire population of peregrine falcons west of the Rockies, are the members of a species living together in a particular locality. One or more populations of a particular kind of plant or animal make up a species (Figure 1.3).

---

1 Predators are generally larger than the organisms they eat, while parasites are smaller than the organisms they eat.

In general, even though they all belong to the same species, the members of a particular population resemble each other more closely than they resemble members of other populations. The reasons are that pairing is more likely to occur between individuals within a population than between those in different populations, and that members of a population are all subject to similar environmental influences upon their direction of evolution.

In some cases, geographic conditions in which various populations of a species live are very different, or the barriers to travel between the localities are great. For example, populations of the same species may live on different islands or in different rivers or on opposite sides of mountain ranges. The populations may then be found to have some very different genetic characteristics (although not enough differences to prevent successful interbreeding if they were given the opportunity). Populations of a species that are unlikely to breed because of geographic factors and that show genetic differences are sometimes defined as *subspecies* or *races*. Dr. Cade's peregrine falcons seem to breed successfully (although he does use some complicated artificial insemination techniques) and so it has been argued that they are all members of one species—or at worst, subspecies—and the executive order does not apply. The Fish and Wildlife Service appears to be unconvinced by this argument, but has agreed to allow an exception to the executive order so the peregrine falcons can still be released.

How Organisms Live Together

Communities

Within a given environment, the various species do not simply act in a random fashion, but are organized into a **community.** This community includes all the living organisms, both plant and animal (including microorganisms), interacting in that particular environment. The term "community" can be applied to a rather small group of organisms, such as those that make up a small pond community, or it can be used to describe a much larger group, such as the eastern deciduous forest community into which Cade hopes to reintroduce the peregrine falcon. This larger community includes populations of small bird species, which are the prey of the peregrine falcons, as well as populations of many other species of large and small animals, and populations of various species of trees and other plants.

Ecosystems

In ecological terms, community refers to a system of living organisms. This living system, plus its nonliving or **abiotic** components and the ecological processes that take place there, make up an **ecosystem.** The nonliving parts of an ecosystem include such things as soil type, amount of rainfall, and sunlight.

Because of the interactions among the organisms in an ecosystem, you might say that an ecosystem adds up to more than the sum of its parts. That is, the system itself has certain ecological properties in addition to the characteristics of the individuals or species making up the community. Examples include the flow of materials or energy through communities. These processes are the subject of the next chapter.

Some systems are not self-sufficient, but depend on neighboring communities for such inputs as organic food materials; an example might be a small stream ecosystem. Others are large and complete

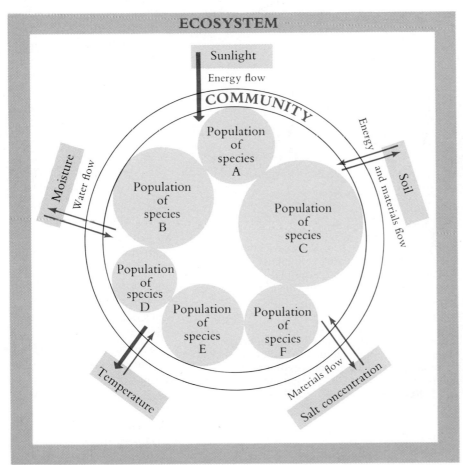

Figure 1.4 A diagrammatic representation of the terms population, community, and ecosystem. The populations (circles) of the various species living together in a specified location are a community. Together with the nonliving features of the area (rectangles) and the ecological processes (arrows) that take place there, they make up an ecosystem. Many ecological processes take place in an ecosystem, such as energy flow and the cycling of water. Only a few are shown in the diagram. (All these processes are discussed in Chapter 2.)

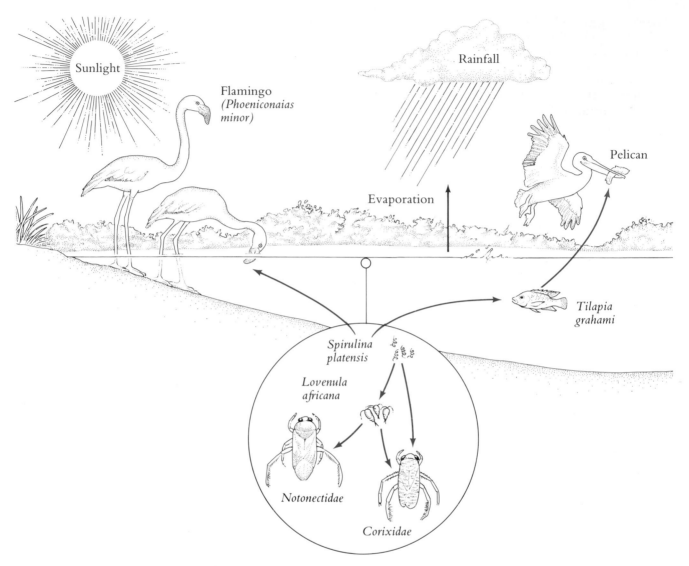

Figure 1.5 A Simple Ecosystem: Lake Nakuru, Kenya. Conditions are such that few species can live here. The living community in Lake Nakuru consists of populations of various species that can withstand the high concentration of sodium carbonate and the low concentration of hydrogen ions (pH 10.5). The main species found are: blue-green algae (*Spirulina platensis*); flamingos (*Phoeniconaias minor*) and fish (*Tilapia grahami*), both of which eat the algae; and fish-eating birds such as pelicans and cormorants. Some of the important nonliving components of this ecosystem, in addition to sodium carbonate, are sunlight and rainfall. In more hospitable environments, such as the eastern deciduous forest community, many more species are found. (Adapted from H. Remmert, *Ecology*. New York: Springer-Verlag, 1980.)

enough to require little more than energy from the sun in order to function as a unit. Examples of ecosystems include a meadow, a stream, a forest, a lake, or any other clearly defined part of the landscape. (See Figures 1.4 and 1.5.)

Abiotic Factors in Ecosystems

The abiotic factors in an environment influence the kind and numbers of organisms found there. Often when we speak of pollution we are concerned about substances or procedures that change these nonliving factors in an ecosystem, and in turn, affect the community of organisms living there. For instance, later we discuss pollutants that can change patterns of rainfall, increase or decrease environmental temperature, affect the oxygen or salt content of water, or change the concentration of inorganic plant nutrients in water. For this reason, we now examine more closely the important abiotic components of ecosystems and briefly consider how they affect the living community.

Light

Sunlight is one of the principal nonliving factors in an ecosystem because green plants use sunlight to produce organic material. This process is called **photosynthesis** and sunlight provides the energy. Almost all living creatures depend on this organic material for food, whether they eat the plants directly or eat animals that eat the plants. (See the supplemental section at the end of this chapter for more on photosynthesis.)

In most ecosystems, sunlight is present in sufficient amounts. Only in the depths of the ocean or deep in inland lakes or in caves does the lack of sunlight limit growth in an ecosystem. How the energy captured in photosynthesis is transferred within ecosystems is discussed in the next chapter.

Moisture

Water and life. The amount of moisture in different environments varies widely from desert areas to lakes and oceans. All forms of life on earth require water to live, and the abundance and quality of water are major factors in determining what kinds of communities will develop in a given environment. In land

environments the amount of moisture is a function of precipitation, humidity, and the evaporation rate. In water environments the types of communities also depend on the availability of water; however, in this case the availability of water means changes in water levels, e.g., changes with the tides. Availability of water may also refer to differences in its salt content, which affects the rate at which water enters or leaves organisms. In Chapter 8 we examine the different kinds of water environments and the specific communities that inhabit them. In this section we will look at the properties of water itself that have directly influenced the development of life as we know it.

Water and temperature. Water is unusual in that a relatively large amount of heat is needed to change its temperature, or to change solid water (ice) to a liquid or liquid water to a gas (water vapor). For these reasons, temperature changes in water tend to occur slowly and variations in temperature are less than in air. This is important for organisms living in water, since it gives them more time to adjust to temperature changes.

Water reaches its greatest density at 3.94 degrees Centigrade. That is, a given volume of water (for example, a 1-cm cube) weighs more at 3.94°C than at any other temperature. Its density decreases as the temperature decreases below this point. If you keep in mind that ice forms at 0°C, you can see that a given volume of ice (at 0°C) is lighter than the same volume of water at 3.94°C. This is why ice floats on cold water. This is a very important property because it prevents lakes from freezing solid. The ice layer floats on top of the lake and insulates the water beneath it. Many aquatic creatures can survive during winter in the water below the ice.

Warm water, being less dense than cold water, also floats on cold water. This is important in managing reservoirs (Figure 1.6) and also in determining the effects of pollutants on lakes, such as the phosphorus in detergents (Chapter 13).

Water as a solvent. Water is the most common solvent in nature. The amount and kinds of nutrients dissolved in water affect the growth of organisms. In a similar way, pollutants dissolved in water, even those that are only slightly soluble, affect organisms in land or water environments. For instance, acid rain is formed when sulfur oxides, produced by burning fos-

Figure 1.6 During the summer in reservoirs such as this, which is used to generate hydropower, the water forms layers. The warm, oxygen-rich water floats on top of cooler bottom water, which may be low in oxygen. If, as is common practice, the cooler bottom water is released through the dam during power generation, stream communities below the reservoir may suffer from lack of oxygen in the water. (U.S. Department of Energy)

sil fuels, dissolve in rain. This acid rain has reduced forest growth in Scandinavian countries and has caused entire populations of sport fish to disappear from some lakes in the Adirondack Mountains of New York State. (See Chapter 21 for more on acid rain.)

Salinity

Salt waters, such as the oceans, are 3.5% salt, or 35 parts of salt for every 1000 parts of water. (Terms such as parts per thousand, parts per million, and parts per billion are explained on page 127). In contrast, fresh waters average 0.05% salt, or 0.5 parts per thousand. Most of the salt in the oceans is sodium chloride, but many other salts are present.

The salt content of water is one of the major factors determining what organisms will be found there. Freshwater organisms, both plant and animal, have a salt concentration in their body fluids and inside their cells higher than that of the water in which they live. Because substances tend to move from areas of higher concentration to areas of lower concentration, water tends to enter and salts tend to leave these organisms. Freshwater organisms have developed mechanisms or structural parts to cope with this situation. In addition, freshwater organisms have evolved so that they contain lower salt concentrations in their bodies than organisms found in salt water.

Some salt-water organisms (for example, marine algae and many marine invertebrates) have a salt concentration in their bodies or cells almost identical to that of ocean water. However, many marine organisms have body fluids with a lower salt concentration than the water in which they live. For these organisms, water tends to leave their bodies or cells and salts tend to enter. Their regulatory mechanisms must solve a different problem from that of freshwater organisms. Bony fish, for instance, have developed ways of excreting salt and retaining water. The main point is that the two environments, salt water and fresh water, provide different conditions for organisms to adapt to, and thus are inhabited by different kinds of organisms.

In addition to salt and fresh waters, there are brackish waters, with intermediate salt concentrations. Such waters occur wherever salt and fresh waters meet—in estuaries, for instance, or where salt water intrudes on fresh groundwaters. Certain organisms are adapted, for all or part of their life cycles, to various intermediate salt concentrations.

Land-dwelling animals and plants tend to lose water to the atmosphere. In this respect they resemble many marine species because during their evolution they have also had to develop mechanisms to conserve water.

The kinds of water communities that develop in salt and fresh water are examined more closely in Chapter 8.

Temperature

Temperature has a profound effect on the growth and well-being of organisms. The biochemical reactions necessary for life are dependent on temperature. In general, chemical reactions speed up 2–4 times for a 10°C rise in temperature. Nonetheless, it is not possible to make sweeping generalizations about the effects of environmental temperature on the distribution of organisms because organisms have developed so many and varied mechanisms to deal with temperature changes.

Warm-blooded organisms, such as humans, are able to maintain a constant body temperature independent of the temperature of their environment. They are called endothermic or homeothermic organisms. Body warmth is a by-product of internal biochemical reactions that produce energy for the organism. In a cold environment warm-blooded animals can retain this body warmth by insulation (blubber, feathers, fur, clothing). In a warm environment the heat is lost by processes usually involving the evaporation of water (humans sweating from the skin or a dog panting and allowing its tongue to hang out). During very cold weather, when animals have difficulty finding enough food to burn to keep their body temperatures at a high level, some of them hibernate. During this time, their rate of energy use falls and so their body temperature falls as well.

The temperature of so-called cold-blooded (ectothermic) animals, as well as plants and microorganisms, varies with that of the environment. However, even in this group of organisms there is a variety of mechanisms to adjust body temperature. Most of these methods would be classified as behavioral methods of regulation. For example, bees can warm their hive by beating their wings. This is such an effective method that bees can live and reproduce in arctic regions. Many insects, snakes, and lizards warm themselves in the sun, taking up a position broadside to the sun's rays during the cool morning hours. Mosquito larvae develop quickly in the uppermost layers of ponds, where the sun's rays warm the water. When temperatures rise, many organisms take refuge in holes or burrows or under rocks. This helps them escape a lethal rise in body temperature or, in the case of desert organisms, prevents excessive use of precious water for cooling. During freezing weather, ectothermic animals and plants may produce antifreeze substances in their cells to prevent them from freezing. Many animals produce glycerol, while plants produce sugars such as hamamelose.

Photosynthesis does not depend on temperature as strongly as other reactions in organisms because it is not just a biochemical reaction, but also involves photochemical (light-driven) reactions. Thus photosynthesis is almost as effective in producing organic material in cold as in warm climates.

Most of the observations we can make about temperature and its effects on organisms are on a large scale. For instance, fewer types of organisms seem able to adapt to conditions in the arctic, where temperatures are far below the biological optimum. Even this seemingly obvious principle is complicated by the observation that not only is the severity of temperature important, but also the variability of conditions. That is, fewer types of organisms are found where temperatures vary widely from day to night or season to season than where temperatures are more constant. However, one thing we can say with confidence is that organisms, during the course of their evolution, have developed mechanisms to deal with temperature as it is found in their environment, whether that is warm or cold, constant or fluctuating. Human actions that change these temperatures can have devastating effects on ecosystems. For example, in Denmark the brown weevil (*Hylobius abieties*) normally takes three years to develop. When the forest is **clear cut** (all trees cut, regardless of size or species), the sun warms the ground more than before and the weevil matures in two years. For this reason, the weevil does much greater damage to forests where clear cutting is allowed. In Chapter 24 we discuss the effects of thermal pollution: the addition of excess heat to water environments.

Oxygen Supply

Both plants and animals use oxygen in the process of respiration, whereby they obtain energy for growth and metabolism. (See the supplemental material at the end of this chapter for more about respiration.) In land environments, oxygen is rarely in short supply. (Exceptions would be in some soils or on high mountain tops.) In water, on the other hand, the supply of oxygen may easily be a problem. The concentration of oxygen in water depends on the rate at which the gas diffuses into water, as well as the rates at which it is produced by the plants living there and used by the plants and animals in the water.

In some lakes the supply of plant nutrients allows the growth of masses of algae, which die, sink to the bottom, and are decomposed by bacteria. This last process can use up all the oxygen in the water. Other desirable organisms then cannot live in this oxygen-poor water. The addition of sewage to natural water environments results in the loss of much of the oxygen present. These effects are discussed more fully in Chapters 12 and 13 on organic wastes, dissolved oxygen, and eutrophication.

Fire

Fire can also be considered an abiotic factor that influences the types of communities in an ecosystem. Some environments are subject to regular natural cycles of fire. In southeastern pine forests, for instance, and also grassy savanna and steppe regions, periodic fires are a natural event.

Trees in forests where fires are regular may have thick bark that enables them to survive fires. The cones of some pines, such as Jack pine (*Pinus banksiana*), release seeds best when heated to a certain temperature. In this way, the seeds are sown at times when other plants that might compete for living space have been eliminated from the area. In fact, in the pine–spruce forests in northern Europe fire actually allows pines to grow. There are certain areas in these forests where the spruce have grown up in dense stands, crowding out the pines, which do not compete well for living space. Although the spruce are easily damaged by fire, once they form a dense stand it is difficult for fire to spread because the short spruce needles pack tightly on the ground and are resistant to fire. However, where pine and spruce are mixed, the forest floor accumulates loose piles of litter shed by the pines. Periodic fires in the mixed pine–spruce forest injure the spruce and allow pines to flourish.

In several cases, it has been shown that the vegetation growing up after a fire has more nutrients, such as phosphorus, potassium, calcium, and magnesium. Animals feeding on this vegetation may be better nourished. When humans prevent these natural fires they are really causing changes in ecosystems that have come to depend on fire for periodic renewal.

Fire has now become an accepted part of forest management (see more in Chapter 34), although the public has been slower to accept this idea (see Controversy 34.1).

Soil Type

The type of soil found in an area is very important to humans because different soils vary widely in their ability to support crops. The most useful soils from this point of view are the grassland and temperate forest soils. Other types, such as desert soil or the soil in tropical rain forests, are not as suited for raising crops. However, scientists looking at the effect of soil type on the kinds and distribution of organisms in an area have concluded that while soil type can have some effects on communities, the communities themselves have a profound effect on the type of soil in an area. To understand this it is necessary to realize that soil is not an entirely abiotic component of ecosystems, but is rather a mixture of living and nonliving materials.

The nonliving part of soil is the finely divided particles produced by the action of weathering on the parent material of the earth's surface. Combined with this is organic material: organisms and their products, which grow, die, and become mixed in. Many soil scientists believe that the kind of soil that develops in a given region is usually governed by two major factors: the climate of the region and the community that grows up there, rather than by the type of parent rock. (Climate as a determining factor in soil type is discussed in Chapter 19.)

As an example of the effect of community on soil type, grassland soil is characterized by its dark color, a result of having a great deal of finely divided organic material called humus. The entire grass plant, including roots, is short-lived. Each year, therefore, large amounts of organic material are added to the soil, where it is quickly decomposed to humus. This humus is slowly decomposed by soil microorganisms to inorganic materials (mineralization). In a forest, on the other hand, the leaf litter and roots are slow to decay to humus but humus is quickly mineralized. Little humus can accumulate in forest soil.

From this discussion you might get the impression that rather than soil affecting an ecosystem, the reverse is true: an ecosystem determines its soil type. To an extent, this is so, but the process takes a long time. In geologically young regions, soil type is more likely to be determined by parent material or some other condition, such as water level, or topographic features such as hills.

Exceptions also occur where climatic conditions are extreme, such as in the desert. Here, small differences in soil composition can make a large difference in the type of community that grows up. The maps of western hemisphere climates (Figure 19.3), soil types (Figure 19.4), and communities (Figure 19.5) are interesting to compare.

In Chapter 5, soil types and their suitability for crop production are discussed again when the different land environments are examined more closely.

Abiotic Factors Working Together

Undoubtedly, many other abiotic components of ecosystems could be mentioned. However, those factors already discussed are generally agreed to be the most important ones.

Although they are discussed separately, it is important to note that abiotic factors act together. Temperature, for instance, almost always acts together with moisture and wind. In order to predict how particular temperatures will affect the kinds of organisms found in a given environment, we need to know about these other factors as well. (Together, these factors describe the climate of an area. See Chapter 19 for a more complete discussion of climate.) Likewise, the development of soil type is an interaction of climate, the organisms that grow up in a given area, and the breakdown of parent rock. All of these factors influence each other.

Habitats and Niches

Before we leave the topic of the living and nonliving components of ecosystems, two useful ecological terms need to be defined. One is habitat; the other is niche.

The physical surroundings in which an organism lives are called its **habitat.** This is where you would go to find that particular organism. For example, the habitat of bloodworms, or chironimids, is the mud at the bottom of lakes, while the habitat of deer mice is temperate zone woodlots. These habitats consist, of course, of a variety of physical factors, such as temperature, soil type, and moisture, but they also contain many other animals as well as plants. Thus, an organism's habitat includes living as well as nonliving factors.

Peregrine falcons prefer to build their nest, or eyrie, on a narrow ledge or a steep cliff. This provides protection for the baby falcons from predators such as owls or raccoons. In at least one case, a skyscraper seems to fill a peregrine's habitat requirements. Scarlet, one of the peregrines released by Dr. Cade, has chosen to roost on a 35-story building in Baltimore, to the delight of its inhabitants.

The term **niche** is somewhat more difficult to define, in part because different people use it in different ways. It can refer to an organism's particular habitat. It can also mean the organism's ecological function, that is, what an organism does in its ecosystem—for instance, what time of day or night it is active, how much sunlight it requires, whether it is a **predator** (eats other organisms) or **prey** (is eaten by other organisms). The niche thus corresponds to an organism's way of life.

It is a general rule that two species cannot occupy the same niche forever. Competition between the species would cause a change in some aspect of the lifestyle of one of them, or force one to move somewhere else, or eliminate one of them. Sometimes two species are found who do live in the same habitat and eat the same food. However, closer examination usually proves that their niche is not the same because some other factor, such as predators or parasites, prevents either of the species alone from using all of the available habitat or eating all that type of food. There is thus room for two separate niches for the two species. A major problem that can arise when a new species is introduced into an area is that it may occupy the same niche as a native species. Competition between the new and native species may eliminate the native species.

Diversity

Ecological Meaning

When ecologists speak of **diversity,** they mean the number of species in a given area compared to the number of individuals. In other words, it is a way of measuring variety in an ecosystem. One of the reasons this is an important measurement is that human interference with a natural ecosystem often results in reduced diversity. Fewer kinds of species are found in a polluted ecosystem than in a similar ecosystem that is not polluted. Measurements of diversity in a given area over time can be an indication of the effects of pollution.

Evolution and Diversity

To understand how human interference can cause reduced diversity, it is necessary to consider, briefly, why there are a variety of species in the first place. According to theories of evolution, organisms of a particular species that are best able to survive and grow in a particular habitat will do so, while others, less able to compete, are eliminated. Organisms can also survive this competition, however, by moving into other habitats or developing a different set of ecological requirements in the same habitat (a new niche). Thus organisms become specialized to the conditions found in a particular subpart of their environment. Eventually, loss of the ability to breed with the original competitors gives rise to new species.

The more diverse the environmental conditions, the more different niches are possible and the more species are found. However, many ecologists believe that in order for this wide diversity to come into being, a long time period of relatively stable conditions is necessary.

It should be emphasized, though, that there are basic biological processes common to all organisms. These processes work best at certain optimum temperatures, light intensities, and so forth. The further conditions are from these optima, the less easily organisms can adapt to the conditions. For this reason, the further conditions are from biological optima, the fewer species are found. To put it another way, fewer species are able to use the available niches in harsh climates. Thus, fewer species are found in arctic regions, on icy mountain tops, and in deserts than in more moderate environments such as temperate or tropical forests.

These three factors—diversity of environmental conditions, harshness of environmental conditions, and the length of time a system has had to evolve—are not the only factors affecting diversity, but they appear to be major controlling influences. Other factors will be mentioned in the next chapter.

Human interference in natural ecosystems often takes the form of reducing the diversity of natural conditions. Thus, clearing a mixed forest to plant a single variety of pines destined for a paper mill reduces the number of niches. The result is a lower diversity of animal and plant species in the new pine forest than in the old mixed forest.

Human interference may also move conditions further from biological optima. For instance, the addition of sewage to a stream reduces the amount of oxygen available in the water. Only a few organisms are adapted to low oxygen concentrations. As a result, the diversity in the polluted stream is much lower than in a normal, well-oxygenated stream. It should be emphasized that the actual number of organisms in the polluted and nonpolluted habitats may be similar, but the organisms in the polluted habitat will mostly be members of one or a few species. It is the number of different species that is reduced.

Peregrines and Diversity

Dr. Cade's program to restore some natural diversity to our ecosystems by reestablishing the peregrine falcon population in the United States is an example of wildlife management techniques at their best and most imaginative. Not only has he developed ways of encouraging the birds to breed in captivity (Figure 1.7), but he and his staff also teach the newly fledged falcons to hunt for themselves in the wild. At about four weeks, when the birds can tear and eat their food, they are taken from the laboratory to old falcon nesting

Figure 1.7 Baby peregrines bred in captivity. In areas where DDT is found in the environment, peregrine falcons lay eggs with thin shells. These eggs are more likely to break before hatching than eggs with normal shells. This is believed to be the reason why there have been no natural populations of peregrine falcons in the Eastern U.S. since the early 1960s. (The Peregrine Fund)

sites or to specially made towers. There they are protected 24 hours a day by attendants who keep away owls and raccoons that would eat the young birds. The peregrines return to the nest site to eat until they are skilled in catching their own meals.

Still, we do not have anywhere near enough researchers to undertake a similar program for other endangered creatures. There are also many reasons why it would not work in other cases. Clearly we must try to prevent species from becoming extinct rather than try to repair the damage once it is done.

In the following chapters, we explore first the dynamics of ecosystems and then the problems faced by the human species as well as those faced by wildlife populations. For, as the fate of the peregrine falcon was changed by human acts, first for ill and then for good, so are human fates and fortunes tied up with those of the other species on this planet.

Photosynthesis

Green plants use the energy of sunlight to manufacture organic materials out of the gas carbon dioxide and water. The organic materials formed include a variety of sugars. **Carbon dioxide** is a chemical compound that can be described as "low energy," while sugars have a great deal of energy stored in their chemical bonds. Thus in photosynthesis the energy from sunlight is turned into, or stored as, chemical energy.

In addition to organic materials, oxygen is produced during photosynthesis. Plants themselves use some of this oxygen, but they generally produce more than they need. All animal life, including human, depends on this excess oxygen for respiration. We also depend for food on the sugars produced during photosynthesis, whether we eat the plants directly or whether we eat animals that eat plants.

In summary, the process of photosynthesis can be written as the following equation:

$$\text{sunlight} + \frac{\text{carbon}}{\text{dioxide}} + \text{water} \xrightarrow{\text{green plants}} \frac{\text{organic}}{\text{material}} + \text{oxygen}$$
$$(\text{sugars})$$

Or, using chemical formulas, photosynthesis can be summarized in this way:

$$6\,CO_2 + 6\,H_2O \xrightarrow[\substack{\text{Green plants} \\ (\text{chlorophyll, enzymes})}]{\substack{\text{Energy from} \\ \text{sunlight}}} C_6H_{12}O_6 + 6\,O_2$$

However, this equation is only a summary. Photosynthesis actually consists of many separate reactions catalyzed by a number of different enzymes (biological catalysts). A variety of different organic materials are formed, here represented by the formula $C_6H_{12}O_6$.

Respiration

Both animals and green plants need energy for growth and other metabolic processes. This energy is obtained from a series of biochemical reactions called respiration. In this series of reactions, an energy-containing organic material such as the sugar glucose is broken down by biological catalysts called enzymes. The energy released is used to drive other reactions in the cell. If oxygen is available, the material is broken down all the way to carbon dioxide, water, and energy. The summary of these reactions resembles photosynthesis in reverse.

$$\text{oxygen} + \frac{\text{organic}}{\text{material}} \xrightarrow{\text{respiration}} \frac{\text{carbon}}{\text{dioxide}} + \text{water} + \text{energy}$$
$$(\text{e.g., glucose})$$

Using chemical formulas, respiration can be summarized as:

$$6\,O_2 + C_6H_{12}O_6 \xrightarrow{\text{enzymes}} 6\,CO_2 + 6\,H_2O + \text{energy}$$

Questions

1. How would you define a species? a population? a subspecies?
2. What could be the purpose of an executive order prohibiting the introduction of exotic species into the country?
3. What is the difference between a community and an ecosystem?
4. Why does a community have more characteristics than just those of the species from which the community is made up?
5. Why is sunlight an important abiotic factor in ecosystems?
6. What special properties does water have in terms of its function in ecosystems?
7. How does salinity affect the distribution of organisms?
8. Briefly note how organisms deal with temperatures that are too high or too low for normal biological functioning.
9. Why do we class fire as an important abiotic factor in some ecosystems?
10. Of what importance is soil type to humans? to ecosystems?
11. Differentiate between an organism's habitat and its niche.
12. What does an ecologist mean by diversity? Of what importance are measurements of diversity?

Further Reading

If you are interested in ecology, you may want to read some of the following materials.

Cade, T. J., *The Falcons of the World*. Ithaca, N.Y.: Cornell University Press, 1982.

Cargo, D. N., and B. H. Mallory, *Man and His Geologic Environment*. Reading, Mass.: Addison-Wesley, 1977.

Gates, D. M., *Biophysical Ecology*. New York: Springer-Verlag, 1980.

Levin, D. A., "The Nature of Plant Species," *Science,* **204** (27 April 1979), 381.

> In this article, Levin states:
>
> "For humans the environment has meaning only when its components can be interrelated in a predictive structure. We try to make sense out of nonsense and put the world into some perspective which has order and harmony."

> The species concept has caused a great deal of argument and confusion. This article is a thoughtful consideration of the problem, as well as a thought-provoking discussion of the human need to impose order where it is not always clear that order exists.

Levin, H. L., *The Earth Through Time*. Philadelphia: Saunders, 1978.

Odum, E. P., *Fundamentals of Ecology*. Philadelphia: Saunders, 1971.

Remmert, H., *Ecology*. New York: Springer-Verlag, 1980.

Watt, K., *Principles of Environmental Science*. New York: McGraw-Hill, 1973.

CHAPTER TWO

Lessons from Ecology: Dynamics of Ecosystems

Trophic Structure
Food Chains and Trophic Levels/Food Webs/Diversity and Stability

The Flow of Materials in Ecosystems
Hydrologic Cycle/Carbon Cycle/Sulfur, Nitrogen, and Phosphorus Cycles/Biological Magnification

Energy Flow in Ecosystems
Productivity/Thermodynamic Laws and Ecosystems

Population Dynamics
Age Distribution Graphs/Population Growth Curves/r and K Selection/Factors Related to Population Density

 r- and K-Selected Populations

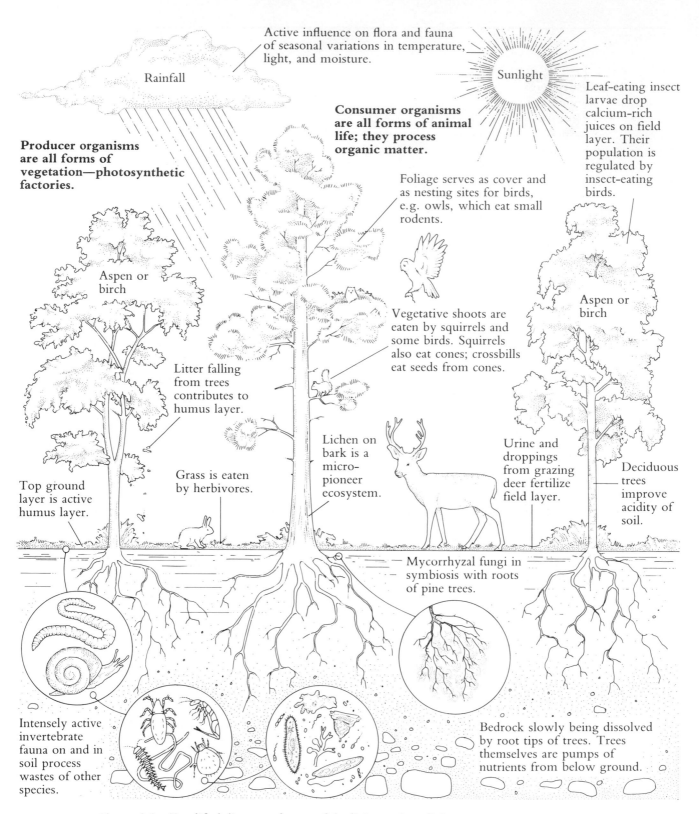

Figure 2.1 Simplified diagram of some of the living and nonliving components of a pine forest ecosystem. (From Darling and Dasmann, in: *Impact of Science on Society*, Vol. XIX, No. 2, 1969. Courtesy of UNESCO.)

In the first chapter we discussed the parts of ecosystems; that is, the living organisms as well as important nonliving or abiotic factors. Figure 2.1 is a simplified picture of a pine forest ecosystem, while Figure 2.2 is a photograph of a pine forest. However, one central point needs to be emphasized: ecosystems are not like photographs, static in time. Instead, growth and change are characteristics of ecosystems. The living and nonliving components of ecosystems act on each other and cause changes. Further, there is a flow of materials and energy through ecosystems. These ecosystem dynamics are the subject of this chapter.

Trophic Structure

Attempts are being made to reestablish peregrine falcons not only in the East, but also in parts of the West from which they have disappeared. For instance, the State of Montana is attempting to reestablish populations of peregrine falcons in abandoned eyries along the face of the Montana Rocky Mountains. This area contains a large enough prey base of smaller birds, such as cowbirds, meadowlarks, and killdeer, for the peregrines to feed on. However, while they are growing and learning to fend for themselves, the peregrines must be protected from predators such as golden eagles, great horned owls, coyotes, and raccoons (Figure 2.3).

Figure 2.2 Pine Forest Ecosystem. Beside the pines, many other organisms are found in pine forest ecosystems. Examples are otters, beavers, turtles, tree frogs, ospreys, and even bald eagles. Mosquitoes, deerflies, and poison ivy are also often found. (Larry E. Morse)

Figure 2.3 In Montana's Rocky Mountains a five-year program has begun to reintroduce peregrine falcons. Here a researcher watches the development of young falcons placed on a cliff-top eyrie and guards against golden eagles, which prey on the young birds. (The Peregrine Fund, William R. Heinrich)

In any particular community, the organisms are linked together in a pattern of preying and being preyed upon—of eating and being eaten. This pattern is the **trophic structure** of the community: the way in which the various organisms in the community obtain their nourishment. Such patterns are often described as **food chains** or **food webs.**

Food chains and trophic levels. Food chains can be diagrammed in a relatively simple, straight-line fashion. Creatures in the chain generally feed on only one or a few species and are preyed on by one or a few other species. Food chains are found where conditions are so harsh that few animals or plant species are able to survive—for instance, on the arctic tundra (Figure 2.4) or in the high carbonate concentrations in Lake Nakuru (Figure 1.5).

The food chain described in Figure 2.4 is a grazing food chain. There are several distinct trophic levels in the chain. Organisms at the bottom of grazing food chains are **producers,** plant species that convert sunlight into food energy by photosynthesis (the grasses, sedges, and lichen in Figure 2.4). Creatures that eat producers are **herbivores** or green-plant eaters (the caribou in Figure 2.4) and they are called *primary consumers.* Next in line are the meat eaters, or **carnivores,** which eat the primary consumers. These are the *secondary consumers* (men, wolves in Figure 2.4). Longer food chains include *tertiary consumers* that eat the secondary consumers, and so on.

At another trophic level are the **decomposers,** or detritus feeders. This group of organisms is extremely important because it feeds on dead organic matter, breaking it down into inorganic and organic materials.

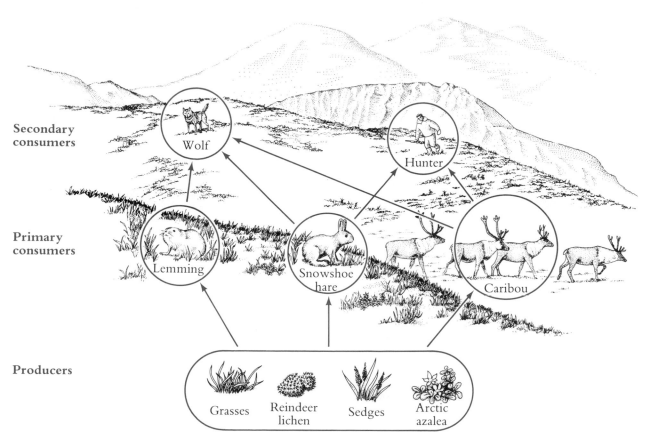

Figure 2.4 Arctic food chains are simple and short because few species are able to live in the harsh and unpredictable arctic climate. One arctic food chain is made up of a few arctic tundra plants; the caribou, lemmings, and snowshoe hares that eat the plants; and wolves and humans.

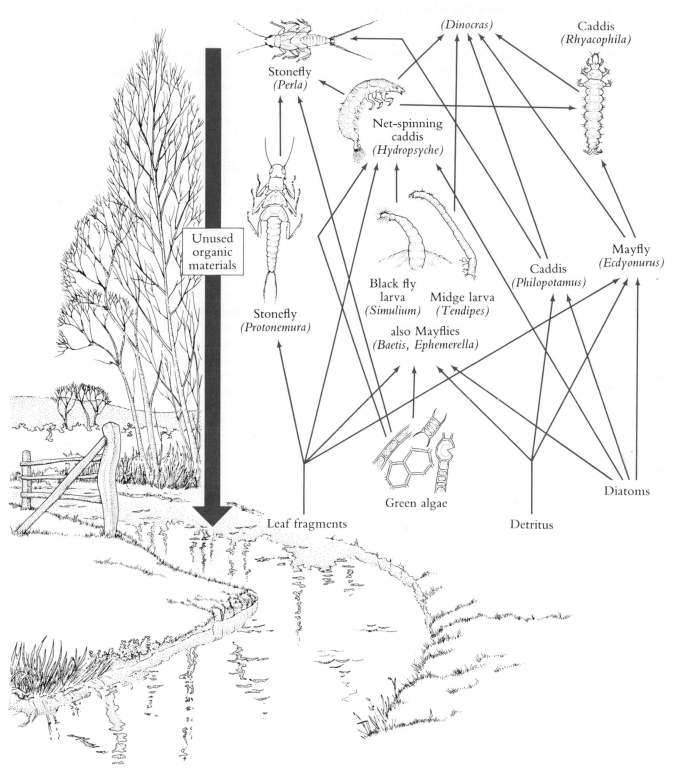

Figure 2.5 Part of the food web in a small stream community in South Wales. Note that the net-spinning caddis is both a primary consumer (eats green plants) and a secondary consumer (eats primary consumers). The complexity of most food webs is emphasized by the fact that this illustration shows only some of the insects involved. Many other creatures—for example, small fish—are a part of this web. [Adapted from J. R. Jones, *Journal of Animal Ecology*, **18** (1949) 142.]

These materials may be used by plants and animals or may inhibit or stimulate other organisms in the ecosystem. Fungi and bacteria are common decomposers, but animal species may also be important.

Food webs. In temperate and tropical climates, where many species interact with each other, trophic structure is best described as a **food web.** Food webs have more organisms and interconnections than do food chains. Figure 2.5 shows an example of a food web from a small stream in South Wales. Food webs are more complicated than food chains and are more difficult to sort into the trophic levels of producer, consumer, etc. One reason is that food webs involve so many species that we cannot always be sure we know all of the creatures in a particular food web. Also, some species occupy more than one position in a web. That is, an organism may eat both green plants and such primary consumers as insects, making it both a primary consumer and a secondary consumer. The net-spinning caddis in Figure 2.5 illustrates just such a complication.

In the small stream food web, the decomposers feed upon materials carried by water from upstream, such as leaves, feces, dead organisms, etc. Many food webs, especially in water, are based on detritus. The large arrow on the side of Figure 2.5 indicates that all of the unused organic material produced at various levels of the food chain (feces, secretions, dead plants or animal bodies, etc.) is either broken down by the decomposers to basic plant nutrients or cycled back up the food chains as decomposers themselves are eaten.

Diversity and Stability

Because of their complicated interrelationships, food webs are often described as more stable than food chains. That is, food webs should remain in balance despite the loss of a species. According to this theory, in a temperate-region field community, for example, the loss of an animal such as a fox, which preys on another species such as rabbits, should not cause a rapid increase in the rabbit population because there are other predators who also eat rabbits. In a food chain, the loss of a predator might lead to an explosive increase in its prey population. However, this point

has not been proven. It should be remembered that because ecosystem webs are so complicated, no ecosystem has yet been completely examined, neither in terms of all the creatures that may be a part of the system, nor in terms of how important the various pathways are. Possibly, most of the food energy in a web flows along only a few pathways, resembling food chains (Figure 2.6). In addition, although on the surface, stability is observed in communities that have many species, it could be due to other factors. For instance, mixed fields of crops do not suffer the massive outbreaks of insect pests often seen in fields with all one sort of crop. In some cases, this happens because insect pests locate the plants on which they feed by means of chemicals given off by the plants, and the insects have trouble locating these plants in a mixed field. The stability of the mixed field ecosystem thus results from a sensory phenomenon rather than from additional predator–prey relationships. As another example, tropical rain forests are among the most diverse ecosystems known. Yet investigators believe them to be one of the systems most easily disrupted by human actions.

We might say that very diverse ecosystems appear to have evolved under fairly constant conditions. Their higher diversity affords many checks and balances that act to keep the system constant as long as there is little outside interference. Changed environmental conditions and new species can have drastic effects on these delicately balanced systems. On the other hand, elastic systems—that is, those that can spring back after an outside influence—may well have evolved under more variable environmental conditions. For instance, the reed communities along lake shores, which are naturally subject to variable conditions such as water-level fluctuations, mowing, and ice movements, appear able to recover relatively quickly from pollution incidents.

The Flow of Materials in Ecosystems

A unifying feature of the flow of materials in ecosystems is that the flow is cyclic. That is, materials flow through systems in a largely circular manner: into the living organisms, back to the abiotic environment, and then into the living organisms again.

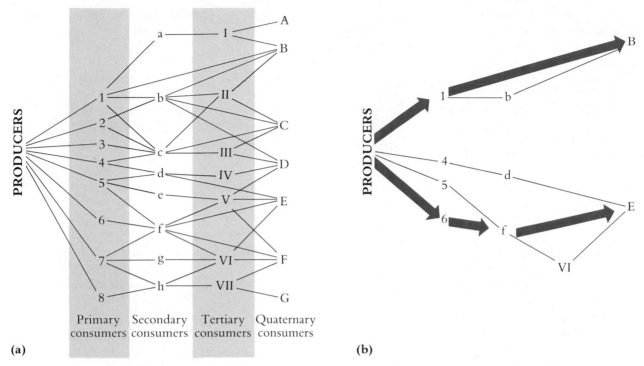

(a)

(b)

Figure 2.6 (a) A hypothetical food web, showing the flow of food or energy through an ecosystem from producers to quaternary consumers. (b) Another view of the same system, in which all pathways accounting for less than 0.001 of the total are left out. Now the significant structure more resembles a set of connecting food chains. (Adapted from Remmert, H., *Ecology*. New York: Springer-Verlag, 1980.)

Hydrologic Cycle

The cycle with which we are most familiar is the **hydrologic cycle.** It consists of three distinct and continuing events: the evaporation of water; condensation and rainfall; and runoff. Water evaporates from the surfaces of lakes, ponds, streams, wetlands, rivers, and oceans. It also evaporates from soil and vegetation and is transpired by plants.[1] This water returns to the earth as rainfall. More water evaporates from the surface of the oceans than returns to the oceans as rain, however. On land, the opposite is true—less water evaporates from soils, vegetation, and surface waters than returns by rainfall. (Except for deserts, where evaporation is greater than rainfall.) The balance, of course, is maintained by runoff, or water flowing from streams, lakes, rivers, and groundwaters to the oceans (Figure 2.7).

The hydrologic cycle can be compared to the process of distillation, in which water is vaporized by heating it in a flask. The water vapor leaves behind in the flask dissolved materials such as salt. This does not mean that rain is completely free of contaminants—rain and snow may become contaminated with gases and particles in the air. (For example, acid rain may form; see Chapter 21.) However, rain is relatively pure. In some areas of the world, with few rivers and little water in the ground for people to utilize, rain furnishes drinking water. In such places, rainfall is collected in cisterns and stored for later use.

The processes in the hydrologic cycle are being modified by human activities. Some of these changes

---

1 Transpiration refers to the release of water vapor from the above-ground parts of plants. Evaporation and transpiration from plants together are sometimes referred to as "evapotranspiration."

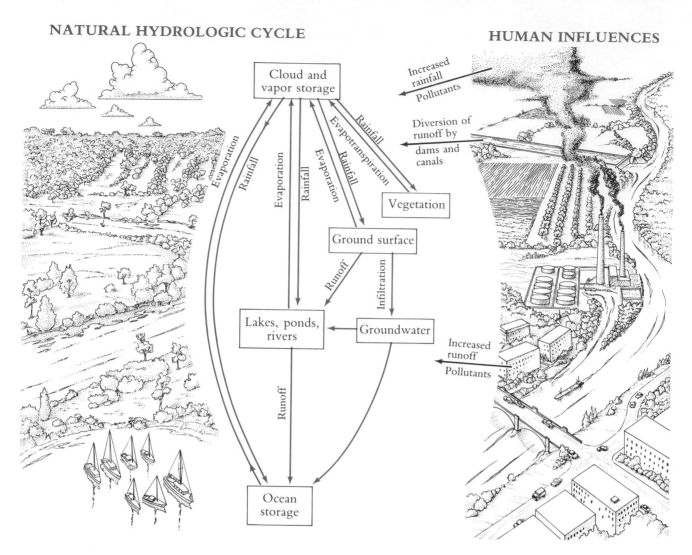

Figure 2.7 The Hydrologic Cycle. Water evaporates from soil and surface waters and is transpired by vegetation. This water returns to the earth as rainfall. In addition, water moves from land areas to the ocean as runoff in streams, rivers, and groundwater flows. Humans influence the hydrologic cycle in many ways. For instance, humans divert runoff, decrease the amount of water held in soil by destroying natural vegetation, increase rainfall via air pollution, and pollute water.

are intentional; others are accidental. For instance, rainfall has increased in industrial areas because water droplets condense more quickly around minute mineral particles in the air. As another example, runoff increases when vegetation is destroyed. Trees, grasses, and other plant covers capture and hold rainfall, allowing it to percolate down through layers of decaying organic materials and rocks in the soil. Too rapid runoff may lead to flooding. Also, where there is less water slowly percolating through the deep soil layers,

the amount of trace minerals dissolved from buried rocks decreases. These minerals are necessary for plant growth, and may have to be added to fertilizer in some areas. On the other hand, the runoff that carries water via streams, rivers, and lakes to the oceans is interrupted by humans at many points and for many purposes. Not only is runoff interrupted, but it is also contaminated with chemical and biological wastes.

These problems are described in greater detail in Part III, Water Resource Problems.

Carbon Cycle

All living creatures on earth consist mainly of water. About 80–90% of living material is water. However, the most important structural material for life forms is carbon. This carbon, which forms the backbone for all organic compounds in living creatures, comes from carbon dioxide in the atmosphere.

As explained in Chapter 1, plants use carbon dioxide during photosynthesis to manufacture some of the organic compounds necessary to life. Carbon from the atmosphere thus enters living systems through photosynthesis at the level of the ecosystem producers. The carbon moves through the ecosystem from one trophic level to another until it re-enters the atmosphere during respiration[2] or until the organisms in which it is contained die. The process of decomposition forms carbon dioxide again as dead material decays (Figure 2.8). This is only a part of the **carbon cycle,** however.

---

2 Respiration is explained in the supplemental section at the end of Chapter 1.

Most carbon involved in the cycle is in the oceans. This carbon (in the form of carbonates) controls, to a large extent, the amount of carbon dioxide in the air. Excess carbon dioxide in the atmosphere can dissolve in the oceans, forming carbonate and bicarbonate ions. In the opposite direction, the oceans can release carbon dioxide to the atmosphere. The oceans thus act to buffer, or keep constant, the carbon dioxide concentration of the atmosphere. Scientists believe this mechanism kept the amount of carbon dioxide in the atmosphere relatively constant until industrialization began. Human activities since then appear to be unbalancing the carbon cycle in several ways. An extra 5.8 billion tons of carbon dioxide are added to the atmosphere each year when fuels are burned at their present level of use. Some scientists believe that the rapid cutting of forests to satisfy human needs for farmland and wood products over the past few years may be decreasing the amount of carbon dioxide used by plants.

Although some of the excess carbon dioxide from burning of fuels dissolves in the ocean, half or more remains in the atmosphere. Measurements show that

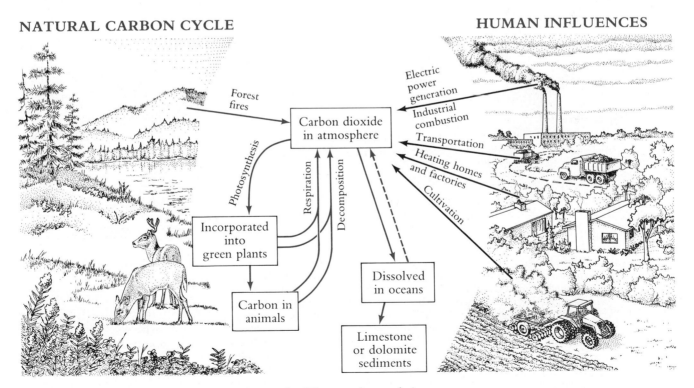

NATURAL CARBON CYCLE　　　　　　　　　　　　**HUMAN INFLUENCES**

Forest fires

Electric power generation
Industrial combustion
Transportation
Heating homes and factories
Cultivation

Carbon dioxide in atmosphere

Photosynthesis

Respiration

Decomposition

Incorporated into green plants

Carbon in animals

Dissolved in oceans

Limestone or dolomite sediments

Figure 2.8　Part of the carbon cycle. (The complete cycle is discussed in Chapter 20 and Figure 20.2.)

the level of atmospheric carbon dioxide has increased steadily since the 1950s. The complete carbon cycle, as well as the possible effects of an increase in atmospheric carbon dioxide, are explored in Chapter 20.

Sulfur, Nitrogen, and Phosphorus Cycles

Three other important materials cycle through ecosystems along with water and carbon: these are nitrogen, phosphorus, and sulfur. All three of these chemical elements are essential components of living matter.

NATURAL NITROGEN CYCLE

HUMAN INFLUENCES

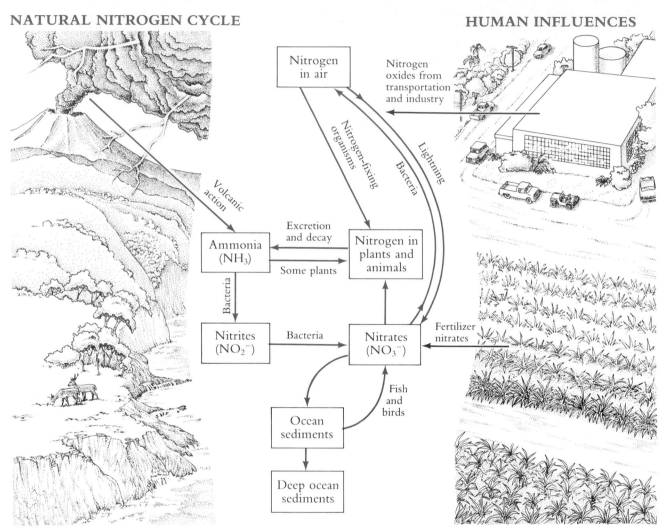

Figure 2.9 The Nitrogen Cycle. Plants and animals generally require nitrogen in the form of nitrates. A small amount of nitrate is formed from atmospheric nitrogen by lightning. However, most nitrogen enters living systems through bacteria and algae that can "fix" nitrogen gas from the air into organic compounds. Biological matter is broken down into ammonia, which is cycled by bacteria through nitrites and back to nitrates. A portion of the nitrates are lost to the cycle when they are buried in deep sea sediments, while nitrogen is added to the cycle by volcanic action. Certain bacteria can "denitrify" nitrates back to nitrogen in the air. Humans influence the nitrogen cycle by adding nitrates in fertilizers and nitrogen oxide gases from transportation and industry.

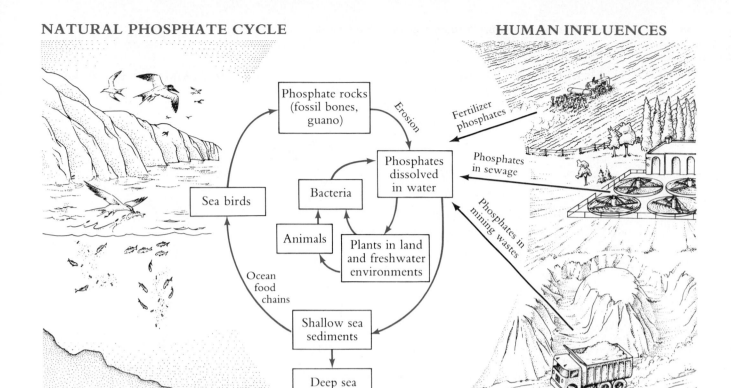

Figure 2.10 The Phosphorus Cycle. Erosion and mining release phosphates from rocks and other deposits to water environments. Phosphates are eventually deposited in shallow or deep-sea sediments. From shallow sediments, some phosphate is recycled back to land environments.

Nitrogen cycle. Nitrogen is present in the atmosphere as nitrogen gas (N_2). The air we breathe is about 80% nitrogen. This nitrogen becomes part of biological matter almost entirely through the bacteria and algae that can "fix" atmospheric nitrogen into organic compounds and nitrates. Legumes, such as clover, alfalfa, soy beans, and locust trees, form little nodules in their roots where such nitrogen-fixing bacteria live. Farmers often fertilize their land naturally by growing these types of crops and then plowing them into the soil.

Nitrogen is lost from the cycle (Figure 2.9) to deep-sea sediments, but this is just about balanced by additions from volcanoes. Humans interfere with the nitrogen cycle by producing large quantities of nitrogen oxide gases, which are normally present in very small amounts. These gases contribute to the formation of smog (Chapter 22). In addition, excess nitrate from agricultural fertilizers can lead to overgrowths of algae, or eutrophication (Chapter 13).

Phosphorus cycle. The sea is a major part of a number of important cycles. One example is the hydrologic cycle; another is the phosphorus cycle, shown in Figure 2.10. Phosphorus enters the environment mainly from rocks or deposits laid down in past ages. Erosion and **leaching** gradually release the phosphorus. Some is used by biological systems, but much of it is washed into the sea, where it settles in the sediments. Phosphorus in shallow sediments is cycled by bottom dwellers through fish and then to birds, which deposit the phosphorus back on land as droppings or guano. But a portion of the phosphorus is not recycled; this is the amount buried in deep sediments.

Large quantities of phosphorus are used as fertilizer, and most of this is eventually lost as deep-sea sediments. At some time, though not in the near future, shortages could result. Phosphorus is an essential nutrient for plants and, along with nitrogen, is often in short supply compared to other nutrients in water. Phosphate, which enters natural waters from sewage,

NATURAL SULFUR CYCLE

HUMAN INFLUENCES

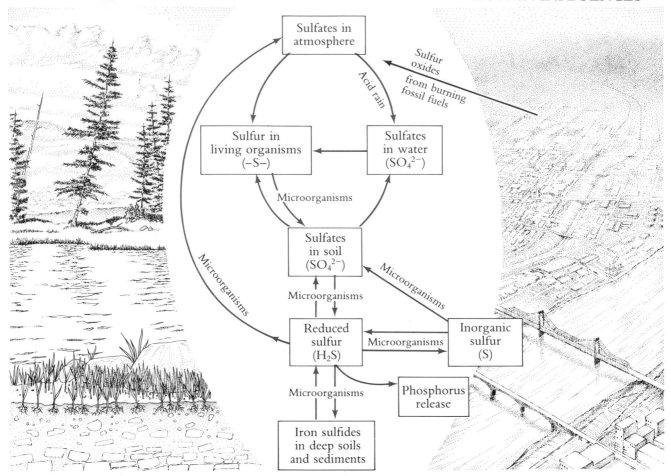

Figure 2.11   The Sulfur Cycle. There is a large pool of sulfur in the soil and sediments, and a smaller pool of sulfur in the atmosphere. Plants and animals use sulfur, as sulfates, from soil, water, and air. This sulfur is then incorporated into organic sulfur compounds, such as sulfur-containing proteins. Sulfur is changed from sulfates to reduced sulfur compounds and back again by a variety of microorganisms in the soil and sediments.

fertilizer runoff, and mining wastes, contributes to the problems of eutrophication (Chapter 13).

Sulfur cycle. In the sulfur cycle, sulfur is changed from one form to another by a variety of microorganisms in the soil and sediments (Figure 2.11). Plants and animals require sulfur as sulfate (SO_4^{2-}) in order to build organic materials such as amino acids and proteins. In the deep sediments and soils, sulfur is changed to iron sulfides at the same time that phosphorus is changed from an insoluble to a soluble form. In this way, cycles of sulfur and phos-

phorus interact.

Human activities, notably the burning of high-sulfur coal to generate electric power, are affecting the sulfur cycle by overloading the atmospheric pool of sulfur with sulfur dioxide, a compound normally present at very low levels. (See the effects this has in Chapter 21.)

Water, carbon, nitrogen, phosphorus, and sulfur are not the only materials that cycle in ecosystems, but from the point of view of their significance to living organisms and their relationship to current environmental problems, they are the most important.

Biological Magnification

Some pollutants not only cycle in the environment, but also tend to accumulate in living organisms. In some cases, the concentration of a pollutant found in living organisms increases as the pollutant is passed up the food chain. This happens when organisms take in a pollutant faster than they eliminate it. Mercury, for instance, may be present at relatively harmless concentrations in water or bottom muds, but may be concentrated to lethal levels in shellfish growing in the water (see Chapter 11). Pesticides such as DDT also are affected this way: they may be almost undetectable in the water, but present in greater and greater concentrations at higher trophic levels. This is the phenomenon called **biological magnification.** We will explain it more fully when we discuss pesticides in Chapter 6.

Energy Flow in Ecosystems

In contrast to the cyclic flow of materials we have been discussing, energy flow on earth is more like a one-way street. The energy in ecosystems comes from the sun and is finally lost as heat to the cosmic reaches of outer space.

The flow of energy through systems is of major concern to ecologists and should be of interest to everyone else too. An understanding of energy flow, and the basic laws that govern it, leads to a better understanding of why natural systems work the way they do. Equally important, such an understanding helps us see the limits of our ability to make changes in our environment without damaging it.

Figure 2.12 is a simplified diagram showing energy and materials flow for a stream ecosystem. The energy from the sun is the driving force. Green plants capture the sun's energy by photosynthesis and use this energy to produce organic materials. The energy stored in these materials serves as the energy source for the primary consumers that eat the plants and for the higher consumers that eat primary consumers, and so on. Even in ecosystems called "incomplete" because they lack the producers that capture sunlight, if we follow the energy source back far enough, it will be found to be the sun. For instance, in the deep sea bottom, where no light penetrates, photosynthesis cannot occur. Instead, the energy source is **detritus.** But this detritus originates from organic materials produced in the upper layers of the ocean, where there is light and photosynthesis.

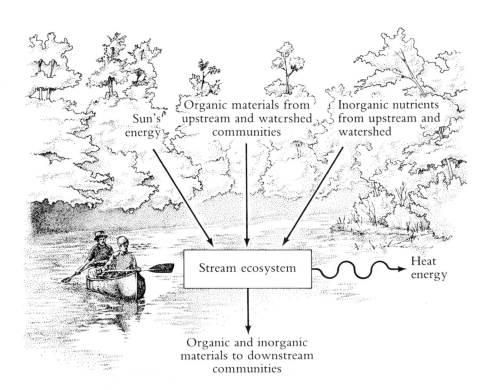

Sun's energy

Organic materials from upstream and watershed communities

Inorganic nutrients from upstream and watershed

Stream ecosystem

Heat energy

Organic and inorganic materials to downstream communities

Figure 2.12 Energy and materials flow for a stream ecosystem. Living systems are open systems: materials and energy flow into the system from its surroundings and out again into the environment.

Productivity

Productivity vs biomass. The amount of organic matter produced in an ecosystem or by an individual in a given time is its **productivity.** Productivity is measured in a variety of ways. It is often measured as the *increase* in weight of biological matter for a given area over a given time period. (This includes all the new plant tissue, roots, leaves, etc., as well as the increased weight of animals during the time specified.) Productivity may also be stated as a number of calories for a given time period, when measuring the energy captured by green plants and then transferred from one trophic level to another in an ecosystem.

In contrast, the **biomass** of an area or ecosystem is the *total* weight of all living material, plant or animal, in that area at a specified moment. For instance, the productivity of a forest ecosystem might be measured during a year's time. The biomass at the end of the year would include not only that year's productivity, but also plant and animal tissue (e.g., tree trunks) produced in previous years.

Primary and secondary productivity. The productivity of an ecosystem can be further divided into primary and secondary production. Primary productivity is the amount of organic material produced by green plants (and some bacteria) using the sun's energy. Secondary productivity is the organic material produced by organisms that are not photosynthetic; that is, fungi, most bacteria, and animals. These organisms consume organic matter from primary producers to gain energy and materials for their own bodies.

Ecological pyramids. Ecologists often draw pyramids to show the amount of energy transferred from one trophic level in an ecosystem to another. Figure 2.13 shows an imaginary and very simplified system in which alfalfa plants in a field use sunlight to produce plant tissue, which is eaten by calves, which, in turn, are eaten by a boy. The first pyramid shows the biomass, or weight of living material, at each level. The second shows the yearly energy transfer in calories through the system.

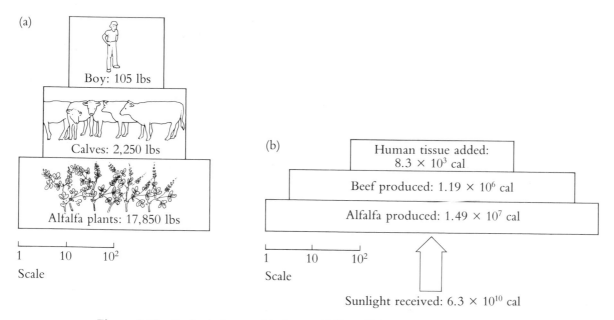

Figure 2.13 Ecological pyramids for an alfalfa–calf–boy system, calculated for 10 acres and one year. (a) The biomass of the living material at each trophic level; (b) the productivity in calories at each level. (Note that both of these are drawn using a logarithmic scale in order to accommodate the large values at the producer level.) (Adapted from Odum, *Fundamentals of Ecology,* 3rd ed. Philadelphia: Saunders, 1971.)

These particular pyramids clearly illustrate how much primary productivity is necessary to provide the yearly protein consumption of just one boy if his requirement is met by meat alone. The implications of this in terms of the world food supply are discussed in Chapter 7.

Gross vs net productivity. Several factors are not at all clear from illustrations such as Figure 2.13. In the first place, we should distinguish between gross and net productivity. Gross productivity is the total amount of organic material produced by plant photosynthesis in a given time period. Part of this, however, is used up by the plants in respiration. The amount left is net productivity. This is actually what we measure, and it is the more important figure because only net productivity is available to the next trophic level.

Availability of produced material. Second, the available organic material is usually not completely consumed by the next trophic level. As a rule, primary consumers utilize only 5–15% of the primary production. There are several reasons for this. One is that some of this primary production is simply not available. In a tropical rain forest, for instance, most of the primary production is in the leaves of the tree canopy high above the forest floor, largely unavailable to browsing animals. Another reason is that dry or cold seasons create a variation in primary productivity: when it is cold or dry, primary productivity drops. The number of animals that can live on primary production is governed by the availability of food at the worst times. There simply are not enough animals in an ecosystem to consume all primary production during periods of high productivity; however, this effect is partly offset by migrations of consumers into areas where productivity is high. Even that portion of the available food actually eaten by animals is not all turned into flesh to be eaten by the next trophic level. Some part of the food is not digestible, and part of the energy in the food that is digested must go to maintain normal vital processes. Finally, in these processes, energy transfer is not 100% efficient because energy is lost as heat (this will be explained later). In general, only 1–10% of the food eaten by animals is used to form new body materials.

Energy budgets. All of these losses are summarized in Figure 2.14, which is a more detailed energy-flow diagram than Figure 2.12. Such a diagram is sometimes called an energy budget, since it attempts to account for all of the energy entering and leaving a system. Even this complicated-looking diagram is not as complex as a real ecosystem. Several things are not shown in the diagram—for instance, small herbivores (plant eaters) probably migrate into or out of such ecosystems from other areas. Further, some of these small herbivores are eaten by secondary consumers, who are eaten by tertiary consumers, and so on. Note that all of the unused organic material eventually enters the food chains responsible for the decomposition of dead material. On land, these are mainly in the soil; in bodies of water, they are usually found toward the bottom.

Thermodynamic Laws and Ecosystems

First law of thermodynamics. Energy budgets follow the first law of thermodynamics, which states that energy can be changed from one form to another, but it cannot be created or destroyed. Thus, all of the energy entering an ecosystem must balance with the amount remaining there plus the amount leaving the system, as is done in Figure 2.14. (Note that the amount of energy consumed by "other herbivores" does not quite balance the amount lost as heat and to decay processes. The remainder represents the "other herbivores" eaten by secondary consumers, which are not shown in the diagram.)

Limits explained by the first law. The first law of thermodynamics explains certain facts about the environment. For instance, it can help us understand limits on food production. Food contains stored chemical energy used by living organisms. When we eat food, it is broken down in our bodies with the release of this stored energy. Some of the energy is stored in body structure; some is lost as heat; and some powers the reactions essential to life.

In order to increase food production, especially in the developed world, farmers use great quantities of synthetic fertilizer. However, this fertilizer is produced by using fossil fuels, both as raw materials and to run the manufacturing process itself. That is, we are increasing the production of one kind of energy—the food energy in plant crops and the animals that eat them—by using up stores of high-energy fossil fuels.

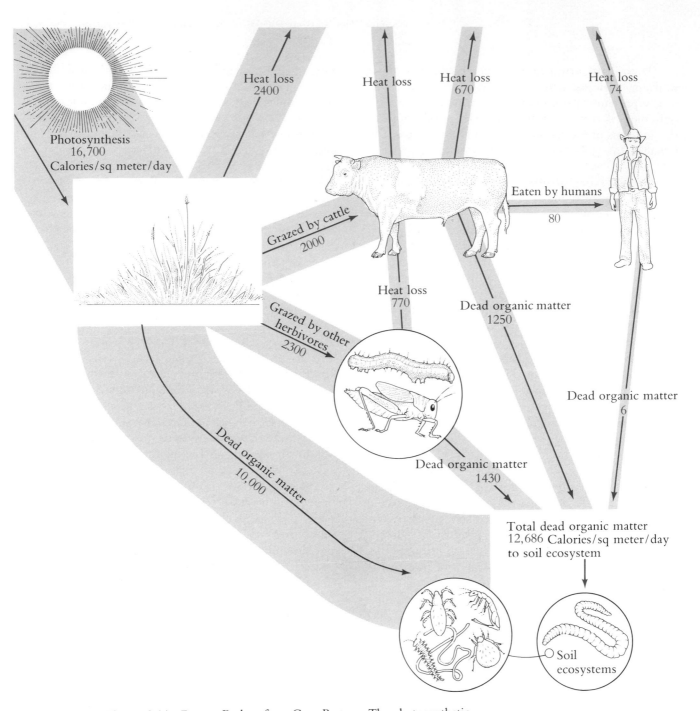

Figure 2.14 Energy Budget for a Cow Pasture. The photosynthetic activity of the grasses captures 16,700 cal/m²/day. The energy flow through the system is shown. Note the large proportion of energy that enters detritus food chains in the soil. Note also the proportion of energy lost as heat at each stage of transfer. Eventually the 12,686 calories that enter soil food chains will also be lost as heat when the organic material containing this energy is completely broken down. Note that not all energy flow is accounted for in this simplified diagram—see text. (Adapted from Remmert, H., *Ecology*. New York: Springer-Verlag, 1980, p. 211.)

In a similar way, synthetic foods cannot be manufactured without using some other energy resource to run the process—energy cannot be created. Of course, that does not mean that no improvements can be made in the efficiency of food production. A major effort involves breeding plants in which more growth takes place in parts useful to humans as food, such as the kernels of a wheat plant rather than the stalk. In plant breeding as in other sciences, however, ecological energy considerations must be taken into account. Remmert[3] gives the example of attempts to breed a plant that can fix nitrogen from the air rather than requiring soil nitrates (which in many farming areas are supplied in the form of expensive fertilizer). In terms of energy costs to the organism, fixing nitrogen costs more than the uptake of nitrate from the soil. This energy must come from some other biological process. Remmert speculates that if the experiment is successful, the plants will be found to be much less productive than the original variety, which would have the effect of cancelling out the cost savings in reduced fertilizer use.

The second law of thermodynamics. While explaining ecologic pyramids, we mentioned that energy-transfer reactions are not 100% efficient—some energy is lost as heat. In this way, living systems obey the second law of thermodynamics. This law was first discovered by physicists attempting to turn heat energy into useful work by machines such as steam engines. They discovered that heat could never be converted 100% into useful work (more on this in Chapter 15). Similarly, in living systems, energy gained from breaking down energy-rich food substances is not all available for use in other reactions—some energy is always lost as heat.

This is one reason why few individuals (such as large carnivores) stand at the top of food chains compared to the vast numbers of individuals at the bottom. The loss of energy at each step in the food chain makes it impossible for more than a few individuals to be supported. This also explains why it is not energy-efficient to eat from the top of food chains, as humans often do. When we eat meat or fish, we eat food that contains only a small part of the energy with which the food chain began.

The large amount of energy lost to ecosystems as heat during energy transfers in the food web must be made up from outside sources (sunlight, detritus) or the whole system would run out of energy and collapse. Materials, such as minerals or water, cycle in the environment, but energy does not. Energy flows into ecosystems and is lost as heat.

Randomness and order. A much broader and sometimes more useful statement of the second law of thermodynamics is that all systems tend to become random, or disordered, spontaneously, that is, on their own. For instance, if you dump marbles onto the floor from a bag, they will spread out spontaneously in a disorderly fashion. The release of air pollutants, such as those formed when coal is burned (Chapters 20 and 21), demonstrates the second law in action. The pollutants, released from a power-plant smokestack, disperse spontaneously in the atmosphere. We actually depend on this phenomenon: because the pollutants disperse, they become more dilute and less dangerous. Of course, this sort of system only works up to a point. As populations grow, power demands and pollutant levels increase. Dispersal or dilution of pollutants becomes a less and less effective solution.

On their own, the marbles we spilled will not regroup and hop back into the bag to restore order in the room. We need to expend energy hunting for them and picking them up and putting them back in the bag. The formation of order is a nonspontaneous process requiring energy, whether someone is cleaning up marbles from a floor or whether an organism is growing by the orderly arrangement of molecules.

When we first look at living organisms, it may seem that here is a system becoming more orderly spontaneously. A growing, living organism, considered alone, might seem to fit this description. However, at the same time that the organism is growing in an orderly fashion, biochemical reactions going on inside it produce heat. This heat, which is given off into the environment, increases the random motion of molecules in the organism's surroundings. Thus the organism plus its environment is still tending spontaneously toward disorder.

Using the second law, we can project some of the effects of population growth. Increasing numbers of people, who themselves live and grow by the production of order, will cause increasing levels of disorder in their environment. Air pollution caused by increased power demands of growing populations is one example of this. Another example is the erosion of soil from

---

3 Remmert, H., *Ecology*. New York: Springer-Verlag, 1980, p. 6.

tightly knit forest ecosystems, after the forest has been cleared to plant crops or after the land has been turned over in the search for minerals.

Although these laws of thermodynamics are important to understanding relationships between organisms and the effects of organisms, including humans, on their environment, most people fail to take them into account. A study of how these natural laws operate with respect to power generation, for example (see Chapters 15 and 24), can help us find ways to minimize our impact on the environment.

Population Dynamics

Up to this point, we have been considering individual parts of ecosystems and some of the processes that go on between parts of the system. Now we have enough information to consider how these processes affect the size and distribution of the various populations in ecosystems—that is, population dynamics.

Age Distribution Graphs

In many natural populations, if we compare the numbers of individuals at various ages, we obtain a graph resembling Figure 2.15(a). In these cases, younger in-

dividuals are present in much greater numbers than older ones. The death rate of these young organisms is very high compared to that in other age groups. Thus, the numbers of individuals that survive to enter the older-age categories are relatively small. This is shown also in the form of a bar graph in Figure 2.15(b).

The shape of the age distribution graph depends on the specific population under study. For instance, the female Pacific herring produces some 8000 eggs each year, of which 95% hatch. Only 0.1%, however, live to maturity. A bar graph of this population would show a very broad base and a very narrow top. On the other hand, species such as elephants and whales, which produce only one offspring and that not every year, would not have as wide a base and would have a larger percent of the population in the adult age categories.

Population Growth Curves

If a population of organisms grows under conditions in which essentially no death occurs before organisms become sexually mature, growth is exponential. A curve of such growth would resemble Figure 2.16(a). An example of exponential growth in natural systems arises when organisms colonize a habitat in which there is no competition. However, it is obvious that

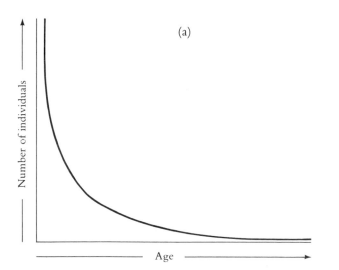

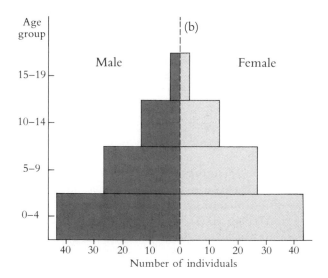

Figure 2.15 Age Distribution Graphs. (a) Generalized curve showing numbers of individuals in a population versus their age group. (b) Bar graph of this population type, characterized by high rates of birth and high death rates in the younger age categories.

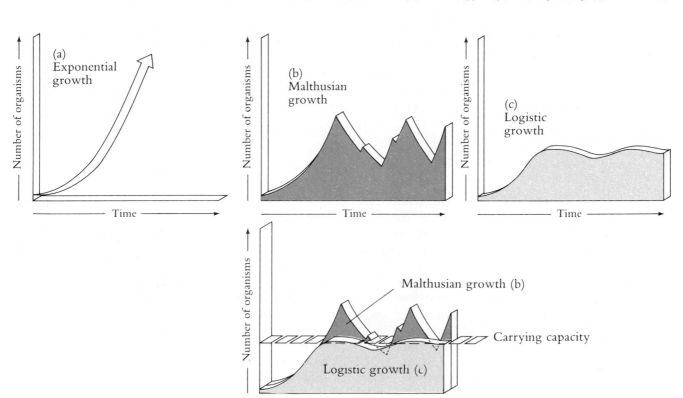

Figure 2.16 Patterns of Population Growth. (a) Exponential growth typical of organisms colonizing newly formed habitats. (b) Malthusian or irruptive growth pattern, in which population growth is limited by disastrous population crashes. (c) Logistic growth, in which the population gradually levels off. In the bottom figure, Malthusian and logistic growth are superimposed on each other for comparison. Also shown is the carrying capacity of an environment, or the largest population that environment can support without being damaged. Note how the population showing Malthusian growth tends to overshoot the carrying capacity and then fall in a series of sharp population crashes. Populations growing logistically, however, level off smoothly around the carrying capacity of their environment. (This is discussed further on page 44.)

this type of growth does not continue as the curve does, because otherwise such an organism would soon cover the earth completely.

Malthus recognized these facts in the late 1800s, and proposed that populations grow until they use up the resources available to them, and then are limited by catastrophes such as famine, disease, and violence. The curve in Figure 2.16(a) thus levels off into a series of fluctuations, as shown in 2.16(b).

Some animal populations do seem to follow this pattern of explosive growth and then collapse. In 1944,

for instance, 29 reindeer were brought to St. Matthews Island in the Bering Sea. By 1963, there were 6000 deer on the island. Immediately thereafter, however, the population suffered a catastrophic decline to less than 50 animals. Such growth is called irruptive or Malthusian.

Other populations do not seem to follow such a growth pattern, but instead show a similar initial exponential increase followed by a smooth leveling off over a period of time, at a level often referred to as the carrying capacity of their environment. (Carrying capacity is discussed more fully shortly.) Organisms with simple life histories demonstrate this type of growth, called logistic growth, very well [Figure 2.16(c)]. As life histories become more complex, however, neither curve may fit the history of a population's growth very well, or a curve may fit only at certain periods of time.

r *and* K *Selection*

Habitats such as the Montana Rocky Mountains or the eastern deciduous forest, where attempts are being made to reestablish peregrine falcon populations, are relatively stable over long periods of time. That is, unless humans intervene, conditions do not change rapidly. (With the notable exception of natural disasters.) Most of the organisms in these ecosystems have adapted in ways that allow them to make the maximum use of their environment without harming it. In other words, these organisms are adapted to the capacity of the environment. A growth curve for such a population of organisms might resemble Figure 2.16(c), the logistic growth curve. There are other habitats, however, which develop rapidly and disappear again—for instance, seasonal ponds that fill with rainwater only during rainy seasons. The growth curve for organisms colonizing such habitats commonly resembles Figure 2.16(a).

Organisms characteristic of stable habitats are often termed K-selected, where K is a symbol for the capacity of the environment. K-Selected organisms are generally long-lived, have few offspring, and are successful in situations where they must compete for scarce resources. A second type of organism, characteristic of fleeting environments, is called r-selected, where the r refers to their high rate of reproduction.[4] In addition to showing a high reproductive rate, these organisms generally develop rapidly. They are not good competitors, however, and can establish themselves only where there is little or no competition. In general, trees and animals with long life spans are K strategists. Small animals and the plants that first colonize habitats after a disaster (pioneer plants) are r strategists. Of course, there is no neat division of all organisms into one or the other type. Many organisms are better classified somewhere between the two categories.

---

4 The origin of the r and K is explained in the supplemental material at the end of this chapter.

Further, even stable habitats have subdivisions that favor r strategists. The bodies of organisms that die in forests, for example, provide a habitat for r strategists (e.g., fly maggots) in an area where K-selected species are usually favored.

Factors Related to Population Density

Carrying capacity and space requirements. Populations have certain space requirements that ensure such things as adequate food supply, space to seed or to raise young, the availability or reproductive partners, and a necessary minimum of stress-producing encounters. The **carrying capacity** of an ecosystem is the maximum-sized population for which the ecosystem can provide these requirements for an indefinite period of time. [Carrying capacity is shown in graphic form in Figure 2.16(c).] We might now ask, what are the factors that determine the carrying capacity of an environment for a particular species and how do organisms in nature adjust to this capacity?

Density-dependent and density-independent factors. Most, if not all, factors that adjust population size and carrying capacity are density-dependent; that is, they may kill off 80% of a dense population but only 10% of a sparse one. An example would be a disease that spreads quickly in dense populations with many contacts between individuals, but which does not make much headway in a sparse population whose individual members rarely come in contact.

Extreme climatic conditions sometimes act as density-independent factors in population regulation. For instance, a severe winter frost in the intertidal zone can cause rock surface to crumble away, removing snails that cling there without regard to the number of snails present. In many other cases, however, climatic factors act in a density-dependent manner. For example, during a flood, a larger percentage of individuals in a sparse population may find refuge (such as higher ground) than in a denser population.

Density-dependent factors controlling populations can work by either decreasing the birth rate in a population or by increasing the death rate. The example given of a disease controlling a population size once it reaches a certain level can be considered a special case of predator–prey relations. The disease is the predator and the host the prey.

Predator-prey relationships. Predator-prey relationships appear to play a part in controlling population size. However, each system must be studied to determine whether the predator is controlling the prey or vice versa. One famous example is the nine-year population cycle of alpine hares and the lynx that prey on them. The alpine hare population peaks in approximately 9-year intervals, and each peak is followed by a peak in the lynx population—but, as the lynx population peaks, the hare population crashes. This was originally interpreted as an example of a species (the lynx) exceeding the carrying capacity of its environment by consuming too much of its food supply, following which the lynx population collapses and the whole cycle starts again. However, in regions where the lynx has become extinct, the hare population still cycles. Whatever the reason for this cycling, it is clear that the hare population was controlling the lynx and not vice versa.

On the other hand, predators much smaller than their prey (some scientists would then call them parasites) may be able to control the prey population. For instance, a European leaf beetle, imported into this country, successfully controls the introduced weed St. John's wort (*Hypericum perforatum*). In areas of great species diversity, several predators may keep the population of a prey species much smaller than it would be under the same environmental conditions with fewer predators. Several species with similar ecological needs can then occupy the same area. This may be a partial explanation for the richness of species on coral reefs and in rain forests.

Food supply and food competition. On the face of it, food supply and the competition for food appear to be a likely limiting factor for populations. However, field researchers have not been able to show that population density varies in any consistent way with food supply. Probably food supply acts with other factors to limit populations.

Abiotic factors affecting population density. Abiotic factors are known to affect population density. That is, the density of a population tends to decrease as abiotic factors such as climate or soil type become less and less optimal for the species in question. As an example, soil type has been found in some cases to be related to the quality of the food available. In Scotland, basaltic soils rich in minerals support a much greater population of hares and red grouse than do mineral-poor granitic soils. Even though the amount of primary production on both soils is similar, the plants are richer in minerals on the basaltic soil and so provide a higher quality food for wildlife.

Autoregulation. In some cases, population size appears to be regulated by factors other than those already mentioned. Scientists have described this situation as autoregulation. The stress caused by overcrowding may be a factor in autoregulation. Stress is normally increased by crowding because there are more aggressive meetings with other organisms. Autoregulation can take several forms; for example, parental care of offspring can improve or deteriorate depending on density, or mass migrations out of a habitat may occur at peak population densities.

Sometimes an actual physical basis for autoregulation behavior is known. That is, some biochemical reaction to crowding affects births or deaths or triggers behavioral patterns. As an example, crowding in tree shrews (*Tupaia glis*) causes an increase in blood urea concentrations. Offspring mature more slowly or not at all. Mothers no longer protect their young by marking them with a glandular secretion, and the young are then eaten by other tree shrews. Population size is kept constant in these ways, even if food is present in excess. If crowding is increased artificially—for instance, by decreasing the size of the cage in which the population is kept—blood urea concentrations rise even higher and some of the animals die. In the field, lemmings have been shown to have similar blood urea concentrations during peak population densities.

The factors mentioned cannot give us a complete explanation of either the carrying capacity of an environment or of how populations adapt to that capacity. In fact, the science of ecology has not advanced to a point at which it could give a complete explanation. However, there are fascinating questions raised by the insights ecology does give us. For instance, we could ask whether humans are r- or K-selected. Considering mechanisms by which animal populations appear able to adjust to the carrying capacity of their environment leads us to wonder what mechanisms might or might not adjust human populations to the carrying capacity of our environment. How do we determine this capacity? Will we even recognize it before we have exceeded it? In the next two chapters, we deal with these and other human population issues and with wildlife resource problems.

✎ *r*- and *K*-Selected Populations

The "*r*" in "*r*-selected" comes from the equation describing exponential growth:

$$\frac{dN}{dt} = rN$$

where *N* is the number of individuals, *t* is time, and *r* is the population's intrinsic rate of increase.

The "*K*" in "*K*-selected" comes from the equation describing logistic growth:

$$\frac{dN}{dt} = r \left(\frac{K - N}{K} \right) N$$

where *N* is the number of individuals, *t* is time, *r* is the population's intrinsic rate of increase, and *K* is the carrying capacity of the environment.

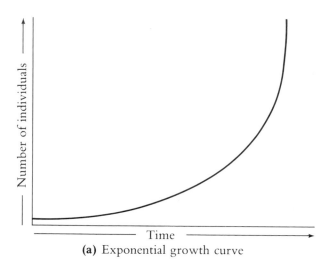

(a) Exponential growth curve

Figure 2.17

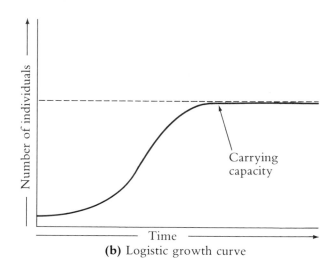

(b) Logistic growth curve

Questions

1. Diagram a simple food chain that might exist in the area where you live. Who are the producers, consumers, decomposers?
2. What is a food web? Are food webs, because of their more complicated system of checks and balances, more stable than food chains? Explain.
3. Describe the hydrologic cycle.
4. How is atmospheric carbon dioxide related to the carbon in living organisms?
5. Why are we concerned about nitrogen and phosphorus cycles?
6. At what point in the sulfur cycle are human influences felt?
7. How does energy flow differ fundamentally from the flow of materials such as carbon, phosphorus, nitrogen, and sulfur?
8. What is meant by biomass? Differentiate between biomass and productivity.
9. What is the main point made by ecological pyramids?
10. How do the first and second laws of thermodynamics work in ecosystems?
11. What is the ultimate source for all energy on earth? Explain, using the idea of energy budgets.
12. Explain what is meant by exponential, Malthusian, and logistic growth.
13. Define the carrying capacity of an environment.
14. List the factors that affect carrying capacity and a population's adjustment to that capacity.

CHAPTER THREE

Human Population Problems

Growing and Changing Populations

Population Growth through History/The Demographic Transition/Population Growth/Rural to Urban Migration

Effects of Population Growth

Consequences in Developing Countries/Consequences in Developed Countries

Limiting Population Growth

The Problem of Time Scale/Methods of Limiting Population Growth

Outlook for Population Growth in the Future

Demography—The Study of Populations

CONTROVERSIES:

3.1: *Family Planning in India—A Lesson on the Meaning of Freedom*

3.2: *The Ethics of Limiting Family Size*

3.3: *Population Growth in Africa*

Growing and Changing Populations

Population Growth through History

When the beings that we call "first humans" lived on the earth, life itself must have been a very uncertain proposition. Humans had to hunt and gather their food, and so were at the mercy of the weather and of fluctuating animal populations. They did not yet grow food or store it for times of drought, unseasonable cold, or poor hunting. There were probably no more than 10 million humans in the world until people learned, about 8000 B.C., to domesticate animals and to grow and store their own food. After this, the human population grew more quickly. Did this growth occur because the death rate decreased as food supplies became more certain? Perhaps not. Some scientists point out that death rates may even have increased as crowded living conditions in stationary communities allowed diseases to spread more easily.

Probably a more important factor was the spacing of births. Studies of present-day hunter–gatherers, such as the Dobe !Kung, show that births are spaced in such nomadic societies so that there is at least three years between children. This is necessary because mothers must carry their infants as the tribe moves about. Also, nomadic peoples do not have soft foods, which are necessary for feeding young children after they are weaned. Thus, children are usually nursed for a relatively long time (five years, in some cases). However, agricultural societies are sedentary; therefore, children need not be carried long distances. In addition, farmers can grow more easily digested foods. These factors would allow for more closely spaced births and would account for population growth as agricultural societies developed. (You may wonder how primitive people could have spaced births at all. The reference by Dumond has some interesting thoughts about this.)

A much greater population growth spurt began just before the start of the Industrial Revolution (Figure 3.1). This growth spurt is a result of decreasing death rates. (The **death rate** is defined as the number of people who die each year per thousand people in a population. Similarly, the **birth rate** is the number of people born each year per thousand people in a population.) Death rates have fallen because of improvements in sanitation, medicine, nutrition, and agriculture. In developed or industrialized countries, food is grown in larger quantities and transported more efficiently. People are healthier because they are better fed.

The Demographic Transition

In the developed nations, death rates fell during the 1800s to the middle of the 1900s. Birth rates generally declined along with the death rates in these countries.

Those countries that have managed to reduce both their birth and death rates are said to have undergone the **demographic transition**. Demographic transition is the name population scientists apply to a pattern of change in birth and death rates. First seen in the industrial nations of Europe and America, the transition consists of a decline in the death rate resulting from the many medical or nutritional advances the nation has come to enjoy. While the decline in death rate is underway, the birth rate, too, falls to a lower and fairly stable value. However, it would be wrong

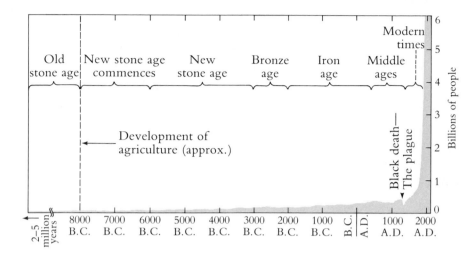

Figure 3.1 World Population Growth through History. There has been a steady increase in the human population since 8000 B.C., when agriculture became a significant way of obtaining food. However, the current enormous population growth spurt began much more recently, just before the start of the Industrial Revolution. (From Elaine M. Murphy, *World Population: Toward the Next Century*, Population Reference Bureau, Washington, D.C., 1981.)

to conclude that birth rates in developed nations fell solely because people became aware of the decline in death rates. There is evidence that birth rates actually began to fall before the medical and sanitary advances that lowered death rates.

Most industrial nations have already undergone this demographic transition. Further changes in birth and death rates in these countries in the near future are likely to be small. The situation is different, however, in the countries labeled **developing**.

Population Growth

Why rapid population growth occurs. In countries all over the world, death rates are being lowered by public health measures. Diseases such as malaria and yellow fever, which are spread by mosquitoes, are prevented by the use of insecticide sprays and by draining and filling the swamps where these insects breed. Cholera, typhoid, and other intestinal diseases are prevented by chlorinating and filtering public water supplies. The pasteurization of milk can prevent the spread of bacterial diseases in one of the most common childhood foods. General food sanitation, made possible by refrigeration and a pure water supply, may also contribute to reducing childhood diseases.

Antibiotics save many children who might once have been victims of diseases such as pneumonia and scarlet fever. Drugs that cure tuberculosis and a vaccine that prevents it are available. Vaccinations can now prevent most children from ever having smallpox, diptheria, tetanus, whooping cough, polio, or measles. In addition, better nutrition and care for pregnant women have improved both the health of mothers and the chances that their babies will survive.

All of these medical and public health advances have combined to cause a dramatic drop in the death rate among infants and small children. The death rate of infants and children has fallen not only in countries that are developed, or industrialized, but also in developing countries where many health and sanitation measures have been instituted. Nevertheless, even though the odds on child survival have increased in these countries, even though more of a couple's children survive to adulthood, there has not always been a corresponding drop in the birth rate. That is to say, in many cases, parents still have as many children as they did when it was necessary to have five or six children in order to be sure of raising two to adulthood.

Components of the growth rate. In developing countries, the birth rate is likely to be as high as 30–45 births per 1000 people in the population each year, while the death rate may be 15 per 1000 per year. In contrast, among the developed countries, birth rates have fallen to 10–15 per year for each 1000 people in the population, while death rates range from 6–13 per year (Figure 3.2).

If we subtract the death rate from the birth rate, we will have the **rate of natural increase**, a number that tells us how fast a population is growing. In countries classified as economically undeveloped, this rate

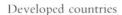

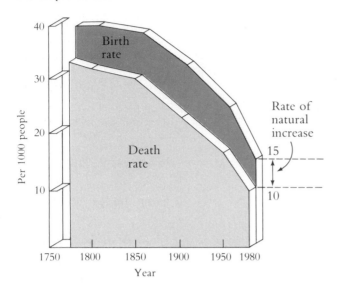

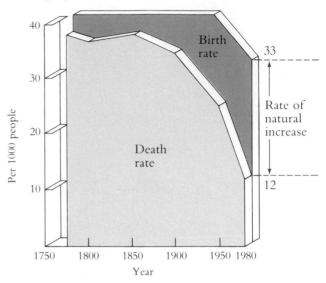

Figure 3.2 World Birth and Death Rates. The rates of birth and death are shown for both developed and developing countries. By subtracting the death rate from the birth rate, we obtain the rate of natural increase. (From Elaine M. Murphy, *World Population: Toward the Next Century,* Population Reference Bureau, Washington, D.C., 1981.)

may range from 15–30 people per year for every 1000 people already in the population. The **growth rate** is a combination of the rate of natural increase and the movement of people into or out of a country (Figure 3.3).

Immigration and emigration. In some developed countries with low birth rates, **immigration** may be a major part of the growth rate. In the U.S., for instance, in 1981, net immigration (immigration minus **emigration**) accounted for almost one-third of population growth.

Throughout its history, the U.S. has received a great many immigrants, a greater number than any other country in the world. Table 3.1 lists the numbers of immigrants from 1880 through 1981, and also shows how immigration has often been a significant factor in U.S. population growth.

Immigration causes serious problems in some undeveloped countries also. In Nigeria, for instance, in 1982, the government issued a decree that all aliens were to leave the country. Hundreds of thousands of workers had entered Nigeria from countries such as Ghana, many illegally, to work in jobs generated by

Nigeria's oil boom. When falling world oil prices dried up the job market, the Nigerian government determined that Nigerians should fill all remaining jobs. (Immigration problems are discussed further on page 59.)

A comparison of growth rates. Growth rates vary in developed countries. In Denmark, Sweden, the Federal Republic of Germany, and Austria, populations are hovering near a zero growth rate. In Italy, Poland, Canada, and the United States, birth rates still exceed death rates. As a group, the developed nations have a growth rate of about 0.6% per year, contrasted with an average of 2% per year for the developing nations.[1]

The world population as a whole is increasing at the rate of nearly 1.7% per year. What does this mean? In 1825, the world population was one billion. One hundred and five years later, in 1930, it had reached

---

1 In 1981, the U.S. had a population growth rate of 1%. The rate of natural increase in the U.S. was close to the average for developed countries; however, net immigration, combined with natural increase, pushed the U.S. growth rate to one of the highest for any developed nation.

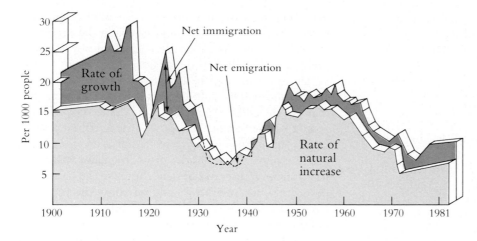

Figure 3.3 Rate of Growth and Rate of Natural Increase. The rate of growth for a country depends on both the natural rate of increase (birth rate minus death rate) and the rate at which people immigrate (move in) or emigrate (move out). These rates are shown for the United States from 1900 to 1981. At the present rate of growth, the population will double in about 87 years. (Source: Population Reference Bureau, 1981.)

two billion. Forty-six years later, in 1976, the world population had doubled again to four billion. In 1982, it was estimated that the world population of $4\frac{1}{2}$ billion would double in 40 years if present birth and death rates continue.

Why rapid population growth continues. In some ways, Westerners find it difficult to understand why birth rates remain high in undeveloped countries even when death rates are falling. The reasons involve differences in the value of children to African or Asian societies compared to their value in Western society. In Western societies today, most of the benefits of raising children could be called psychological. The pleasures of parenting, the feeling of passing on values to another generation, the fulfilling of other people's expectations (e.g., grandparents)—none of these bring any economic reward to the parents. Most of these psychological rewards are gained by having one or two children, although a third child may be wanted if the first two are the same sex. This, combined with the high cost of raising children in Western society today and with effective, readily available contraceptive methods, has made the two-child family more and more the accepted norm.

The situation is different in most developing nations. In many parts of rural Africa, labor is a scarce resource and land is not. The family that has many working members can farm more land and become wealthier. As one writer notes of Mali,

. . . except for recently introduced plows and carts, the physical factors a man needs to pro-

Table 3.1 Levels and Rates of U.S. Immigration, 1880–1981

Time period	Average annual immigration	Percent population growth due to net immigration
1880–1889	520,000	41
1890–1899	350,000	28
1900–1909	860,000	38
1910–1919	560,000	17
1920–1929	420,000	14
1930–1939	40,000	—
1940–1949	120,000	11
1950–1959	250,000	12
1960–1969	320,000	17
1970–1979	420,000	22
1980	800,000	33
1981	600,000	30

Large waves of immigration from the 1880s through the 1920s were composed of Europeans. The large increases in 1980 and 1981 were made up mainly of Southeast Asian, Cuban, and Haitian refugees. These numbers all represent legal immigration. Illegal immigration has become an increasingly large problem in the U.S., possibly reaching 500,000 a year.

Data from: U.S. Population, Population Bulletin, vol. **37**, No. 2, Population Reference Bureau (June 1982)

duce food are readily available. Land is abundant, "capital goods"—hoes and oxen—are cheap, and the wood and iron for making them are abundant. The crucial factors of production are the family's labor, of course, and "technology."[2]

While in Uganda:

> . . . extensive labor is needed for herding . . . A moderately prosperous herd owner with say 100 to 150 cattle, 100 sheep and goats and a few donkeys needs about six herd boys ranging from 6 to 25 years of age to maintain a herd by himself. A man with many cattle but few sons must herd together, and share the yield of his stock, with a man who has few cattle and many sons.[3]

Further, in some parts of Africa—for example, the Upper Volta and the Sudan—private ownership of land does not exist. A person owns the crops grown on land, but no one can own the land itself. Land-use priorities are often given out according to the number of people in a family who can work on the land.

Also important is the fact that in many developing countries the definition of family differs from that in the West. In parts of Africa, a father, his sons, and their wives and children may live together, or at least work together, under the father's direction. The economic unit of one man, his wife, and their children is often not the usual one. In fact, in parts of Africa, a man may have more than one wife. In situations like this, property rarely belongs to a husband and wife together, and may not be passed down from father to son. By contrast, European peasants felt the need to limit family size to avoid splitting inherited land among too many heirs. In parts of Africa where land is not held privately or where a father's property is not divided among his sons, factors that help limit family size are missing.

In some Indian states, the problems of inheritance are seen as the problems of the next generation. The father has the right and the duty to work for the maximum benefits for himself and the family he heads. This

may very well mean more children now to provide more labor and produce more goods in the present.

African and Asian families are often not as child-centered as families in the West. Parents may be expected to eat more than a proportionate share of food and use more of the family resources for clothing, etc., than are the children. A child's duty is to work for the benefit of the parents. In some African tribes, a son's loyalty and labor belong to his father even after he has a family of his own. Besides providing labor to free the parent from household or field chores so the parent can do other work, working children allow the parent to have some leisure time. Younger children even allow older children more freedom.

In developing nations children can be useful at as young an age as five years (Figure 3.4). If they cannot do an adult's labor in the fields, they can begin to help take care of younger brothers and sisters and do some of the household chores. In developing countries, such as Bangladesh, marketing and preparing meals for a family may be a full day's work for one person in the

Figure 3.4 Child Worker. This Venezuelan boy is one of what the International Labour Organization estimates as at least 52 million child laborers in the world today. (UNICEF photo.)

---

2 W. Jones, "The Food Economy of the Ba Bugu Ijoliba, Mali," in *African Food Production Systems*, P. F. McLoughlin, Ed. Baltimore: Johns Hopkins University Press, 1970.

3 R. Dyson-Hudson and N. Dyson-Hudson, "The Food Production System of a Semi-Nomadic Society: The Karimojong, Uganda," in *African Food Production Systems*, P. F. McLoughlin, Ed.

family. Children can be trained to help with meal preparation chores. One writer estimated that a man living alone in Bangladesh would spend 90% of his working time on household chores—tasks women and children would otherwise do.[4]

Children also provide a sense of security against old age or ill health, especially where there are no government old-age programs. In at least one sense, children are better than pensions or government social security since the benefits they promise will not be eaten away by inflation. Even when parents must invest resources in children during the years of child rearing, they can expect to reap some return when the children are grown. This enables parents to even out their own lifetime consumption—they spend more when they are young, vigorous, and able to work, and draw on this investment when they are old or infirm. In Africa, parents who can expect their children to support them when they are old tend to respond to arguments that fewer children are needed as medical care improves. Thus, they have fewer children. However, African parents living in cities, who seem to feel less certain that the old values will hold, tend to have even more children because it is then more likely that at least a few will provide for their parents' old age.

In developed and in certain undeveloped nations, parents can make choices about how they will educate their children. They may educate all of them, splitting the family's resources; they may educate only the brightest, sending that one child as far up the educational ladder as possible; or they may educate none of them. In countries with enough economic development and where social classes are not rigid, parents may see a benefit in producing only a few, well-educated children. These children can be expected to make something of themselves and add to the family's prestige and wealth. In poorer countries, such as Bangladesh, parents may not have this choice—their resources are often so limited that even one child cannot be educated. In such a situation, the only wealth available to a family is a large number of children.

To summarize, in many developing nations, children provide parents with wealth, with leisure, and with security in old age. These factors, which promote large families, are largely absent in industrialized societies.

---

4 M. Cain, "The Economic Activities of Children in a Village in Bangladesh," Paper presented at the IUSSP International Population Conference, Mexico City, 1977.

Most nations that have managed to slow their growth rates are also characterized by increased economic development and by what is called social mobility. That is to say, jobs are available to those who are willing and able to work, and further, there are no barriers of class or custom to prevent a family from bettering itself. In most of these countries, parents cannot benefit from their children's labor because of laws against child labor and because children are often required to be in school. In a climate such as this, parents seem to limit family size in order to gain extra income by working. A few well-educated children are seen as a better investment than a large family.

Rural to Urban Migration

Although many people are concerned about the rapid growth of the world's population, another equally far-reaching phenomenon is occurring in this century, which affects both the environment and how people interact with one another: namely, the way in which society is organizing itself. We are in the midst of transition from a mainly rural society to a mainly urban society. Although at one time most people lived on farms, soon most will live in cities. In the early 1900s, only one-quarter of the world's population lived in cities. By the year 2000, it is believed that 60% will live in cities.

Cities have certain advantages. They allow for greater division of labor, providing more jobs. Cities offer more specialized occupations for people to find work—for example, taxicab or rickshaw drivers, policemen and firemen—jobs that may not even exist in the hinterlands. Even crime and begging provide ways to sustain life in the cities. A more concentrated population can also be more easily provided with health and welfare services. These factors are important in drawing people to cities.

Another reason for the rural to urban migration is that the land available for farming is limited. That is, as the population in rural areas increases, there will not be enough land to support the entire population in farming or in other traditional rural occupations. In both developed and developing nations, fertilizers, pesticides, and modern farming equipment (when available) make it possible for fewer people to till the same area of land. Employment opportunities are then fewer and people move to the cities to support themselves.

Figure 3.5 Sidewalk Dwellers. These typical sidewalk-squatter dwellings of children in Bombay contrast starkly with the modern high-rise apartment buildings that house a fortunate few. (UNICEF photo.)

In developing nations, poor people who move to cities to find work often find conditions of extreme squalor. In the major cities of India, many people must live on sidewalks, making their home on nothing more than the pavement (Figure 3.5). The water and sanitation conditions of these wretchedly poor people are far worse than in the rural areas from which they came. Cardboard and tar-paper shacks seem a step up to people who have so little. Sanitation conditions are a modest improvement in communities of shacks, but they remain extremely primitive, with no running water and only ground disposal of waste. Intestinal diseases are rampant in these unsanitary conditions, and tuberculosis is spread by such close contact. For most people in developing nations, the move to the city is a misery often comparable to, and in some ways worse than, the misery left behind.

In the wealthier, developed countries, poor people who move from rural areas to the cities may find noise, dirty air, violence-ridden schools, crime, readily available drugs, and, too often, ugly surroundings.

These problems of cities in the developing and developed world are in part due to the speed of the transition from rural to urban society. Laws and social customs have not evolved quickly enough to cover situations in which enormous numbers of people are crowded together into cities. Perhaps new customs and laws will in time help solve some of these problems. As an example, wise land-use programs may be used to redirect the rural–urban migration. Such programs may preserve open space, limit "urban sprawl," prevent strip development, or confine industrial operations.

In the U.S., the trend of rural to urban migration reversed in the 1970s. Instead, more people began to move out of cities and into rural areas. That is not to say that many returned to farming—the farm population decreased a further 25% during this period. Rather, people seem to want a more rural life-style. Improvements in transportation and communications, as well as movement of industry to rural locations, made it possible to live in the country and still enjoy some of the benefits of urban life.

Effects of Population Growth

To a large extent, the problems caused by rapid population growth differ depending on whether a country is a developed one, such as the United States and the countries of Western Europe, or whether it is a developing one, such as the countries in Latin America.

Consequences in Developing Countries

Food production. In developing countries, the most pressing need is often simply to produce enough food to feed the rapidly growing numbers of people. Government programs must tend heavily in this direction to prevent starvation. In fact, the percentage of people in a country involved in the actual labor of growing food gives us two insights. It tells something about how developed a country is, because as a country develops, fewer and fewer people are needed to grow food for the rest. In addition, this percentage also points out how important food production is in developing countries. China-watchers estimate that fully 70% of the Chinese people are involved in agricultural labor. In the United States, one farmer's labor feeds 59 people; on a worldwide basis, one farmer's labor feeds only 5 people. Resources are employed that could be used in many other ways.

Investments in food production are also important for the stability of the government. A hungry population is a discontented one, and discontent can flare into revolution and overthrow of the government. This possibility sometimes leads governments to act in ways that are not economically wise, but will help them stay in power. For instance, food prices are often kept at a low level in large cities. However, this policy results in low prices for the farmer, which in turn dis-courages increases in food production. Farmers will not want to raise more food if they do not receive a reasonable price for their efforts. Low farm prices also mean that farmers cannot invest in modern farming methods that would increase their crop yields. Thus, countries that most need more food may, in some ways, be hindering greater food production (Figure 3.6). In this manner, population pressures can force governments to slant their policies in ways that are not in the long-term best interest of that country. The world food situation is covered in more detail in Chapter 7.

Health consequences. Because so much money must be spent on food production in developing nations, little is left over for health programs. Among

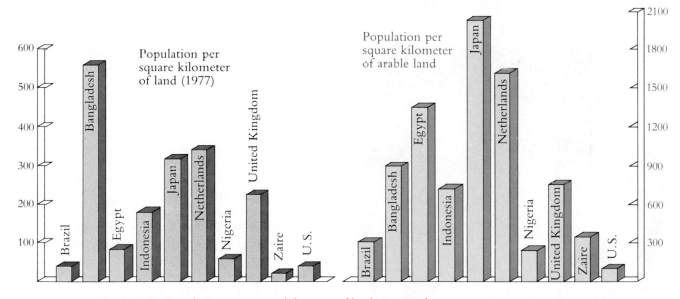

Figure 3.6 Population per square kilometer of land compared to population per square kilometer of arable land (1977). When the population size of various countries is compared with their land area, Bangladesh leads with the highest density of people per square kilometer of land in the world. Interestingly, the Netherlands is second. However, when population density is measured in people per square kilometer of arable land (land that can be farmed), the picture changes. Now Japan leads the world in terms of population density. The Netherlands is still second, but Bangladesh is in a somewhat better position with respect to its ability to grow enough food to feed its people. Actually, population density is not always the important measure when population problems are considered; sometimes the rate of growth of a population is more important. A sparse but rapidly growing population can strain a nation's resources, while a densely populated but nongrowing nation may be able to provide for its citizens relatively well. (Source: Population Reference Bureau, 1977.)

women who have many children, the rates of maternal illness and death are higher in developing countries. The babies born to these women tend to suffer from malnutrition (Figure 3.7). The lack of sufficient food in childhood leads to poor health and decreased mental ability.

A developing nation's annual investment in health activities must cover salaries of physicians and nurses, medicines, rents, equipment, etc. Yet, in most developing countries, the total annual investment in health activities, when divided by the population, comes to the figure of only $1 to $2 per person per year. Obviously, not much health care can be obtained for this sum.

Ill health has its effect on productivity. Burdened by ailments of many sorts, individuals are not likely to be able to pull themselves and their families out of poverty by their own labor. In this way, the expanding

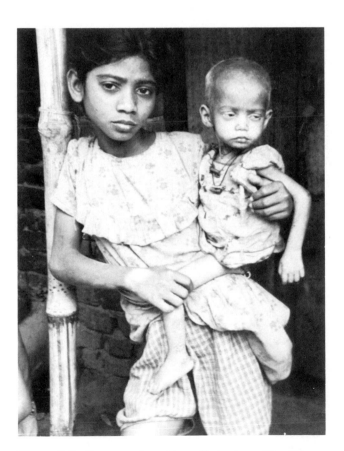

Figure 3.7 Severe malnutrition affects about 10 million children under age five in developing countries. (UNICEF/Abigail Heyman.)

populations of developing nations diminish the health resources available to each person and so reduce each person's potential to succeed.

Social and economic consequences in developing nations. Not only is little money available for health programs, but education is also likely to be slighted. The government is often unable to increase educational services at the same rate at which the population is growing (Figure 3.8). Despite actual increases in school enrollment, the number of illiterate people in the world rose between 1950 and 1965 because the population grew so fast. As increasing numbers of young people go out to look for jobs each year, their low educational levels and the scarcity of jobs force them into unskilled positions at best. People who are forced to accept these low-paying jobs have limited upward social mobility and very little opportunity to improve their economic lot even by very hard work. These factors lead to the development of two groups in a developing nation: one rich and well-fed, one poor and ill-nourished. The rich are likely to remain rich and privileged. The poor and the children of the poor have little hope to escape their fate.

At a national level, as population grows, the per capita income generally decreases. That is, the number of people to be divided into the Gross National Product is larger each year, so the result, which is the per capita income, is smaller. There are two ways of looking at this particular fact. Some experts feel that the lower per capita income decreases savings, cuts down on the amount of capital available for investment, and prevents people from buying manufactured goods, thus retarding industrialization. Others point out that in many undeveloped countries, the capital and most of the income is in the hands of a privileged few anyway. They argue that, unless population programs are combined with some form of redistribution of this wealth, population reduction among the poorer people will not help improve their economic status at all.

Environmental effects in undeveloped countries. Increased agricultural development needed to feed a growing population can cause increased soil erosion, contamination of soil and water with fertilizers and pesticides, and the destruction of wildlife habitats and areas of natural beauty. Erosion, pollution, loss of species, and loss of land—all result from unchecked population growth.

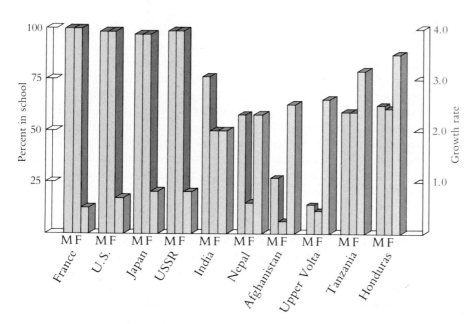

Figure 3.8 School enrollment in 10 selected countries compared to their growth rate. The graph shows the percent of girls and boys age 6–11 enrolled in school (in black) compared to the rate of natural increase for that country (in color). (Data from 1982 World Population Data Sheets, Population Reference Bureau, and UNESCO Statistical Yearbook, 1981.)

We catalog these events because they matter to us. Our perspective is influenced, however, by the mountain of plenty on which we stand. Well-fed, well-clothed, well-educated, we can afford the luxury of seeking quality in our environment. Our concern with environment in the developing countries has a certain hollowness to the people of these nations.

It is not hard to understand why wise land-use policies, pollution control, or protection of endangered wildlife are very low on the list of priorities in developing nations. Where children and parents are perpetually hungry, often sick, without adequate clothing—where there is little hope for people to better themselves—the immediate future of the environment does not really matter very much. Unless these other, primary needs are met, there will be little enthusiasm about seeking a quality environment.

Nonetheless, certain elements of the environment we seek are of long-term importance to people everywhere. Without good agricultural methods, such as proper irrigation techniques, careful use of pesticides, and control of erosion, the productivity of the soil will eventually be lost and food production will fall. Without adequate treatment of water supplies and of wastewaters, diseases will continue to ravage and weaken the population and periodic epidemics of chemical poisoning will occur. (See, for instance, the reference by Agarwal et al. on environmental problems in India.)

The list of long-term problems could continue. The point is that, even in developing nations where the most pressing needs are the most basic human ones—food, shelter, and good health—thought must be taken for the environment. Unless the environment remains healthy, it cannot long support a human population.

Resources and technology in developing countries. Since there is little advanced technology in developing countries (which is, after all, why they are called undeveloped), there is little demand for the resources that fuel an industrialized society. Thus, scarcity of resources is not a problem faced by developing countries. The one serious exception to this is materials used to produce fertilizers. Developing nations need large amounts of fertilizer to produce food; but fertilizers need nitrates (which are produced from natural gas), phosphates, and energy for their production. For this reason, high world oil prices affect not only the developed parts of the world, but also undeveloped countries such as India. Except for agricultural development, technology has little to offer undeveloped nations. "Modern technology has increasingly been directed toward doing better or more cheaply or in a greater variety of ways those things that most people in the poorer countries cannot afford to do at all."[5]

5 *Rapid Population Growth Consequences and Policy Implications*, Volume I, Summary and Recommendations, National Academy of Science. Baltimore: Johns Hopkins University Press, 1971, p. 31.

Consequences in Developed Countries

Resources and technology. In developed countries, by contrast, technology and the scarcity of resources are subjects of primary concern with respect to population growth. The energy crisis has made many people aware that we are running out of such resources as natural gas and oil. Shortages are also predicted by the turn of the century in many minerals, among them copper, tin, magnesium, and zinc.

Looking back in history, it is hard to find instances in which we have actually run out of some critical resource. What happens, commonly, is that scarcity causes a price rise that eventually forces a switch to some other resource. Thus, a scarcity of wood in England in past centuries led to the burning of coal. Whether this mechanism, which economists call "substitution," can be counted on to continue to work in the face of increasing population growth is not clear. For instance, will technology provide us with solar or fusion power before we run out of uranium, oil, and gas? We are at least gambling with the possibility of future resource shortages.

The illustration of the switch from wood to coal fires in England brings up another problem. By the 1900s, coal fires were responsible for a level of air pollution that made London almost uninhabitable at certain times of the year. Air pollution incidents, such as the London fog of 1952 that caused 4,000 deaths, were primarily due to pollutants derived from the burning of coal.

Pollution is one of the most serious problems we face in developed countries. We need more fertilizer and pesticides to grow ever increasing quantities of food. We need more petroleum to produce the gasoline needed to get increasing numbers of people to ever

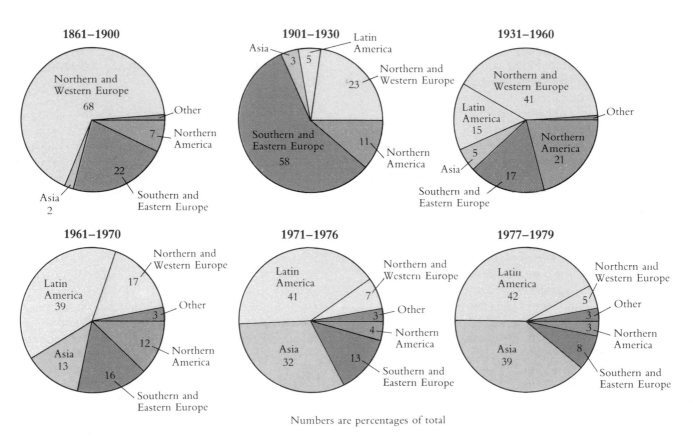

Numbers are percentages of total

Figure 3.9 U.S. Immigrants by Region of Birth, 1861–1979. (from: Gouvier, L. F., *Immigration and Its Impact on U.S. Society*, Population Trends and Public Policy Series #2, Population Reference Bureau, 1981.)

Figure 3.10 Crowded Campground in South Tyrol. In developed countries, rapid population growth leads to overcrowding of recreational and wilderness areas. In a real sense, some of our freedom of choice is lost. (Werner H. Muller/Peter Arnold, Inc.)

more distant places of work. This results in increasing levels of smog. More people need more housing and more farm products. Housing and farming use up forested lands and lead to increased soil erosion. More people generate more sewage and solid wastes, which either foul natural waters or demand technological solutions for their disposal. And more people consume more products whose manufacture generates chemical waste products. Although technology may be able to solve some of these problems, more and more of our money and effort go to maintaining the quality of life. Like the Red Queen in *Through the Looking Glass*, we run in order to stay in the same place.

Immigrant issues. Because immigration makes up such a large portion of U.S. population growth (33% in 1980; 30% in 1981), special problems may arise. Figure 3.9 illustrates how the regional origins of U.S. immigrants have changed over the years. Although the earlier European immigrants usually adopted American customs and language quickly, parts of the recent hispanic immigrant population desire to retain their own language and customs. This has raised fears of separatist difficulties, such as those between French-speaking Quebec and the rest of Canada. However, the Hispanic immigrants themselves come from a variety of cultural backgrounds (Puerto Rican, Cuban, Mexican), and so these fears may be unjustified.

Immigrants are concentrated in certain states. In 1980, only six states accounted for 70% of the 4.5 million permanent resident aliens (those intending permanent settlement in the U.S.). This concentration can impose a severe burden on those states for social services (schools, hospitals) to immigrants.

Limiting choices. As population grows, we limit not only our choices about spending society's money, but we also limit personal choices. For example, whole areas of the country (such as the Eastern Seaboard) are becoming urban. Recreational areas are threatened by throngs whose presence may even destroy the values of the area (Figure 3.10). Furthermore, as population increases, available open space decreases or becomes more costly. Wilderness turns into overcrowded parks, and parks degenerate into outdoor slums. Our lives become increasingly regulated by rules that maintain social order.

Scientists have studied animals under crowded conditions, and some of their results are worth noting. For instance, rats in crowded cages begin to lose their normal patterns of social behavior. They neglect their young and sometimes resort to cannibalism. Some rats become overly aggressive, while others withdraw from the community. Sexual behavior becomes abnormal. Monkeys also show some of this same behavior when they are crowded together.

In the wild, animal populations, too, occasionally reproduce very rapidly. The needs of the expanded population may exceed the food supply or living space available. In such cases, starvation, disease, and reduced fertility will lead to a decline in the animal population. Some scientists think the mass migrations of lemmings is a response to rapid population growth.

Can the results of these studies be applied to human populations? Probably they cannot be applied directly. Nevertheless, the catastrophes that occur in animal communities under stress from rapid growth and crowding warrant consideration.

Perhaps humans can survive under relatively crowded conditions. On the other hand, we may prefer not to see the changes such crowding will cause. If we wish to influence the future, we must make decisions about population growth. Otherwise the choice will have been made for us.

Limiting Population Growth

The Problem of Time Scale

Programs designed to reduce population growth have one very serious and built-in problem: it takes some 10–20 years before the results of any such program are seen. The reason is that fertility-reduction programs have an effect only on the number of babies currently being born; they can obviously have no effect on the group of children, age 0–15 years, already born. These children will grow to adulthood and marry and have children themselves during the next 10–20 years.

In most countries, the size of this group (0–15 years) is larger than that of any other age group in the population. If these people in the 0–15 age category only have enough children to reproduce themselves, they will still swell the size of their country's population. (See page 68 for more details.) Thus, most countries would continue to grow in population for a time even if birth-control programs immediately reduced the birth rate to about 2.1–2.3 children per couple. (This is the birth rate that provides for each person to reproduce himself.[6]) Even at a figure of 2 children per family, the United States population will grow in size

until the middle of the 21st century, especially if net immigration continues at high rates. Of course, at a figure of 3 children per family, growth would be even faster.

The fact that population programs take such a long time to make their effects felt is a great handicap to effective action. It takes five to six years before decreased school enrollments are seen, 15 years before a reduction in the labor force occurs, and 15–19 years before stabilization begins in the amount of food needed. A government may thus seem to be spending a great deal of money, for many years, with very little obvious effect. Because of this time gap, it is important to try to separate birth-control programs from politics and politicians. Politicians and political movements or ideologies have a way of changing or going in and out of favor in less than 10 or 20 years.

In China, the government has considered this lag time in population-reduction programs and has decided that even the 1981 fertility rate of 2.3 children per couple would not slow population growth quickly enough. So the government has adopted a "one couple, one child" policy. Economic incentives, such as higher pensions, larger personal land allotments, health care, and schooling benefits, are offered to families who pledge to have only one child. In addition, strong social pressure is brought to bear on couples to formally sign a one-child pledge. Some experts are unsure whether this policy will work in rural areas where children are a source of labor. There may also be problems with minority groups who are suspicious that it is a way of further decreasing minority numbers.

Methods of Limiting Population Growth

Birth control information. When a government wishes to limit its population growth, usually the first policy adopted is providing people with information and materials for birth control. In many cases, abortions are made legal and thus easier for the poor to obtain. Although several religious organizations object to both abortion and to the distribution of birth control information, there is usually also a great deal of popular support for this type of policy. In most situations, people will at least tacitly agree that couples should be able to limit their family size if they wish to do so. For instance, in mostly Roman Catholic Co-

---

6 The number is not exactly two because some people die before reaching reproductive age. The slightly-greater-than-two figure allows for this mortality.

lombia, despite initial opposition by the Church, wide government distribution of contraceptives has helped reduce the birth rate from 3.4% to 2%.

A policy of providing access to information and materials on birth control can be justified on grounds of humanitarianism. Fewer children will, in general, be better fed and cared for, since a family's resources will not be spread too thin. These children may even be healthier, since repeated, closely spaced pregnancies have been shown to be harmful both to mothers and to babies. Furthermore, if only wanted pregnancies occur, mothers have no need to resort to abortions, which, like any operation, carry some risk. For this reason, many anti-abortion groups support contraceptive programs. The cost to society of unwanted children who may become wards of the state in one way or another is also reduced.

However, even if the government is willing to provide birth control services, there is no guarantee that people will see a need to use them. As previously mentioned, there are a number of reasons why couples may choose to have more children than is considered ideal for society as a whole. A massive program of public education and contraceptive distribution in Pakistan from 1965 to 1980 failed almost completely due to social factors. The failure was partly administrative; information about the program failed to reach a large part of the population. But, in addition, a demand for contraceptives was not generated in this society.

If birth control information and services prove ineffective in slowing the expansion of a nation's population, the type of policy to be implemented next is not at all clear. A great deal of controversy has arisen, in fact, over what types of policies should be adopted. Three general types of policies have been suggested.

Coercive measures. The first group of policies involves coercive measures, in which the government takes active control of individual reproduction. In such a situation, a license might be issued to allow a couple to have a child. Forced sterilization programs might be carried out, as they were in parts of India in the 1970s.

There are a number of valid objections to these policies. Who would decide, and on what basis, who deserves to have children? The tenets of individual freedom would be lost in such programs. It should also be pointed out that, at this time, we do not have the technical capability to carry out such a policy.

There is no magic chemical that could be put in water to prevent fertility, without other serious effects. Not enough doctors and nurses are available for forced, large-scale contraceptive or sterilization procedures. From a purely practical standpoint, family limitation programs, if they are to succeed, must be made to seem desirable to everyone. (See Controversy 3.1 for India's experience.)

Rewards and incentives. The second group of policies offers various economic rewards for not having children or metes out penalties to those who do have children. One suggestion in the United States was to remove welfare benefits or income tax deductions for dependent children after the second child. Other suggestions have included fines for having more than a specified number of children, and educational or other bonuses for families with few children. The main objection to these policies is that the extra children, who had no control over whether or not they were born, suffer along with the parents. Other children in the family likewise suffer.

Despite these drawbacks, such policies are, in fact, being followed in several countries and are planned in others, such as Bangladesh. State employees in certain Indian states are not eligible for housing loans and land grants if they have more than three children. Singapore levies increasing fines for the birth of a third, fourth, fifth, or sixth child. Maternity leave is not allowed for the birth of a third or subsequent child, and hospital fees rise with the number of children in a family. On the other hand, sterilization after two children is rewarded with education and employment for the children and extra vacation time for the parent. Foreigners who wish to marry Singapore nationals must agree to be sterilized after the birth of their second child.

Methods that preserve freedom of choice. The third group of policies do not put burdens on innocent children. These policies also allow for some preservation of an individual's freedom of choice, and they often have some other value to society in addition to the reduction in population growth.

In many developing countries, children may be sent to work and so represent to the parents the possibility of additional family income. The passage and enforcement of laws prohibiting child labor can make additional children less desirable by removing their income potential. Parents must then feed and clothe the

Family Planning in India—
A Lesson on the Meaning of Freedom

Among the nations of the developing world, India has had one of the fastest-growing populations. At 300 million just after World War II, India doubled its population by the mid-1970s. Efforts to limit population have been underway there since the early 1950s. In fact, India was the first nation to create a national program to limit population through family planning. Even into the 1970s, however, the results were barely noticeable. In 1976, it was estimated that only 17.5 million couples out of 103 million in the reproductive age groups were using contraceptives.

India's program up to that time was traditional: sterilization or birth control methods and devices were made available to those who asked for them. Although no one was forced to seek sterilization, incentives were offered. For a time, a man could obtain cash or a transistor radio (a coveted article) in return for having a vasectomy.

In April, 1976, a new policy was adopted emphasizing sterilization by vasectomy and, in many ways, compelling men to submit to the operation. Mrs. Indira Gandhi, prime minister of India, was quoted as saying, "We must act decisively and bring down the birth rate. . . We should not hesitate to take steps which might be described as drastic. Some personal rights have to be kept in abeyance for the human rights of the nation: the right to live, the right to progress." The government's goal: a reduction in the birth rate from 35 per thousand to 25 per thousand by 1984. It is of use to note that the Indian government had declared a state of emergency in June, 1975, and that a number of civil rights, including free speech, had been either suspended or decreased since that time. The intensive program of sterilization was begun in this political climate. Although the government did not mount a national campaign of compulsory sterilization, individual states were encouraged to do so.

The State of Maharashtra passed laws calling for compulsory sterilization for the father of three living children and compulsory abortion of a pregnancy leading to a fourth child. Incentive payments to those who submitted to vasectomies were a part of the law, as were payments to informers. In Delhi, the capital, and in the states of Punjab and Haryana, laws were passed withdrawing vital government benefits and services from married men with two children who did not submit to vasectomy. Loss of subsidized housing, loss of free medical care and loans, and even loss of employment were possible if a man did not comply with the sterilization laws.

Widespread abuses, in which people—especially poor people—were compelled to submit, were reported as the state governments attempted to fill sterilization quotas. Riots broke out in Northern India over the issue. In the final months of Mrs. Gandhi's rule, sterilizations numbered a million per month. Then, in early 1977, Mrs. Gandhi's government fell, as the voters sent her and her Congress Party from office. It is not clear if her government fell because of her near-dictatorial rule and her suspension of civil liberties or because of the highly visible, highly controversial program of sterilization she promoted. Many regarded the sterilization program as the largest factor in her defeat. The government that succeeded hers, led by Prime Minister Desai, abandoned the use of force and punishment for failure to be sterilized, presumably because the issue remained highly charged politically. In 1977–78, sterilizations fell to 11% of the previous year's total. Although she was reelected in January 1980, Mrs. Gandhi has generally kept a low profile on population control efforts.

This episode raises some basic questions about birth control policies. Mr. Kaval Gulhati asks, in an article in *Science*:[*]

> Should they stand by and wait for economic development and family planning programs to motivate contraception? Or should they take the destiny of the people in their hands, and force a fertility decline?

Mr. Gulhati asked this question before Mrs. Gandhi's government was rejected. In light of her government's defeat, the question might be expanded to ask:

> Can the government of a free people successfully undertake a compulsory program of population control and itself survive?

---

[*] K. Gulhati, "Compulsory Sterilization: The Change in India's Population Policy," *Science,* **195** (25 March 1977), 1300.

child for a longer time, during which he cannot work. Such laws benefit children by allowing them time for schooling and, depending on what work they would have done, possibly improving their health (Figure 3.11). The state of Kerala in India has a per capita income below the average for all of India, yet it devotes 39% of its budget to education and 16% to health and family planning services. Almost all children in Kerala attend at least the primary school grades. Birth rates in Kerala have declined from 37 per thousand in 1966 to 25 per thousand by 1978.

Recall from earlier discussion that children represent security for parents in old age. If old age pensions, such as the Social Security program of the United States, are created in developing nations, parents may be able to give up the idea that they need many children. Bangladesh is planning Social Security benefits but they are to be awarded only to parents of two or fewer children, in order to further emphasize the desirability of small families.

Another reason many parents have more than two children is to ensure that they will be able to raise at least two children to adulthood. The region with the world's highest birth rate also has the world's highest death rate.[7] Continuing reductions in infant and child diseases should eventually lead to some reduction in births.

Education, besides helping the children themselves and society as a whole, has several benefits related to population control. Educated people tend to marry later and have fewer children. This is related to the fact that education allows people to move upward in economic and social status. Upward mobility is hindered by large numbers of children to feed, clothe, and care for. Educated women tend to limit their families to allow themselves the satisfaction of a career and additional family income.

Improvement in the status of women may have the same effect of limiting population growth. If the need to care for young children prevents mothers from taking advantage of new or better job openings, they seem to limit their family size to take advantage of those opportunities. In the United States, the educational achievements of women, as well as men, have been steadily growing. In addition, the percentage of women who work has been increasing. These factors

Figure 3.11 Young, nimble-fingered children toiled long hours at low pay in the 19th-century factories and mines of today's more developed countries. (U.S. Library of Congress; photo by Lewis Hines.)

have probably contributed to the downward trend in the U.S. birth rate. On the other hand, in parts of Africa and Asia where many family members can help care for young children, or where uneducated and unskilled babysitters are easy to find, an increase in the number of women in the work force does not have much influence on the birth rate.

Another way to influence population growth is to control housing availability. In addition, the government can require a period of military or other national service for all young people. This action tends to increase the age at which marriage occurs, reducing the span of years in which a couple can conceive.

These population-control methods all allow individuals some freedom of choice on the number of children they plan to have. They open other areas of satisfaction and they also make people aware of the additional costs that children bring. In 1982, it was estimated that each child costs a middle-income American family $85,000 in direct costs and another $55,000 in lost opportunity costs (mostly income the mother forgoes while taking care of the child). These methods

---

7 The region is Africa, which in 1982 had a birth rate of 46 and a death rate near 17 per 1000 per year.

The Ethics of Limiting Family Size

. . . Any choice and decision with regard to the size of the family must irrevocably rest with the family itself. . .

**United Nations Universal Declaration
of Human Rights (1967)**

Freedom to breed will bring ruin to all.

Garrett Hardin

Many difficult decisions are involved in the control of population growth. One of the most difficult is whether couples should have the complete freedom to decide how many children they wish to have. That is, should governments set and enforce upper limits of family size?

In 1967, the United Nations passed the Universal Declaration of Human Rights, which stated:

The Universal Declaration of Human Rights describes the family as the natural and fundamental unit of society. It follows that any choice and decision with regard to the size of the family must irrevocably rest with the family itself, and cannot be made by anyone else.

In the past few years, however, a few nations have decided that this concept is unworkable. Thus, in Bangladesh, Singapore, China, and India, governments have begun to use both economic threats and bonuses designed to limit population size.

Demographer Garrett Hardin believes that the decision to limit family size cannot be left to the individual. He says that such a decision is similar to one which faced cattle owners in 18th-century England. At that time, some pastureland was held in common. Hardin argues that there was a great advantage to a herdsman to add more cattle to his herd, if he could, since the pastureland was free to all. If everyone did that, however, the pasture would be overgrazed and thus ruined. Hardin argues that, in a sense, our welfare-oriented society is like a commons. Individuals can choose to use up more than a fair share of resources,

both social and environmental, by having more than their fair share of children.

Hardin writes:*

Perhaps the simplest summary of this analysis of man's population problems is this: the commons, if justifiable at all, is justifiable only under conditions of low-population density. As the human population has increased, the commons has had to be abandoned in one aspect after another.

First we abandoned the commons in food gathering, enclosing farm land and restricting pastures and hunting and fishing areas. These restrictions are still not complete throughout the world.

Somewhat later we saw that the commons as a place for waste disposal would also have to be abandoned. Restrictions on the disposal of domestic sewage are widely accepted in the Western world; we are still struggling to close the commons to pollution by automobiles, factories, insecticide sprayers, fertilizing operations, and atomic energy installations.

. . . I believe it was Hegel who said, "Freedom is the recognition of necessity."

The most important aspect of necessity that we must now recognize, is the necessity of abandoning the commons in breeding. No technical solution can rescue us from the misery of overpopulation. Freedom to breed will bring ruin to all. At the moment, to avoid hard decisions many of us are tempted to propagandize for conscience and responsible parenthood. The temptation must be resisted, because an appeal to independently acting consciences selects for the disappearance of all conscience in the long run, and an increase in anxiety in the short.

The only way we can preserve and nurture other and more precious freedoms is by relinquishing the freedom to breed, and that very soon.

Do you agree with Hardin? Should the decision on family size be made by each couple, or should governments have a voice in it? If you feel that population growth should be slowed, should appeals to conscience be tried first? What did Hardin mean when he said "an appeal to independently acting consciences selects for the disappearance of all conscience in the long run"? Suppose such appeals don't work—can you think of economic measures, such as the use of taxes, that could influence family size? Suppose neither appeals nor economic measures are effective—are penalties justifiable?

★ Garrett Hardin, "The Tragedy of the Commons," *Science,* **162** (13 December 1968), 1243.

---

provide other means of achieving the same benefits obtained in large families. Through such mechanisms, population growth might be reduced without coercion. Evidence shows that, under the right circumstances, people can become aware that having more than two children is not in their own best interests. Countries offering the greatest possibility of upward movement in economic and social status have shown the most dramatic decreases in birth rate. Thus, the ideal of bringing population under control may be achieved by means that, at the same time, help people reach a satisfying standard of living.

The main problem in devising ways of reducing population growth is, of course, the fact that the issue is not some abstract scientific problem solvable by dollars and technology. Instead, it is a problem that touches people directly. We can program computers to provide projections of population, but only people's behavior, plans, and aspirations can determine what the future will be. People's needs and feelings affect how they react to incentives and plans. People also exert a degree of control over policies that affect them. In the United States, for instance, population control policies must be seen as desirable to the majority of people before they have any chance of being put into action.

Outlook for Population Growth in the Future

Late in the 1970s, it became clear that population growth rates in developing countries were decreasing slightly. Is this a sign that developing countries are beginning to undergo a demographic transition in which birth rates will come into line with lowered death rates? In fact, two trends are occurring. First, there is a small but real decrease in world population growth rates. Second, there is a general and important decline in the fertility of women in the less developed countries. Demographers define the **fertility rate** for a nation as the number of live births a woman would be expected to have during her childbearing years. The decline in fertility, reflected in the observed slower rates of growth, is a dramatic signal of possible changes on the horizon.

According to the U.S. Bureau of the Census, the rate of world population growth fell from 1.98% per year in the period 1965–1970 to 1.88% in the period 1975–1977 and to 1.7% by 1982. While the decline is small, it is significant because it represents a reversal in the pattern of growth seen in recent history—populations now seem to be growing more slowly. It gives hope that future rates of growth will fall still further as family planning becomes an integral part of total planning in the less developed nations. Exceptional declines in growth rate have occurred in Sri Lanka, the Philippines, the Republic of Korea, Thailand, and Colombia. Only African countries seem to be running counter to the trend of declining growth rates.

The second trend, a decline in fertility, is the underlying explanation for the decrease in growth rates. On a worldwide basis, the average total fertility rate fell from 4.6 in 1968 to 3.9 live births per woman in 1982. Fully 80% of the world's population is in countries with declining fertility rates.

Table 3.2 Decline in Fertility in Some Less Developed Nations[a]

Country	Fertility (live births per woman)		Percent decline	Population in 1982 (millions)
	1968	1975–1980		
Bangladesh	6.98	6.3	10	93.2
Colombia	6.54	3.8	42	25.6
India	5.67	5.3	8	713.8
Indonesia	6.46	4.7	27	151.3
Mexico	6.59	4.8	27	71.3
Pakistan	6.84	6.3	8	93.0
Peoples Republic of China	4.20	2.8	33	1000.0
Thailand	5.86	3.7	27	49.8

a Data from 1982 World Population Data Sheet, Population Reference Bureau.

Table 3.3 Mean Number of Children Desired in Selected Countries

Country	Mean number desired
Kenya	6.8
Jordan	6.3
Sierra Leone	6.1
Paraguay	5.1
Costa Rica	4.7
Dominican Republic	4.6
Guyana	4.6
Mexico	4.4
Philippines	4.4
Malaysia	4.4
Panama	4.2
Pakistan	4.2
Fiji	4.2
Venezuela	4.2
Bangladesh	4.1
Colombia	4.1
Indonesia	4.1
Jamaica	4.0
Nepal	3.9
Peru	3.8
Sri Lanka	3.8
Thailand	3.7
Haiti	3.6
Korea, Rep. of	3.2
New Zealand	3.0
Turkey	3.0
Spain	2.8
Taiwan	2.8
Great Britain	2.6
Czechoslovakia	2.4
Belgium	2.3
Japan	2.2
Hungary	2.1

(From Mary M. Kent and Ann Larson, *Family Size Preferences: Evidence from the World Fertility Surveys*, Population Reference Bureau, Washington, D.C., 1982.)

In a study of 113 less developed countries,[8] 95 nations had seen fertility declines from 1968 to 1975. Table 3.2 lists the decline in fertility in a number of the larger less developed nations. Using the new estimates of fertility, the world's population in the year 2000 may be 6 billion rather than the 6.3 billion predicted by the United Nations only a few years earlier.

Statistics such as these provide genuine hope that world population is coming under control. This by no means erases the problem of population growth. On the contrary, it appears to show that rapid population growth is a problem that more effort and money can affect. Studies from the cooperative international effort known as the World Fertility Studies (WFS) show that the mean number of children wanted by women in many developing nations is still well above replacement level (2.1 to 2.5, depending on a country's death rate). Table 3.3 shows some of these family size preferences.

A close look at recent declines in fertility rates in developing countries suggests that two equally important factors were at work. One was an improvement in economic conditions within some of the countries. In many ways, as we have seen, better economic and social conditions cause parents to feel that it is in their own best interest to have fewer children. The second important factor in many countries was a strong family planning program. In countries such as Mexico and

8 "World Population 1977: Recent Demographic Estimates for the Countries and Regions of the World," available from U.S. Bureau of the Census, Population Division, Washington, D.C. 20333, January 1979.

Population Growth in Africa

No programme will work in Africa if it is imported.

Professor Mungai, University of Nairobi

In 1982, the World Fertility Study noted fertility declines of at least one child per woman for the undeveloped countries studied in Asia, Latin America, and the Middle East. Some of these declines even occurred in countries where population reduction was considered very difficult—for instance, the nearly one child per woman decline in Pakistan. But no decline was found in Africa.

Africa as a region has the highest birth rate and also the highest death rate in the world. Moreover, Western concerns with population reduction are not generally shared by African government leaders. Louis Ochero wrote in the Population Reference Bureau's *Newsletter* of October 1981:

> The advocacy of family planning as synonymous with or contributing to population control, considered by the advocates as a basic requirement for economic development, was almost universally rejected in black Africa. In some areas these were viewed as thinly veiled efforts at genocide, or at best an unrealistic approach to the real issue of development.
>
> The obstacle to the spread of family planning practices in Africa is not unconnected with their foreign origin. Most of the personnel propagating family planning in Africa are either expatriate or foreign-financed. This has generated a suspicion that family planning is foreign-supported for political and even racial reasons.

As Professor Mungai of the School of Anatomy, University of Nairobi, warned in his address to a seminar in 1975:

> "No programme will work in Africa if it is imported. . ."

There is a great deal more to the control of population growth through reducing the birth rate than only the conception control which is dispensed at the family planning clinics. Experience has shown that the birth rate of the traditional societies often falls effectively only under the impact of urban industrial development, education, improvement of health and the lowering of the death rate. Propaganda for birth control or government-sponsored family planning programs does not seem to have played an important role in lowering the birth rate of the developed nations either. The unjustifiably exaggerated emphasis on family planning only and its identification with population growth control has annoyed and irritated the intelligentsia in the developing countries and has aroused opposition to family planning. The less intellectual fail to see the reasons for "downgrading" their existing traditional methods of child spacing.

The most important factors which have caused the lowering of the birth rate in the advanced countries have been the same as have been responsible for the raising of the standard of living of the people living there. Industrialization, urbanization, rural development, the spread of education, better health services, raising the age of marriage for women can be specially mentioned. No amount of exhortation for population control can take the place of these factors. There is already a strong tendency in developing countries for women to get married late. It is expected that this trend will continue and will also become more widely accepted. Along with this, the rising status of women will be a major factor affecting population growth.

What do you think Mr. Ochero meant when he said there is a suspicion in Africa that family planning is "foreign-supported for political and even racial reasons"? Do you think this is an accurate perception of the policies of developed nations toward family planning aid to undeveloped nations?

Do you think wealthier countries should increase their aid to poorer countries, such as those in Africa whose populations are growing rapidly? If so, what kind of aid should be given?

Do you think developing countries will lower their fertility rates in the same way the developed countries did a century ago?

Colombia, as well as the Indian state of Kerala, a strong commitment to population control on the part of the government, backed up with a commitment of money and personnel, appears to have been a major factor in the success of their programs.

Zero population growth. Twenty-four countries have actually reached a fertility rate of replacement-level or lower; that is, 2.1 births per woman or lower.[9] In some of these countries, such as Sweden,

the population has reached a stage where births are equal to or less than deaths each year. These countries face new problems—those brought about by zero population growth (ZPG). Perhaps the most serious problem is the increasing proportion of older people in the population who must be supported by an increasingly smaller labor force. (See the following section on demography for an explanation of why the proportion of elderly people is increasing in ZPG societies.)

Sweden has a comprehensive welfare system. Costs for support of the elderly have been rising much more rapidly than other welfare costs. How Sweden will deal with this as well as other problems of ZPG is of interest to countries such as the U.S., which, at present fertility levels, is beginning to face these same problems.

---

9 Australia, Austria, Belgium, Canada, Cyprus, Denmark, Federal Republic of Germany, Finland, France, German Democratic Republic, Hungary, Italy, Japan, Luxembourg, Malta, Netherlands, Norway, Portugal, Singapore, Sweden, Switzerland, United Kingdom, United States, Yugoslavia.

❧ Demography—The Study of Populations

To understand how we can influence what will happen to a population in the future, let us take note of the people in two age ranges: those 0–15 years, who are not yet old enough to have children, and those 15–35, who are at the childbearing age. We can see that knowing the size of the total population is not as important as knowing how many people are in these younger age groups, since people in the older age groups are no longer likely to have children. These older people are thus not able to affect the future size of a country's population.

Because the 0–15 age group in most developing nations is larger than any other age group, the country's total population will grow even if people now in this group only reproduce themselves on a one-to-one basis. In order to have the total population stay at the same level as it is today, parents have to average fewer than two children per couple. This is not a likely situation. Most people feel that they have at least the right to reproduce themselves. China's attempt to foster a one-child per couple policy runs directly counter to these feelings; how successful it will be remains to be seen.

Because changes in the near future are the result of the age structure existing today, population programs started right now will have a 40–60 year lag time before they can result in stable populations. This is true

even if the programs are immediately and successfully implemented. It has been calculated that the United States' population will take 60–70 years to stabilize and reach the state of zero population growth, even though the fertility rate is now at slightly less than two children per woman. Actual experience has shown that, in most countries, population programs are not immediately successful, but take a number of years to become accepted.

Figure 3.12 shows the age structure of the U.S. population from 1900 to 2000 as a series of population profiles. Note how the large group of babies born in 1955–1959 forms a bulge in the population profile. The females in this large age group will be in their childbearing years through the 1990s. Even if their fertility rate remains at less than replacement level (1.8 for instance), the large actual number of children born to these women will cause a growth in the U.S. population.

Certain features of populations show up very well on bar graphs. For example, many developing countries show a characteristic "pinched profile." In these cases, a decreased death rate in younger portions of the population, brought about by better health care, has swollen the younger age groups compared to the older ones. The profile has a wide base and a very narrow top (Figures 3.13 and 3.14). Compare the profile in

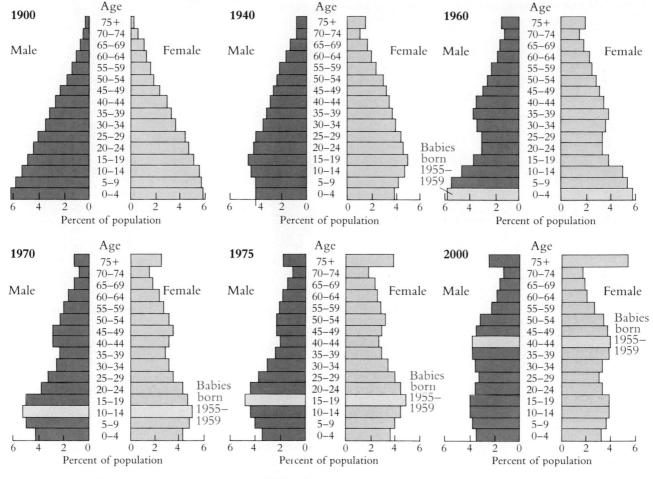

Figure 3.12 The population of the United States at various times is shown here in the form of bar graphs or population profiles. From these graphs you can see how a large group of babies born during one period (1955–1959) will swell the size of other age groups in later years. (Source: Population Reference Bureau, 1976.)

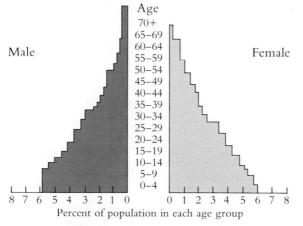

Figure 3.13 Population of India in 1951. (Source: W. S. Thompson and D. T. Lewis, *Population Problems*, 5th ed. New York: McGraw-Hill, 1965.)

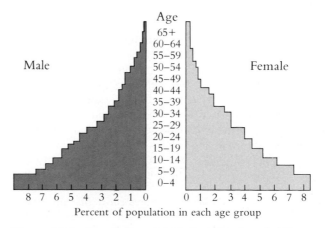

Figure 3.14 Population of India in 1970 ("pinched" profile). (Source: *Population Bulletin*, Population Reference Bureau, 1970.)

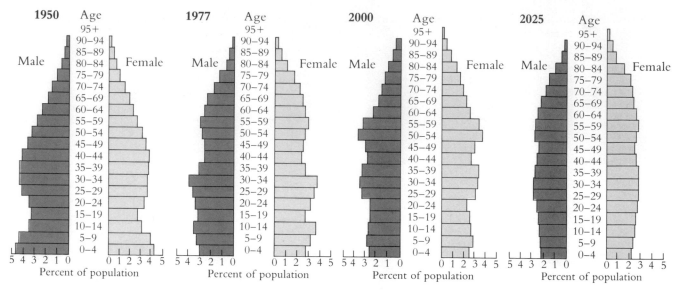

Figure 3.15 Population of Sweden, 1950–2025 (projected). (Source: Population Reference Bureau, *Population Bulletin*, No. 35, 2 June 1980, page 28.)

Figure 3.14 to the one of India in 1951 (Figure 3.13). Note how the younger age groups in 1970 are very large compared to the older groups. This is the result of better infant and child survival as health care improved.

Figure 3.15 shows a series of profiles for Sweden from 1950 to 2025 (projected). Notice the obelisk shape of the graph, which is characteristic of countries that have achieved replacement-level fertility. Distribution among the age classes is more uniform. Notice also how the proportion of the population in the older age groups increases with time, while the proportion of people in the labor force shows a relative decrease. This can lead to serious social and economic problems.

Questions

1. What is meant by the terms **birth rate, death rate, rate of natural increase, fertility rate**?

2. Although the United States fertility rate has reached replacement level or lower, the U.S. population is still growing. At current birth, death, and immigration rates, the population will double in about 90 years. Explain why the U.S. population will continue to grow.

3. Let us assume that a doubling of the U.S. population is unacceptable. Outline a program of population growth reduction for the U.S. Explain what methods you consider reasonable and workable in this country.

4. In the latest U.S. census, figures show that migration from rural areas to cities has begun to reverse for the first time in this country's history. However, three out of four Americans still live in a large city or its suburbs. What do you see as the advantages and disadvantages of living in a large city? What are the advantages and disadvantages of living in rural areas? Do you feel that, on the whole, the world trend toward increasing urbanization is good, bad, or inconsequential?

5. What is meant by **demographic transition**? Why do you think this transition has occurred in developed nations but not yet in undeveloped ones?

6. List the problems caused by rapid population growth in developing nations. Compare them to the problems caused by rapid population growth in developed nations.

7. Some experts have called high fertility a result of poverty as well as a cause of poverty. What does this mean?

8. Can you support the proposition that developing countries should spend some of their desperately scarce funds on protecting the environment?

9. Population reduction policies that penalize families with more than a certain number of children can be viewed as unfair to innocent children. Are policies that reward small families similarly unfair? Explain, using examples such as housing loans, employment and education bonuses, old age social security, etc.
10. How can economic development affect population growth?
11. Why is the status of women (education, employment opportunities) an important factor in fertility levels?
12. How many children do you want to have? What advantages do you see to having children in the U.S. today? What disadvantages do you envision? If you were a poor farmer living in Bangladesh, what advantages and disadvantages might there be? How many children would you want to have if you lived in Bangladesh?
13. What major problem does a country face when it has stopped growing?

Further Reading

Bachrach, C. A., "Old Age Isolation and Low Fertility," *Population Bulletin,* **7** (1), January 1979.

Once we achieve a nongrowing population, what changes can we expect? Will there be large economic and social changes? This article examines a future that we should, perhaps, prepare for now.

Day, L. H., "What Will a ZPG Society Be Like?" *Population Bulletin,* **33** (3), Population Reference Bureau (June 1978) Also, "Sweden Faces Zero Population Growth," *Population Bulletin,* **35** (2), June 1980.

Intercom, *The International Newsletter in Population,* Population Reference Bureau Inc., Washington, D.C.

The Population Reference Bureau, 1337 Connecticut Avenue, N.W., Washington, D.C. 20036, publishes a monthly newsletter as well as many position papers, teaching aids, wall posters, and population data sheets. Memberships, which entitle the owner to almost all the publications (a real wealth of information), are available to teachers and students at reduced rates. The writing is clear and at a level intended for the general public.

Revelle, R., A. Khosla, and M. Vinovskis, *The Survival Equation.* Boston: Houghton Mifflin Co., 1971.

Revelle presents a thoughtful, humanitarian point of view about rapid population growth and possible solutions to the problems it causes. The book is clearly written and organized. Well worth reading for the arguments against coercion in birth control programs and the arguments that, under the right circumstances, all peoples can see that control of rapid population growth is in their best interests.

"U.S. Population: Where We Are; Where We're Going," Population Reference Bureau, Vol. 37, No. 2, June 1982.

This volume has an especially good discussion of immigration and its effects on U.S. population growth and composition.

"World Population and Fertility Planning Technologies: The Next 20 Years," Congress of the United States, Office of Technology Assessment, 1981.

This report from the OTA examines the various available contraceptive methods as well as prospects for new methods.

References

Agarwal, A., et al., *The State of India's Environment,* Washington, D.C.: International Institute for Environment and Development, 1982.

Canadian Census Data: Available from Publications Distribution, Statistics Canada, Ottawa K1A OT6.

Curtin, L. B., "Status of Women: A Comparative Analysis of Twenty Developing Countries," Reports on World Fertility Survey, No. 5, Population Reference Bureau, June 1982.

Dumond, D. E., "The Limitation of Human Population: A Natural History," *Science,* **187** (28 February 1975), 713.

Faucett, J. T., et al., "The Value of Children in Asia and the United States: Comparative Perspectives," Papers of the East–West Population Institute, No. 38, July 1974.

Lightbourne, R., et al., "The World Fertility Survey: Charting Global Childbearing," Population Reference Bureau, Vol. 37, No. 1, March 1982.

Robinson, W. C., et al., "The Family Planning Program in Pakistan: What Went Wrong?" International Family Planning Perspectives, Vol. 7, No. 3, September 1981.

Visaria, P., and L. Visaria, "India's Population, Second and Growing," Population Reference Bureau, Vol. 36, No. 4, October 1981.

Ware, Helen, "The Economic Value of Children in Asia and Africa: Comparative Perspective," Papers of the East–West Population Institute, No. 50, April 1978.

This paper (and others) are available in single copies from East–West Population Institute, East–West Road, Honolulu, Hawaii 96822.

CHAPTER FOUR

Protecting Wildlife Resources

Is a Little Fish Worth More than a Big Dam?

The Value of Species
*Practical Value in Medicine and Agriculture/Value as Part of
Food Webs and the Gene Pool/Intrinsic Value of Species*

How Species Become Endangered
*Hunting/Habitat Alteration/Competition with Introduced
Species: The Special Case of Islands/Pesticides and Air
Pollution/Collection of Plants*

Protecting Wildlife Resources
*Wildlife Protection Laws/Wildlife Management
Techniques/Preserving Endangered Species Worldwide*

**Endangered Ocean Mammals versus
Endangered Native Cultures**

CONTROVERSIES:

4.1: *The Most Important Environmental Problem:
Extinction of Species*

4.2: *Do We Need Flexibility in Laws Protecting
Endangered Species?*

4.3: *Endangered Species versus Human Health
Benefits*

4.4: *The Ethics of Hunting Whales*

Is a Little Fish Worth More than a Big Dam?

In the free-running waters of the Little Tennessee River lived a tiny fish called the snail darter (Figure 4.1). When this small member of the perch family was first discovered in 1973, virtually all the snail darters in the world lived in this one place. In 1966, seven years before anyone even knew the snail darter existed, Congress had authorized the Tennessee Valley Authority (TVA) to build the Tellico Dam and reservoir across the Little Tennessee River. Builders had half-completed the dam before the snail darter was discovered. Since the snail darter cannot breed in the still waters of a reservoir (it needs free-running water), completion of the 116-million-dollar Tellico Dam threatened to destroy the entire newly discovered population at a blow.

In the same year the snail darter was discovered, Congress passed the Endangered Species Act. An **endangered species** has so few living members that it is in danger of becoming extinct in the near future. The act states, in part, that actions of federal government agencies may not "jeopardize the continued existence of endangered species and threatened species or result in the destruction or modification of habitat of such species which is determined to be critical."

The Tellico Dam was three-quarters finished in 1975 when the Secretary of the Interior listed the snail darter as an endangered species. Clearly, by destroying the breeding grounds of the snail darter, the Tellico Dam, if completed, would violate the Endangered Species Act. Several environmental groups sued to have construction stopped and the case went all the way to the Supreme Court. In 1978, with the dam 90% complete, the Court ruled that the project did

Figure 4.1 Snail Darter. This endangered fish stood in the way of completion of the $116-million Tellico Dam. The controversy over the fish and the dam led the Tennessee Valley Authority to rethink the whole idea of placing a dam on the Little Tennessee River, which provided one of the few remaining stretches of good cold-water fishing in the region. The controversy also caused Congress to rethink the Endangered Species Act, which protects even tiny fish from large construction projects. (For the final results of the controversy, see p. 81.) (NYT Pictures.)

indeed violate the Endangered Species Act and so must be either stopped or changed.

Was the protection of a small population of 3-inch (7.5 cm) fish really what Congress had in mind when it passed the Endangered Species Act? As one writer put it: "Undoubtedly many members of Congress were thinking thoughts of brown-eyed creatures and soaring winged things when they cast their vote, and now are finding themselves confronted with a Pandora's box containing infinite numbers of creeping things they never dreamed existed."[1]

---

1 C. Holden, "Endangered Species: Review of Law Triggered by Tellico Impasse," *Science*, **196** (27 June 1977), 1427.

What is the value of a species? Why should we attempt to save species from extinction? Can we say that some species are more worth saving than others?

Scientists estimate that probably about 10 million species exist in the world, but ecologists have so far discovered and described only some one and one-half million. However, the discovery of new species is becoming a race against extinction. In prehistoric times, one species was lost perhaps every 1000 years. Between 1600 and 1950, the rate increased to around one species lost every 10 years. Today, one species becomes extinct every year. By the end of this century as many as one million species may be lost, most in the tropical forests.

The Value of Species

Practical Value in Medicine and Agriculture

Lost species can be viewed as lost opportunities. Animals and plants provide us with drugs, with foods, and with raw materials for industry. Twenty-five percent of prescription drugs dispensed in the U.S. contain plant extracts that cannot be synthesized. These drugs include the tranquilizer reserpine as well as a variety of antibiotics, pain killers, and drugs used to treat heart disease and high blood pressure. Vincristine, a drug extracted from a tropical periwinkle, is used in the successful treatment of Hodgkins Disease, a cancer that strikes 5000–6000 Americans each year. However, only 5000 plant species have been investigated for useful drugs. Scientists feel that another 5000 useable drugs could be found among the 500,000 species of plants believed to grow in the world.

Agricultural researchers have found uses for a number of organisms. For instance, an important technique in farming is the use of biological controls, which involves using one species to prevent another from harming crops. As an example, certain wasps can successfully prevent the sugar-cane borer from destroying whole fields of cane. Another technique of modern agriculture involves the cross-breeding of various plant species to develop crops with higher yields or resistance to disease, drought, or heat. Every time we allow a plant or animal to become extinct, we run the risk of losing a possibly helpful organism. (In some

parts of the world, the spread of modern hybrid grain varieties threatens the survival of traditional varieties, many of which have useful characteristics. This problem is discussed again in Chapter 7.)

Value as Part of Food Webs and the Gene Pool

The loss of a particular species or group of species may have far-reaching effects on the community in which the species live. Complex food webs are common in temperate and tropical climates; however, since only a relatively few webs have been thoroughly studied, we usually lack the knowledge to predict what the effect will be of the extinction of any particular plant or ani-

Figure 4.2 An Endangered Plant—Persistent Trillium (*Trillium persistens*). Concern about endangered plant species has come much more slowly than concern about endangered animals; yet the two are so intimately related that they cannot be conserved separately. Many examples are known of animals driven to the brink of extinction because a particular plant on which they feed or under which they shelter became scarce. It has been estimated that for every plant species that becomes extinct, 10–30 other species of insects, higher animals, and plants may also face extinction (Dr. Peter Raven, Missouri Botanical Garden, reported by Natural Resources Defense Council). (U.S. Department of the Interior, Fish and Wildlife Service; photo by Dr. John Freeman, Auburn University).

Figure 4.3 Sea otters were almost wiped out by fur trappers in the 18th and 19th centuries. Now, due in part to laws such as the Marine Mammal Protection Act, sea otter populations are recovering. In fact, they are threatening to avenge themselves in the process, if not on man himself, then on species dear to his gastronomic heart: abalone, Pacific lobster, and crabs. From a few individuals discovered near Monterey, California, in 1938, otters have increased to almost 2000 animals, ranging along a 150-mile (240-km) stretch of coast. Unfortunately, this same stretch of coast is also famous for seafoods such as abalone, a shellfish marketed for $8–10 a pound. Commercial fishermen are demanding that otter herds be limited in size to prevent further depredation of the profitable fishing industry. On the other hand, ecological studies show that the sea otter is a vital member of the shore community. By feeding on species such as sea urchins, otters protect seaweeds such as kelp from being overgrazed. In turn, kelp beds are at the bottom of the food webs that sustain species such as harbor seals and bald eagles. (Dr. Daniel Costa, Joseph M. Long Marine Laboratory, University of California, Santa Cruz)

mal species. Many rare insects, snails, and birds depend on a particular kind of plant for food or as a place to live. If the plant becomes extinct, the animal is also likely to become extinct (Figure 4.2). In another case, a predator that normally keeps some pest under control might be lost. As an example of this effect, in some areas sprayed with DDT, red spider mites have gone out of control and damaged crops. This occurred because DDT killed the ladybugs that usually eat the mites, leaving the mites (which were not affected by the DDT) free to reproduce in enormous numbers. (See also Figure 4.3.)

Wolves are threatened by humans in part because their role in food webs is not understood. Wolves kill animals like deer for food, but tend to select the sick, the weak, and the old. They can thus help keep a herd of deer healthy and of a size proper to the available food supply. Humans, hunting competitors of the wolf, pride themselves in taking the finest deer, thus lowering the quality of a herd!

The loss of species can cause other, less obvious problems. Genes, the hereditary material contained in the nucleus of living cells, determine the characteristics of the organism of which they are a part. Genes are also the means by which characteristics from two parents combine and are passed to their offspring. The science of genetic engineering is in its infancy, yet it appears as though in the future scientists may be able to transfer, from one plant to another, desirable characteristics such as disease or drought resistance, insect resistance, drug-producing ability, or higher protein content (Figure 4.4). Reducing the number of species in the world reduces the size of the gene pool.

Intrinsic Value of Species

Besides all the practical reasons we might give, there are also philosophical arguments in favor of preserving as many species as possible. Any species lost is gone forever. If we fail to do what is in our power to prevent these losses, we make a choice not only for ourselves but also for future generations. We are saying that they will never see the same living creatures we can see; they will never enjoy the diversity we enjoy. It may not even be a question of enjoyment—having evolved in the midst of such diversity, humans may require it to maintain their own mental health.

Figure 4.4 A kind of wild corn (*Zea diploperennis*) was recently discovered on a hillside in Mexico. This wild corn is perennial—that is, it doesn't need to be planted each year, as cultivated corn must. In addition, the corn is resistant to various viruses and grows well in wet soil. These are all desirable characteristics to transfer to cultivated corn by traditional plant breeding techniques or by genetic engineering. The new species has so far been found only in this one location. The discovery was made just before the hillside was to be plowed up. (World Bank Photo; Larry Daughters)

All these reasons, of course, consider other species only from the viewpoint of their usefulness to humans. Henry Beston wrote:

> Remote from universal nature, and living by complicated artifice, man in civilization surveys the creature through the glass of his knowledge and sees thereby a feather magnified and the whole image in distortion. We patronize them for their incompleteness, for their tragic fate of having taken form so far below ourselves. And therein we err and greatly err. For the animal shall not be measured by man. In a world older and more complete than ours they move finished and complete, gifted with extensions of the senses we have lost or never attained, living by voices we shall never hear. They are not brethren, they are not underlings; they are other nations, caught with ourselves in the net of life and time, fellow prisoners of the splendour and travail of the earth.[2]

_____

2 From Henry Beston, *The Outermost House.* Copyright 1928, 1949, 1956 by Henry Beston; copyright © 1977 by Elizabeth C. Beston. Reprinted by permission of Holt, Rinehart and Winston.

How Species Become Endangered

Hunting

The reason that comes most quickly to mind for the disappearance of species is probably hunting. Hunting has contributed to the loss of a number of animals, especially vertebrates. (Vertebrates are animals that have a backbone.) In certain well-managed wildlife populations, hunting need not harm the population—in fact, it can contribute to its welfare, most notably in cases where a population threatens to grow too large for its habitat. Unregulated hunting, however, has contributed to the loss of species. The buffalo of the American plains was hunted almost to extinction in the 1800s. Trainloads of hunters came for the sport, often carrying home no more than a buffalo head to mount as a trophy. In Africa, game officials have stopped or limited the hunting of many big game species lest these animals cease to exist except in zoos. (See Figure 4.5 and Controversy 4.4.)

Habitat Alteration

Hunting is not the main problem faced by most endangered species. The majority of these are threatened with a loss of their **habitat**, the area in which they grow, breed, seek food, and find shelter. As human populations grow, they require more houses, roads, and shopping centers. Forests are cut down; marshes, estuaries, and bays are filled in; and land is overturned in the search for coal. All of these processes reduce the land or food supply available to various animals and plants. In a sense, humans are increasing their own habitat at the expense of the habitats for other creatures.

In some cases habitat destruction is a result of game-management procedures such as burning or flooding, done to make areas more attractive to game species. Populations of elk, pronghorn antelope, white-tailed deer, and mule deer have all increased greatly as a result of such management techniques. In the process, however, the habitat becomes unsuitable for many other, nongame species.

Tropical rain forests. Although destruction of all types of habitats is occurring, the problem is most critical in tropical rain forests. Each year, an area equal in size to Great Britain is clear-cut or otherwise degraded. At present rates of destruction, in 20 to 30 years these forests will not exist in their present form. Yet two-thirds of the 3–10 million organisms believed to exist on earth are found in the tropics, most in forests.

In addition, a major portion of the nutrients in tropical ecosystems are in the biomass. Thus, when forests are cleared, impoverished soil remains. Left without cover, the soil is subject to massive erosion with resultant flooding and loss of productive capacity.

Human population growth has often been blamed for much of the loss of tropical forests. Population increases in developing countries can lead to increased wood gathering and "slash-and-burn" agriculture by native peoples. In this type of agriculture, farmers cut down trees and raise crops for a few years. Then, as the soil becomes depleted of nutrients, the farmers move on to new plots and cut down more trees. Some experts feel the blame is misplaced, however, stating that such subsistence farming is responsible for only 10–20% of forest loss. Large-scale cattle raising projects and military road building in Brazil, as well as the demand for tropical woods from Brazil, Africa, and Southeast Asia, destroy a much greater portion of tropical forests (Figure 4.6).

Special niches. Many endangered plants are living links with past ages, remnants of species that flourished in an earlier age and climate. They now exist in special niches along river banks, in bogs, marshlands, and barrens. Others are found on isolated mountain faces or in valleys or areas where glaciers never reached. These plants are rare precisely because they are adapted to grow in their present location. They can only be preserved if their habitat is protected.

Figure 4.5 African elephant herds have decreased dramatically in the past ten years, due mainly to poachers seeking ivory tusks. In 1970, approximately 5,000,000 elephants lived in Africa; by 1980, only 1,300,000. In Uganda's Kabalega National Park, poachers reduced a herd of 9000 elephants to 160 terrified beasts who kept moving night and day, unable to find refuge. In Asia, the elephant is also endangered—only about 15,000 remain. Here the problem is loss of habitat as increasing areas are farmed. (Judy Rensberger/NYT Pictures)

The Most Important Environmental Problem: Extinction of Species

> Despite concern in the U.S. over pollution, it is about the least important aspect of environment.
>
> *Lee M. Talbot*

Lee M. Talbot, an ecologist at the Council of Environmental Quality, points out that pollution "is about the least important aspect of environment" because it is, in most cases, reversible. But changing land use, such as leveling forests or filling in wetlands, eradicates entire habitats and causes some species to be lost to the world forever. Plants and animals that may now be regarded as dispensable may one day emerge as valuable resources.*

Russell Train, President of the World Wildlife Fund, agrees:

> I have spent most of my time over the past several years working on a variety of pollution problems—air, water, and chemical among others. As I review these efforts, I am struck by the fact that the real "bottom line" is the maintenance of life on this earth. Time is running out rapidly on the natural systems of the earth, and particularly on the survival of species. The loss of genetic diversity which threatens everywhere and the resulting biological impoverishment of the planet have grave implications for our long-term future.
>
> We need nothing less than a comprehensive program worldwide to preserve and protect representative ecosystems. . . .
>
> We human beings are relative newcomers on the face of the earth, but we now possess the power of life or death over our fellow creatures. While the scientific and economic arguments for the maintenance of species are compelling, it seems to me that we have an overriding moral responsibility to help preserve the other forms of life with which we share the earth.†

Do you agree that loss of habitats of wild animals and plants is the most serious environmental problem

Figure 4.6 Tropical Cloud Forest, near Xalapa, Veracruz, Mexico. Tropical cloud forests are found at the cloud line on tropical mountainsides. Many epiphytes (air plants, which do not have roots in soil), such as orchids and bromeliads, flourish in these forests. Like lowland tropical forests, these cloud forests are often cleared for agricultural development. In the region where this photograph was taken, much forest land has been cleared for coffee plantations. (Larry E. Morse)

we face? If not, what do you think is the most serious problem?

---

* Lee M. Talbot, Council on Environmental Quality, in *Science* **184** (10 May 1974), 646.
† Quoted from "Letters to the Editor," *Science* **201** (28 July 1978), 324.

Competition with Introduced Species: The Special Case of Islands

A number of species have become endangered or are already extinct because they were unable to compete with new species introduced into their habitat. Island ecosystems are especially fragile in terms of species loss due to competition with introduced species.

Of the 161 birds that have become extinct since the year 1600, 149 lived on islands. On some islands—for instance, the Hawaiian Islands—many species have evolved fairly recently. These islands exhibit a variety of different habitats due to differences in soil, rainfall, and elevation. During the evolution of native plants, competition from other species was different from what happened on the continents, allowing many new species to survive. In Hawaii, 97% of the native species are endemic (found nowhere else). Unfortunately, many of these species have been unable to compete with plant species or survive predation by animal species introduced in modern times. Imported mouflan sheep threaten both the mamane tree and the honeycreeper, a bird dependent on the tree as a food source. Mongooses were brought to the American Virgin Islands to control rats, but instead attacked other native species. Of the native plant and animal species of the Hawaiian Islands, 36% are in danger and more than 10% may already be extinct.

Pesticides and Air Pollution

Many habitats that are otherwise undisturbed are poisoned by air pollutants, acid rain (see Chapter 21), or pesticides. Pines in the mountains near Los Angeles are injured by smog from the city. The large-scale use of pesticides in agriculture places further stress upon many endangered species. For instance, birds in the raptor group, which includes hawks and falcons, are affected by the use of DDT. About 20–30 years ago, these birds began laying eggs with very thin shells, so thin that they crack before they can hatch. Thinning is believed to be due to DDT, a pesticide once used extensively in the U.S. DDT is now banned in the U.S., largely because of the effects it has on certain bird species. (See Figure 4.7. See the reference by Grier for actual correlations of eggshell thickness and DDT products in the environment.)

As part of a pest-control program in the American West, attempts were made to kill coyotes, foxes,

Figure 4.7 Brown Pelicans. Their population along the California coast decreased steadily in the 1950s and 1960s as the birds accumulated high levels of the pesticide DDT. Since 1970, however, the pelicans seem to be staging a comeback. In that year, a plant manufacturing DDT stopped discharging wastes through a sewage outflow near Los Angeles. The number of breeding pairs of pelicans has risen from 1000 pairs in the 1960s to 5000 pairs in 1981. However, on another front, almost the entire population of brown pelicans in Louisiana died in 1975, apparently from pesticide poisoning. The new problem seems to have been triggered by high tides in the Mississippi, which caused the birds to use body fat, stored for times of stress. Body fat, however, is also where small amounts of pesticide, eaten in fish, are stored. The birds may have gotten a poisonous dose of pesticides from their own body tissues. The brains of all birds tested contained a lethal dose of the pesticide endrin. (Dept. of Interior, Sport Fisheries and Wildlife/Photo by Luther C. Goldman.)

and wolves by using poisoned baits. These methods had a severe effect on populations of endangered species, among them the bald eagle, that also take the bait. The eagle, a symbol of national pride, is in danger not only from pesticides, but also from loss of habitat and from hunting by farmers who believe eagles kill their livestock.

Collection of Plants

Certain plants, especially cacti, orchids, and carnivorous plants, are so desirable that they have been collected almost to extinction. Dealers in Texas and Mexico dig up huge piles of cactus and truck them to market, selling them to collectors and for use in southwestern landscaping schemes. Half the cacti go as far afield as Europe and the Far East.

Living creatures must certainly change as environmental conditions change. Species unable to adapt to new conditions die out and new species evolve to take their place. Dinosaurs and flying reptiles no longer live on the earth, but other creatures have arisen that did not exist when the dinosaurs did. Humans, however, are speeding up the rate of change to the point where species cannot evolve quickly enough to replace those that are lost. One-half of the world's extinct mammals died out within the last 50 years.

Protecting Wildlife Resources

The protection of wildlife can be approached in a variety of ways. Many laws have been passed in the U.S. to help preserve native wildlife. The development of specific wildlife-management techniques has brought some endangered species back from the brink of extinction and has increased populations of several species of game animals. A third approach looks more closely at why species are becoming extinct at such a rapid rate all over the world. This kind of analysis has produced suggestions for economic incentives to help preserve endangered species worldwide.

Wildlife Protection Laws

Refuge system. In the United States, during the early 1900s, Congress began to set aside areas of wildlife habitat, or refuges, to help protect endangered wildlife. The development of the National Wildlife Refuge System is explained in the general context of land preservation in Chapter 34.[3]

Plant species, especially, can best be saved by setting aside part of their natural habitat as a preserve. A few individuals of a species in botanic gardens are not

enough to ensure reproduction and species survival. The first refuge intended to save endangered plants was purchased in 1980. This is the Antioch Dunes in California, home of the endangered Contra Costa wallflower and the Antioch Dunes evening primrose. A number of animals are also protected in refuges. The trumpeter swan, for instance, flourishes in Red Rocks Lake Refuge in Montana.

Many wildlife experts point out, however, that refuges must be large areas, measured in thousands of square kilometers. Smaller areas may not be able to support certain species, often precisely those most endangered. For example, large predators such as wolves or the big cats must roam vast areas to find food. In addition, larger reserves are able to buffer species from border pressures, such as pollution or human disturbance.

Because the animals and plants living in a particular area depend upon each other, we must attempt to save whole communities rather than just single species. With this in mind, the United Nations Educational, Scientific and Cultural Organization (UNESCO) has begun to identify "biosphere reserves" or "ecological reserves." This is meant to be a network of protected samples of the world's major ecosystem types. The reserves must be large enough to support all the species living there, buffer them from the outside world, and protect their genetic diversity. In this way, the reserve allows both growth and evolution, and acts as a standard against which human effects on the environment can be measured.

Besides laws establishing preserves, stronger laws are necessary to limit the spraying of pesticides near game preserves or near the other habitats of endangered species.

Hunting and fishing laws. Among the first federal laws dealing with wildlife passed in the U.S. were those that taxed hunting and fishing equipment and required permits for these sports. The money collected has been used to buy land for wildlife refuges. Hundreds of millions of dollars have been raised for this purpose (a fact that must be taken into account by those who oppose hunting). It has been suggested that horticultural items could be taxed to provide a similar fund for plant preservation.

Endangered Species Act. In 1966, Congress passed an endangered species law. This legislation was

---

3 Pages 649–650 deal specifically with the wildlife refuges.

designed not only to protect wildlife, but to determine just how extensive is the problem of disappearing wildlife. It directed that a list be made of endangered species. Estimates must be obtained on numbers of individuals of a species as well as the range over which the species is found.

In 1973, this law was made much stronger by a series of amendments. The new law recognized that we may want to protect species that face extinction in the U.S., though not worldwide. It also established a new category called "threatened species." **Threatened species** are those not now endangered but whose populations are heading in that direction. By recognizing this fact early, perhaps we can do more to save them. A further important change was that a new category—endangered plants—was added to the endangered species list. In addition, the new amendments directed that federal agencies could not undertake projects that threaten endangered species or their habitat. Although this particular provision drew little notice when the amendments were passed, it became the basis for the conflict between the snail darter and Tellico Dam. As written, the law allowed no weighing of benefits versus costs if the extinction of a species was involved. (See Controversy 4.2.)

In 1978, as a result of the snail darter controversy, the act was again amended to make it more flexible when in conflict with government projects. Now a committee must first decide whether those in charge of a project have considered all reasonable alternatives. If conflict still exists, another committee, composed of the Secretaries of Interior, Agriculture, and the Army; the chairman of the Council of Economic Advisors; and representatives of the state in which the project is located, will rule on whether a disputed project gives benefits that clearly outweigh preserving an endangered species. At its first meeting, in January 1979, this committee ruled against the completion of the Tellico Dam. Although millions of dollars of public money had already been spent on the project, this was ruled not sufficient reason to allow a species to be exterminated, even though the species involved had no sport or commercial value. The committee noted further that the threat to the snail darter pointed to other environmental problems that completion of the Tellico Dam would cause, problems that only came to public attention because of the fight over the snail darter. Many acres of fertile farmland would be covered with water when the dam was completed, decreasing a vital

agricultural resource. In addition, a recreational resource would be lost: the last free-flowing stretch of the Little Tennessee River. Third, land of historical value, the ancestral homeland and grave sites of the Cherokee Indians, would be flooded. This case illustrates how endangered species, even seemingly insignificant ones such as the snail darter, act as barometers for environmental problems in general.

Jimmie Durham, a Cherokee Indian leader, noted in testimony before a House committee that many people were making fun of the snail darter. "I would like to ask why it is considered so humorously insignificant," he asked. "Because it is little, or because it is a fish?"[4]

Although it seemed that the snail darter and the stretch of free-flowing river in which it lived were now safe, this was not so. In September, 1979, supporters of the dam tacked onto an energy-development bill an amendment authorizing completion of the dam "notwithstanding the provisions of" the Endangered Species Act. This bill was signed by the President, to the dismay of many environmentalists.[5]

In 1982, the Endangered Species Act was reauthorized for another three years. In addition, several amendments designed to improve procedures for the protection of endangered plants were added. Before 1982, only about 70 U.S. plant species had been listed as endangered (compared to about 225 animal species), despite studies showing that almost 3000 U.S. plant species are in danger (Table 4.1).

Several factors have combined to make the listing of plants a long and time-consuming procedure. Plant taxonomy is not as well advanced as is animal taxonomy, nor has as much field work been done. Thus it is not always clear whether plant species are rare or just little known. Further, before the 1982 amendments, the Endangered Species Act required that a plant's critical habitat be defined when a plant was listed as endangered. This is the area in which the plant is found; any areas to which it could be moved if necessary; and a buffer zone if needed. Field work to determine this critical habitat has not been done for most plants in jeopardy. Finally, several analyses were required before a species could be listed. An economic analysis of

4 *The New York Times*, 26 June 1978.
5 Since 1979, three small populations of snail darters have been found in remote locations by TVA zoologists. There are still too few snail darters, however, to remove the species from the endangered list.

the effect of setting aside a critical habitat was required by the Endangered Species Act, and economic and regulatory analyses were required by other acts and executive orders (Regulatory Flexibility Act, Paperwork Reduction Act). Staff and budget cuts at the Fish and Wildlife Service have slowed progress in this kind of analysis for both plants and animals.

The 1982 renewal specifically addressed these problems, noting that the only criteria for listing a species should be biological—that is, it is either in danger of extinction or it is not. Economic and other criteria are not to be used. Further, critical habitat is now to be defined as well as possible at the time of listing. This is an area in which citizen participation can be

Table 4.1 U.S. Plants in Danger

State	Persuasive evidence for extinction	On endangered species list[a]	Vul-nerable[b]	Endemics	State	Persuasive evidence for extinction	On endangered species list[a]	Vul-nerable[b]	Endemics
Alabama	2	1	90	17	Missouri	0	0	26	3
Alaska	0	0	33	30	Montana	2	0	12	1
Arizona	6	5	167	115	Nebraska	0	0	3	1
Arkansas	0	0	25	3	Nevada	3	1	132	70
California	30	15	761	662	New Hampshire	0	1	7	0
Colorado	5	4	53	32	New Jersey	0	0	19	0
Connecticut	0	0	10	1	New Mexico	2	5	50	22
Delaware	0	0	12	0	New York	0	1	24	1
District of Columbia	0	0	1	0	North Carolina	5	2	74	10
					North Dakota	0	0	3	0
Florida	6	2	214	154	Ohio	0	1	15	0
Georgia	7	3	88	8	Oklahoma	0	0	15	4
Hawaii	177	6	798	797	Oregon	11	2	167	88
Idaho	2	2	47	19	Pennsylvania	0	0	16	2
Illinois	1	0	27	2	Rhode Island	0	0	4	0
Indiana	1	0	20	0	South Carolina	0	2	47	2
Iowa	0	1	8	0	South Dakota	0	0	2	0
Kansas	0	0	8	0	Tennessee	2	2	62	11
Kentucky	1	0	33	4	Texas	10	9	219	192
Louisiana	2	0	18	3	Utah	5	8	169	124
Maine	0	1	12	3	Vermont	1	1	10	1
Maryland	0	0	16	0	Virginia	1	1	42	3
Massachusetts	1	0	11	0	Washington	2	0	78	48
Michigan	0	0	19	1	West Virginia	0	0	16	1
Minnesota	0	0	13	3	Wisconsin	0	1	16	2
Mississippi	0	0	22	0	Wyoming	1	0	13	8
					Total	283	61[c]	2882[d]	2448

The first column lists plants for which the U.S. Fish and Wildlife Service has persuasive evidence of extinction or maintenance only in cultivation. The second column gives the numbers of plants actually protected by listing as endangered species according to the Endangered Species Act. The third column shows the numbers of plants either formally proposed for listing or those being considered for proposal (candidate plants). A useful comparison to this third column is the number of vulnerable plants in a state which are found nowhere else in the U.S. (endemics); these are enumerated in the fourth column.

a   As of 1982.
b   Either formally proposed for listing or candidates for proposal.
c   Since this table was prepared, three plants have been moved from the vulnerable to the formally listed category, making a total of 64 listed plants.
d   Plus 117 species in Puerto Rico and U.S. Virgin Islands.

Do We Need Flexibility in Laws Protecting Endangered Species?

No compromise is possible when the problem is stated in terms of the question "Do organisms have the right to exist?"

Wayne Grimm, National Museum of Canada, quoted in Science, 196 (24 June 1977), 1428

Are you going to do anything to get the snail darter off our backs?

Unidentified Alabama Congressman to Interior Secretary Cecil Andrus, quoted in Science, 196 (24 June 1977), 1427

When the Endangered Species Act halted construction of the Tellico Dam because a completed dam would have wiped out the endangered snail darter, many congressmen began to feel that the act was "inflexible." That is, the act had no provision for considering the value of a project compared to the value of an endangered species. People began thinking of cases in which, they felt, a project or action could have more value to humans than the continued existence of a species. Senator William Scott of Virginia argued:

Suppose a bird of some endangered species was in front of an intercontinental ballistic missile. . . . They could not release that missile. To me that would be a ridiculous offense. . . . Any commander worth his salt . . . would go ahead and release the missile, but he would be disobeying the law and would be subject to a fine of $20,000 and imprisonment for up to a year.★

Others argued that the act was actually working well. They pointed out that the Tellico Dam was the only project ever halted by the law and one of only three to go to the courts at all. (Some 5000 possible problems were solved by consultation with the Fish and Wildlife Service.)

But aside from questions about how well the original act worked is the question of whether flexibility is desirable. Do you feel there are instances in which a project could have more significance to humankind than the survival of an endangered species? Can you think of an example? Or do you think that no species should become extinct because of a construction project, however beneficial to humans? Can you justify this view?

★ Quoted in *Science* 201 (4 August 1978), 427.

very effective. Several states have private conservation groups devoted to plant protection. Almost one-half the states have groups such as the California Native Plant Society, or botanical clubs such as those in Michigan and Maine, which work to determine the rarity and distribution of their state's plant species. Federal funds and advice are sometimes available through state governments to local society botanists. Their work helps in assembling the data needed before a plant can gain official endangered-species status.

Local garden clubs are encouraged by the Fish and Wildlife Service to "adopt-a-plant" (by protecting its habitat in the wild rather than by moving it to a garden). The FWS is willing to provide help in choosing those plants most in need of attention.

The Endangered Species Act does not control the effects of nongovernmental projects. Private citizens are not bound by law to consider the effects of projects, such as housing developments, on endangered species. Several states do have programs to encourage wildlife conservation on private lands, providing cost-sharing, free or low-cost materials, and free labor. Most of these programs involve farmland and are meant to conserve animal rather than plant species.

The 1973 amendments to the Endangered Species Act served one further purpose, which was to ratify the Convention on International Trade in Endangered Species of Wild Fauna and Flora (CITES). The treaty sets up a system of permits for both exporting and importing threatened and endangered species, or products made from them. Trade in nearly extinct species is practically prohibited, while strict controls are set for other endangered or threatened species. Other countries, including Canada, have signed CITES.

Wildlife Management Techniques

A variety of special techniques have been developed to preserve species in danger of extinction or to increase the range of animals considered very desirable (i.e., those that people like to hunt). In some cases, animals may be transferred from their natural habitat to a similar area where they were not previously found. This has been done mainly with nonendangered game species like Canada geese. The wild turkey, which has been introduced in a number of areas, now occupies more territory than it did during Colonial times.

When the judgment has been made that a species will not survive on its own, even if given a fair chance, eggs may be collected and hatched in captivity or breeding programs can be instituted at zoos. The animals can in some, but not all cases, be successfully reintroduced into the wild. Sea turtles, which by instinct run to the sea after hatching and later return to their birthplace to lay eggs, never seem to get their bearings right if hatched in captivity. They swim off into dangerous waters and fail to return to suitable beaches for successful egg laying. On the other hand, about half of the whooping cranes alive today were hatched and reared in captivity.

In some cases management procedures on preserves are so successful that limited hunting can again be allowed. One hundred years ago, the American bison lived in herds so huge it sometimes required several hours to pass by a herd. Fifty years ago there were only a few hundred bison left. Within the past few years, there have been enough bison again to allow some hunting.

Preserving Endangered Species Worldwide

In order to understand the third approach to saving endangered species, we need to consider a little bit about how economics of the marketplace influence decisions people make. Most people would agree that other creatures have a right to survive on the earth. People rarely intentionally set out to wipe out other species. Yet, as a number of experts have shown, our economic system is set up in such a way that we tend to do just that.

Species, like air and water, are, in a sense, a common resource. That is, we all stand to benefit from having a wide variety of plants and animals on the planet. None of us, however, "owns" any particular wild species and so no one is directly responsible for the survival of any particular species. Individuals can easily see the benefits gained from hunting tigers, capturing apes, or building a housing development on some other creature's habitat. Much harder to keep in mind are the benefits, to all of us, of having a great variety of species, since these benefits (esthetic or moral) are long-term and less visible (medical uses, agricultural). In other words, the short-term, visible benefits go directly to the individuals involved, while the losses are mainly long-term and are spread over

CONTROVERSY 4.3

Endangered Species versus Human Health Benefits

> Surely it is not beyond our scientific ingenuity to find alternative methods.
>
> **F. B. Orlans**

> It is not consistent with the genius of the American people to restrict the progress of scientific knowledge.
>
> **A. S. Packard, Jr., and E. D. Cope,
> American Naturalist, 17, 175 (1883).**

The use of animals in scientific research was once opposed mainly on the grounds of possible pain and cruelty. Scientists went to great lengths to assure the general public that the animals used in experiments never feel pain.

Now, however, new ethical concerns are being raised. Are research animals housed in such a way that social and behavioral needs are met? That is to say, normally social animals like chimpanzees should not be kept in individual, isolated cages because this would be a form of mental cruelty.

To go even further, should an animal with a dwindling population be used in research at all, even if humans stand to benefit greatly? N. Wade writes:

> . . . production of the [hepatitis] vaccine may well pose a fatal conflict between the interests of mankind and those of chimpanzees. Chimps are the only species, other than man, in which the safety of the vaccine can be tested . . . if the chimpanzees are protected—the species is already classified as threatened—it may prove impossible to safety test and hence to manufacture the vaccine. Yet even in developed countries, where the disease is comparatively rare, hepatitis B takes a heavy toll. In the United States 15,000 cases were reported in 1976. The true incidence was probably 150,000 according to the Center for Disease Control, of which probably about 1500 cases ended in death. . . .
>
> . . . officials deny that their chimpanzees would be captured inhumanely. "The method of capture is generally by locating a group of chimpanzees, surrounding them with a number of people and chasing them. The juveniles would usually tire first and these were captured by hand," a Merck official told the Federal Wildlife Permit Office. . . .
>
> ". . . Totally impossible unless you had big nets," says Jane Goodall. "Utterly fanciful. . . . Given the sort of habitat where wild chimpanzees are found, no human being could keep up with a wild chimpanzee, much less run it to the ground. . . . I can only conclude that someone is seeking to conceal the actual but less humane method of capture used—that is, shooting the mother to recover the young, which is the standard method used in Africa."[*]

F. B. Orlans adds:

> . . . a way must be sought to solve this conflict in a manner that is not detrimental to the chimpanzees. In the past, alternative methods of producing other vaccines (notably that for polio) have been found so that animal lives are spared . . . the ethical concerns for elimination of inhumane killing (in Wade's words, "to capture a chimpanzee: first shoot the mother") and for preservation of this dwindling species of animal are overriding.[†]

Do you feel that a clear human need should outweigh the need to preserve an animal species?

[*] N. Wade, "New Vaccine May Bring Man and Chimpanzee into Tragic Conflict," *Science* **200** (2 June 1978), 1027.

[†] F. B. Orlans, Letter to the Editor, *Science* **201** (7 July 1978), 6.

society as a whole. As a result, we are as wasteful of the resource of species as we have been of air and water, other resources that no one owns[6] (Figure 4.8).

Because these sorts of effects are not an intended result of people's actions, economists call them *externalities*, or spill-over effects. People who undertake more economic activities are likely to cause greater spill-over effects. People in developed countries use the most raw materials, which are often obtained by disturbing the habitat of creatures in less developed countries, and are, in this sense, most responsible for loss of species. Several suggestions have been made to remedy this situation.

In the first place, corporations operating overseas could be required by law to determine the effects their operations have had on animal and plant species in foreign countries. Policies could be worked out to allow tax credits for conservation measures or fines for damaging operations.

Under the National Environmental Policy Act (NEPA), U.S. government agencies are required to examine their actions abroad for possible ecological harm. International public organizations, such as the U.S. Agency for International Development or the United Nations agencies, could be required to write environmental impact statements for their projects. Even if it were decided, because of great benefits to humans, to go ahead with plans that would damage habitats of other species, the costs would be recognized and weighed, avoiding much unnecessary damage. The U.S. has signed the Convention on Nature Protection and Wildlife Preservation in the Western Hemisphere. Because of this convention, special plans were developed during Pan American Canal negotiations by the U.S. and Panama to protect endangered vegetation on Barro Colorado Island.

In some cases, developed countries should help pay the costs of saving species in undeveloped countries, either because the countries in which the endangered species are found are too poor to undertake protection, or because their people are too far from achieving minimal human needs, in terms of food, shelter, and health care, to be able to choose wildlife protection over exploitation of natural resources. Tanzania, which has the largest wildlife population in Af-

Figure 4.8 Devil's Hole Pupfish (*Cyprinodon diabolis*). There are 12 species of pupfish in various locations around the U.S. Important to science because of their ability to withstand extremes of temperature and salt concentration, the fish have run afoul of humans in several instances. In 1976, the U.S. Supreme Court ruled that the Endangered Species Act prohibited ranchers from pumping so much underground water for their ranches that the water level was lowered in Devil's Hole, home of the endangered Nevada pupfish. The fish have been marooned there since the last glacier receded, leaving much of Nevada desert country. Not so lucky, California's Tecopa pupfish was recently declared extinct. Thirty years ago, builders of a bathhouse diverted waters from the thermal pools and springs in which the fish lived. Unable to adapt to life in the new swift stream, the fish finally died out. The bathhouse is no longer in use. (U.S. Fish and Wildlife Photo/Jack E. Williams)

---

6 These principles were first clearly applied to environmental problems by Garrett Hardin in his "Tragedy of the Commons," *Science,* **162** (13 December 1968), 1243.

---

rica, has asked that other countries contribute to the cost of guarding wildlife from poaching. Needed equipment, such as surveillance helicopters for game wardens, is beyond the reach of many developing countries. That such measures can help is shown by the success of Project Tiger to preserve the Bengal tiger in India. This program, supported by international conservation groups, has increased the tiger population by 65%. The key feature of the project is preserves, surrounded by buffer areas, in which humans are not allowed. Poaching has been eliminated by laws and by payments made to farmers who lose stock to the tigers.

In the final analysis, the fate of other species is a barometer for the fate of the human species. If the

Figure 4.9 Humpback Whale. This whale is "breaching," or leaping out of the water. See the supplemental material at the end of this chapter for a discussion of the plight of this endangered giant. (Ken Balcomb/World Wildlife Fund)

outlook is not sunny for the survival of other creatures on this planet, surely human survival is in for a stormy time (Figure 4.9).

> I heard the song
> Of the world's last whale
> As I rocked in the moonlight
> And reefed the sail.
> It'll happen to you
> Also without fail
> If it happens to me
> Sang the world's last whale.[7]

We would do well to be guided by the maxim attributed to Aldo Leopold:

> The first requirement of intelligent tinkering is to save all the pieces.

---

7 "Song of the World's Last Whale" by Pete Seeger. Copyright 1970 by Stormking Music, Inc. All rights reserved. Used by permission.

Endangered Ocean Mammals versus Endangered Native Cultures

Eskimos and the Sale of Products from Sea Mammals

In recent years, the U.S. government has twice been forced to look for a middle ground that would allow the survival of both endangered sea mammals and equally endangered native cultures. The issues involved will come up again in other forms and in other parts of the world. As human populations expand and as their needs for living space and food grow, pres-sures on wildlife species will increase. In some cases, animals and plants may become endangered because their habitats are turned to human uses. In other cases, hunting, for food or other reasons, may bring species to the point of extinction. In the final analysis, choices will have to be made between human needs and the survival of species. The following examples, drawn from the experiences of Alaskan Eskimos, illustrate how difficult such choices can be.

In 1972, Congress passed the Ocean Mammal Protection Bill. This law protects arctic fur seals, walruses, and whales, among other ocean mammals, from commercial hunting, which has reduced some species almost to extinction. However, during hearings, before the bill was passed, it became clear that a total prohibition of the hunting of sea mammals or the sale of products made from sea mammals would threaten something else that was endangered: native Eskimo culture.

The following testimony is excerpted from hearings before the Subcommittee on Oceans and Atmosphere of the Committee on Commerce of the United States Senate in the spring of 1972 (before the Ocean Mammals Protection Bill was passed):

Statement of John Henry: I don't have no written statements. . . . So, I'm going to make my own—what people thinks. As people thinks, I'm Eskimo, just like Indian or Aleut, all same. We are from Norton Sound on the coast area, where we could hunt on the sea and on the ground. We have two ways to hunt. To feed ourself from the sea and from the ground. Even we don't have no money. Sometimes we have hard times. We can't purchase from the ground, we can't purchase from the sea. But we are Eskimo, we can't write a check to purchase something that we could buy. Just sit down and write a check to purchase something for our children. . . .

Sea mammals. It's our living and I born on sea mammals. I born, we didn't have much money. My dad didn't have much money. He hunt and hunt, just the same as like other peoples in the other villages. He try to support us from the hunting, that's all. . . . No other resource we have, just from his hunting. Just like other villages does. And ancestor, from our ancestor I think we born like them. I will be like that, because my ancestors I know. I was born without money. I will be an ancestor for my own folks without money. . . .

Statement of Myrtle Johnson: I am a resident of Nome, and was born and raised in the village of Golovin. . . . I will draw from people I know personally to paint an imaginary picture based upon the truth.

John and Mary are in their mid-50s. John works part-time seasonally in casual labor, and they care for four foster children plus one grandchild. John is a successful seal hunter. He also catches Beluga. If he had a chance, he would join a crew and hunt black whale as well. This would bring him a share of meat and muktuk— something his family otherwise must buy at \$3 per pound or receive as a gift that will obligate him to return in some other form.

John's family depends heavily upon sea mammal meat and oil all year long. The spring greens are stored in fresh seal oil, some of this being sent to family and friends living in the cities. The meat is dried, frozen, or in other ways preserved for the time when it is not so plentiful. . . .

The skins will end up as mukluks, parkas, parka-pants, and vests for the family, as well as surplus skins or garments to be sold or traded to others. Both raw and stretched, skins are a source of income and add to the internal welfare system of their village.

Not all men in a village hunt seal. Some are working or cannot hunt for other reasons. In every village a certain amount of sea-mammal meat, oil, and skins are purchased by nonhunters. This stretches their food dollars. A portion of the village sea-mammal harvest is always provided for the poor or the aged in a fashion that is in keeping with village traditions.

Beef and pork, seldom carried at a village store, must be priced 25% to 50% higher than at the outlets with jet flight service. Thus, the importance of local meat harvest cannot be underestimated. Without the meats and oils, many families would be without this basic food in their diet. . . .

Caucasians speak of bread as the staff of life. For coastal Eskimos the seal represents the single basic food staple with all the meanings others may associate with bread. . . .

To take away our Eskimo bread is to deny the men and women of the villages the right to work to meet their needs as they see them and can mean the final end of our way of life before we are fully adapted to the modern world.

As a result of the hearings, the final act allowed exemptions for subsistence hunting (hunting for food or clothing) and the sale of the native handicrafts[8] (Figure 4.10).

---

8 When the bill was renewed in 1981, the subsistence exemption was changed to one for rural users, rather than specifically mentioning Eskimos.

Subsistence Hunting and the Bowhead Whale

In 1976, even the Eskimo right to subsistence hunting was called into question.

For centuries, humans have hunted whales for oil and meat. In the 1700s and 1800s, the right and bowhead whales were the main targets. These were the "right" whales because they float after they are killed and because they are slower than other species. Now these whales, along with the blue whale, the humpback, and the gray whale, are near extinction.

The great blue whale is the largest animal now living on the earth. Although they swim in the oceans, as fish do, whales and their smaller relatives, the dolphins, are mammals. Whales are warmblooded and suckle their young. They are also social animals, living in herds and possibly even in families. Within the past

Figure 4.10 Eskimo craftsmen carve and sew materials from ocean mammals into useful and beautiful objects. Above is an example of a parka sewn from seal skins. (Dobbs Collection/Alaska Historical Library.)

few years, scientists have realized that whales can "talk" to each other. Some beautiful records have been made of the singing sounds they make under water.[9]

Whaling was once a dangerous occupation. The small crew of men, with their harpoons and in their fragile boat, and the huge whale were relatively well matched. Mostly the men won, but sometimes the whale did, too. Modern methods, however, have tipped the balance far in favor of men (Figure 4.11).

Modern non-Eskimo whaling fleets consist of fast killer boats, which kill the whales with harpoon guns, and large factory ships, which process the whales at sea. Whales are located by sonar or small planes, then driven by high-frequency sound waves until they are exhausted. A harpoon carrying a grenade, which explodes within the whale, is used to kill them. With these methods, modern whaling fleets have killed more whales in this century than in the 16th, 17th, 18th, and 19th centuries together. Within the past 50 years, the numbers of blue, humpback, and gray whales have been reduced from hundreds of thousands to a few thousand. The California gray whale, whose breeding grounds are now protected by the Mexican government, has begun to recover in numbers, perhaps even to reach a population size near that of the 1800s. The bowhead whale, on the other hand, has been protected for 35 years, but does not seem to be increasing in numbers.

About 30 nations belong to the International Whaling Commission, which sets yearly whaling quotas (the numbers of each type of whale that may be caught in the various oceans). Nations that have whaling fleets, such as Russia and Japan, and those that no longer have fleets, such as the U.S., Canada, and South Africa, send delegates to the Commission meetings. Over the past few years, quotas have been slowly reduced as new scientific data came to light on the population sizes and reproductive ability of whales. However, quotas also declined because, in a number of cases, whalers were simply unable to find enough whales to fill them.

For many years, conservationists have worked to pass a ten-year moratorium on all whaling, allowing time for the study of whales and their survival needs, before they are unintentionally hunted to extinction.

---

9 One record of these sounds is *Songs of the Humpback Whale,* Capitol Records, Inc., Hollywood and Vine Streets, Hollywood, Calif.

Conservation groups such as the Animal Welfare Institute have conducted a vigorous campaign to support the moratorium. In addition, they urge that people boycott goods from Japan and Russia, major whaling countries. (See Controversy 4.4.) In 1982, the IWC finally approved a moratorium, to begin in 1985–1986. However, long before this, in 1946, the IWC treaty noted that the bowhead whale was the most endangered whale species. The commission at that time forbade all but subsistence hunting.

United States Eskimos had been catching about 10–15 bowheads per year since 1946. However, in the 1970s, the catch increased. Further, the number of whales struck but lost steadily increased (Table 4.2). It is estimated that at least half of these later die. The IWC was concerned about the effect this increased hunting would have on an already severely endangered species. In 1976, the commission called for a total ban on subsistence hunting of the bowhead.

Why did Eskimo hunting of the bowhead whale increase? The answer is tied up with economics and the impact of outsiders on native culture. R. Storro-Patterson writes:[10]

> Correlated with the increase in the Eskimo harvest of bowhead whales is the growing number of whaling crews:
>
> Number of Whaling Crews—Spring Hunt
> | 1971 | 1972 | 1973 | 1974 | 1975 | 1976 |
> |------|------|------|------|------|------|
> | 25 | 27 | 45 | 46 | 75 | 86 |
>
> The rise in the number of whaling crews may in turn be significantly related to the availability of jobs and money for Eskimos. Traditionally, to be a whaling captain was not only a matter of great prestige, but great difficulty. One had to either inherit the equipment, marry to obtain it, or gain sufficient wealth to purchase it. The latter option was seldom possible. Recently, however, such construction projects as the Alaskan pipeline, and oil drilling operations, as well as the Alaskan Native Claims Settlement have changed this. An ambitious Eskimo can save $9,000 for an "outfit" plus another $2,000 for provisions for a whaling crew.

The Eskimos were outraged by the ban, as were many other people who sympathize with the native Alaskans' difficult fight to survive and preserve their culture in a relatively hostile natural environment.

Table 4.2 Alaskan Eskimo Bowhead Harvest[a]

	1973	1974	1975	1976	1977[b]
Landed	37	20	15	48	26
Killed, lost	0	3	2	8	2
Struck, lost	10	28	26	37	77
	47	51	43	93	105

a Source: R. Storro-Patterson, "The Bowhead Issue," *Oceans* (January 1978), p. 63.
b The 1977 figures represent the spring hunt only.

> "Hunting the bowhead is what keeps our communities together. Our people depend on the bowhead, first for food and, just as importantly, for the survival of our culture," said Dale B. Stotts, an official of Alaska's North Slope Borough, which includes a number of whaling towns and villages along the coast of the Beaufort Seas.[11]

W. H. DuBay wrote, in defense of the Eskimo hunters:

> . . . People outside Alaska don't seem to realize that the Eskimos are the residents of the Arctic and that, were it not for their aggressiveness in protecting their Arctic homeland from those who would destroy it, there would be no effective environmental safeguards at all operating in the U.S. or Canadian Arctic.
> . . . The issue is not just the best way to manage the preservation of the bowhead whale, but also the great question of subsistence hunting rights of American native peoples and the basic human right to eat what you have to eat in order to survive in your own environment.[12]

Environmental groups were torn between their sympathy for the natives and the need to protect whales. Conservationists felt that if the U.S. filed a formal objection to the IWC decision, which would cancel the ban, other nations, such as Japan or Russia, would feel free to file objections to quotas the IWC had set on other whale species:

> "For the first time, this has put the United States in the position of being the affected party in restricting whaling," said Dr. Roger Payne, a

---

10 R. Storro-Patterson, "The Bowhead Issue," *Oceans* (January 1978), p. 64.

11 Quoted by B. Rensberger in *The New York Times,* 5 October 1977.

12 W. H. DuBay, Letter to the Editor, *The New York Times,* 15 November 1977.

ONTROVERSY 4.4

The Ethics of Hunting Whales

Whaling is an affront to human dignity.

Sir Peter Scott, The New York Times, *8 June 1982, p. A25.*

In a word, we are dissatisfied with the meeting. We are trying
to avoid taking extreme actions.

Shigeru Hasui, *Managing Director, Japanese Whaling Association,*
at the 1982 meeting of the International Whaling Commission

Most of us would agree that it would be terribly wrong to hunt whales to extinction. But to go a step further, should we "harvest" them at all?

At the 1982 meeting of the International Whaling Commission, a moratorium on commercial whale hunting was passed. The moratorium will not start until 1986 and will be reviewed in 1990. Japan was not at all happy with the decision of the IWC.

Whale products are used to manufacture fertilizers, transmission fluids, and animal feeds, although other materials could be used to make all of these items. Whale meat is eaten in Japan and the Soviet Union. The Japanese claim that although whale meat forms only a small part of the Japanese diet (possibly as little as one percent of the protein), it is eaten mainly by the poor, who do not have many other choices.

On the other side is the evidence that whales are able to communicate with each other, that at least some whales travel in family groups, and that whales have large brains. (The ratio of brain weight to body weight in dolphins—if the blubber layer is ignored—is 2.0, compared to 1.9 for humans.)

How should we balance these factors, as we deal with this "other nation" with which we inhabit the earth? Should there be any commercial whaling?

Do you think the U.S. should continue to ask for an exemption for Alaskan Eskimos to hunt endangered bowhead whales, even though this makes it more difficult for us to support a total moratorium on whaling at IWC meetings?

Figure 4.11 Whaling with a hand-thrown harpoon in a small boat, as is still done in some areas, is a difficult and dangerous business. The whale might overturn the boat, with tragic results for the whalers. (Courtesy American Museum of Natural History, Dr. F. Rainey, photographer) In contrast, gigantic modern whaling ships, such as this Soviet factory ship, ensure there is little danger to the whalers. They are frighteningly efficient at killing whales, however. (Matt Herron/Black Star)

whale specialist with the New York Zoological Society. "If the United States files an objection to the I.W.C. position, this country could lose all its credibility in whale conservation."

"The original exemption on the bowhead was to guarantee what are known as the aboriginal rights of the Eskimos," Dr. Payne said. "The Eskimos aren't aborigines any more and along with modern hunting weapons there have to go some controls."[13]

The U.S. government finally decided to ask that the Eskimos be given a quota on the number of bowheads they could catch. At an emergency meeting in December 1977, the IWC voted a quota of 12 whales landed or 18 struck, whichever came first.

The Eskimos objected that this meant that some people must go hungry in the coming year. Nonetheless, they agreed to obey the quota and set up a self-

governing system to do so. A further quota was granted in 1982 for 17 whales each year for a period of three years.

Perhaps the necessary balance has been found for the moment. It seems unfair that the Eskimos, as they attempt to protect their cultural heritage, must shoulder one more burden, that of nurturing back to health a whale population plundered by commercial whalers in the beginning of this century. But it also seems unfair that the bowhead whale, so close to extinction, should not have all possible chances to recover. "Surely the whales are also a 'special interest group' and need an advocate to defend their right to survive."[14] As a *New York Times* editorial noted,[15] "if Eskimo culture needs the bowhead, it must be saved; killing off the few that remain would not long satisfy either stomach or spirit."

---

13 B. Rensberger, *The New York Times,* 29 September 1977.

14 R. Ellis, Letter to the Editor, *The New York Times,* 11 November 1980.

15 *The New York Times,* 1 October 1977.

Questions

1. How would you explain the value of preserving a wide variety of species, both plant and animal?
2. What do we mean by the terms *extinct, threatened, endangered,* when they are applied to species?
3. Briefly list the main reasons why plant and animal species become extinct or endangered. Which is the most important reason?
4. Do you believe hunting should be permitted? Give your reasons for or against and any special requirements you would like to see enforced.
5. How does the fact that wildlife species are usually viewed as a common resource, in the same way that water or air are considered common resources, contribute to the rapid loss of species today?
6. Recount some of the ways in which attempts are currently being made to preserve species from extinction. What further measures could be taken?

Further Reading

Animal Welfare Institute, P.O. Box 3650, Washington, D.C. 20007.
　This conservation organization is one of the most active in campaigns to save whales. If you write to them, they'll put you on a mailing list for current information and also send directions for staging your own "Save the Whales" campaigns.
Canadian Nature Federation.
　Information on endangered species in Canada and Canadian wildlife laws can be obtained from this organization at: 75 Albert Street, Ottawa, Ontario, K1P 6G1.

Elias, Thomas S., "Rare and Endangered Species of Plants—The Soviet Side," *Science* **219**, 7 January 1983, p. 19.
　In 1980, the USSR passed an endangered animal protection law. Although some 2000 plant species are in need of some protective measures and about 200 are in danger of extinction, plants are not yet protected by a specific law in the USSR. This paper contrasts U.S. and USSR approaches to the protection of endangered species.
Endangered Species Technical Bulletin, Vol. VI, No. 1, Department of the Interior, U.S. Fish and Wildlife Service.

This bulletin and the December 15, 1980, Federal Register list those plants considered in danger in the U.S. and defines the problems they face. If you want to know which plants in your area need protection, these are the places to look. A similar list of vulnerable animal species was published in the Federal Register, December 30, 1982. More information is available from: The Director (OES), U.S. Fish and Wildlife Service, Dept. of the Interior, Washington, D.C. 20240.

Graves, William, "The Imperiled Giants," *National Geographic*, p. 752 (December 1976).

This beautiful pictorial essay, and the following essay by V. B. Scheffer, explore the life history of whales as well as their often ill-fated relationship with humans.

Mowat, F., *People of the Deer*. New York: Pyramid, 1968.

Mowat, F., *The Desperate People*. Toronto: McClelland and Steward Ltd., 1975, Volumes I and II.

These three books tell of the vanishing people, the Ihalmiut, who were the Eskimos of Canada's inland region. The story tells of the delicate balance between humans and wildlife, which once enabled the Ihalmiut to live and flourish in a harsh climate. The books will give you an understanding of the plight of native peoples and endangered animals in an increasingly technological world.

Plumwood, V., and Routley, R., "World Rainforest Destruction—The Social Factors," *The Ecologist*, **12**, No. 1 (1982), p. 4

The social, political and economic factors involved in rain forest destruction are thoroughly explored in this article. Its conclusions are not the same as those in the National Geographic article on India and wildlife (below).

Putnam, John J., "India Struggles to Save Her Wildlife," *National Geographic*, p. 299 (September 1976).

India provides one of the best examples of how an expanding human population exerts intolerable pressures on wildlife populations in developing countries. The almost unresolvable conflicts between human and animal needs are highlighted in this perceptive article.

References

A Legislative History of the Endangered Species Act of 1973 as amended in 1976, 1977, 1978, 1979, and 1980, Committee on Environment and Public Works, U.S. Senate, February 1982.

Franklin, J. F., "The Biosphere Reserve Program in the United States," *Science*, **195**, 262 (21 January 1977).

Grier, J. W., "Ban of DDT and Subsequent Recovery of Reproduction in Bald Eagles," *Science*, **218**, 1232 (17 December 1982).

International Wildlife Conservation, Hearing before the Subcommittee on Resource Protection, Committee on Environment and Public Works, U.S. Senate, November 7, 1979.

MacBryde, B., "Why Are So Few Endangered Plants Protected?" *American Horticulturist*, October/November 1980, p. 29.

A good review of U.S. laws relating to plant conservation.

Marine Mammal Protection Act Amendment, Report No. 97–228, 97th Congress, 1st session, House of Representatives, 1981.

Myers, N., "An Expanded Approach to the Problem of Disappearing Species," *Science*, **193**, 198 (16 July 1976).

Rare Plant Conservation: Geographical Data Organization, L. E. Morse and M. Henifin, eds. New York Botanical Garden (1981).

Citizens interested in helping with studies necessary before plants can gain official endangered species status can get an idea of the type and quantity of needed data from this publication.

The Evolution of National Wildlife Law, Council on Environmental Quality, 1977.

The World's Tropical Forests: A Policy, Strategy and Program for the United States, U.S. Interagency Task Force on Tropical Forests (1980).

Wildlife and America, Howard P. Brokaw, ed., Council on Environmental Quality (1978), U.S. Government Printing Office, 1978-0-235-920.

(WFP photo by T. Page)

ART TWO

Resources of Land and Food

Soil as a Resource

The United States is truly blessed with soil resources. This country contains the largest area of prime farmland in the world, some 346 million acres. Prime farmland is land with the proper combination of soil characteristics, climate, and water supply for agriculture. With proper management, such land can produce high crop yields year after year. This land, as well as pastureland, rangeland, forest, and nonprime cropland, makes up a huge and valuable resource. It is, of course, valuable to humans for food production, but it is also a basic part of wildlife habitats and recreational resources.

Yet the soil that forms the very basis of these resources is threatened in several ways. The main threat to agricultural lands is soil **erosion**, the loss of soil to winds or water flow. In certain areas, **salinization**, the accumulation of salts in soil, is a problem. Salinization often occurs when irrigation water evaporates, leaving behind the salts it contains. For this reason, it is a problem in dry climates such as the American Southwest, where a great deal of irrigation water is used. In other areas, the structure of the soils has been hurt by traditional farming methods that compact soil with heavy equipment, or by recreational vehicles run over delicate desert soils. Acid rain leaches needed minerals from both agricultural and forest soils, especially in the Northeast. Finally, we are paving over some 3 million acres (1.2 million hectares) a year of our soil, almost 1/3 of it prime farmland, as cities and suburbs grow to accommodate a growing U.S. population.

These problems are not unique to the U.S., of course. Erosion, for instance, is a serious problem in many countries today, especially in areas where tropical forests have been cleared for crops. Similarly, acid rain affects forests and croplands in a number of European countries.

Desertification

All over the world are areas where groundwater tables are declining as people pump water out faster than it can be naturally replaced, or where erosion or salinization are destroying soil resources. In some of these places—for instance, in the African Sahel and parts of the American Southwest—vegetation has vanished.

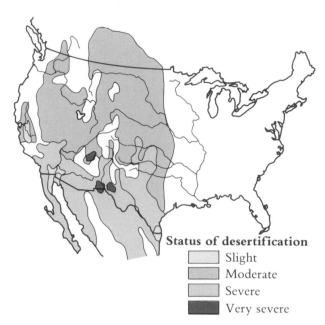

Status of desertification

- Slight
- Moderate
- Severe
- Very severe

Figure A Status of desertification in hot arid regions. (Source: United Nations)

Such areas are said to be undergoing **desertification** (Figure A). Although it may appear that the desert is creeping along and engulfing more and more land, this is not what actually happens. Desertification is a term applied to the severe degradation of land, to the point where it resembles a desert. It results from overuse by humans, sometimes in combination with variations in climate such as droughts. Desertification is the ultimate human insult to land and soil resources.

From 1968 to 1973, in the Sahel region of West Africa, a severe drought, along with past patterns of overuse, combined to form new deserts on the southern fringe of the Sahara. Thousands of people and animals starved as a result. The problem is not new, however. As long ago as 300 B.C., once-fertile lands along the coast of Turkey became deserts because of overuse by humans.

In this part of the text we examine some of these threats to soil resources and food production. In Chapter 5, soil structure and processes of soil formation are explained. We also describe various natural plant communities. In the next chapter we discuss erosion, a major soil problem. An equally important environmental problem associated with food production is pesticide use, also covered in Chapter 6. In Chapter 7, the world food situation and prospects for the future are reviewed.

We first look at the scope of the food problem. Is there a food supply crisis? How can farmers, especially those in developing countries where population growth is most rapid, grow more food? But these questions deal with only part of the story. Social and political realities control, to a very large extent, how much food is actually produced and where it goes (Figure B). In Chapter 7, we attempt to give a feeling for these additional factors that affect our ability to provide enough food for all the world's people. Then we summarize the optimistic and the pessimistic views of future world food supplies. This last chapter also deals with the issue of world grain reserves, because here may be the solution to both unnecessarily high food prices and catastrophic famine.

Water resources, which along with soil resources determine an area's food-producing ability, are covered in the next part, in Chapter 9, while acid rain and its effects on soil fertility are noted in Chapter 21. The last two chapters in the book, 33 and 34, detail and summarize problems of urbanization and recreational use of land resources.

Figure B Feeding the world's people involves many problems—some technical, but many economic, social, and political as well. These school children are fed in a special government program to improve nutrition and also attract children to school. (WFP photo/P. Morin)

CHAPTER FIVE

Lessons from Ecology:
Land Habitats and Communities

Succession and Climax

A Tall-Grass Prairie Park/Succession/The Climax Concept/Succession and Climax in Natural Ecosystems/Diversity and Succession/Energy Use and Succession/Climax Community Characteristics

World Biomes

Tundra/Forests and Grasslands/Deserts

Soils and Ecosystems

Soil Types/Soil Formation/Soil Type and Agriculture

Influences on Succession

Succession and Climax

A Tall-Grass Prairie Park

When the first American pioneers moved westward, they found vast expanses of grassland: the American prairies. Their covered wagons, moving through seemingly endless seas of waving grasses, were called prairie schooners. These prairies once covered more than a million square miles (2.6 million km²), from southern Canada all the way to the Gulf of Mexico, and from the Eastern forests to the Rockies (Figure 5.1). Prairie grasses range in size from short grasses a few inches (10 cm) high to tall grasses such as big blue stem, which grows taller than a human. At one time, tall-grass prairie covered 400,000 square miles (1 million km²), but now barely 1% of that remains. The rest, along with 50% of the short-grass prairies, has been farmed or paved over to make roads, housing developments, and shopping centers.

Figure 5.1 The American prairie grasslands once occupied a million square miles across the entire mid-section of the U.S. Tall-grass prairie, where grass grew more than 6 feet (1.8 meters) tall, covered the easterly portion and blended into a mid-region of mid-height and mixed-height grasses. Farthest west was the short-grass prairie. (Source: Save the Tallgrass Prairie, Inc.)

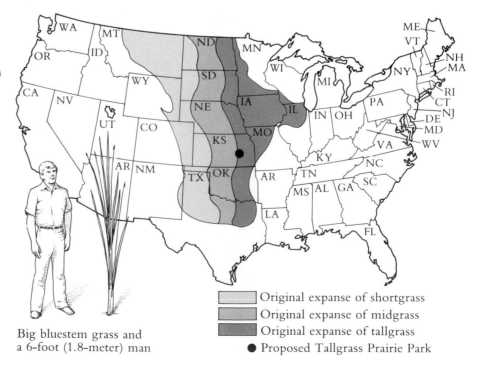

Big bluestem grass and a 6-foot (1.8-meter) man

☐ Original expanse of shortgrass
☐ Original expanse of midgrass
☐ Original expanse of tallgrass
● Proposed Tallgrass Prairie Park

Figure 5.2 A coyote roams through tall-grass prairie in spring, before the grass has reached its full height. (U.S. Fish and Wildlife Service/E. P. Haddon) "There was only the enormous, empty prairie, with grasses blowing in waves of light and shadow across it, and the great blue sky above it, and birds flying up from it and singing with joy because the sun was rising. And on the whole enormous prairie there was no sign that any other human being had ever been there." (Quote from L. I. Wilder, *Little House on the Prairie*. New York: Harper and Row, 1935, p. 40.)

Tall-grass prairie forms a stable ecosystem composed of characteristic grasses: big bluestem, Indian, little bluestem, and switch grass, which grow to heights of 3–8 feet (1–2.4 meters). Many wildflowers are mixed in with the grass and 80 species of mammals, including deer, bobcats, and badgers, find a home there (Figure 5.2). In addition, 300 bird species and well over 1000 kinds of insects inhabit the prairie. However, once the grass is plowed up, it may be difficult or impossible to reestablish the system in any reasonable length of time. Several conservation groups are working to establish a Tall-Grass Prairie National Park before there is nothing at all left to save.

Succession

The prairie grasslands are part of an ecosystem that is characteristic of that given area and also stable in time. That is, the kinds and proportions of the various plants and animals do not change much in time. Not all communities have this sort of stability.

Consider the fate of an abandoned farmer's field in a temperate climate such as the northern U.S. The first year after the farmer gives up the field, it fills with annual weeds, the kind of weeds that grow up each year, produce seed, and die. Perennial weeds, which are dormant during the winter months, start to appear the second year. Eventually, small shrubs and trees grow among the weeds and then take over the field. As the shrubs and trees grow larger and begin to shade the ground, the weeds that once grew in the field no longer find conditions suitable for growth. The kinds of animal species found in the field change as the vegetation alters. This rather orderly process is called **succession**. During succession, each community of plants and animals actually changes the environment in which it lives so that conditions favor a new community, one usually involving larger species. The changes may involve such factors as the temperature and acidity of the soil or the amount of water and sunlight available. For example, pines may be the first trees to grow in a field. Pine branches eventually shade the ground, however, and pine seedlings do not do well in shade. Deciduous tree seedlings, such as oak and hickories, which can grow in shade, may then take over the forest floor and supplant the pines themselves.

Succession occurs not only in abandoned fields but in all kinds of habitats. For instance, after a volcano erupts, tiny plants called algae begin to colonize the cooled lava. Over the years, a series of communities will replace each other until communities similar to those inhabiting the area before the volcano erupted appear. This succession, which starts from bare rock, is called *primary succession*. Primary succession also occurs in ponds and lakes (Figure 5.3). The term *secondary succession* is used to describe the process by which the community originally present in the abandoned farm field is reestablished.

The Climax Concept

In many areas, a community finally appears that is not supplanted by another one, as long as major climate changes do not occur. This last community does not change conditions to make them unsuitable for itself. Such a community may survive, perhaps for centuries, until other factors, such as climate, change, or until disease or human activities are introduced. This relatively stable community is called the **climax community**. Major areas of the world have characteristic, or regional, climax communities. Regional climax communities are called **biomes**. They appear to be determined mainly by temperature and rainfall within a

Figure 5.3 Lake Succession. With time, lakes and ponds may begin to fill up with sediments and plant debris. The waters become more and more shallow and less and less clear. (This is the process of eutrophication.) Eventually, the lake or pond fills so completely that it becomes a bog or marsh. Following this, a succession of land plants appears until the normal climax for the region establishes itself. Succession starting from an aquatic environment, such as a lake or pond, or from a terrestrial environment, such as an abandoned field, tends to lead to the same climax vegetation in a given area, although the path the succession takes may be very different. This supports the idea that the climate of a given area determines what the climax vegetation will be. The pond in the photo is a good distance along the road to becoming a bog. The margins are filled with spongy accumulations of plant material. Only a small area of open water remains. (The Nature Conservancy—Bruce Lund)

given area, although other factors, such as soil type and soil age, can also influence the kind of vegetation that grows up (Figure 5.3).

Succession and Climax in Natural Ecosystems

This concept of succession leading to a climax vegetation is a clear and straightforward idea. Unfortunately, it does not perfectly fit natural ecosystems. Thus, systems occur in which two or more climaxes appear to alternate. An example is Neusiedler Lake on the border between Austria and Hungary, which seems to appear and disappear in long cycles, perhaps 80 years or more. One theory holds that reeds in the lake grow until they occupy the entire lake (which is shallow). At this point the reeds use up the entire amount of water in the lake. The dry lake bed can no longer support the reed ecosystem, which collapses, only to reappear after the lake fills again.

The main point is that there may be long cycles of change even in climax ecosystems. It is important to take this into account when humans are planning changes in ecosystems, lest the natural and human changes add together and have an unexpectedly greater result. For instance, seeds are not produced every year by forest trees such as beech, spruce, and maple, but rather at irregular intervals depending on the climate. Passenger pigeons were probably dependent on seeds of this type. Some ecologists believe that large flocks of passenger pigeons were the result of a series of good seed years. The birds may have become extinct because humans killed so many birds that flocks were too small to survive the following series of poor seed years.

A second problem with the climax concept is that succession sometimes appears to linger at a particular stage rather than advance to the climax. Intermediate succession communities can be very vigorous and sometimes resist replacement by climax species. In some areas of Central Europe, for instance, dry grasslands do not move toward the expected climax beech forest. Regular mowing by humans or grazing by animals, which removes the shoot tips of small trees, halts succession at the grassland stage.

As another example, when tropical rain forests are cleared, only a poor secondary growth forest returns, not the original rain forest. This is due in large part to the fact that nutrients in tropical rain forests are tightly locked into cycles within the living organisms. There is no slow cycle of decay in the soil and eventual reuse of nutrients, as exists in temperate-region forests. Thus when tropical forests are logged, most of the ecosystem nutrients are removed in the tree trunks, leaving little for a new forest to grow up on. If the tropical rain forest climax regenerates at all, it is on a time scale that has little relevance for humans.

However, if the concepts of succession and climax are not applied with too much strictness, they can be useful in describing events in natural systems.

Diversity and Succession

Generally, during succession, the number of species increases in each successive community. The first communities that colonize a lava flow, sand dune, or field are simple: they contain few species. As more plants appear, niches arise for more species of insects and other animals. These creatures, in turn, feed on plants and create pressures for species of resistant plants to flourish, and so on, through the various stages of succession. In this way, diversity increases as succession goes forward. However, diversity of species is not greatest in the climax community. Rather, diversity seems to peak sometime before the climax community appears.

Energy Use and Succession

Another community measure that increases during succession is energy use. As different groups of plants succeed each other, taller species grow up. The amount of biomass, or the weight of living material, both plant and animal, increases. Thus more and more biomass is produced for the same energy input (i.e., sunlight) as succession progresses. However, energy use, too, peaks before the climax community occurs.

Climax Community Characteristics

In climax communities, other factors seem to be more important than those leading to species diversity and increasing biomass. An example is the increasing size of individuals. This gives species the ability to store nutrients or water against seasonal shortages. This and other factors lead to increased competition between species and a loss of species in the climax community.

World Biomes

The world's climates and the factors that influence them are discussed in Chapter 19.[1] The following sections describe the major communities of plants and animals, or biomes, that are characteristic of different climatic regions.

---

1 See the map of world biomes, Figure 19.5, and world climates, Figure 19.3.

Tundra

In the far north, the top few inches of soil thaw only in the summer. Below this thin layer, the soil remains permanently frozen year round. The frozen layer is called **permafrost** and the whole area is known as the **tundra biome**. Roots cannot penetrate the icy layer, and so vegetation is limited to mosses, lichens (such as reindeer moss), grasses, and small woody plants (Figure 5.4). Overall, growth on the tundra is very slow. For this reason, when the tundra is disturbed, as by tire tracks, the scars do not heal. In the summer, the tracks fill with water and mud and in the winter they freeze to ice. This particular problem was of much concern during the planning of the Alaskan pipeline, which crosses so many miles of tundra.

Forests and Grasslands

Below the tundra is an area called the **taiga**, or **northern coniferous forest biome**. Although this biome is cold in winter, when summer comes the soil thaws completely—there is no permafrost (Figure 5.5).

Farther south are several biomes, depending on rainfall and temperature patterns. In areas with 30–60 inches (75–150 cm) of rainfall, moderate temperatures, and definite winter and summer seasons, the **deciduous forest biome** occurs. This includes most of the eastern United States, central Europe, and parts of Asia (Figure 5.6).

Where rainfall is too sparse to support trees, grasslands develop. Grasslands need 10–30 inches (25–75 cm) of rain a year, compared to forests, which require 30–60 (75–150 cm). The taller grasses generally need more rainfall and deeper, richer soils than the short or mid-height grasses (see Figure 5.7). Grasslands also occur in tropic areas where rainfall may total 40–60 inches (100–150 cm). However, in these areas the rain comes all in one season, followed by a dry season.

Areas of the tropics where little seasonal variation in temperature occurs but much rain falls (60–90 inches [150–225 cm] a year) are generally covered by tropical rain forests. Many species of both plants and animals live in these forests. In fact, tropical rain forest biomes (and coral reefs) contain the greatest diversity of life found in any biome; that is, the largest number of different species per square meter are found there (see Figure 5.8).

Figure 5.4 Arctic Tundra. (Pro-Pix/Monkmeyer Press Photo
Service) "Across the northern reaches of this continent there lies a
mighty wedge of treeless plain, scarred by the primordial ice,
inundated beneath a myriad of lakes, cross-checked by innumerable
rivers and riven by the rock bones of an older earth. They are cold
bones into which an eternal frost strikes downward five hundred feet
beneath the thin skin of tundra bog and lichens which alone feel
warmth under the long summer suns; and for eight months of the
year this skin itself is wrinkled by the frosts and becomes part of the
cold stone below. . . .

Yet of all things that it may be, it is not barren. During the brief
arctic summer it is a place where curlews circle in a white sky above
the calling waterfowl on icily transparent lakes. . . . It is a place
where minute flowers blaze in microcosmic revelry, and where the
thrumming of insect wings assails the greater beasts, and sets them
fleeing to the bald ridge tops in search of a wind to drive the unseen
enemy away. And, not long since, it was a place where the caribou
in their unnumbered hordes could inundate the land in one hot flow
of life that rose below one far horizon, and reached unbroken past
the opposite one." (From Farley Mowat, *The Desperate People*.
Boston: Little, Brown and Co., 1959, p. 21.)

(a)

(b)

Figure 5.5 Northern Coniferous Forest Biome. This biome is made up of conifers such as spruce, fir, and tamarack, along with some deciduous trees such as birch. Bears, wolves, moose, squirrels, lynx, and many other mammals are common, as are many species of birds in the summer. (a) A spruce–pine forest in a mountainous region. (b) Alaskan taiga. (Photographs: U.S. Department of the Interior, Bureau of Land Management.)

Figure 5.6 Deciduous Forest Biome. In different areas, the mixture of deciduous trees (trees that shed their leaves each year) making up the forest will vary. For instance, in the North Central U.S., the forests are predominantly beech and maple, while in the West and South, the forests are composed mainly of oak and hickory. Deer, bear, squirrels, and foxes roam the woodlands, while many bird species—such as woodpeckers, thrushes, and titmice—find a home here. In the fall, the leaves turn color and then fall to the ground. (National Park Service photo)

Figure 5.7 Grassland Biome—native big blue stem and wildflowers in the Flint Hills of eastern Kansas. This 1000-acre (405-hectare) tract of virgin tall-grass prairie was donated to Kansas State University by the Nature Conservancy and is near the site of a proposed national park. Many grazing animals live on the grasslands. In the U.S., buffalo and pronghorn antelope were once numerous. Burrowing rodents such as gophers and prairie dogs are found, along with their predators—coyotes, kit foxes, and badgers. (Nature Conservancy)

Figure 5.8 Tropical Rain Forest Biome. Typically, three layers of vegetation occur. A sprinkling of tall trees, which lose their leaves during dry seasons, pokes up through a canopy of broadleaved evergreens and plants. Below this is a third layer of plants that flourish wherever there is a hole in the canopy. Lower levels of the forest are very humid and have a constant temperature. Climbing vines and plants growing on other plants are common. So many species are found in tropical rain forests that it may be hard to find two trees of the same species over an area of several acres. Animals and insects are likewise numerous and diverse. (World Wildlife Fund—U.S./Russel A. Mittermeier)

Figure 5.9 Desert Biome. Desert plants have adapted to survive long periods without water. Some have few, small, leathery leaves, which can be dropped during the dry season to reduce water loss—examples are sagebrush and mesquite. Other plants can store quantities of water in their tissues—these are the cacti and the euphorbias. A third group includes the annuals, which spring up, flower, and set seed in the short period after a rain. Desert animals have similarly adapted to the conditions of their environment. Many are active only in the cooler night periods, burrowing into the ground during the day to remain cool. (National Park Service Photo; Richard Frear)

Deserts

In areas where rainfall is 10 inches (25 cm) per year or less, **desert biomes** occur. Only organisms adapted to long periods of drought and widely varying day–night temperatures can survive these conditions. Because there is little or no vegetation to moderate the sun's energy, days on the desert tend to be very hot, while as soon as the sun goes down, night temperatures plummet (see Figure 5.9).

The biomes described here are the major biomes found in the world. Many sub-types exist, as does a variety of biomes covering smaller areas. (See Figure 19.5 for a map of world biomes.)

Soils and Ecosystems

Soil Types

If we dig a trench vertically down into a particular kind of soil, along the sides of the trench we can see several layers, or horizons. The arrangement of these horizons is called the soil profile (Figure 5.10). This profile differs for soil found in various biomes.

The top layer, or A horizon, varies in depth from less than an inch (2.5 cm) in tropical rain forests to several feet in some grasslands. This layer, often called **topsoil**, contains plant roots, fungi, microorganisms, and a wide variety of soil insects and other burrowing animals. Also found here are dead and decaying parts of plants and animals. In land ecosystems, this is where the chief turnover of organic matter occurs. Here all the unused organic materials from the various trophic levels are recycled and broken down, first to humus and eventually to inorganic materials. **Humus** is an organic substance that is broken down relatively slowly. It is not a plant food; however, it helps retain water in soil and to keep soil loose, or friable. These are important qualities for soil fertility.

Inorganic substances, formed from decomposition in the topsoil, filter down into the second soil layer, or subsoil. Finally, we come to the third layer, or parent material, on which the process of soil formation began.

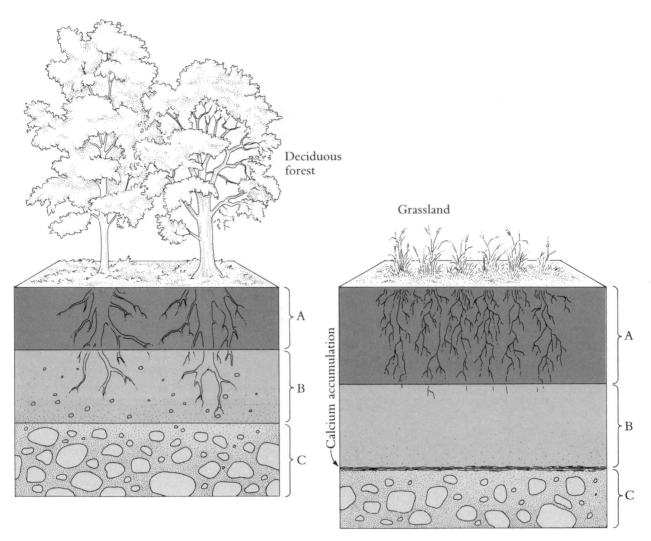

Figure 5.10 Soil Profiles. This is how soil would appear on the sides of a trench cut into the ground. The upper layer, or A horizon, is topsoil and contains plant and animal debris, which is being turned into humus by soil microorganisms. In the B horizon, most of the organic material has been changed into its inorganic components. The third, or C, horizon is parent material; that is, soil characteristic of the rock from which it was formed. Climate and vegetation act on parent rock material over the ages to form soil characteristic of a given area.

Soil Formation

Many soil scientists believe that the type of soil that eventually forms in an area is dependent only on the climate of the region. Although plant and animal materials and the parent rock contribute the substances from which soil is formed, climate determines the process of soil development. According to this theory, the rock from which the soil is originally derived is not important except in early stages of soil formation, when it has an effect on the type of vegetation that arises. In strong support of this theory is the fact that world climate maps closely match maps of world soil types (see Figures 19.3, 19.4, 19.5).

Temperature and precipitation are the two climatic factors of greatest importance in soil formation. As an example of how climate affects soil type, consider hot, humid climates. The warm temperatures and moisture speed the processes of decay, while heavy rainfall causes soluble materials to leach rapidly out of the topsoil layers. You would expect the result to be a topsoil having little organic material and not much in the way of soluble plant nutrients, either. This is exactly what is found in hot, humid tropical rain forests. Some of these soils are so poor in humus and minerals that they are almost white.

Soil formation is a very slow process, taking place over thousands of years in temperate regions. Some temperate-region soils are millions of years old. The tragic processes of erosion, which allow soil to wash away, are not naturally reversible within the length of human life spans. (The time span involved in soil formation is discussed on page 112; erosion is described on page 110.)

Soil Type and Agriculture

Grassland soil is typically black and rich in finely divided, organic humus. Plant nutrients such as calcium, magnesium, and potassium are abundant. Such soils are very valuable for agriculture. Temperate soil forests have less humus, and nutrients tend to leach out more easily, but these soils too can be successfully farmed if fertilizers and lime are added. Other soil types are not at all as suitable for agriculture.

Desert soils tend to be coarse and high in salts or lime, since little water is available to leach them out. Further, evaporation draws salts up to the surface where they can form a crust on the surface, called hardpan or caliche. In the tropics, with high temperatures and abundant rainfall, soils have little humus, because it rapidly decomposes to inorganic materials. Other materials, such as silica, are rapidly leached out by the heavy rains, leaving high concentrations of aluminum, magnesium, and iron oxides in the soil. In some areas, when tropical rain forests are cut down and the soil is laid bare, the iron enrichment of certain soil layers can cause the formation of **laterite**. The result is a soil so hard that it has been cut up into building blocks. Some of these blocks have lasted for 400 to 500 years in Southeast Asian temples. Obviously such soils can no longer be cultivated.

Influences on Succession

The discussion of succession ignored two factors that may have a strong influence on the type of vegetation in a specific area. One factor is, of course, human activities. Wherever we live, we humans turn the land to our own uses, cutting down or burning those species for which we have little use and encouraging those species we need. The way humans manage forests for the production of wood provides an example.

We noted earlier that climax communities do not produce the greatest biomass per unit of energy input. Productivity is not as great in climax ecosystems as in some earlier stages of succession. In order to increase production of food or wood, humans will try to keep an area at a younger successional stage. These younger stages, however, are less balanced and less able to maintain themselves than an older climax stage. Large inputs of energy and ingenuity are needed to maintain them. For instance, in some areas, paper companies plant vast tracts of forest land with softwood tree species that are fast-growing and suitable for pulp. Such stands are much more subject to severe insect damage than the normal, mixed climax community. For this reason, quantities of pesticides must be used to protect the trees. Manufacturing and applying pesticides, of course, requires energy. Pesticides can also have many adverse effects upon the environment, some of which are detailed in Chapter 6.

Grasslands are used by humans to graze cattle or to grow corn and wheat. Up to a point, these are reasonable uses. It is not strange that the American farmer has done well farming ploughed grassland, since the two plant communities, grains and grasses, involve similar ecosystems. Grasslands, however, can be overgrazed (Figure 5.11). In tropical grasslands and in tropical rain forests, most of the plant nutrients are cycling in the plants themselves and do not spend much time in the soil. Thus, removal of too much vegetation by overgrazing removes almost all the nutrients, leaving soils that no longer support plant life.

Similar to the problem of overgrazing is that of overuse of slash-and-burn farming in areas of tropical rain forest. In this type of agriculture, the farmer cuts and burns the natural vegetation before planting his crops. Such plots can only be used for a few consecutive years and then must lie fallow for 10 or even 30 years to regain their fertility. As populations increase in these areas, farmers are tempted to reuse a plot too soon.

Figure 5.11 Desertification. Overgrazing strips land of its protective vegetation. Dust storms can then pick up soil particles and blow them away. This is one way in which once-fertile areas can come to resemble deserts. (USDA—Soil Conservation Service)

A special influence humans have on vegetation is the result of strip mining. If the stripper is not careful to save the topsoil and spread it out on top again after he fills in his mining trenches, a wasteland can result. The top layers of soil, which contain the nutrients cycling from dead plants and animals to living ones, are the most important from the point of view of supporting growth. When the strip miner turns the layers over, the deeper nutrient-free layers are turned uppermost. This, in addition to the acidity produced by coal wastes, gives rise to a soil in which almost nothing can survive. Only imported topsoil or the slow, centuries-long processes of soil formation and washing away of acid, will restore the original vegetation to land treated in this manner. Erosion, as described in Chapter 6, can produce similar results because the topsoil is washed away by rains. These are examples of how growing human populations, with their increasing needs for food, fiber, and mineral resources, can unbalance their environment.

On the positive side, people have turned deserts into gardens by adding the missing factor—water. Here again there are problems, however. The rapid loss of water by evaporation leaves behind minerals. Unless this is accounted for in irrigation schemes, minerals can accumulate until nothing will grow.

The other factor to which little attention has been given is fire. Many ecosystems are adapted to the natural occurrence of periodic fires. African grasslands, for instance, are composed of fire-tolerant species because fires are common during the long dry season. The topic of fire and forest management is covered in Chapter 34.

Questions

1. Describe what is meant by succession.
2. What is the main characteristic of a climax community? What are some of its other characteristics compared to earlier stages of succession?
3. What are the major factors influencing which climax community grows up in an area?
4. UNESCO has a program to identify and preserve examples of all the world biomes. What is a biome? Of what possible use could it be to preserve an example of each type?
5. Do soils influence what vegetation grows up in an area or vice versa? Explain.
6. What kinds of ecosystems have soils that are best for agriculture? Why?
7. Explain how clearing forest land for farming is an example of increasing food production at the cost of unbalancing the environment.

CHAPTER SIX

Soil Erosion
and Pesticides Usage

Soil Erosion
*Kinds of Water Erosion/Environmental Effects/Effects of
Wind Erosion/Methods of Controlling Erosion/Encouraging
Soil Conservation*

Pesticides and Food Production
*The Need for Pest Control/Early Use of Pesticides/Biological
Magnification/Persistence/Resistance/The Organic Farming
Solution/The Integrated Pest Management Solution/Pest
Control Experts*

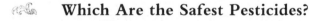 **Which Are the Safest Pesticides?**

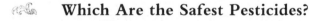 **How Much Is a Part per Billion?**

CONTROVERSIES:

6.1: *Should the U.S. Sell Dangerous Pesticides
Overseas?*

6.2: *Organic Gardening Economics*

Soil Erosion

Storm clouds gather and a brisk wind stirs the poplars in the hedgerow. A tiny drop of rain falls on the freshly plowed earth. Then another drop, and another, and the storm begins in earnest. Each drop hits the earth like a miniature meteor—soil particles spray in a circle around the drop and a small crater is formed. As more droplets reach the earth, their arrival rate begins to exceed the rate at which the water can be absorbed into the soil. Puddles form, then rivulets. The looser and finer soil, the best soil, is picked up by the rivulets and carried downslope. The velocity of the rivulets increases; more and more soil is carried in suspension. Rivulets join to form temporary streams, which flow in gullies, and the rapidly flowing water, en route to a true stream, cuts sharply into the earth.

This event is not isolated or uncommon—it is any rainstorm on a plowed field. Multiplied by thousands of farms or millions of acres, it results in as much as four billion tons of soil being carried off each year by water erosion in the U.S., about three-quarters of it from farmland. Wind may scour away another billion tons of soil. A report by the Soil Conservation Service in December, 1980, estimated that as much as one-half of all cropland in the U.S. was being eroded so fast that its productivity would suffer. Such losses as these must have consequences. What are the effects of erosion on the land and water? How can erosion be slowed?

Kinds of Water Erosion

Soil scientists divide water erosion into three components: sheet, rill, and gully erosion. In sheet erosion, surface water moves downslope in a wide flow; it may roll soil particles with it, bump them along, or completely suspend them in the water, depending on the size of the particles and the velocity of the water. Since sheet erosion occurs over wide areas in a field, it may not be noticeable at first, and awareness of it may dawn only after much damage has been done.

In rill erosion, in contrast to sheet erosion, the water joins into rivulets (Figure 6.1). Flowing at higher velocities through narrow miniature valleys, the water has greater erosive power. However, because rivulets are small and can be plowed up with the rest of the soil and smoothed over, they are also easy to miss in farming. You may be able to see large rills forming in a hill of fresh earth at a construction site.

Gully erosion is not so easy to miss as sheet or rill erosion; the rivulets join into such large flows that tractors cannot smooth them over (Figure 6.2). Gullies resemble ordinary streams with soil banks, except that water flows in them only after rainstorms; at other times they are dry. The banks of the gully are usually only loose soil, which the water erodes away. Earth overhangs may be exposed by the water that undercuts the soil banks; since they are structurally weak, it is common for the overhangs to collapse into the gully, accelerating the erosion process.

Figure 6.1 Severe Rill Erosion in the State of Washington. A fine tilling of the summer's cover crop, combined with poor growth in the fall of the wheat crop, left this steep slope in a highly eroded condition. The county crew that came to clear the road was shooed off by the farmer, who told them, "I'll haul the soil back up the hill." (USDA, Soil Conservation Service; Photographer: Fred Wetter.)

Figure 6.2 Gully Erosion in Montana. This gully cuts deeply through a hay meadow. The farmer will have to do significantly more plowing to prepare his land—now in two fields instead of one. (USDA, Soil Conservation Service; Photographer: D. J. Anderson.)

Wind also erodes soil, lifting and transporting the finer particles and organic matter. Such erosion leaves behind the less desirable coarser soils, which, because they retain water less well, are themselves more likely to erode.

All these forms of erosion carry away soil nutrients and plant residues, essential components of a productive earth. Pesticides are also absorbed onto the soil particles and can later dissolve in waters into which the soil is carried.

Environmental Effects

It is well to remember that erosion is a natural process. It has been observed for centuries; it even has beneficial effects. The rich delta lands built up at the mouths of the Mississippi, the Rhine, and the Nile, all exist because soil was carried off the land in enormous quantities. The richness of the river deltas is a two-edged sword, however. If the delta soils are rich, the soil that remains on the land upstream must have been robbed of some of its value. This, in fact, is one of the principal concerns we have about soil erosion: the decreasing productive capacity of the soil.

Although erosion is a natural process, the massive extent to which it is presently occurring is the result of human activities. The farming of more land, only in this century made productive by irrigation; the grazing of cattle; the construction of buildings and highways; and the activities associated with mining—all disturb the soil and accelerate erosion. In addition, the pavement that now overlays so much of America contributes to faster erosion. Because water cannot be absorbed by paving materials, it collects and may run swiftly off the pavements. Its increased volume and velocity will erode and suspend soil, carrying it into streambeds.

Rate of soil formation versus soil loss. Soil is formed in a slow and continuous process. The rate of new soil formation is about one inch (2.5 cm) of topsoil every 100–1000 years. However, the rate varies widely, depending on climate, vegetation, soil type, and land use.

Soil erosion is measured in tons of soil lost per acre (metric tons/hectare, or t/ha). To understand what rates of soil loss mean, we need to know a few facts. About 6 inches (15 cm) of topsoil is usually cultivated in modern agriculture. This 6 inches weighs approximately 1000 tons/acre (2245 t/ha). If soil erodes at about 15 tons/acre/year (33 t/ha/year), about 0.10 inch (0.25 cm) of soil is lost each year and the whole plow layer in 60–70 years. Experts estimate that, depending on soil types, only 1–5 tons of soil can be lost per acre each year (2.2–11 t/ha) without decreasing productivity in the long term.

Loss of topsoil in the U.S. The three billion tons of soil eroded from U.S. farmland each year come from individual farms, where erosion could be more or less than some average figure. From sloped land that is unprotected and has been freshly plowed, under an intense rain, annual erosion losses could reach 60–100 tons per acre (132–220 t/ha). On a national basis, cropland erosion losses average about 4–5 tons per acre (9–11 t/ha) per year, but averages are deceiving. Only 6% of the nation's cropland accounts for some 43% of the soil erosion (Table 6.1).

Table 6.1 Serious Erosion Problems in the U.S.

Area and location	Severity of problem
The Palouse (Hilly area covering parts of Washington, Oregon, and Idaho.)	Runoff from rain and snow causes erosion rates of 50–100 tons/acre (110–200 t/ha).
Southeastern Idaho	Summer rainstorms and late winter rains cause erosion rates of up to 16 tons/acre (35 t/ha) on slopes.
Texas Blackland Prairie	Gently rolling, highly erodable land. Losses higher than national average.
Southern Mississippi Valley	Row crops grown without soil conservation practices experience losses of 20 tons/acre (44 t/ha).
Corn Belt States	Some of the highest average rates in the country. Iowa: 9.9 tons/acre (22 t/ha); Illinois: 6.7 tons/acre (15 t/ha); Missouri: 10.9 tons/acre (24 t/ha).
Aroostook County, Maine	Potatoes grown on slopes of up to 25%. Upper 2 feet (0.6 m) of soil gone. Some sloping fields lose as much as 160 tons/acre/year (352 t/ha/year).

(Source: America's Soil and Water Trends, U.S. Department of Agriculture, Soil and Water Conservation Service, December 1980)

Such losses create numerous problems. The productive capacity of the soil may decrease, depending on the depth to which the topsoil extends. One figure cited is that a loss of 2 inches (5 cm) of topsoil will cause the yield per acre (measured in bushels, let us say) to decrease by 15%. The loss of 6 inches (15 cm) of topsoil will make most cropland much less productive, if not unsuitable for farming. For example, losses of this magnitude in Tennessee have reduced corn yields by 42%. These decreases in productivity may occur even if the farmer tries to make up for lost nutrients by adding more fertilizer. Loss in productivity is not only due to loss of the more fertile topsoil, but is also due to changes in the soil structure. When topsoil has been carried off, the remaining soil may not absorb water as well. Hence, more runoff is likely to occur, making less water available to the crop. Since the subsoil is generally less permeable, less water can be stored between rains, making the likelihood of damage from droughts greater.

The less desirable subsoil remaining is also more difficult to till; rills make plowing more difficult yet; and gullies may block plowing entirely, resulting in the complete loss of portions of the land. Not only is the land lost, but the fields are cut into smaller blocks by the gullies; hence, the time and effort to cultivate an area increase. It is akin to having to mow the grass on an acre of land (about 200 feet by 200 feet or 61 m by 61 m) in blocks that are 10 feet (3 m) on a side, instead of making cuts that are 200 feet (61 m) long. A final effect of water erosion on farmland is the damage to seedlings. These immature plants may be dislodged and carried off, or they may be buried by sediment deposited on them.

Effects on streams, reservoirs, and aquatic life. Soil particles carried in suspension and bumping along the bottom of streambeds will eventually settle when the velocity of the stream decreases sufficiently. As sediments accumulate on the bottom of the stream channel, the carrying capacity of the stream decreases. (The carrying capacity is the maximum flow rate the stream can support without overflowing its banks.) At the same time, runoff will increase from eroded areas because of a loss of absorptive capacity of the soil. Thus, more water will be coursing down a stream channel that is unable to handle it. Compounding the problem is the fact that the volume of runoff is increased again by the volume of the soil particles carried

in suspension.[1] Flooding by streams flowing through badly eroding areas is more likely.

Not only are stream channels filled up by sediments, but reservoirs fill up as well. These structures, which are used to store water for irrigation, for hydropower, and for water supply, bring the flowing waters to a standstill. When this occurs, most of the sediments, which were in suspension, drop down. The build-up of sediments in reservoirs decreases the quantity of water available on a reliable basis. Large reservoirs are actually designed with extra capacity to hold the sediments. Of course, this added capacity drives up ths cost of the reservoir. A 1968 survey of nearly 1000 reservoirs in the U.S. showed about 40% of them filling at over 3% per year. These were most often smaller reservoirs, less than 100 acre-feet.

Besides using up the capacity of rivers and reservoirs, sediments carried in suspension affect aquatic life. Sediments may bury the habitat of bottom feeders. The turbid water allows less sunlight to enter, hindering the growth of aquatic plant species.

Finally, we must list among the effects of water erosion the poor water quality caused by the substances carried into our streams in runoff. Fertilizers and animal wastes use up vital oxygen in the water and destroy the habitat of aquatic life. Pesticides also enter the water through erosion. In fact, the Commission on Pesticides of the Department of Health, Education, and Welfare, estimated that erosion is the most common route of pesticide entry into water.

Effects of Wind Erosion

Wind erosion has its own set of effects. Of course, its effect on the productivity of the soil is similar to the impact of water erosion. The lighter, more granular soils, with high organic content, are scoured away by the wind, leaving a coarser, less absorbent subsoil exposed. The impact of wind erosion on human activities, however, is far different from that of water erosion.

Dust storms, when the winds lift and carry soil particles on a massive scale, are fearful indeed. One episode is recorded in which nearly 1300 tons of parti-

---

1 One extreme case of erosion has been documented on the Yellow River (Hwang Ho) in China. At flood stage, the flow in that river has been as much as 50% sediment by weight. The yellow color of the river is due to the color of the soil particles eroded from the uplands.

Figure 6.3 In the Midwest suffered a series of severe dust storms caused by winds blowing soil off fields during dry weather. So many dust storms hit the Southern Great Plains that the area became known as the "Dust Bowl." (Reprinted by permission from R. P. Beasley, *Erosion and Sediment Pollution Control*. Copyright 1972 by Iowa State University Press, Ames, Iowa 50010.)

cles were airborne per cubic mile of air (455 t/km^3). Annual soil losses due to wind erosion have been measured at levels up to seven tons per acre. People and livestock have difficulty breathing in dust storms and develop respiratory and eye ailments. Crops, especially seedlings, cannot survive such conditions. In the 1930s, the Midwest went through such a severe series of dust storms and blowing soil that the Southern Great Plains were nicknamed the "Dust Bowl" (Figure 6.3). Dust storms seem to come in bunches. For instance, there were 120 storms recorded near Dodge City, Kansas, in 1936–1937. Another cluster of storms struck the area about 20 years later. Apparently some combination of wind, weather, and soil conditions triggers these storms.

Although the storms are distinct and noteworthy events, blowing soil is common. Its effects are the same as those of dust storms, though slower and less dramatic. Scour of plant leaves by soil particles may reduce crops. If plant roots arc uncovered, the crop will not survive. Soil is stripped of its richer portions.

Methods of Controlling Erosion

Perhaps the most important single activity that brings about soil erosion is farming. When soil is turned over, exposing a surface with no cover—indeed, with no protection whatsoever—the possibilities for erosion are greatly increased. One set of alternatives to control erosion, then, concentrates on how soil is tilled.

Reduced tillage farming. The fastest-spreading soil conservation technique today is undoubtedly reduced tillage farming. In this method, some sort of crop cover is left on the soil at all times. The soil is never plowed and harrowed to turn under old crop residues and break up soil lumps. Instead, herbicides (which kill vegetation) are applied to kill weeds or the previous crop. Then the new crop seeds are planted, using special equipment that inserts the seeds directly into the old stubble or opens only a tiny furrow for the seed. For instance, rye may be planted in a field in the fall and then killed by a herbicide in the spring before corn is planted in the field. The remains of the rye crop are left on the field to reduce soil erosion by decreasing the impact of raindrops and slowing the flow of water across the field (Figure 6.4).

Reduced tillage farming can decrease soil erosion to almost zero. In the past 10 years, the number of acres in the U.S. planted by reduced tillage has increased dramatically from about 35 million to almost 100 million acres (14 to 40 million hectares). (Compare this to a total U.S. cropland of 413 million acres [165 million hectares].)

Developing countries also might profit from this technique. This is especially true in tropical climates, where the topsoil layer is very thin and quickly eroded. Although reduced tillage is now primarily used by large farmers in developing nations, small farmers could benefit from the reduced plowing labor involved. The energy involved in plowing fields with hand tools often limits the amount of land a farmer can plant.

There is a major environmental problem associated with reduced tillage farming, however. This is the large amount of pesticides necessary to kill weeds and plant pests that are ordinarily controlled by plowing. Of course, farmers using conventional techniques may already be using pesticides. Further, the reduced runoff from reduced tillage fields also decreases the runoff of pesticides. However, it is a problem that needs to be studied further. (Pesticides are discussed in more detail later in this chapter.)

Figure 6.4 Reduced Tillage Farming: Soybeans Growing in Grain Stubble. Reduced tillage farming, in which crops are planted into the residue of a previous crop, has several advantages. Soil erosion is reduced almost to zero. In addition, water is conserved, as runoff and evaporation decrease. Less expensive machinery is needed and about 10–20% less energy is used, because the tractor makes only one or two passes over a field. This last feature also helps reduce the packing down of soil by heavy machinery. On the negative side, insects and plant diseases may be more common because the nonplowed seedbed provides a more favorable habitat. Soil temperatures are lower because of the plant cover. This is a negative feature in some areas, such as the northern U.S., but can be an advantage in the tropics. In general, greater management ability on the part of the farmer is needed, at least until the reduced tillage technique is completely adapted to the particular farming area. (USDA—Soil Conservation Service; Photographer: L. C. Harmon)

Crop rotation. On plowed land, the method of plowing and the type of crops planted can be planned to reduce erosion. When crops are rotated, by planting cash crops such as wheat alternately with grasses or legumes, the soil gains protection from erosion and may be nourished as well. Unfortunately, the high price of wheat in the late 1970s drew many farmers away from wheat–fallow rotations and wheat–sorghum–fallow rotations[2] to continuous planting of wheat; no intermediate plantings were used to firm the soil and restore its nutrients. The problem of making farmers act in their long-term interests as opposed to short-term interests is a basic one that we shall address in a moment.

Contour plowing, terracing, and shelter-belts. An important factor in the severity of erosion is the slope of the soil: the greater the slope, the more severe erosion is likely to be (other factors such as rainfall and soil types remaining the same). This has led to the development of soil management practices such as contour plowing or terracing. Such methods are designed to interrupt, divert, and absorb the flow of water in its path down a field (Figures 6.5 and 6.6).

Two other major influences on the extent of erosion—the frequency and amount of rainfall—are not generally controllable by human activity.

We conclude by noting that methods involving the use of vegetation can protect the soil not only from water erosion but from wind erosion as well. Protection against wind erosion, however, may also involve planting trees such as willow or poplar, two fast-growing species. Arranged in long lines, called shelterbelts, the trees decrease the wind speed downwind from the belt. Earth banks, wooden fences, and rock walls may also be useful in checking wind erosion.

Encouraging Soil Conservation

One would think that soil conservation is in the farmer's best interests. It may be, but only in the long run. In the short run, the farmer must survive to plant another year. We indicated a moment ago that crop rotations in growing wheat may be abandoned in the quest for immediately higher profits. Soil conservation

---

2 A fallow crop is one that nourishes the soil. Sorghum, although grown as a grain, is a grassy plant.

Figure 6.5 This aerial view of a field of peach trees in New Bridgeton, New Jersey, resembles nothing so much as a tufted bedspread. Yet the pattern actually reflects good soil conservation practice. Between the rows of peach trees are strips of clover sod, which simultaneously enrich the soil and help to anchor it in place. (USDA, SCS photo by Clarence Deland)

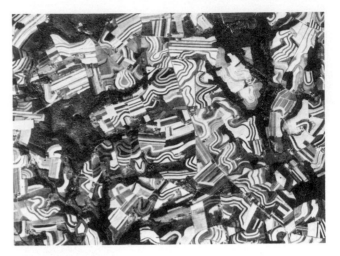

Figure 6.6 Conservation Farming in Pennsylvania. One of the first Soil and Water Conservation Districts in the U.S. was founded in Lancaster County, Penn., in 1937. From a wasteful pattern of straight-line furrows that climbed up hillsides, the pattern of cultivation evolved to its present state. This aerial view shows furrows following the contours of the hills and strip cropping (growing alternate strips of the main crop and a ground cover, such as hay or grass), two soil conservation practices that check erosion and hence preserve the soil. (USDA, SCS photo by Don Schuhart)

measures that involve restructuring the land cost money. And land taken out of production is land that will not produce a crop and an income. Furthermore, the loss in fertility of the soil due to erosion can temporarily be overcome by liberal doses of fertilizer.

The U.S. Department of Agriculture (USDA) has sponsored soil conservation programs since the middle 1930s, when the Dust Bowl era dramatized how very fragile our soil resource is. Recent studies have shown, however, that the programs are not working. A 1977 study by the U.S. General Accounting Office investigated conditions and practices on 283 farms, chosen on a random basis. The farms, which were in the Midwest, Great Plains, and Pacific Northwest, consisted of 119 farms that had had conservation plans prepared by the Soil Conservation Service of the USDA and 164 farms that had not. Of the 283 farms, 84% were losing soil by erosion at an annual rate of five tons per acre (11 t/ha) or more; experts believe a loss of five tons per acre each year cannot be sustained without a loss of productivity, even by deep soils. Of the 119 farms with conservation plans, fewer than half were actually using them.

Although the USDA does have a cost-sharing plan (with about $200 million to share each year) to stimulate conservation measures by farmers, many of the uses to which this money has been put turn out to be

measures to improve the yield of crops rather than stabilize the soil. A more radical proposal, one that definitely limits the farmer's freedom of action, would link farm benefits to the soil conservation measures put into effect. That is, the farm owner would only be able to participate in the price support, crop insurance, and farm loan programs if a defined program of soil conservation had been or was being carried out on the farm. Precedents exist for such a linking of programs—the Federal Housing Administration will only insure loans for houses that meet a minimum standard of quality. The U.S. Department of Transportation allocates federal money only to states that enforce a safety measure, the 55-miles-per-hour speed limit. Thus, the linking of conservation programs with farm benefit programs is not unthinkable.[3] The more appropriate question is whether such a concept could ever be legislated, given the many farm states in the nation and their representation in Congress. It might take another Dust Bowl.

---

3 More on this in T. Barlow, "Solving the Soil Erosion Problem," *Journal of Soil and Water Conservation,* **32**(4) (July–August 1977), 147.

Pesticides and Food Production

The Need for Pest Control

Each year an estimated half of the world's critically short food supply is consumed or destroyed by insects, molds, rodents, birds and other pests that attack foodstuffs in fields, during shipment and in storage.[4]

Experts believe that if the insects and diseases that attack crops such as cereal grains in the field could be controlled, some 200 million extra tons of grain would be available each year. This amount of grain would feed one billion people. Moreover, if we could stop the pests, such as rats and insects, that eat or destroy food once it is harvested, there could be as much as 25% more grain available to the world's hungry people, without any increase in the amount of food actually grown (Figure 6.7). Estimates say that in parts of India, as much as 70% of stored foods are lost to pests each year.

The challenge, and it's not a simple one, is to solve these pest problems without harming people or the environment. In India, stored grain is often illegally treated with the pesticide DDT to kill insects. As a result, people in India have more of this poisonous and carcinogenic chemical in their body fat than people in any other country in the world.

The use of pesticides creates many environmental problems. This section focuses on the use of pesticides in growing food. The problem of pesticides in water is covered in Chapter 11, while the effects of a specific pesticide, DDT, on wildlife are noted in Chapter 4. In addition, the possible effects of pesticides in food on human health are considered in Chapter 31.

Early Use of Pesticides

Any material used to kill pests is called a **pesticide**, and usually any organism that competes with humans for food, fiber, or living space is called a pest. The use of chemicals to kill pests is not a new idea. For centuries, farmers have used minerals such as arsenic, lead, and mercury or natural plant substances such as pyrethrum (obtained from a daisy-like plant) to kill insects and other pests. However, in 1945 a new era in pest control began—DDT came into widespread use

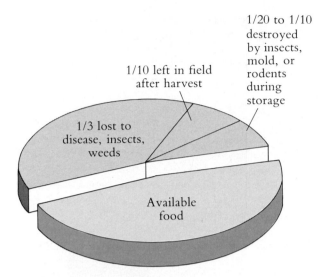

Figure 6.7 Crop Loss in the Field and in Storage. In the U.S., more than one-third of the crops grown are lost to insects, disease, and rodents in the field and during storage. One-third is lost in the field, another 10% is left in the field after harvest, and 5–10% is lost during storage. In other parts of the world, these figures are even higher. On average, as much as one-half of the world's food crop is lost to disease, insects, and rodents during growing, shipping, and storage.

to control the lice and fleas that plagued the armies of World War II. DDT was the first widely used synthetic, or laboratory produced, pesticide. In the years that followed, many other synthetic pesticides were introduced. In a number of cases, spectacular increases in crop yields resulted where the new pesticides were used. By 1967, half of all of the pesticides used were, like DDT, in the chemical family of chlorinated hydrocarbons.

Slowly, however, evidence began to accumulate that the synthetic pesticides were not an unmixed blessing.

Biological Magnification

Problems with pesticides arose in three main areas. In the first place, certain pesticides[5] tend to accumulate in living organisms. In some cases, pesticides not only accumulate in organisms to levels greater than are found in the environment, but concentrations keep in-

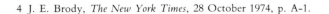

4 J. E. Brody, *The New York Times*, 28 October 1974, p. A-1.

5 Mainly, chlorinated hydrocarbons such as DDT and mercury-containing compounds.

creasing as they move up food chains. This is the effect known as **biological magnification**.

DDT is an example of a pesticide that is magnified biologically. DDT is soluble in fat. When an organism ingests DDT—whether from water, from the detritus of plants that have been sprayed, or from insects that have eaten the plants—the DDT becomes concentrated in the organism's fatty parts. DDT is lost very slowly from these fatty tissues. If another creature in the food web eats the first organism, the consumer will be eating a concentrated dose of DDT (Figure 6.8). Organisms at the top of their food chain (e.g., humans or predatory birds such as eagles and falcons) are eating foods with a much higher level of DDT than is generally present in the environment. The DDT has been biologically magnified. One effect of such DDT levels on birds is that they lay eggs with shells that are much thinner than normal. These thin shells break easily and so are no protection for the developing chick inside the egg. In this way, the Eastern peregrine falcon failed to reproduce and so became extinct in the 1960s.

Persistence

The second area of concern involves the length of time pesticides remain in the soil or on crops after they are applied. The chlorinated hydrocarbons, such as DDT, and pesticides that contain arsenic, lead, or mercury, are all known as persistent pesticides. This means that they are not broken down within one growing season by sunlight or bacteria. The half-life of DDT, for instance, may be as long as 20 years. (That is, after 20 years, only one half of an amount of DDT used has been broken down to simpler compounds.) Mercury and arsenic are never really broken down—they are moved around or buried in muds.

Persistent pesticides can build up in the soil if a farmer applies them year after year. In some orchards where lead arsenate has been used to control insects, arsenic in the soil has built up to a level that kills the fruit trees. The long life of persistent pesticides is a major factor in the process of secondary contamination, whereby foods that were never treated with pesticides still become contaminated. For example, DDT sticks to soil particles after it is applied. In many areas, dust storms have picked up this contaminated dust and blown it, literally, around the world. Rain washes this

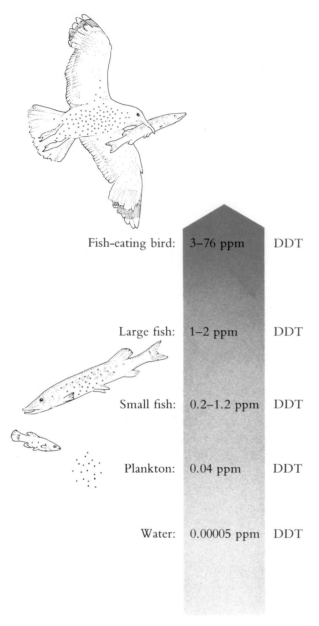

Fish-eating bird:	3–76 ppm	DDT
Large fish:	1–2 ppm	DDT
Small fish:	0.2–1.2 ppm	DDT
Plankton:	0.04 ppm	DDT
Water:	0.00005 ppm	DDT

Figure 6.8 Biological Magnification. In 1967, fish-eating birds from a Long Island salt marsh estuary contained almost one million times more DDT than could be found in the water. At each step of the food chain, DDT was concentrated, as organisms consumed and absorbed more DDT than they were able to excrete. If you are not familiar with the term parts per million, see page 127. (Adapted from W. Keeton, *Biological Science*, 3rd ed. New York: W. W. Norton, 1980.)

dust out of the air in places where pesticides and even farming are practically unknown. Thus seals, penguins, and fish in the Antarctic all show traces of DDT in their fat. One author points out that during the heyday of DDT use, in the 1960s, 40 tons of DDT were dumped on England every year in rainfall. In this way, pastures and crops of all kinds receive an unintended and unwanted treatment with DDT.

Resistance

Biological magnification and persistence are not the only problems associated with the use of pesticides. A third serious problem is that pests can become **resistant** to pesticides—the pesticide will no longer kill them. This can come about because of mutations occuring in some of the enormous numbers of insects hatching each season. **Mutations** are changes in an organism's inherited genetic material, the material that determines the organism's characteristics.

A pesticide may kill off most of the insects in a certain area. However, a few that have mutated so that they have slightly different characteristics than the others, may survive. These few can repopulate the sprayed area, passing on their pesticide resistance to their offspring. In some cases, pesticides themselves may cause the development of resistance by stimulating the production in pests of enzymes that act to break down pesticides. Those insects that have been stimulated to produce the largest amounts of the enzymes will be most likely to survive and reproduce. In several generations, insects with sufficient enzyme to break down any reasonable amount of pesticide will appear. This sort of effect has led to the development of houseflies resistant to all of the major types of pesticides used.[6]

The commonly used "broad-spectrum" pesticides, which kill a wide variety of insects, make the development of resistance simpler. They ensure that the newly resistant insects will have little competition for food, and also kill off many parasites and predators of insect pests. In fact, predators may receive a higher dose of pesticide than their prey because of biological magnification. Resistant insects are thus given a good chance to establish themselves, free from normal controls. In a number of cases, pesticides have created new

pests in this manner. For instance, the ladybug normally controls a type of citrus scale insect. If an orchard is sprayed with DDT, however, the ladybugs are killed while the scale insect is not. The scale insect can then multiply unchecked and cause severe damage to the citrus crop.

In some cases, extensive use of pesticides has left farmers worse off than before. For instance, cotton has long been grown in Peru in the Cañete Valley. Before World War II, insects were controlled with arsenic and nicotine sulfate. Yields of cotton were about 470 bushels per acre (42.5 kl/ha). In 1949, crops were severely damaged by cotton bollworms and aphids, causing growers to try the new chlorinated hydrocarbon insecticides such as DDT and toxaphene. At first, it seemed to be a successful move. Yields nearly doubled. However, by 1955, the picture had changed. Not only had insect pests become resistant to synthetic insecticides, but new pests were appearing because the insecticides had killed off the beneficial insects that preyed on pests. Chemicals were unable to control insect populations and, by 1956, the cotton-growing industry had collapsed.

A similar sequence of events occurred in the Rio Grande Valley in Texas. By 1968, cotton pests were resistant to all available insecticides. Despite as many as 15–20 treatments with all possible combinations of pesticides, many fields were totally lost due to insect damage (but see Figure 6.9).

Even the successful public health uses of pesticides are threatened by insect resistance. In many countries, walls inside houses are sprayed with DDT and dieldrin. For months after the spraying, mosquitoes that land on the walls are killed. In this way, diseases like malaria and yellow fever, which are spread by mosquitoes, can be controlled. Worldwide, malaria is the largest single cause of death and debilitation. In the 1950s, campaigns to eradicate malaria were begun. Based mainly on spraying with chemical pesticides such as DDT, the campaigns were at first successful. However, the mosquitoes became resistant to DDT and then to propoxur, the spray that replaced DDT. Now cases of malaria are once more on the rise. In 1955, when eradication programs began, 250 million cases were reported and 2.5 million people died. By 1965, only 107 million cases were reported, but in 1976 the number rose to 150 million. Furthermore, the malarial parasite is becoming resistant to drugs used to treat the disease itself.

---

6 Which pesticides are safest to use? See p. 127 for one view.

Should the U.S. Sell Dangerous Pesticides Overseas?

States have the responsibility to insure that activities within their jurisdiction or control do not cause damage to the environment of other states.

United Nations Conference on the Human Environment, Stockholm, 1972

In a world of growing food interdependence, we cannot export our hazards and then forget them. There is no refuge.*

D. Weir and M. Shapiro

It only results in a loss of American jobs.†

Lewis A. Engman, President, Pharmaceutical Manufacturers Association

The U.S. presently sells overseas some pesticides that are banned in the U.S. (except for public health emergencies), including DDT, Aldrin, Dieldrin, Heptachlor, and Chlordane. In 1976, several environmental groups successfully sued to prevent any sale of these unregistered pesticides that involved the use of money from the U.S. Agency for International Development. Chemical companies can still ship unregistered pesticides and other countries can still buy them, however, as long as they don't use U.S.A.I.D. money. Ten percent of U.S. pesticide production consists of unregistered or banned products destined for overseas markets.

One problem with these sales is that agricultural workers in undeveloped countries may be poisoned by such pesticides. (A lack of pesticide regulation, combined with a lack of education amongst the laborers, makes even the use of U.S.-approved pesticides dangerous.) The World Health Organization estimates that half a million people are poisoned each year by pesticides, most of them in undeveloped countries.

Another problem is that the greatest proportion of pesticide use in undeveloped countries (50–70%) is on crops destined for export to developed nations such as Japan and the U.S. A 1979 report by the U.S. General Accounting Office stated that a large portion of the food imported into the U.S. may contain unsafe pesticide residues.

What do you think? Should we allow sales to other countries of pesticides we restrict severely in our own country? Do we have the right to make this decision for other people? Do we have the responsibility? What about the argument that if we do not sell pesticides to other countries, someone else will?

---

* Weir, D. and Shapiro, M., "The Circle of Poison," *The Nation*, 15 November 1980.
† *Science*, **211**, 27 March 1981.

Table 6.2 Pesticides in Americans

	Percent of people with detectable amount	No. of people examined	Mean value found (in parts per million)
DDT in blood	99	6000	—
Other halogenated pesticides in blood	4–14	6000	—
Petachlorophenol in urine	79	4000	0.007
Other halogenated compounds in urine	4–7	4000	
Carbamates in urine	2–4	4000	~0.01–0.04
Organophosphates in urine	6–12	4000	~0.02–0.03
DDT in adipose tissue	99.9	785	3.6
PCBs in adipose tissue	31	785	1–3
Dieldrin in adipose tissue	95.5	785	0.12
Chlordane in adipose tissue	95	785	0.13

As part of a continuing survey of the health of Americans, blood, urine, and adipose (fatty) tissue samples are being checked for traces of pesticides. This table highlights some of the findings of pesticides in human tissues and fluids as sampled in the studies.

Data from: Health and Nutrition Examination Survey (HANES) and Human Adipose Tissue Survey, National Center for Health Statistics, Pesticides Monitoring Report, U.S. EPA, Exposure Evaluation Division, Volume 1, Nos. 1 and 2, 1980.

Thus, to summarize, synthetic chemical pesticides have not proven to be as easy to use nor as beneficial as was originally hoped. Insects and other pests, such as rats, have shown the ability to develop resistance to chemical pesticides, while pesticides that are persistent can accumulate from year to year in the soil and in the muds at the bottom of lakes and rivers. Fat-soluble pesticides absorbed from water, grass, or crops become concentrated in the milk and body fat of many animals, including humans (Table 6.2), at the level of parts per million or parts per billion. (Do you understand what is meant by parts per million or parts per billion? If not, see the supplemental material at the end of this chapter.)

The Organic Farming Solution

If pesticides are now seen as a somewhat tarnished miracle, what might be a better strategy in growing food and fiber? One possibility is to apply no synthetic pesticides at all, and use only manure or special crops for fertilizer. This is, of course, organic farming. Once the only possible type of agriculture, organic farming became associated with health-food faddists and back-to-nature cultists after the development of synthetic pesticides and fertilizers. Often, now, organic farming is more charitably viewed as a fine ideal for the home gardener, but not a reasonable goal for large-scale farmers. In 1971, Earl Butz, then Secretary of Agriculture, said in an interview, "Before we go back to an organic agriculture in this country, somebody must decide which 50 million Americans we are going to let starve or go hungry." Nonetheless, evidence is accumulating that in some situations, and when properly carried out, organic farming may be a match for what is now called conventional farming—i.e., using synthetic pesticides and fertilizers (see Controversy 6.2). The USDA estimates that if no pesticides at all were used in the U.S., farmers would lose some 70% of their crops to pests, but other experts do not agree with this figure. Professor David Pimentel of Cornell University counters that the use of crop varieties resistant to pests, and other nonchemical techniques—such as changing the kind of crops grown in different areas—could cut total losses to 16%. Certain crops, such as cabbage, potatoes, and some fruits, might suffer heavier losses.

One study of organic farming in the corn belt seemed to show that yields of corn and soybeans were almost as good as on conventional farms (corn yields were about 10% lower and soybean yields about 5% lower). However, wheat yields were as much as 25% lower on organic farms. When energy cost savings on organic farms were taken into account, profits were almost the same for both kinds of farming.

The Integrated Pest Management Solution

A second possibility is to use small amounts of synthetic pesticides in combination with a variety of other techniques to control pests. This solution is called IPM, for Integrated Pest Management. The farmer practicing IPM chooses from a variety of pest control methods (Table 6.3).

Table 6.3 Integrated Pest Management Techniques

Cultural practices (e.g., plowing, disposing of crop wastes at the end of the season, crop rotation)

Use of resistant plants

Biological control by parasites, predators, and pathogens

Reproductive control by releasing sterile insects, chemical sterilants, or incompatible pest strains

Use of hormones to interfere with developmental stages of pests

Use of pheromones or other behavioral chemicals

Selective use of pesticides

Cultural techniques and resistant crops. Before synthetic pesticides were invented, farmers used many cultural practices to reduce insect damage. Some of these old techniques and some newly developed ones are very useful. A book published in 1860 lists this control for apple worms:

> The insect that produces this worm lays its egg in the blossom-end of the young apple. That egg makes a worm that passes down about the core and ruins the fruit. Apples so affected will fall prematurely and should be picked up and fed to swine. This done every day during their falling, which does not last a great while, will remedy the evil in two seasons. The worm that crawls from the fallen apple gets into crevices in rough bark, and spins his cocoon, in which he remains till the following spring.[7]

Plowing alone can destroy up to 98% of the corn earworm pupae that winter in the soil. Rotating the kind of crops grown in a field each season, or mixing crops in a field rather than growing large fields of the same kind of crop year after year, can prevent the buildup of pests. Destroying crop residues after harvest destroys the winter home of pests such as the boll weevil. By growing certain kinds of hedgerows around fields, a farmer can increase the number of predatory insects in a field, leading to better control of the plant-eating insects that damage crops (Figure 6.9).

Some plant varieties have been bred so that they are resistant to major insect pests. In some cases, this has been accomplished by breeding plants that look or taste unattractive to pests. Wheat resistant to Hessian fly has been available for 30 years. As with pesticides, however, new insect mutants may develop and begin to attack the resistant species, making it necessary to again breed new crop varieties.

Monitoring and treatment. In integrated pest management, the farmer must carefully monitor his crops to watch for the beginnings of insect attacks. Traps baited with a variety of substances lure insects and give the farmer early warning of a pest invasion (Figure 6.10). Only when something must be done to prevent serious crop damage does the farmer or his advisor choose a pest control treatment. This need not

7 J. H. Walden, *Soil Culture*, Saxton, Barker and Co., 1860, p. 22.

Organic Gardening Economics

You would never have believed it—we outyielded our neighbors by 100% or better on everything.

K. C. Livermore, *Nebraska organic farmer*

. . . The result of abandoning the use of pesticides? . . .
Silent Autumn.

ChemEcology, *January 1978*

Although most agriculturists remain unbelievers, a small but growing number of people believe that even large farms can be run, profitably, without synthetic pesticides and fertilizers. The high prices of fertilizer and pesticides, due partly to the energy crunch, has led many farmers to experiment with using fewer synthetic chemicals.

K. C. Livermore, a Nebraska farmer, grows alfalfa, oats, soybeans, and corn on 260 acres (104 hectares). He says:

We've done much better without chemicals. We hurt some at first when we switched over because we had to get the soil back in balance, get the poisons worked out of it. But in our fourth year there was a big turnaround and now we're outyielding our "chemical neighbors" by far.

Mr. Livermore says about insect and weed problems:

. . . We don't have an insect problem like our chemical neighbors do. We don't have an altered plant. Our plants are natural and healthy. They pick up antibiotics from the soil, which turns insects away as nature intended. And we have insects, like ladybugs, which fight off the enemy insects. Ladybugs thrive on our farm.

Also, as soon as you get natural, healthy soil, there isn't any weed problem. Nature puts in weeds to protect the soil. Weeds grow down in the soil and pick up trace minerals, and as they die they deposit these minerals on the soil's surface.

And when you have your soil in balance, weeds just don't grow as fast and you don't grow as many of them. Another thing is that when we used chemicals we had a clotty soil.

Now it will run through your hands just like flour at times. Earthworms and other life in the soil are alive and can loosen it. It's easy to push the weeds right over when we cultivate.★

Other farmers, such as Mike Shannon, who farms 30,000 acres (12,000 hectares) in San Joaquin, California, use a minimum of pesticides. His S–K ranch actually owns a pesticide supply company and a crop-dusting service, but has cut its own pesticide usage by two-thirds. "It costs money," he says. "We have the planes, but I'd rather not touch them." The S–K ranch is advised by Richard Clebenger, a man trained in integrated pest management—a strategy that attempts to use as little pesticide as possible. Clebenger admits, "There still are a lot of farmers . . . who can't sleep right unless they've given their fields a good spray." Still, his business is now worth $400,000.

Dow Chemical, on the other hand, has published a book, *Silent Autumn*, that describes the plagues of Biblical times as well as a variety of famines caused by uncontrolled pests in recent years. The book emphasizes the idea that careful use of pesticides is necessary to feed people in an ever more crowded world.

For the experimenting organic farmers, though, K. C. Livermore sums it up:

We'd like to see this thing get turned around. . . . We'd like to see the wildlife and the birds back here like it was in the 1940s and 50s. Is that a profitable way to farm? You bet it is. We use one-fourth less input and get as much or more back than anybody else. That should be real easy to calculate in your mind. . . .

★ All quotes from *EPA Journal*, pp. 24–25 (March 1978).

Figure 6.9 The Boll Weevil. Cotton growing has resumed in the Rio Grande Valley in Texas as a result of IPM methods. Short-season cotton, which matures before boll weevil populations become large; cultural techniques, such as destroying crop residues; and the use of small amounts of insecticides, control pests without destroying beneficial insects or causing pest resistance to develop. (USDA photo; Lewis Riley, photographer)

be a chemical spray in all cases, since a number of methods of biological pest control are known.

Biological controls and use of natural biochemicals. Investigations of how insects reproduce and how they interact with each other have identified several points at which humans can interfere in order to reduce insect damage to crops. For example, insects produce chemical substances called pheromones. Some pheromones are sex attractants, which help insects find mates; other pheromones mark paths to food sources. If large amounts of a pheromone can be obtained, the pheromone can be used to bait traps. This has been done with the gypsy moth, whose sex-attractant pheromone has been artificially synthesized. Traps baited with pheromone can be used either to catch insects and kill them with a poison or to give early warning of a pest's appearance in the field.

Insects and weeds are sometimes imported from other countries without their natural parasites, preda-

tors, or diseases. Such pests may then spread unchecked and cause great damage. Successful attempts have been made to control many of these undesirable immigrants, as well as some native pests, by introducing parasites, predators, and diseases. For instance, a tiny wasp, *trichogramma*, controls cotton bollworms by parasitizing bollworm eggs. Two kinds of leaf-eating beatles have been able to keep Klamath weed, once a serious problem in California, under control. (See also Figure 6.11.) In such cases, of course, it is necessary to determine in advance that the new species will not begin to eat crops or kill desirable insects if it runs out of pests.

Bacteria that attack only pest species also can be used as a control measure. Bacteria causing milky-spore disease will kill Japanese beetles and can be applied much like a synthetic insecticide.

Certain kinds of insects can be grown in large numbers and then sterilized by radiation or chemicals.

Figure 6.10 A scout collects insects from a trap in a field. This trap serves as an early warning of insect pests for farmers in the area. (USDA photo)

Figure 6.11 Gypsy Moth Caterpillar and Parasitic Fly. In this country, gypsy moths do much more damage than in their native Europe and Asia, where parasites keep their numbers low. However, none of the moth's natural enemies accompanied them here. Gypsy moths strip the leaves from millions of acres of trees (especially oaks) in the Northeast and central Michigan every year. The moth infestations seem to be spreading west and southward. At one time, 12 million acres (4.9 million hectares) a year were sprayed with DDT to control gypsy moths. The fly shown here is one of the parasites scientists are studying to find a biological control for the moths. (USDA photo; Murray Lemmon, photographer)

Sterile males of these species, when released, mate in the wild with normal females, but the females then lay eggs that do not hatch. Screw worms have been eliminated from a Caribbean island, Caracoa, and controlled substantially in the southwestern United States by the release of sterile males. Mediterranean fruit flies are partially controlled in California and Florida by the release of sterile flies. Not all insects can be grown artificially in large enough numbers, or sterilized easily enough, to use this method.

Economics of IPM. For a number of crops, integrated pest management is cheaper than conventional techniques, which involve spraying with chemicals at regular intervals. Even though farmers must pay to have their crops monitored, cotton, apples, and citrus fruits all can be grown more economically using IPM methods. The savings are mainly due to the use of less chemical spray, although sometimes yields are increased also. In one example, 600 acres (243 hectares) of California tomatoes were being sprayed with chemicals four or five times a season, at a cost of $20–30 per acre ($50–75/hectare). In spite of this, fruit worms continued to damage the crop. A company specializing in IPM was able to control the fruit worms, as well as reduce costs to $8–10 an acre ($20–25/hectare) and, during the second year of the program, no chemical sprays were necessary at all. As the costs of fertilizer

and pesticides go up due to fossil-fuel shortages, the savings in such IPM programs will become more and more attractive.

Pest Control Experts

Understandably, farmers are slow to change their farm practices. Their income depends on the crops they raise, and so they want to be sure of the effectiveness of a new method before they risk using it. Nonchemical pest control methods tend to be more complicated to use than chemical sprays, and they usually do not have the immediate and obvious "knockdown" effect that sprays have.

Farmers and experts who are used to reducing pest populations as close as possible to zero may be uncomfortable with one of the basic ideas of IPM: that pest populations should be kept just below the level at which they cause economic injury, in order to maintain predators and parasites in the ecosystem. Furthermore, farmers get a good deal of advice from pesticide manufacturing companies, which are hardly disinterested parties.

These factors, as well as the fact that there are so many pest control techniques from which to choose, have led to the need for people specifically trained in pest control, as well as for "scouts"—people who monitor fields for the presence of insect pests. A num-

Table 6.4 Relative Toxicities of Selected Pesticides[a]

Common name	Toxicity to rat[b]	Toxicity to fish[c]	Longevity[d]	Bioaccum-ulation[e]	Total
Chlorinated hydrocarbons					
Aldrin	3.2	3.9	4.0	3.1	14.2
DDT	2.7	3.7	4.0	4.0	14.2
Dieldrin	3.1	3.9	4.0	3.0	14.0
Endrin	3.5	4.0	4.0	2.8	14.3
Lindane	2.7	3.4	4.0	1.5	11.6
Mean value	3.0	3.8	4.0	2.9	13.7
Organophosphorus chemicals					
Dichlorvos	3.0	2.7	1.3	1.0	8.0
Disulfoton	3.9	3.3	1.1	1.0	9.3
Malathion	1.8	3.2	1.1	1.0	7.1
Parathion	3.6	3.3	1.3	1.0	9.2
Phorate	4.0	3.7	1.1	1.0	9.8
Mean value	3.3	3.3	1.2	1.0	8.8
Carbamates					
Carbaryl	2.1	2.4	1.1	1.0	6.6
Carbofuran	3.6	2.9	1.4	1.0	8.9
Mean value	2.8	2.6	1.2	1.0	7.6
Triazines					
Ametryn	1.8	2.4	1.2	1.0	6.4
Atrazine	1.7	2.0	3.6	1.0	8.3
Prometone	1.7	2.5	4.0	1.0	9.2
Mean value	1.9	2.3	2.5	1.0	7.7
Organic acids					
Dalapon	1.5	1.5	1.1	1.0	5.1
Dicamba	1.8	1.0	1.3	1.0	5.1
Picloram	1.1	2.4	3.6	1.0	8.1
2,4-D	2.4	1.4	1.1	1.0	5.9
2,4,5-T	2.5	2.8	1.4	1.0	7.7
Mean value	2.0	2.1	1.6	1.0	6.7
Miscellaneous					
Captan	1.0	3.0	1.0	1.0	6.0
Copper sulfate	1.7	3.0	4.0	1.0	9.7
Ethylmercury chloride	3.3	2.4	4.0	3.0	12.7

a Data used with permission of Dr. Jerome B. Weber, North Carolina State Univ.

b How poisonous a pesticide is to rats (based on the LD50, the amount that will kill half of a group of test animals) is a good indication of how toxic it is to mammals in general. On a scale of 1 to 4, 4 is the most toxic.

c How poisonous a pesticide is to fish (based on the LC50, the concentration that will kill half of a group of test fish) is a good indication of how toxic it is to other aquatic life. On a scale of 1 to 4, 4 is the most toxic.

d How long a pesticide persists in the environment is indicated by how long it remains in the soil after it is applied.

readily broken down:	15 weeks or less
moderately broken down:	15–45 weeks
slowly broken down:	45–75 weeks
persistent:	75 weeks or longer

On a scale of 1 to 4, 4 is the most persistent.

e Bioaccumulation can be measured by exposing a test species, such as oysters, to the pesticide for a length of time and then measuring how much accumulates in the oysters. On a scale of 1 to 4, 4 represents the highest accumulation.

ber of small firms have already begun offering pest management services. The Department of Agriculture is attempting to work out certification programs to ensure that persons calling themselves pest control experts actually have the necessary training. Several university programs are being developed in pest control.

They will eventually produce people knowing the advantages and disadvantages of chemical sprays, cultural practices, and biological controls. These trained specialists can then help to choose, for each situation, the combination of methods that will best solve a pest problem.

Which Are the Safest Pesticides?

In order to say whether or not a pesticide is safe, a number of questions have to be asked. How poisonous is the pesticide to humans and to wildlife? How long does the pesticide persist in the environment? Does it accumulate in living organisms? Jerome Weber, an agricultural scientist at North Carolina State University, has developed a simplified rating scheme to determine pesticide safety on the basis of the answers to questions such as these (Table 6.4). According to this table, the all-around most dangerous pesticides are the chlorinated hydrocarbons and mercury compounds, such as ethylmercury chloride. It is easy to see why the EPA has forbidden or restricted the use of these chemicals in the U.S.

Organophosphorus pesticides are very toxic to rats and humans and are therefore dangerous for the people who use them. But, since they do not accumulate in the environment or in living organisms, their total hazard is somewhat less than that of chlorinated hydrocarbons. Herbicides, such as the triazines and organic acids listed in the table, are widely used in agriculture to kill weeds and to clear roadsides. These compounds score the lowest, generally 7 or lower.

Among the factors this rating scheme does not take into account are the potentials these chemicals have as mutagens (causing changes in heredity), carcinogens (causing cancer), or teratogens (causing birth defects). Such properties are more difficult and expensive to determine than the four factors considered, but are equally important in the long run.

How Much Is a Part per Billion?

A number of environmental problems are defined in terms of parts per million, parts per billion, or even parts per trillion of pollutant chemicals or particles in a given quantity of the resource (in this case, water). These are very small numbers, indeed. But are the numbers so small they can be safely ignored? This, of course, depends on the frame of reference. The following tables attempt, in two different ways, to put these small numbers into proper perspective.

The December, 1976, issue of *ChemEcology* carried a table compiled by Dr. Warren B. Crumett of the Dow Chemical Company, which was designed to show just how small trace concentrations are (Table 6.5).

Table 6.5 Trace Concentration Units

Unit	1 part per million (ppm)	1 part per billion (ppb)	1 part per trillion (ppt)
Length	1 inch/16 miles	1 inch/16,000 miles	1 inch/16,000,000 miles (A 6-inch leap on a journey to the sun.)
Time	1 minute/2 years	1 second/32 years	1 second/320 centuries
Money	$1/$10,000	$1/$10,000,000	$1/$10,000,000,000
Weight	1 oz. salt/32 tons potato chips	1 pinch salt/10 tons potato chips	1 pinch salt/ 10,000 tons potato chips
Volume	1 drop vermouth/ 80 fifths of gin	1 drop vermouth/ 500 barrels of gin	1 drop vermouth/ 25,000 hogsheads of gin
Area	1 sq. ft./ 23 acres	1 sq. ft/36 sq. miles	1 sq. in./250 sq. miles
Quality	1 bad apple/ 2000 barrels	1 bad apple/ 2,000,000 barrels	1 bad apple/ 2,000,000,000 barrels

Feeling that Dr. Crumett may have overstated his case, Colin Chriswell of Iowa State University wrote (*ChemEcology,* November 1977):

> Even a minute concentration can add up to a sizable total amount of material when contained in a large sample. For example, consider the amount of material that would be involved in the amount of water used in the Los Angeles area (3,000,000 acre-feet/year) if some selected contaminants were present at a part per billion level.

Substance	One ppb in 3,000,000 acre-feet/year would amount to enough to:
Lead	Cast 1,000,000 bullets
Chromium	Plate 50,000 car bumpers
Mercury	Fill 4,000,000 rectal thermometers
Phenols	Produce 250,000 bottles of Lysol
Herbicide	Kill all the dandelions in 100,000 lawns
Insecticides	Fill 5,000,000 aerosol cans of bug killer
Gold	Run the federal government for nearly 20 minutes or support 50 average families for eternity

A few numbers from the biological world (Table 6.6) might also be worth comparing to those in Table 6.5:

Table 6.6 Effects of Trace Concentrations on Living Organisms

Parts per million	Parts per billion	Parts per trillion
If there is 1 part per million oil in the water, $\frac{1}{2}$ of the exposed Dungeness crab larvae will be killed	At levels of 20 parts per billion in their blood, humans show symptoms of mercury poisoning	Brook trout cannot grow properly or reproduce at levels of toxaphene over 39 parts per trillion

Questions

1. Explain what erosion is and what its major causes are.
2. Summarize the effects erosion has on the environment.
3. What basic principles are involved in controlling erosion?
4. What do you see as the advantages and disadvantages to controlling erosion by reduced tillage farming?
5. Why do you think farmers do not seem particularly interested in erosion control?

6. What are the major environmental problems associated with the use of pesticides?
7. If pesticides cause so many problems, why do farmers use them at all?
8. Explain the basis of integrated pest management.
9. If you were a farmer, would you be an organic farmer, would you use integrated pest management, or would you use chemical pesticides whenever you felt you had an insect problem? Why?

Further Reading

Desertification of the United States, Council on Environmental Quality, U.S. Government Printing Office, Washington, D.C., 20402 (1981).

This clearly written publication details the causes of desertification (soil erosion, overgrazing, overuse of water resources, salinization, etc.) in the U.S.

Phillips, R. E., et al., "No-Tillage Agriculture," *Science*, **208** (6 June 1980), p. 1108.

No-tillage agriculture is a fast-spreading solution to soil erosion problems on cropland. The pros and cons are carefully detailed in this paper.

Carson, R., *Silent Spring*. Boston: Houghton Mifflin, 1962.

Graham, F., Jr., *Since Silent Spring*. New York: Faucett Publications Inc., 1970.

The classic book by Rachel Carson began a public controversy over the effects of pesticides that continues today. Carson was eventually proven correct in many of her claims, although, as detailed in F. Graham's book, they were attacked violently when her book was first published.

EPA Journal, **4** (March 1978).

Most of this issue is devoted to a series of articles on methods and promise of integrated pest management.

Pimentel D., ed., *World Food Pest Losses and the Environment,* AAAS Selected Symposium 13, 1978, American Association for the Advancement of Science, Washington, D.C.

This volume contains a series of papers on food production and pest control. Both the hazards of synthetic pesticide usage and the promise of IPM are examined.

Oelhaf, R. C., *Organic Agriculture*. New York: Halsted (Wiley), 1979.

Safe Pest Control Around the Home, New York State College of Agriculture and Life Sciences, Cornell Bulletin #74 (1979). Obtainable from Mailing Room, 7 Research Park, Cornell University, Ithaca, N.Y., 14853.

References

National Agricultural Lands Study, National Association of Conservation Districts, Washington, D.C. (1981).

Perkins, J. H., *Insects, Experts and the Insecticide Crisis*. New York: Plenum, 1982.

Adkinson, P. L., et al., "Controlling Cotton's Insect Pests: A New System," *Science,* **216**, 2 April 1982, p. 19.

Lockeretz, W., et al., "Organic Farming in the Corn Belt," *Science,* **211**, 6 February 1981, p. 540.

Batra, S., "Biological Control in Agroecosystems," *Science,* **215**, 8 January 1982, p. 134.

Report and Recommendations on Organic Farming, U.S. Department of Agriculture (1980), U.S. Government Printing Office, Washington, D.C. 20402.

Bottrell, D. G., *Integrated Pest Management*, Publication 041-011-0049-1 (1980), U.S. Government Printing Office, Washington, D.C. 20402.

Larson, W. E., et al., "The Threat of Soil Erosion to Long-Term Crop Production," *Science,* **219**, 4 February 1983, p. 458.

CHAPTER SEVEN

Feeding the World's People

Is There Really a Food Crisis?

Are People Dying from Starvation?/Children and Poor Nutrition/Estimating Calorie Requirements/The Food Crisis in Historical Perspective/Distribution and Wastage Problems

How Can We Grow More Food?

Traditional Farming versus the Modern American Farm/Fertilizer, Energy, and Food Production/Irrigation and Salinization/The Supply of Arable Land/Fishery Resources

The Green Revolution

High-Yield Grains/Green Revolution II

Social, Economic, and Political Aspects of Food Production

The Pessimist Looks at the Future

Population Growth May Outstrip Food Supply/Practical Limits to a Green Revolution

The Optimist Looks at the Future

What Technology Promises/Appropriate Technology

Are Grain Reserves a Solution?

History of the Idea/Farmers and Grain Reserves/An International Grain Reserve

CONTROVERSIES:

7.1: *The Lifeboat Ethic*

7.2: *Not Triage, But Investment in People*

Is There Really a Food Crisis?

A friend recently said that he had attended a dinner publicizing the food crisis. When someone asked if a dinner was really an appropriate way to highlight this particular problem, he replied that only rice and tea were served—and every third person got nothing to eat.

Is this an accurate reflection of the world food situation? Is there really a food crisis? Some people say yes; others argue no.

Are People Dying from Starvation?

It is undoubtedly true that too many people in the world do not get enough to eat, by any standard. Areas of chronic starvation exist where day after day, people eat fewer Calories or less protein than they need. (Calories are a measure of the heat or energy content in food. Human food requirements are measured in kilocalories, and the term is written with a capital c, "Calories.") Over the long term, this leads to both mental and physical crippling. Times of actual famine also occur when people die in large numbers. In Bangladesh, in 1974, the aftermath of the war for independence, combined with serious flooding, led to mass starvation. In the years prior to 1978, the African Sahel region was struck by a drought that led to famine for the wandering tribes inhabiting the area.

But problems arise in estimating how many people are starving or malnourished, and in defining what a reasonable standard of nutrition should be. In truth, people do not commonly die of starvation. In most cases, a person who has too little food is more likely to develop diseases than someone who eats well. Furthermore, the undernourished victim is more likely to die of diseases from which a well-nourished person could recover. Typhus, cholera, smallpox, plague, influenza, tuberculosis, and relapsing fever all commonly strike people weakened from lack of food. This means that the actual cause of death is listed as something other than starvation, even though starvation is clearly an underlying or major cause of death.

Children and Poor Nutrition

Children suffer from two diseases specifically related to malnutrition. **Kwashiorkor** occurs when a child's diet may have enough Calories but too little protein. A major symptom of the disease is edema, or swelling of the abdomen (Figure 7.1). Kwashiorkor usually occurs after children are weaned. Marasmus, which involves a shortage of both protein and Calories, occurs when infants less than one year old are fed overdiluted formula from unsterile nursing bottles. Severe diarrhea almost always occurs in both diseases. Children with marasmus and kwashiorkor have higher death rates and may also suffer permanent brain damage.

If mothers nurse their babies, infantile marasmus can be prevented because human milk provides needed protein for the child. In addition, because human milk is a completely sanitary food source, chances are less that the child will be infected by diarrhea-causing organisms. Even kwashiorkor is uncommon if a child is nursed through the second year or longer. Furthermore, nursing has an incomplete but significant contraceptive effect. Women who nurse are *less likely* to conceive again while nursing. This can have a positive effect in terms of population control where other birth control methods are not available or are unacceptable.

Unfortunately, in some developing countries, companies that produce infant formulas have engaged

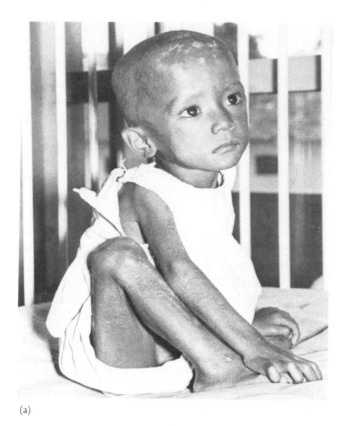

(a)

(b)

Figure 7.1   (a) Suffering from acute malnutrition, or kwashiorkor, a small boy sits on his hospital bed in Djakarta. (UNICEF photo by Jack Ling) (b) Infantile marasmus. (UNICEF photo by Lynn Millar)

in campaigns to sell parents on formula feeding of infants. In many of these countries, the cost of infant formulas is very high compared to workers' incomes. In some cases, commercial formula, if fed at the proper strength, would cost one-fourth to one-third of a worker's income. Thus, the formula is often overdiluted. The high cost and the lack of necessary sanitary conditions for preparing formula mean that in many parts of developing countries, a switch from breast feeding to bottle feeding or earlier weaning will cause infant and child health to suffer.

Estimating Calorie Requirements

Estimates of the number of people receiving less food than they need are complicated by a lack of agreement on how many Calories are needed. For instance, people living in warmer climates may need fewer Calories. Even within a region, Calorie requirements probably vary by as much as 50% from person to person.

Another problem is that nutritional estimates, based on what people eat in developed countries, may be too high, since people there are generally regarded as overfed. In addition, economists who estimate the food available tend to take into account only those crops moving through the marketplace—for instance, cereal grains, or those animals that can be counted easily, such as beef cattle. This method probably underestimates how much people get to eat, at least in rural areas, since family gardens and other local sources of food do exist.

On the other hand, sometimes the food supply in an area has simply been divided by the number of people there. This method takes no account of the fact that income is unevenly divided, making purchasing power uneven. Starvation can exist in a land of plenty if some people cannot afford to buy food. This situation exists in all countries, undeveloped or developed, including the U.S. Even within a family, food distribution may be unequal. In some cultures, men in general or the breadwinner in particular, eat first and the rest of the family shares what is left.

The latest Food and Agriculture Organization Survey estimated that 25% of the people in undeveloped countries are malnourished (over four million people, mostly women and children).[1] The people lack one or more nutrients (such as protein) needed in a healthy diet. A smaller percentage is undernourished; that is, they do not get enough Calories to maintain their body weight with normal activity. Most undernourished people, of course, are also malnourished.[2]

The Food Crisis in Historical Perspective

Periodic famine has always existed. The concept of "food crisis" seems to imply something more. Are we reaching the point in terms of population growth where we simply cannot increase food production to meet the demand created by new mouths to feed? Might large segments of the world population starve to death in a nightmarish solution to overpopulation?

Since World War II, just that sort of prediction has been made and then withdrawn several times. Perhaps something may be learned from past history of the world's food supply. After World War II, when public health measures reduced death rates in many undeveloped countries and population growth rates began increasing, many writers predicted that worldwide famine was on the way. This feeling was bolstered by Food and Agriculture Organization reports, which gave the impression that one-half to two-thirds of the world's population was malnourished. This was probably an overstatement. Statistics on food supplies were scanty and unreliable, and high values were used for the number of Calories required in a healthy diet.

Developing countries were actually making slow but steady progress toward feeding their populations until 1965–1966. In those years, India suffered two droughts that markedly decreased her food output. Because India is home to such a big chunk of the world's population (one out of every six people in the world is Indian), these crop failures loomed large in totals of world food production. Many people felt that the world famine had arrived.

Immediately after this, however, favorable weather and the introduction of high-yield grain varieties in India and other parts of Asia reversed the trend. World food production rose and a mood of optimism prevailed through 1971. The Green Revolution had begun. In the U.S., the government paid farmers not to produce grain to prevent prices from falling too low. However, in 1972, the Russian grain crop failed. The Soviet Union and eastern European countries purchased 28 million tons of grain, reducing grain stocks in the U.S. and Canada to a 20-year low and raising grain prices all over the world. Demands from other countries also rose, in part due to rising incomes in countries such as Japan and in Western Europe that enabled more people to afford meat. Increased meat production means grain consumption by increasing numbers of cattle.

About this same time, the Peruvian anchovy fishery collapsed as a result of changes in the normal water temperature and overfishing. Harvests that had totaled 12.3 million metric tons in 1970 fell to 3 million metric tons in 1973. Visions of feeding the world from the oceans began to fade in the hard light of reality.

Then the weather demons struck again. In 1974, drought and early frosts caused the worst growing season in the U.S. in 25 years. Grain harvests were 20% below expected levels and grain reserves hit new lows. In addition, food prices hit new highs (Figure 7.2). World famine was again predicted.

There were certainly areas of famine in the 1970s, but the predicted world food crisis did not occur. Indeed, since 1975, favorable weather in most areas of the world has led to a new build-up of grain stocks. By the end of the 1978–1979 growing season, world reserves totalled about one-fifth of the year's consumption. Many experts now agree that a world food crisis does not have to materialize in the near future. This does not mean we have no problems, however. Most of these problems revolve around the issue of fair distribution of available food.

Distribution and Wastage Problems

Even when there is a reasonable surplus of food, the world remains troubled by uneven distribution. One-half of the 1979 grain reserves were in one country, the U.S.

---

1 This number does not include people living in Asian centrally planned nations—for example, China.

2 See the reference by Poleman (1981) for a clear review of the problems, methods, and assumptions underlying such estimates.

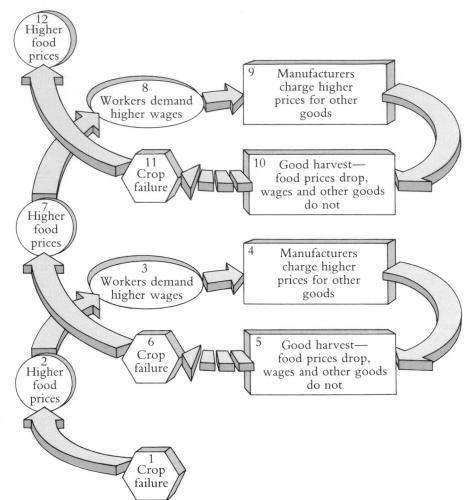

Figure 7.2 The "ratchet effect" of rising food prices. When food prices rise, they cause an increase in wage demands and also in the prices of other goods. When food prices drop, however, wages and other prices do not follow suit. Thus, if food prices lurch from high to low and back again, they have the effect of "jacking up" the cost of living.

While food production in developing countries has increased almost 120% since the 1950s, population has doubled. This growth ate up most of the food production gains, leaving only a minimal improvement in per capita food production. In the developed nations, on the other hand, food production increased by 100% but population increased by only 1/3. Thus per capita food production in developed nations grew at a rate 15% better than in undeveloped nations (see Figure 7.3).

One problem, then, is transportation: getting food surpluses to where the needy people are. Another problem involves decreasing the portion of crops lost to insects and disease in the field, in storage, and in transportation. As noted in Chapter 6, some 25% more grain could be available if such losses were prevented. The highest losses occur in developing coun-

tries, which can least afford them (see Figure 6.7).

A third problem is damping the price fluctuations of grain on the world market. These fluctuations are caused mainly by bad weather, which decreases harvests in an unpredictable fashion. Because an increasing part of the world's grain harvests are spoken for in long-term contracts between the wealthier nations, any decreases in harvest seriously affect small, poor nations. These nations do not have the money or political clout to compete for grain when world supplies are small and prices high. Fourth is the problem of changing income distributions so that the countries and the people presently too poor to purchase enough food can afford a good diet. The social, political, and economic problems involved are enormous and will be discussed later. First, we look at the prospects for increasing the world's food supply in absolute terms.

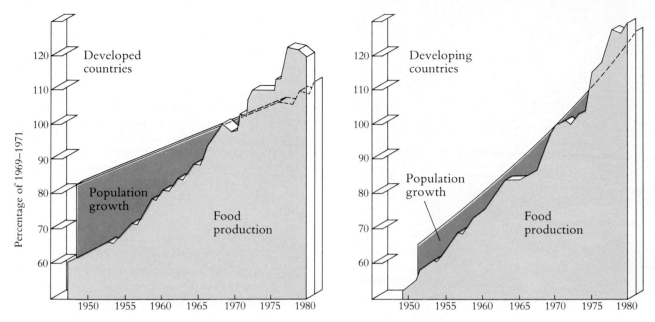

Figure 7.3 Food Production and Population Growth in Developed and Developing Nations. Although food production has increased at similar rates over the last three decades, the two groups have not enjoyed similar per capita gains. In developed countries, food production has increased twice as fast as population. In developing countries, population increases have absorbed more than four-fifths of the gain in food production, leaving little improvement in per capita output. (From Barr, T. N., "The World Food Situation and Global Grain Prospects," *Science, 214*, 4 December 1981, p. 1087)

How Can We Grow More Food?

Traditional Farming versus the Modern American Farm

The Igorot. In the central highlands of the Philippine island of Luzon are the villages of a fierce mountain people. The Igorots were once noted as head hunters, but their fame now rests on their outstanding ability to farm the steep mountain slopes of the island's central region. In the incredibly short period of a few hundred years, these people have built a series of terraces on the mountainside, by hand labor with only a few primitive tools (Figure 7.4). On these terraces, using no artificial fertilizers, they grow rice in amazingly high yields. A single hectare of land (2.47 acres) yields a whole year's supply of rice, their main food, to a family of five.

The high yields are due partly to the enormous amount of labor the Igorots put in. On the terraces, rice is grown by water culture. The terraces must be weeded continually to keep the water-retaining walls from crumbling. Another important factor appears to be nitrogen-fixing algae that grow in the water along with the rice. These algae provide nitrogen, which fertilizes the rice crop, while the rice provides carbon dioxide and some necessary shade for the algae. The Igorots have developed an environmentally sound and self-renewing agricultural system that supports them well. Along with the agricultural system, a social system has developed. Villages are organized into work groups that construct irrigation canals, terrace new slopes, plow the fields, and harvest crops, far more efficiently than solitary workers could (Figure 7.5). From these work groups the whole social structure of the villages has grown.

Figure 7.4 Igorot rice terraces are marvels of engineering and skill, practiced by a people with no modern tools or equipment. (Photo by Charles Drucker)

They sponsor rituals, and they help to resolve disputes; they form political parties and in times of trouble they become military units. They enter into virtually every phase of the communities' activities. Perhaps most importantly, they provide the Igorots with a stability of association, continuity in their personal relationships that in our own affluent, mobile society is becoming ever more rare and valued. The interdependence of the working groups has made the Igorots so culturally conservative that almost a century of contact with Western Society has resulted only in superficial changes in their way of life."[3]

The wet-culture of rice is a special type of farming. All over the world, farmers grow crops such as this, developed over long periods of time by methods requiring much manpower but little or nothing of modern agricultural technology. Such crops are usu-

---

3 C. B. Drucker, "The Price of Progress in the Philippines," *Sierra* (October–November 1978), p. 22.

Figure 7.5 Among the Igorots, work groups are much more than temporary labor gangs: they form the basis for society itself. Here work groups transplant rice seedlings. (Photo by Charles Drucker)

ally well adapted to the climate, soil type, and availability of water where they are grown. But in many cases, these farming operations are at or near the subsistence level. The farmer grows enough food for his own family and some to trade for the other goods needed to live. What hope does technology offer for bringing these farmers to a point where they can grow enough food for the rapidly growing urban populations in their countries? Before answering this question, it will be helpful to describe the modern American farm.

Modern American farms. There appear to be two kinds of U.S. farms. Some 20% of the farms produce 75% of the food and fiber; the rest yield less than $20,000 per year in sales. Presumably these latter farms are worked only part time. Profitable farms are big. This may mean 600–800 acres (240–320 ha) in the Corn Belt, where one family can work this much land alone with the proper machinery and a small amount of help at harvest time. A dairy farm must have 40 cows per worker to see a profit because, again, a great deal of machinery is involved (Figure 7.6).

Still, except for certain types of farming, such as raising broiler chickens, large corporations are not farm owners. Most farming does not lend itself to corporate management, since managers would have to put in long hours and have knowledge of optimum farming techniques for each separate locality.[4] Nor is farming as profitable, in terms of return on investment, as many other business ventures.

---

4 About 2.1% of farms are owned by corporations, but almost 90% of these are family corporations.

Figure 7.6 An American farm family in 1887 in front of their sod home. Although transformed by modern agricultural technologies, the family farm remains the characteristic unit in the Corn Belt and Wheat Belt. Nonetheless, the look of the farm has changed quite a bit since that time. In 1850 there were 1.5 million farms in the U.S., with an average size of 196 acres (78 ha). By 1920, 6.5 million farms existed, averaging 149 acres (60 ha) each. By 1959, only 3.7 million farms were counted and the average size was 303 acres (121 ha). However, there are indications that this trend is now reversing, at least in the Northeast. The number of small farms is growing and may increase by 18–20% during the 1980s. (USDA photo)

The next few sections detail some of the agricultural developments—such as synthetic fertilizers, irrigation schemes, and high-yield crops—that have increased U.S. food production to a point where North American farms produce most of the food exported in the world today. We will also examine the promise of similar techniques in developing countries.

Fertilizer, Energy, and Food Production

Energy in farming. With few exceptions, the American farmer uses large amounts of artificial fertilizer. In fact, 30–40% of increased U.S. productivity in recent years has been attributed to increased use of fertilizer. Nitrogen fertilizers, which account for half of all fertilizer in the world, are produced mainly from fossil fuels such as natural gas and coal. Energy is also needed for the manufacturing process itself. Shortages of fossil fuels have raised the cost of fertilizers over threefold in recent years, which, in turn, has contributed to the higher cost of food.

An especially unfortunate result of higher fertilizer prices is that poorer countries can now afford less fertilizer—yet these areas are precisely where it would do the most good. The reason is that as more and more fertilizer is added to a field, the increased crop yield per pound of fertilizer begins to decrease. The field is, so to speak, saturated with fertilizer. Most large farms in developed countries, such as the U.S., fall into this category. In contrast, farms in undeveloped countries, where little or no fertilizer has been used in the past, show a much larger increase in yield for a smaller amount of fertilizer. In terms of the total world food supply, then, the distribution of fertilizer is uneven.

Energy is used in modern agriculture not only to produce fertilizer, but also to run the machines that plant and harvest crops and in the after-harvest processes: handling, storage, processing, and distribution (Figure 7.7).

Meat versus grain. One further point should be made in relation to energy and agriculture. Most people prefer to eat meat if they can afford it. Yet, as the second law of thermodynamics predicts, meat is an inefficient way of obtaining food energy because of the energy transfers involved. As a general rule, an animal eats 3–10 pounds of grain to produce one pound of meat. A much more efficient use of food resources is

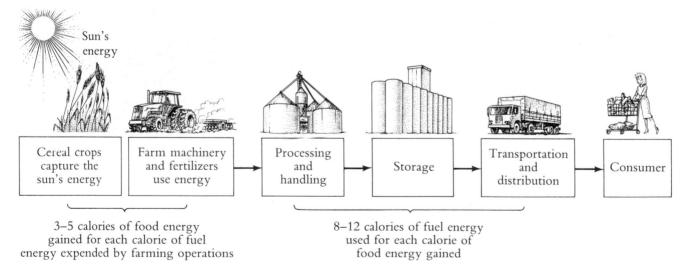

3–5 calories of food energy
gained for each calorie of fuel
energy expended by farming operations

8–12 calories of fuel energy
used for each calorie of
food energy gained

Figure 7.7 Energy costs for producing food in a developed country such as the U.S. Farming itself shows a net gain of energy because more of the sun's energy is captured by plants than is used by farming procedures. After this, however, processing, handling, storage, and distribution of food use up this energy gain and more, so that by the time food reaches the consumer, 5–7 Calories of fuel energy have been used to produce each Calorie of food energy.

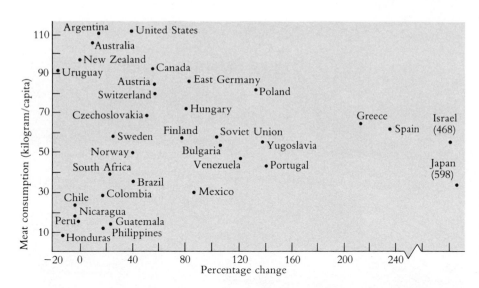

Figure 7.8 Change in meat consumption between 1961 and 1980. As people earn higher incomes, they spend more money on meat. The actual meat consumption per person is shown on the vertical axis, while the percent change from 1961 to 1980 is shown on the horizontal axis. Argentina and the U.S. lead in the amount of meat eaten per person, but the greatest changes have taken place in countries such as Israel and Japan. (From Barr, T., *Science,* **214**, 4 December 1981, p. 1087)

for people to consume grain directly, as is now done in most undeveloped nations (Figure 7.8).

Animals and animal products still have an important place in improving agricultural systems of developing nations, however. For instance, farmers can use animals to pull machinery such as plows or harvesting equipment. In many parts of the world, this means that a small family can tend the same amount of land as used to be tended by a larger family without animal help. This could lead to improved diet or excess crops for sale. The animal must be fed, of course, but it can graze on land that is not as suitable for human food crops (too steep, too cold, too rocky), or fed a second crop harvested at a different time from the main human food crop. In any case, animals and animal feed are generally easier to come by in developing nations than tractors and gasoline.

Animal products such as milk and meat also can provide the farmer with year-round income, as well as some small insurance against starvation in bad crop years.

Irrigation and Salinization

In many areas, notably the American Southwest, fertile lands that receive little rainfall are successfully farmed after the development of irrigation systems. Some 15% of the world's farmland is irrigated, and 30% of the world's food is produced on these acres. Again, energy is a necessary component of the system. In most cases, fossil fuel is needed directly for pumping or draining systems.

In many areas, however, irrigation brings hazards along with benefits. In tropical countries, irrigation canals serve as a breeding place for the snails that carry the disease schistosomiasis. This is a nonfatal disease, but one that takes a great toll of strength from people living in areas where the disease did not exist before irrigation (Figure 7.9).

Figure 7.9 Irrigation project in Khuzestan, Iran. Canals such as this are good breeding spots for schistosomiasis, a disease spread by a snail that lives in slow-moving water. Children playing in the water are quickly infected. (Paul Hebert, 1975)

Figure 7.10 Salinization. In California's Imperial Valley, soil salinity levels on irrigated land pose a major problem to crop production. Areas of this cotton field show evidence of crop failure due to the high salt content in the soil. (USDA—SCS photo by Tim McCabe)

Furthermore, irrigation water evaporates to some degree, leaving behind salts that were in the water. In some areas, including parts of the American West, this has left soils so salty that crops will no longer grow there. Soils can also become waterlogged if drainage systems are not included in irrigation plans (Figure 7.10). Irrigation schemes may even have adverse social effects, such as causing overpopulation in newly irrigated areas.[5]

Some scientists have expressed concern that within the next hundred years, irrigation may become so widespread that the flow of major rivers into the sea may completely stop. The Nile and other rivers entering the Mediterranean are most likely to suffer that fate. The environmental effects of such an occurrence are unknown, but might include such things as decreased fisheries, as the flow of nutrients from land into sea is stopped. Water as a scarce resource is discussed in Chapter 9.

The Supply of Arable Land

There is no shortage of land physically able to grow crops. In fact, the earth may have as much as twice again the amount of arable land as is now farmed. Some important qualifications must be made, however. In the first place, most of the land that is easily cropped is already being farmed. Using the remaining land would require a great deal of energy and labor for such tasks as clearing forests, building irrigation systems, or transporting crops to distant markets. Furthermore, the land that is considered most fertile is generally already in cultivation. Some of the land not now being farmed is only marginally fertile. Thus, if new land is farmed, expected yields may not be as great as on present farmland. However, to offset this last point, land can be improved by farming if the proper techniques and energy are available. Indian farmers cultivate almost the same number of acres as U.S. farmers, but produce only 40% as much crops, partly because of a lack of fertilizers and machinery.

In many other countries, farmers keep their fields fertile by shifting cultivation every few years. When

---

5 For more on this, see the reference by E. Barton Worthington.

the land is allowed to lie fallow for a few years or a few decades, natural processes such as nitrogen fixation renew essential plant nutrients. In addition, populations of insect pests and weeds, which take a heavy toll of the crop, die down during the fallow period. These same ends can be achieved to a large extent if artificial fertilizers and pesticides are available. Apparently, even tropical forest land can be farmed continuously if planters take care to prevent erosion and use suitable fertilizers or nitrogen-fixing crops.

In Africa, much of the fertile uncultivated land lies in areas infested with tsetse flies (which transmit trypanosomiasis, or sleeping sickness, to most breeds of cattle). Although the tsetse fly cannot yet be controlled over these large areas, conceivably it might be restrained in the future. Experts caution that clearing and planting or grazing this land would lead to severe erosion and desertification problems unless great care is taken.

Plans for clearing new lands for farming, however, put food production experts on a direct collision course with those concerned about preserving endangered plant and animal species. In South America, as well as in Africa, a large part of the land that could be farmed is now under forest cover and provides habitats for large numbers of animals and even larger numbers of plant species. The concept of preserving a wide diversity of species in the world comes into conflict with the vision of providing enough food for everyone. As human populations increase until more land is necessary for food production, other desirable uses, such as species protection, are edged out of the picture.

In the U.S., many acres of fertile farmland fall victim to suburban development each year. This problem is examined in Chapter 33.

Fishery Resources

At one time humans looked to the oceans as a vast reservoir of food, waiting only for the technology to harvest and turn it into tasty dishes. However, reality, in the form of declining fish catches and even the complete collapse of some fisheries,[6] has caught up with this notion. The seas hold a finite food resource. If we overexploit this resource, it appears that commercial enterprises will no longer be worthwhile and some species may be driven to extinction.

How much protein, then, could we harvest from the sea on a sustained basis—that is, year after year? Estimates vary from 60 million metric tons to 400 million tons, depending on the figures and assumptions used. The current world catch of marine fish is 50–60 million metric tons per year (Table 7.1).

Increased yield estimates are sometimes based on the total biomass available in an area. This, however, includes species not now practical to catch (e.g., too small) or use (e.g., not accepted by consumers). Current fisheries usually exploit only one or a few species in an ecosystem. Increased yield estimates are also sometimes based on simple extrapolation of current trends, assuming that productivity limits would not apply for some time. This, however, ignores biological limits such as the population sizes needed for reproduction or interactions between populations in predator–prey systems, or the lack of a pollution-free environment.

In fact, the total world marine catch has remained relatively constant over the past 10 years, despite improved technology and increased fishing activity. This suggests that many currently harvested species are fully exploited and in some cases are decreasing in abundance. Rather than increasing the harvest, technology will be hard put to keep it from decreasing in the near future.

Although people have from time to time suggested harvesting other marine protein sources, such as plankton, current harvest and processing technology are lacking. Antarctic krill (Figure 8.9) might provide a significant harvest in the near future (perhaps 10 mil-

---

6 The California sardine fishery collapsed in the 1950s, the North Sea herring fishery in 1969, the West Africa/Nambia pilchard fishery in 1970, and the Peruvian anchovy fishery in 1972.

Table 7.1 Major Species Categories of World Marine Catch in 1976

	Millions of metric tons
Diadromous fish (sturgeon, salmon, shad, etc.)	1.45
Marine fish	55.10
Crustaceans (lobster, shrimp, crab, etc.)	2.01
Molluscs (oysters, clams, squid, etc.)	3.05
Aquatic plants (brown, red, green seaweeds)	1.29

lion metric tons per year). However, it is not known what effect harvesting krill or phytoplankton, which are very important parts of ocean food chains, could have on other fishery harvests. (See Chapter 8 for a comparison of productivity in ocean ecosystems and land ecosystems and Chapter 24 for a discussion of "farming" the sea, or mariculture.)

The Green Revolution

High-Yield Grains

Perhaps the biggest hope technology has held out to farmers in developing nations is the promise of high-yield grains. New varieties of wheat and rice have been developed that far outproduce traditional varieties. Many new wheat and rice varieties are shorter in stature than older varieties and so channel more energy into seed production than into stem growth. Further, they often mature more quickly than older varieties, allowing a farmer to plant and harvest up to three crops a year. Using the new varieties, a farmer can grow more grain even without increasing the amount of land under cultivation. This development has become known as the "Green Revolution."

In practice, however, difficulties remain. The new grains are very responsive to fertilizers—in fact, the best yields depend on the use of large amounts of fertilizer. In addition, the new varieties need more water than older varieties to reach their full potential, and in some cases they are more susceptible to disease and insects. All this means that a farmer must have the technical knowledge of how to use the new varieties and, often, the money to spend on necessary supplements such as fertilizers, pesticides, and irrigation mechanisms. Without these supplements, the new grain varieties may be no better and are sometimes worse than the older ones. Further, the intensive farming encouraged by the Green Revolution can cause serious erosion, salinization, and pesticide pollution unless farmers are educated to deal with such problems.

This is not to say the Green Revolution is without value. Even without optimum amounts of fertilizer, the new varieties often yield substantially more than older varieties. In a period of six years, from 1965 to 1971, farmers in the northwest area of India achieved spectacular increases in food grain production, in some cases averaging almost 10% more grain per year. A combination of high-yield grain, more fertilizer, and good rainfall led India to announce in 1971 that she soon expected to be self-sufficient in food. Although the failure of the monsoon rains in 1973 and 1974 caused a setback, India in 1978–1979 had a carry-over of almost 25 million tons of grain at the time of the next harvest. In short, however, the new varieties are best suited for certain environments where water, energy, education, and capital are not severely limited. Three-quarters of the rice and half of the wheat sown in developing nations still consists of locally adapted varieties.

Green Revolution II

New advances in agricultural production will need to take place in the areas of breeding crop varieties or developing methods that produce high yields on marginal lands. Agricultural researchers are looking for new varieties that are tolerant of salt or aluminum, can survive swamping during monsoons, or have natural defenses against plant pests (Figure 7.11). Scientists hope that many of these advances will come from techniques such as genetic engineering or the use of chemicals to regulate plant growth. These relatively new kinds of biotechnology are sometimes labeled Green Revolution II.

Genetic engineering, which involves such things as the transfer of genes governing desirable traits between different species, is perhaps farthest from practical application at the moment. Still, it holds the promise of advances such as major crop plants that fix their own nitrogen, or the transfer of a gene that makes plants resistant to herbicides.

Newly discovered plant-growth regulators may eventually be able to increase crop yields or, conversely, prevent weed growth. Other newly found chemicals seem to stimulate natural plant defenses against insect damage and so may eventually be used in reducing pesticide needs.

One unfortunate result of Green Revolution I was the virtual disappearance of local varieties of crop plants in some areas, when all the farmers switched over to new hybrid varieties. As Green Revolution II gets underway, we can see even more clearly the need to preserve as many different species of plants as possible. For example, some of the traits that will be used to

Figure 7.11 Triticale–A New Grain. Above right is triticale, a cross between wheat (left) and rye (middle). Triticale gives better yields than wheat, corn, or soybeans on marginal land, especially acid soil. (USDA photo)

produce improved crop varieties by genetic engineering will probably be found in wild plants and locally adapted varieties of crop plants.

Social, Economic, and Political Aspects of Food Production

Technology offers farmers in developing countries a great deal of hope for increasing food production. Whether the farmers can take advantage of this help depends in large part on political and economic conditions. For instance, before small farmers in a developing country can take advantage of many kinds of improved technology, credit must be available so they can purchase seeds, pesticides, and other equipment. Large banks, however, are usually wary of lending money to small farmers. Some of these farmers may turn to local money-lenders, friends, or relatives. The energy crunch is apparently making the credit problem worse, since prices for fertilizers, fuel, and related goods are rising. In the long run, price increases will be a curb on the success of the Green Revolution.

In the long run, too, land reform may be necessary in countries where a few wealthy landowners hold most of the political power. Such landowners are able to obtain new technological information for themselves fairly easily and see no need to make such information available to the small farmer. Yet the small farmer must have help, especially education in the technology of using high-yield grains, if the developing countries are to significantly increase agricultural production.

Another political restraint on agricultural production in the past has been that politicians have often tried to hold down food prices. This is often done to keep living costs down so that the large numbers of city dwellers will continue to vote for the politicians in power. The effect of this policy is to remove any incentive for farmers to grow more food than their family needs. Market prices simply remain too low to justify the work and investment involved in more crops. As an extreme example, a careful study in Mali showed that it cost a farmer 83 Malian francs to produce a kilo of rice, but the government paid the farmers only 60 francs per kilo. As a result, farmers smuggled rice across the borders into countries where they could get 108–128 francs per kilo for it.

In the Sahelian states (e.g., Mali, Niger, Upper Volta) it has been a practice to assure a job to any high school graduate. Such a policy has led to vast increases in the number of civil servants, whose salaries can be kept very low by keeping food prices at low levels. This has acted to slow food production because the farmers see no reason to grow more crops at current prices.

The management of irrigation schemes, too, can have important effects on crop production. Managers are often political appointees, living in the cities, too far away from the scene to know how to manage the water resource effectively. Public water managers have tended to focus on avoiding disagreements about the proper distribution of water, rather than making hard decisions on what is really the most efficient way to use the available supplies. Furthermore, the water itself is usually undervalued. This is shown by the success of private water distribution schemes in countries such as India, even when they compete directly with public water supplies. That is, farmers are willing to pay higher prices for additional water from private sources even when some public water is available.

The Pessimist Looks at the Future

Population Growth May Outstrip Food Supply

As the history of the world's food supply shows, predictions about the ability of agriculture to feed the world are usually colored by the current stock of food on hand. When harvests are bad and grain surpluses fall, the pessimistic view prevails. Experts make calculations showing that we simply cannot boost agricultural production to a rate significantly higher than the population growth rate. When harvests are good and surpluses accumulate, the optimistic point of view catches hold and experts predict that a good diet for everyone is just around the corner.

Historically, pessimists can point to the fact that food production has just barely kept ahead of population growth (Figure 7.3). With so many of the world's

Figure 7.12 In many countries, food production barely keeps up with rapid population growth. Unexpected bad weather can cause crop failures that lead to famine, when the margin between the amount of food produced and the amount of food consumed is small. The United Nations Food and Agriculture Organization operates WFP (World Food Programme) to provide emergency famine relief and to help increase food production in developing nations. Almost half of WFP's resources go to projects to foster land improvement, help new settlers, and raise better livestock. (WFP photos: left photo by J. Bradford; right photo by B. Bhansali)

Figure 7.13 A farmer in India irrigating his field. Agriculture in undeveloped countries uses much human-power and few machines. If farms are mechanized to increase agricultural production, will people be put out of work? Some experts think not. They point out that labor demands may actually increase when high-yield grains and machinery such as tractors, threshers, and harvesters are used because these enable farmers to farm more intensively, planting several crops a year. (William Borders/NYT pictures.)

people already undernourished, keeping up is not good enough (Figure 7.12). Population growth in developing countries continues, although there are signs that growth may be leveling off in some.

Practical Limits to a Green Revolution

Furthermore, the pessimists warn us that high-yield grains themselves are susceptible to disease and insects. As these varieties become more common, the stage may be set for a crop failure similar to the potato blight that devastated 17th-century Ireland. A great variety of seed types in some measure protects us from widespread crop failure due to disease or insect pests. In semi-arid regions, intensive agriculture without safeguards against erosion and the increasing use of irrigation that leaves salty residues in soil, bring areas closer to desertification. This state, which has been loosely defined as human-induced barrenness of land, is often blamed on drought or the "march" of the desert to engulf nearby semi-arid land. But desertification is, in fact, a result of human stresses. Natural events such as droughts are only the final straw. Unless great care is taken to prevent overgrazing, erosion, salinization, and compaction of the soil, the resultant desert-like area will no longer grow enough food for the human population dependent on it.

In terms of increasing the actual supply of food, the Green Revolution is held back by the high price for energy, which prevents developing nations from obtaining the fertilizer and fuel necessary to grow high-yield grain varieties successfully. Even when farm machinery and the fuel to run it are available to farmers, problems will arise if people in rural areas are put out of work by machines (Figure 7.13).

In socio-economic terms, as incomes rise in some countries and people can begin to afford meat, grain is diverted to feed meat-producing animals and food-purchasing power shifts further away from poorer countries. One result of the energy crisis has been to separate the undeveloped countries into two groups: the wealthy undeveloped OPEC countries, who are the oil exporters, and all the others, who have no oil. The oil-exporting countries are able to buy the food needed to upgrade the diet of their people. For this reason, they stand in a somewhat different position with respect to possible food shortages than the other undeveloped countries.

The Lifeboat Ethic

> For posterity's sake we should never send food to any
> population that is beyond the carrying capacity of its land.
>
> **Garrett Hardin***

> This obscene doctrine . . .
>
> **Roger Revelle**

> It is just as obscene to let people die in the future as it is to let
> them die now.
>
> **J. D. Martin, Sociologist,**
> **Lakehead University, Ontario, Canada**

Garrett Hardin has proposed an analogy, related to population growth and the supply of food in the world, that is known as the "lifeboat ethic." In this analogy, so many people crowd onto a lifeboat that it sinks and all are lost. If fewer people had been in the boat, the argument goes, those few might have reached shore safely.

Roger Revelle stated in an editorial in *Science*:†

The specter, unseen by some and ignored by others, looming over the World Food Conference this week in Rome is the continuing rapid population growth of the world's poor countries. Some scientists and publicists have seriously advocated a "lifeboat ethic," saying that nations which do not *compel* human fertility control (by what means is never stated) are endangering the survival of our species—hence they should be starved out of the human race by denying them food aid. This obscene doctrine assumes that men and women will not voluntarily limit their own fertility when they have good reasons and the knowledge and means to do so.

The sharp decline in birth rates during the past decade in a dozen developing countries belies the assumption. But one thing is clear from this experience: environmental changes can bring down birth rates only if they affect the people who have the children—the great mass of the poor who now have little hope for a better life.

Many people sent letters to the editor in reply to this editorial. J. D. Martin, a member of the sociology department at a Canadian university, wrote:

The idea of letting people die, Revelle says, is "obscene." Well, if so, it is just as obscene to let people die in the future as it is to let them do so now.

. . . non-Western populations will keep growing until we can't feed them, even at great cost to the quality of our soils.

We Westerners brought it on ourselves, by saving lives through medical skill and humanitarian generosity. . . . The millions of lives saved by our medical help became the hundreds of millions of lives that are due to be lost in famines.

. . . Shall we impoverish the West in order to make the problem even worse, and in the process weaken both our land and theirs?

I think not; this is the essence of the "lifeboat ethic" which Revelle criticizes. Let too many people into a lifeboat and all will sink. The same may be true of our spaceship called Earth.‡

F. A. Cotton, a member of the chemistry department at Texas A & M University, supports Martin's position:

Nobody can look without horror on the prospect, let alone the actual spectacle, of fellow human beings starving to death. It is a monstrous thing, but we live in an age of monstrosities—some still latent but imminent, unless actively forestalled—and it is literally necessary to consider not only relative degrees of monstrousness, but the fact that some monstrosities are qualitatively more ghastly than others. Overpopulation and starvation are interdependent monstrosities, but of a qualitatively different nature. I believe that the former is far more dire than the latter.

If a quarter of the people in the world starved to death next year, the human condition, in the larger sense, would not be basically or permanently changed. After a few generations, this calamity would leave no basic imprint on our collective consciousness, any more than did the deaths of one-fourth of the people in Europe in the great plague of the 14th century.

However, if the population of the world goes on increasing at the present rate for much longer, the human condition will be basically and catastrophically altered, in an irreversible way.§

What do you think of Martin's argument that "we Westerners brought it [a food crisis] on ourselves" by

providing medical aid and knowledge to developing countries? Should we have denied them this aid in the first place? Do you agree with Cotton that worldwide starvation can be equated with the deaths from plague in the 14th century? Do you agree with Revelle that the lifeboat ethic is an obscene one, or do you feel that the horrors of overpopulation call for such a drastic response?

After you have thought about these questions, you might want to read *The New York Times* editorial by Barbara Ward, reprinted in Controversy 7.2. Another way of looking at this issue is to ask: Is the nutrition problem really a population problem, as Revelle's critics contend, or is the population problem, at least in part, a nutrition problem? Could better nutrition help people in undeveloped countries solve population problems themselves? Beverly Winikoff [in *Science*, **200** (26 May 1978) p. 895] considers some of the ways these questions can be answered.

* *Atlantic Monthly*, **247**, No. 5 (5 May 1981), p. 60.
† *Science*, **186** (15 November 1974), p. 589.
‡ *Science*, **187** (21 March 1975), p. 1029.
§ *Science*, **187** (21 March 1975), p. 1030.

The Optimist Looks at the Future

What Technology Promises

Optimists support their point of view by noting that in most undeveloped countries the Green Revolution has barely started. New grain varieties and attention to crops such as the root vegetables, on which relatively little research has been done, promise great increases in yields. Both traditional plant-breeding techniques and the new biotechnology methods of Green Revolution II promise better grain varieties for even the people who must farm on land that is less than ideal. The United States itself could at least double agricultural production if the returns were great enough—and here lies a major **if**. If population growth can be slowed, if incomes can rise, then the historic problem of poor nourishment can be solved. In the simplest terms, people are undernourished because they are poor. With jobs and reasonable incomes, they could buy food, and, according to most experts, the world's capacity for producing food could probably meet this demand.

The paradox is that developing countries must find a way to provide more jobs in rural areas so that people there will not migrate to the cities (where there are no jobs either), while modernization of agriculture may cause reduction in the number of people required to run farms. At the same time, agricultural and other development schemes are threatening to destroy the social and economic fabric on which present incomes depend.

One example of the problem of modernization is the Philippine government proposal to build four dams in the central mountains of Luzon. The dams are intended to provide hydroelectric power for a variety of Philippine development schemes. They would also flood as many as a dozen Igorot villages, submerging thousands of hectares of laboriously built rice terraces and forcing thousands of these mountain people to move: to cities where jobs are scarce; to a government resettlement area where they would have to begin building all over again; or to other Igorot villages where overcrowding could destroy the carefully worked out social fabric.

Appropriate Technology

Rather than simply promoting a wholesale transfer of modern agricultural technology, many experts now recommend a careful study of what people in developing nations need and can use in the way of technology. This idea has come to be known as "appropriate technology." For instance, farm machines do not have to be labor-saving to be useful. In many developing countries, farmers must wait for the rainy season before plowing. Machines to help farmers plow the hard caked earth allow them to have their crops planted in time to take advantage of the monsoon rains.

Amulya Reddy, of the Indian Institute of Science, points out that technology in developing nations must meet several criteria. It must focus on what the people themselves need most; it must be environmentally sound; and most important, it must encourage self-reliance for those who use it. In terms of food production, this means that studies are needed to determine the actual practices and needs of farmers in developing nations. The technology then developed is more likely to be adopted and is more likely to improve production without causing severe environmental and social problems (Figure 7.14).

Reddy warns against trying to develop gadgets and devices for Indian villages while sitting in a lab in Cambridge, Massachusetts. They will never work, because one has to understand the whole social ecology as well as the physical circumstances of a village to know what people need and will use. Besides, the locals have to be involved every step of the way. "This is considered the obvious thing to do for the urban architect," notes Reddy, but somehow with poor people the idea of investigating their habits and preferences and doing test marketing falls by the wayside. "Then they don't like what we have done and we say they are stupid."[7]

There are really, then, two categories of food problems. First, can we actually produce enough food to provide an adequate diet for everyone? This question is one of technology, with certain added social and political features. The answer, for the near future, is probably yes. Second, can we ensure that everyone actually receives enough food? The problems in this category may be more difficult to solve. Based on social inequalities and economic variables, these problems and their solutions penetrate to the very core of whether humans, across the world, will care and take responsibility for each other (see Controversy 7.1).

One way that has been suggested for the short-term relief of famine and long-term relief from rising food prices is the establishment of grain reserves.

7 Holden, C., "Pioneering Rural Technology in India," *Science,* **207** (11 January 1980), 159.

Figure 7.14 In developing nations, new, "appropriate technology" solutions to farming problems hold out the hope of increasing yields without displacing workers. This human-powered tractor is used in Philippine rice paddies, along with an easily built and repaired rice thresher designed to thresh Philippine rice. (Courtesy of J. K. Campbell, Cornell University, Department of Agricultural Engineering)

Are Grain Reserves a Solution?

History of the Idea

. . . Let Pharoah proceed to appoint overseers over the land, and take the fifth part of the produce of the land of Egypt during the seven plenteous years. . . . That food shall be a reserve for the land against the seven years of famine which are to befall the land of Egypt, so that the land may not perish through the famine.[8]

So spoke Joseph to the Pharoah of Egypt some 3600 years ago. In ancient China, too, we are told, the Confucians created a "constantly normal granary." In the modern era, Henry Wallace, Secretary of Agriculture for President Franklin D. Roosevelt and Presidential aspirant in the 1940s, was the leading advocate of grain reserves. Though Wallace had been urging an "ever-normal granary" since 1912, not until the 1930s, when farms were being destroyed by dust storms and drought was he able to create a government agency that would store grain. Even then, the publicly announced objective behind government purchasing of grain stock was to stabilize farm incomes. The stabilizing of farm incomes was a goal for two reasons. First, the political power of farmers made a program of insured farm incomes important. Second, preventing farmers from going out of business because of losses in bad years did help make the supply of food more secure. Wallace's program used crop insurance, paid for by a contribution of a portion of the crop itself; limited direct payments; acreage allotments; and price-support loans which, if defaulted on, did not require repayment. Although farm incomes stayed up, output grew and the stocks of grain grew, bringing criticism of the program.

Criticism ended, however, when World War II broke out: the stocks of wheat, cotton, and corn became a vital military resource, as the United States undertook to aid its allies with badly needed contributions of food. By the close of the period of reconstruction after the war, all current U.S. farm production was being consumed and the stocks were gone. Stocks began to build again as farm productivity grew to new levels and government purchases at price-support levels continued.

In 1954, Congress passed Public Law 83-480, a temporary measure now become permanent, which aimed at reducing the growing stocks in a way that benefitted both poorer nations and American foreign policy. Under this law, the government was allowed to sell some of its surplus stocks to nations needing grain. Emergency relief could also call forth shipments from the overflowing U.S. cupboard. Famines and natural disasters were among the situations to which the U.S. could respond with food aid. Voluntary relief organizations and governments were the recipients who then distributed the food. Under 83-480, the U.S. participates in the World Food Program. We have been a consistent and generous supporter of this program, a UN-FAO initiative that supplies food aid for emergency situations and to nations carrying out development projects.

To restrain grain production further and to dampen the build-up of stocks, Congress established the Soil Bank in 1956. Under this short-lived program, farmers were paid for the acres they kept *out of production.* Public criticism of possible abuses led to an early end of the program in 1960, before it had been fully tried. By 1961, stocks of feed grain and wheat had reached nearly 1.5 billion bushels and the federal government was paying out a million dollars a *day* in storage costs. Congress felt the need to direct the Department of Agriculture to dispose of some of its stocks.

Surpluses and contributions of food aid continued through the 1960s, but the early 1970s brought a confused situation with alternating years of surpluses and shortages. The combination of a lower-than-trend production of grain, heavy export demand for wheat, and the inflated price of wheat led the government to dispose of nearly all its grain stocks by mid-1973, and the stocks were not rebuilt. High prices brought about by the high level of demands, as measured against current production, brought welcome profits to farmers. If the government had been selling grain from stockpiles, lower market prices and lower farm profits would have resulted.

In the early 1980s, bumper harvests caused grain prices to farmers to drop. The lower prices obtained per bushel, combined with high interest rates and fuel costs needed to produce the crops, pushed many farmers close to bankruptcy. In 1982, farm prices for wheat were as much as 40% below the average estimated cost of production. Farm interests again looked to the government for help.

---

8 Genesis 41:34, 36.

Not Triage, But Investment in People

Barbara Ward, an economist and writer, wrote the following editorial in *The New York Times*:*

Now that the House of Representatives has bravely passed its resolution on "the right to food"—the basic human right without which, indeed, all other rights are meaningless—it is perhaps a good moment to try to clear up one or two points of confusion that appear to have been troubling the American mind on the question of food supplies, hunger, and America's moral obligation, particularly to those who are not America's own citizens.

The United States, with Canada and marginal help from Australia, are the only producers of surplus grain. It follows that if any part of the world comes up short or approaches starvation, there is at present only one remedy and it is in Americans' hands. Either they do the emergency feeding or people starve.

It is a heavy moral responsibility. Is it one that has to be accepted?

This is where the moral confusions begin. A strong school of thought argues that it is the flood tide of babies, irresponsibly produced in Asia, Africa and Latin America, that is creating the certainty of malnutrition and risk of famine. If these countries insist on having babies, they must feed them themselves. If hard times set in, food aid from North America—if any—must go strictly to those who can prove they are reducing the baby flood. Otherwise, the responsible suffer. The poor go on increasing.

This is a distinctly Victorian replay of Malthus. He first suggested that population would go on rising to absorb all available supplies and that the poor must be left to starve if they would be incontinent. The British Poor Law was based on this principal. It has now been given a new descriptive analogy in America. The planet is compared to a battlefield. There are not enough medical skills and supplies to go round. So what must the doctors do? Obviously, concentrate on those who can hope to recover. The rest must die. This is the meaning of "triage."

Abandon the unsavable and by so doing concentrate the supplies—in the battlefield, medical skills; in the world at large, surplus food—on those who still have a chance to survive.

It is a very simple argument. It has been persuasively supported by noted business leaders, trade-unionists, academics and presumed Presidential advisers. But "triage" is, in fact, so shot through with half truths as to be almost a lie, and so irrelevant to real world issues as to be not much more than an aberration.

Take the half truths first. In the last ten years, at least one-third of the increased world demand for food has come from North Americans, Europeans, and Russians eating steadily more high-protein food. Grain is fed to animals and poultry, and eaten as steak and eggs.

In real energy terms, this is about five times more wasteful than eating grain itself. The result is an average American diet of nearly 2,000 pounds of grain a year—and epidemics of cardiac trouble—and 400 pounds for the average Indian.

It follows that for those worrying about available supplies on the "battlefield," one American equals five Indians in the claims on basic food. And this figure masks the fact that much of the North American eating—and drinking—is pure waste. For instance, the American Medical Association would like to see meat-eating cut by a third to produce a healthier nation.

The second distortion is to suggest that direct food aid is what the world is chiefly seeking from the United States. True, if there were a failed monsoon and the normal Soviet agricultural muddle next year, the need for an actual transfer of grain would have to be faced.

That is why the world food plan, worked out at Secretary of State Henry A. Kissinger's earlier prompting, asks for a modest reserve of grain to be set aside—on the old biblical plan of Joseph's "fat years" being used to prepare for the "lean."

But no conceivable American surplus could deal with the third world's food needs of the 1980's and 1990's. They can be met only by a sustained advance in food production where productivity is still so low that quadrupling and quintupling of crops is possible, provided investments begin now.

A recent Japanese study has shown that rice responds with copybook reliability to higher irrigation and improved seed. This is why the same world food plan is stressing a steady capital input of $30 billion a year in third-world farms, with perhaps $5 billion contributed by the old rich and the "oil" rich.

(What irony that this figure is barely a third of what West Germany has to spend each year to offset the health effects of overeating and overdrinking.)

To exclaim and complain about the impossibility of giving away enough American surplus grain (which could not be rice anyway), when the real issue is a sustained effort by all the nations in long-term agricultural investment, simply takes the citizens' minds off the real issue—where they can be of certain assistance—and impresses on them a nonissue that confuses them and helps nobody else.

Happily, the House's food resolution puts long-term international investment in food production firmly back into the center of the picture.

And this investment in the long run is the true answer to the stabilizing of family size . . . the whole experience of the last century is that if parents are given responsibility, enough food and safe water, they have the sense to see they do not need endless children as insurance against calamity . . .

Go to the root of the matter—investment in people, in food, in water—and the Malthus myth will fade in the third world as it has done already in many parts of it and entirely in the so-called first and second worlds.

It may be that this positive strategy of stabilizing population by sustained, skilled and well-directed investment in food production and in clean water suggests less drama than the hair-raising images of inexorably rising tides of children eating like locusts the core out of the whole world's food supplies.

But perhaps we should be wise to prefer relevance to drama. In "triage," there is, after all, a suggestion of the battlefield. If this is how we see the world, are we absolutely certain who deserves to win—the minority guzzlers who eat 2,000 pounds of grain or the majority of despairing men of hunger who eat 400 pounds? . . .

Is this the battlefield we want? And who will "triage" whom?

★ *The New York Times*, 15 November 1976.

Farmers and Grain Reserves

Farm interests argue against a national grain reserve first on the basis that for the past three decades the U.S. has consistently been able to export grain and still meet domestic demands. The argument that our national grain demand can surely be met, is only partly true. In the presence of heavy demand for export, grain prices rise and U.S. consumers have to pay the higher price. Further, the same high grain prices that annoy the affluent citizens of this country are a deadly blow to individuals in developing nations.

In nations where perhaps 60% of the average family's budget is spent on food, a price rise for wheat or rice can make the difference between poor nutrition and malnutrition. The need for a grain reserve becomes more important as a means to dampen price increases for people in the developing nations, not simply as a source of emergency grain.

It is at times of shortage and high prices that a grain reserve is most needed and that farmers argue most strongly against it. The reason is that a grain reserve would dampen prices and profits in times when production levels are down in relation to demand. At present, farm interests are largely unwilling to forego profits at such times, even in return for protection of profits when production relative to demand is high. The protection is discounted and dismissed because farmers see the near future as a time when all their products will continue to be absorbed on the world market.

When production exceeds demand, causing prices to fall, farmers want the government to buy up surplus so that prices fall no further. Thus, the ideal time, from a political point of view, for discussion of building a grain reserve is when farm prices are low due to unusually high levels of production.

In the United States, government purchases of grain for a federally held reserve will serve to protect farmers' income in times of production excesses. What would happen if developing nations were to build grain reserves from their own harvests? Such reserves could make a very big difference in the nutritional levels of the population and could also aid the national economy of a developing country by eliminating the need for food imports, thus preventing a deficit in the balance of trade. The World Bank, through the use of loans, and the FAO are encouraging individual nations to build grain reserves, but the costs of purchase and storage are considerable.

An International Grain Reserve

Another possibility is less costly and more efficient, but more difficult to achieve politically. An international grain reserve, not belonging to a particular country, could be created. Proposals for such a reserve were made in the mid-1970s. In the early 1980s, a similar international bank was started for commodities such as tin, rubber, and cocoa. In this case, the intention was to protect less developed countries, whose economies often depend on only one crop or mineral, from severe swings of the marketplace.

Resistance to an international grain reserve is great, but the benefits are large. Consider how an international agency operating a grain reserve might function. Two prices might be used to trigger the op-eration of the reserve. A price for grain falling below a preset low price would signal the agency to buy grain. The low-price signal to buy serves two functions. First, low prices are the time to set grain aside; holding or investment costs for the agency are kept low in this way. Low prices also mean an abundant harvest relative to demand, indicating that now is the time to buy and store grain without driving the price up significantly. Second, purchase of the excess harvest acts as a price support for grain. Farmers would find that a good portion of their excess harvest is thus absorbed without the price falling still lower. Thus, the presence of a grain reserve helps to ensure the income and prosperity of farmers in times when too much good weather works against them.

In similar fashion, when the price for grain rises above some predetermined high limit, the agency would offer its grain on the international market. The grain would be available for purchase by those poor nations for whom higher prices are a burden. The price at which the grain is sold essentially puts a lid on the international price of grain: who would pay more than the high-signal price if grain can be purchased at that price from the international agency? True, farm profits are dampened when the agency decides to sell its grain, but the diminished profits are not the normal profits that keep a farm in business, but the profits built of hardship and malnutrition for the poorest. The farmer selling grain at the grain reserve price still makes a sturdy profit.

Joseph seems to have had quite an idea.

Questions

1. If people die from too little food every day in the poorer countries, how can we say people rarely die of starvation?
2. What special effects does poor nutrition have on children?
3. Suppose you were in charge of a program to encourage breast-feeding in developing countries. On what arguments would you base your campaign?
4. Why is it sometimes said that the food crisis is a problem of distribution rather than of supply?
5. What is the Green Revolution or Green Revolution II? What problems are tied to the Green Revolution?
6. Are the resources from the sea inexhaustible? Why or why not?
7. Imagine you are a farmer in a developing country. Describe the values and drawbacks you see in synthetic fertilizers, a local irrigation scheme, high-yield grains, a government plan to clear the forest near you for the planting of crops (you would share in the increased farmland), and gasoline-fueled machinery for planting and harvesting crops.
8. Explain how economic and political problems can be as important in terms of food production in developing countries as introducing new technologies.
9. What is "appropriate technology"? Give an example.
10. How would grain reserves help even out fluctuations in food prices?
11. What effect could this evening out have on the inflation-

ary spiral?

12. Suppose you are an American farmer. Are you in favor of a grain reserve? Why?

13. Are you optimistic or pessimistic about future world food supplies? Why? Explain why you do or do not feel the answer to this question is very important.

Further Reading

Eicher, C. K., "Facing Up to Africa's Food Crisis," *Foreign Affairs Quarterly*, **61**, No. 1 (Fall 1982).

Lele, Uma, "Rural Africa: Modernization, Equity and Long-Term Development," *Science*, **21** (6 February 1981), p. 547.

Franke, R. W., et al., *Seeds of Famine*. Montclair, N.J.: Allanheld, Osmun (1980).

Food and population problems in Africa are perhaps the most severe in the world. Twenty-two of the poorest countries in the world are African. Population growth continues at high rates while food production languishes. Africa possesses almost half of the idle farmable land in the world, yet the prevalence of sleeping sickness and other diseases currently limit its use. These fascinating articles detail why the green revolution seems to have bypassed Africa and what social and economic barriers appear to prevent progress in food production. They also give some history of the region and the colonial influences that have contributed to Africa's problems.

"Harvesting the Sea," *Oceanus*, **22**, No. 1 (Spring 1979).

This entire issue is devoted to fishery resources and food from the sea.

Intercom: The International Population News Magazine. Population Reference Bureau, Inc., 1337 Connecticut Avenue, N.W., Washington, D.C. 20036.

This publication carries a capsule on the world food situation each month, prepared by FAO. Good for up-to-the-minute information.

Winikoff, B., "Nutrition, Population and Health: Some Implications for Policy," *Science*, **200** (26 May 1978), 895.

Popkin, B., et al., "Breast-Feeding Patterns in Low-Income Countries," *Science*, **218** (10 December 1982), 1088.

In these articles, the authors discuss the interrelationship between the food and population problems, and also advantages and social aspects of breast-feeding in developed and undeveloped countries.

References

Worthington, E. B., "The Greening of the Desert: What Cost to Farmers," *Civil Engineering—ASCE* (August 1978), p. 60.

This article points out the social effects of irrigation projects, often overlooked by planners and engineers.

Plucknett, D. L. and N. J. Smith, "Agricultural Research and Third World Food Production," *Science*, **217** (16 July 1982), p. 215.

Poleman, T. J., "Quantifying the Nutrition Situation in Developing Countries," Food Research Institute Studies, Vol. XVIII, No. 1 (1981), p. 1.

Swaminathan, M. S., "Biotechnology Research and Third World Agriculture," *Science*, **218** (3 December 1982), p. 967.

Impacts of Technology on U.S. Cropland and Rangeland Productivity, U.S. Congress Office of Technology Assessment, August 1982.

Ward, G. M., et al., "Animals as an Energy Source in Third World Agriculture," *Science*, **208** (9 May 1980), p. 570.

Desertification in the U.S., Status and Issues, Working Review Draft, U.S. Department of Interior, 1980.

National Park Service Photo; Richard Frear

PART THREE

Water Resource Problems

Human survival depends upon a number of natural resources. Water is certainly one example; air is another, and energy resources a third. The importance of plants and animals as wildlife resources was discussed in Part I, while land and food resources were examined in Part II and land resources will be taken up again in Part VIII.

The next chapters are concerned with the resources air, water, and energy. We begin with water, in part because the environmental movement first took shape around efforts to protect water supplies. More than 100 years ago, people realized that water could carry disease. Because of that recognition, the profession of Environmental Engineering, or, as it was then called, Sanitary Engineering, grew up. The environmental movement, as a visible phenomenon, had begun.

Chapter 8 examines water from the ecologist's point of view. Water is a resource with unique properties, essential to all life on earth. It is a basic factor in the growth of natural communities and human civilizations. The chapter describes the various kinds of wa-

ter habitats and the organisms that live there, using New York's Hudson River to illustrate several of the major kinds of water habitats.

In Chapter 9, with the Colorado River and the Ogallala Aquifer as examples, we detail how people have managed their water resources and put water to use. We examine also the deep-rooted arguments over water transfers and the use of water for development. Water is seen as a critical element for the growth and maintenance of human society.

Chapters 10–14 examine water problems from the point of view of human needs. When is water pure and when is it safe? What are the substances that contaminate water supplies and how do we remove them? How do we handle wastewaters and what effects do they have on natural waters?

The first of these questions deals with *the water we drink.* It involves basic facts we should know about the safety and quality of our supply of drinking water.

The impurities that influence the safety of a water supply for drinking fall into three broad classes (Figure A). *Inorganic chemicals* are one class; included in it are

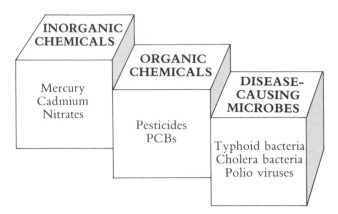

Figure A The three classes of impurities in drinking water.

the ions arsenate, nitrate, fluoride (at high levels), and other chemicals that can have adverse effects upon our health. *Organic chemicals*, a second category, may also be dissolved in the water; some of these compounds have been linked to cancer. Finally, water may contain *microorganisms (microbes)* that cause diseases such as typhoid and cholera. Fortunately, these diseases are only a distant memory to most of us in the United States. Once widespread in this country, they still occur commonly in nations that have not yet treated their water supplies.

In addition to being concerned with the safety of water supplies, we are concerned with other water characteristics, such as clarity, odor, and taste. Water may also be investigated for its "hardness." Hardness is caused by the presence in solution of compounds such as calcium carbonate and magnesium carbonate. Hardness decreases the effectiveness of soap and makes washing clothes, dishes, and people more difficult. In Chapters 10 and 11 we explain the possible hazards in drinking water and then look at water treatment, the methods by which we attempt to make water safe for drinking.

The effectiveness of current water treatments, however, depends in part on the amount and kinds of wastes that we allow to contaminate water we may later wish to drink. The remaining chapters in this part deal with this other kind of water resource problem: problems caused by *the water we waste*. Wastewaters flow from cities and industries, from mining operations, from farming and rural homes. These wastes are treated in different ways but are generally disposed of in the same way: into the nearest river or lake or into the ocean.

The problem with wastewaters is not just that they might contaminate our drinking water supplies. Wastewater can be fairly easily chlorinated so that it does not carry large quantities of disease-causing organisms, and the water we drink, even if it comes from a highly polluted source, can be treated until it is not only healthy but pleasant to drink. Thus, our insistence that wastewater be treated before it is released is not focused only on human health problems. We must also consider how the wastewater will affect the natural waters into which it flows.

If organic pollutants are not removed from wastewater, they can set up a chain reaction that robs water of the oxygen normally present. Further, certain chemicals, such as pesticides, may be directly poisonous to aquatic organisms. Fish and other aquatic creatures may not be able to live under these conditions, and other less desirable species may take over. In addition, certain inorganic elements in wastewater, such as phosphorus and nitrogen, cause excessive growths of the microscopic green water plants called algae. These "blooms," as they are called, form unpleasant scums and mats over the surface of lakes.

Water pollution control is the term given to methods of cleaning up wastewaters so that they can be released to natural waters without causing problems. What are these methods? Surprisingly, biological processes are used to purify the water.

It should be noted that several wastewater pollutants are covered in detail elsewhere in the book: acid mine drainage (along with its cause, coal mining) is discussed in Chapter 16. Oil pollution is described in Chapter 23. Water pollutants that can cause cancer (e.g., asbestos) are covered in Chapter 30.

CHAPTER EIGHT

Lessons from Ecology:
Water Habitats and Communities

The Hudson River

Water—A Limiting Factor for Life
Availability of Water/The Watershed

Water Habitats
Freshwater Habitats/Estuaries/Marine Habitats

**How Lake and Reservoir Water Can Become
Oxygen-Poor**

The Hudson River

High in the Adirondack Mountains of New York State, the Hudson River begins as a small, clear lake, Lake Tear of the Clouds. Flowing first as a brook and then as the Opalescent River, the waters run south, joined by many other streams. At the town of Newcomb, still in the mountains, the Opalescent officially becomes the Hudson. Just above Troy, the Mohawk River joins the Hudson, which is now one of the mightiest rivers in the U.S. Three hundred and fifteen miles from its origin, as it passes New York City, the Hudson meets the Atlantic Ocean in New York Bay, which is the Hudson's estuary (Figure 8.1).

The communities of organisms living in various parts of the Hudson change as conditions in the river change. Small swift streams that feed the river provide homes for different species than those in the slower-moving river itself. Organisms living near the point where the river meets the ocean tides must be able to live in varying concentrations of salt. The ocean itself is a very different environment from the freshwater river, ponds, and lakes.

Organisms living in the Hudson River have many problems with which to contend. For years, a number of cities have dumped raw or poorly treated sewage into the river. Industry has also been guilty of using the Hudson as a sewer. But before we deal with the problems of polluted waters, let us look at life in natural, unpolluted waters.

Figure 8.1 View of the Hudson River today from near Bear Mountain. (Courtesy of New York State Dept. of Commerce)

Water—A Limiting Factor for Life

Availability of Water

We pointed out in Chapter 1 that water is a necessity for all forms of life on earth. As noted in Chapter 5, in land habitats the abundance of water (a function of rainfall, humidity, and the evaporation rate) determines the kinds of communities that develop. In water environments, the availability of water changes with changes in water levels—for instance, with tides. It also changes according to differences in the salt content of water, which affect the rate at which water enters or leaves organisms. In water environments, as well as land environments, the type of community that develops depends on this availability of water.

The Watershed

Surface waters such as lakes, rivers, and oceans are highly visible features of the environment. We can easily see that they provide different habitats for living organisms than do land areas. Not as easily seen is that the two kinds of habitats, land and water, are tied together by the cycling of energy, water, and nutrients through the environment. For instance, the Hudson River is not a self-contained system. Energy comes from the sun. Organic and inorganic nutrients wash into the Hudson, by erosion and stream flow, from the river's banks and from land bordering all those streams flowing into the Hudson. Even the river water itself cycles through the hydrologic cycle—some water leaves as evaporation and some water enters the river as rainfall and stream flows. Pollutants, too, reach the river not only directly but also from the land areas surrounding it.

We can see, then, that the functioning unit is not simply the river itself, but also the whole land area that drains into the river. This area is the **watershed**. In terms of understanding and maintaining the quality of natural waters, the whole watershed is the ecosystem that must be studied or managed.

Water Habitats

Water habitats are usually differentiated on the basis of salt content (i.e., saltwater versus freshwater habitats) and whether a current is present or absent (for instance, streams with swift-flowing waters versus still lakes or river pools)(Figure 8.2). The **turbidity** of wa-

(a)

(b)

(c)

Figure 8.2 The three main types of water habitat. (a) In freshwater habitats, the water is either still, as in lakes and ponds, or moving, as in streams and rivers. (Yellowstone Lake, National Park Service Photograph) (b) In marine habitats, such as oceans and seas, the water moves continuously as a result of various currents. The salt concentration is, of course, much higher than in fresh water. (U.S. Coast Guard) (c) Estuaries are partly enclosed bodies of water where salt water meets fresh water. An example is Chesapeake Bay, shown here. (Office of Tourist Development, Maryland Department of Economic and Community Development)

ter, which is a measure of its clarity, is also important. Turbidity affects how far sunlight can penetrate in water. Green plants can only live in the water zone into which sunlight reaches because they need sunlight for photosynthesis.

Freshwater Habitats

The most important physical characteristics of freshwater habitats involve: temperature; turbidity; current (or lack of current); and the amount of dissolved materials, including solids, such as nitrate and phosphate salts, and gases, such as oxygen and carbon dioxide.

Lakes. Most organisms living in lakes and ponds or in the quiet pools of streams are adapted to life in still waters. In the shallow water zone along the shore, light reaches all the way to the bottom. Here live rooted water plants and floating algae, as well as a variety of animal life (Figure 8.3).

Farther from shore is the area of open water. The upper layer, which light penetrates, is home to minute plants and animals called **plankton**, as well as to fish. Plankton are microscopic, drifting organisms found in lake waters as far down as light penetrates. Plankton species are, in general, unable to move against currents, but drift along with water movements. Plant species, called **phytoplankton**, include many species of algae; animal species are known as **zooplankton**. Phytoplankton are important producers in the lake ecosystem. They capture the sun's energy and turn it into organic compounds that form the basis of many of the lake's food chains. Zooplankton feed on phytoplankton, and so are primary consumers in the lake ecosystem. Along with the phytoplankton in the open

waters, plants in the shore zone are producers in lakes and ponds. However, phytoplankton have an enormous rate of reproduction. Thus, although they are small in size compared to rooted plants, phytoplankton species are the more important producers in aquatic systems.

Living in the deeper water layer and on the bottom, where not enough light penetrates for photosynthesis to occur, are organisms that live on dead organic matter. Bacteria, fungi, small clams, and bloodworms all "re-process" organic matter, which is then carried by currents or swimming creatures back to the other lake zones. Thus, the deep zone houses the detritus feeders in lake and pond communities.

Lakes often stratify, or have layers of different temperatures. In summer, a warm upper layer floats on top of a cold layer. Organic materials produced by photosynthesis in the upper layer filter down into the lower layer, where decomposition reactions take place. This decomposition uses up oxygen, lowering the oxygen content of the lower layer drastically. Because oxygen is sometimes in short supply in deep lake waters (see the supplemental material at the end of this chapter for a discussion of this problem), many organisms living there have adapted to low oxygen concentrations or even no oxygen at all.

While some consumers such as frogs and snakes live along the shore line, other consumers such as fish may range over all three zones, depending on the season and availability of food. Figure 8.3 summarizes some features of lake communities.

Ponds. Ponds differ from lakes in that the shore zone is relatively large and the open-water zone is comparatively small. Often, ponds are too shallow to have a layer of water that light does not reach. Thus,

---

Figure 8.3 (opposite page) Simplified Lake Community. In the shallow waters along lake shores grow rooted water plants such as cattails, water lilies and muskgrass. These, along with algae, are the shore-zone producers. Also found there are consumers such as pond snails, dragonfly nymphs, and sunfish. On the surface balance waterstriders and backswimmers. In the open-water zone, algae and dinoflagellates are producers. They are found as deep as sunlight penetrates. These producers are eaten by zooplankton such as copepods and rotifers, which are eaten by small fish who are, in turn, eaten by larger fish. The deep zones and bottom muds are inhabited by creatures that live on detritus: bacteria and

fungi, chironomid larvae, tubifex. (a) cattails (*Typha*); (b) water lilies (*Nymphaea*); (c) muskgrass (*Chara*); (d) amphipod (*Gammarus*); (e) pond snail (*Heliosoma*); (f) dragonfly (*Aeschna*); (g) backswimmer (*Notonecta*); (h) frog (*Hyla crucifer*); (i) copepod (*Canthocampus*); (j) waterstrider (Gerris); (k) sunfish (*Centrarchidae*); (l) small-mouth bass (*Micropterus dolomieui*); (m) algae (*Schenedesmus, Navicula, Anabaena*); (n) dinoflagellate (*Euglena*); (o) cladocera (*Diaphanosoma*); (p) rotifer (*Asplanchna*); (q) copepod (*Cyclops*); (r) walleye (*Esox*); (s) bacteria and fungi; (t) chironomid larvae (*Tendipes*); (u) clam (*Sphaeriidae*); (v) *Tubifex*.

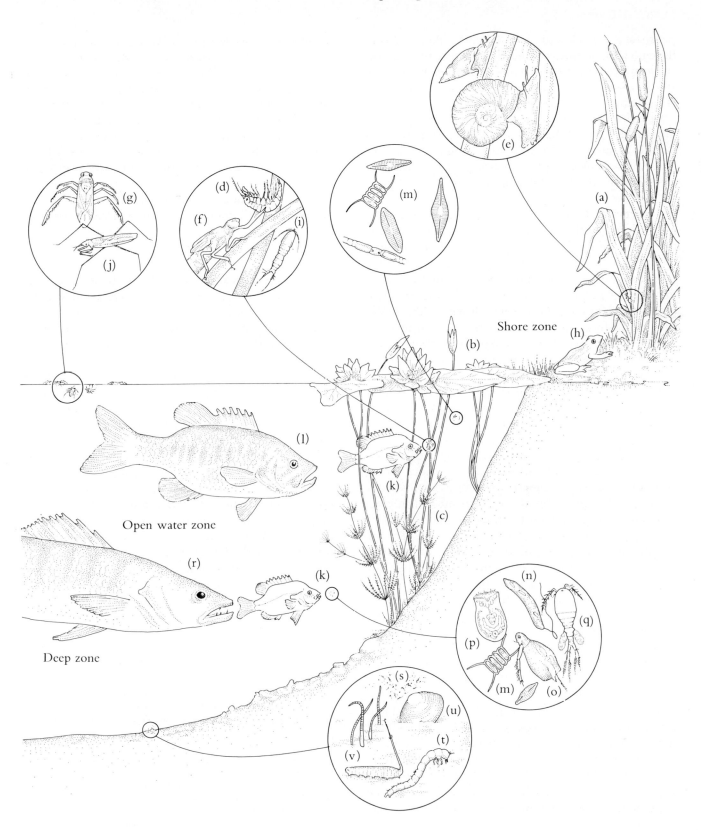

Shore zone

Open water zone

Deep zone

photosynthesis takes place at all depths. Ponds usually have no temperature stratification because they are too shallow to prevent thermal currents from mixing the waters. Some ponds dry up during part of the year, creating particular stress on their communities. Organisms living there must have a dormant stage to survive the dry period. For instance, fairy shrimp lay eggs capable of surviving for months or longer in dry soil. Some organisms can live both on land and in water—for example, amphibians such as frogs.

Rivers and streams. Three features of the environment in rapidly flowing waters are very important to understanding the types of organisms that live there: the presence of a current; the high oxygen concentration; and the source of nutrients. A current is one of the main factors making life in a stream or river different from life in lakes and ponds. However, the difference is not found in all parts of these environments. Streams have pools or areas of quiet flow where organisms find similar habitats to those in lakes. In addition, a lake shore, where waves keep water moving, provides organisms with a habitat similar to a rapidly flowing stream or river. There are thus two types of stream or river communities: those in flowing water and those in quiet water.

A major, and very understandable, feature of organisms living in moving water is that they usually have some way of hanging on to surfaces such as rocks or stream bottoms. Some are cemented firmly to stones or other objects in the flowing waters. Others have hooks, suckers, or sticky undersides. Stream creatures also have streamlined bodies to reduce resistance to flowing water and are often flat so they can crawl under rocks to escape the pull of the current.

Because of the current, rapidly flowing streams or rivers usually have a high oxygen content. The waters, moving and tumbling over rocks, become well mixed with air and so absorb a great deal of oxygen. Organisms living in rapidly flowing water are used to these high oxygen concentrations. When pollutants that use up oxygen in water are added to streams, the clean-stream organisms cannot survive the low oxygen levels. The stream communities found in polluted and clean water are contrasted in Chapter 12.

A large part of the nutrients in streams and rivers either wash or fall into the water from the banks and surrounding watershed. Plant nutrients, such as nitrate and phosphate, and organic material, such as leaves on

which detritus feeders live, all enter the stream from its watershed. Stream organisms have adapted to this constant flow of fresh nutrients and also to the removal of their waste products by the current. Waters with currents thus provide an environment fundamentally different from the still waters of ponds and lakes.

Organic materials and plant nutrients are not the only materials washed into rivers and streams from watersheds. Pesticides or industrial wastes in groundwaters may also wash in. For this reason, when river and stream pollution problems arise, we must, in examining possible solutions, take into account not only the stream itself but also the land surrounding the stream.

The Hudson as a sewer. Partly because currents usually carry wastes downstream, rivers and streams have always seemed to humans to be especially handy and inexpensive ways to dispose of wastes. For years, the Hudson River was used as a sewer for many communities along its banks. This practice resulted in a series of typhoid epidemics around 1890 in communities that not only dumped sewage but also drew drinking water from the Hudson (Chapter 10). The latest story about pollution in the Hudson River, however, goes back only to 1975, when it was discovered that as a result of industrial sewage, river fish contained dangerous levels of toxic chemicals called PCBs. PCBs are lethal in very small amounts (as little as 10 parts per billion) to many insect larvae and to some small fish.[1] Large fish, at the top of their food chains, were found to have PCB concentrations well over the legal limit for human foods (Figure 8.4).

As a result of laws such as the Toxic Substances Control Act of 1976, the manufacture and disposal of PCBs are now controlled much more closely. For the Hudson River, however, this comes too late. (See more on PCBs in Chapter 11.)

Estuaries

As the Hudson winds its way to the Atlantic Ocean, it reaches at last an area where its fresh waters mix with the salt waters brought by ocean tides. In the Hudson, salt and fresh waters meet in New York Bay, and

---

1 For some comparisons that may help you understand this number better, see the supplemental material at the end of Chapter 6 (pages 127–128).

Figure 8.4 *Shad Fishermen on the Shores of the Hudson River,* by Pavel Petrovitch Svinn (1787–1839). Henry Hudson anchored his ship *Half Moon* off this point as he returned from exploring the river that was later given his name. But Indians fished for shad in the Hudson long before Hudson arrived in 1609. Fishing for some species of fish is no longer allowed in the Hudson. Unhealthy levels of the chemicals known as PCBs are now found in river fish. Despite the long passage of time and the pollution, some uses of the river have not changed much. Page 225 shows a picture of men fishing for shad on the Hudson River today. (The Metropolitan Museum of Art, Rogers Fund, 1942)

some salt water moves as far up river as the city of Troy.

River mouths, salt marshes, bodies of water behind barrier islands, and coastal bays—such as New York Bay—are all estuaries (Figure 8.5). **Estuaries** are coastal bodies of water partly surrounded by land but still having an open connection with the ocean. In these areas, fresh water drains from the land and mixes with tidal currents of salt water. Estuaries, especially marshes, are often looked upon as wasteland, best dredged or filled. But this is a real misunderstanding of the role estuaries play.

Productivity. Estuaries are very productive systems, and are generally more fertile than either the neighboring ocean or the fresh waters that flow into them. (See Table 8.1, page 169.) The reason is that nutrients are easily trapped in estuaries. The nutrients are trapped in a physical sense, first as sediments settle out from river inflows and then by the action of the tides and fresh water flow (Figure 8.6). Nutrients are also trapped in a biological sense because they are recycled rapidly by a network of producers, consumers, and detritus feeders. Unfortunately, pollutants are also recycled and sometimes biologically magnified in estuaries, so the effect of toxic materials such as DDT can be more serious than in a river or the ocean.

Another factor contributing to the fertility of estuaries is tidal action. Tides, which cause the water in

estuaries to flow back and forth, make it possible for organisms like the oyster, which feeds by filtering sea water, to sit and have their food brought to them. In the same way, their wastes are removed.

Furthermore, estuaries provide good habitats for a variety of producer organisms, from the large rooted grasses like eel grass, turtle grass, or salt marsh grass, to the tiny floating plants or bottom-dwelling algae. In fact, often more organic material is produced than can be recycled in the estuary itself. The excess nutrients flow out into the ocean and fertilize these waters. Good fishing is the result.

Estuaries serve yet another important purpose: providing a nursery for many ocean species, such as shrimp. The larvae, or immature stages, of these species find protection and food in the estuary. Some fish, such as salmon or Hudson River shad, which live in salt water but return to fresh water to breed, require estuaries as places to rest during their journey. Thus, when estuaries are unthinkingly filled in, the effects fall not only upon the creatures that spend their whole lives there, but also upon many ocean species that use estuaries or the food produced there.

Marine Habitats

Geographic features of the sea bottom. Actually, the Hudson River does not disappear into the

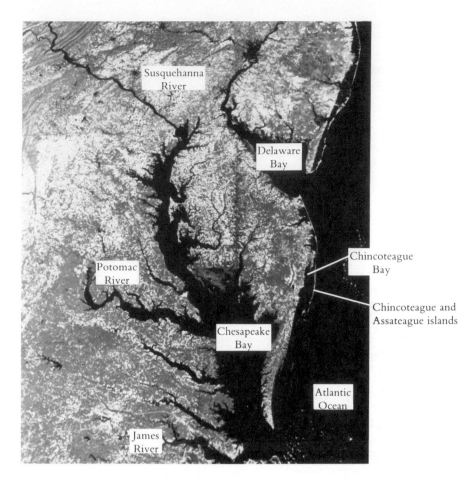

Figure 8.5 Estuaries. Chincoteague Bay is an estuary formed by the mainland on one side and the barrier islands, Chincoteague and Assateague, on the other. Chesapeake Bay is a huge estuary where fresh water from many large rivers—such as the James, the Susquehanna, and the Potomac—mix with salt waters from the Atlantic Ocean. (NASA)

Atlantic Ocean without a trace. Three miles (5 km) southeast of Ambrose Lightship, an underwater channel begins and runs 130 miles (210 km) along the ocean bottom. Geologists believe that at the end of the last ice age the shore was 150 miles (230 km) farther into the ocean than it is now. They speculate that the roaring Hudson, fed by melting ice, cut a channel through

this land on its way to the sea. Hudson Canyon is a natural wonder. At some points it is 7 miles (11 km) deep, deeper than the Royal Gorge of the Colorado River. But it now lies hundreds of feet below the surface of the ocean.

Besides Hudson Canyon, many interesting geologic features appear below ocean waters. Looking at the sea in profile, as if it were cut in half from top to bottom, several well-defined areas can be seen (Figure 8.7). For some distance, the ocean floor slopes gradually away from the land. This area is called the continental shelf. The floor then drops off sharply (continental slope) and again levels off, into the continental rise. Finally, the floor drops off once more to a level plain, the abyssal plain, 6000–15,000 feet (2000–5000 meters) below the surface. Toward the middle of the ocean, a series of ridges lie scattered across this abyssal plain.

Currents. Unlike lakes, where layers of water may remain still for long periods, the water in the sea moves continually. Currents are caused by a variety of forces, such as temperature differences, differences in

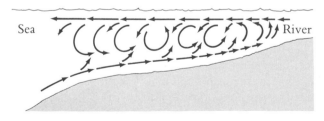

Figure 8.6 Mixing Currents in Estuaries. Fresh water, which is lighter, tends to float on the heavier sea water. As one rolls over the other, mixing currents are set up that tend to recirculate nutrients. (Adapted from E. P. Odum, *Fundamentals of Ecology.* Philadelphia: Saunders, 1971, p. 354.)

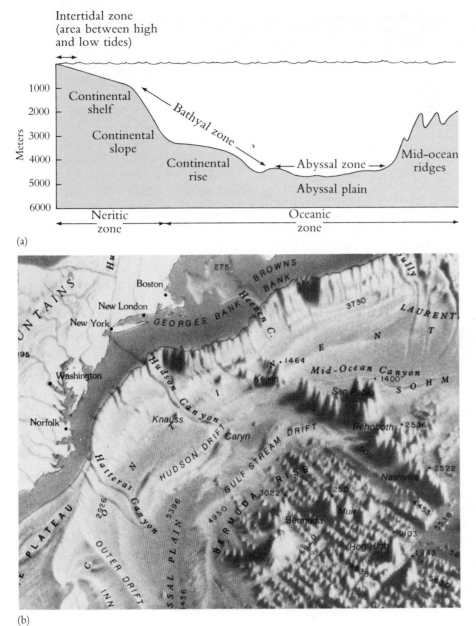

Intertidal zone
(area between high
and low tides)

Continental
shelf

Continental
slope

Bathyal zone

Continental
rise

Abyssal zone

Abyssal plain

Mid-ocean
ridges

Meters
1000
2000
3000
4000
5000
6000

Neritic
zone

Oceanic
zone

(a)

(b)

Figure 8.7 (a) The seas in cross section. The shallow water of the continental shelf is called the neritic zone, including the area where the tides cover and uncover the shore twice a day. The rest of the open sea is called the oceanic zone. The bottom of the sea along the continental slope and rise is called the bathyal zone, while along the deeper plain it is called the abyssal zone. As in fresh water, light penetrates only the top layer of water. All below this layer is in darkness. (b) Panorama of ocean bottom. Note the Hudson Canyon shown in the upper left-hand corner of the picture. (Panorama of ocean bottom showing Hudson Canyon is from World Ocean Floor Panorama by Bruce C. Heezen and Marie Tharp, 1977. © Marie Tharp)

salt content, the rotation of the earth, and winds. Because of these mixing currents, even the deep parts of the oceans have a constant supply of oxygen in the water. In some areas, along steep coastal slopes, winds continually blow the surface water away from shore, allowing cold bottom waters, rich in nutrients, to rise to the surface. This is called **upwelling.** The areas where this occurs—for instance, along the coast of Peru—are the most fertile in the seas. In general, although life is found in all areas of the sea, the major commercial fisheries are all located on or near the continental shelf. The reason is that many ocean food chains are based on the microscopic green plants, which grow best in areas of coastal upwelling.

Communities. Figure 8.8 shows some of the species found in the coastal zone. The communities that live in the intertidal zone consist of organisms specially adapted to the periodic absence of water when the tide goes out. (Some scientists are concerned that these organisms might be endangered by the development of tidal power: the use of the tides to generate electricity. We shall examine this more fully in Chapter 25.) In the open ocean, species have adapted to

Figure 8.8 Some of the organisms (not drawn to scale) found in communities in the region of the continental shelf. Phytoplankton species, including diatoms, dinoflagellates, and microflagellates, are the producers, which form the base of the food chains in the coastal zone. "Seaweeds," algae adapted to hold on to rocky shores as the tide washes over them, are also producers in this zone. Tiny zoo-plankton such as copepods, pteropods, and jellyfish (medusae), as well as the shrimp-like krill (*Euphausids*) feed on the producers and are fed on by fish, squid, and sea mammals such as whales. Sea birds, large fish such as tuna and swordfish, and humans are top consumers in these food chains. Also found in this region are the larvae or immature stages of many marine species. Buried in the bottom muds or clinging to rocks are the worms, clams, snails, crabs, and bacteria that form the benthos, or bottom dwellers, which feed on detritus. (Adapted from John D. Isaacs, "The Nature of Oceanic Life," *Scientific American,* September 1969.)

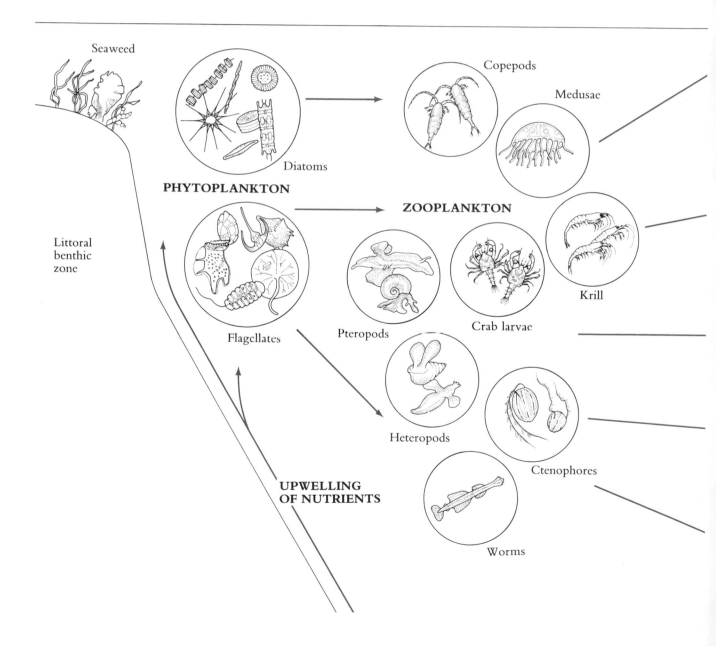

Seaweed

Copepods

Medusae

Diatoms

PHYTOPLANKTON

ZOOPLANKTON

Littoral benthic zone

Krill

Flagellates

Pteropods

Crab larvae

Heteropods

Ctenophores

UPWELLING OF NUTRIENTS

Worms

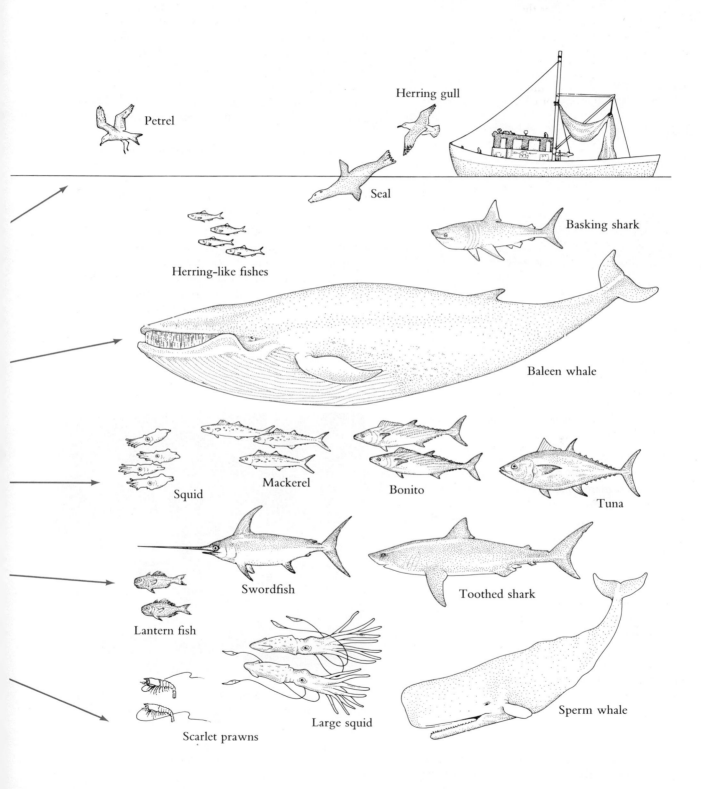

Petrel

Herring gull

Seal

Basking shark

Herring-like fishes

Baleen whale

Squid

Mackerel

Bonito

Tuna

Swordfish

Toothed shark

Lantern fish

Large squid

Sperm whale

Scarlet prawns

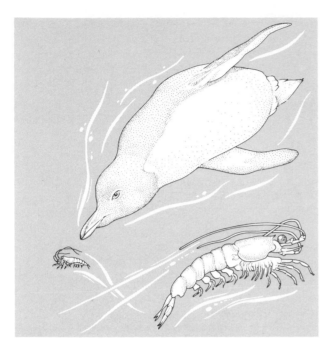

Figure 8.9 Shrimp-like krill (*Euphausia superba*) are a vital link in ocean food chains between the microscopic producers and larger creatures such as whales. However, nations hungry for protein are now looking at the 5-cm-long krill (shown here being chased by a penguin) as a possible food source for their human populations. Krill can be made into a sort of shrimp paste. What effect massive harvesting of krill would have on ocean communities is unknown.

life far from shore. In the top layer of water, where light penetrates, drifting microscopic plants and animals live. Microplankton species such as *chloramoeba* are the main producers in the open oceans. The shrimp-like krill (Figure 8.9), as well as small zooplankton species, feed on microplankton. Larger fish and sea mammals such as the baleen whales range over both the open ocean and the shore areas in search of krill and zooplankton, on which they feed. Oceanic birds such as petrels, albatrosses, and frigate birds feed on the open oceans except during breeding time, when they fly to land.

The sunlit zone in the open oceans does not support as much life per square meter as the light zone in coastal areas does. However, the oceans cover 70% of the earth's surface and much of this is open ocean. For this reason, the photosynthetic organisms in the open ocean are very important in world oxygen and carbon dioxide balances.

We know relatively little about communities in the deep zones of the ocean. Only recently have a

number of facts come to light—quite literally, since one of the main physical characteristics of the deep sea regions is relative darkness. Sunlight does not reach these regions (although they are not completely dark). Since there is not enough light for photosynthesis, organisms depend on producers in the top layer of water for organic nutrients. The major portion of the organic matter reaching the deep ocean zones is probably composed of fecal pellets from zooplankton on the surface.

Diversity in deep sea communities. Much of the deep sea bottom is covered with thick layers of mud. In some areas, this appears to be formed into topographic features such as ridges and cliffs. In and on the bottom muds of the plains live many species of worms, clams, and crustaceans (Figure 8.10). These organisms live in an area that maintains a very stable temperature and energy supply. Although the organisms living here are small, an enormous variety of them exists. In fact, the diversity is comparable to that found in tropical rain forests and in coral reefs, two other areas noted for their great variety of species. These habitats are similar in that they are physically stable (that is, little or no changes in physical conditions take place over long time periods) and have a long evolutionary history. High diversity is character-

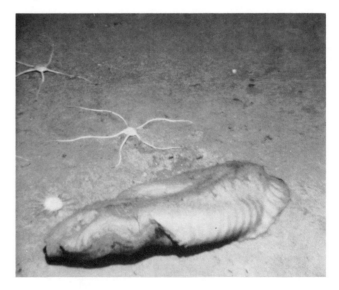

Figure 8.10 Life on the Deep Ocean Bottom. Although a few large species are visible on the surface, most of the organisms in the deep sea are found in the mud. Here, a holothurian, or sea cucumber (*Holothuroidea*), an urchin (probably *Lytechinus*), and brittle stars (*Ophiuroidea*) share the sediment surface. (Dr. Fred Grassle, Woods Hole Oceanographic Institute.)

istic of ecosystems that have not been disturbed for long periods of time.

Productivity of marine versus land ecosystems. Table 8.1 gives a comparison of the estimated productivity of the various ecosystems on earth. The estimates in the table are of gross productivity rather than net productivity; that is, the amount of produced material used by organisms themselves in respiration has not been subtracted out. Because of the loss due to respiration, as well as for a variety of other reasons,[2]

only one-third or less of the gross productivity in an ecosystem is available for humans to harvest.

Table 8.1 underscores the fact that compared to most land ecosystems, the productivity of the open oceans is relatively low. Some researchers believe this is primarily due to a lack of nutrients in the open oceans. However, especially in warm oceans, it is probably also due to the fact that the primary producers are nannoplankton. These extremely small plankton species are only used as a food source by very specialized small planktonic animals. As we explained in Chapter 2, only about 10% of the available energy is passed from one trophic level to another; thus the addition of another level at the bottom of the trophic pyramid greatly decreases the production measured at higher levels.

---

2 For example, unavoidable losses to insects and diseases, and the fact that not all parts of plants or animals are, or are considered, suitable for human consumption.

Table 8.1 Primary Productivity in Various Ecosystems

Ecosystem	Area (million square km)	Gross primary productivity (kcal/square m/year)	Total gross production (10^{16} kcal/year)
Marine ecosystems			
Coastal zone	34.0	2,000	6.8
Areas of upwelling	0.4	2,000	0.2
Open ocean	326.0	1,000	32.6
Estuaries and reefs	2.0	20,000	4.0
			43.6
Natural land ecosystems			
Deserts and arctic tundra	40.0	200	0.8
Grasslands	42.0	2,500	10.5
Dry forests	9.4	2,500	2.4
Moist temperate forests	4.9	8,000	3.9
Coniferous forests	10.0	3,000	3.0
Tropical rain forests	14.7	20,000	29.0
			49.6
Cultivated land ecosystems			
No energy inputs (no fertilizer, pesticides, etc.)	10.0	3,000	3.0
Fuel-subsidized agriculture	4.0	12,000	4.8
			7.8

Adapted from Odum, E. P., *Fundamentals of Ecology*, 3rd ed. Philadelphia: W. B. Saunders, 1971, p. 51.

At the moment, humans eat near the top of ocean food chains, thus decreasing even further the amount of productivity available to them. At one time, scientists thought that plankton might be harvested from the ocean and used as a human food source; however, productivity in the open oceans is too low to make harvesting plankton a reasonable alternative. That is, too much energy would need to be invested in running the fishing boats and processing the plankton compared to the amount of food energy obtained. In addition, of course, we would have to find ways to make plankton appealing to humans as a food. It is largely due to low productivity that once widely held hopes of feeding the world from the oceans are doomed to disappointment.

In addition, the primary producers in the ocean are important to the world's oxygen and carbon dioxide balance. Schemes to harvest primary producers in the ocean could seriously interfere with this balance.

Certain coastal marine areas are suitable for **mariculture**. Aquatic species such as clams, lobsters, and certain fish can be raised in underwater farms (see Chapter 24). However, such schemes are probably not practical in the open oceans because of the low nutrient concentrations there. These plans are also endangered by pollution of coastal waters with sewage and hazardous wastes.

In the following chapters, various factors are discussed that can disrupt the functioning of aquatic ecosystems and make waters unfit or unpleasant for human uses.

How Lake and Reservoir Waters Can Become Oxygen-Poor

To understand how decreased quality of lake and reservoir waters comes about, we need to discuss the effects of temperature on the water in a lake reservoir. An annual cycle occurs in the temperature profile of such a body of water.

For convenience, we describe the cycle as beginning in late fall, when the weather has begun to cool. The autumn and winter winds transfer their energy to the lake or reservoir in the form of waves, which mix into the body of water. The waters are well-mixed during this period. Thus, a fairly uniform character in terms of quality and temperature of the water exists at all depths. That is, if we were to measure the temperature at all depths, we would find that, within limits, the temperature of the water is roughly the same everywhere. Such a picture continues through the winter months.

In the spring, however, sunlight warms the waters nearest the surface and warmer stream inflows enter the cold body of water. Above 4°C, warm water is less dense than cooler water[3] and thus floats above it.

Hence, the new warmer inflows tend to form layers near the surface of the reservoir. Similarly, surface waters that have been warmed by sunlight tend to remain near the top. The wind's force at mixing is now less effective because the warmer, more buoyant water, though pushed into the interior, rises again toward the surface.

Essentially, three layers of water form at successively warmer temperatures. The warmest layer, called the epilimnion, forms at the top; it stays at the top because it has the lowest density. The coldest layer, called the hypolimnion, lies at the bottom; it has the highest density. Between these layers, a sharp temperature transition occurs; this middle layer is called the thermocline. This formation of layers of differing density and temperature is known as stratification.

Stratification of water into temperature layers is one of two phenomena that combine to cause a decrease in the quality of lake and reservoir water. The other phenomenon is the growth of algae in relatively clear waters. While some algae are growing, others are dying. The dead algae sink to the bottom layers; there, microorganisms in the water consume the dead algae as food. The microbes also remove oxygen as they grow and maintain themselves. Since the water in the reservoir has stratified and little mixing occurs be-

---

3 At temperatures at and not far from freezing, cooler water becomes less dense, and more buoyant. You know this from your observation that ice accumulates first on the top of a pond rather than in its depths.

tween the upper and lower layers, once oxygen is removed in the bottom layer, it is not restored quickly. Thus, the bottom waters of a stratified lake or reservoir may become very low in oxygen. These are the waters, we must point out, that are commonly released to downstream uses. In the section on Organic Water Pollution (Chapter 12) we see that waters that are low in oxygen do not support desirable aquatic life.

As colder weather arrives, the upper layers become cooler and cooler until they approach temperatures just less than the temperature in the bottom layers. At that point, the layering becomes unstable, and the layers "flip." This event is called the "fall overturn," and the lake or reservoir now becomes relatively well-mixed; that is, the water at all depths is of about the same temperature. The well-mixed character is sustained through the winter months.

Questions

1. Why is the watershed of a lake or river considered the important ecosystem, rather than just the river or lake itself?
2. What major factors define water habitats? Describe some of the major ways in which organisms have adapted to these habitats.
3. Why are estuaries important ecosystems?
4. What is the importance of plankton to fresh and salt-water communities?
5. What is a possible explanation for the high diversity found in deep sea communities?
6. Do you think the oceans are a vast untapped resource of food to feed the hungry nations of the world? Why or why not?
7. What are the advantages and disadvantages of eating krill instead of whale meat?

CHAPTER NINE

Water Resources

How We Obtain Fresh Water

Surface Water and Reservoirs
Uses of Reservoirs/Environmental Impact of Reservoirs

Groundwater Resources

Extending Water Supplies without More Reservoirs

Water Resources Decisions
The High Plains and the Ogallala Aquifer/The Colorado River and Conflict in the Southwest/Grand Plans

CONTROVERSIES:

9.1: *Should Water Be Transferred to the High Plains?*

9.2: *Has the West Had Enough Water Projects? And Enough Growth?*

In more and more places on the earth, we are finding that fresh water is not present in the amounts needed for growing communities. Wells are being driven deeper; water pipelines and aqueducts are being built; dams are being constructed—all with the goal of capturing more water.

Although in the chapters to follow we focus on the *quality* of the water we consume, in this chapter we investigate how people obtain adequate water supplies and whether the *quantity* of available water is sufficient. We do not mean to suggest that water quality and water quantity are separate issues; they are, in fact, interlocked. When engineers and planners ask if adequate water resources are available, they really are asking: "Do we have available a sufficient quantity of water of the quality that we need?"

How We Obtain Fresh Water

We draw water in two fundamental ways: either from wells, tapping underground sources of water called aquifers; or from surface flows—that is, from lakes, rivers, and man-made reservoirs. Other means exist for producing potable (drinkable) water. In some wealthy or technologically advanced areas, *desalination*—the desalting of sea water by such means as distillation, for example—can make even ocean water fit to drink. In water-poor regions of the world without wealth, cisterns collect rainwater for human use. The extent of water production by such exotic means is negligible, however. Generally, we rely upon fresh surface water and groundwater.

We can examine the functions of these two water sources by comparing the man-made above-ground structures known as reservoirs with the below-ground water-filled spaces we know as aquifers.

Reservoirs extend the amount of water available to communities. Without a reservoir, the maximum steady uninterrupted rate of water withdrawal from a flowing river, as measured typically in millions of gallons per day, can be no greater than the lowest daily flow rate occurring in the river during the time of operation. Any larger quantity of water could not be withdrawn day after day. However, a water-supply dam thrown across a river backs up the water flow behind its face, forming the reservoir, while releasing through its gates enough water to sustain the flow downstream from the dam. The reservoir's large volume captures high flows as well as low flows. A city or town extending its water-supply conduit to the reservoir can draw water steadily from it week after week, month after month, without interruption.

The above-ground reservoir is a human invention to steady the flow of fresh water through time, collecting high stream flow from one season to make water available in another season. The aquifer, in contrast, is a natural underground reservoir, where water temporarily resides on its route to lakes, streams, rivers, or oceans. Storage is temporary because groundwater flows toward the ocean, just as surface water does. An above-ground reservoir is a large empty space when not filled with water. An aquifer is not necessarily so empty. It may consist of free-flowing water, as in an underground cavern or stream, but it may also be water that simply fills the spaces between particles of sand and gravel. Aquifers can be huge, extending for hundreds of miles. The volumes of water in such aquifers are enormous. The volume of water stored in the

Ogallala Aquifer, in the states of the High Plains, probably is comparable to the volume of Lake Huron.

Water from a surface reservoir and water from an underground aquifer are likely to be quite different in quality. In the surface reservoir, water will contain sediment picked up by the river filling the reservoir. Some of the sediment particles settle to the bottom of the reservoir, depending on how long the water is detained there, but particles do remain in the water drawn for human use. To remove these particles, it is common to filter reservoir water prior to use.

Organic matter that can cause the removal of oxygen is likely to be present in surface water, having arisen from both municipal and agricultural discharges. Nitrogen and phosphorus compounds enter surface water from these same sources. Algae often find the reservoir environment inviting—they find the nutrients and sunlight they need for growth. Because reservoirs receive "fresh" sources of pollution, water from a reservoir, if used as drinking water, ought to be fully treated. Treatment is required to remove undesirable tastes, color, and odor; to make the water clear; to make it free from hazardous chemicals; and to destroy any disease-causing organisms. However, not all reservoir water destined for human consumption is fully treated, even in the United States. Filtration is excluded on occasion for surface waters of good clarity, but it is a dangerous step to exclude.

Water drawn from aquifers, on the other hand, is likely to be far clearer than water from surface reservoirs, especially if the aquifer has not been drawn on for long nor extensively depleted. While water from an aquifer is likely to have a higher content of dissolved minerals, it will not have algae in it, since no sunlight has illuminated the water for many years. Because of the way water reaches the aquifer—that is, because it is likely to have filtered through thick layers of soil—its bacterial and viral content is likely to be lower than that in surface reservoir water. The odor of hydrogen sulfide is common in groundwater, due to the bacterial degradation of organic material that takes place in groundwater in the absence of oxygen. But there are exceptions to these generalizations about quality; groundwater can become polluted by chemicals and by microorganisms. In fact, failure of treatment or lack of treatment of groundwater is a problem that threatens many communities in the United States.

In a number of sections of the country, groundwater from large aquifers may seem more reliable than surface water. The same steady flow may be drawn from groundwater, month after month and year after year, whereas the local surface stream flow may be highly variable. This is especially true in the arid southwest United States, where desert river beds are often dry but flows swell to flood proportions after rainstorms. Large underground reservoirs may seem to provide highly reliable yields in these areas, but the reliability may be only temporary, on the order of, say, tens of years. The yield appears so reliable because the water is drawn from an enormous underground reservoir that may have taken hundreds of years to fill. If the rate of withdrawal from the aquifer is far greater than the natural rate of recharge or refilling, the groundwater resource will be steadily depleted.

Depletion of an aquifer by too rapid withdrawal is one threat to groundwater; eventually, the aquifer will be unable to sustain the rate of withdrawal. We will discuss several areas of the country where withdrawal rates exceed rates of natural recharge and where, as a consequence, the aquifers are steadily going down. Depletion of an aquifer can lead to other problems. As wells go deeper to reach water, the waters they draw are from older portions of the aquifer, and are likely to have a higher mineral content. Eventually the water can become so salty that it is undrinkable, and not useful for irrigation even though still available. In ocean coastal areas where groundwater is in use, salt water may move in (intrude) to fill some of the void spaces created by drawdown of an aquifer. Again, the water can become undrinkable as salt concentrations build up.

Chemicals and bacteria can threaten groundwater supplies. Chemicals from landfills are seeping through the soil into the groundwater reservoir that provides water for the city of Tucson, Arizona. In 1964, bacteria entered the aquifer supplying a portion of the water in Los Angeles, and because of inadequate chlorination, an epidemic of diarrheal illness struck the city. Once polluted, groundwater may remain polluted through many lifetimes because aquifers may be recharged very slowly, often over hundreds of years. Groundwater supplies are extremely fragile, then, relative to the entry of pollutants.

Because aquifers are recharged so slowly, it would seem wise to discourage any long-term industry that uses groundwater at a rate higher than the natural rate of recharge. Irrigated agriculture using groundwater much faster than its rate of recharge would be an activity to avoid. But in the High Plains, in southern Arizona, and in the San Joaquin Valley of California, irrigation with groundwater has been going on for many years. We will discuss the serious impact of these activities shortly.

Surface Water and Reservoirs

Uses of Reservoirs

Reservoirs are of two types: single-purpose and multiple-purpose. Single-purpose reservoirs provide only a single function, such as community water supply. Their operation is relatively straightforward—for instance, water-supply reservoirs release only the amount of water that is needed. Multiple-purpose reservoirs may serve a variety of purposes, including

community water supply;

irrigation;

navigation;

recreation;

hydroelectric power generation;

flood protection; and

low flow augmentation and environmental flow-by.

Community water supply includes water for drinking and washing, for industrial purposes, and possibly for watering lawns. Irrigation water is designated for the growing of crops; its use is often highly seasonal, with large requirements during the hot seasons. The capability of river navigation may be maintained by steady water releases through the year. Recreation such as boating, picnicking, and the like is served by keeping a relatively constant volume of water in the reservoir so that shorelines do not fluctuate extensively. Hydroelectric power generation requires both steady releases and high water levels. Flood protection requires that the reservoir be kept as empty as possible. Low flow augmentation and environmental flow-by imply that releases are made during times of low flow to protect the quality of the water and the species that inhabit it. Such releases dilute wastes, thereby decreasing their total demand for oxygen from the water. Such releases may also flush out salt water from estuaries, maintaining the proper environment for the native estuarial species.

The operation of reservoirs serving these multiple functions is far more complex than that of a single-purpose reservoir because a number of these purposes are in conflict. A reservoir that serves only a water supply function should be kept as full as possible. If the purpose of a reservoir is flood control only, it should be kept as empty as possible so the largest flood flows can be caught and then released more slowly. The operation and purposes of a reservoir can significantly influence its impact on the environment—and reservoirs have a strong impact on the environment.

Environmental Impact of Reservoirs

As flowing waters slow down, which indeed they must as they are brought to a halt by a reservoir, the sediment suspended in the turbulent waters settles out and drops to the bottom of the reservoir. Downstream from the reservoir, the clear water released into the river erodes earth from the river bank at a faster rate than the free-flowing river would have, as if to make up for the sediment lost in the reservoir. Increased erosion downstream from a reservoir is a common occurrence.

The bottom of the reservoir becomes coated with the sediments transported from upstream areas (Figure 9.1). This blanket of sediments is exposed to view periodically as the level of water in the reservoir rises and falls in response to inflows and releases. The sediments gradually build up and, unless dug out occasionally, begin to consume the storage volume of the reservoir. That is, a reservoir created to store water for water supply or to catch water for flood control can gradually lose its effectiveness unless the solids deposited in its banks are excavated from the site.

Large deposits of sediment may be prevented, in part. Although erosion and sediment transport are natural and continuing events, farming, road development, construction of homes, and forest cutting all accelerate erosion processes by exposing fresh and unanchored soil. Careful management of the soil (see Chapter 6) can help reduce the burden of sediment carried by streams and thus help prevent the rapid deposition of sediment in reservoirs.

The unsightly mounds of sediment that become visible during times of low reservoir storage are one reason why many individuals express distaste for dams. Another reason is more basic: valued lands are lost to view and to use forever. Valued animals and

Figure 9.1 A silted reservoir near Ithaca, New York. This reservoir has been drained because a new community water supply has been created on a nearby lake. The sediments, accumulated over some 50 years of operation, were consuming a large part of the storage volume of the reservoir. Consequently, a reasonable water level in the reservoir looked like a larger volume than was actually stored.

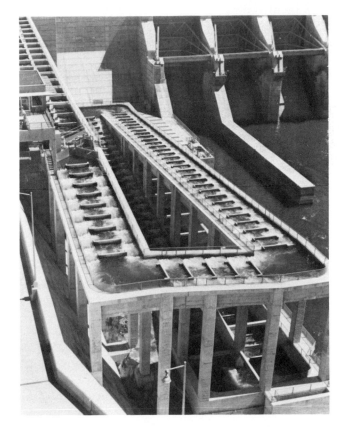

Figure 9.2 The Long Road Home. This fish ladder is on the John Day Dam in Oregon. (Photo courtesy of the U.S. Army Corps of Engineers.)

plants are also destroyed. Not only land species are lost; fish that inhabit the stream may also be barred. Where once they swam upstream to spawn, the reservoir wall now blocks their path.

Reservoirs may be built so that fish are still able to pass upstream. Fish ladders consist of concrete steps down which a stream of water flows (Figure 9.2). The ladder is so constructed that fish can jump from one step up to the next, and eventually to the reservoir and upstream.

The geologic history of a region is often revealed in the layers of rock through which a river has cut its course. This history may be lost when a dam is thrown across a river. And the sheer majesty of the river's power demonstrated by the gorge may be replaced by a quiet pool of water. Each potential reservoir site has its own characteristics. It may include a mature forest or possess a cluster of rare wildflowers. The reservoir may flood out valuable farm lands, or displace families or Indian tribes from ancestral homes. Graveyards and tribal burial grounds may be forever obliterated. It is not always clear that a just compensation can be found for the people uprooted from such an area, for their loss is more than economic.

A reservoir may have still other impacts in addition to these. During certain times of the year, the quality of water within the reservoir and the quality of the water released from it may be surprisingly poor. During the summer and early fall, the lower layers of

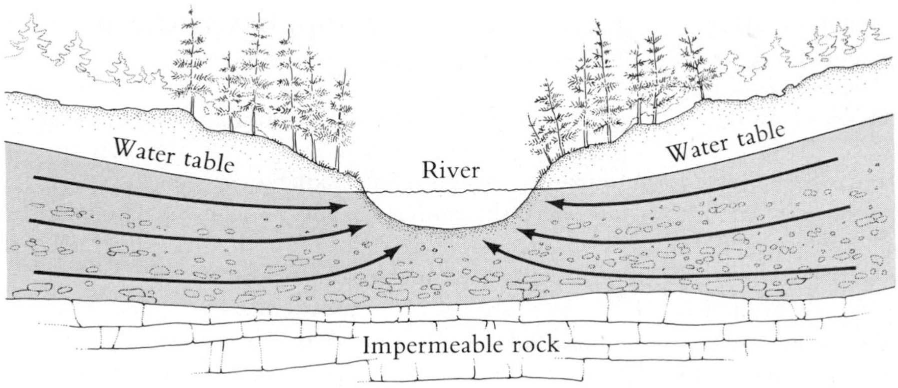

Figure 9.3 Groundwater helps to maintain stream flows. Especially in dry seasons, the relative contribution of groundwater to stream flow may be substantial.

water in a reservoir can become very low in oxygen. This low oxygen condition is caused by a combination of two phenomena. The first is a lack of mixing of the reservoir waters during the summer and early fall. The second phenomenon is the consumption by bacteria of the dead algae in the bottom layers of the reservoir. The consumption causes a removal of oxygen from the water. If this oxygen-poor water is released, fish and other aquatic creatures downstream from the reservoir may be harmed.

Groundwater Resources

Groundwater serves a more limited set of functions than does surface water. In many cities, groundwater provides community water supplies. Especially in rural areas, where the cost of extending the water distribution system is very high, people turn to wells to deliver their water needs. Groundwater is also used for irrigation of crops, a very common practice in farming areas that are short of surface water or where the building of irrigation canals is very costly. Experts estimate that groundwater provides the water for about half the population in the United States. Surprisingly, groundwater may also supply approximately a third of the total water used for irrigation in the country. In the western states, groundwater supplies just over half of the irrigated acreage.

Groundwater also serves another important function, but a relatively unseen and unappreciated function. Groundwater flow contributes to and often sustains the summertime flow of streams and rivers—which themselves may be used for water supply (see Figure 9.3).

Groundwater may seem an almost infinite resource in some places. In the conterminous U.S., about 25% of the water that falls as precipitation each year becomes groundwater. Indeed, of the world's fresh (nonsaline) water, groundwater resources far exceed surface water resources (see Table 9.1). The appearance of enormous resources is deceiving, however, because groundwater has collected over hundreds and thousands of years at relatively slow rates. Rapid withdrawal of groundwater is not matched by rapid inflows, but by the same slow, steady infiltration that occurred in the past. And groundwater resources below 0.8 kilometer are often too laden with salts to be useful for water supplies.

Table 9.1 Freshwater Resources of the World

	Volume (thousands of cubic km)
Fresh water in lakes and inland seas	125
Fresh water in streams, rivers, etc. (average)	1.25
Groundwater within 0.8 km of surface	4,200
Groundwater between 0.8 km and 4.0 km depth	4,200
Fresh water in glaciers and icecaps	29,000

(Adapted from H. Bouwer, *Groundwater Hydrology*. New York: McGraw-Hill, 1978.)

Groundwater use provides a number of advantages to consumers. First, there may be cost advantages. Since groundwater may be located very near its point of use, savings in piping and possibly pumping costs can occur. Second, the firm yield that can be sustained over a long period through both dry and wet seasons may be very much greater than the comparable yield of a surface water reservoir. This advantage, however, may prove to be illusory if the aquifer is depleted by a consistent overdraft. Third, groundwater is often of a fairly consistent and reliable quality. In fairly pristine settings, it is not usually subject to bacterial, viral, or chemical contaminations. Minerals and hydrogen sulfide do occur in groundwater with enough regularity to make people alert to this possible compromise in quality.

What is the safe yield of an aquifer? The safe yield of a surface water reservoir depends on the flows that enter the reservoir. The same is true of an aquifer. No more can be withdrawn year after year from an aquifer than the annual recharge rate of the aquifer—*unless* the user, the remover, is willing to draw down the volume in the aquifer. This is happening in a number of parts of the country: withdrawal rates are exceeding recharge rates and the aquifers are being depleted.

One aquifer in the southern High Plains region of Texas has a volume five times that of Lake Mead on the Colorado River. However, the rate of recharge of Lake Mead is, on average, 20 times greater. Though the aquifer has a seemingly larger volume against which to draw, its withdrawal limit is 1/20 that of Lake Mead, based on its rate of recharge—unless the users are willing to see the aquifer drawn down and eventually depleted. In fact, in desert basins, rainfall only rarely recharges an aquifer. During most years, evaporation draws water up from the surface into the atmosphere. Only during extremely wet years will enough water be present for some of it to recharge the aquifer.

Finally, in this brief description of groundwater and its uses, we should mention that groundwater is often used in conjunction with a surface water supply. We call this blending of surface and groundwater *conjunctive use*. Because surface water supplies are more "flashy"—that is, because they exhibit greater variability than groundwater supplies—the groundwater may be used to "fill in" the shortage periods. Supplies are then "firmed" at a higher level without extensive drawdown of an aquifer.

Extending Water Supplies without More Reservoirs

In the past, water supply engineers would look at growing water demand and reach for new water sources to meet it. Dams, aqueducts, pipelines, and the like were the traditional means to meet demands for new water supply. Now, more than ever before, this approach is running into difficulty.

As more people express concern for the environmental alterations caused by reservoirs, new water projects are becoming less popular than they once were. New water projects also are having trouble getting off the ground because of increasing competition and conflicts over who owns the water. Extensive growth in water use makes the increasing conflict fairly certain to continue as more and more water sources are allocated.

Even though fewer new water sources are available, in many cases it is still possible to meet growing demands. One obvious means of accommodating more users with the same basic water system is to encourage people to conserve water. A higher price for water is one way to get people looking for ways to conserve. People, industry, and agriculture can all find ways to save water if given the incentive to do so.

Another way to meet growing water demands without new sources is by the interconnection and joint use of water sources. We mentioned that groundwater can be used to "firm" or fill in the water availability from a reservoir. Groundwater fills up the inflow gaps, stabilizing supply at a higher level without extensive use of the groundwater resource.

The modern engineering discipline of water resources systems analysis has found methods for taking independent but interconnected supplies and managing them in a fashion that increases the system yield above the yield that occurs with independent operation. That is, the component reservoirs of a system can reliably deliver more water when their releases are synchronized than the sum of the safe yields for the individual reservoirs. For example, in the Metropolitan area of Washington, D.C., sixteen new reservoirs had been proposed by the U.S. Army Corps of Engineers to meet the demands of the growing capitol district. However, with *only one* of the proposed reservoirs built, a study of the water supply system showed that the needs of the area could be met to the year 2020 if the releases from the existing reservoirs were coordinated properly in time.

In 1981, several New Jersey communities ran short of water due to a severe drought that was fairly general throughout the northeastern United States. The state called in a consultant, who reminded officials that during World War II, interconnections had been built between the major New Jersey water sources as a means of preventing possible disruption of water supplies. He also reminded them that fiercely independent communities had torn up those interconnections after the war, rather than share their water. If the connections had been left intact, areas with surplus water could have helped shortage areas through the drought. The consultant, Dr. Abel Wolman, knew of the interconnections because he had caused them to be built during the war, some forty years previously. Interconnections and integrated system management are two options that can secure adequate water supplies for the future without requiring new sources.

Water Resources Decisions

Across the country, many water resource projects have been undertaken, such as dams, canals, hydropower installations, dredging, and water transfers. In many areas, new projects are being considered to meet emerging needs. Decisions on these projects are complex because water is so highly valued and so many people claim it. Also, water resource projects are often very expensive, and who should pay for them or how the cost burden should be allocated is not always clear. In addition, water resource projects often have significant environmental effects that must be weighed.

We have selected for discussion two areas of the country where decisions are currently being considered. In both cases, problems are in the offing that require resolution and clashes are occurring over policy, economics, and environmental values.

The High Plains and the Ogallala Aquifer

In parts of Texas, Oklahoma, New Mexico, Kansas, Colorado, and Nebraska, the pumps that draw groundwater to irrigate crops are drawing mostly air. Irrigation with groundwater is proceeding at so rapid a pace in these states that the aquifers are being drawn down faster than they are being recharged. One aquifer in particular that extends through these states, the Ogallala Aquifer, has been particularly hard hit. However, other aquifers of lesser extent are also being used in local areas that have suffered from groundwater overdraft for irrigation. (See Table 9.2 and Figure 9.4.)

Table 9.2 Approximate Withdrawals and Remaining Storage in Portions of the Ogallala Aquifer by State (1977)

	Approximate annual withdrawal[a] (millions of acre-feet)[b]	Remaining in storage (millions of acre-feet)[b]	Withdrawal-to-storage ratio (fraction withdrawn annually)
Colorado[c]	1.0	75	1/75
Nebraska[d]	1.5	450	1/300
Kansas	2.9	750	1/250
Oklahoma	0.6	60	1/100
New Mexico	0.75	30	1/40
Texas	8.0	270	1/30

a Withdrawal figures do not take account of natural recharge.
b An acre-foot is the volume of water covering one acre to a depth of one foot.
c Excludes Baca and Prowers Counties.
d South of the Platte River.
(Data from B. R. Beattie, "Irrigated Agriculture and the Great Plains," *Western Journal of Agricultural Economics,* **6,** No. 2, December 1981.)

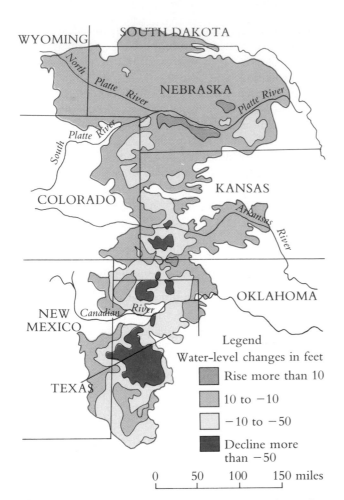

Figure 9.4 The Ogallala Aquifer in the High Plains of the United States. The aquifer holds a volume of water roughly equivalent to that of Lake Huron. The extensive use of the aquifer for irrigation of farms has depleted its volume by almost 23% since the 1950s. Changes in water level are shown from the period before development up to 1980. (Source: U.S. Geological Survey Yearbook, 1982)

In the area near Lubbock, Texas, in Floyd County, where agriculture relies on the Ogallala, the level of the water table has been falling 5 feet per year during the decade of the 70s, and has fallen 200 feet (61 m) since 1940. (Water table is the term hydrologists use to describe the level of the aquifer closest to the surface. See Figure 9.3.) In Grant and Finney Counties, Kansas, the water table has fallen over 100 feet (31 m) since 1940. In response to these declines in the water table, wells have been drilled deeper to tap the vital water. And the water table falls still further. A falling water table indicates that the withdrawal rate exceeds the natural rate of recharge of the aquifer. In total, the Ogal-

lala Aquifer that irrigates the farms of the High Plains may have declined in volume by 23% since the mid-1950s. Not all areas have been so hard hit. The problem is spotty, but indicates more widespread problems to come.

The Ogallala Aquifer covers an area the size of California; it includes southwestern Nebraska, eastern Colorado, western Kansas, Oklahoma, Texas, and eastern New Mexico (see Figure 9.4). The aquifer holds a volume of water roughly equivalent to that of Lake Huron. The aquifer underlies a region of relatively low rainfall, about 12–22 inches (30–56 cm) per year—normally not enough to support extensive irrigated agricultural development. Perhaps only ½–3 inches of the rainfall in the region penetrates down to the water table and recharges the aquifer. The groundwater resource, while it does renew itself, does so on such a slow time scale that it is essentially being "mined." Hydrologists estimate that the annual overdraft—the amount by which the withdrawals exceed the natural recharge—is on the order of the annual flow of the Colorado River, a figure you can verify when you read the next section.

The Ogallala Aquifer is not uniform throughout the region. In Nebraska, the water table is not far from the surface, and the aquifer is as much as 1000 feet (300 m) thick. Although irrigation is decreasing the water table in Nebraska, it will be a long time before falling groundwater levels threaten agriculture there (see Figure 9.5). In the irrigation area northwest of Lubbock, Texas, the aquifer is 100–300 feet (31–92 m) thick. To the south of Lubbock, the aquifer thickness falls to 25–150 feet (8–46 m). In the Texas portion of the High Plains, the pumpage rate in the mid-1970s ranged from 5–10 million acre-feet per year depending on the local precipitation, causing sharp drops in the water table. More than 65% of the irrigated acreage in the Texas High Plains is watered by groundwater, and percentages similar to this extend up into Kansas.

The agricultural base of the High Plains is extensive. The region produced more than 50% of the nation's beef cattle in 1980, and important quantities of wheat, cotton, corn, and other grains (Table 9.3). The value of its agricultural product was estimated at $4.2 billion in 1977, of which about $3 billion came from irrigated acreage. The impact of irrigation on yield may be seen from the fact that High Plains corn yields increased from about 50 bushels per acre in 1968 to just over 100 bushels per acre in 1977. During the same

Table 9.3 High Plains Agricultural Production, 1977, and Projections for 1985

		Percent of national production	Percent of production from irrigated acreage
wheat	1977	16.4	18.9
	1985	13.4	11.9
corn	1977	13.1	93.9
	1985	13.1	96.9
sorghum	1977	39.7	57.4
	1985	36.8	54.3
cotton	1977	24.9	66.3
	1985	31.2	72.4

(Source: *Six-State High Plains Ogallala Aquifer Regional Resources Study,* High Plains Associates, A report to the U.S. Department of Commerce, July, 1982.)

period, average U.S. yields went from 85 to 95 bushels per acre. The entire High Plains region has 13.1 million acres of irrigated land, for which 20 million acre-feet of water are withdrawn—both from surface reservoirs and from groundwater.[1] The 13.1 million acres

---

1 An acre-foot is the volume of water covering one acre to a depth of one foot.

of irrigated cropland in the High Plains region constitutes about 10% of the cropland acreage in the six-state region.

The Ogallala Aquifer has been described by hydrologists as an "egg carton"; that is, they believe that the aquifer has large compartments with small connections rather than being a single large volume of water. This description suggests that the drawdown of the aquifer in one area (and one compartment) shouldn't have much immediate effect on the water table in an adjacent compartment. If this is the case, high use rates in Nebraska should not soon imperil groundwater supplies in Texas—which is fortunate, since Texas is threatening its own supplies without outside help.

What happens if groundwater pumping continues at the same rate in these areas and the aquifer continues its decline? New wells will have to be drilled and pumps lowered still further. Since the water will have to be pumped from a lower level, electrical energy costs will increase. Consequently, farmers will need higher prices for their produce to make up for the increased irrigation costs. If the farmers can't get these higher prices, they may have to go back to dry-land farming without irrigation, the kind of farming in which success of the crop depends on rain—rain that doesn't always arrive in time.

Another factor, besides energy costs, could drive the farmer back to dry-land farming: as the wells go

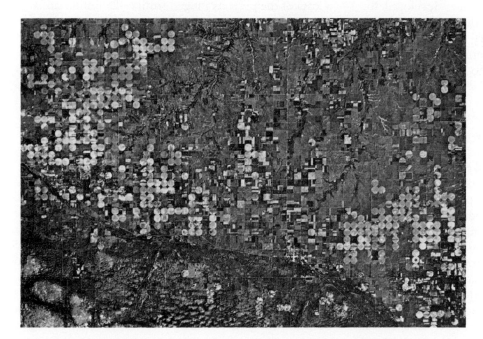

Figure 9.5 Center-pivot irrigation systems in Nebraska, as shown by a Skylab photo. Each circle in the photo represents the area swept out by the rotating arms of the center-pivot irrigation apparatus. Most of the increasing amount of Nebraska irrigation uses groundwater. (U.S. Geological Survey)

deeper, the "sweet" water starts to decline. Water drawn at deeper depths is older water, water that has been dissolving minerals from the soil over perhaps a thousand years. Such mineral-laced waters are said to have a high salinity. If the salt content is high enough, the water will not aid the growth of crops, but can even harm the soil and plants. When a well starts to draw saline water, the end of groundwater irrigation is at hand.

The problem of the Ogallala, however, is now clearly drawn and understood. In Colorado and Kansas, state legislatures have passed laws to limit the number and spacing of new wells, though enforcement is in doubt. Enforcement requires on-site inspection of farmers' well meters. Farmers can be an independent breed, and water is their lifeblood—they might be expected to resist an inspector's "intrusion."

Nevertheless, farmers realize that the water they began drawing in the 1950s cannot continue to be used at so rapid a rate. Many farmers have taken steps to cut their use of water. One such step is the installation of small dams on streams on the farm property, creating ponds that can be used as a source of irrigation water. Another step is to spray-irrigate at lower water pressures, thereby using less water. Also, if irrigation is scheduled in the cooler hours, water can reach the soil and the plant with less evaporation loss.

One novel way to hold rainfall as well as irrigation water is the Texas diker. The device, developed for cotton growing, is run through the fields and forms small dams, 2.5 inches high, every seven feet in every other furrow. Two of the small dams form a miniature reservoir and are said to be able to trap up to a three-inch rainfall within their borders without loss.

Still another form of irrigation could have a significant impact on water use. "Trickle" irrigation can deliver water to certain kinds of growing crops with high efficiency. In trickle irrigation, a pipe with multiple outlets is placed on the ground and extended parallel to a row of crops. Water "trickles" from the pipe into the soil near the roots of the growing plants. Water is not wasted on earth that lacks crops and the water does not evaporate so readily because it is not sprayed into the air.

These steps to conserve irrigation water are making a difference. In 1981, a water conservation official in Lubbock, Texas, estimated a reduction of one-third in pumping rates over a five-year period. However, the reduction only buys time; the water table is still falling, though at a slower rate.

If farmers go back to dry-land farming, their productivity declines. With smaller crop yields, their income and financial stability are threatened. Prices of wheat, corn, and other food basics could rise. It has been estimated that 40% of the irrigated acres in Colorado would have to be returned to dry-land farming by 2020 if groundwater pumping continues at its present rate. That translates into about 1.3 million acres. In Texas, a 1.4-million-acre decline in irrigated cropland is projected to occur between 1975 and 2000—about a 30% reduction. The situation is less serious in other states relying on the Ogallala because use began later in those states and the water volumes stored in the aquifer locally are greater. But it is only a matter of time before the effects of declining water supply are felt.

In an effort to determine how to sustain the agricultural productivity of the region, the federal government commissioned a six-million-dollar study of the High Plains Region. Released in 1982, the study reviewed steps at the farm level as well as regional solutions to the problem of declining water tables. Among the regional solutions considered was a system of dams, canals, pipelines, and pumping stations to bring "surplus" water hundreds of miles uphill from the Quachite, Sulphur, Sabine, Red, Arkansas, White, and Missouri Rivers to the High Plains (see Figure 9.6). Costs of the projects ranged from 6–25 billion dollars, depending on the scale of the projects envisioned. These are the costs of moving "surplus water," but it is not clear that there is any "surplus water" that states are willing to make available (see Controversy 9.1).

The Colorado River and Conflict in the Southwest

The Colorado River is both the mighty carver of the Grand Canyon and the lowly watering trough of much of the southwest United States. Rising in the Rocky Mountains of Colorado and Wyoming, fed by the Green and San Juan Rivers, the Colorado winds majestically south through 1400 miles of gorge and plain to end as a mere trickle into the Gulf of California. On its way, the waters of the Colorado are drawn off time and again to irrigate the arid plains of the Southwest. Along its route, deserts become verdant valleys with some of the most valued agricultural production in the

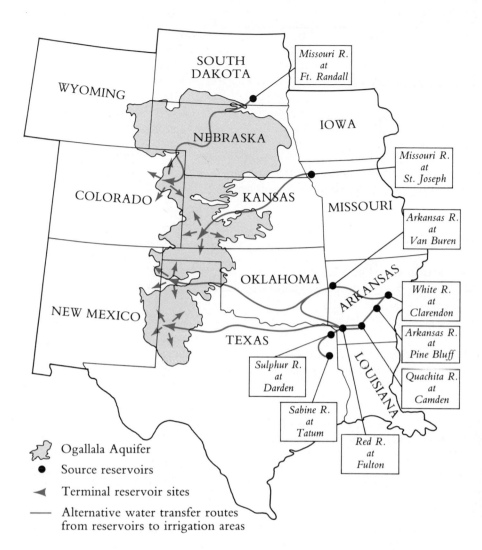

Figure 9.6 Interstate water transfer options assessed by the Corps of Engineers. (Source: Adapted from Figure 5, Review Draft, Water Transfer Elements of High Plains—Ogallala Aquifer Study, January 1982, U.S. Army Corps of Engineers.)

world. Though its annual flow is only on the order of the Delaware River in the East, the inhabitants of much of the Southwest have come to depend on the Colorado. They have thrown huge dams across it, built canals to transport its water, fought over rights to the river's water, allocated its flow, and gone to court over it. The Colorado, one of the most highly controlled rivers in the world, gives life to the Southwest.

The flow of the Colorado River is committed to many users over the Southwest—indeed, overcommitted—by a quirk of fate. At the time that the compact on the allocation of the river was signed in 1922, the region had just experienced a number of very wet years, years with stream flows that have not been seen since. Believing the future would resemble the past,

the states that signed the Colorado River compact in 1922 allocated among themselves a total annual quantity of water that is no longer generally available.

To the Upper Colorado Basin, above Lee Ferry, Arizona, the agreement allocated water consumption of 6.7 billion gallons per day on an annual basis, or 7.5 million acre-feet per year. To the Lower Basin was allocated another 6.7 billion gallons per day on an annual basis. In 1944, a treaty promised Mexico an additional 1.5 billion gallons per day. The total of these promises is 14.9 billion gallons per day over a year. The average daily inflow to the Colorado in both the Upper and Lower Basins in an average year is *now* estimated at 14.3 billion gallons per day. Thus in about half the years, the flow is expected to be less than this

CONTROVERSY 9.1

Should Water Be Transferred to the High Plains?

. . . there is no way realistically and economically that we can sustain the kind of development we have, short of bringing in new water.

*Morgan Smith**
Commissioner of Agriculture of Colorado

The bottom line is that most of the High Plains area is over-developed and its water is over-appropriated.

Keith Lebbinl,
Scott City, Kansas

Is there any point in leaving the water [in the Ogallala Aquifer] underground simply to enjoy knowing that it's there?

Anonymous Hydrologist†

. . . massive investment to augment the declining Ogallala is not economically efficient—not now or in the forseeable future.

Bruce R. Beattie,‡
Professor of Agricultural Economics,
Montana State University

Now that the over-appropriation of water has been recognized, the question remains as to what to do. Farmers are implementing conservation measures, but these steps only delay the day of reckoning. It is a matter of either going back to dry-land farming when the aquifer is finally exhausted (or no longer usable), or bringing in more water. The water would come via pipeline from new dams on the Arkansas and Missouri River Basins. Costs of construction could range from 6–25 billion dollars.

Are water transfers the answer? If so, who should pay for the massive projects? Or should the land go back to dry-land farming? (Remember the possible impact on crop yields, food supply, and prices of a return to dry-land farming.) Do you think ground-water should be used for irrigation, and if so, under what conditions?

---

* *The New York Times,* August 11, 1981, p. B4.
† *Science 81,* **2,** No. 5, June 1981, p. 35.
‡ B. R. Beattie, "Irrigated Agriculture and the High Plains," *Western Journal of Agricultural Economics,* **6,** No. 2, December 1981.

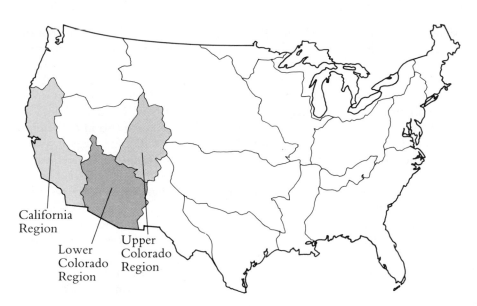

Figure 9.7 Water resources regions: California, the Lower Colorado, and the Upper Colorado (as defined by the Water Resources Council, 1978). Areas outlined represent other water resource regions.

number; in the other half it will be greater. Even in an average year, then, the Colorado is overcommitted.

Fortunately, though the water is committed, not all users consume their full allocations at this time. The time is approaching, however, when the Lower Basin will have more users on line than water to go around, as we will describe shortly.

Upper Colorado Region. In our discussion of the Colorado, we will use the water regions defined by the Water Resources Council in their 1978 National Water Assessment. The region of the Upper Colorado Basin, which includes portions of northern Arizona, northern New Mexico, Colorado, Utah, and Wyoming (see Figure 9.7), produces most of the flow of the Colorado River. Today, the Upper Basin is not consuming its guaranteed share of water from the Colorado.

The annual inflow to the Colorado in the Upper Region, before losses, averaged 14.0 billion gallons per day from 1906 to 1975. The outflow from the region in a year when this average inflow occurs, and with the 1975 level of development in the Upper Basin, is about 10.0 billion gallons per day. Thus, some 4.0 billion gallons per day evaporate or are consumed from the Colorado in the Upper Basin, based on the 1975 level of development. Most of this consumption occurs in irrigation.

The Upper Basin has rights to an additional consumption of 2.7 billion gallons per day. By the year 2000, consumption is projected to increase by 1.1 billion gallons per day, which would decrease the flow into the Lower Basin to 8.9 billion gallons per day in an average year. At least for an *average* water year, even by the year 2000, the Upper Basin will not bump up against its limit of consumption and will deliver sufficient water to meet the requirements of the compact agreement.

One problem arises because the Lower Basin is already consuming virtually all 10 billion gallons per day of the water that the Upper Basin is letting through. A second and potentially more immediate problem is that the virgin (undepleted) inflows into the Upper Colorado (averaged over a year) have fallen as low as 5.0 billion gallons per day (1934), even though flows rose as high as 21.4 billion gallons per day in 1917. Low flow years are likely to happen again.

The Upper Colorado Region had developed as agricultural land about 1.4 million of its 66 million acres by 1975; all of this cropland was irrigated. Over 7 million more acres in the region have potential for productive agriculture if irrigation water can be made available. The region also has an abundance of coal and the world's largest known resources of oil shale. The exploitation of the oil shale to produce oil may eventually place heavy water demands on the Colorado

River. According to one report, a shale oil industry producing 1,300,000 barrels per day at about the year 2000 is estimated to require 0.09–0.13 billion gallons per day of water.

The water that returns from irrigation to the Colorado River is richer in dissolved minerals from its tour of farmland. These minerals add to an already substantial burden of minerals in the river. The base load of minerals is largely of natural origin, but evaporation from reservoirs concentrates the salts, and irrigation return flows make another contribution. In the Upper Colorado above Lee Ferry, Arizona, salinity has reached no more than about 500 milligrams per liter and the water is still usable for all purposes. But by the time the waters of the Colorado reach Mexico, the salinity level has reached 1500 mg/l—too salty for normal irrigation.

Erosion brings substantial volumes of sediment into the river, and sediment is slowly filling the reservoirs on the river. Because of sediment accumulation, the capacity of these reservoirs to store water for irrigation and for hydropower production is decreasing. The ability of these reservoirs to protect against floods is decreasing as well.

A number of reservoirs are located along the Colorado and its tributaries, designed to even out the varying flows of the river. The reservoirs both protect against floods and provide water for irrigation, for municipal use, and for industrial use. The reservoirs also must supply the water committed to the Lower Colorado Region by the 1922 compact. Glen Canyon Dam, 17 miles upstream from Lee's Ferry, releases the required water to the Lower Colorado Basin.

The construction of dams has made possible the development of many areas in this region. Without the water, development would likely not have taken place. Further economic development might require more dams, and the region possesses a number of unspoiled rivers. However, these streams may qualify for designation as Wild Rivers, and would have to remain free of dams if so designated. Thus, a conflict appears in the offing between the preservation of wilderness areas in the region and the further development of its economic potential.

Lower Colorado Region. While the Upper Colorado Region produces most of the water, the Lower Colorado and the California Regions use it most extensively (see Figures 9.7 and 9.8). The Lower Colorado Region is defined by the drainage area of the Colorado River south of Lee Ferry, Arizona; this is the historic dividing point of the Upper and Lower Basins agreed on in the Colorado River Compact of 1922. The major state in the Lower Colorado Region is Arizona, but small portions of New Mexico, Colorado, Nevada, and eastern California are also in the region.

Beginning in the Lower Colorado Region, major diversions of the Colorado from its main stem draw water off to cities and agriculture. Water for Las Vegas, Nevada, comes from Lake Mead, the reservoir formed by Hoover Dam. Water for Los Angeles and San Diego flows via the Colorado River Aqueduct from Lake Havasu, the reservoir formed by Parker Dam; this aqueduct supplies 75% of the needs of these cities. Irrigation water for the lush Imperial Valley of California flows through the All American Canal System, and is drawn from the reservoir created by Imperial Dam. And water from the Colorado will soon be available to the desert cities of Phoenix and Tucson, Arizona. Taken together, the civil works projects on the Colorado River are massive; their operation literally changes the face of the land.

The Central Arizona Project, scheduled to deliver water by 1985, will supply the needs of Phoenix and Tucson and the irrigation needs of the agricultural valley in which they lie (see Figure 9.9). For Tucson, the water is meant to replace the groundwater upon which the city has relied, and which has been both polluted and drawn sharply down. Groundwater withdrawal in Tucson is lowering the water table 4–10 feet (1.2–3 m) per year; land has subsided in the nearby desert, apparently because of the withdrawal, and fissures have broken the earth.

The water supplied by the Central Arizona Project will have the possible effect, even in average years, of decreasing the quantity of water available to Lower California. California's rich agricultural output may be threatened by this tightening of supplies, and California has responded with a new and controversial plan. The plan would import additional water to Southern California from the water-rich northern portion of the state. We will discuss California's response to the Central Arizona Project in more depth in a moment.

Irrigated agriculture in the Lower Colorado Region has developed on about 1.3 million of the area's 99 million acres, but another 35 million acres could be converted to agricultural production if irrigation water were available. Crop yields per acre on the lands under

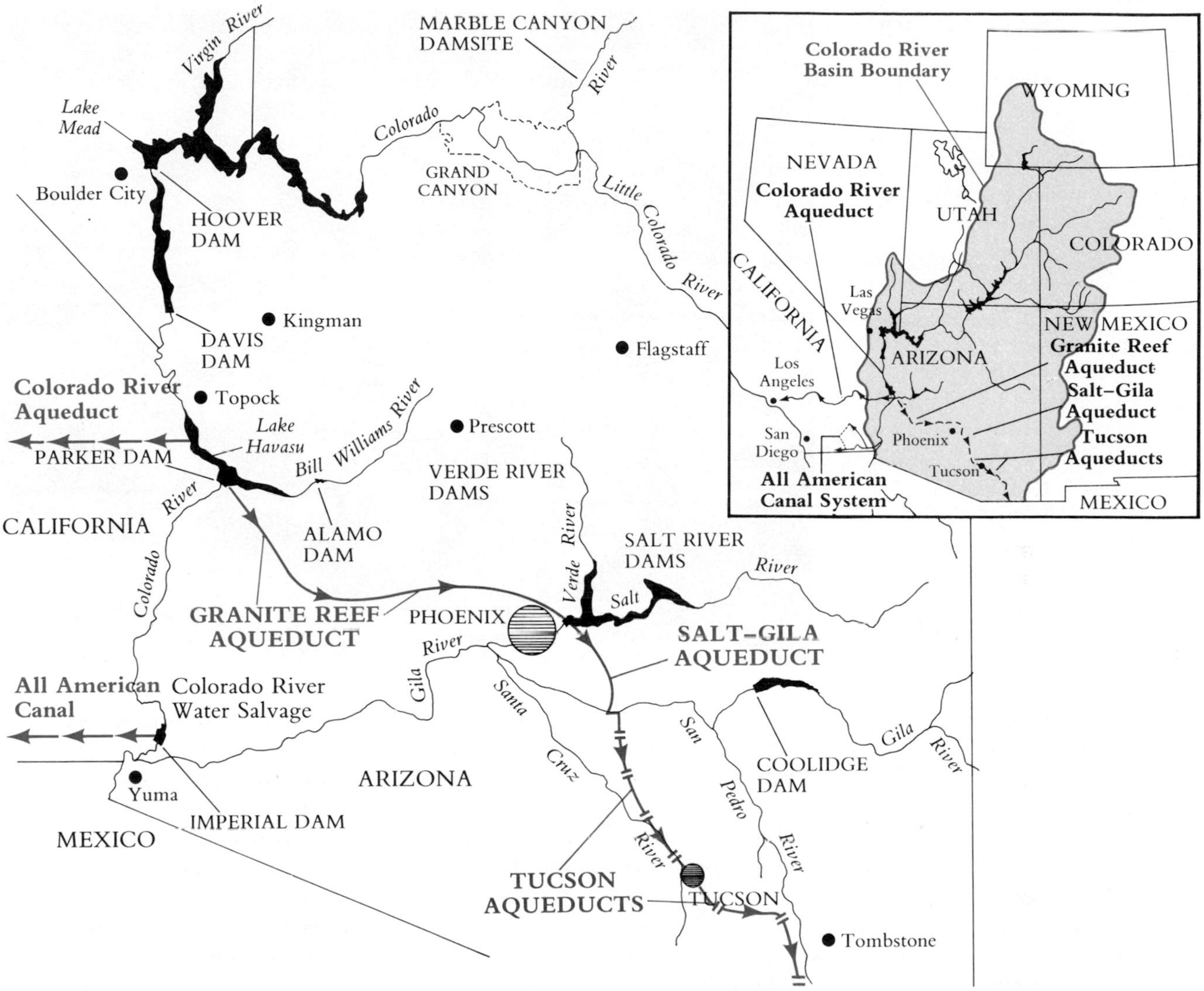

Figure 9.8 The Lower Colorado River Basin is the scene of intense conflict over the rights to water. Los Angeles, San Diego, and the farmers of the Imperial Valley have been drawing off Colorado River water for several years. Now, however, Phoenix and Tucson are claiming their share of the water, and these cities will soon have the means, through the Central Arizona Project, to physically take the water. (Source: U.S. Department of the Interior, Bureau of Reclamation.)

irrigation already are among the highest in the nation for most crops, suggesting that irrigation water would make this a most rich agricultural area. Many portions of the Lower Colorado Region are nearly frost-free for the entire year, adding even more potential for agricultural production. Cotton is now the most important crop in the region, but vegetables are also produced in abundance. About 50% of the irrigated acreage in the region is currently supplied by groundwater.

Large dams on the Colorado in the Lower Colorado Region include Hoover Dam, Parker Dam, and Davis Dam. In addition to generating electricity, the

Figure 9.9 The Central Arizona Project, under construction. (Photo courtesy of the Metropolitan Water District of Southern California.)

dams are used to control the ravages of the Colorado River, for the Colorado was once a wild and untamed river. A flood of the Colorado in the early 1900s flashed across Southern California; the water that the flood left in the Salton Basin was the origin of California's Salton Sea. Hoover Dam and the additional dams in the Upper Colorado Region have now effectively ended the rampages of the Colorado.

Of the average 14.0 billion gallons per day entering the upper Colorado River, 4.0 billion gallons are lost, leaving 10.0 billion gallons per day for the Lower Colorado Region. This quantity is more than the compact requirement of 6.7 billion gallons per day (7.5 million acre-feet) and should remain so through the end of the century. Another 0.3 billion gallons per day enter the Colorado in the Lower Colorado Region, making available a total of 10.3 billion gallons per day. Leaving the Lower Colorado Region, however, the river carries barely 1.5 billion gallons per day into the Republic of Mexico, and the water is highly saline. This 1.5 billion gallons per day is the precise amount required by the 1944 treaty with Mexico.

The difference between 10.3 and 1.5 billion gallons per day, some 8.8 billion gallons per day, represents water consumed in irrigated agriculture in the Lower Basin; water evaporating from reservoirs; and water exported to Nevada and California. When the Central Arizona Project begins operation in 1985, the new demand for water from the lower Colorado is expected gradually to increase up to 1.07 billion gallons per day (1.2 million acre-feet per year). To continue supplying the required treaty water to Mexico, some quantity of use is likely to have to give way. The probable loss is a portion of California's water withdrawals. California has been drawing water from the Colorado at a rate of about 4.5 billion gallons per day, which is more than its court-ordered allotment of 3.9 billion gallons per day.

Thus, the Central Arizona Project may spell problems for California and her rich agriculture. The debate in California concerning what to do when the Central Arizona Project begins sending water east to Phoenix and Tucson is sharp. Politics plays a role; continued growth and economics are argued; environmental protection is an important factor; and equity among portions of the state is an issue.

The California Region. California is so blessed with mild climate and fertile soil that it leads the nation in the production of nearly 50 crops. In the era before irrigation, cattle and sheep ranching and grain farming were the main agricultural activities. Now its leading crops are cotton, grapes, hay, citrus fruits, and tomatoes. Its coastal valleys produce most of the country's spinach, brussels sprouts, avocados, and artichokes. In total, the nine million acres under cultivation in the state produce a yearly crop valued at about $10 billion.

If California is blessed with climate and soil, water is its Achilles heel. Nearly 97% of those nine million acres must be irrigated to produce their abundant yields. Water, much of it from the Colorado River, has helped convert the deserts of California into one of the most productive agricultural regions of the world. Population growth in the South, such as Los Angeles and San Diego, has also been made possible by water imported from the Colorado River.

In 1981, California imported 4.5 billion gallons per day (5.1 million acre-feet for the year) of water from the Colorado, a quantity of water equivalent to the domestic needs of a city or region of 40 million people. California has also harnessed its own rivers: nearly 1100 state and local reservoirs exist in Califor-

nia, and about 150 Federal reservoirs. Ten of California's reservoirs can each store more than a million acre-feet of water. But the reservoirs are not always sufficient; a severe drought occurred in 1977, and it was Colorado River water that carried California through.

The locations of California's water resources and her agricultural output do not coincide. About 80% of the water needs of the state lie south of Sacramento; yet 70% of the stream flow that could be captured for use occurs north of Sacramento. Water now flows from north to south via the California Aqueduct and the Delta–Mendota Canal (see Figure 9.10). These canals drain water from the southern portion of the Sacramento–San Joaquin Delta to the southern portion of the state. The San Joaquin (pronounced San Wakeen') River and the Sacramento River together feed the estuarial island-studded region known as the Sacramento–San Joaquin Delta, which itself feeds into San Francisco Bay, a body of salt water.

The draining of the fertile delta region to supply water to Southern California has allowed salt water

Figure 9.10 *California Aqueduct. This aqueduct is the main means of water transfer from California's water-rich north to its water-poor but populous south. (Courtesy of California Department of Water Resources)*

from San Francisco Bay to intrude into the normally less saline estuary. This intrusion is changing the ecological balance in terms of the species that can thrive in the estuary. An overdraft of groundwater for irrigation in the San Joaquin Valley has also contributed to the intrusion of salt water from the bay. The region most affected is the southern portion of the estuary, for which the San Joaquin is the principal source of fresh water.

California relies on groundwater in addition to its reservoirs. Altogether, about 48% of its water withdrawals come from groundwater. Summed over the state, groundwater is presently being used to the extent of 2.2 billion gallons per day in excess of the annual rate of recharge. Many of the aquifers in the state cannot be used because they produce water too saline (salty) for crops or domestic consumption.

Irrigation is the largest use of water in California. Of the 39.6 billion gallons per day of fresh water withdrawn in 1975, irrigation drew 34.6 billion gallons. And of these 34.6 billion gallons, 24.3 billion gallons were lost—the water filtered into the soil and into groundwater; evaporated; or transpired from plants.

During dry years, which seem to have become the norm for the Colorado River, the new demand on the Colorado by the Phoenix–Tucson complex could seriously affect the ability of California to sustain its agriculture, its industry, and its population.

Of the 4.5 billion gallons per day (5.1 million acre-feet per year) that California has been drawing from the Colorado (1981), only 3.9 billion gallons per day are California's rightful share, according to a Supreme Court decision (Arizona vs. California, 1964). Of this 3.9 billion gallons per day, the irrigation districts of Southern California rather than the cities have rights to the first 3.4 billion gallons per day (3.85 million acre-feet per year) under their water-supply contracts with the Secretary of the Interior. Of the 1.1 billion gallons per day (1.2 million acre-feet per year) being drawn by the Metropolitan Water District serving Los Angeles and San Diego, the District has rights to only 0.5 billion gallons per day. The remaining 0.6 billion gallons per day being drawn by the District is at risk when the Central Arizona Project begins to draw Colorado River water. Obviously, California must adjust, either by using water more efficiently or by adding new water sources.

While some people see greater efficiency as the answer, others advocate a new canal from the Sacra-

mento River in the northern portion of the Sacramento–San Joaquin Delta. The Peripheral Canal, as it has come to be known, would link the Sacramento River to the California Aqueduct, supplying water for ultimate transport to Southern California and agricultural use. Costs would run between 3 and 20 billion dollars, depending on whether you believe advocates or critics of the canal.

As the earth-walled Peripheral Canal swings south in an arc around the delta, it would release portions of its flow into the delta at a number of points. The increased freshwater flow into the southern portion of the delta would flush out some of the salt water that has intruded from the Bay. This flushing of the delta has been used as a selling point to those who, for other reasons, oppose another transfer of water from North to South. Nonetheless, a 1981 statewide referendum in California has said "no" to the new canal. The "no" vote may stimulate some rather painful adjustment in Southern California, as growers and cities watch the faucet being tightened down on water from the Colorado River.

Grand Plans

Perhaps the grandest plan ever seriously advanced for providing water for the West was the NAWAPA concept advanced by the Ralph M. Parsons Engineering Company in the early 1960s. NAWAPA stands for North American Water and Power Alliance; as the name implies, the plan was truly continental in scale.

The concept would draw "surplus" water from the Fraser, Yukon, Peace, and Athabasca Rivers in Alaska and Canada. New canals and tunnels, as well as rivers, would be used to distribute the water to the Canadian Plains, to the Western and Midwestern United States, and to Northern Mexico. Water for the Western U.S. would need to be pumped over the Rocky Mountains, but hydropower projects in the NAWAPA system would provide such power as well as additional electric generation. Water would be delivered to the Great Lakes as well via a new waterway from the Canadian Rocky Mountains to Lake Superior.

The NAWAPA system was to have over 4.3 billion acre-feet of water storage and was estimated to cost about $80 billion at the time of discussion in 1964. According to a 1967 version of the plan, total delivery

CONTROVERSY 9.2

Has the West Had Enough Water Projects?
And Enough Growth?

In normal years, we'll have some trouble. In drought years, we'll have a disaster.

David Kennedy, [*]
Metropolitan Water District (of Southern California)

The best cure for a threatening water shortage is not necessarily more water; savings in water use, or transfer of water to less consumptive, higher-yield applications, . . . may offer better solutions.

Committee on Water [†]
of the National Research Council

Unless the Lower Colorado River Basin Project as planned is allowed to proceed, the nation will not keep faith with the people of the Pacific Southwest.

Craig Hosmer, [‡]
former Congressman from California

The Commission believes that since the West has been won, there is no reason to provide additional interest-free money for new irrigation development in the 17 Western states. . . .

National Water Commission [§]

Kennedy says that more water is needed for Los Angeles and San Diego, but the Committee on Water argues for more intelligent use of water. Former Congressman Hosmer expresses the view that the West is "owed" more water, and the National Commission replies that the debt has been paid.

Together, these arguments are a miniature picture of the debate on Western water policy during the past two decades. They represent the "no new water projects" philosophy and the "we need the water" philosophy.

In the West, population growth, industry, and agriculture go hand in hand with water supplies. Each new supply seems to make more demand possible. Many preservationists ask for an end to the building of water projects. They are trying not merely to protect free-flowing rivers, fish and wildlife, gorges and native vegetation. They are also calling for an end to growth in the West. The West has grown enough, many of them say.

Has the West grown enough? Are water projects, canals, pipelines, and dams only the pets of politicians? Or do they serve and supply existing real needs? Are growing demands inevitable and should those demands be met as long as possible? Are more water projects the best use of limited federal and state money? Where should the line be drawn, or should even the grandest plans be considered for development?

---

[*] *Newsweek*, May 31, 1982, p. 33.
[†] Quoted from *Water and Choice in the Colorado Basin*, Committee on Water, National Research Council, National Academy of Sciences, 1968.
[‡] *Ibid.*
[§] Press Release of the National Water Commission, June 14, 1973.

of water by the system was to be 180 million acre-feet (m AF) per year, of which 136 m AF was destined for the United States, 25 m AF for Canada, and 20 m AF for Mexico. While Congress held hearings to consider the NAWAPA concept, no action was ever taken on the projects specific to its plan.

Still another grand plan was proposed by the Los Angeles Department of Water and Power in the early 1960s. The Department suggested tapping the Columbia River in Washington and the Snake River in Idaho for water for the Colorado River System. The water would flow through Nevada by pipeline or canal into Lake Mead behind Hoover Dam, from which it could be distributed. The plan drew strong opposition from the governors of Washington and Idaho and was never pursued.

Such water transfers as these have never reached fruition. Yet plans continue to evolve even now as new demands for water arise in the West. The water in the Great Lakes is not far from the area where energy developments are taking place. Wisconsin is reported to have been approached by coal transport companies in Montana for water from Lake Superior. The water would be used for pumping crushed coal in coal slurry pipelines to other sections of the U.S. Texas is considering pumping water from the Mississippi River for its use. South Dakota has already agreed to sell Missouri River water to a coal transport company in Wyoming that would use the water to supply a coal slurry pipeline from the Powder River Basin coal fields. The sale has produced dismay from the states into which the Missouri flows. Huge diversions of water are still possible.

Questions

1. Compare the seriousness of a toxic chemical leaking into a river used for water supply and the same toxic chemical leaking into an aquifer used for water supply, in terms of the long-term impact.
2. What potential problems can arise from persistent overdraft of groundwater? Give at least two examples.
3. To what extent is the Ogallala Aquifer in Texas influenced by drawdowns of the aquifer in Nebraska? Why?
4. How can water resources be extended or increased without building new dams?
5. Describe the physical and ecological impact of a dam in place across a river.

Further Reading

The Nation's Water Resources, 1975–2000, Second National Water Assessment by the U.S. Water Resources Council, Volume 4: Lower Colorado Region, Upper Colorado Region, California Region, 1978.

Beattie, Bruce R., "Irrigated Agriculture and the Great Plains," *Western Journal of Agricultural Economics*, **6**, No. 2, December 1981, p. 287.
　　Takes a bit of economist's jargon to penetrate, but otherwise fun to read.

Boslough, John, "Rationing a River," *Science 81*, June 1981, page 26.
　　A very readable article with many color pictures of the Colorado.

Water and Choice in the Colorado Basin, a report by the Committee on Water of the National Research Council, published by the National Academy of Sciences, Publication 1689, Washington, D.C., 1968.
　　Technical, but readable, highly iconoclastic.

Fradkin, Philip, *A River No More: The Colorado River and the West*. New York: Knopf, 1981.
　　The big picture of the Colorado, including historical development. The book may be a bit preachy in tone.

Shoji, Kobe, "Drip Irrigation," *Scientific American* (November 1977), p. 62.
　　Fairly technical, but still accessible treatment of a highly efficient method of irrigation.

Kohrl, William, *Water and Power*, University of California Press, 1982.
　　This is a very readable story of the political power struggle in which Los Angeles secured one portion of its water supply from the Owens Valley of California.

CHAPTER TEN

Purifying Water

Learning from Past Mistakes
The Discovery that Water Can Carry Disease/Filtration Is Found Effective/Chlorination Is Found to Kill Microbes/Virus Diseases Are Spread by Water/A New Threat to Water Supplies

How Can We Tell if Water Is Unsafe to Drink?

Water Treatment
Taste and Odor Removal/First Chlorination/Removal of Particles and Colors/Filtration to Remove Microbes/Final Chlorination/Chlorine and Chlorinated Hydrocarbons

Testing for Indicator Organisms

A Simple Test for Free Chlorine

Ozonation to Purify Water

CONTROVERSIES:

10.1: *Should Water-Supply Reservoirs Be Used for Recreation?*

10.2: *Is There Such a Thing as a Zero-Risk Environment?*

In this chapter we examine how people learned to protect drinking-water supplies from contamination by bacteria and viruses. We also discuss the methods of testing water for such contamination. Finally, we describe the modern processes for purifying drinking water.

Learning from Past Mistakes

The Discovery that Water Can Carry Disease

As, it seems, with all science, the first notion that water could carry disease occurred to the ancient Greeks. The physician Hippocrates, the ancient innovator of medical ethics, advised that polluted water be boiled or filtered before being consumed. Despite Hippocrates' recommendations, only 150 years ago most people were still not aware that human diseases could spread by water. For this reason, many communities dumped their raw sewage into the same lake or river that they, or other communities, used for drinking water.

Typhoid and Asiatic **cholera** are two diseases spread by water polluted with human wastes. These diseases attack and infect the intestinal tract of humans. Bowel discharges of infected individuals contain the pathogens (disease-producing microorganisms) that spread the diseases. If these bowel discharges enter a water source, there is a high probability that new infections will occur among those who drink the water. Typhoid, besides being spread by water polluted with the pathogen *Salmonella typhi,* may also be spread by human carriers of the disease. In the past, typhoid was spread principally by polluted water.

In the United States today, most cases of typhoid are due to contaminated food. The food was probably contaminated in its preparation by an asymptomatic carrier of typhoid (an individual who carries the pathogens but exhibits no symptoms of the disease). One famous case involved a woman who was a cook for a family on Long Island in the summer of 1906. Typhoid infections occurred among the members of this family and an investigation followed. Investigators found that typhoid had occurred in all the families for whom the woman had cooked in the past several years; Mary, the cook, was a chronic carrier of typhoid. The state of New York eventually found employment for "Typhoid Mary" in a laundry during the remainder of her life to ensure that she would not endanger others.

Chronic carriers of cholera, unlike typhoid, are not common. The disease is spread mostly via the water route. However, in the midst of an epidemic, individuals may temporarily carry the pathogen without showing signs of the disease. Such people are capable of transmitting the disease through contact and food-handling.

During the period 1850–1900, people gradually became aware that diseases such as typhoid and cholera are associated with polluted water. Although this is common knowledge today, it was not common knowledge in 1854, when an epidemic of Asiatic cholera struck hundreds of residents in the London parish of St. James. The pump on Broad Street in St. James served a large number of families and industries in the area. Its water was regarded by many, in fact, as being even better than that of other wells. Nevertheless, in the 17-week epidemic of Asiatic cholera that occurred in that year, over 700 of the 36,000 residents of the St. James district died. Although the death rate from chol-

● ● Location of pumps.

····· Location of fatal
      cholera cases.

——— Boundary of equal
      distances between
      Broad Street pump
      and other pumps.

(After the original
map by Dr. John Snow.)

Figure 10.1   Asiatic cholera and
the Broad Street pump, London
1854. (Source: Sedgwick's *Principles
of Sanitary Science and Public Health*,
revised and enlarged by S. Prescott
and M. Norwood. New York:
Macmillan, 1948.)

era was over 200 deaths per 10,000 population in St.
James Parish, the two districts adjacent to St. James,
Charing Cross and Hanover Square, had cholera death
rates during the 17-week period that were much lower:
only 9 and 33 per 10,000, respectively. Clearly, some
special circumstance was responsible for the greater
number of cholera deaths in St. James.

In the hope of discovering the cause of the outbreak, an inquiry committee was appointed to study
conditions in the parish. The committee looked at the
population density in the parish, the weather conditions, and the cleanliness of the houses. They also studied the cesspools, the sewers, and the water supply.
The committee, it must be concluded, was fundamentally unaware of the cause of the epidemic. Picture, if you will, a gentleman in white wig and frock,
delivering the following pronouncement, in the name
of the committee:

> . . . a previous long-continued absence of rain
> . . . ; a high state of temperature both of the air
> and of the Thames [River] . . . ; an unusual
> stagnation of the lower strata of the atmosphere,
> highly favorable to its acquisition of impurity
> . . . ;

. . . their combined operation, either by favoring a general impurity in the air or in some
other way, concurred in a decided manner, last
summer and autumn [1854], to give temporary
activity to the special cause of that disease . . .[1]

Though still unaware of the cause, however, they recognized an unusual circumstance:

> But, as previously shown in the history of
> this local outbreak, the resulting mortality was
> so disproportioned to that in the rest of the me-
> tropolis, and more particularly to that in the
> immediately surrounding districts, that we must
> seek more narrowly and locally for some pecu-
> liar conditions which may help to explain this
> serious visitation.

One member of the inquiry committee, John
Snow, began an investigation of the deaths in the area
near the Broad Street well (Figure 10.1). His report of

---

1 These citations are from *Report on the Cholera Outbreak in the
Parish of St. James, Westminster, during the Autumn of 1854*, presented to the Vestry by the Cholera Inquiry Committee, July
1855. London: J. Churchill, 1855.

the epidemic in St. James Parish revealed that 73 of the first 83 deaths in the epidemic occurred among individuals whose closest source of water was the Broad Street pump. Interviews indicated that about 90% of these 73 deaths had occurred among people known to drink from the well either constantly or occasionally. The ten additional deaths in the first stage of the epidemic occurred in households nearer to some well other than the Broad Street well. However, it was found that five of the ten households were in the habit of sending to the Broad Street pump for their water. Dr. Snow writes in the Report of the Inquiry Committee:

> I had an interview with the Board of Guardians of of St. James's Parish on the evening of Thursday, 7th of September, and represented the above circumstances to them. In consequence of what I said the handle of the pump was removed on the following day . . .[2]

In the spring of 1855, the Reverend Mr. Whitehead, a member of the Inquiry Committee, discovered that residents of the house at No. 40 Broad Street had had an unidentified disease prior to the epidemic. He noted further that their dejecta had been disposed of in a cesspool not far from the well of the Broad Street pump. The brickwork of the cesspool was found to be defective in the sense that

> the bricks were easily lifted from their beds without the least force; so that any fluid could pass through the work . . . into the earth . . .[3]

The main drain leading from the house was in a similar state. The cause of the epidemic had been discovered.

These were the circumstances surrounding the 1854 epidemic at St. James. From the investigations at St. James, we had learned that cholera infections could be transmitted through a water supply. It took 70 years more, however, before engineers had perfected ways of purifying water to guard citizens from disease.

Still other notable epidemics were to occur, many of them contributing to a growing awareness of the link between water supplies and disease. In the spring of 1885, the mining town of Plymouth, Penn-sylvania, had been drawing its water from two sources. One was a mountain stream of seemingly high quality; the other was the Susquehanna River, into which Wilkes-Barre, upstream from Plymouth, was discharging its wastes. In April of that year, an epidemic of typhoid struck the town, affecting 1100 of the town's 8000 inhabitants; 114 people died.

Several explanations were advanced for the epidemic. Some people blamed the generally poor sanitary condition of the town for the epidemic. The sudden onset was attributed to the warm rays of the April sun falling upon the dirt in the town and thereby releasing poisonous vapors. Dr. L. H. Taylor, however, who investigated the outbreak, noted that Plymouth was no dirtier than neighboring towns and no dirtier in that year than in previous years. "Thoughtful minds," he tells us, "turned to the water supply as furnishing the true cause of the invasion."[4]

It was the mountain stream, in fact, that carried the typhoid pathogens and brought the epidemic to the town. An inhabitant of a house located on the stream banks had contracted typhoid in late December of 1884. His illness persisted for three months, and during this period his dejecta were thrown or deposited on snowy or frozen ground. From there, a rain or thaw had carried them down the slopes into the stream that fed the town's reservoir. An apparently pure mountain stream, trusted for its quality, had caused the epidemic.

Dr. Taylor drew two conclusions from the epidemic: first, that the dejecta of typhoid fever patients must be carefully disposed of and disinfected; second, that water companies "should be compelled to remove from the banks of their streams and reservoirs not only all probable, but all possible sources of pollution,"[5] a warning that many water managers carefully heed to the present time.

Many areas of the United States experienced typhoid in this era. In Massachusetts, in 1890, an epidemic was traced to the Merrimac River, which supplied water to the numerous manufacturing towns of the region. In that same year, an epidemic of typhoid occurred in New York State, and again the path of

---

2 Report of Cholera Inquiry Committee.
3 Report of J. York, Secretary and Surveyor to the Cholera Inquiry Committee.

4 First Annual Report, State Board of Health and Vital Statistics of Pennsylvania, 1886.
5 *Ibid.*

infection followed a river, this time the Mohawk and Hudson Rivers. The cities of Schenectady, Cohoes, West Troy, and Albany all experienced typhoid epidemics. Eighty years later, we find this same stretch of the Hudson River as the route of contamination of the chemicals known as PCBs.

Filtration Is Found Effective

By 1872, **filtration** through sand beds was being recognized as a way to make water clear and relatively safe for drinking. In that year, Poughkeepsie, New York, became the first city in the United States to install sand filters to purify its water. The filters in Poughkeepsie were still in operation over 100 years later, in 1982.

The effectiveness of the sand filter at removing disease-causing organisms was demonstrated dramatically in 1892. Altona and Hamburg, sister cities on the Elbe River in Germany, shared a common boundary, but did not have the same water supply. Hamburg's water was drawn from the Elbe above the two cities. In contrast, Altona drew its water from a site downstream from *both* cities. Thus, the supply of Altona was potentially polluted by the sewage of both cities. To produce water of sufficient quality, Altona began to filter its water supply through beds of sand. Hamburg, however, drew its water directly from the river without any treatment.

The cholera epidemic that struck Hamburg in the fall of 1892 killed more than one citizen in 100 and made nearly twice that number ill. In Altona, the cholera death rate and case rate was less than 1/7 of Hamburg's, and the difference was correctly attributed to the filtration of its water. The cholera cases that did occur in Altona are explained by noting that many residents of Altona worked in Hamburg. The protection provided by sand filtration was clear, and cities moved quickly in the coming years to install filters in an effort to prevent such epidemics. Hamburg was among the first to take the step.

During the early part of the 20th century, the city of Chicago was discharging its municipal wastes into Lake Michigan. It was also drawing its water supply from the same source via an intake pipe extending into the lake, just as it does today. Not surprisingly, ty-phoid fever was endemic in Chicago in this era. One epidemic, in 1885, was caused by heavy rains washing wastewater far out into the lake, out to the water intake. The death toll from cholera, typhoid, and other diseases is thought to have reached 90,000 people, an eighth of the city's total population.

The city recognized the connection between its high rate of typhoid fever and its water supply, but rather than purify its water supply mechanically by filtration, Chicago chose to divert its sewage from the lake. To accomplish this, the Chicago River, which carried the city's sewage and which had emptied naturally into Lake Michigan, was diverted into a canal. The Chicago Drainage Canal, as it is called, was created to carry the river's flow and the sewage of Chicago all the way to the Mississippi River, over 350 miles distant. This step, of course, polluted the Mississippi, and eventually Chicago was forced to treat its sewage before it could be discharged into the Drainage Canal.

Typhoid and cholera are not the only bacterial diseases transmitted through polluted water. Paratyphoid, dysentery, and other diarrheal diseases can also be spread by bacteria in water. The failure of the water treatment system of Detroit, Michigan, led to an epidemic involving 45,000 cases of bacterial dysentery in 1926. Nor are bacteria the only microorganisms that spread disease. A form of dysentery referred to as amoebic dysentery is caused by an amoeba, a one-celled animal living in water. This disease is accompanied by a diarrhea that brings severe weakness. Chicago suffered an epidemic of amoebic dysentery caused by *Entamoeba histolytica*. Visitors to the Chicago World's Fair of 1933 who stayed at two of the city's first-class hotels were the principal victims. The water-supply mains of the hotels were discovered to have been built with cross connections with the hotel sewer lines.

Chlorination Is Found to Kill Microbes

Filtration was not the only process devised to purify water. A chemical "sterilization" using chlorine and chlorine compounds was also investigated in the early part of the century. Experiments with **chlorination** using calcium hypochlorite (also called bleaching powder) were conducted in the United States in 1896 by

Should Water-Supply Reservoirs Be Used for Recreation?

Without strict controls over recreation, waterborne diseases become a serious problem.

American Water Works Association, 1971

No amount of properly planned recreation can generate enough contamination of reservoir water to pose any technical difficulties in treatment.

President's Council on Environmental Quality, 1975

The American Water Works Association issued the following statement on the use of water-supply reservoirs for recreation:*

The American Water Works Association supports the principal that water of highest quality be used as source of supply for public water systems. Since each water utility is responsible for its product, determination of type and extent of recreational use of impounding reservoirs shall be vested in the water utility.

Water utilities of the United States and Canada provide water of the highest quality in the world. To improve further upon this record, each water utility must continue to recognize its responsibility to deliver a safe and appealing product to its consumers. A growing demand for use of reservoirs for recreational purposes, however, may make this responsibility more difficult to carry out in the future. In some areas, uncontrolled use of reservoirs already has resulted in deterioration of water quality. Without strict controls over recreation, water-borne diseases could become a serious problem. . . .

Research on the effects of recreation on water quality should be expanded. Reservoirs used for recreation must receive close supervision. In all cases, water from reservoirs used for recreation must be treated, with the degree of treatment to be determined according to state or provincial and local utility laws, regulations, and policies. . . .

Public water-supply reservoirs should not be used for recreation if other surface waters are available. Bodily-contact sports such as swimming should not be allowed in public water supply reservoirs. For the purposes of this policy statement, reservoir means an impoundment reservoir subsequent to which water receives treatment before consumption. Distribution reservoirs from which water is supplied directly to the public require the strictest of controls and under no condition should be used for recreation.

The recreation referred to as permissible under certain circumstances is presumed to include boating, fishing, picnicking on shores, etc. The Association takes the position that water-supply reservoirs should not be used for recreation activities unless there are no other possibilities in the area. Essentially this is a weak statement. It admits that recreation on reservoirs is not so serious a threat to water quality that it must be prevented at all costs. It recognizes that such practices already exist. The only strong statement is the directive against swimming in public water-supply reservoirs. In fact, however, a number of reservoirs in the Midwest and West are used for swimming.

The President's Council on Environmental Quality disagrees with the American Water Works Association.† It advocates (and attempts to justify) more, not less, recreation, including swimming in larger reservoirs. The council argues that modern water treatment processes are fully adequate to safeguard water supplies.

If your community is served by a reservoir, find out whether recreational activities are permitted in the watershed area or on the reservoir. Do you think it wise to prevent any public use of the reservoir and watershed area? Can you support your opinion? If recreational uses other than swimming are permitted, what precautions would you recommend to go along

with such use? That is, what design features would you incorporate in the recreational area? And what activities would you restrict?

---

* Recreational Use of Domestic Water Supply Reservoirs. A statement adopted by the Board of Directors on 26 January 1958, and revised 25 January 1965, and 13 June 1971.
† President's Council on Environmental Quality, *Recreation on Water Supply Reservoirs*, September 1975. This pamphlet can be obtained from the U.S. Government Printing Office, Stock No. 041-011-00027-1.

George Fuller in Kentucky. In the following year, sodium hypochlorite was successfully introduced to the water supply of Maidstone, England, to arrest a then-occurring epidemic of typhoid. The process was first used on a continuous basis in Middlekirke, Belgium, in 1902. In England, the Metropolitan Water Board of London began continuous chlorination of a portion of its supply two years later, using sodium hypochlorite.

In 1908, because its supplies were polluted, the East Jersey Water Company, a private firm that furnished water to Jersey City, was ordered by the courts to make its water pure and safe. Calcium hypochlorite was introduced to the water in order to eliminate the bacterial contamination. The court agreed that the bacteria had indeed been destroyed and issued the following opinion:

> From the proofs before me of the constant observations of the effects of this device, I am of the opinion and find that it is an effective process, which destroys in the water the germs, the presence of which is deemed to indicate danger, including the pathogenic germs, so that the water after treatment attains a purity much beyond that attained in water supplies of other municipalities. The reduction and practical elimination of such germs from the water was shown to be substantially continuous. . . . I do therefore find and report that this device is . . . effective in removing from the water those dangerous germs which were deemed by the decree to possibly exist therein at certain times.[6]

The adoption of chlorination in many cities, now sanctioned by law as a means to purify water, proceeded rapidly. Today, nearly all municipal water supplies in the United States use chlorination to disinfect their water supplies, although liquid chlorine has replaced calcium and sodium hypochlorite. Calcium hypochlorite still remains a valuable tool to employ on a temporary basis for disinfection when disasters such as floods or earthquakes strike, or in times of war. The United States Public Health Service still tells travellers how to disinfect water by using bleach: four drops of a bleach mixed into each quart or liter of water is sufficient to disinfect water within 30 minutes.[7]

The use of liquid chlorine instead of calcium or sodium hypochlorite is based upon a variety of reasons. For one, it is more effective in destroying bacteria. In addition, it is easily and safely stored in large quantities and its introduction into the water is easily regulated. We will have more to say about the specific process of chlorination when we discuss the current methods of water treatment in more detail.

Since those early times, chlorination has been properly regarded as one of the most important tools scientists have to ensure the safety of drinking water. It has prevented illnesses and deaths in untold numbers around the world. Nonetheless, there is growing reason to question its side effects on human health, and we will explore this issue in the last section of this chapter.

---

6 George Johnson, "The Purification of Public Water Supplies," U.S. Geological Survey Paper No. 315, 1913.

7 *Health Information for International Travellers*, 1979, USDHEW, Public Health Service, Center for Disease Control, Atlanta, Georgia.

Virus Diseases Are Spread by Water

Viruses, too, can cause waterborne epidemics. Viruses, the simplest of living organisms, cannot reproduce themselves as bacteria can. Instead, they invade plant, animal, or bacterial cells and use the reproductive capability of these cells to produce more of their own kind. Smaller than even bacteria, viruses are very difficult to detect in water samples and require highly specialized equipment for detection.

Our knowledge of the potential for waterborne viral epidemics has come only recently. Scientists believe that certain polio epidemics have resulted from water supplies contaminated by sewage. Cases have also been reported in which viral respiratory diseases were spread by swimming in contaminated pools. The only proven waterborne epidemics caused by a virus, however, have involved the hepatitis virus. In hepatitis, the victim suffers an inflammation of the liver. Symptoms include a feeling of weakness, nausea, and fever. The liver is unable to perform its normal function of breaking down certain pigments in the blood; these pigments build up in the body and turn the skin and the whites of the eyes a yellow color. This condition is called jaundice.

Most hepatitis epidemics have occurred in small private water supplies or resulted from the consumption of raw clams or oysters grown in sewage-polluted water. One large-scale epidemic, however, was recorded in New Delhi, India, in 1956. In this outbreak, 50,000 people were infected with hepatitis from the city water supply, which had become contaminated with sewage. Interestingly, the water supply had been chlorinated before distribution to the public. During the hepatitis epidemic, no increase in waterborne bacterial illnesses such as cholera or typhoid was observed. This points up a difference between bacteria and viruses: bacteria are much more easily killed by chlorine than viruses are, although chlorination is effective against viruses if properly applied. In developing countries, where water supplies are often polluted, the presence of a group of viruses known as ECHO viruses contribute to chronic diarrheal disease.

Waterborne diseases in epidemic proportions have continued to occur in the U.S. throughout this century, although with decreasing frequency. The cities struck by these epidemics are typically small and have insufficient treatment of their water supplies. Although such epidemics are becoming infrequent in the United States, they continue to occur where water supplies are not treated or where treatment systems fail or where the water distribution systems may become contaminated.

In 1980, just over 20,000 people in the U.S. were affected in 50 outbreaks of diseases traced to water supplies. In only 22 of the 50 outbreaks was the causative agent identified. In 30% of the cases where the causative agent was identified, the responsible agent was a little-known human parasite, *Giardia lamblia*.

A New Threat to Water Supplies

Although awful waterborne epidemics seem to be past in the United States and Canada, as well as the other industrial nations, a new threat has only recently surfaced. Outbreaks of a disease with diarrhea, cramps, and weight loss have occurred in recent years, typically in smaller communities with less advanced treatment facilities. The disease does not occur until one to eight weeks after infection and may last two to three months. The disease is giardiasis and is caused by the protozoan *Giardia lamblia*. The delay in the onset of the disease occurs while the parasite population is building up to the level that causes symptoms. Long thought to be a cause of disease only in hunters, backpackers, and others who drank untreated water, the *Giardia* parasite is now known as the cause of many waterborne outbreaks in the United States.

The flagellated protozoan *Giardia lamblia* has been recognized since the 1800s, having first been described by the famous Dutch microbiologist Leeuwenhoek. The parasite has been found in about 4% of the population of the United States. Surprisingly, the same parasite can be found in beavers, coyotes, cattle, dogs, and cats. This fact explains why other warm-blooded animals can serve as a "reservoir" of infection for humans. In fact, an outbreak of giardiasis in Camas, Washington, in 1976, has been associated with the presence of infected beavers living upstream from the water supply.

Although *Giardia* may have been one of the responsible agents in an epidemic among U.S. residents in an apartment building in Tokyo in 1947, the first recognized outbreak among U.S. citizens occurred in 1970 when over one hundred people in two U.S. tourist groups in Leningrad, U.S.S.R., became ill with the disease. Thereafter, many more cases have occurred among U.S. travellers who consumed tap water in Leningrad. Giardiasis was also observed in U.S. travellers returning from the Spanish island of Madeira in 1976. A significant but unknown fraction of the resident population of these places is likely to have had giardiasis and have recovered from the disease.

In the United States, more than 7000 cases of giardiasis have occurred from 1965–1977; the more significant outbreaks, along with causes, are listed in Table 10.1. From these outbreaks, two obvious causes stand out: the absence of any water treatment and the failure of treatment processes. Also plain to see is the evidence that chlorination alone may be inadequate to destroy the parasite. In three of the eight cases shown in Table 10.1, filtration did not accompany the chlorination process. Its presence might have prevented the outbreaks. The explanation for the importance of filtration and the possible failure of chlorination lies in the nature of the *Giardia* protozoan. The parasite passes out of the intestine in the form of a cyst, a resistant form of the protozoan. Chlorination, unless precisely practiced in terms of contact time and water temperature, may fail to destroy the cyst. Then only filtration stands between the delivery of pure or contaminated water.

The lesson is that a clear water supply, though it appears to be clean enough without filtration, may not be adequately treated by chlorination alone. The water may appear pure due to the destruction of the coliform bacteria (the organisms that indicate the presence of sewage contamination), but the *Giardia* cyst may survive. The lesson seems plain, but numerous water supplies in the United States still do not include filtration in their treatment scheme. Among these are New York City with its Catskill Mountain and Croton reservoirs and Boston with its Quabbin reservoir, where input is only partially filtered.

Cases of giardiasis have continued to occur since the last outbreak noted in Table 10.1. A number of the outbreaks occurred in Colorado, where the parasite seems to be very common. Whether the parasite is spreading among the populace is unknown; hence such a spread cannot be pointed to as the cause of an increasing number of outbreaks. The possibility that the parasite might be spreading, however, should give pause to cities with "chlorination-only" water supplies.

Table 10.1 Waterborne Outbreaks of Giardiasis in the United States[a]

Location	Date	Cause	Number of cases
Aspen, Colorado	Dec. 1965– Jan. 1966	Well-water supply contaminated by sewage; chlorination only	123
A camp, San Juan area of Utah	September 1972	Water was not treated at all	60
Meriden, N.H.	June–Aug. 1974	Water was only chlorinated	78
Rome, N.J.	Nov. 1974– June 1975	Water was only chlorinated	4800
Camas, Wash.	May 1976	Chlorination and filtration were inadequate	600
Berlin, N.H.	April 1977	Filtration inadequate	750
Hotel, Glacier Park, Montana	July 1977	Surface water supply not treated	55
West Sulfur Springs, Montana	July 1977	Surface water supply was only chlorinated	246

a Only outbreaks with more than 50 cases are listed here.

How Can We Tell if Water Is Unsafe to Drink?

Three major classes of disease-causing organisms are found in water: bacteria, viruses, and protozoans. The maladies they spread range from severe diseases, such as typhoid and dysentery, to minor respiratory and skin diseases.

Examining water in order to isolate all the possible disease-causing organisms in it would be a difficult, costly task. To avoid such exhaustive examination, special tests have been devised. These tests detect the presence of a group of bacteria that, if present, are very likely to be accompanied by disease-producing microorganisms. If this group of bacteria is absent or if the level of these bacteria in a water sample is sufficiently small, it is doubtful that the less numerous disease-causing organisms are present.

The bacteria for which scientists test are nonpathogenic; that is, with rare exception, they do not cause disease. These bacteria are found naturally in the intestines of warm-blooded animals, including humans, and are known as **coliforms**. Although coliforms rarely cause disease, their presence in water indicates that the water may have been contaminated with raw (untreated) sewage. To drink such water is, obviously, unwise.

Thus, the procedure of examining water for the presence of disease-causing organisms does not involve looking for the pathogens themselves; it is sufficient to show that water has in some way been contaminated with sewage or not treated enough to kill all the coliform bacteria in it. Now let us look again at the careful wording of the court's opinion on chlorination (p. 199). We can see that the judge was referring to the coliforms when he referred to organisms "the presence of which indicate danger." The coliforms are often called "indicator organisms."

The federal Environmental Protection Agency has recently drawn up drinking water regulations that set forth limits to coliform bacteria. Although the standards have actually existed for nearly half a century, they could not be enforced by the federal government unless the water moved in interstate commerce (for example, bottled water). According to these new regulations, if more than four coliform bacteria are present in a 100-milliliter sample taken from water destined for human consumption, the supplier must notify the state. Further, the supplier must take action

Table 10.2 Activities Permitted for Various Levels of Coliform Bacteria in Water

Coliform level	Activity permitted
1 coliform or fewer per 100 milliliters of water	Water safe for drinking
4 coliforms or more per 100 milliliters of water	State must be notified and corrective action taken
2300 coliforms or fewer per 100 milliliters of water	Swimming is allowed
10,000 coliforms or fewer per 100 milliliters of water	Boating is allowed

that will decrease the coliform count to less than one in each 100 milliliters of water.

Four coliforms per 100-milliliter sample is the point at which corrective action for drinking water is required. In contrast, water with more than 2300 coliforms per 100 milliliters is considered unsafe to swim in. Boating is allowed in water with up to 10,000 coliforms per 100 milliliters (but don't fall in!). (See Table 10.2.)

In 1969, 969 public water supplies were examined to see how pure the tap water really is in the United States. Somewhat surprisingly, 12% of the water supplies were found to exceed the permissible limits of coliform bacteria. Figure 10.2 shows that the water-supply systems in violation of coliform standards were generally in smaller communities. For instance, 79 of the 120 systems found in violation were in communities of less than 500 people, and 108 of the 120 systems served communities of 5000 people or fewer.

A 1982 study of rural water supply in the United States focused on private well-water supplies rather than community systems. From a random geographical sample of about 3000 rural households, nearly 30% of the water supplies exceeded the allowable bacterial count.

Water Treatment

Communities treat their water to remove both disease-causing organisms and harmful chemicals. In addition, water purification is designed to make water

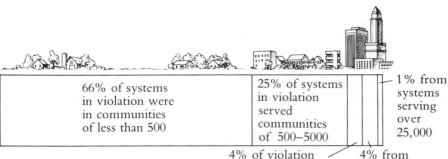

66% of systems in violation were in communities of less than 500	25% of systems in violation served communities of 500–5000		1% from systems serving over 25,000

4% of violation from systems serving 5,000–10,000

4% from systems serving 10,000–25,000

Figure 10.2 Distribution of violations of coliform standards among communities of various sizes. [Data from: J. McCabe, *et al.*, "Survey of Community Water Supply Systems," *Journal of the American Water Works Association,* **62,** 670 (1970).] A study of rural households completed in 1982 showed that 29% of the homes surveyed had evidence of bacterial contamination of their water supplies, suggesting that not much change or improvement occurred in the 1970s.

pleasant to drink. Not all cities purify their water in the same way because of the presence of different substances in their basic supply. We will describe a set of processes that are fairly typical among cities and indicate the order in which these processes are commonly arrayed.

Taste and Odor Removal

Tastes and odors that exist in the untreated supplies of some cities may be offensive to consumers of the drinking water and hence require removal. The causes of such odors and tastes may include:

1. Hydrogen sulfide gas dissolved in the water. This condition is most common in water drawn from wells.

2. Dead or decaying organic matter stemming from algae or aquatic weeds. Water from reservoirs may contain such materials.

3. Chemicals used in agriculture to control weed and insect pests and chemicals from industrial wastes.

The methods of controlling tastes and odors may be classified in two groups: (1) methods to *prevent* growth of algae and aquatic plants; (2) *treatment methods* to remove tastes and odors once they occur. To prevent algae and other aquatic plants from growing, chemicals may be added to the water while it is in reservoirs. Herbicides were once considered for this task, but have now been rejected. The one herbicide that was thought to be unobjectionable (but never used) is now known to cause cancer. This is the compound Lindane.

Copper sulfate is a commonly used substance for algae control. It may be slowly dumped from boats crisscrossing in the reservoir to ensure good distribution in the water.

In the treatment plant itself, tastes and odors may be removed by aeration (exposure to the air) or by chemical treatment. Water taken from a river or lake or from community wells is often aerated during treatment. The water is sprayed into the air from banks of fountains or allowed to flow over racks in a thin film. You may have seen the spraying step at a local water distribution reservoir. The aeration accomplished by spraying or by the rack method removes dissolved gases (e.g., hydrogen sulfide) from the water. With the removal of hydrogen sulfide, a substantial control of odor may be achieved. (Hydrogen sulfide gas is what we smell when an egg becomes rotten.)

First Chlorination

After aeration, chlorine gas is added to the water to kill disease-causing organisms. This step, called **chlorination**, occurs again after all other treatment steps are completed. More detail on chlorination is provided later in this chapter.

Removal of Particles and Colors

Coagulation–Settling. The minute suspended particles that give color to water will neither dissolve nor settle rapidly from the water; particles behaving this way are termed *colloidal*. We can conduct an experiment to illustrate the concept of a colloid. A teaspoonful of rich earth (humic soil) is mixed into a gallon jug of clear water. Note how the passage of light is restricted through such water. Observe the yellow-brown color imparted to the water by the particles.

These particles can be removed by a process called **coagulation**.

In the first step of coagulation, alum or ferrous sulfate is added continuously and rapidly mixed into the stream of water flowing through the plant (Figure 10.4). When the chemicals are added, a precipitate, called a "floc," forms, consisting of mineral particles that are insoluble in water. A floc particle resembles a feathery cloud in miniature. The floc then passes through a flocculation basin, where the floc is thoroughly mixed into contact with the suspended particles.

Next, the water is pumped slowly through a **settling tank**. Here the water is detained long enough to allow most of the floc or precipitate to settle to the bottom of the tank. In the process of settling, the "blanket" of floc will capture and drag down a large share of the particles suspended in the water. The precipitate is scraped away from the bottom by a scraping bar to prevent build-up.

The water leaving the settling tank has had much of its turbidity (cloudiness) removed. In part, these preceding steps protect the effectiveness of the next step, filtration through beds of sand, by removing relatively large particles that could clog the filter.

Carbon treatment. In many water-treatment plants, a small amount of activated carbon particles may be added along with the alum or ferrous sulfate in the first step of the coagulation process. Colloidal particles bind to the carbon. When the carbon is removed in a later step, much of the colloidal material responsible for color is carried with it. Not only is color improved when the carbon particles are removed, but tastes and odors are decreased as well because many dissolved organics have been bound to the carbon particles and are removed from the water.

Contacting the water with powdered activated carbon particles is a very popular method of removing tastes and odors, dating to the 1930s. In that era, filters using activated carbon as the filter medium were used in Bay City, Michigan. In contrast, in recent years the practice has been to add a small amount of carbon in the first step of the coagulation process. Interestingly, the filtration of water through beds of granular carbon is now being recommended as part of water treatment by the Environmental Protection Agency as a means of removing low-level but important contaminants such as pesticides and other resistant organics. Some of these compounds are now viewed as potential causes of cancer. Thus, we may see carbon filtration re-introduced as a common water treatment method. (Also see the last section of this chapter.)

Filtration to Remove Microbes

After the floc has settled out, the water stream passes downward through beds of sand. From top to bottom, the sand placed in the filter bed gets larger and larger. The construction of a typical sand filter is shown in Figure 10.3.

The removal of the larger particles achieved by the coagulation step prevents the sand filter from clogging rapidly. The sand filter does cause a further removal of particles. However, the primary purpose of the filter is to capture and retain bacteria, viruses, and other

Figure 10.3 A Sand Filter. Particles are removed when water is passed through the filter. The process eliminates most disease-causing bacteria and protozoa from a community water supply. When the filter becomes clogged with particles, water from the clear well is pumped up through the filter. This reversed flow dislodges particles and cleanses the filter for further use.

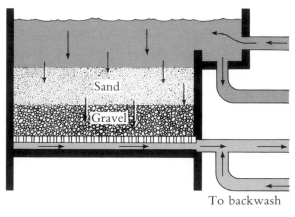

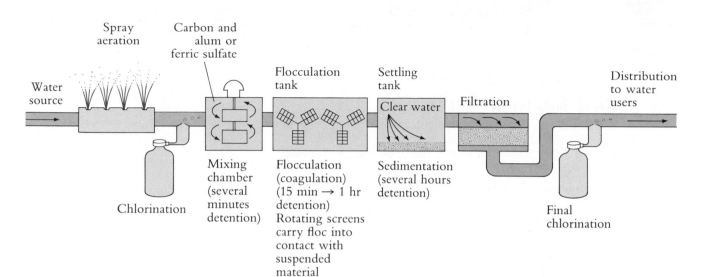

Figure 10.4 A typical water-treatment plant scheme.

organisms. Most disease-causing microorganisms are thus prevented from entering the system that distributes clean water to the city.

Periodically, the sand beds must be washed clean if they are to retain their effectiveness at filtering out microorganisms. Some of the filtered water is saved in a "clear well" for this purpose. The washing proceeds by passing the clean water up through the filter to dislodge particles retained in the spaces between the grains of sand. The dislodged particles are now suspended in the upward-flowing wash water, which is then dumped back in the river. Once the filter is clean, it is ready for continued use. The dirtier the water entering the filter, the more frequently backwashing is needed.

All of these treatment steps—aeration, chlorination, coagulation/settling, and filtration—are shown schematically in Figure 10.4.

Final Chlorination

Even though the sand filter is very effective at removing bacteria and viruses, the removal of these microorganisms is not certain. An additional step, a second step of chlorine addition, finally destroys any microorganisms remaining after sand filtration. Chlorine also reacts with any ammonia that may be present in the water. Chlorine is added beyond the level required to kill all microorganisms and also beyond the amount required to react with the ammonia present in the wa-

ter (Figure 10.5). This results in "free" or unreacted chlorine in solution. If free chlorine is found to be present when the water is tested, the water may be

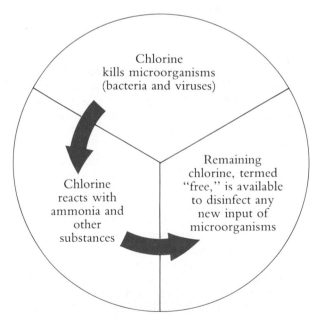

Figure 10.5 Chlorine is the favored disinfectant for water supplies. Chlorine kills microorganisms and reacts with ammonia. Any chlorine in excess of the amount needed to disinfect and react with ammonia will remain "free" in solution, protecting the water supply from any new sources of pollution en route to the consumer.

Is There Such a Thing as a Zero-Risk Environment?

We are dealing with a time bomb with a 25-year fuse.

Dr. Robert H. Harris,
Environmental Defense Fund

Crash programs . . . to meet "yesterday" deadlines are not understandable at all.

A. E. Gubrud,
American Petroleum Institute

Several environmental groups were very unhappy with how the Environmental Protection Agency set about fulfilling (or not fulfilling) the toxic pollutants portion of the 1972 Federal Water Pollution Control Act—so unhappy, in fact, that in 1975 they sued the EPA in federal court. The petition charged the EPA with failure to require that water supplies be checked for the presence of heavy metals such as cadmium and with failure to require proper removal of pesticides, viruses, and asbestos.

"We are dealing with a time bomb with a 25-year fuse," said Dr. Robert H. Harris, at a news conference at which the petition was discussed. He said Americans could not afford to wait to see what would happen, as they had done with cigarette smoking.

Dr. Harris noted that it often took many years for a cancer to develop. Therefore, he argues that exposure of a population to cancer-causing substances might not be reflected in an obvious rise in the cancer rate until 25 years or so after the exposure began. . . .

Dr. Harris, who is associate director of the Environmental Defense Fund's toxic chemicals program, said traces of known and suspected cancer-causing chemicals had been found in water samples from several major cities. . . .

Dr. Harris said federal regulatory agencies had done "virtually nothing" to protect the public from cancer-producing chemicals in water supplies.★

The suit was settled by an agreement between the EPA and the environmental groups, which was approved by a judge of the District Court of Appeals. The agreement required the EPA to start a huge program of experiments on 65 specified toxic pollutants. The agency also agreed to begin studies of the available methods of control and economic effects of controlling these pollutants. Many industry groups were upset by the agreement. As one industry spokesman put it:

. . . (A) disturbing trend is emerging, ever more clearly. In the past, the scientific and technical communities have experienced difficulty enough in responding to "technology-forcing" legislation and regulation. Now, however, as a result of the Toxics Effluent Guidelines Settlement Agreement, the scientific and technical communities are being asked not merely to accelerate development of toxic pollutant controls, but to improve the whole state of toxic pollutant knowledge by several orders of magnitude. It is extremely doubtful whether all of the combined talents in government, industry, and academia could achieve that end in the time allowed by the Settlement.

The occasional acceleration of a government program in response to a clear and present environmental danger certainly is understandable. Crash programs in the absence of such danger to meet "yesterday" deadlines are not understandable at all. They arise not in response to specific, scientifically demonstrated needs, but from the belief in some quarters that *all* people can be protected at *all* times from *all* real or suspected environmental dangers—the belief, in short, that a "zero-risk" environment is attainable.

Zero-risk is not attainable in the environmental area, any more than it is attainable in any other area of human endeavor. I know of no health specialist who would maintain that there are absolute thresholds, other than zero, below which no health risks will exist for anyone.

We must continue to work hard and systematically to improve the quality of our environment and eliminate true threats to the public health. But crash programs arising out of a kind of national hypochondria not only can yield dubious and harmful results, but also are extremely wasteful.†

Do you feel that this is the type of question that can be settled in the courts, in the full glare of public view? Are we requiring too much of industry in this case? Can we depend on their good-faith attempts to achieve pollution control without these kinds of legal threats? Can we achieve a "zero-risk" environment? Should we try?

---

* Dr. Robert H. Harris, quoted in *The New York Times*, 18 December 1975.
† A. E. Gubrud, Environmental Affairs Director, American Petroleum Institute, at the Federal Water Quality Association Conference on Toxic Substances in the Water Environment, Washington, D.C., 28 April 1977.

regarded as "safe" because any new contamination by bacteria will be prevented.

A simple, quick, and inexpensive test can determine whether unreacted or free chlorine is in water. One reason that chlorination is so favored as a means to disinfect public water supplies is that this "leftover" or residual chlorine remains and a quick, simple test for it exists. When the test shows that free chlorine is present, one can be confident that any newly entering microorganisms will be killed. (The simple test for free chlorine is described on p. 211.)

One alternative to chlorination is disinfection with ozone. (The advantages of ozonation are described on p. 211.) The ozonation procedure, however, does not leave any substance or by-product remaining in solution that can be measured quickly to check whether disinfection has been successful. In one sense, the lack of any residual or by-product is good, but from the standpoint of judging the water's safety, ozonation has a disadvantage. To check the safety of ozonated water requires the millipore direct count of coliforms, which takes 18 hours. To someone interested in discovering if water is safe, a day's delay seems long (Table 10.3).

Chlorination, in terms of its effects on humans, has recently come under attack because chlorine acts on hydrocarbon compounds dissolved in water. Some of the chlorinated hydrocarbons produced from this reaction are potential causes of cancer.

Table 10.3 A Comparison of Chlorination and Ozonation

Process	Substance used	Effect	Objectionable reactions, by-products	Check on water safety
Chlorination	Chlorine gas	Kills microorganisms	Chlorinated hydrocarbons, taste and odor substances	Immediate check: safe if free chlorine is present
Ozonation	Ozone gas	Kills microorganisms	None	Must run standard tests for coliforms (minimum 18 hours)

Chlorine and Chlorinated Hydrocarbons

In 1967, the Environmental Protection Agency began to study why drinking water drawn by the city of New Orleans from the lower Mississippi River smelled and tasted "fishy" or "oily." The source of the problem seemed fairly clear. A number of petroleum, chemical, and coal-tar products industries discharge wastes into the Mississippi. Thus, it was no surprise that the 1972 EPA report on the problem stated that chemicals in the wastes from these industries were responsible for the tastes and odors in New Orleans drinking water.

The report also noted, however, that chlorination of drinking water might be adding to the problem because chlorine, in the form used to disinfect drinking water, can react with natural or pollution-related organic materials to form chlorinated hydrocarbons. Some of these chlorinated hydrocarbons are known to cause cancer.

Since that report was issued, this last aspect of the problem has caused the greatest controversy. Chlorinated hydrocarbons and other organic compounds have now been found in many water supplies. In one survey of 129 municipal drinking-water plants that chlorinate water, chlorinated organic compounds at levels of a few parts per billion were found in the finished water of almost every plant. Chemicals found included chloroform and carbon tetrachloride, both known carcinogens. In the experiments that showed these chemicals to be carcinogenic, however, the doses of chloroform and carbon tetrachloride were much higher than humans could get by drinking New Orleans water. The problems involved in deciding whether the low levels in water are hazardous are explained in Chapter 29.

It is unlikely, however, that we will stop chlorinating drinking water. The immediate risk of incurring epidemics of waterborne bacterial diseases such as typhoid fever and viral diseases such as hepatitis or polio is too great.

Instead, the focus has been on removing organic compounds from drinking water. Organic compounds in water may stem from industrial wastes, from urban wastes, and even from natural sources. Pesticides are organic compounds also, and these may occur in the runoff from farmland. Many of these organic compounds are known to be toxic or carcinogenic. Not only has attention been given to removing organic compounds, but water-treatment methods are being changed as well in an effort to reduce the formation of chlorinated hydrocarbons.

The EPA is now requiring many water suppliers to reduce the level in drinking water of a group of chemicals called trihalomethanes, a group that includes chloroform. Trihalomethanes are one class of compound that may be formed when chlorine reacts with hydrocarbons dissolved in water. Reduction of trihalomethane levels can be accomplished in several ways.

1. The source of raw water can be changed to a less polluted one, decreasing the amount of organic material able to be chlorinated to trihalomethanes.

2. Adjustments can be made in the water-treatment scheme, such as changing the point at which chlorine is added.

3. Other methods of disinfection, such as the use of ozone, can be tried.

Finally, because there are too many different possible organic contaminants to be able to set standards for each one, the EPA has required the use of granulated active-carbon filters to remove most of the chlorinated hydrocarbon compounds.

Cities that can show, by sampling and monitoring, that they draw water from unpolluted sources (unpolluted deep wells or managed watersheds, for instance) do not have to treat water for removal of toxic organics. Initially, the regulations were written to apply only to large water supplies (those servicing over 75,000 people). Smaller water supplies are being phased in more slowly as administrative, technical, and economic problems are worked out. The new regulations have met a good deal of opposition from the people supposed to put them into effect.

The other way of attacking the problem is, of course, to reduce the amount of toxic organic material added to natural waters in the first place. This is part of the focus of efforts in the control of toxic substances, discussed in Chapter 32.

🦠 Testing for Indicator Organisms

Two methods are in use to test water for the presence of coliform indicator organisms. In the older tests, water samples are added to a special broth containing lactose sugar (Figure 10.6). A number of test tubes are prepared in this way. Relatively few bacteria besides coliforms are able to use lactose sugar as their energy source. Therefore, if gas forms in the broth within 48 hours (an indication of bacterial activity), it is considered evidence that at least one coliform organism is present. Such samples are called "presumptive positive." These samples are then grown on a special agar (which is a gelatin-like substance), to which the chemicals eosin and methylene blue have been added. These chemicals cause colonies of coliform bacteria to take on a special appearance: they become dark green and have a "sheen" to them in addition to having a dark center. If coliform colonies appear on the agar, the sample is "confirmed positive."

A simpler method is now in use that gives information more directly and more quickly than the test-tube procedure. A measured sample of water is filtered through a special filter membrane. This membrane is made by the Millipore Company and is called a Millipore filter. It is manufactured with pores or holes so minute in size that bacteria cannot pass through them, although water can. The filter itself, on which the bacteria are retained, is then added to a medium containing the nutrients that coliform bacteria need to grow. On that medium, each coliform bacterium captured on the membrane filter divides and forms a colony of bacteria. These colonies are visible to the naked eye; they can be identified and counted in order to estimate the level of pollution in the water (see Figure 10.7).

Determining the presence of viruses in water is not as simple as detecting the presence of bacteria. Most viruses are too small to be trapped by a filter and cannot be grown in a simple broth. However, just as water is not routinely examined for the disease-causing bacteria themselves, water is not commonly checked for disease-causing viruses either. Again, the presence of coliform bacteria in a water sample is taken as evidence that disease-causing viruses as well as bacteria are likely to be present in the water.

Bacteria and viruses are not the only kinds of pollutants that occur in water, however. Poisonous metals and a variety of organic chemicals can also pollute water supplies. These problems are subjects of the next chapter.

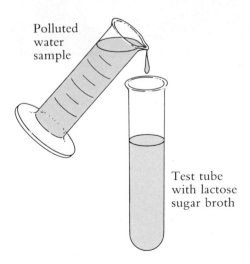

Polluted water sample

Test tube with lactose sugar broth

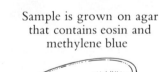

Sample is grown on agar that contains eosin and methylene blue

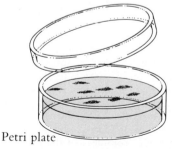

Petri plate

Figure 10.6 Testing for the coliform indicator organisms.

Presumptive test
If gas forms within 48 hours, at least one coliform was present. The test is then called "presumptive positive."

Confirmed test
If dark green, shiny colonies grow on the agar, the sample is termed "confirmed positive."

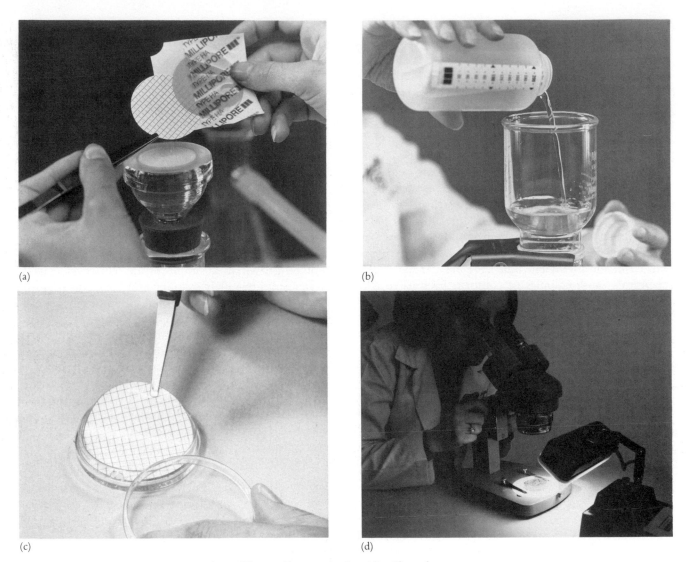

(a)

(b)

(c)

(d)

Figure 10.7 Using the Millipore-filter method to identify and count coliform bacteria. (a) A sterile Millipore membrane is placed on a filter resting on the base of a funnel. (b) A glass top has been added to the funnel base and the sample of water to be tested is poured into the funnel. The water is drawn through the funnel by applying vacuum suction. Coliform bacteria are too small to pass through the filter and are retained on its surface. (c) The membrane filter is removed from the funnel and placed in a petri dish. The petri dish contains a medium rich in nutrients that favor the growth of coliform bacteria. The dish is covered and placed in a constant-temperature "incubator" at 35°C for 24 hours. If coliforms were captured by the filtering step, "colonies" of the bacteria will grow up around each bacterium lodged on the filter. (d) After 24 hours of incubation, the coliform colonies are counted under a low-power microscope. In this photo, the cover of the petri dish has been removed and the bottom of the dish with the medium and colonies has been placed in the microscope field. (Photographs courtesy of the Millipore Corp., Bedford, Massachusetts.)

A Simple Test for Free Chlorine

Chlorine dissolves in water to produce hydrochloric and hypochlorous acids (both very dilute) and free chlorine. Both free chlorine and hypochlorous acid act as disinfectants. That is, they are capable of destroying disease-producing microorganisms. However, they also react with substances dissolved in the water.

Ammonia and hydrogen sulfide are two common chemicals dissolved in water that react with hypochlorous acid and chlorine. These reactions use up the disinfectant potential of chlorine. Thus, when these two substances are present, more chlorine may be required to achieve disinfection.

A simple test for the presence of free chlorine involves the use of starch and iodine. Add a small amount of starch plus a few crystals of potassium iodide to water from a swimming pool or to chlorinated tap water. Free chlorine oxidizes the iodide to free iodine (chlorine is simultaneously converted to the chloride ion). Free iodine in the presence of starch produces a characteristic blue-purple color. Presence of the color indicates that free chlorine is available to oxidize the iodide ions. The starch–iodine test indicates whether or not free chlorine is present in sufficient concentration to destroy the pathogenic bacteria. It gives a very swift and very reliable answer to the question: is the disinfection process operating properly— that is, is the water safe to drink?

Ozonation to Purify Water

The ozone process, like the chlorination process, is a simple contacting of water with gas. Ozone gas is a powerful oxidizing agent, which destroys bacteria and viruses. In contrast to chlorination, in which chlorine may combine with hydrocarbons, ozone may actually destroy some of the hydrocarbon compounds by oxidation. Also, potential taste and odor problems created by chlorination are avoided when ozone is used. The combination of chlorine reacting with phenol in water produces an undesirable taste and odor, but ozone does not. The creation of chlorinated hydrocarbons is also avoided with ozone. Furthermore, ozone is effective in removing color.

Ozonation of water supplies is practiced in a number of European cities, but chlorination is still the process most water-supply engineers choose in the United States. In the United States, ozonation is used by a number of companies that produce bottled water. They choose this method of disinfection to avoid the taste chlorination gives to water, a taste which the consumer may associate with a municipal water supply. The city of Whiting, Indiana, does use ozone for purifying its water supply. Their water supply, drawn from Lake Michigan, contains phenols, which would react with chlorine to give the undesirable taste and odor mentioned earlier.

Questions

1. Name three classes of organisms that can affect the safety of a water supply for human consumption. Give examples of each class.
2. The spread of certain diseases to epidemic levels can be described in terms of a "cycle." Describe the role of water in the spread of diseases such as typhoid and cholera.
3. The engineering solution to the spread of infectious disease has been to break this cycle. List several ways in which this can be accomplished.
4. List three diseases, other than typhoid and cholera, which are spread by polluted water. Indicate the type of microorganism involved in each of the diseases you list.
5. What is the function of sand filtration in supplying water for human consumption? Does this process alone ensure safety against infectious disease?
6. What is the function of chlorination in supplying water for human consumption?

7. Explain what is meant by "indicator organism." Discuss the role of coliform bacteria in monitoring water quality.

8. Where does your drinking water come from? Has your water been tested for safety recently?

9. Complete the following table, which summarizes the procedures or steps in treating water to make it safe to drink and the reason for these steps.

Water Treatment

Procedure	Reason for this procedure
Spray aeration	Removes dissolved gases like hydrogen sulfide
Addition of activated carbon	
Coagulation with alum or ferrous sulfate	
Sand filtration	
Chlorination	

10. How can chlorination of drinking water cause a possible cancer problem?

11. Should we stop chlorinating drinking water? What are the possible solutions for this problem?

12. Describe briefly two methods used in testing a water sample for the presence of coliform indicator organisms. Are these methods suitable tests for detecting virus contamination? Explain your answer.

Further Reading

Drinking Water and Health, Safe Drinking Water Committee, National Academy of Sciences, Washington, D.C., 1977.

This publication is the result of the definitive and comprehensive study undertaken by the National Academy of Sciences at the request of Congress when it passed the Safe Drinking Water Act of 1974.

National Interim Primary Drinking Water Regulations, Office of Water Supply, Environmental Protection Agency, EPA 570/9-76-003, 1976.

This document, which replaces the Public Health Service Drinking Water Standards of 1962, contains the standards for all contaminants that influence the safety of water supplies. The publication contains, in addition to the standards, background material that explains the basis for the values specified for the standards. The drinking water standards provided in this document became effective in 1977, but will probably be modified, in time. One new standard was issued in 1979 for trihalomethanes. Material on this standard can be found in *Federal Register*, Vol. 44, No. 231, November 29, 1979, p. 68,624.

Sedgwick's Principles of Sanitary Science and Public Health, rewritten and enlarged by S. Prescott and M. Horwood. New York: Macmillan, 1948.

The classic public health text. It is easy reading, and especially absorbing for those interested in the historical roots of the environmental/public health movement.

Standard Methods for the Examination of Water and Waste Water, 13th ed., prepared and published jointly by the American Public Health Association, American Water Works Association, Water Pollution Control Federation, 1980. Available from the American Public Health Association, 1790 Broadway, New York, N.Y. 10019.

This book provides step-by-step descriptions of all laboratory tests used to assess the quality of water and wastewater. The tests "are the best available and the generally accepted procedures for the analysis of water and wastewater."

Craun, G., "Outbreaks of Waterborne Disease in the United States: 1971–1978," *Journal of American Water Works Association,* July 1981, p. 360.

Outbreaks of waterborne disease continue into the present. This article reviews the most recent outbreaks with explanation of the causes.

Green, Phillip, "Infectious Waterborne Diseases," Geological Survey Circular 848-D, 1981. Available free from Distribution Branch, Text Products Section, U.S. Geological Survey, 604 South Pickett Street, Alexandria, Va 22304.

The list of disease-causing organisms spread via the water route is far longer than has been indicated in this chapter. This little booklet is remarkably complete; virtually all the viruses, bacteria, and parasites that infect humans are listed here.

"National Statistical Assessment of Rural Water Conditions," Report to the Environmental Protection Agency, J. Francis, Principal Investigator, 1982.

The study provides a picture of private household water supplies in rural America. These are the supplies typically drawn from wells.

Waterborne Transmission of Giardiasis, W. Jakubowski and J. Hoff, eds., U.S. Environmental Protection Agency, June 1979, EPA-600/9-79-001. Document available from National Technical Information Service, Springfield, Va 22151.

The articles collected in this volume form a virtual textbook on giardiasis. Well-written and clear.

Miller, G. W. and R. G. Rice, "European Water Treatment Practices," *Civil Engineering, ASCE* (January 1979), p. 76.

The arguments for using ozone rather than chlorine are given in this paper, along with the experiences of European water-supply managers, who use ozone regularly.

Water Quality and Treatment, 3rd ed. New York: American Water Works Association, Inc., 1971.

Water Treatment Plant Design. New York: American Water Works Association, Inc., 1969.

Both of these sources are somewhat technical in nature, intended principally for environmental engineers and chemists.

Contamination of Groundwater by Toxic Organic Chemicals, Council on Environmental Quality, January, 1981. (For sale by the U.S. Government Printing Office.)

A review of incidents and potential for groundwater contamination.

CHAPTER ELEVEN

Chemicals in Drinking Water

The Problem of Chemicals in Drinking Water

Inorganic Chemicals in Drinking Water
Mercury/Cadmium/Nitrates in Drinking Water/Sodium in Drinking Water

Organic Chemicals in Drinking Water
PCBs/Phthalates: An Environmental Problem in the Making?/Pesticides in Drinking Water

How Safe Is U.S. Drinking Water?
Safe Drinking-Water Laws/U.S. Drinking-Water Problems

National Interim Primary Drinking-Water Standards

Why Nitrites Are Poisonous

CONTROVERSIES:

11.1: *Minimata Revisited: Mercury*

11.2: *The Economics of Environmental Improvements: PCBs*

11.3: *Shaping Up Utility Companies: Safe Drinking Water Laws*

The Problem of Chemicals in Drinking Water

In 1980, the town of Rockaway, New Jersey, discovered the chemical trichloroethylene (TCE) in town drinking water at 2–3 times the recommended maximum levels. Pushed by the New Jersey Department of Environmental Protection, town officials looked around for possible solutions to the problem. Faced with drilling new wells, buying water from a nearby town, or treating their own well water to remove the TCE, officials found a fourth option. Apparently TCE was mainly seeping into one well and contaminating the whole water supply. Officials began pumping water 20 hours a day from the most contaminated well into nearby Beaver Brook. This approach seemed to solve Rockaway's problem—TCE values dropped to only trace levels. New Jersey DEP however, was not happy with the fact that the TCE-contaminated water flowed into Boonton Reservoir, Jersey City's water supply! Rockaway Mayor William Bishop declared that this wasn't a problem because TCE is volatile and thus evaporates from Beaver Brook. "We're giving the water to Boonton Reservoir, but we're giving the TCE to God."[1]

This story points up some of the problems we now face in providing drinking water that is not only safe from bacterial contamination (as described in Chapter 10), but also pure, in the sense that it contains no harmful chemicals.

In the past, most attention has focused on chemical contamination of surface waters, which has been a serious problem. Unwanted chemicals find their way into surface waters from many sources. To a young nation, with an apparently limitless supply of fresh water, it seemed reasonable to dispose of industrial waste chemicals into waterways. Mining wastes have been allowed to leach or flow unchecked into nearby rivers. Pesticides and other agricultural wastes are washed by rain into nearby rivers and lakes. Rain also washes a variety of contaminants from the air into water. Further, in the winter, the salting of icy roads leaves chemicals that wash into the nearest watercourse at the first thaw. Inevitably, some of these chemicals find their way into lakes or rivers from which drinking water is taken.

However, almost half of all Americans use groundwater for drinking (see Chapter 9 for a discussion of groundwater resources). This water has been considered a relatively pure source, safe from a number of the problems that plague surface water sources. Shallow groundwaters tend to be pure because soils and soil microbes filter out or degrade many pollutants, such as harmful bacteria or materials that make water turbid. However, these processes do not remove most synthetic organic chemicals such as TCE. Organic contaminants are often volatile and so may eventually evaporate from surface waters, but they are trapped in groundwater. In addition, once groundwaters filter down to deeper levels, no further cleansing of pollutants takes place. Groundwater aquifers, once contaminated, can remain contaminated for hundreds or even thousands of years.

At present, the greatest chemical threat to groundwater supplies is posed by industrial and municipal waste landfills, many of which are unlined and sited in permeable soil near aquifers or wells. (Landfill problems are described in detail in Chapter 32.) In certain

_____
1 *Civil Engineering—ASCE,* September 1981, p. 67.

areas, mines and facilities for petroleum and natural gas production are important sources of groundwater pollution. In other areas, septic tanks pose a threat. Even the chemicals used to clean septic tanks are a hazard because the major solvents in them are organic chemicals such as trichloroethane.

In this chapter, we discuss some of the most serious chemical pollutants of drinking water drawn from groundwater and surface water sources. In the last section we look at how well U.S. utilities are doing in supplying their customers with safe drinking water. A summary of current EPA standards for drinking water is found in the supplemental material at the end of the chapter.

Inorganic Chemicals in Drinking Water

Mercury, cadmium, and nitrates are examples of inorganic chemicals that contaminate water supplies. These chemicals are discussed individually in the following sections.

Arsenic and lead are also inorganic chemicals that may be found in drinking water. Most of the arsenic used in this country (80%) is used in agricultural pesticides and defoliants. Arsenic is also found in smoke from burning coal. Thus, surface waters near coal-fired power plants or waters in agricultural areas may be subject to arsenic pollution. In some areas, arsenic from rock contaminates groundwaters to such an extent that they are not suitable as drinking-water sources.

Lead from leaded gasoline is a common air pollutant that is washed down by rains into surface waters. The lead problem is covered in Chapter 21.

Iron and manganese may also contaminate waters that serve as community water sources. In most cases, high levels of these metals are a result of **acid mine drainage** (see Chapter 16).

Mercury

A strange disease. In 1953, people living in the region around Minimata Bay in Japan began to suffer from a mysterious nervous disease. Symptoms included a narrowing of their field of vision and a lack of coordination. One unusual aspect of the epidemic was that animals and birds seemed affected as well as peo-

ple. Because of this, public health officials were led to suspect that an environmental poison was causing the epidemic. The symptoms were, in fact, those of mercury poisoning.

Mercury has long been known to be a poison. The expression "mad as a hatter" originated in the times when many workers in the felt-hat industry, who were exposed to high mercury levels in their work, suffered from mental problems. In mild cases, mercury causes symptoms that mimic mental and emotional disorders: insomnia; inability to accept criticism; fear; headache; depression; and generally exaggerated emotional responses.

In Minimata, some 120 people were affected and 46 died before investigators realized that people and animals were being poisoned by eating mercury-contaminated fish and shellfish from the bay. The origin of the mercury was a plastics factory located on a stream leading into Minimata Bay. Although mercury is toxic to fish as well as people, mercury levels in the water were not high enough to prevent fish and shellfish from living there. Two natural processes changed the factory's waste product from a trace chemical in the environment to an epidemic-causing pollutant.

In the first place, a transformation of the mercury was taking place. There are several forms of mercury. Elemental mercury (used in thermometers) and inorganic salts of mercury (for instance, mercuric chloride) are excreted relatively quickly from the body. Much more poisonous are the alkyl mercury compounds, like methyl mercury and ethyl mercury. These compounds are excreted very slowly from the body—perhaps only 1% of the total amount present is removed each day. Although a great deal of mercury that finds its way into natural waters is in the form of inorganic mercury, the mercury found in fish is almost always the more toxic methyl mercury. Studies have shown that bacteria in the bottom muds of lakes and rivers and in the slime on fish themselves are capable of transforming inorganic mercury to methyl mercury. Some of the mercury released into Minimata Bay was in the form of methyl mercury, but a good deal more methyl mercury was apparently formed by bacteria in the bottom muds.

In the second place, **biological magnification** (see Figure 6.8) increased mercury concentrations in fish and shellfish to levels many times greater than those found in the bay water. Fish and shellfish in the bay concentrated the methyl mercury to levels that were toxic to humans who ate the seafood.

The possibility of these two processes, transformation of a substance in the environment and concentration or magnification by living organisms, must always be taken into account when a decision is made about the hazard of a particular chemical.

Mercury in water. Mercury enters natural waters from many sources. The English River in Ontario, Canada, has been contaminated with mercury discharged from a chlorine–caustic soda plant. Mercury levels in fish caught in the river have been reported as high as in fish from Minimata Bay. Furthermore, cats fed on fish from the river have shown symptoms of mercury poisoning. Since fish from the river form a major portion of the diet of Indians living in two nearby reservations, another Minimata-type epidemic may soon be revealed. (See Controversy 11.1.)

In the U.S., scientists have estimated that chlorine–caustic soda plants released $\frac{1}{4}$–$\frac{1}{2}$ pound (100–200 g) of mercury for each ton of caustic soda (sodium hydroxide) produced until the early 1970s. Today, strict U.S. laws prohibit the discharge of mercury by industry. However, in areas where mercury was formerly discharged—for instance, near pulp and paper mills or chlorine–caustic soda plants—mercury in the bottom mud often still contaminates the water and the organisms living there (Table 11.1).

Lakes and rivers may take 10–100 years to cleanse themselves of mercury after all additions of mercury have stopped. Fishing restrictions have been set in many states because fish are concentrating mercury dumped into the water many years ago. The following story is in many ways typical of how mercury contaminates water sources and aquatic life.

From the 1930s to about 1950, DuPont manufactured acetate, using mercuric sulfate, in a plant at Waynesboro, Virginia. The plant was located on the South Fork of the Shenandoah River. It appears that mercury, spilled from the plant over 25 years ago in the 1930s and 1940s, is contaminating the river. Writing in *Science,* Luther J. Carter noted:

> Not all of the Shenandoah River is contaminated, only the South Fork, which many regard as the best of it . . .
>
> For the canoeist or the float fisherman (the South Fork is famed for its small-mouth bass fishing), the scene is ever-changing but is always good and sometimes spectacular, especially when the winding river turns toward the steeply rising slopes of the Massanutten. Along with the rest of the Shenandoah, the South Fork has long been a prime candidate for consideration as part of the national system of wild and scenic rivers.
>
> Although the Shenandoah was known to have some water quality problems, especially overfertilization from the runoff from farmland and other sources, it was not until this spring that state officials got word that part of the river might be heavily polluted with mercury.[2]

Seeping from the abandoned plant site, the mercury seems to have settled into crevices and nooks on the limestone bottom of the river, and is slowly dissolving into the water. Bass caught more than 77 miles (124 km) downstream from the plant contain over twice the legal limit for mercury.

Mercury in air. Mercury is not only a water pollutant, but also an air pollutant from sources such as coal-fired power plants and mercury-ore refining plants. The fossil fuels (coal and oil) contain mercury, which is released to the air when they are burned. Estimates say that some 5000 tons of mercury may be added to the air each year from fossil fuels. A significant amount of this airborne mercury is washed out of the air by rainwater.

The mercury level in the air in cities tends to be much higher than in rural areas, perhaps due to the greater use of fossil fuels in cities. For instance, in rural areas, mercury levels of 0.003–0.009 micrograms (μg) per cubic meter of air have been measured. In Chicago, levels average 0.01 μg per cubic meter and in New York City, values of 7–14 μg per cubic meter have been found. Mercury-ore processing plants and

Table 11.1 Mercury Found in Organisms Living Upstream and Downstream from a Paper Mill[a]

	Living in stream above the paper mill (parts per million)	Living in stream below the paper mill (parts per million)
Sowbug	65	1900
Burrowing alderfly	49	5500
Caddis fly	—	1700

a From A. G. Johnels, et al., *Oikos,* **18** (2), 323 (1967).

2 Luther J. Carter, *Science,* **198** (9 December 1977), 1015.

Minimata Revisited: Mercury

> The autopsy evidence of Minimata Disease in two domestic cats fed mercury-contaminated fish from the lower English River . . . is a significant indication of the hazard to Indians and others who also eat fish from the same waters.*
>
> *T. Takeuchi et al.*

In Dryden, Ontario, a chlorine–caustic soda plant discharges 3.3 g of mercury per metric ton of chlorine produced. Between 1962 and 1970, 9000–11,000 kg of mercury were discharged into the Wabigoon River on which the plant is located.

Fish caught in this river system show mercury levels comparable to those found in Minimata Bay. Further, two cats fed on fish from the river were found to have symptoms of Minimata Disease, or mercury poisoning. Concern has been voiced because two Ojibway reservations are located nearby—the Grassy Narrows and the White Dog Reserves on the lower English River, which is part of the same river system. The Indians eat a diet high in fish. In fact, government studies have found that many of the Indians have excessive mercury levels in their blood. In addition, a Japanese group that came to study the Indians found signs of Minimata-like mercury poisoning, such as visual problems and lack of coordination.

One group of scientists has written:

(V)ery little has been published even though 26% of the 110 individuals tested at Grassy Narrows had mercury blood concentrations in excess of 100 ppb.† One individual's mercury blood level was reported to be 385 ppb, while the mean value for the group was about 70 ppb. . . . Furthermore, since the highly contaminated fish were discovered in 1969, no provincial or federal government agency has attempted to establish, by conducting autopsies on Indians who died of unknown causes at either Grassy Narrows or White Dog Reserves, whether either chemical or pathological evidence of Minimata Disease existed. Since cats develop the symptoms of methylmercury poisoning more readily than human beings, they serve as indicator organisms for this disease. Thus, the autopsy evidence of Minimata Disease in two domestic cats fed mercury-contaminated fish from the lower English River in the vicinity of the Grassy Narrows and White Dog Reserves is a significant indication of the hazard to Indians and others who also eat fish from the same waters.†

Why do you think so little has been published or done about this problem? What steps could the government take? What effects would these steps have on native culture?

---

* T. Takeuchi, F. M. D'Itri, P. V. Fischer, C. S. Annett, and M. Okabe, "The Outbreak of Minimata Disease (Methyl Mercury Poisoning) in Cats on Northwestern Ontario Reserves," *Environmental Research* **13,** 215 (1977).

† Humans generally begin to show signs of mercury poisoning when blood levels reach 20 parts per billion.

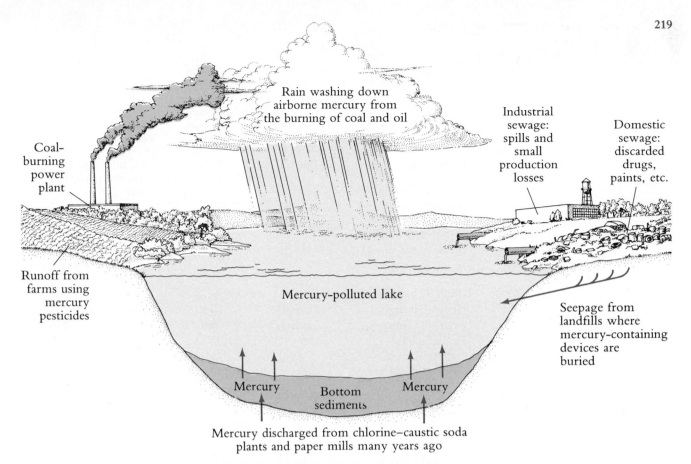

Coal-
burning
power
plant

Rain washing down
airborne mercury from
the burning of coal and oil

Industrial
sewage:
spills and
small
production
losses

Domestic
sewage:
discarded
drugs,
paints, etc.

Runoff from
farms using
mercury
pesticides

Mercury-polluted lake

Seepage from
landfills where
mercury-containing
devices are
buried

Mercury   Mercury

Bottom
sediments

Mercury discharged from chlorine–caustic soda
plants and paper mills many years ago

Figure 11.1   Mercury Pollution of a Lake. In this imaginary lake, mercury enters the water from industrial and domestic sewage and in seepage from landfills where mercury batteries, switches, and other devices are buried. Rainwater washes mercury-containing pesticides into the lake and also washes down mercury in the air from the burning of fossil fuels. Furthermore, mercury washed into the lake many years ago has settled into the bottom muds, where it is slowly being converted to toxic methyl mercury by bacteria and is then entering food chains. Unfortunately, although this particular lake is imaginary, the processes shown are not. Examples of all of them can be found in many areas of the country. In developing countries, too, mercury pollution is becoming a problem. Djakarta Bay, in Indonesia, was reported in 1980 to be seriously polluted with mercury, and 85% of the seafood eaten in Djakarta comes from this bay.

chloralkalai plants have added significant amounts of mercury to the air in the past. However, new standards are reducing air pollution from these sources to much lower levels.

Mercury in industrial and consumer products. About six million pounds of mercury are used each year in the U.S., in electrical devices, thermometers, fungicides, dental fillings, drugs, and paints. Although three-quarters of this mercury could be recycled, at least half is not recycled. That is, it finds its

way into the environment and eventually into natural waters.

Mercury could be called a "permanent" pollutant in that, once released to the environment, it appears to be cycled from the air to water, to organisms living in the water, to human food supplies, and perhaps to humans themselves, in seemingly endless cycles. Many years go by before environmental mercury becomes covered with such thick layers of mud at the bottom of lakes or the ocean that it becomes harmless (Figure 11.1).

Table 11.2 Mercury Concentrations inside Buildings[a]

Location	Date	Mercury concentration (nanogram/ cubic meter)[b]	Remarks
Home 6, bedroom	10 Dec. 71	262	New home, painted with latex-base paint 30 days before
Home 7, living room	29 Dec. 71	1560	New home, painted with latex paint 7 days before
Home 7, bedroom 1	29 Dec. 71	1560	New home, painted with latex-base paint 7 days before
Home 7, bedroom 2	29 Dec. 71	3070	
Home 8, bedroom	9 Nov. 71	12.9	Home paneled
Home 8, kitchen	9 Nov. 71	5.0	
Office building 1, room 1	9 Dec. 71	172	Painted with latex-base paint, June 1970
Office building 1, room 2	9 Dec. 71	245	Painted with latex-base paint, June 1970
Office building 1, room 3	9 Dec. 71	203	Painted with latex-base paint, June 1971
Office building 1, room 4	9 Dec. 71	116	Paneled office
Office building 2, room 1	17 Dec. 71	183	
Doctor's room 1	13 Dec. 71	4950	Mercury thermometer broken in the past
Doctor's room 2	13 Dec. 71	5680	Mercury thermometer broken in the past
Doctor's room 3	13 Dec. 71	4550	Mercury thermometer broken in the past
Dentist 1, room 1	23 Dec. 71	5550	Mixing area for silver amalgam
Dentist 1, room 2	23 Dec. 71	5030	
Dentist 1, room 3	23 Dec. 71	4770	
Dentist 2, room 1	28 Dec. 71	1295	Inactive for previous 4 days
Dentist 2, room 2	28 Dec. 71	1135	
Dentist 2, room 3	28 Dec. 71	1160	
Washington, D.C.	2 Feb. 72	3.25	Potomac River at Key Bridge
San Francisco	19 Jan. 72	3.14	8 km south of San Francisco on the beach
Dallas	11 Jan. 72	3.38	16 km southwest of Dallas

a From R. S. Foote, *Science,* **177,** 513 (11 August 1972).
b Some recommended maximum levels for exposure to mercury in the air are: 40-hour work week, 50 micrograms per cubic meter; 24-hour exposure, 300 nanograms per cubic meter. A nanogram (ng) is 1/1000 of a microgram.

Closer to home, mercury used as a preservative in latex paints may be a hazard. Tests in houses recently painted with latex paints show that mercury levels in the air can remain for at least a week at five times the level considered safe (Table 11.2). The same table shows that broken thermometers can also be a local source of high mercury levels if the mercury is not cleaned up properly. Studies of mercury concentrations in people show that dentists seem to be at a high risk of mercury poisoning. Small amounts of mercury may be spilled during the mixing of the amalgam used to fill teeth. (The amalgam, once mixed, is considered safe, however.) Once on the floor, the mercury slowly vaporizes. Dentists would do well not to carpet their offices, because the carpet acts as a trap for spilled mercury. Spilled mercury should be covered with powdered sulfur and then damp-mopped up. The area can also be coated with commercial hairspray.

Because of its toxicity and tendency to accumulate in living organisms, the standard for mercury in drinking water is set at 0.002 mg per liter. Fish with more than 1 part per million mercury are considered unsafe for human consumption.

Cadmium

Environmental sources. In the Fuchu area of Japan, an unusual disease occurs, called *itai-itai,* or "ouch-ouch" disease, because it is so painful. The victims' bones become brittle and break easily. Even walking can be painful. The cause of the disease has been traced to contamination of the Jintsu River, the area's water supply, with waste from the Kamioka mine. Cadmium is believed to be the contaminant responsible for the disease. Cadmium levels in the rice grown in the district are at least three times higher than in unpolluted areas. In several other prefectures in Japan, cadmium from mines or refineries pollutes water supplies.

The natural occurrence of cadmium in water in more than minute amounts is almost unknown. In the past, detectable levels were usually the result of contamination from mining or industrial wastes. Cadmium is used in the manufacture of such products as paints, alloys, light bulbs, pesticides, and nuclear reactor parts. Wastes from electroplating plants have contaminated groundwater in the United States. Experts now worry that cadmium-containing products burned at dumps release cadmium to the air, while those buried in landfills are contaminating groundwaters.

Cadmium's toxic effects. The main effect of cadmium in the body appears to be on the kidneys, where it accumulates. In addition to kidney disease, cadmium may contribute to high blood pressure.

In fish, cadmium accumulates in the liver, kidneys, and other organs. The flesh is usually low in cadmium. Products made from whole fish, however, can have quite high levels of cadmium. (Fish paste has been reported with up to 5.3 parts per million of cadmium.) Shellfish may also accumulate cadmium, as do wheat and rice.

Smokers seem to face a special risk from cadmium, which is found in tobacco products (1.5–2 micrograms per cigarette). Autopsies on smokers have shown twice the amount of cadmium in the kidneys, liver, and lungs of smokers as compared to nonsmokers. (The topic of chemicals in cigarette smoke is covered more thoroughly in Chapter 31.)

Once a dose of cadmium is absorbed by the body, either from the digestive tract or from the lungs, an extremely long time may pass before it is excreted—perhaps 10–30 years before one-half of the cadmium absorbed is lost. Some scientists feel that cadmium levels found in the kidneys of the general population in the U.S. already approach half the level known to be toxic, and express concern that increasing environmental levels will endanger health.

Cadmium may be removed by water-softening treatments used at drinking-water plants. On the other hand, drinking water that is relatively acidic and high in oxygen content may corrode pipes easily and pick up cadmium after leaving the treatment plant. The Environmental Protection Agency has set a limit of 0.010 mg per liter for cadmium in drinking water.

Nitrates in Drinking Water

A special hazard to infants. In the late 1940s, doctors noticed that certain infants, rushed to the hospital because they were listless and suffering from a blueish coloration of the skin, would recover spontaneously while they were in the hospital. Strangely, they would turn blue again when they returned home. These babies lived in rural areas and were fed on formulas made with well water. Doctors became suspicious that something in well water was making the babies ill.

Thousands of such cases have been reported and are now known to be caused by nitrates. Only infants seem to be endangered. Parents and older children drinking the same water remain unaffected. Nitrates contain the chemical grouping NO_3^-. Water containing more than 10 mg of nitrate per liter is considered unsafe for drinking, mainly because it is likely to be poisonous to infants.

Why are only infants affected? Apparently, in some infants the stomach does not yet produce enough acid to prevent the growth of bacteria that convert nitrates (NO_3^-) to the highly poisonous nitrites (NO_2^-). (If you want to know more about why nitrites are poisonous, see the supplemental material at the end of this chapter.) In older children and adults, who produce the proper amound of acid, bacteria that convert nitrates into nitrites cannot grow in the stomach.

Sources of nitrate contamination. In rural areas, poorly built wells can be contaminated by nitrates from nearby septic systems. Wells have also been contaminated by fertilizers that have back-siphoned into the wells during spraying or diluting of the fertilizers. In some areas, well water is naturally high in nitrates. In other places, groundwater has become contaminated with nitrates from industrial or agricultural sources. In parts of Israel, for instance, some wells are increasing in nitrate concentration by 2 mg or more per liter per year. Plans for recharging groundwater supplies with treated wastewater, which is high in nitrates, will increase this problem.

Some evidence exists that Vitamin C, from orange or tomato juice, can prevent nitrite poisoning in infants, perhaps by combining with the nitrites. Actually, however, only a part of the nitrates we receive come from water. Nitrates are also found in many vegetables, and both nitrates and nitrites are added to cured meats, such as hot dogs and bologna. This is discussed in Chapter 31.

Sodium in Drinking Water

Concern about the amount of sodium in drinking water is a relatively recent development. Scientists first noted in 1957, in Japan, that in communities with soft water, middle-aged men seemed to be at greater risk of dying from heart attacks than in communities with hard water. Since then, this effect has been confirmed in several other countries. It is still not clear what chemistry involving hard and soft water is responsible for this phenomenon.

There is some evidence that either calcium or magnesium in hard water may have a protective effect. On the other hand, sodium is present naturally in high concentrations in some soft water and the main methods of water softening, as well as some water treatment processes, add sodium to water. Medical evidence strongly points to excessive sodium intake as an aggravator, if not a cause, of high blood pressure. Studies have shown that children living in areas where water is softened have higher blood pressures than children in non-water-softened areas. Thus, the sodium in or added to softened waters may well be a health hazard and may contribute to cardiovascular disease.

Utilities could switch treatment processes to decrease sodium addition. For instance, a hydrogen-ion exchange softening process, which adds no sodium to water, could be used. Plumbing for households with home water softeners can be changed so that cold-water taps carry only unsoftened water (except for the tap to the utility area).

There are no national standards for sodium in drinking water at present, although some states have a sodium standard. In areas with naturally high sodium waters, sodium values may reach 300–1700 mg sodium per liter. Water softening in exceptionally hard water areas may result in water with 200–300 mg sodium per liter. The National Academy of Sciences has estimated that keeping sodium concentrations in drinking water at 100 mg/liter or less would be of benefit to the U.S. population. The American Heart Association has suggested a standard of 20 mg/liter to protect people with heart or kidney ailments.

Organic Chemicals in Drinking Water

Phthalates, PCBs, and **pesticides** are typical **organic chemicals** that find their way into drinking water. All are, or were, widely used chemicals. Water contamination, as well as contamination of air and food, results from this widespread use. Another important category of organic chemicals in drinking water is known as **chlorinated hydrocarbons.** Evidence is accumulating that some of these possibly carcinogenic chemicals are actually formed during the water purification procedure, chlorination. This especially difficult problem was noted in Chapter 10. The following sections detail a few typical organic chemical drinking-water problems.

PCBs

In 1968, health officials became aware of yet another curious disease spreading in Japan. Victims developed acne and brown blotches on their skin. In addition, they often complained of severe weakness and eye discharges. The disease was at first named Yusho, or rice-oil disease, because it seemed to result from eating a certain brand of rice oil. Japanese public health detectives finally discovered that the rice oil was contaminated with polychlorinated biphenyls (PCBs). But before the cause was found, over 2000 people were affected and 16 had died.

The abbreviation PCBs actually refers to a whole family of similar chemicals. Because they do not burn readily, they are widely used to transfer heat from one material to another, as in rice oil manufacture. The Japanese tragedy was caused by leaking pipes, which allowed PCBs to flow into the rice oil.

PCBs are used in electrical devices, such as transformers and capacitors, where a spark could ignite a flammable material. PCBs have also been used in a wide variety of other ways: as solvents for paint and ink, in pesticide sprays, and in plastics. In fact, so much PCB has been manufactured and used, we now find small amounts of contaminants in water all over the world (Figure 11.2). Like mercury, PCBs are concentrated by fish and other water-dwellers (Table 11.3). Even people retain PCBs—some 40–45% of the American public has 1 part per million or more PCBs in their fatty tissues, mainly as a result of PCBs in food.

Although the effects of large amounts of PCBs are known (in the Japanese epidemic, people ate an average of two grams of PCBs in a short time), the effect of small amounts of PCBs ingested over a long period of time is not as clear. People exposed to PCBs at work have developed nerve, skin, and liver ailments.

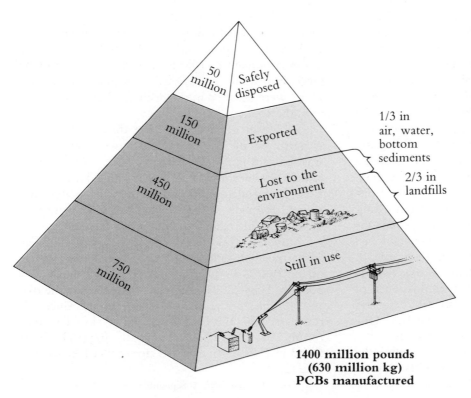

1/3 in air, water, bottom sediments

2/3 in landfills

**1400 million pounds
(630 million kg)
PCBs manufactured**

Figure 11.2 The Fate of PCBs. Of the 1.4 billion pounds (630 million kg) of PCBs manufactured since 1929, some 450 million pounds (203 million kg) are now in the environment. About 750 million pounds (350 million kg) are still in use. Some of what is in use can be expected to enter the environment. Since PCBs are broken down very slowly, environmental levels of PCBs will probably decrease very slowly.

Table 11.3 Concentration of PCBs in Cayuga Lake Trout as a Function of Age

Age of fish (years)	Mean PCB concentration (ppm)
1	1.3
2	2.1
3	1.8
4	3.2
6	7.0
7	6.3
8	11.7
12	15.5

Data from Bache, C. A., et al., *Science,* **177** (29 September 1972), p. 1192.

In the laboratory, PCBs cause birth defects and cancers in animals (possible at levels as low as 100–300 parts per million in their diet).

PCBs in the environment. By 1970, enough concern had been expressed about the contamination of the environment with PCBs that the only U.S. manufacturer, Monsanto Chemical Co., announced it would sell PCBs only for use in closed systems; that is, systems in which no PCBs would be released to the environment. Allowable uses included the manufacture of devices such as capacitors and transformers but did not include PCBs in paints, inks, or plastics.

Soon after these restrictions went into effect, studies showed PCB concentrations in some rivers to be dropping. For instance, in rivers flowing into Green Bay, Wisconsin, levels dropped from 0.15–0.45 parts per billion to almost nothing. Many people heaved a sigh of relief—unfortunately, prematurely. In August, 1975, the New York State Commissioner of Environmental Conservation issued a warning about eating salmon and bass from the Hudson River or Lake Ontario, because the fish had been found to contain well over the 5 parts per million of PCBs considered safe. In fact, federal researchers sampling fish for toxic materials notified the New York State Commissioner of Environmental Conservation that striped bass and salmon taken from the Hudson River and Lake Ontario had as much as 37 parts per million PCB.

Further study showed more rivers, including the St. Lawrence, to be contaminated with PCBs. A wide variety of fish from these waterways contained over the allowable limit of 5 parts per million PCBs. However, fish from the Hudson River still had the distinction of having higher PCB levels than fish from anywhere else in the country.

Many people paid no attention to the warning, however. "I had a striped bass for lunch yesterday," said Andrew Garzel, co-owner of *Sweets* (a Manhattan restaurant). "It was delicious. I'm going to have more. Our bass come from Provincetown, Massachusetts, and what happens in the Hudson doesn't affect the fish we serve here."[3] In fact, bass tagged in the Hudson River have been caught off the Massachusetts coast. The Hudson River is a major spawning ground for the entire East Coast from Massachusetts to Delaware. Thus, pollution of the Hudson has widespread effects.

Where were the PCBs coming from? A check of manufacturing plants on the river showed that two General Electric plants, making capacitors, had permits from the federal government to discharge PCBs into the river. Further investigation showed that General Electric had been discharging PCBs into the Hudson for almost 20 years, possibly as much as 48 pounds (22 kg) per day.

Meanwhile, the U.S. Environmental Protection Agency announced that at least 12 other plants in the United States were also dumping PCBs into waterways or municipal sewage systems. Russell Train, then EPA Administrator, said that PCBs "are present in our environment to a far greater degree and at higher levels than we have previously thought." He said that PCBs were detected in the drinking water of cities all over the country.

As a result of hearings held by the New York State Department of Environmental Conservation, G.E. was judged guilty of violating state water-quality laws. The Chairman of the hearings called the situation a regulatory agency failure as well as corporate abuse, since G.E. made no secret of what it was doing from the responsible federal and state agencies.

Some 500,000 pounds (225,000 kg) of PCBs are still in the Hudson River, 80% within 10 miles (16 km) of the two G.E. plants. Between 5,000 and 10,000 pounds (2250–4500 kg) of PCBs move down the river at 4–10 miles per year (6–16 km per year). Since February of 1976, the Hudson has been closed to most commercial fishing (Figure 11.3). In 1971, PCBs found in Lake Michigan fish closed that lake to com-

3 *The New York Times,* 9 September 1975.

Figure 11.3 Hudson River Fishermen. Closing a river to commercial fishing can mean the loss of thousands or even millions of dollars to the industries depending on those fishing grounds. It also means a loss of livelihood to the fishermen involved. (Edward Hausner/NYT Pictures)

mercial fishing. The Wisconsin Department of Natural Resources is still waiting for the PCBs to degrade so the fisheries can resume, but the department admits it may take decades.

In October, 1976, New York State and G.E. agreed to a compromise settlement in which the company contributed seven million dollars to a Hudson reclamation program and a research and testing program for PCB substitutes. In 1982, New York State began work on a plan of dredging to remove the worst PCB deposits from mud in the Hudson. The state estimates that total cleanup would cost over $200 million, and so will attempt only a partial cleanup at an estimated cost of $25 million.

PCB control—still a problem. The United States Congress passed an amendment to the 1976 Toxic Substances Control Act that banned the manufacture of PCBs in the U.S. after January 1979. The Environmental Protection Agency followed through by forbidding any discharge of PCBs into U.S. waters. In addition, rules were made to phase out the use of PCBs, even in totally closed systems.

How much of a problem remains then? Although PCBs are no longer allowed in new manufacturing systems, some 750 million pounds (350 million kg) of PCBs are still in use in electrical equipment, mostly in capacitors and transformers used by electric utility companies (Figure 11.4). These devices may leak while in service; they will contaminate natural waters if improperly disposed of; and capacitors can explode, scattering PCB-containing oils in the environment.

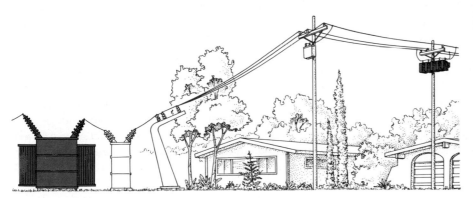

Figure 11.4 PCB-Containing Transformers and Capacitors. Although PCBs can no longer be used in the manufacture of electrical equipment, many old PCB-containing transformers and capacitors (shown in color) are still in service. In many cases, utilities do not know whether older equipment contains PCBs or not. (However, most capacitors made before 1977 are believed to contain over 500 ppm PCBs.) Leaks from this equipment, as well as improper disposal of old transformers and capacitors, are responsible for continuing pollution of the environment with PCBs.

Expert opinion appears to be that PCB-containing capacitors and transformers could not all be replaced immediately because not enough high-grade oil manufacturing capacity is available to accomplish this. Instead, methods are being developed to remove PCBs from existing equipment and dispose of PCBs safely and easily. Incineration at sea at high temperatures is probably the safest disposal method at present.

Meanwhile, a continuing series of contamination incidents is being reported by the press. In some cases, soil and water are reported contaminated by exploding capacitors or abandoned transformers. In other cases, PCBs find their way into human food supplies. A 1979 incident in Billings, Montana, involved a PCB-containing transformer that was accidentally knocked off a pole and into the production system of a meat-packing plant. In this particular incident, PCBs eventually contaminated food shipped to 17 states (see Figure 31.7). The result was a loss of some 10 million dollars to food industries. The problem of accidental contamination of food by PCBs and other materials is discussed in Chapter 31. The Department of Agriculture is attempting to prevent further accidents by requiring meat, poultry, and egg-processing plants to replace all PCB-containing equipment.

Phthalates: An Environmental Problem in the Making?

In the past, environmentally harmful chemicals have been identified because they caused a problem: people became sick, fish were killed, trees died. However, we are becoming more sensitive to the possibility of harm. We look more suspiciously at chemicals that are likely to become widespread in the environment, whether or not we have evidence that they are dangerous, because these are the chemicals in a position to cause widespread damage.

A case in point is that of phthalates (thal´ ates).[4] Phthalates are chemicals used in polyvinyl chloride (PVC) plastics. Phthalates change PVC from a hard, glass-like material into a flexible plastic. Phthalate-containing plastics have been incorporated into automobile interiors, clothing, food wraps, and medical products, such as the plastic bags that hold blood for transfusion (see Figure 11.5). Phthalates are not actually attached to the plastic. They fill spaces between the plastic molecules and act, in a way, as lubricants. Because they are not bound to the plastic, they can and do migrate out into the materials with which the plastic comes in contact. Because phthalates have a tendency to migrate and because of the widespread use of plastics, we might have predicted that phthalates would become a widespread environmental pollutant. In fact, this is just what has happened. Plastics burned in incinerators contaminate air, plastics buried in landfills seep into groundwater and run off into lakes and rivers. Very low levels of phthalates are even detectable in water samples taken from the open oceans and the Gulf of Mexico (0.005–0.090 micrograms per liter). For organisms living in water, phthalates appear to be concentrated up to 13,000 times over their concentration in water.

The next logical question is: what effects do phthalates have on living creatures? Over the short term, phthalates do not appear to be highly toxic chemicals, even in large doses. However, we do not know the effect of small doses, such as are now found in water, over long periods of time. Some experiments have shown that phthalates interfere with reproduction in water creatures, for example, water fleas, guppies, and zebra fish, even at levels as low as 3 micrograms per liter (a level found in some rivers). Higher doses of phthalates cause reproductive problems in mice. In 1982, the Public Health Service announced that in one experiment, two phthalate compounds were found to cause liver cancers in rats. The study will have to be repeated and expanded to determine whether these chemicals are truly carcinogens and if they pose a serious threat at their present levels in the environment.

Does this group of chemicals pose a long-term threat to humans, perhaps through direct exposure via water, food (food wraps), or blood transfusions? Or do we face a situation in which subtle changes in food webs will occur as certain aquatic creatures accumulate phthalates and/or suffer reproductive problems? The questions are serious, but the answers may be a while in coming.

---

4 For information on the chemical structure of phthalates, see the Environmental Health Perspectives volume mentioned in the References.

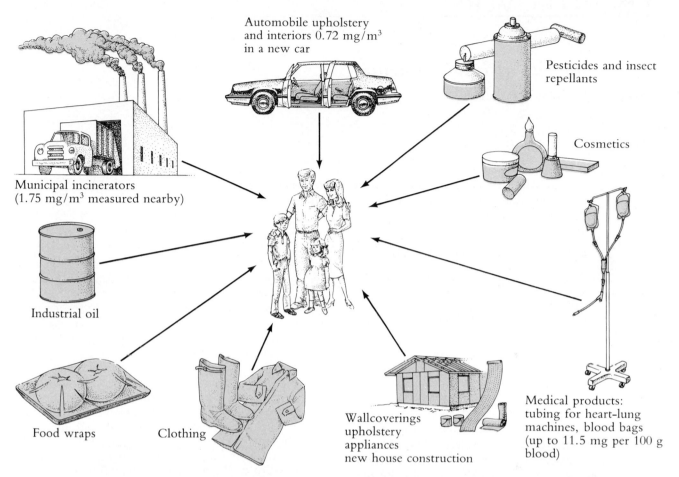

Automobile upholstery and interiors 0.72 mg/m³ in a new car

Pesticides and insect repellants

Cosmetics

Municipal incinerators (1.75 mg/m³ measured nearby)

Industrial oil

Food wraps

Clothing

Wallcoverings upholstery appliances new house construction

Medical products: tubing for heart-lung machines, blood bags (up to 11.5 mg per 100 g blood)

Figure 11.5 Phthalates: A Widespread Environmental Contaminant. Phthalates have been used in a variety of products for home and industry. Although phthalates are relatively nontoxic (humans could probably survive a daily dose of 3–480 mg of phthalates; workers in plastics industries begin to show symptoms of poisoning at about 2–70 mg of phthalate per cubic meter of air), long-term effects are unknown.

Pesticides in Drinking Water

Sources. Pesticides, which were discussed in Chapter 6, are chemicals designed to kill insects and other pests. These chemicals are widely used. In fact, they have become part of our daily lives. A study by the U.S. Environmental Protection Agency showed that DDT, a pesticide once very popular in this country, can be found at levels of a few hundred parts per trillion in many U.S. waters, including drinking water supplies. Other studies have shown that Americans have detectable levels of DDT and other pesticides in their fatty tissues (Table 6.2).

How do pesticides find their way into the water we drink? There are several ways. Pesticides run off farmlands during rains, either dissolved in the rainwater or attached to dirt particles. Rainfall itself may be contaminated with pesticides that have evaporated or remain in the air after the spraying of crops or woodlands. Industrial plants manufacturing pesticides may accidentally spill or intentionally discharge pesticides into nearby rivers and lakes. An example of some of these processes can be seen in the story of how two related pesticides, Kepone and Mirex, came to contaminate such major fisheries and water sources as the James River and Lake Ontario.

The Economics of Environmental Improvements: PCBs

> For the state to require zero discharge would put us on the "same old self-destruct course" of creating an unfavorable climate for industry here.
>
> *A top aide to Governor Carey,*
> *quoted in* **The New York Times,** *30 January 1976*

> Tommyrot and balderdash!
>
> *Ogden Reid,*
> *quoted in* **The New York Times,** *18 February 1976*

Efforts to improve the environment often appear to run head-on against economic interests. For instance, when the New York State Commissioner of Environmental Protection, Ogden Reid, ordered General Electric to stop all discharges of PCBs into the Hudson River immediately, he was quickly overruled by the governor of New York, who gave G.E. a whole year to reduce its PCB discharge. Governor Carey thought that the stricter ruling might cause General Electric to move to a more lenient state. New York would lose tax money and jobs.

"For the state to require zero discharge of PCB's," one of the Governor's top aides said . . . , would put us on the "same old self-destruct course" of creating an unfavorable climate for industry here.

"Furthermore," this aide said, "after all those years of discharging 30 pounds a day, no one can tell me there's one bit of difference between zero and 3.5 ounces."

"It's a totally phony, symbolic issue," he added.

Mr. Reid said it was "tommyrot and balderdash" that New Yorkers had to choose between jobs and a healthy environment. "There are alternatives to PCB's and we believe it is possible to work this out so that G.E. can continue," Mr. Reid said.

He also criticized those who say G.E.'s proposal to discharge no more than 3.5 ounces a day by 1977 is reasonable. "That amount would impair water quality and we estimate that minnows would carry over four parts per million of PCB's as a result," Mr. Reid said. "Striped bass that eat the minnows would bioaccumulate those PCB to perhaps 11 parts per million. Saying that a little of PCB's don't hurt is like saying you're a little bit pregnant."

It will undoubtedly cost G.E. more money to discharge absolutely no PCBs. However, this is not the only cost involved in the case. How many other costs can you think of that result from the polluting of the Hudson? Who pays these costs?

Suppose G.E., unable or unwilling to reduce its discharge to zero, planned to move out of the state instead. Do you feel that New York State should then compromise its standards? Why?

Kepone: A case in point. In June, 1975, an ailing employee of the Life Science Products Corp. in Hopewell, Virginia, was finally persuaded by his wife to see a doctor. The doctor thought that the dizziness and trembling the man suffered might be due to a chemical to which he was exposed at work. When a blood sample was sent to the Public Health Service Center for Disease Control in Atlanta, it was indeed found to contain a high concentration of the pesticide Kepone, which was manufactured at the Hopewell plant. Thus began a series of inquiries into an incident that a United States attorney later called "the environmental disaster of the decade."

A state official who visited the Hopewell plant found piles of Kepone dust and contaminated water. He also found seven workers ill enough to require immediate hospitalization. Over 100 people, including wives and children of plant employees, were found to have Kepone in their blood. Thirty people were hospitalized and found to be suffering from tremors and visual problems. Further, some of these people may be permanently sterile and all are threatened with the possible development of liver cancer, since Kepone is known to cause such tumors in experimental animals.

But this was only the tip of the iceberg. Studies detected Kepone in the air and water as far as 40 miles (64 km) away from the factory site. In addition, Kepone was found first in fish and shellfish in the James River, on which Hopewell is located, and then in fish and oysters in Chesapeake Bay, into which the James River empties. Eventually, traces of Kepone were found in bluefish along hundreds of miles of the Atlantic coast. Commercial fishing and oyster harvesting were banned in the James River and parts of the Chesapeake. Millions of dollars were lost by those who usually fish these waters for a living.

Is it possible to fix the blame for what happened? Life Science Products was a small company started by two former employees of Allied Chemical Corp. The smaller company manufactured Kepone from raw materials supplied by Allied and then sold the finished product exclusively to Allied. Officials of Allied Chemical said the smaller company knew that Kepone was a hazardous substance. Workers at Life Science said they were never told it was dangerous. Scientists calculated that workers received daily doses of Kepone over a period of 15 months that were within a factor of five of the amount producing cancers in experimental rats and mice exposed for $1\frac{1}{2}$–2 years (Figure 11.6).

Figure 11.6 The former Life Science Products Corporation in Hopewell, Virginia. Nothing is left on the site today. The plant has been razed to the ground, by order of the Virginia State Health Department, and pieces of it were buried in a special lined pit. (Photo by R. S. Jackson, M.D., Virginia State Department of Health)

Wastes from the Life Science plant found their way into Hopewell's sewage treatment plant and caused sewage digesters there to fail within a few months after the plant began operations. Kepone, which was marketed as a control for ants, roaches, and the banana-root borer, had killed the bacteria in the digesters. These bacteria normally oxidize waste materials (this process is explained in Chapter 14). The city of Hopewell asked the U.S. Environmental Protection Agency for information on how to handle the wastes. State agencies advised that the wastes should be specially treated before being run through the plant. This recommendation was apparently ignored.

Oysters, shrimp, and fish accumulate Kepone 425–20,000 times over the level found in the water in which they live. Kepone is toxic to some water creatures at very low levels—for instance, shrimp fail to reproduce properly at 0.2 part per billion. Kepone is also slowly degraded in the environment. As a result of the discharges from the Hopewell plant, fish and shellfish taken from the James River in 1976 were found to contain 0.1–1 part per million Kepone. (Safe levels were considered to be 0.1 ppm at that time, but since been raised to 0.3 ppm.)

Allied Chemical Corp. was eventually fined eight million dollars for polluting the nation's waterways, under the Water Pollution Control Act of 1972. Another $5.3 million was paid to the city of Hopewell and the State of Virginia for repairs on the sewage treatment plant and patrols on the river to ensure the fishing ban. Cleaning Kepone out of the river doesn't seem to be technically feasible at the present time. The river may eventually bury the Kepone in natural sediments.

The Mirex problem. Meanwhile, an almost identical pesticide, Mirex, was being used to control the fire ant in several southern states. This species of ant first entered the United States in the late 1950s from Argentina. It resembles ants found here except that it can deliver a painful sting, similar to that of a bee or wasp. The sting is not fatal except to certain people who develop an allergy to it. Fire ants are obviously a serious nuisance in residential areas. They also build large earthen hills, which can interfere with farming.

Mirex was used against fire ants starting in 1962. It was mixed in small amounts with ground corncobs and soybean oil to form a bait, which the ants carried back to their nest, thus killing the whole colony. Unlike aldrin and dieldrin, which had been used previously on fire ants, Mirex at first seemed to be harmless to other creatures in the area. The Department of Agriculture began an eradication campaign in which the bait was dropped from airplanes in an attempt to wipe out the fire ant.

Evidence began accumulating, however, that Mirex was not as harmless as was originally thought. It was found to cause cancer in mice, to be toxic to crabs and shrimp, and to last longer in the environment than DDT. The Environmental Protection Agency joined the fight against Mirex in 1973 when it began hearings on the pesticide. Despite mounting evidence of hazard, the Department of Agriculture and the State of Mississippi were reluctant to give up their goal of eradicating the fire ant. When Allied Chemical Corp., as a result of increasing pressure brought to bear by environmental groups, decided to discontinue Mirex production, the State of Mississippi took over the Allied plant and bought its inventory. Possibly as a result of the Kepone disaster and its publicity, however, a compromise was reached between the EPA and the State of Mississippi in which the use of Mirex bait to control fire ants was to be phased out by June 1978.[5]

---

5 The Mississippi Fire Ant Authority received permission to use a related pesticide, Ferriamicide. This is a mixture of Mirex and a few other chemicals, supposed to help degrade Mirex more quickly in the environment. Probably this was, in large part, a political solution. In any case, use of Ferriamicide was prevented, pending further tests, by a suit brought by environmental groups.

Before that happened, a study was carried out which detected Mirex in the fatty tissue of about 25% of the people living in the area sprayed with Mirex. Thus, food and water must have become contaminated either directly (by spraying) or indirectly. An example of the latter might be the contamination of milk from cows grazing in pastures onto which Mirex had drifted.

This was not the whole story, however. Mirex was also found in fish in Lake Ontario, which is certainly a long way from the southern states where the fire-ant program was carried out. This time the source was found to be the Hooker Chemical plant, located on the Niagara River, which flows into Lake Ontario. Hooker was the supplier for technical-grade Mirex to the Mississippi plant.

Thus, one pesticide spray program resulted in the absorption and concentration of the substance by organisms, including humans, in the sprayed area and also in the industrial contamination of waters where the pesticide was being manufactured, far from the sprayed area. In cases where a possible hazardous chemical is to be used or manufactured, only extreme care can prevent the material from contaminating the environment and concentrating in food chains. The quantities of Kepone and Mirex released from Life Science at Hopewell and by the fire-ant eradication program will remain in the environment and as part of food chains for many years to come.

How Safe Is U.S. Drinking Water?

Safe Drinking-Water Laws

In December, 1974, President Nixon signed into law the Safe Drinking Water Act. Several unsuccessful attempts had been made in other years to pass such a law, but it required some hint of an environmental disaster to provide the final push that propelled the law through Congress. In this case, the push was the publicized discovery of a number of possibly carcinogenic substances in the drinking water of New Orleans. (For more on this discovery, see Chapters 10 and 29.)

The Safe Drinking Water Act directs the Environmental Protection Agency to establish regulations ensuring the purity and safety of United States drinking water. Although each state is expected to be responsible for setting and enforcing its own standards, the federal government, under the law, can step in if the

CONTROVERSY 11.3

Shaping Up Utility Companies: Safe Drinking-Water Laws

> No one knows at this point how much time, money and effort, or blood will have to be devoted to . . . this provision . . . and certainly no one knows how many utility personnel it will drive to an early retirement or to another profession.
>
> **John M. Gaston,**
> **California State Department of Health**

One provision of the Safe Drinking Water Act is of special interest to both consumers and to the people who run water treatment plants. This is the requirement that after June, 1977, water treatment agencies must notify their customers if the water they supply does not meet all of the standards.

Some concern has been voiced that the public will become unnecessarily alarmed if they are told their drinking water does not meet standards. However, John Gaston, a Senior Sanitary Engineer at the California Department of Health, sees the provision as a way of forcing utilities to "shape up." He provides the following data from the California water supply system (only water supplies with at least 200 connections are included). Assuming the notification provision had been in effect in 1976, this is what would have happened.

In the first eight months of 1976, 72 water utilities would have notified the public a total of 80 times. One unfortunate utility would have notified four out of the eight months of record in 1976. This notification would have been because of bacteriological quality failure, and would have included television, radio, newspa-

per, and direct mail notification. Sixty-six utilities would have notified the public a total of 102 times because they failed to take enough bacteriological samples. One utility didn't take enough samples five out of the eight months of record.[*]

In fact, a General Accounting Office Study in 1982 found that utilities were actually notifying the public in only about 11% of the cases required by law.

Certainly the public should be notified of emergency health hazards such as the 70-ton spill of chloroform that travelled down the Ohio River in February, 1977. (The city of Cincinnati takes its drinking water from the Ohio.) But do you think consumers really want to know every time their water suppliers "forget" to take a sample to check for bacteria? Do you think the new provision will improve the record in terms of water supplies meeting the drinking-water standards? Have you ever been notified by your utility that your water did not meet standards?

---

[*] John Gaston, "Consumer Notification—Public Awareness or the Smoking Pistol," *Journal of the American Water Works Association* (November 1977), p. 574.

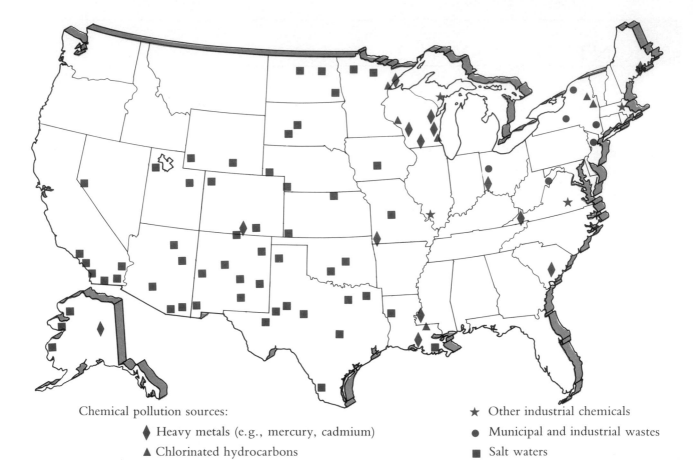

Chemical pollution sources:

◆ Heavy metals (e.g., mercury, cadmium)

▲ Chlorinated hydrocarbons

★ Other industrial chemicals

● Municipal and industrial wastes

■ Salt waters

Figure 11.7 Chemical Drinking-Water Problems in the U.S. The squares indicate areas where salt waters threaten groundwater drinking-water sources. The dots represent areas where chemicals have contaminated drinking-water sources. (Source: U.S. Water Resources Council and the Conservation Fund.)

states' regulations are not at least as stringent as federal regulations, or if the state does not enforce its regulations. This is a major change. In previous years, only suggested guidelines were set by the U.S. Public Health Service. The federal government had no power to regulate any water supply except that used by interstate carriers such as buses or trains. Forty-eight states now have approved drinking-water programs and so have been granted full control of their drinking-water safety, or "primacy," by EPA.

U.S. Drinking-Water Problems

The old system of drinking-water regulation simply did not work well enough. This is shown by the following quotes from a 1970 study by the Department of Health, Education and Welfare:

Thirty-six percent of 2600 individual tap water samples contained one or more bacteriological or chemical constituents exceeding the limits in the Public Health Service Drinking Water Standards.

Seventy-seven percent of the plant operators were inadequately trained in fundamental water microbiology; and 46 percent were deficient in chemistry relating to their plant operation.

Seventy-nine percent of the systems were not inspected by State or county authorities in 1968, the last full calendar year prior to study. In 50 percent of the cases, plant officials did not remember when, if ever, a State or local health department had last surveyed the supply.

Is the new system working? An ongoing study by The National Research Council reported in 1980 that 30–40 states still had serious drinking-water problems (Figure 11.7). Problems included bacterial contamination of water supplies; chemical contamination with toxic waste chemicals; and intrusion of salty water into groundwater drinking-water sources (see Chapter 9 for a discussion of this last problem). Nearly every state east of the Mississippi reported problems, as did several nonindustrial Western states.

In a 1982 study of rural households, most of which draw their water from individual wells, almost one-third were found to have bacterial contamination of their drinking water. Surprisingly, high levels of cadmium, mercury, lead, and selenium were also found.

Problems do still exist, then, but the focus has changed somewhat. Bacterial contamination is still a problem. (Twelve states report that a portion of their drinking-water supplies still do not meet standards for bacteriological quality. Some improvement in this area has been noted in the case of supplies that serve large numbers of households.) However, contamination of groundwater drinking supplies with organic chemicals and heavy metals from industrial and municipal wastes is definitely a new focus of concern.

Table 11.4 summarizes one final drinking water issue—fluoridation. Fluoride is a trace chemical added to water to help prevent tooth decay. Although the American Dental Association is in favor of adding fluoride to drinking-water supplies, many people still feel that fluoride may be harmful. We may, in fact, be reaching optimum levels of fluoridation in the U.S. because fluoridated water use contributes fluoride to canned drinks and other food items. See the Further Reading section for more on this.

The drinking water standards set by the Environmental Protection Agency are discussed in the supplemental material at the end of this chapter.

Table 11.4 Percentage of State Populations Using Fluoridated Water (1975)

State or district	Percent of population using fluoridated water	State or district	Percent of population using fluoridated water
Alabama	31.5	Missouri	42.0
Alaska	42.7	Montana	26.7
Arizona	31.2	Nebraska	45.8
Arkansas	38.2	Nevada	3.0
California	22.0	New Hampshire	13.3
Colorado	83.8	New Jersey	21.4
Connecticut	79.4	New Mexico	63.8
Delaware	39.5	New York	66.1
District of		North Carolina	45.4
Columbia	98.4	North Dakota	50.8
Florida	35.8	Ohio	41.4
Georgia	41.4	Oklahoma	63.3
Hawaii	6.4	Oregon	10.7
Idaho	33.5	Pennsylvania	46.2
Illinois	86.2	Rhode Island	66.3
Indiana	61.2	South Carolina	52.8
Iowa	61.9	South Dakota	61.5
Kansas	51.5	Tennessee	66.9
Kentucky	51.0	Texas	58.7
Louisiana	23.2	Utah	2.4
Maine	40.6	Vermont	37.0
Maryland	68.1	Virginia	51.2
Massachusetts	21.6	Washington	39.8
Michigan	76.1	West Virginia	50.7
Minnesota	71.5	Wisconsin	62.2
Mississippi	24.7	Wyoming	21.3

(Source: Morbidity and Mortality Weekly Report, U.S. Department of Health, Education, and Welfare, 1977)

🐚 National Interim Primary Drinking-Water Standards

How Standards Are Set

Standards dealing with microbiological contamination of drinking water are fairly easy to set. We have methods of detecting such contamination and know approximately what levels of contamination cause outbreaks of disease. Setting standards for chemicals in drinking water is not an easy task. In some cases, we do not have sufficient evidence about a chemical to know if there are safe levels. For instance, a number of possibly carcinogenic chlorine-containing organic chemicals can be found in water. Their long-term effects at low levels are unknown.

In other cases, chemicals may be harmless, or even necessary, to life in trace amounts, but toxic in larger quantities. This is true of copper, for example, which is an essential element in trace amounts in human and animal diets, but in larger amounts, has been responsible for epidemics of sheep poisoning in the Netherlands. Large amounts of fluoride are poisonous; however, fluoride is added to water in trace amounts to reduce tooth decay.

Furthermore, not all chemicals can be detected by a simple, inexpensive, and reliable test. The chlorine-containing organic chemicals fall into this category. In addition, standards are meant to ensure that drinking water is safe when it leaves the water treatment plant. However, water may pick up metal contaminants from the pipes in which it is transported. Soft waters with low pH (relatively acidic) are especially corrosive. Studies have shown that, depending upon the type of pipes it travels through, water can pick up cadmium, chromium, cobalt, copper, iron, lead, manganese, nickel, silver, and zinc between the time it leaves the treatment plant and the time it reaches the consumer.

Despite these problems, in order to protect consumers, the Environmental Protection Agency has set standards limiting the levels of certain chemicals in drinking water (Table 11.5). With respect to organic chemicals, standards have been set only for six pesticides and for total trihalomethanes.[6] EPA has published "suggested no adverse response level" (SNARL) documents for a variety of other organic compounds. Even though these are not legal standards, some states and communities have adopted them as levels above which they will close wells.

The standards set by the EPA are interim standards; that is, they will be revised as new information becomes available. The EPA has also begun research in areas where information is needed.

---

6 Trihalomethanes are chlorinated organic chemicals formed during one of the processes of water purification, chlorination (see Chapter 10). The standard, 100 parts per billion, applies only to drinking-water supplies serving communities of 10,000 or more people.

🐚 Why Nitrites Are Poisonous

Nitrite (NO_2^-) is an oxidizing agent, and can change the iron atom in the blood protein hemoglobin from Fe^{2+} to Fe^{3+}. The changed hemoglobin is called methemoglobin. Hemoglobin gives blood its red color and carries oxygen from the lungs to all other parts of the body. Hemoglobin can only function if the iron atoms tucked into the center of the heme groups are Fe^{2+} rather than Fe^{3+}. If enough hemoglobin is changed from hemoglobin (Fe^{2+}) to methemoglobin (Fe^{3+}), the victim is said to have methemoglobinemia and may suffocate.

Table 11.5 National Interim Primary Drinking-Water Standards

Contaminant	Maximum Allowable Level (milligrams per liter or ppm)	Comments
Inorganic Chemicals		
Arsenic	0.05	
Barium	1.00	Mainly a problem in agriculture—barium is toxic to plants.
Cadmium	0.010	
Chromium	0.05	
Lead	0.05	See Chapter 21 for a discussion of lead.
Mercury	0.002	No standard had been set previously.
Nitrate	10.00	This standard is reduced from 45 mg per liter.
Selenium	0.01	
Silver	0.05	
Fluorides	1.4–2.4	Depending on the average daily maximum air temperature.
Organic Chemicals[a]		
Chlorinated hydrocarbon pesticides		
Endrin	0.0002	
Lindane	0.004	
Methoxychlor	0.1	
Toxaphene	0.05	
Chlorophenoxyl pesticides		
2,4-D	0.1	
2,4,5-TD	0.01	
Total Trihalomethanes	0.1	Applies only to community drinking-water systems serving more than 10,000 people
Turbidity		The presence of particles in the water, or turbidity, is both esthetically undesirable and also can interfere with disinfection, since particles can combine with chlorine and use it up. The standard is set at a maximum of one turbidity unit.[b]
Microbiological contaminants		No more than four coliform bacteria per 100 ml may be found when the water is sampled. (See Chapter 10 for a discussion of this standard and what it means.)
Radioactivity		
Naturally occuring		
Radium 226 and 228	5 pCi/l[c]	
Gross alpha-particle activity	15 pCi/l[c]	
Man-made		
Average annual dose	4 millirem/year or less[d]	

a Although the Safe Drinking Water Act was passed in large part due to public concern about organic chemicals like chloroform in drinking water, it was finally decided that no good, simple methods are available to test for them. For this reason, no standard was set for many organic chemicals. Rather, the EPA will do expensive and sophisticated tests on samples from certain water supplies. At the same time, research is being carried out to find a simple test for the chemicals of most concern.

b For an explanation of this term, see Standard Methods for the Examination of Water and Waste Water, American Public Health Association, American Water Works Association and the Water Pollution Control Federation, Thirteenth Edition. Available from the American Public Health Association, 1790 Broadway, New York, NY 10019.

c pCi/l = picocurie per liter, a standard unit used to measure radioactivity.

d A millirem is a measure of the dose of radioactivity received by a human, weighted to account for the biological damage that a particular form of radioactivity can do to various organs (see Chapter 30).

Questions

1. Explain why the mercury concentration found in the waters of Minimata Bay was much lower than those found in the fish and shellfish of the bay. What does this imply for people living in that region?

2. Suppose you are a public health official in charge of deciding whether to allow a manufacturer to sell a new chemical, which is supposed to kill nuisance plants in lakes. List three questions you would want answered before you decide whether the chemical seemed safe to add to the water environment.

3. What does it mean to recycle mercury? Name some uses of mercury in which it recycles. What are some uses of mercury in which it does not recycle? Why can't the mercury be recycled?

4. Explain how the cadmium concentration of municipal drinking water can depend, in part, upon the characteristics of the water itself.

5. In order to discover how much of a particular pollutant people are exposed to, all possible sources of exposure must be examined. List all the ways you can think of that people in Fuchu were exposed to cadmium once it contaminated their water supply.

6. If a septic system is located too close to a well or uphill from a well, the wastes from the septic system may contaminate the well. Even if no disease-causing bacteria or viruses occur in the contaminated well waters, problems may still occur if a young infant drinks the water. Explain. Why are high nitrate levels in drinking water only a problem with respect to young infants?

7. What are the main sources of sodium in drinking water? Why might high concentrations of sodium be a problem?

8. PCBs have been found in water all over the country, but only in trace amounts. Nowhere have sampling studies found enough PCBs to be immediately poisonous to humans. Give three reasons why PCBs at these low levels are still a serious problem.

9. Should General Electric be legally responsible for cleaning PCBs from the Hudson River, in addition to paying fines for exceeding allowed discharge levels? In a broader sense, should industry be responsible for cleaning up environmental pollution caused by illegal discharges? by accidents? by employee negligence? by "acts of God"? (Oil companies are legally obligated to clean up oil spills that occur during drilling for oil in waters off the coast *whatever the cause of the spill*.) Is this an acceptable burden on a particular industry? Are there other ways to achieve the clean-up?

10. People who fish on the Hudson River have effectively been deprived of their jobs because G.E. dumped PCBs into the river. Should they be recompensed? Who should help them and why?

11. PCBs are found in almost every fluorescent light fixture, as well as in some 895 million capacitors scattered all over the country. (It has been estimated that 250 million pounds of PCBs are in capacitors now in service.) If special plans are not made and followed for the disposal of these items, the PCBs they contain will almost certainly find their way into dumps, landfills, or incinerators and then into the nation's waters.
    a. Diagram how the PCBs in a fluorescent light fixture might end up in a large lake.
    b. The EPA is looking for ideas on how to interrupt this flow of PCBs into the environment. High-temperature incineration destroys PCBs. Can you think of ways to encourage people to dispose of PCB-containing items appropriately? (*Note:* Because of the smallness of the items and their wide distribution, a simple law requiring proper disposal probably would not be very effective.)

12. Compared to chemicals like DDT, which is both very poisonous and a cancer-causing substance, phthalates are not very toxic. Why, then, should we worry if they become widespread in the environment?

13. Trace a possible route of the phthalates in a piece of food wrap entering the ocean environment.

14. Although at one time enough Kepone or other toxic material was in the Life Science plant effluent to actually overcome a worker at the Hopewell sewage treatment plant, the levels of Mirex that leached or were sprayed into local ponds and rivers during the fire ant program were generally low, not enough to kill fish directly. Why then was the Mirex still a problem in water? (You will probably think of more than one reason.)

15. Briefly diagram three routes by which Mirex could have found its way into the fatty tissues of Mississippians.

16. How could wives and children of workers at the Life Science plant have had Kepone in their blood? (Assume they didn't visit their husbands or fathers at the plant.)

17. What is the source of your drinking water? Is it a protected source?

18. Contrast groundwater versus surface water as a source for drinking water. What problems are each subject to?

19. Not all trace minerals are undesirable. Some give water the flavor to which we are accustomed. Try the taste of distilled water, which has none of these minerals: catch the steam from a teakettle on a piece of aluminum foil (be careful—steam burns are painful) and let the drops run into a glass. Compare this to a glass of water straight from the tap.

Further Reading

Evans, R. J., J. D. Bails, and F. M. D'Itu, "Mercury Levels in Muscle Tissues of Preserved Museum Fish," *Environmental Science and Technology,* **6** (10), 901 (1972).
One of the unusual aspects of environmental mercury contamination is that fish near the top of saltwater food chains, like tuna or swordfish, seem to have dangerous levels of mercury even in areas where no human sources are found. This paper gives an introduction to the literature on the subject of mercury in food fish.

Zim, M. H., "Allied Chemical's $20-Million Ordeal with Kepone," *Fortune* (11 September 1978), p. 82.
It is fascinating to read Allied's view of the Kepone incident. The effect of this disaster on corporate policy has been far-reaching.

Safe Drinking Water Act, Report from the Committee on Interstate and Foreign Commerce, House of Representatives, 93rd Congress, Second Session, #931185, 1974.
The report summarizes and explains the provisions of the Safe Drinking Water Act. In addition, it provides some background about why the act was needed.

Leverett, Dennis H., "Fluorides and the Changing Prevalence of Dental Caries," *Science,* **217** (2 July 1982), p. 26.
Even health professionals strongly in favor of fluoridation feel that fluoride levels in water supplies must be watched carefully. This paper gives many background references on fluoridation and also discusses what we should do now that fluoridation seems to have proven a successful method of prevention for tooth decay.

References

Friberg, L., et al., *Cadmium in the Environment II, 1973;* prepared for the U.S. Environmental Protection Agency; available from National Technical Information Service, Report #EPA-R2-73-190.

Wallace, R. A., W. Fulkerson, W. Shults, and W. Lyon, *Mercury in the Environment: The Human Element,* Oak Ridge National Laboratory Report ORNL-NSF-EP-1, Oak Ridge, Tennessee, 1971.

Kaiser, K. L., "The Rise and Fall of Mirex," *Environmental Science and Technology,* **12** (May 1978), p. 520.

Cutshall, N. J., et al., "Man-Made Radionuclides Confirm Rapid Burial of Kepone in James River Sediments," *Science,* **213** (24 July 1981), p. 440.

Shuval, H., and N. Gruener, "Epidemiological and Toxicological Aspects of Nitrates and Nitrites in the Environment," *American Journal of Public Health,* **62**(8) (1972), p. 1045.

Environmental Health Perspectives #1, NIH 72-218, U.S. Department of Health, Education and Welfare, April 1972. [PCBs]

Gaffey, W. J., *Epidemiology of PCB's,* Monsanto Co., 1981.

Environmental Health Perspectives, experimental issue #3, U.S. Department of Health, Education and Welfare, Public Health Service, National Institute of Health, January 1973. [Phthalates]

Giam, C. S., et al., "Phthalate Ester Plasticizers: A New Class of Marine Pollutant," *Science,* **199** (27 January 1978), p. 419.

Burmaster, D. E., and R. H. Harris, "Groundwater Contamination: An Emerging Threat," *Technology Review,* July 1982, p. 51.

Lauch, R. P., and T. J. Sorg, "The Occurrence and Reduction of Sodium in Drinking Water," *Journal of the American Water Works Association,* May 1981, p. 256.

States Compliance Lacking in Meeting Safe Drinking Water Regulations, Report to the Administrator of EPA by the U.S. General Accounting Office, March 3, 1982.

Drinking Water and Health, National Research Council, National Academy Press. Volume I (1977): Health Effects of Microbiological, Radioactive, Particulate, Inorganic and Organic Contaminants; Volume II (1980): Chlorination and Alternative Means of Disinfection; Volume III (1980): Health Effects of Trihalomethanes and Also Relationship between Water Hardness and Cardiovascular Disease; Volume IV (1982): Chemical and Biological Problems Related to Water Distribution Systems and Also Health Effects of Additional Inorganic and Organic Contaminants.

CHAPTER TWELVE

Organic Wastes and Dissolved Oxygen

Why Organic Wastes Are Pollutants
A Simple Experiment/Oxygen in Natural Waters

Stream Health and Dissolved Oxygen
What Happens at a Sewage Discharge/Clean-Water Zone/Zone of Decline/Damage Zone/Zone of Recovery

Organic Wastes: How to State Their Levels and How to Measure Them

Why Organic Wastes Are Pollutants

Water pollution means different things to different people. To some, it means the presence of toxic chemicals; to others, the presence of disease-causing bacteria. To still others, pollution is seen in floating weeds and algae. To many engineers and scientists, however, water pollution is indicated by organic wastes. Such wastes can come from industry, from farming, and from cities. These organic wastes are composed mainly of carbon, hydrogen, oxygen, and nitrogen. Oxidation of the carbon, hydrogen, and nitrogen in these wastes is responsible for many of the unpleasant conditions that exist in polluted rivers and lakes.

A Simple Experiment

To understand what organic water pollution is, consider the following experiment. If we mix and dissolve sugar and gelatin in an open bottle of water, the resulting solution is clear. The solution does not appear to us to be "polluted." A fish should be able to live in this water (after any chlorine has been allowed to diffuse out), at least for a time. Yet, the contents of the bottle become clouded in less than a week. Dissolved oxygen, which fish need to live, is used up, and the fish dies from lack of oxygen.

The organic wastes, in this case, were sugar and gelatin. Together, they caused a condition in which fish and many other water creatures cannot live. That condition is a lack of oxygen, resulting from the breakdown of the organic materials (sugar and gelatin) by bacteria. In the presense of oxygen, these materials are broken down to simpler substances by bacteria, a process known as biochemical oxidation. If enough species of bacteria are present, the carbon (C) in the sugar and gelatin is oxidized to carbon dioxide (CO_2). Furthermore, the hydrogen (H) in the sugar and gelatin is oxidized to water (H_2O). Last, the nitrogen (N) in the gelatin is eventually converted to nitrate ions (NO_3^-). All of these reactions utilize oxygen. The net effect is to reduce the amount of oxygen dissolved in the water. (For measuring the potential for oxygen removal of organic wastes, see p. 243.)

Oxygen in Natural Waters

When water is well mixed in the presence of air, the water dissolves a maximum amount of oxygen, called the *saturation concentration*. This is the number of milligrams of oxygen that can be dissolved in one liter of water before the water can hold no more. The saturation concentration achieved by thorough mixing is between 8 and 9 milligrams of oxygen per liter of water, depending on the temperature. In summer, at higher water temperatures, the saturation concentration falls as low as 8 milligrams per liter. In winter, at colder water temperatures, the saturation concentration is nearer 9 milligrams per liter.

If the concentration of organic material in a sample of water is sufficiently large, oxidation of the organics by bacteria and protozoa may use up all the oxygen in the sample. That is, the concentration of oxygen in water may be reduced to zero. When there is no oxygen in natural waters, we say the water is in an **anaerobic condition.** Under anaerobic or very low oxygen conditions, normal aquatic life, such as fish species, die away. The few species that do survive and thrive in any numbers are those especially adapted to these low-oxygen conditions.

In low-oxygen conditions, the normal bacterial population that uses oxygen also dies away. In its place, a population of bacteria grows up that lives on sulfur instead. The sulfur atom occurs in organic wastes and is structurally similar to oxygen, except that it has one more ring of electrons. Sulfur then takes the place of oxygen in the "oxidation" reaction, producing, for instance, hydrogen sulfide (H_2S) instead of water (H_2O). The well-known smell of rotten eggs is the smell of hydrogen sulfide.

To summarize, oxidation of organic wastes by bacteria leads to a loss of oxygen from water. An environment in which oxygen is absent possesses no fish and only a few specially adapted animal species. In this anaerobic environment, specialized bacteria produce, among other end products, the foul-smelling gas hydrogen sulfide.

Stream Health and Dissolved Oxygen

What Happens at a Sewage Discharge

Now that we know how organic pollution can alter the dissolved oxygen concentration in a body of water, we can investigate how aquatic life responds to such pollution. When organic substances from the waste outfall of a community or industry enter a river or stream, the concentration of dissolved oxygen in the stream decreases. This is due to the oxidation of the organic material by bacteria and protozoa. Natural mixing of the stream with the air does tend to replace the removed oxygen, but not immediately. Instead, a

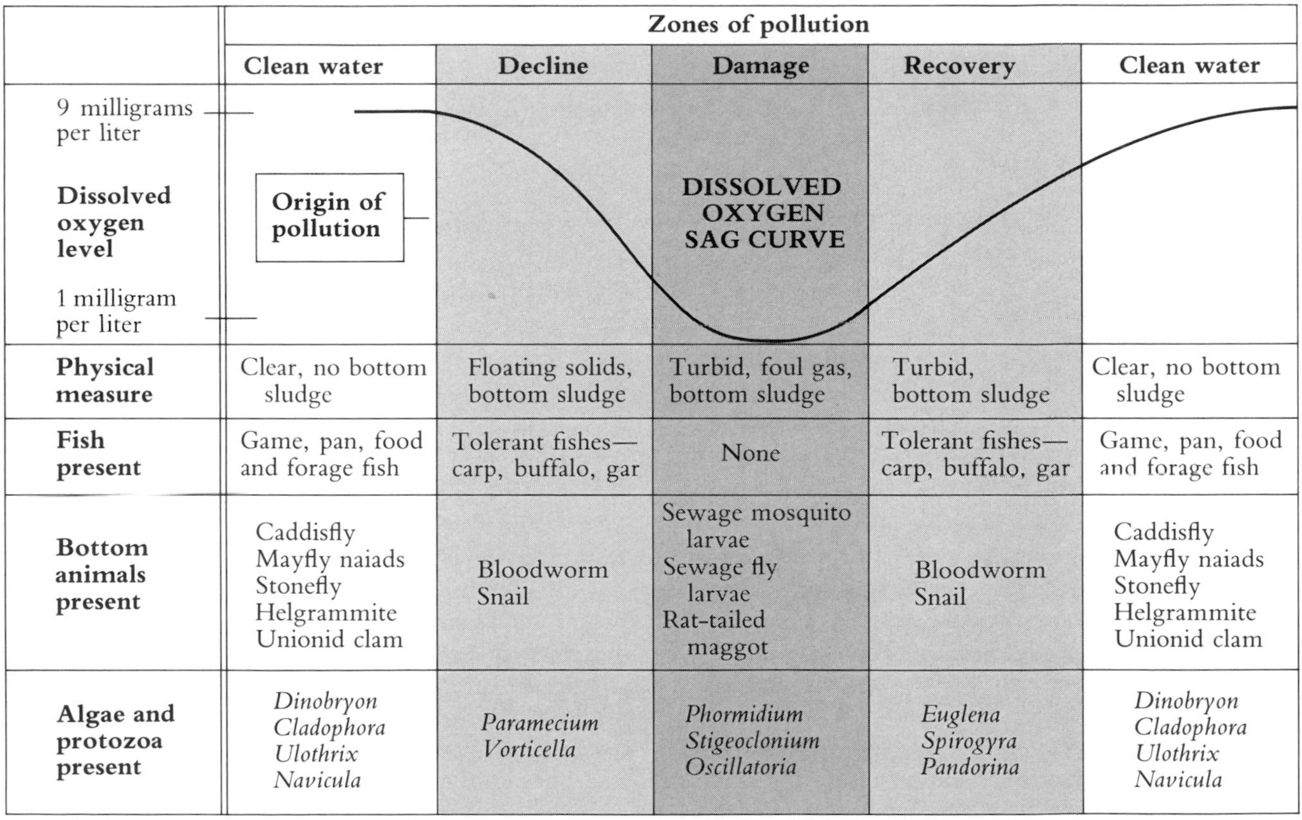

	Zones of pollution				
	Clean water	**Decline**	**Damage**	**Recovery**	**Clean water**
Dissolved oxygen level (9 milligrams per liter to 1 milligram per liter)	Origin of pollution		DISSOLVED OXYGEN SAG CURVE		
Physical measure	Clear, no bottom sludge	Floating solids, bottom sludge	Turbid, foul gas, bottom sludge	Turbid, bottom sludge	Clear, no bottom sludge
Fish present	Game, pan, food and forage fish	Tolerant fishes—carp, buffalo, gar	None	Tolerant fishes—carp, buffalo, gar	Game, pan, food and forage fish
Bottom animals present	Caddisfly Mayfly naiads Stonefly Helgrammite Unionid clam	Bloodworm Snail	Sewage mosquito larvae Sewage fly larvae Rat-tailed maggot	Bloodworm Snail	Caddisfly Mayfly naiads Stonefly Helgrammite Unionid clam
Algae and protozoa present	*Dinobryon Cladophora Ulothrix Navicula*	*Paramecium Vorticella*	*Phormidium Stigeoclonium Oscillatoria*	*Euglena Spirogyra Pandorina*	*Dinobryon Cladophora Ulothrix Navicula*

Figure 12.1 The zones in a polluted water course. (Adapted from K. Mackenthun, *Toward a Cleaner Aquatic Environment*, U.S. Government Printing Office, Washington, D.C., 1973)

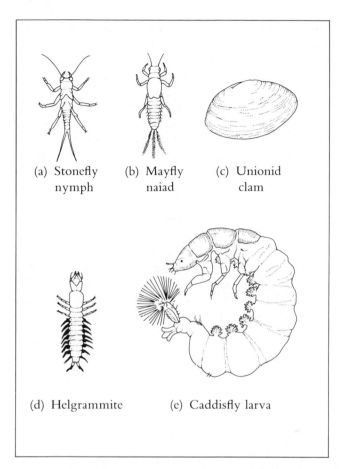

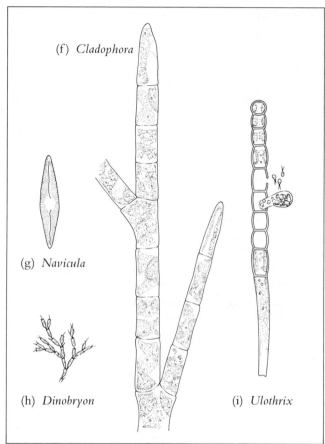

Figure 12.2 (a)–(e) Clean-water animals associated with streambed; (f)–(i) clean-water (sensitive) algae.

competition arises between the oxygen-depleting forces (the oxidation of organic wastes) and the oxygen-restoring forces (the mixing of water with air). This competition produces a typical picture of oxygen concentrations along the stream. The profile is shown in the top of Figure 12.1.

The four basic sections of the stream are:

1. a clean-water zone (high dissolved oxygen) upstream from the pollution source;

2. a zone of decline (falling level of dissolved oxygen);

3. a zone of damage (relatively constant and low level of dissolved oxygen);

4. a zone of recovery (rising dissolved oxygen).

Of course, if more than one sewage outfall occurs along the stream or river, the zone of damage may extend for many miles. Such conditions may occur along rivers in major urban and industrial areas.

Clean-Water Zone

The clean-water zone upstream from the waste discharge may sustain fish, mayflies, clams, stoneflies, and other species (Figure 12.2). These species require oxygen in water in order to survive. If the oxygen concentration falls, these sensitive species are among the first to disappear. Such fish as trout, bass, salmon, and minnows are among these sensitive species. While the needs of various fish species vary, biologists generally believe that dissolved oxygen levels of 5 milligrams per liter or better are necessary to ensure survival of fish.

Zone of Decline

A zone of decline follows the introduction of organic wastes. Species that can survive at these somewhat lower levels of dissolved oxygen are referred to as "intermediately tolerant." They are pictured in Figure 12.3. Solids from the wastewater outfall may cloud the water.

Damage Zone

The damage zone, which follows the zone of decline, is one in which dissolved oxygen is almost gone. When dissolved oxygen has fallen to very low levels, only a few species are capable of survival, and the numerous species that characterized the clean stream have disappeared. In their place arise a group of organisms referred to as "pollution tolerant" because of their ability to survive under conditions of low dissolved oxygen. One such organism is the sludgeworm, which consumes sludge and thrives in water with as little as $\frac{1}{2}$ milligram of oxygen per liter. Another is the rat-tailed maggot, a resident in the sludge of stream bottoms; the maggot breathes by a long tube that reaches to the water surface. The maggot is the larva of the drone fly. Its numbers may so increase that they cover the bottom of the streambed in a waving red sheet. Another resident of this zone is the bloodworm. This organism, like the sludgeworm, consumes sludge as its diet. These species are illustrated in Figure 12.4.

In the clean-water zone, many species exist side by side, and each is moderately represented in terms of numbers of organisms. In the damage zone, only a very few species survive, but these can be present in enormous numbers. If, in the damage zone, one fails to find large numbers of organisms of the pollution-tolerant species, it is likely that some chemical poison in the sewage may be preventing their increase.

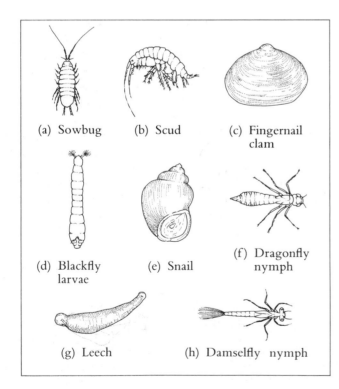

(a) Sowbug (b) Scud (c) Fingernail clam

(d) Blackfly larvae (e) Snail (f) Dragonfly nymph

(g) Leech (h) Damselfly nymph

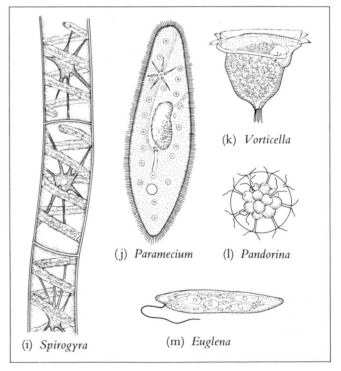

(k) *Vorticella*

(j) *Paramecium* (l) *Pandorina*

(i) *Spirogyra* (m) *Euglena*

Figure 12.3 Intermediately tolerant aquatic species.
(a)–(h) Intermediately tolerant animals associated with streambed;
(i)–(m) intermediately tolerant algae and protozoa.

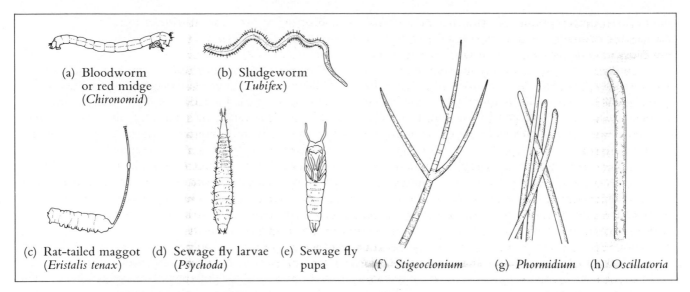

(a) Bloodworm or red midge (*Chironomid*)

(b) Sludgeworm (*Tubifex*)

(c) Rat-tailed maggot (*Eristalis tenax*)

(d) Sewage fly larvae (*Psychoda*)

(e) Sewage fly pupa

(f) *Stigeoclonium*

(g) *Phormidium*

(h) *Oscillatoria*

Figure 12.4 Pollution-tolerant aquatic species.
(a)–(e) Pollution-tolerant animals associated with the streambed;
(f)–(h) pollution-tolerant algae.

Zone of Recovery

A zone of recovery follows the damage zone. Here the water is likely to be clear, allowing sunlight to penetrate. The oxygen is on its way up to more reasonable concentrations. With the clearing of the water and recovery of dissolved oxygen, algae may begin to grow. Their presence can result in fluctuation of the oxygen content in the water. During daylight hours, the algae produce oxygen as a by-product of photosynthesis. But at night, respiration and decomposition of algae remove oxygen from the water, yielding a low oxygen concentration. These algae-caused swings in dissolved oxygen may even deplete the stream's oxygen to such an extent that normal aquatic life, which requires oxygen, may not be re-established. Beyond the zone of recovery, the species of the clean-water zone are found again.

Organic Wastes: How to State Their Levels and How to Measure Them

It may have occurred to you to ask how we determine the amount of organic wastes in a stream. This knowledge could help us predict the amount of oxygen that would be used up in oxidizing the organic substances. It would also help us predict what species will be present in the water.

Unfortunately, to measure the quantity of every organic substance in a waste stream is an exhausting and perhaps impossible job. Each substance present would first have to be identified, already an enormous task. Next, once a substance is shown to be present, the quantity of that substance has to be measured. Then, if we also know the oxygen consumption when bacteria oxidize one gram of each substance present, we could predict how much oxygen would be removed by oxidation of all the wastes in a liter of the polluted water.

A much easier way to measure the concentration of organic wastes in water was proposed in England near the turn of the century. The method determines not only the concentration of organics, but also reveals the amount of oxygen that all the organic wastes, acting together, will eventually remove from the water. The number reported is the **Biochemical Oxygen Demand (BOD).** This is the amount of oxygen that would be consumed if all the organics in one liter of polluted water were oxidized by bacteria and protozoa. It is reported in number of milligrams of oxygen per liter.

Suppose we are studying a sample of polluted water discharged as a waste stream from a community. The sample is reported to have a BOD of 120 mg per liter. This means that bacteria and protozoa oxidizing all the organic substances in one liter of the water would consume 120 mg of oxygen. Now suppose that 50 milliliters (0.05 liter) of the polluted water were mixed with 950 ml of clean water. The sample has been diluted to 50 parts in 1000 parts, or to 1/20th of the mixture. The BOD of the mixture will be (1/20) × 120 or 6 mg per liter.

This dilution is similar to what happens when the flow of sewage from a community enters a watercourse that has a substantial water flow. Suppose the BOD in the discharge is the 120 mg per liter we were discussing. Assume further that the community empties 1 million gallons of waste a day into the stream and that the clean stream is flowing at 19 million gallons per day past the outfall. The BOD of the mixture is diluted to 1/20th of the discharge value, or 6 mg/liter. This number, 6 mg/liter, is the amount of oxygen that can be removed from every liter in the combined flow of stream and wastewater. Given that clean water can only possess up to about 9 mg of oxygen per liter, the oxygen level of the mixture can be reduced to a very low level. How low it will actually go depends on many factors, but especially on the rate of oxidation by bacteria and on the natural rate of restoration of oxygen to the stream.

Thus, the BOD value tells a biologist the capability of the polluted water to deplete the oxygen resources in the stream. This is a valuable indicator of pollution, since the lack of oxygen is responsible for fish kills and also causes foul odors and populations of undesirable pollution-tolerant organisms. The BOD does not explain what organic substances are present in the water nor in what quantity they are present. Nonetheless, it provides a rapid and important insight into the maximum pollution damage the wastewater could cause.

Measuring the Concentration of Organic Wastes

The specific procedure for determining the value of Biochemical Oxygen Demand (BOD) involves a number of steps. First, a carefully measured sample of the polluted water is diluted in a much larger, measured volume of unpolluted water. A 300-ml bottle is specially made for this test. The unpolluted water has previously been well shaken in the presence of air so that the water has absorbed as much oxygen as it can. This water is saturated with oxygen. The water may also have been "seeded" with microorganisms known to break down organic wastes in the presence of oxygen. If the BOD of polluted *river* water is to be measured, however, the needed microorganisms are likely to be present in the polluted water sample already. The mixture of polluted water and clean water is poured into the bottle, completely filling it. The bottle is then closed to the air by insertion of a glass stopper. This closure prevents any new oxygen in the air from entering the bottle. Two bottles are commonly prepared in this way from samples of the same polluted water.

One of the bottles is set aside in the dark at 20°C and taken out at the end of five full days. The darkness is necessary to prevent the growth of algae. Algae may contribute oxygen to the water as a by-product of photosynthesis. Such oxygen would interfere with the measurement of BOD and so is avoided. One bottle is not set aside, but is examined immediately to determine the quality of oxygen dissolved in it (Figure 12.5).

Several methods exist to measure dissolved oxygen, including a chemical test and a test done by electrodes. The chemical test requires special solutions and procedures. The electrode test only requires the insertion of a probe (electrode) into the water. However, one must first ascertain that the electrode is calibrated—that is, it reads the appropriate value—and this requires prior testing.

Suppose the quantity of dissolved oxygen in the bottle not set aside but measured immediately was 7.5 mg. Suppose further that after five days, the quantity of oxygen in the bottle stored in the dark was 6.0 mg. No oxygen has been added to the second bottle from any other source. Therefore, the difference between the initial quantity of oxygen and the quantity present after five days must be the oxygen removed by microorganisms, 1.5 mg.

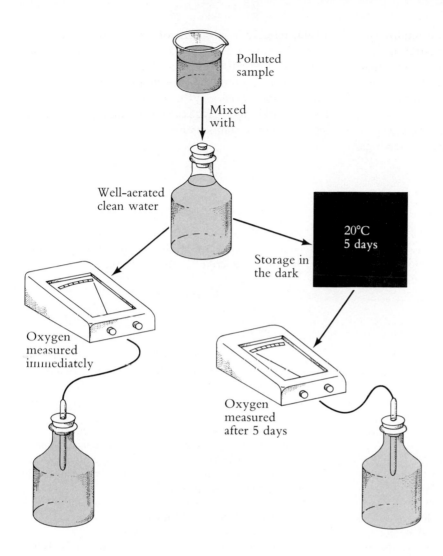

Polluted
sample

Mixed
with

Well-aerated
clean water

20°C
5 days

Storage in
the dark

Oxygen
measured
immediately

Oxygen
measured
after 5 days

Figure 12.5 Measuring the concentration of organic wastes (Biochemical Oxygen Demand).

Suppose a 10-ml sample of polluted water was diluted initially. Since 10 ml of this polluted water caused the removal of 1.5 mg of oxygen, 150 mg of oxygen could be removed by a 1-liter sample (1000 ml) of the polluted water. The BOD (Biochemical Oxygen Demand) of the polluted water is 150 mg per liter.

By the end of five days, the rate of oxygen removal has become very slow. Only a little more oxygen will be removed each day after that. The five-day BOD then is the number commonly used to measure the oxygen-consuming ability of polluted water.

Chemical Oxygen Demand (COD)

The BOD is an important measure of organic water pollution because it tells us how much oxygen will ultimately be removed from the water by biological oxidation of the wastes. Unfortunately, it takes five days to get such a reading, and sometimes an estimate is needed more quickly.

To meet the need for a quick measurement of organic waste concentration, water scientists have developed a chemical test that does not rely on microorganisms to do the oxidation. Instead, the test uses potassium dichromate and sulfuric acid to oxidize the organic material. Nearly all the organic matter present is oxidized by this mixture, even organic material that microorganisms could not oxidize. Thus, the test generally reports a greater concentration of oxidizable organics than the BOD test does.

Test results are reported as the Chemical Oxygen Demand (COD). The units of COD are the same as BOD, namely, milligrams of oxygen per liter. As an example, suppose we had a sample of polluted water with a COD of 100 mg/liter. The organics in one liter of this water would be expected to remove a bit less than 100 mg of oxygen from a stream. The removal is less than 100 mg because the COD test oxidizes organics that microorganisms cannot.

Questions

1. When organic wastes are added to natural waters, they may cause fish to die, even though the wastes themselves are not directly poisonous to fish. How do you explain this?

2. Describe how sewage changes the stream communities found downstream from the outfall. What are the zones that can be described in such a stream?

Further Reading

Hart, C., Jr., and S. Fuller, *Pollution Ecology of Freshwater Invertebrates*. New York; Academic Press, 1974.
A research volume with papers by a number of investigators, each dealing with a particular group of invertebrates. Includes methods of using organisms as indicators of water pollution.

Kemp, L., W. Ingram, and K. Mackenthun, eds., *Biology of Water Pollution*, Federal Water Pollution Control Administration,* U.S. Department of Interior, 1967.
This is a collection of biologically oriented papers from journals treating stream pollution, biological waste treatment, and public health as influenced by stream pollution. The classic paper "Stream Life and the Pollution Environment" by Bartsch and Ingram is reprinted here. This article alone makes a copy worth obtaining from the library.

Mackenthun, K., *Toward a Cleaner Aquatic Environment*, U.S. Environmental Protection Agency, 1973, available from U.S. Government Printing Office, Stock No. 5501-00573.
This is virtually a textbook on the biology of water pollution. It is interesting, well-written, well-illustrated, and is rarely too technical for the novice to follow. Directions for identifying algae associated with water pollution are provided.

Mackenthun, K., *The Practice of Water Pollution Biology*, Federal Water Pollution Control Administration,* U.S. Government Printing Office, 1969.
Another of Mackenthun's well-written and well-illustrated texts.

---

* The organizations marked by an asterisk are not separate government entities, but evolved through administrative change to become part of the Environmental Protection Agency.

CHAPTER THIRTEEN

Eutrophication

Feeding Lakes

How Water Becomes Eutrophic
Natural versus Artificial Eutrophication/Limiting Nutrients/Nitrates and Eutrophication/Phosphates and Eutrophication

How Can Eutrophication Be Controlled?
The Detergent Phosphate Problem/Sewage Treatment as a Solution/Controlling Other Phosphate Sources/Helping Eutrophic Lakes Recover

Lake Erie: A Case History

Other Problems with Detergents: Foaming Waters and Biodegradability

CONTROVERSY:

13.1: *The Scientist and Social Responsibility: Optical Brighteners*

Feeding Lakes

In 1974, Russell Train, Director of the U.S. Environmental Protection Agency, warned a meeting of the New Hampshire Lakes Region Clean Water Association that nutrients that cause lakes to age rapidly are increasing in U.S. waters. What did he mean by this? How can a lake age? What nutrients are overfeeding lakes?

Carbon, oxygen, hydrogen, nitrogen in nitrates, and phosphorus in the form of phosphates are some of the elements, or "foods," needed by plants for growth. Smaller amounts of many other substances, such as iron, calcium, and copper, are also required.

Lakes with large amounts of these necessary plant nutrients are called **eutrophic** (from Greek *eu,* well, and *trophé,* nourishment). Relatively high concentrations of plant nutrients in eutrophic lakes allow huge amounts of water plants, such as algae, to grow. This over-population of algae makes a lake unpleasant to swim in. Some kinds of algae, which grow in long strands, wind themselves around boat propellers, making boating impossible. The water in such a lake tends to be scummy, cloudy, or even soupy green. A rapidly growing population of algae is called a bloom. The algae in a bloom may wash onto the lake shores in storms and high winds, and die and decay there, producing bad smells. Sport fish have trouble living in a lake with blooms because the algae use up most or all of the oxygen in the lake at night, leaving none for fish. Such a lake is clearly not the kind people enjoy using for recreation or drinking water. Cities or industries that want to use eutrophic lake water find they must filter it first to remove the algae, thus increasing costs.

An **oligotrophic** lake (from Greek *oligo,* poorly, and *trophé,* nourishment) is one with low levels of plant nutrients. Such lakes have only small populations of algae, and remain clear and sparkling. They are fun to swim in or boat on, delightful to look at, and should provide good drinking water. Lake Tahoe in Nevada is an example. (See also Figure 13.3.)

How Water Becomes Eutrophic

Natural versus Artificial Eutrophication

Many lakes are oligotrophic when first formed. However, over the centuries, silt, debris, and dissolved nutrients accumulate in these lakes, filling in their originally deep basins and turning the lake waters into a nutrient-rich soup. In this way, after a few thousand years, a lake can change naturally from an oligotrophic lake to a eutrophic lake. This process is called aging.

Not all lakes age naturally. Some lakes, in the normal course of events, maintain their oligotrophic characteristics. However, any beautiful, clear lake can be made eutrophic if large amounts of plant foods are added. The most common way this happens is when city or industrial sewage is dumped into lakes. The sewage from a city or town is rich in nutrients, such as nitrates and phosphates, that plants need for growth.

Limiting Nutrients

Some controversy exists about which plant nutrients are most responsible for causing the premature aging of U.S. lakes, rivers, and coastal waters. In general, experts agree that phosphates and nitrates are the most likely culprits. Although there are many other necessary nutrients, these two are noteworthy because, together or singly, they are usually the **limiting factors**

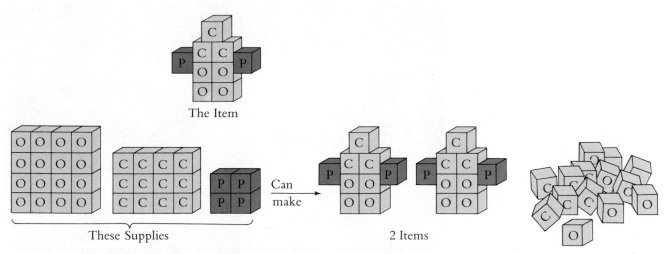

Figure 13.1 The Limiting Factor Concept. Suppose in order to make up an item you needed 4 blocks of O, 2 of P, and 3 of C. If you had 16 blocks of O, 4 of P, and 12 of C, you could only make 2 of the item because after that you would run out of P. In this case, P is the limiting factor. No amount of extra O or C blocks will make up for having no more P.

in natural, unpolluted waters. A limiting factor is the nutrient present in the smallest amount compared to the amount needed for growth (Figure 13.1). That is to say, water plants such as algae will grow until all available nitrogen or phosphorus is used up. Unnatural amounts of phosphorus or nitrogen can allow more than the normal amount of algae to grow.

Nitrates and Eutrophication

In most U.S. coastal waters and in lakes such as Tahoe, which are fed by steep mountain streams, a low concentration of nitrates seems to be the factor that normally limits plant growth. Nitrates can enter natural waters from several sources. Sewage from cities or animal feedlots are two major sources, since human and animal wastes are about one-half nitrates. Fertilizer, which runs off croplands or suburban lawns during a rainstorm, also contains a large amount of nitrates.

In addition, several natural sources of nitrates exist. Volcanic eruptions and lightning can change nitrogen gas in the air into nitrates. Just as legumes found in land ecosystems can "fix" atmospheric nitrogen into a form usable by other plants, some blue-green algae found in aquatic systems can turn nitrogen from the air into nitrates. The process is called nitrogen fixation. These algae, which are usually considered the worst nuisance plants, thus do not depend on an outside source of nitrate.[1] Finally, waterfowl, which feed on the shores of lakes and ponds and then fly over the water, are the source of what one wildlife expert termed "bombed in" nitrates.

Phosphates and Eutrophication

In contrast to nitrogen, phosphorus in the form of phosphates is found naturally in waters in only trace amounts. Most pollutant phosphorus comes from human activities. Phosphate mines pollute certain areas of the country, such as central Florida. Fertilizer runoff contains large amounts of phosphate. Domestic sewage is high in phosphates, too, with one-half coming from human wastes and the other half from detergents.

Animal feedlots, where cattle are grouped together in large pens to be "finished," or fattened for

---

1 There is some evidence that decreasing the amount of nitrates in water, without decreasing the amount of phosphate as well, may even favor the growth of the undesirable blue-green algae because they can fix nitrogen while other water plants cannot.

Figure 13.2 The role of phosphorus in lake eutrophication is dramatically illustrated in this photograph of Lake 226 in northwestern Ontario, Canada. The lake has two basins. Phosphorus, carbon, and nitrogen were added to the far basin (upper part of photo). An algal bloom covered the basin in two months. The lower basin was treated at the same time with the same amounts of nitrogen and carbon, but no phosphorus. This basin remained clear and sparkling. (Photograph courtesy of D. W. Schindler.)

market, are sources of both nitrates and phosphates. Although growth of algae in coastal waters and some wilderness lakes is limited by low concentrations of nitrates in the water, growth in most U.S. lakes and rivers is believed to be limited by the amount of phosphate present (Figure 13.2).

How Can Eutrophication Be Controlled?

The Detergent Phosphate Problem

Since phosphate is the nutrient that limits plant growth in most U.S. waters, and because detergents are a major source of phosphates, it is natural to ask whether we wouldn't be better off banning the use of phosphates in detergents in an effort to stop eutrophication.

Detergent composition. The chemical present in the largest amount in phosphate detergents is the phosphate builder. Dry detergents usually contain sodium tripolyphosphate, or STPP, as a builder. Liquid

detergents may contain sodium or potassium phosphates. Builders are necessary because water often contains calcium and magnesium ions. Water that has more than 75 milligrams per liter of calcium and magnesium (measured as calcium carbonate) is called hard water. The calcium and magnesium ions combine with soap to form a hard precipitate. This precipitate does not make suds and does not dissolve grease or dirt. In a similar fashion, calcium and magnesium ions can tie up the surfactant, or dirt-dissolving, molecules in detergents. Builders "complex" or tie up these calcium and magnesium ions. Without builders, manufacturers would have to include a great deal of relatively expensive surfactant in their detergents to be sure their product performed well in hard-water areas. In addition, builders help keep dirt from reattaching itself to clothes.

Phosphate bans. Many scientists feel that a complete ban on phosphates in detergents would remove about half of the phosphates in sewage. In a number of areas, a ban on or a reduction in the amount of phosphate allowed in detergents has been tried. One such locality is the area around Onondaga Lake in New York. For many years, the lake had been the dumping ground for sewage from the city of Syracuse. As a result, blooms of scum-forming algae were covering the lake every summer. In July, 1971, areas around Onondaga Lake passed laws limiting the amount of phosphates allowed in detergents. Two years later, the blooms were noticeably not as bad as in previous years. There is now more oxygen in the lake water than there was before.

This example brings up a serious problem, however. If manufacturers cannot put phosphates in detergents, what can they use instead? In some areas, water is naturally soft enough (that is, it has little or no calcium or magnesium) for soap to be used. Water can also be made soft by passing it through a water softener or ion exchanger. However, in most houses the water is hard enough so that some sort of detergent builder is necessary.

The choice of a chemical builder to replace phosphates must be made carefully. Just as people did not realize a few years ago that phosphates could harm the environment, we have no way of knowing whether a substitute for phosphates might not be harmful. If a single chemical were chosen, the possibility arises that we could be adding as much as two million pounds of some new material to the waters, every year.

At the present time, phosphate-free detergents contain:

1. Carboxymethyl cellulose, an organic carbon compound that may present a problem because it may not be biodegradable.[2]

2. Carbonates and silicates, in the form of sodium metasilicate and sodium carbonate. Both combine with calcium and magnesium ions to give insoluble precipitates. Large amounts of carbonates and silicates, however, can make phosphates more soluble. Vast amounts of phosphates are already captured in the muds of lake bottoms. Any substance that helps mud phosphates dissolve into lake water is not a good ingredient in a phosphate-free detergent.

3. Borates, in the form of sodium tetraborate decahydrate, or borax. However, waters having more than one part per million of borax are toxic to most plants.

---

2 "Biodegradable" is another term, along with "phosphate-free," that you often see on detergent labels. The meaning of biodegradable is explained on p. 255.

Although many states and localities either enacted or considered bans on phosphates in the early 1970s, at least 20 of these bans have now been revoked. Indiana, New York, and several other states and cities still do not allow phosphates in detergents, but many other areas have settled for a limit on the amount of phosphate (8.7% phosphorus, which is equivalent to 50% phosphate). Phosphate-free detergents, which once accounted for 14% of the market, now have about 3–4% of sales. The main reason for the failure of the antiphosphate campaign has been the lack of a suitable substitute.

Sewage Treatment as a Solution

However, another way to attack the phosphate problem does exist. Phosphates can be removed from sewage by a method called **tertiary treatment** of wastewater. In this method, phosphates are precipitated out of sewage before the treated water is released into lakes or rivers. (See Chapter 14.) About 80–90% or even more of the phosphates in sewage can be removed, and the cost is not great. Why then shouldn't we allow the use of phosphate detergents and remove the phosphate from sewage by tertiary treatment? One reason is that at present, not everyone in this country is served by a sewage treatment plant. Some have septic tanks, others have only treatment ponds. Only 1% of the people in this country are served by sewage treatment plants that have facilities for phosphate removal. This particular solution, then, is tied up in the economics and politics of building and improving sewage treatment plants. Tertiary treatment is a solution that seems ideal and should be worked for; but it also must be viewed as still in the future, while eutrophication of lakes is occurring right now.

Meanwhile, a combination of solutions seems called for. People living in areas served by a sewage treatment plant that removes phosphates should use a phosphate detergent. Those of us who live in soft-water areas or have water softeners could use soap. The rest of us can use a detergent that has no phosphates or the very least amount of phosphates that will get clothes clean. Finally, we should very strongly support the building or equipping of sewage treatment plants to remove phosphates in our area.

The Scientist and Social Responsibility: Optical Brighteners

> We must be sure that a full-sized, hungry, four-footed wolf, with teeth, is coming before we start crying out about it . . .
>
> **B. J. Kilbey and G. Zetterberg**

> . . . I prefer not to give the product the benefit of a doubt . . .
>
> **B. Gillberg**

In 1971, Bjorn Gillberg stated that he had found evidence that optical brighteners used in modern detergents caused mutations in yeast. Some evidence exists that chemicals causing mutations are also likely to cause cancer. (See Chapter 29.) However, scientists in another laboratory could not repeat his experiment. B. J. Kilbey and G. Zetterberg wrote:

> We do not think that our experiments indicate unequivocally that no danger exists from optical brighteners. The data are insufficient at present for this conclusion to be drawn . . . However, we must be sure that a full-sized, hungry, four-footed wolf, with teeth, is coming before we start crying out about it. For environmental biologists, this means doing all in our power to be sure that the right experiments are done, positive results are reproducible, and any artifacts of method are excluded . . . If we startle the public too many times with sensational claims that are later retracted, we run a real risk of losing our most valuable ally if and when a real crisis comes.[*]

Gillberg replied:

> I have not made any sensational claims about brighteners; the only thing I say in my paper is that I consider it of importance to carry on with genetic studies of brighteners against the background of my results. Research has now begun in other laboratories that should have been undertaken before the brighteners were released on the market. The benefits of a product must of course always be weighed against the risks it may create. In such a situation I prefer not to give the product the benefit of the doubt if there are some questions raised. Questions have been raised about these compounds, and I believe that it is my social responsibility to tell my fellow citizens.[†]

Who is right? At what point should the public be told that some doubt has been cast on the safety of a product that is in common use? Must we be careful not to "cry wolf" too many times to avoid producing boredom about environmental dangers, or is it important to give people all available information as soon as it comes to light?

---

[*] B. J. Kilbey and G. Zetterberg, "Optical Brighteners," *Science,* **183** (1 March 1974), 798.
[†] B. Gillberg, "Optical Brighteners and Social Responsibility," *Science* **184** (13 September 1974), 901.

Controlling Other Phosphate Sources

Treatment of sewage to remove phosphates not only helps solve the detergent phosphate problem, but also removes phosphates from other sources, such as human wastes. What, then, remains? The main problem is nutrients that enter natural waters in small amounts but from many sources. This is called **non-point-source pollution** because it is usually hard to pinpoint exactly where the phosphates or nitrates are coming from. The total amount of nutrients from non-point sources can be very high. In some areas, it is equal to or greater than the amount from sewage treatment facilities.

The runoff of fertilizers from agricultural land is one example of non-point-source pollution. Poorly working septic systems, drainage from cattle feedlots, and city storm drains, which contain nutrients from lawn fertilizers, pet wastes, and other sources, are more examples.

Nutrients from manure and chemical fertilizers, contained in agricultural runoff, can be controlled by good farming practices, such as strip cropping, terracing, and careful timing of fertilizer application so that fertilizer is not immediately washed away by rains.[3] Feedlot operations may need to install sewage treatment equipment to prevent animal wastes from enriching nearby streams.

Some communities have adopted the solution of piping municipal sewage and storm waters around threatened lakes and into some other body of water. For instance, sewage that once flowed into Lake Washington in Seattle now is routed around the lake and into Puget Sound. Although Lake Washington is slowly recovering from its gradual slide into a eutrophic state, we need only think about Puget Sound to realize that this is a temporary "solution." Other solutions to non-point-source pollution are discussed in Chapter 14.

Helping Eutrophic Lakes Recover

Once lakes or ponds are eutrophic, simply preventing the addition of any more phosphates is often not enough to help them become clear and sparkling again. The reason is that phosphates settled into the bottom muds can, in many cases, dissolve back into the waters. Thus, phosphate levels may remain high for years, even with no new additions. Ways of combatting this problem have met with varying success. In some cases the flow of water through a lake can be increased. This helps wash nutrients and algae out of the lake. Low-phosphate water from Seattle's sewage treatment plant is used to increase the flow through Green Lake in Eastern Washington.

Chemicals such as aluminum, iron, and calcium have been used to precipitate phosphate out of lake waters. The cost of this sort of treatment is about $150–300 per hectare (depending on how polluted the lake is) and results last about three years. Fly ash from power plants has also been used to precipitate phosphate. In addition, the fly ash can seal off the lake sediments, preventing phosphates from dissolving back into the water. This also provides a way for power plants to dispose of fly ash. However, the ash is known to contain heavy metals, which could be toxic to aquatic life.

Oxygen can be added directly to the oxygen-poor lower layers of eutrophic lakes. This allows fish to take up residence there again and also helps prevent phosphate release from lake muds. (Release is more likely to occur under conditions of little or no oxygen.) This method has been used in Europe for years and has been tried successfully in several lakes and reservoirs in the U.S.

In some cases, there seems to be no alternative but to actually dredge the phosphate-rich sediments out of a lake (Figure 13.3.). This is, however, the most expensive method of lake restoration.

The National Clean Lakes Group is a coalition of interested parties that has convinced Congress to allot funds to communities wishing to restore eutrophic lakes. This is basically a continuation of the Clean Lakes Program run by the Environmental Protection Agency from 1976 to 1982. In order to be eligible for a grant, the lake must be available to the public; it must be classed as fresh water; and the applicants must show that their plan for lake restoration is likely to produce long-lasting benefits. Almost 100 lakes were helped under the EPA's Clean Lakes Program between 1976 and 1982 (Figure 13.4). More information can be obtained from EPA regional offices.

---

3 Strip cropping and terracing also reduce the damage done to farmland by erosion. These techniques are explained in Chapter 6.

(a)

(b)

Figure 13.3 The town of Vaxjo, Sweden, stopped pouring sewage into overexploited Lake Trummen in 1958. (a) By 1969, the unrecovered lake contained no oxygen, no fish, no underwater vegetation, and was of little use to humans. The main problem was the rapidly increasing black muddy sediment caused by decaying plankton. Starting in 1970, the sediment was sucked out and put into settling ponds; runoff water was cleansed of phosphorus and returned to the lake. (b) This is how Lake Trummen looks now, a revitalized recreational asset. (Photographs by S. Bjork)

Lake Erie: A Case History

The Great Lakes—Superior, Michigan, Huron, Erie, and Ontario—lie between the United States and Canada. Combined, they represent the largest reservoir of fresh water in the world. However, because they are the site of large population centers in both the U.S. and Canada (Duluth, Green Bay, Milwaukee, Chicago, Gary, Toledo, Cleveland, Erie, Buffalo, and Toronto, among others), the lakes are becoming more and more polluted. In fact, Lake Erie, the smallest and shallowest of the lakes, has been described as "dead." This really is not the right term to use—the lake still supports a large fish population. What has changed, however, are the kinds of fish in the lake. When the lake area was first settled in the 1700s, Lake Erie was home to large numbers of commercial and game fish. There are now no longer any blue pike, sauger, or

Figure 13.4 Volunteers used this truck to pull stumps out of the lake bed as part of an EPA-sponsored clean lakes project. (Photo by Harold Woodworth)

native lake trout. Only a few lake herring, sturgeon, whitefish, and muskellunge are left, along with some walleye and northern pike. Instead the fishery is composed of yellow perch, white bass, channel catfish, freshwater drum, carp, goldfish, and rainbow smelt.

Intense commercial fishing was responsible in part for the change in species. The introduction of new species into the lake also had unwanted effects. Sea lamprey, which feed on desirable food and game fish, found their way into Lake Erie sometime before 1921 via the Welland Canal. More serious is the invasion of rainbow smelt, which began about 1931. The smelt became very abundant, preying on the young of desirable fish such as lake trout, blue pike, and lake whitefish.

However, eutrophication of the lake has also played a large part. The amount of algal growth in the lake has increased 20 times since 1919. When these algae die and decay, the level of oxygen in the water plummets, especially in the cooler bottom waters. Not only does this destroy the summer habitat of many preferred fish species, but it also destroys important fish-food species, such as the burrowing mayfly. The kinds of algae in the lake have also changed. Blooms of blue-green algae are now common. In contrast to green algae, many species of blue-green algae are not eaten by fish.

In addition to providing a large part of the nitrates and phosphates that have led to the eutrophication of Lake Erie, sewage inflows have added toxic chemicals such as mercury, PCBs, and Mirex. These chemicals accumulate in fish and have led to severe restrictions on the kinds and sizes of fish that can be kept or eaten when caught in the Great Lakes.

In 1972 and 1978, the U.S. and Canada agreed on plans to clean up the Great Lakes. A central feature of the plans is building sewage treatment plants to reduce the amount of phosphate entering the lakes. Canada met the first deadline (31 December 1975) for phosphorus reductions, but the U.S., for political and economic reasons, did not. Although the amount of phosphorus added to the lake has decreased 75% in the past 10 years, the bottom waters of Lake Erie's central basin still become oxygen-poor in the summer months. If stated goals of lowering phosphorus inputs from 20,000 metric tons/year to 11,000 metric tons/year are met, it seems likely that the lake will recover. However, this will require not only sewage treatment but also reduction of at least 1/3 in the amount of phosphate that runs off farmland in Lake Erie's drainage basin. As one Canadian official, annoyed at the lack of U.S. progress, put it, the present situation is "like mixing a glass of clean water with a glass of dirty water, you end up with dirty water."

In summary, one might say that reports of the death of Lake Erie have been exaggerated. However, if we do not make progress in reduction of sewage inflows to all the Great Lakes, both because of the nutrients and the poisons they contain, the lakes may become effectively dead for human purposes.

Other Problems with Detergents: Foaming Waters and Biodegradability

In the early 1950's, many people noticed a strange sight. Brooks and streams, rivers, and even lakes were beginning to foam. Wherever water tumbled over stones or waterfalls, wherever winds rippled the surface, accumulations of bubbly froth built up. The explanation for this odd circumstance was not hard to find. Huge amounts of detergents (close to two billion pounds per year) were being used to wash clothes. The dirty wash water ran down millions of drains and into nearby streams, lakes, and rivers. Sometimes it first passed through sewage treatment plants and sometimes it did not. In fact, it made no difference whether it did or not. The chemical in detergents that made them foam was not removed by sewage treatment plants. The material, called alkyl benzene sulfonate (ABS), has several special properties. The property that makes it useful in washing clothes is that it can help greasy dirt dissolve in water. That is the good part. The not-so-good part is that its long, branched shape makes the molecule hard for bacteria to break down. Treatment plants depend upon the use of bacteria. Because bacteria do not have the necessary catalysts (called enzymes) to destroy ABS compounds, these molecules began to accumulate in the early 1950s wherever wash water ran into natural waters. When the concentrations of ABS were high enough, foam formed, just as it did in washing machines and dishpans.

This problem, fortunately, could be solved. Detergent manufacturers now use surfactant molecules having straight, rather than branched, side chains. Such surfactants are known as linear alkyl sulfonates or LAS. These compounds are called biodegradable because bacteria are able to digest them, and so the foam has almost disappeared from our rivers, lakes, and streams.

Questions

1. The amounts of nitrates that lead to algal blooms are smaller by a factor of at least 10 than the amounts that cause nitrate poisoning in infants. Contrast the sources and effects of nitrates at low and high levels.
2. What is hard water? Why is it a problem? Why are phosphates added to detergents?
3. What problems are caused by too much phosphate in natural waters? Would you call phosphate a poisonous chemical?
4. Why must we be careful about what kind of substitute chemicals are used in detergents instead of phosphates?
5. Does your sewage treatment plant remove phosphates? (You could call and find out.) Does your home have soft water? What would you recommend that your family use to wash clothes? Why?

Further Reading

Eutrophication of Surface Waters: Lake Tahoe's Indian Creek Reservoir, U.S. Environmental Protection Agency, Ecological Research Series, 1975, EP 1.23.

One community's answer to the problem of lake eutrophication was to build a whole new reservoir, Indian Creek, to hold the area's wastewater so that it would not have to be dumped into Lake Tahoe. This report details some of the problems they encountered.

Pirie, N. W., "Water Weed Uses," *Water Spectrum,* Summer 1980, p. 43.

The author puts forth the idea that water weeds should be looked on as a resource rather than as a nuisance. Such an approach could control eutrophication in some cases and in others eliminate it as a problem entirely. See also the article on Lake Erie in the same issue.

References

Phosphates in Detergents and the Eutrophication of America's Waters, 23rd report by the Committee on Government Operations, U.S. Congress, Washington, D.C. (1970).

Reiger, H. A. and W. L. Hartman, "Lake Erie's Fish Community: 150 Years of Cultural Stress," *Science,* **180** (22 June 1973), 1248.

Historical perspective on the effects of civilization on the Great Lakes.

Schindler, D. W., "Eutrophication and Recovery in Experimental Lakes: Implications for Lake Management," *Science,* **184** (24 May 1974), 897.

This article provides an entry into the literature on the limiting-nutrient controversy.

Wentz, Dennis, *Lake Classification—Is There a Method to This Madness?* U.S. Geological Survey, 848-B, 1981, p. B15.

CHAPTER FOURTEEN

Water Pollution Control

The Difference Between Water Treatment and Water Pollution Control

Controlling Water Pollution
Point Sources versus Non-Point Sources

Water Pollution Control at Point Sources
Combined Sewer Systems/Primary Treatment/Secondary Treatment/Digestion/Tertiary Treatment/ Chlorination/Control of Water Pollution in Rural Areas/The Federal Water Pollution Control Act/The "Settlement Agreement"

Control of Water Pollution from Non-Point Sources

Other Ways of Treating Wastewater

Tertiary Treatment—How It Works

Problems from Chlorination of Sewage

Control of Storm Waters

CONTROVERSIES:

14.1: *Can You Save a River Without Removing the Pollutants?*

14.2: *Economics and Environmental Regulation: Pretreatment Standards*

In Koln,[1] a town of monks and bones
And pavements fang'd with murderous stones
And rags, and hags, and hideous wrenches;
I counted two and seventy stenches;
All well defined, and several stinks!
Ye nymphs that reign o'er sewers and sinks,
The river Rhine, it is well known,
Doth wash your city of Cologne;
But tell me, Nymphs! What power divine
Shall henceforth wash the river Rhine?

Samuel Taylor Coleridge

The Difference Between Water Treatment and Water Pollution Control

The term "water treatment" should be distinguished from the term "water pollution control." These are very different activities. Water treatment and water pollution control are the two sides of the coin of water quality. Taken from a lake, river, or well, water must be made both safe and desirable to drink. *Water treatment* is the process that purifies water for the consumer, whether in the home, in business, or in factories. Safety is primarily achieved by destroying disease-causing microorganisms. Palatability is achieved by removing tastes and odors and by making the water clear.

Water pollution control, on the other hand, is directed toward restoring the quality of water that the consumer has used. Organic wastes and bacteria that have been added during use of the water must be re-

moved. The organic wastes are removed primarily to prevent depletion of oxygen in the waters that will receive the waste. The relation of organic wastes to the level of dissolved oxygen was discussed in Chapter 12. In addition to removing organic wastes, water pollution control facilities also aim to destroy disease-causing microorganisms to prevent disease from occurring among those who later use the water.

The relative positions of water treatment and water pollution control are indicated in Figure 14.1.

Controlling Water Pollution

Point Sources versus Non-Point Sources

Many ways are used to combat water pollution, but the methods chosen depend on the origin of the pollution. We can distinguish two general origins of pollutants. First are those pollutants that arise typically in urban areas and enter water courses from a single pipe; we say such pollutants come from a **point source.** Point-

---

1 The modern city of Cologne, Germany.

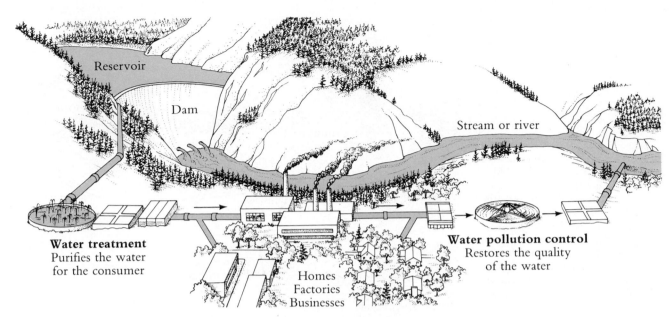

Figure 14.1 Water Treatment and Water Pollution Control. Note that water treatment precedes use of the water by the community. Water pollution control, on the other hand, follows use by the community. Water treatment "prepares" the water for use; water pollution control "repairs" it after use.

source pollutants arise when domestic wastes and industrial wastes are collected by sewers; the sewers carry the wastes to a treatment plant. After the wastewater has been treated, it is discharged to a waterway. We call such pollution "point source" because it occurs at a distinct and identifiable point. The concentrated runoff from large feedlot operations may be considered a point source of pollution.

In contrast are pollutants that come from relatively rural areas; these pollutants enter water bodies in runoff from the land. The runoff occurs in numerous rivulets and streams, and reaches the river at many, many points. We can think of the runoff as being spread out along the waterway, and we call such runoff with its burden of pollution a "non-point source."

The methods of controlling pollution from point sources and non-point sources are very different. To point-source pollution, we have traditionally applied the "technical fix"; the technical fix consists of chemical and biological processes that remove contaminants from the wastewater. Non-point-source pollution, however, is not so easy to control, for it does not necessarily lend itself to technical methods. Instead, changes in land use and in agricultural practices are

needed. Changes are also needed in mining practices and in timber growing and harvesting. In this chapter we discuss the technical solutions that apply to point-source pollution; we also discuss some of the changes in land management needed to control pollution from non-point sources.

Water Pollution Control at Point Sources

Combined Sewer Systems

When wastewater enters a wastewater treatment plant (also called a water-pollution control plant), it passes through a rack and a coarse screen that together prevent large objects from passing into the plant. The large objects could include such items as boots, cloth, branches, etc. Such objects would seem to be unexpected in municipal wastewater, but they are quite common. Their presence is a clue to a very large problem that most communities in the United States continue to face: namely, the drain system that carries storm waters and the drain system that carries domestic wastewater are combined.

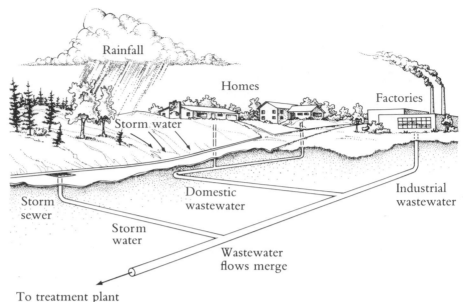

Figure 14.2 The Combined Sewer System of Most U.S. Cities. The runoff from a rainstorm enters a city's sewers and there mixes with the wastes from homes and industries. A treatment plant built for normal wastewater flow will not be able to handle the combined flow. Much of the combined flow will be diverted past the treatment plant and will enter a water body without treatment.

Storm-water flows, which may transport large objects such as those we mentioned, mixes with municipal wastewater during and after a rainfall. The combined flow of storm water and wastewater would overwhelm a treatment plant and must be diverted past it. The concept of combined sewers is illustrated in Figure 14.2. The problem posed by combined sewers is one of continuing concern and is not open to easy, low-cost solutions. (This widespread problem is discussed on p. 273.) To build a storm sewer system separate from the wastewater system would cost billions of dollars in each of a number of U.S. cities.

Primary Treatment

After the wastewater has passed through the rack and screen, the flow moves into the grit chamber. Here very large particles of earth, such as gravel, drop out of the moving stream. The flow next moves to a settling tank, also commonly called a "clarifier," where the organic solids characteristic of wastewater slowly settle out. Settling is a purely physical process; no chemical or biological reactions take place. Settling removes organic solids from wastewater by slowing the velocity of the waste stream. As the velocity decreases, the tendency of the water to keep particles in suspension decreases, and organic particles slowly settle to the bottom of the tank. The sludge of organic solids re-

moved by a "settler" may account for 35% of the organic matter in the wastewater from a typical city. This initial settling is referred to as **primary treatment.**

Secondary Treatment

The processes that follow the settling tank are designed to remove organic materials that are dissolved (as opposed to suspended) in the wastewater. This process step, the removal of dissolved organics, is referred to as **secondary treatment.**

One process, known as **activated sludge** treatment, uses microorganisms in a large, well-aerated tank to break down dissolved organic substances. A steady stream of wastewater from the settler enters one end of the activated sludge tank and exits at the other end. The flow, on exit, is richer in microbes and poorer in dissolved organics (Figure 14.3). About 80–85% of the dissolved organics in wastewater can be removed by the activated sludge process.

The microbial population that grows steadily in the tank uses the organic substances in solution for growth and energy. Thus, the organics are removed from solution and are converted within the tank either to more microbes or to the end products of biological oxidation. In the presence of oxygen, the end products include carbon dioxide and water.

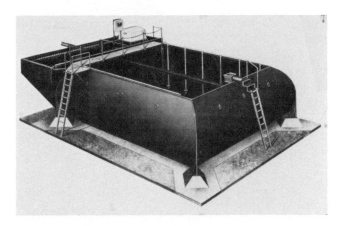

Figure 14.3 The Package Activated Sludge Plant. Often used for subdivisions, these small activated-sludge plants are sold as prefabricated steel package plants. The process flow diagram is essentially the same as in the larger system, except that these small plants typically have no primary treatment and aerate the raw wastewater for a 24-hour period, rather than the 6–8 hours used in conventional plants.

The microbial solids produced in the activated sludge tank are suspended in the waste flow leaving the tank. This waste flow is passed through another settler, which removes the solids. These solids from the second settler flow in a slurry that is 98–99% water; this flow is combined with the solids from the primary settling tank. The combined solids are then treated in a device called a digester (see Figure 14.5). We shall describe the digester shortly.

A portion of the solids recovered from the flow through the activated sludge tank is returned to the activated sludge tank as "seed." These solids mix with the waste stream from primary treatment and provide an initial bacterial population to start the mixture fermenting in the tank.

Another common technology used in secondary treatment to remove dissolved organics is the **trickling filter.** Here the wastewater is distributed over and passed down through a bed of rocks in a large concrete basin. A slime of microbes grows on the surface of the rocks during long operation of the filter. The microbes in the slime remove dissolved organics for growth and energy, just as the microbes in the activated sludge process do. The slime slides off the stones a little at a time and into the wastewater. The result is a waste stream with solid materials in suspension and much of the dissolved organics removed, just as occurred in the activated sludge tank. About 80–85% of dissolved organics may be removed by a trickling filter. The solids are removed in a settling tank, and these solids are combined with those from the primary settling step and pumped to the digester. The trickling filter may be as large as 200 feet (60 meters) in diameter at large installations (Figure 14.4).

Digestion

The process of waste digestion converts the sludge of organic solids to a stable material; the process takes place in a large, heated tank appropriately called a **digester.** The tank is closed to the air so that oxygen does not reach the wastes. Specialized microbes, which live at high temperatures and do not require oxygen,

Figure 14.4 Trickling Filter. Wastewater is distributed over rocks and trickles down through the rock bed. A slime growing on the rocks removes organic material from the wastewater.

grow in the tank. These microbes convert the wastes to stable end products, including the gases methane and hydrogen sulfide. The methane gas is often burned to provide the heat needed to keep the digester at the proper temperature. A stable, nondegrading sludge is the result. This stable sludge is then dried, either in greenhouses or on vacuum drums.

The disposal of sewage sludge differs across the country. For many years, Philadelphia has been dumping its sludge in the ocean; the dumping has been done lately only on an interim basis. Many cities bury their dried sludge in landfills. Others burn it in a special unit and bury only the ash in landfills. Some communities spread the sludge on cropland, but heavy-metal contaminants in the sludges make such use unwise. The city of Milwaukee, Wisconsin, packages its sludge as a commercial fertilizer. Milorganite, as the product is

known, might be called "the sludge that made Milwaukee famous."

The primary treatment (settling) and secondary treatment (activated sludge or trickling filter) remove organic substances. Together they may remove up to 90% of the organic wastes in the water (Figure 14.5). Primary treatment is common at many communities in the United States. Secondary treatment is coming into much wider use.

Tertiary Treatment

Most of the organic material in wastewater is removed by primary and secondary treatment. The processes that follow the secondary treatment step are designed to remove the plant nutrients responsible for "overnourishing" lakes and rivers. Taken together, these

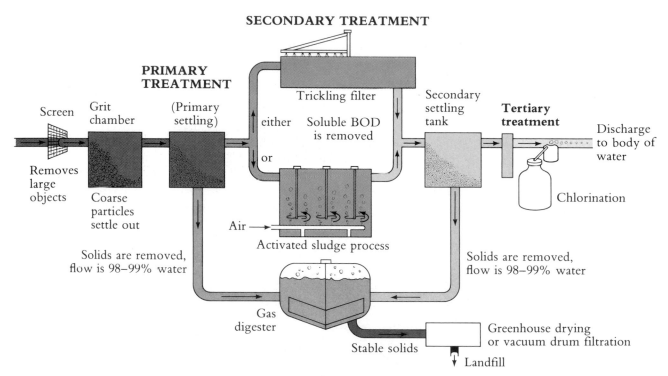

Figure 14.5 Water Pollution Control Processes. Three distinct sets of processes are used in sequence. "Primary treatment" removes suspended solids. "Secondary treatment" removes dissolved organics. "Tertiary treatment" is designed to remove phosphorus, nitrogen, and resistant organic compounds. Digestion, not part of any of these three processes, is the means by which organic solids are degraded and stabilized.

Can You Save a River Without Removing the Pollutants?

We can easily and without spending huge sums of money give back to the river the oxygen which it lacks in the summer due to pollution and a lower water flow.

Chairman of the City Council, Paris, France

In addition to the methods of water pollution control discussed in this chapter, there are two quite different ways to improve the quality of rivers. These options deal with the rivers themselves and are not treatment processes in the usual sense. The first is called "low-flow augmentation" because additional water from an upstream reservoir is released to dilute the downstream concentrations of dissolved organics. The dilution is intended to add oxygen to the stream, thereby reducing the effect of the oxygen-demanding wastewater. For example, the wastewater from a particular community might be able to reduce the dissolved oxygen in the average river flow during August to 3 milligrams per liter. If the flow is augmented sufficiently, the dissolved oxygen in the increased flow may only fall to 5 milligrams per liter.

Among environmental engineers, low-flow augmentation is humorously called "the dilution solution to pollution." It is not really pollution control, however; it is a method of water quality improvement. The same quantity of organic wastes are still entering the watercourse—only their effect is different. Until the early 1970s, the water quality improvement brought about by flow augmentation could be counted as a benefit when the Corps of Engineers proposed a reservoir project to Congress, but this is no longer the case.

In the same category as flow augmentation is the process known as "in-stream aeration." This, too, is a means of improving the dissolved oxygen content of a river without removing dissolved organics. We know that dissolved organics, through the action of bacteria, remove dissolved oxygen from a stream. We also know that the flowing water of the stream captures and dissolves oxygen from the air. Why not help the natural process by bubbling in compressed air? The process is known to work, and aquatic life can be restored to the river by raising the dissolved oxygen level.

The City Council of Paris, France, plans to offer the Seine River, which flows through the city, such a "face-lift." The chairman of the City Council remarked, "We can easily and without spending huge sums of money give back to the river the oxygen which it lacks in the summer due to pollution and a lower water flow." The chairman did not mention that the pollution will still be present.

Is in-stream aeration or flow augmentation a better way than pollution control to restore the quality of a river? Is this the kind of process that should be considered in planning the clean-up of a river? Or should clean-up mean only the removal of wastes?

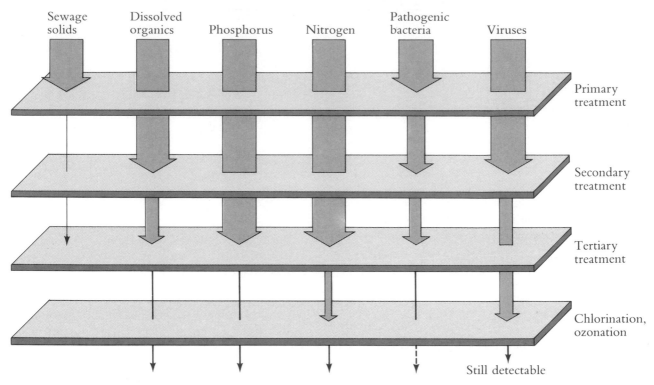

Figure 14.6 Removal of Contaminants by Sewage Treatment. The relation between each stage of treatment and the level of removal of contaminants is shown by this diagram. If an arrow meets a treatment plane with an arrowhead and changes thickness in moving through that treatment plane, the level of the contaminant has been reduced by that stage of treatment. The relative reduction in the thickness of the arrow indicates approximately the extent of the reduction. For instance, sewage solids are nearly all removed in primary treatment. Dissolved organics, on the other hand, are not altered by primary treatment, but are significantly reduced by secondary treatment. Nitrogen and phosphorus are not generally removed by primary or secondary treatment, but are much reduced by tertiary treatment, and so on.

processes, which follow secondary treatment, are known as **tertiary treatment.** Tertiary treatment, in contrast to primary and secondary treatment, is quite uncommon at present.

The serious effects of over-enrichment (eutrophication) of lakes and rivers with plant nutrients have only been recognized in the last few decades. Thus, methods to control eutrophication have not yet been widely applied. At present, probably only a few dozen such installations are operating in the U.S. Nevertheless, there is promise that the processes in tertiary treatment will be added to sewage treatment plants in many areas.

The main purpose of tertiary treatment is to remove compounds containing the elements nitrogen and phosphorus. Thus, nitrogen compounds, such as ammonia, nitrate ions, and nitrite ions, are removed by one set of processes. Phosphate ions, found in both human wastes and detergents, are removed by another process. Tertiary treatment may also include removal by carbon adsorption of those resistant organic substances not removed by secondary treatment. These processes are very expensive. (Tertiary treatment is discussed in more detail on p. 271.)

Chlorination

Bacteria that inhabit the human intestine find the sewage treatment plant cold comfort indeed. Their natural habitat is considerably warmer, so that a gradual reduction in these bacteria is noted as the wastewater moves through the treatment plant. Nonetheless, considerable numbers of bacteria still exist in the waste stream even after secondary treatment. Thus, wastewater leaving the sewage treatment plant is commonly chlorinated in the U.S. in order to destroy disease-causing microorganisms (pathogens). Where the sewage treatment plant was merely a torturer to these bacteria, chlorination is the executioner.

In contrast to bacteria, viruses from the human intestine die away more slowly in the treatment plant. Primary treatment reduces their numbers little, if at all. Secondary treatment via the trickling filter may reduce viruses by 40%, but secondary treatment via the activated sludge process can reduce their numbers by up to 98%. The final step of chlorination reduces their numbers once again. Yet even after chlorination, one can still find live viruses from the human intestine in the waste stream.

The health of people swimming or boating on the water downstream from the treatment plant is protected by destroying the pathogens in the wastewater. The process of chlorination is inexpensive compared to other methods of disinfecting wastewater. Furthermore, the effectiveness of the disinfection step is easily checked by using the test for free chlorine to see if chlorine remains in the wastewater.

Although chlorination is in wide use at waste treatment plants, people do recognize that its presence in water is harmful to fish, even at low levels. Chlorine is also known to react with hydrocarbons to produce some substances that are suspected of causing cancer. (See supplemental material at the end of this chapter.)

The approximate reduction in pollutants, as well as the reduction in bacteria and viruses, by each step in the total treatment scheme is indicated in Figure 14.6.

The devices and processes discussed so far are the wastewater treatment methods used most commonly in American cities today. However, other processes and concepts are being studied or used on a small scale. These include land application of wastewater, physical–chemical treatment, and other processes.

Control of Water Pollution in Rural Areas

Septic systems. In rural areas and small subdivisions where there are no wastewater treatment plants, or no sewers to connect to treatment plants, households most often treat their wastes by using **septic tanks.** The septic tank is the successor to the cesspool. The older device, the cesspool, was a structure resembling a well. It was lined with stones or concrete blocks, with no mortar placed in the joints between the stones or blocks. Raw, unsettled sewage was disposed of directly into the cesspool (Figure 14.7). Groundwater contamination was a distinct possibility.

The septic tank, in contrast to the cesspool, is a watertight tank, constructed of either metal or concrete (Figure 14.8). Depending on the tank size and the size of the home it serves, the tank may hold sewage

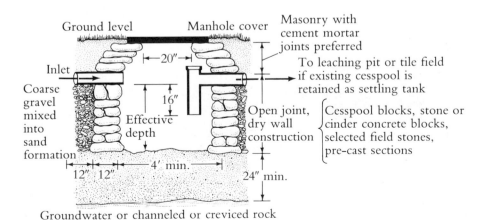

Figure 14.7 Leaching Pit and Cesspool Details. One of the first devices used to dispose of wastes, the cesspool is still a familiar concept. It is now recognized, however, that the cesspool is inadequate and dangerous in disposing of wastes. The potential contamination of groundwater is the principal objection to the use of the cesspool. (Adapted from J. Salvato, *Environmental Sanitation,* Copyright © 1958 by John Wiley & Sons, Inc., N.Y. Reprinted by permission.)

Figure 14.8 Cross Section of a Typical Concrete Septic Tank. The septic tank is the device that replaced the cesspool in rural areas. Closed to the soil by its concrete or metal sides, the device provides a modest amount of treatment to the wastewater from a home. Solids settle out in the tank and are degraded there; while a relatively clear overflow passes out to the leaching field. (Adapted from *Cleaning Up the Water,* Maine Department of Environmental Protection, 1974)

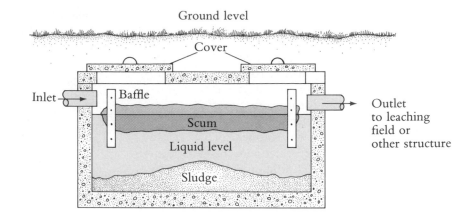

flows for anywhere from half a day to three days. During this time, solids settle to the bottom of the tank, where they are degraded by bacterial action.

The overflow from a septic tank is a liquid from which some solids have been removed. The overflow passes to a distribution box that "distributes" the water to a buried **leaching field.** The leaching field consists of perforated pipes of clay or plastic pieces, and the wastewater seeps into the soil from the joints between the pieces (Figure 14.9). Microorganisms in the soil decompose the wastes. Lush grass may often be observed above the "arms" of the leaching system. The leaching system helps disperse the wastes and decreases the possibility of groundwater contamination.

If a family uses both a septic system and a well, it is good practice to keep the water supply far away (several hundred feet) and preferably uphill from the waste disposal system. Furthermore, the soil into

which a leaching field is placed should be a fairly porous one. (Porosity can be determined by a percolator test.) The soil should not have fractured or creviced rock beneath the surface, for such features could potentially channel the wastes to the groundwater. The top of the water table should be at least several feet below the pipes. Housing lots in suburbia are often required to be very large in order to handle the required leaching field for septic tank wastes.

Lagoons and ponds. One device that finds fairly wide use in smaller communities and also in industry is the **waste stabilization lagoon,** or oxidation pond. The lagoon is essentially a wide shallow pond. It is typically 2–4 feet (0.7–1.2 m) deep, and raw wastewater discharges directly into it. Its use depends on the availability of low-cost land and the character of the surrounding area. Odors may make its use a nuisance

Figure 14.9 Typical Layout of a Septic Tank and Leaching Field. The outflow from the septic tank is distributed to a buried leaching field, consisting of clay pipes with spaces between them. The wastewater enters (leaches into) the soil from these spaces, and soil bacteria degrade the wastes further. (Adapted from *Cleaning Up the Water,* Maine Department of Environmental Protection, 1974)

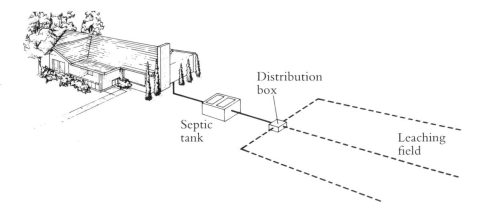

in populated areas.

Most ponds are designed to retain about two weeks worth of flow from the community. Larger ponds can hold nearly two months of flow. Solids settle out as the wastewater flows slowly through the pond. The solids decompose in the absence of oxygen at the bottom of the pond. Near the surface, however, bacteria oxidize organic materials in the presence of oxygen. Weeds and other aquatic plants, including algae, may grow in the pond, making use of nutrients from the wastes. In addition, the algae supply oxygen to the system as a by-product of their photosynthesis. The oxygen helps keep the system operating with minimal odors. Sometimes the lagoon is aerated; that is, air is bubbled into the water, much as in the activated sludge process. The lagoon thus can act as a combination of primary and secondary treatment. In fact, no primary treatment is usually given to the waste entering a lagoon.

In 1978, the contents of a 36-acre (14-hectare) sewage lagoon in West Plains, Missouri, leaked into the ground. Numerous subterranean streams carried the sewage over many miles and contaminated groundwater in the region. Over 750 cases of diarrheal disease were associated with the leak. The incident illustrates a hazard associated with waste lagoons and septic systems: the possibility of leakage to the groundwater through channeled rock.

The Federal Water Pollution Control Act

Congress has attempted to control point-source water pollution by a variety of laws (see also Chapter 32), but the most basic rules are found in the Federal Water Pollution Control Act. This law was passed in 1972 and further amended in 1977. The laws were designed to go into effect in a stepwise fashion; some provisions will not take effect until 1987 or later. Basically, the laws divide pollutants into three classes. For wastes designated as *toxic,* industries must use the Best Available Technology (BAT) in their treatment before the wastes are released into natural waters. For pollutants listed as *conventional,* such as BOD, dischargers must use the best conventional technology. Anything else falls into a third class, *unconventional* pollutants. These pollutants must meet BAT standards, although waivers are possible.

If even the Best Available Technology would not protect certain waters, stricter standards (e.g., no discharge at all) must be imposed. Further, special standards called *pretreatment standards* are imposed for wastes that will pass through sewage treatment plants. (See Controversy 14.2.)

A special permit from EPA or the state is required for discharge of any pollutants into navigable waters. This permit would, of course, require that the discharge meet the appropriate toxic, conventional, or unconventional treatment standards. For ocean discharges, EPA was required to set special standards.

The "Settlement Agreement"

EPA was also supposed to decide which pollutants fall into what category. But in 1976, EPA was sued by a coalition of environmental groups on the basis that the agency was "foot-dragging" in deciding which pollutants were toxic and what discharge standards should be. Only nine pollutants had been classified as toxic by that time. A settlement agreement was finally reached in which EPA would develop standards for 65 listed pollutants on a strict time schedule, and no later than 1983. Nonetheless, the issuance of standards still proceeds slowly.

Control of Water Pollution from Non-Point Sources

By non-point sources of water pollution, we mean those many streams and rivulets that may carry pollutants off the land into major bodies of water, such as rivers or lakes. Non-point-source substances include organic wastes, nitrogen compounds, phosphorus compounds, and sediment or soil particles.

Dairies and livestock farms produce animal manure which, unless disposed of properly, can end up in streams and rivers. The use of feedlots, as opposed to grazing steers on pastureland, can result in a very concentrated source of pollution. The average steer generates about 15 times more wastes than an average person; the hog twice as much. In total, farm animals may be producing 10 times the wastes of the U.S. population. To control pollution from feedlots and dairy farms, waste treatment devices such as lagoons may be installed at the larger operations.

Economics and Environmental Regulation: Pretreatment Standards

Should we not ask the EPA to show us that the benefits of its regulations are at least as great as the costs?

R. W. Crandall

Conventional estimates of gross national product are . . . meaningless so long as they fail to account for the cost of . . . uncorrected levels of pollution.

R. D. DuBoff

The Environmental Protection Agency is now requiring industries that have been discharging toxic wastes into sewer systems to set up pretreatment systems to prevent the wastes. In this way, materials such as toxic organic chemicals will be removed before they can reach municipal sewage treatment plants. The measure should help reduce toxic organics in drinking water by stopping the discharge of these materials into natural waters that might be used as sources of drinking water. In addition, toxic wastes often "knock out" sewage treatment plants by killing microbes that are a vital part of treating sewage. An instance of this occurred in Louisville, Kentucky, when chemicals dumped into a sewer knocked out the city's sewage treatment plant for 45 days, costing the city five million dollars.

Of course, pretreatment is not without cost to industry. Robert W. Crandall, a Senior Fellow at the Brookings Institution, took issue with the EPA's regulations in a letter to *The New York Times,* 3 July 1978.

. . . Jorling is quoted as saying that he did not know how much these industrial standards would cost, but that it would possibly be in the "low billions" of dollars. This rather cavalier assessment of costs is not uncommon among environmentalists who tend to view G.N.P. as a stock which does not change with increasing regulation. Perhaps someone should point out to

Mr. Jorling that there are not many "low billions" of dollars in the entire industrial output of the economy. His assurance that the costs of those latest regulations are in the "low billions" of dollars may be likened to the executioner's assurances to the condemned that he will only have to fall "a few feet" with the noose around his neck.

Recently a colleague, Edward Denison, measured the effect of environmental expenditures by business upon economic growth and found that by 1975 they had reduced economic growth by more than 10 percent of its recent average. The total cost of these regulations in 1975 was in the "low billions" of dollars—$9.6 billion, to be exact. If a single set of new regulations will cost a few billion more, what can we expect when a large number of new air quality standards are promulgated in the next few months? Before we proceed, should we not ask the E.P.A. to show us that the benefits of its regulations are at least as great as the costs?

This was answered by Richard B. DuBoff, an economist at Bryn Mawr College, on 10 July 1978 in *The New York Times:*

Perhaps, to borrow Mr. Crandall's own language, someone should point out to him, and to

Mr. Denison, that their thoroughly conventional estimates of gross national product are overstated, or meaningless, so long as they fail to account for the costs of present, uncorrected levels of pollution—of air, water and industrial inputs (the cotton "brown lung" disease but the latest case). Even a nonradical economist like Paul Samuelson warns students who use his textbook that "clearly we must adjust for any such 'bads' that escape the G.N.P. statistician whenever society is both failing to prevent pollution and failing to make power (or water or air or cotton) users pay the full costs of the damage they do." Once we make these adjustments, we see that net economic welfare grows more slowly than (conventionally measured) G.N.P.

Mr. Crandall's reproof of "environmentalists who tend to view G.N.P. as a stock which does not change with increasing regulation" is another fine example of the tendency of orthodox economists to leap to policy "tradeoff" conclusions without any probing of underlying economic relationships (let alone their broader political context).

Should industry be required to pretreat wastes? Who will pay for this, eventually? In what ways do we pay if industries do not pretreat wastes?

How does this controversy illustrate the way air and water have been treated in the past as free resources?

Growing crops can lead to water pollution because of the fertilizer applied to them. Fertilizer is likely to include compounds containing both phosphorus and nitrogen, the two elements most likely to cause eutrophication. In addition to fertilizer, decaying vegetation and soil particles from erosion may enter streams in runoff. Erosion occurs from cropland, from grazing land, from logged areas, from unreclaimed mining areas, and from construction sites. Soil particles may be suspended in the flowing water, resulting in a murky stream. On a weight basis, in fact, sediment from erosion is the number-one water pollutant in our streams and rivers.

Control of pollution arising from farmland requires more careful application of fertilizers and pesticides. Contour plowing on small hills and terracing on steeper hills also helps decrease runoff. (See discussion of erosion in Chapter 6.) In the last decade, many farmers have gone to "conservation tillage" or "no till farming" as labor- and fuel-saving devices. These methods involve use of a herbicide, which kills weeds before they can sprout from the ground. Such ground does not need to be turned over in the spring to eliminate weeds; hence much less erosion occurs. However, these herbicides may be washed into streams along with other pest-control chemicals applied to the crop. Many chemicals used to control insects and plants have been found to be cancer-causing substances. Therefore, it is important that the herbicides selected for use have as little effect on humans as possible.

Septic systems can give rise to non-point-source pollution. Urban runoff that does not enter storm sewers but flows from gulleys or culverts into streams is a contributor as well. The runoff from storm drains on highways carries metals such as the lead from leaded gasoline. Roads and shopping centers, with their vast impermeable areas, create conditions in which storm water runoff can occur very rapidly and hence with considerable erosive power. Because of the rapid build-up of flow from highways and shopping centers, stream channels can be severely eroded and the sediment increased in the flowing water.

The magnitude of non-point-source pollution is now thought to be larger than was once suspected. The National Commission on Water Quality has projected that when all industrial and municipal sources have installed secondary treatment, the amounts of nitrogen and phosphorus from non-point sources will then match or exceed the quantities from the treated wastewater of industry and cities.

✍ **Other Ways of Treating Wastewater**

Physical–Chemical Treatment

All the secondary treatment processes we have discussed utilize microorganisms to remove dissolved organics from wastewater. There are defects in this biological treatment of wastes, however. Probably the most important problem is the sensitivity of microorganisms to toxic pollutants. An industry may discharge a waste that will poison the microorganisms. As the waste flows through the treatment plant, the special populations of microbes, which had been thriving in ordinary wastewater, may be partially or wholly killed. When this happens, the effectiveness of the entire plant is reduced to the removal levels achieved by primary treatment alone. A whole new population of microbes must grow up to renew the effectiveness of secondary treatment.

In the last decade, much progress has been made on *"physical–chemical treatment"* of wastewater. These processes remove the same substances as are removed by a plant with conventional primary, secondary, and tertiary processes.

Physical–chemical treatment begins as conventional treatment does, with a rack and screen. The next step is the addition of a chemical such as lime or alum to create an insoluble mineral phosphate precipitate. The resulting solid particles, as well as the usual suspended solids, then settle out together in a settling basin, much as in the usual primary treatment step.

The wastewater is then passed through a column containing activated carbon, where dissolved organics are adsorbed onto the surface of the carbon particles. The organic removal achieved in this step matches that of the best biological treatment processes. Eventually, the carbon particles become saturated with organics; then the column is removed for cleaning and replaced by another. Once cleaned, the first column can be placed in service again.

Physical–chemical treatment plants may soon be in service in a number of areas, including the cities of Niagara Falls, New York, and Cleveland, Ohio.

Land Application of Wastewater

One concept of waste treatment being studied at government-supported projects is land application of wastewater. Land application has advantages and disadvantages. Wastewater contains phosphorus, nitrogen, and organic substances. It is a reasonable fertilizer for crops, and, of course, it is a source of moisture. Though its use for irrigation seems attractive, there are other issues to be considered.

First of all, we know that primary treatment ought to precede any application of wastewater to the land. If it did not, the system that distributes the water could be clogged by the solids in the wastewater. Furthermore, without primary treatment, a very large portion of the bacteria and viruses might be applied to the land. Some researchers feel that secondary treatment is not needed before applying wastewater; they assert that the land treatment will remove dissolved organics. Others feel that applying wastewater to the land is an alternative method only for tertiary treatment. The tertiary processes, you will recall, are designed for phosphorus and nitrogen control and for removal of resistant organics.

Three methods of land application have evolved, each designed to accomplish a different objective. In the process referred to as *overland flow,* wastewater is applied to sloping land that has been stabilized by grasses. Gravity carries the water down the slope, across the soil, and through the grasses to a runoff collection ditch. The water does not penetrate the soil deeply, and at the base of the slope it joins other surface runoff. Because the removal processes are biological in nature, areas where freezing temperatures occur regularly are not good candidates for overland flow treatment. The process does not remove much phosphorus and hence, may not be a good substitute for tertiary treatment.

Irrigation with wastewater, in contrast to the overland flow process, is designed to fertilize crops and provides a high degree of treatment. Irrigation could be thought of as an example of water re-use. Probably irrigation with wastewater is best suited for the water-short western states. Muskegon County in Michigan and the city of Lubbock, Texas, have large-scale spray irrigation installations that apply wastewater to the land.

The third method of land application, *infiltration–*

percolation, is primarily a method for recharging groundwater; a crop need not be involved. Treated wastewater is applied to the soil by pumping it into basins, from which the water percolates into the soil. Its use is most appropriate where wells are depleting low-salt-content groundwater and where waters with high salt content are replacing them. Treatment is primarily by filtration through the soil.

Nassau County on Long Island, New York, plans to "polish" its sewage to drinking-water quality and then recharge it to the ground instead of discharging it into the ocean, in this way maintaining its reserves of fresh groundwater. Groundwater recharge is practiced also in Israel and the Netherlands. In the Netherlands, the dirty water of the Rhine River is recharged into the sand and then drawn out before it is used by the city of Amsterdam.

Applying wastewater to the land requires a number of conditions. First, land with no better uses (such as housing, industry, etc.) must be available at a distance not far from the treatment plant. If the land is too distant, the electrical energy needed to pump the wastewater will make the project too costly. Second, citizens must be convinced of the value and safety of the project. It would be untruthful to suggest that applying wastewater to the land is known to be safe in all circumstances. The Water Pollution Control Federation issued a policy statement in 1974 on land application of wastewater. It stated that "present knowledge of the effects of large scale use of land disposal is too meager to justify widespread use for man-made wastes."

In 1976, a report from Israel studied the frequency of infectious diseases in agricultural communities (kibbutzim) that spray-irrigated their crops with wastewater. Investigators found that the infectious disease rate was about four times higher in these areas than in communities that did not irrigate crops with wastewater. Droplets containing bacteria and viruses were probably carried in the air by the wind and caused the increase in disease in the communities.

Another concern is that long-term irrigation with river water or wastewater can cause salts to build up and eventually make the land unfit for agriculture. The Imperial and Central Valleys of California are experiencing a build-up of salts in the soil. A much earlier example occurred in the region of Iraq once known as the Fertile Crescent. Watered by the Tigris and Euphrates Rivers, the land was irrigated to stimulate agricultural production. The term "Fertile Crescent" is no longer used to describe this relatively barren area.

Attention must also be given to the possibility that nitrates may appear in the groundwater unless removed from the wastewater before it is applied. The presence of nitrate ions in high enough concentrations in drinking water can bring about the disease methemoglobinemia. (See page 221.)

∿ Tertiary Treatment—How It Works

Primary and secondary treatment are effective methods of removing sewage solids and dissolved organics. Chemical processes are needed, however, to remove several undesirable chemicals in wastewater. These substances are nitrogen, as ammonia or as ammonium ion, nitrate or nitrite ions, and phosphorus in the form of phosphate ion.

$$NH_3 \qquad NH_4^+ \qquad NO_3^-$$
ammonia gas $\qquad$ ammonium ion $\qquad$ nitrate ion

$$NO_2^- \qquad PO_4^{3-}$$
nitrite ion $\qquad$ phosphate ion

These substances contribute to the eutrophication of natural water, a process characterized by excessive algal growth. (The biological basis of eutrophication is described in Chapter 13.)

The processes used to remove these chemicals are often referred to as tertiary treatment when they follow the secondary stage of treatment. Sometimes the processes for chemical removal can actually be included within the primary and secondary treatment stages. When this is done, the added processes may be referred to as "advanced waste treatment."

Phosphate removal is accomplished by chemical precipitation and settling. Chemicals such as ferrous and ferric salts, or aluminum salts or lime, are added to the waste stream. When one of these chemicals is mixed well with the waste stream flowing from secondary treatment, a precipitate of solid matter forms. For example, the calcium ion from lime (calcium oxide) combines with the phosphate ion in solution to produce solid particles of calcium phosphate. These particles are removed by allowing them to settle out of the waste stream. Engineers have suggested that the addition of chemicals to "bring down" phosphate could also take place just before the primary settling tank. This would allow collection of both the organic solids and the phosphate precipitate to take place at the same time.

After the phosphate is precipitated and settled out, any remaining suspended matter may be removed by filtration. Beds of sand, crushed coal, garnet, and gravel are used in a number of layers as the filter materials for this step. Because of the several components of the filter bed, the process is referred to as "multi-media filtration."

The removal of nitrogen is much more difficult. There is no common insoluble nitrate or nitrite compound that can be settled out; nor is there any common insoluble ammonium compound. Hence, there is no way to remove nitrogen by precipitation in the treatment plant.

Removing nitrogen that occurs as ammonia (NH_3) is important for a number of reasons. First, ammonia is harmful to fish. Second, ammonia combines with the chlorine that was added to destroy harmful bacteria and viruses. Reactions with ammonia tie up chlorine in substances known as chloramines and hence decrease chlorine's effectiveness. Third, ammonia is oxidized by bacteria in the stream, and this oxidation process may remove large amounts of oxygen from the water. Thus, we have many reasons to remove ammonia.

The ammonium ion may be removed as ammonia gas by a process known as "gas stripping." In the process, wastewater is exposed to air in a metal tower packed with small inert plastic shapes. The water is allowed to run down over the plastic surfaces while air passes upward through the column. Ammonia moves from the water into the upward-moving air stream.

Ammonia may also be removed by biological treatment, by passing the wastewater through an aerated tank with a specialized population of microbes. These microbes convert the nitrogen in ammonia and in nitrites to the nitrate ion. The process is somewhat unstable because of the need to maintain the proper type of bacteria growing in the tank at all times.

Removal of nitrate ions, the most oxidized form of nitrogen, is also desirable. Although nitrate ions may have been present initially in the wastewater, they could also have been produced by the oxidation of ammonia and nitrite ions. Nitrate ions should be removed because they increase eutrophication, but an even more compelling reason exists for removing nitrate and nitrite ions. If high-nitrate water is used for drinking, a blood disorder in infants, methemoglobinemia, could result. (See page 221.)

Nitrate removal may be achieved by specialized populations of microbes that convert nitrate to nitrogen gas and water. Methyl alcohol must be present for this reaction, which produces carbon dioxide, water, and nitrogen gas. Again, keeping the proper population of microbes growing in the tank is a problem.

A final procedure in tertiary treatment is to pass the waste stream through a tower containing particles of activated carbon. The activated carbon adsorbs dissolved organic substances that were not removed in any previous processes. The organic materials "stick" to the surfaces of the particles. This "carbon polishing" step restores the wastewater to such an extent that people have considered re-using this water, after it has been filtered and disinfected, in public water supplies.

One of the earliest successful applications of tertiary treatment was at Lake Tahoe, Nevada, in the late 1960s. The clarity of this beautiful lake was threatened by algal growths caused by phosphorus and nitrogen compounds from the community's wastewater. Scientists recognized that the usual primary and secondary treatment steps could not stop the degradation of the lake. Hence, a tertiary treatment plant was opened at Lake Tahoe in 1965. The plant removes phosphate and ammonia and "polishes" the water by passing it through carbon columns to remove resistant organics. The wastewater—now clear, colorless, and odorless—is chlorinated before it leaves the plant. Over 95% of the dissolved organics are removed by the combination of primary, secondary, and tertiary processes at Tahoe, and the same portion of the phosphorus has disappeared as well. Nitrogen removal at Tahoe ranges from 50–95%.

The wastewater from the plant is never discharged to the lake. Instead, the water is pumped to Indian Creek Reservoir, some 27 miles distant, a reservoir created expressly to receive the plant's wastewater.

✍ Problems from Chlorination of Sewage

Though chlorination is commonly used at most treatment plants, there are other ways to disinfect wastewater. There are also reasons to consider these ways. At the same time, environmental engineers are reluctant to abandon chlorination. They understand how well chlorination has served in the past to purify water and prevent disease.

Nevertheless, ecological problems do occur as a result of chlorination. Chlorine and the compounds it forms with ammonia are toxic to fish. Some fish can be poisoned even at concentrations of chlorine as low as 0.002 mg/liter (2 parts per billion). Suppose a wastewater were being discharged into a river and suppose further that the available chlorine (free and reacted with ammonia) were 2 mg/liter. This is a fairly common level for available chlorine in such discharges. The wastewater would need to be diluted 1000 times to decrease the concentration in the receiving body to less than 0.002 mg/liter.

We mentioned in our discussion of drinking water that chlorinated hydrocarbons could form from the reaction of chlorine with hydrocarbons. These persistent and toxic substances, which have been detected in public water supplies, are suspected of being carcinogenic (cancer-causing) agents. They are discussed in greater detail in Chapter 29.

The chlorine, once added, can be removed from wastewater by contacting the water stream with sulfur dioxide gas, leaving sulfate and chloride ions in solution after the process is complete. The sulfur dioxide will even strip chlorine away from the compounds it forms with ammonia, but will not remove any chlorinated hydrocarbons that may have formed. Removal of chlorine by passing the wastewater through towers containing activated carbon is also being investigated. Such towers would also be useful in removing chlorinated hydrocarbons.

An alternative to chlorination of municipal wastes is ozonation. Ozone, though a powerful disinfectant, disappears quickly in contaminated water. No protection is left behind to destroy new microorganisms that may enter the water later. Chlorine does provide this continued protection, but it is just this continued presence that also makes it an ecological problem. In contrast to chlorine, which reacts with ammonia and hydrocarbons, ozone does not seem to form toxic by-products when it reacts with substances in water.

In the United States, only about twenty small communities use ozone to disinfect wastewater, but its popularity is increasing. Paris, France, however, does disinfect its vast quantities of wastewater with ozone. Ozone also seems to be effective at destroying some of the resistant organic compounds that persist through the last stages of treatment. Ozone is manufactured at the treatment plant in most applications, but its high costs continue to make chlorination popular.

✍ Control of Storm Waters

Runoff entering storm drains is simply water that was not absorbed into the ground. The flow of water from storms may, at times, be as much as 100 times the rate of domestic wastewater flow (called dry-weather flow), which averages 100 gallons per day per person. Since storm waters can so greatly exceed the flow of domestic wastewater, sewage treatment plants are designed only for domestic flow (as opposed to 101 times larger). When a storm does occur and storm waters enter the sewers, the two flows mix. However, the treatment plant can handle only its usual volume, the dry-weather flow. The remainder must overflow to the stream or river and is left untreated.

Storm water carries many contaminants, including substances washed from the streets by rain. Such substances as lead (from the lead in gasoline), particles of earth, dog feces, leaf litter, and unburnt fuels may be captured in the flow before it enters the sewers. Once the flow enters the sewers, it may scour up organic solids (including bacteria) deposited there by the slower-moving dry-weather flow. Thus, storm water is not at all clean; it is carrying metals, organics, and possibly pathogenic bacteria into the watercourse.

There are four ways to alleviate this problem: (1) The sewer systems can be separated (at a cost of billions of dollars). (2) The streets may be washed and/or swept on a regular basis. This action reduces the contaminants washing off from streets, but does not alter the scouring up of sediments in the sewer. (3) Very large storage tanks can be inserted in the system to hold the large flows; the tanks then release these volumes gradually to the treatment plant. The tanks may also serve as settling basins so that organic solids settle out of suspension while the water waits for treatment. (4) Facilities may be built to treat combined sewer overflows. Of course, an alternative may include a mixture of these basic options.

Even when separate systems are built, some storm water may enter the sanitary sewer during storm events. Such entry will occur through illegal connections of downspouts and basement floor drains and through leakage into the sanitary sewer system. The joint flow of wastewater and storm water in this separate sanitary sewer system may still be too great for a sewage treatment plant, again requiring bypass of the plant. Engineering devices may, however, be used to route the "first flush" of storm water flow through the treatment plant because of the high level of pollutants it has scoured up from the sanitary system.

While all four of the preceding options exist, in most cities little has been done to deal with the problem of combined sewers because of the enormous expense involved.

Questions

1. Complete the following chart, which lists the steps or procedures in treating sewage before it is released to natural waters, and the reasons for those steps.

Water Pollution Control

Procedure	What is done	Reason
Primary treatment	Settling, removes solid materials	Removes 35% of organic materials, which would otherwise use up oxygen in natural waters
Secondary treatment	Trickling filter removes: Activated sludge removes:	
Tertiary treatment or advanced waste treatment	Phosphate removal: Nitrogen removal:	
Digestion of sludge from primary and secondary treatment		

2. What is the difference between water treatment and water pollution control?

3. How is the sewage from your home treated? Does it go to a sewage treatment plant or into a septic tank or cesspool? If it goes into a septic tank or cesspool, summarize what happens in a few sentences. If it goes to a sewage treatment plant, does the plant give only primary treatment or does it have secondary and/or tertiary treatment equipment too?

Further Reading

Environmental Pollution Control Alternatives: Municipal Wastewater, U.S. Environmental Protection Agency, Technology Transfer, EPA-625/5-76-012.

This is a profusely illustrated, well-written document that deals with most primary, secondary, and tertiary processes for waste treatment, including European practice and emerging technologies. Since it does not go into process design, the reading remains nontechnical at all times.

Fair, G., J. Geyer, and D. Okum, *Water and Wastewater Engineering,* 2 Vol. New York: Wiley, 1968.

These two volumes are basic texts in environmental engineering programs and so are very complete and explanatory. They do, however, become technical in the sense of giving mathematical treatment to many subjects.

Berger, B., and L. Dworsky, "Water Pollution Control," *EOS,* **58** (1) (1977), 16.

The legislative and administrative history of water pollution control efforts by the federal government are covered. It was written by two very knowledgeable people who have served in the government in this area. Nontechnical.

The following three paperbacks, all with the same basic title, total about 200 pages. While there is some technical engineering material in them, on the whole, they are quite readable since they were designed to educate a general audience.

Land Treatment of Municipal Wastewater Effluents, Environmental Protection Agency, Technology Transfer Seminar Publication, 1976: (1) Design Factors—I; (2) Design Factors—II; (3) Case Histories.

Crites, R. and C. Pound, "Land Treatment of Municipal Wastewater," *Environmental Science and Technology,* **10** (June 1976), 549.

This article describes the three land application methods and reviews several examples of land application in the U.S.

(Department of Energy)

ART FOUR

Conventional Sources of Energy: Resources and Issues

In the fall of 1973, the United States came abruptly to the end of an era. This era was marked by the growth of cities, sprawling across the landscape; by record levels of auto ownership and use; and by the spread of interstate highways, slicing the nation. Transit and train travel dropped dramatically during this period. It was a time in which it was fashionable to erect glass buildings with windows that did not open. Insulation was largely ignored. It was, in short, an era of cheap energy.

The sudden increase in oil prices, which occurred in 1974 after the Arab oil embargo, was a clear notice to the United States that the era of cheap energy was coming to an end. Inflation, combined with a decreased energy resource base in the United States, contributed to the death of the era. A more fundamental reason, however, was the realization on the part of a cartel of oil producers that the industrial world was feeding at their trough.

Western Europe, the United States, and Japan were consuming far more energy than they themselves were producing. The industries of these nations require oil in vast quantities. Without oil, the entire industrial structure of the Western world would be threatened. As a result of this realization, the cartel set a new price for oil. The new price was high enough to make the industrial world aware that it had unwittingly become dependent on a group of supplier nations capable of draining off their money in huge gulps.

The embargo in 1973–74, followed by the six-fold price increase in the spring of 1974, was the first signal. In 1973, the United States was paying only two dollars per barrel for foreign oil. By 1981, the United States was paying an average of $37 per barrel for its imported oil. The price of oil dropped by almost a third in 1982, however, as a world-wide recession cut oil use drastically.

These are signals that we cannot ignore. The United States and the rest of the industrialized world must learn, and are learning, to conserve energy and need to discover and utilize new energy sources.

With the embargo and following price rise, we

entered a new era—the era of energy conflict, in which energy users are squeezed by energy producers. Such conflict logically precedes a time of scarcity. Since our energy resources such as coal, oil, natural gas, and uranium are limited, energy use will eventually strike up against the barrier of limited resources; this will be the era of scarcity. Although it has not happened yet, we now have warnings of the scarcity to come, warnings communicated by the higher prices we are paying for energy of all kinds. What has happened can be seen as fortunate in the sense that we have received an early warning of problems still only glimmering on the horizon.

The embargo and meteoric price rise were the results of a reduced energy resource base in the industrial world and an enormous energy base in certain less developed countries. The reality of the global resource situation has now brought us squarely up against the issue of energy resources, and it has happened a great deal sooner than most people would have predicted.

The reason the cartel could raise the price of oil so high so quickly is that the cost of finding and producing new oil in the industrial world is much higher than the price the producing nations had been charging for their oil. There was room for a price increase. Furthermore, in the short term, the industrial nations had no option but to pay the price, since new oil of their own could not be discovered and produced quickly enough to take the place of oil from the cartel.

In a way, the drastic price rise illustrates a classic argument of resource economists. These economists have said that resources are never actually used up, but simply become so costly that substitute resources begin to be found. Resource economists envision that as more of a resource is consumed, additional quantities become more difficult for suppliers to find. To continue to sell the resource at a profit when it is increasingly costly to find and produce, the supplier must charge a higher price. As the price of the resource is pushed higher by continued use and gradual depletion, users become interested in seeking substitutes that are less costly. Substitutes can even replace the original resource and become more widely used than the resource they displace.

Whale oil, shale oil, and coal oil are examples of energy resources from earlier years that were eventually displaced by a cheaper substitute. These oils were used to light lamps in colonial days and into the middle 1800s. It was in the last half of the 19th century that kerosene, a liquid derived from petroleum, replaced these oils because of its lower cost.

The harder it is to obtain a resource, the higher its price. This notion is so basic that the very statement of reserves depends crucially on the price of the resource in the market place. Oil in the North Sea, which cost $5–$6 per barrel in the middle 1970s, was not even counted as a reserve in the early 1970s, when the world price of oil was $2 per barrel.

The main point to be learned from events taking place from 1974 to the present is that the energy crisis is the result of the geographic and political separation of producers and suppliers. The crisis is not yet one of global demand butting up against global supply. Such a crisis may one day come to pass, but the likelihood of its happening is now far less, for we have had a clear warning notice nailed on our collective door.

In Part IV, we discuss conventional ways of generating power and the conventional energy resources we regularly employ. We describe the fundamental limit in the efficiency of converting heat energy to work or to electricity. This limit, imposed by the second law of thermodynamics, explains the source of one of our basic inefficiencies in energy use. Ways to improve the efficiency of electric generation are also noted.

With this background in our use of energy, we go on, in Chapter 16, to an investigation of conventional sources of energy. Coal, the troublesome, dirty fuel we are so rich in, is explored first. Its abundance and uses are contrasted to the host of environmental problems it creates. From the impact on the land of strip mining to the impact on humans of deep mining and power generation, there are numerous points at which coal can degrade the environment.

Oil and gas, two fuels cleaner than coal but in shorter supply, are discussed in Chapter 17. More than most other resources, oil and gas are surrounded by economic, political, and conservation issues that must be aired. What are our current sources of these fuels and what are the patterns by which we consume them? Strong clues on how we should conserve oil and gas will emerge from the answers to these questions. New sources of oil and natural gas are also being sought. For instance, oil-bearing shale exists in abundance in the Rocky Mountains, but the environmental consequences of developing this resource are very great. Natural gas, too, is available from many alternate sources.

In Chapter 18, we discuss uranium resources, which, upon enrichment in the component known as uranium-235, can serve to power nuclear electric plants. Nuclear energy, however, poses basic questions to society about safety, risk, and the spread of nuclear weapons. It is a resource that, if fully developed, will demand the most rigid control and security. The "nuclear option" raises grave issues, to say the least.

In the next major part of the book, Part V, the effects of these conventional energy sources on air and water resources is discussed in greater detail. For example, we explore how our pattern of consumption of various fossil fuels is at the root of most of our air pollution difficulties. Our energy use also produces global water pollution problems. While the transport and production of oil pollute the oceans, the disposal of waste oil provides an unnoticed but very large source of the oil that reaches our waters. Not only does oil pollution threaten aquatic species, but waste heat from power plants, dumped into rivers and oceans, upsets the natural ecological balance, upon which a diversity of species depends.

We seek new sources of fossil fuel to extend our resource base, but we can also seek new sources of energy not linked to fossil fuels—energy from tides, winds, the sun, and the earth's heat. These new sources, which we refer to as natural sources, not only extend our resource base, they also have far less impact on the environment. Many air and water pollution problems are simply swept aside by the use of these natural sources.

In Part VI we explore alternatives to conventional energy resources. Do they provide a solution to our energy crisis, or do they themselves have undesirable environmental effects? Finally, we look at methods of energy conservation, since here, at least, we seem to have a solution guaranteed to do no harm to the environment.

CHAPTER FIFTEEN

How We Use Energy Resources

Energy Conversions and the Second Law of Thermodynamics

Electric Power

Generation, Use, and Impact/Fusion—Electric Energy in the Future?/Fuel Cells and the Hydrogen Economy

The Uses of Energy and the Fuels that Supply It

New Developments in Conventional Electric Generation

CONTROVERSY:

15.1: *Should Electric Companies Be Publicly or Privately Owned?*

Energy Conversions and the Second Law of Thermodynamics

We commonly utilize energy resources in three fundamental ways. We may release heat energy by burning fossil fuels and use this energy directly as heat: heat for homes, schools, factories, and shops. Or we may convert this heat energy into work, using refined oil to drive machinery and power automobiles, trucks, trains, planes, and ships. Last, we may convert the heat energy from burning fossil fuels or from the fission of uranium into electricity and then use this transformed energy either for heat or to do work. Falling water is also commonly used to generate electricity. Electricity, in effect, acts as a middleman between energy resources and final use (Figure 15.1). Just as the presence of a middleman in a market creates higher prices, the use of energy in the form of electricity brings a markup as well—in this case, an energy loss or penalty.

The practice of converting resources into electrical energy arose for many reasons. In some cases, it was not possible to make effective use of certain energy sources except to convert them to electricity. Before

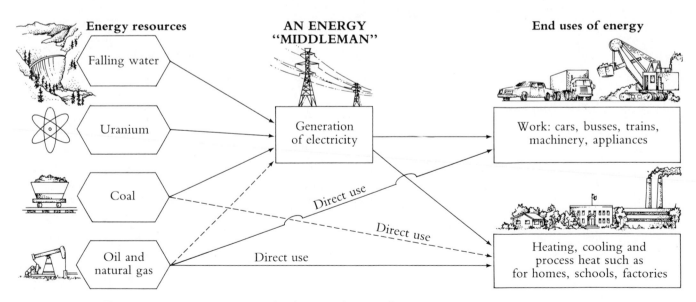

Figure 15.1 Common Routes for the Use of Energy Resources. Electricity is an energy middleman that takes the basic energy resource, processes it, and transfers it to the final user. Like any middleman, it demands that a toll be paid.

the era of electricity, the energy from falling water (hydropower) could only be used to drive machinery at the waterfall itself. Textile mills, grain mills, and lumber mills all operated at waterfalls for a time in early manufacturing history. Hydropower could be used in no other way until it could be used to generate electricity, which then made it possible to operate machines at sites distant from the falling water. Uranium, likewise, is not generally used to fulfill energy requirements except in the form of electric energy. Then, however, like hydropower, it can be used not only to drive machinery, but also to produce heat for homes (an inefficient use), for hot water, and for other uses.

The fossil fuels, in contrast to falling water, were first recognized as sources of light and heat and not as sources of energy to power machinery. Wood and coal, and often dried peat, were burned for heat in homes and commercial buildings, and coal was used to provide the heat needed to manufacture iron. "Carbon oil" from coal was used in lamps. Not until the invention of the steam engine in the 1700s was the potential of fossil fuels to drive machinery discovered. By the early decades of the 1800s, railroad locomotives were being built in the U.S. that burned coal to generate the steam for power. And by the early decades of the 20th century, coal was burned to provide steam to generate electricity, although the production was very inefficient at the time.

Thus, fossil fuels began to assume a new role in the industrial world. From merely providing heat and light, they came, through the use of engines and electricity, to be sources of power for industrial machinery and for transport, as well.

These historical events color our present attitudes. The use of coal and other fossil fuels to generate electric power for machines is very appealing. Since electric power can be transmitted from place to place, machinery can be located far from the site of electricity generation. Electric heating of homes is also appealing: no fuel is consumed in the home; no ashes and smoke are produced there. But there is a penalty for using energy in the form of electricity, far beyond the extensive pollution we so easily ignore at the point of electric generation. This penalty is an enormous loss of efficiency—an enormous waste of the heat energy in the fossil fuels.

Figure 15.2 The inefficiency of generating electricity: the example of oil.

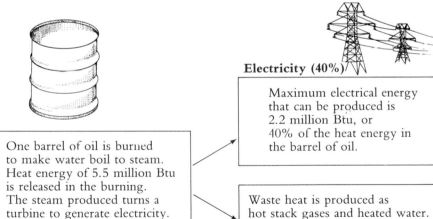

One barrel of oil is burned to make water boil to steam. Heat energy of 5.5 million Btu is released in the burning. The steam produced turns a turbine to generate electricity.

Electricity (40%)

Maximum electrical energy that can be produced is 2.2 million Btu, or 40% of the heat energy in the barrel of oil.

Waste heat is produced as hot stack gases and heated water. Heat energy of 3.3 million Btu or more is wasted, equal to 60% of the heat energy in the barrel of oil.

Waste heat (60%)

As a consequence of the second law of thermodynamics, there is an inherent limit to the amount of heat energy that can be converted into work and hence into electricity. The reverse is not true; that is, other forms of energy can be converted completely into heat. This limit to the efficiency of converting heat into work was not an obvious concept to the early designers of engines that used heat for power. An engineering text[1] tell us that:

> . . . the low efficiencies encountered in the conversion of . . . heat to work were at first attributed to incorrect design of the mechanism employed, or to practical difficulties in operation. Even after careful improvement, however, low efficiencies prevailed. When extensive efforts at improvement proved fruitless, it was finally inferred that there existed some underlying natural limit to the conversion of heat to work.

In fact, most of the possible design improvements in the efficiency of generating electric power from steam have now been made. The modern coal-fired power plant, which produces steam to turn a turbine for generating electricity, has reached 40% efficiency. That is, 40% of the heat in the coal that is burned is converted into electric energy. Oil-fired power plants push up into this range as well. Conventional approaches to electric power generation cannot be expected to increase this efficiency by very much (Figure 15.2). Nuclear electric power plants also produce steam to turn a turbine, and these have attained efficiencies of only 30–32%. That is, only 30–32% of the heat of fission is converted to electrical energy. Attempts to increase this conversion percentage have yet to succeed.

The generation of electricity, then, which is a convenience, and a very great convenience, to be sure, is wasteful of the heat energy of the fossil fuels coal, oil, and natural gas. It is especially wasteful when the electric energy is only converted back into heat at its point of use. Because so much of the energy we use is in the form of electric energy, we need to discuss more completely the sources and uses of electric power, as well as the impact upon the environment that accompanies this use.

---

1 H. C. Weber and H. Meissner, *Thermodynamics for Chemical Engineers,* 2nd ed. New York: Wiley, 1957.

Electric Power

A flick of the finger and we have

> light;
> sound;
> visual communication;
> heat;
> hot water;
> refrigeration; and
> air conditioning;

all from electric energy.

A flick of the finger and we also have

> river valleys buried in sediment;
> sulfur oxides, acid rain, acid mine drainage;
> oil pollution;
> particulate matter;
> nitrogen oxides and smog;
> strip mining;
> thermal pollution;
> radioactive wastes; and
> plutonium for bombs;

all from electric energy.

Generation, Use, and Impact

Electric energy is one of the most easily used forms of energy. One electric line enters a house, and the energy to light the home, cook and bake, warm water, and run machines and appliances is all at our fingertips. The first list above gives us an idea of how widely useful electric energy is. The second list reminds us that the clean fuel at the point of use is anything but clean at the point of mining and at the point of generation.

The use of electricity for household tasks is steadily increasing. Electrical appliances are becoming more widely available as time goes on, as the graph in Figure 15.3 shows.

Because electricity is such a convenient form in which to use energy, it is expected to capture more and more of the energy market. According to the Energy Information Administration, the quantity of electric energy used in the U.S. will increase by almost 53% from 1980 to 1995, even though total energy use will increase by only about 22% in that same period (Figure 15.4).

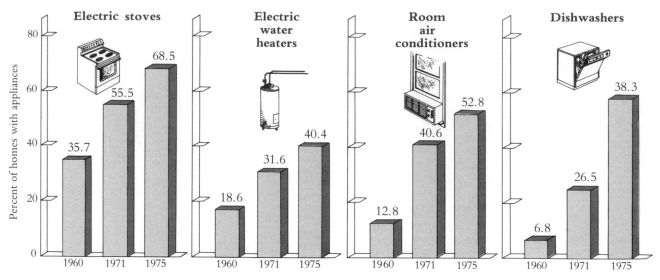

Figure 15.3 Growth in Percentage of Homes with Certain Electric Appliances. (Note that the number of homes increased by about 25% from 1960 to 1971.) (Sources: "Energy and the Environment: Electric Power," The President's Council on Environmental Quality, August 1973; and *EPRI Journal*, December 1977.)

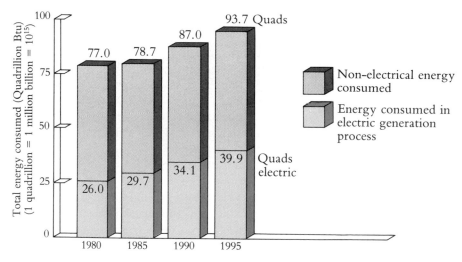

Figure 15.4 Projected Growth in Total Energy Consumption and Growth in Energy Consumed in Electric Generation in the U.S. The energy consumed in electric generation (the areas shown in color) is calculated from the consumption of electric energy in end uses, using the assumption that electrical energy consumed *in end uses* = 30% of the energy consumed in generating the electricity. That is, 70% of the energy consumed in generating electricity is lost, either wasted to the environment or lost in transmission, and only 30% is left for end uses. (1980 figure is reported; other numbers are projections. Source: Energy Information Administration.)

It should be noted that the quantity of energy *consumed* for electric generation at power plants is not, in fact, the quantity of electrical energy *generated* at power plants. Nor is it the quantity consumed in end uses such as heating water. It is, instead, the far larger quantity of energy that was released in the burning of coal or fissioning of uranium at the power plant to produce the electrical energy used in application and lost in the process of transmission. The quantity of electrical energy actually put to use is about 30% of the energy consumed for purposes of electrical generation. This is a result of the inherent inefficiency of electric generation, a consequence of the second law of thermodynamics. Now read this paragraph again to be sure you understand this subtle point.

The energy consumption numbers of Figure 15.4 are given in quadrillions of Btu.[2] A quadrillion Btu, or Quad as it has come to be known, is equal to the heat energy in 172 million barrels of oil or in 44.4 million tons (40 million metric tons) of coal. In 1980 in the United States, one quad could have powered 12 million cars for one year or heated 20 million well-insulated homes for a year.

To understand how fuels are used to generate electricity, we need to describe the operation of a typical "thermal" electric plant (Figure 15.5). By thermal electric plant, we mean a power plant that utilizes heat to produce steam, which turns a turbine. A coal-fired plant is a thermal electric plant; so is one that burns oil

---

2 The Btu, or British Thermal Unit, is a unit of heat energy. One Btu will raise the temperature of 1 pound of water by 1°F.

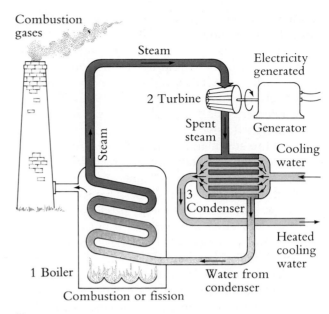

Figure 15.5 Elements of a typical thermal electric power plant.

or gas. A plant that utilizes the heat from nuclear fission is also a thermal electric plant. The principle of operation in all three is the same.

Heat is produced by combustion (or fission). That heat is used to boil water to steam. The steam, at high temperature and pressure, is used to turn a turbine and generate electricity. The turbine's rotation of an armature in a magnetic field induces the flow of electric current. The steam that leaves the turbine has its pressure and temperature much reduced. This "spent"

Beware the Meter

(Following is an excerpt from an editorial that appeared in
The New York Times or, as it was spelled then, *The New-York
Times,* on June 11, 1881.)

Now, there is not the least reason to believe that electricity exerts any better moral influence than gas, and we know that when a number of reasonably Christian men form themselves into a gas company they immediately become pirates of the most merciless and extortionate character.

Why should we look for better things from the electric light companies? They expect to have us at

their mercy, and they will be as merciless as the gas men.

We shall have electric meters in our cellars that will be as mendacious and unprincipled as the gas meters, and the moment we refuse to pay for ten thousand feet of electricity which we have not used our lights will be cut off and we shall be left to candles and kerosene.

steam is converted back to water by a condenser, through which cooling water flows.

The water used to condense the steam has now been heated. It may be returned to the body of water from which it was taken, or alternatively, it may be passed through towers to be cooled and used again in the condenser. The water produced when the steam was condensed now re-enters the boiler to be again converted to steam and circulated to the turbine.

A coal-powered station wastes 60 units of heat to the environment for every 40 units converted to electricity; it works at 40% efficiency. That is, 1.5 units of heat are wasted per unit of energy produced. A nuclear plant, at 30% efficiency, wastes 70 units of heat for every 30 units converted to electrical energy. Thus, 2.33 units of heat are wasted per unit of energy produced at the nuclear power plant. A nuclear plant wastes to the environment about 55% more energy than the coal plant per unit of electric energy produced.

The problem is compounded for a nuclear plant because all its waste heat is transferred to the condenser cooling water. A coal-fired power plant, in contrast, directs only 75% of its waste heat to the cooling water. The remainder of the heat from the coal-fired plant goes up the smokestack.

Electric energy is generated at fossil-fuel power plants, at nuclear power plants, and at hydroelectric plants. The combination of plants from which we draw electricity has been constantly changing for a variety of reasons. During the late 1960s, coal-fired power plants were being converted to residual oil in an effort to reduce sulfur dioxide levels in urban areas. With the onset of the 1973 oil embargo and a massive oil price rise, coal plants, with better pollution control or burning lower-sulfur coal, began to replace the oil-fired plants. A program to convert to coal because of its more certain availability was carried out in an effort to reduce oil imports. In the late 1960s and into the late 1970s, new nuclear plants were being added quickly to the stock of generating stations. But as resistance to nuclear power grew, the number of orders for new nuclear stations dropped to only several plants per year in the late 1970s. Thus, the combination of plants, as we show it for 1980 (Figure 15.6), is actually a snapshot of an evolving system. By the time you read this, oil's share of electric generation will have fallen. Gas use will also fall as natural gas is diverted to use in heating homes. Coal use will increase not only because

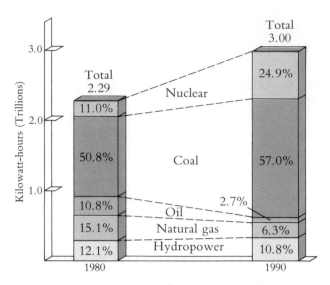

Figure 15.6 Fuels for Electric Generation in the U.S., 1980 and 1990. 1980 column is reported data; 1990 column is a projection. (Numbers may not add due to independent rounding.) (Source: Energy Information Administration.)

of conversions from oil to coal, but because much of the new generating capacity will be in coal-fired units. Nuclear use will have grown because the nuclear units already under construction will have come "on line" in this period. Hydropower will lose a portion of its market share as the market expands because few new large hydroprojects are possible.

That electric generation is not a clean source of power is vividly shown in Table 15.1. Here we can see that emissions from electric generation account for a high proportion of the annual emissions of three major air pollutants. The wide range of environmental insults caused by electric power generation will become clearer in the chapters that follow.

Fusion—Electric Energy in the Future?

In the midst of recurring energy shortages that foreshadow an era of scarcity to come, the question occurs again and again, "What about fusion?" Fusion is not only a specific atomic event that scientists hope to harness to produce electric power; to many, it has become the name for the "dream of limitless energy." Is fusion one piece of the answer to energy-supply problems of the future? How soon could it contribute to our basic needs for energy? We need to describe fusion to approach answers to these questions.

Table 15.1 Sources of Selected Air Pollutants[a]

	Particles, million tons per year	Sulfur oxides, million tons per year	Nitrogen oxides, million tons per year
Electric generation	4.5[b]	16.0	7.1
Residential and commercial combustion	0.6	2.9	1.1
Industrial combustion	2.0	3.2	2.7
Other industrial processes	14.4[c]	7.5	0.2
Transportation	0.8	1.1	11.2
Miscellaneous	12.8	0.4	2.4
Total	35.1	31.1	24.7

a From *Energy/Environment Fact Book,* March 1978; USEPA and USDOE EPA-600/9-77-041 The contribution of electric generation to the production of hydrocarbons and carbon monoxide is negligible.
b If not under control, this amount could be multiplied by a factor of 10.
c Grinding, spraying, demolition, construction, mining, cement plants, grain handling, etc.

While nuclear fission refers to the *splitting apart* of the nuclei of large *heavy* atoms with a release of energy, nuclear fusion refers to the *combining* of the nuclei of certain *light* atoms, such as two atoms of deuterium, an isotope of hydrogen. This combination, which takes place constantly in the sun and stars, releases vast quantities of heat energy. The notion is that the fusion reactor of the future would capture fusion heat and convert it to electric power.

The fuel for fusion reactors is deuterium, an isotope of hydrogen with one proton and one neutron in its nucleus. (Ordinary hydrogen has only one proton and no neutrons in its nucleus.) Deuterium is so abundant in the surface waters of the earth and the fusion reaction is so energetic, that if the energy from deuterium reactions were fully extracted, only eight pounds of water (about a gallon) could supply the energy equivalent of 300 gallons of gasoline. Unfortunately, it is not clear that fusion can ever be exploited because of our inability to somehow "house" the fusion reaction. Thus, cost estimates for fusion really have no basis, for we are not at a stage where we can say that fusion will work.

Containment is needed for a fusion reaction because fusion takes place only at the temperature of a star, 100 million degrees Celsius. Since no substance is known that remains solid at this temperature, scientists have devised several novel concepts to "hold" the fusion reaction. These include the famous *tokamak* concept in which the fusion reaction is held in place by powerful magnetic fields. Another concept that does not use magnetic confinement is laser fusion, in which converging laser beams focus on a near microscopic pellet of "fuel." The heat from the laser beam produces the temperatures needed for fusion. Many nations are now conducting fusion experiments.

The environmental safety of the future fusion reactor is an often-mentioned advantage for fusion. In fact, there may be a significant quantity of radioactive waste to dispose of. Most fusion reactions being considered now produce high-speed neutrons, and the neutrons bombarding the walls of the reaction vessel will make the wall radioactive and a waste to be disposed of somehow.

When active research on fusion began in the late 1950s, the first fusion power was estimated to be 20 years away. Today, nearly a quarter of a century later, scientists are still saying "20 years away." It makes sense to pursue dreams; dreams and ideas are the forerunner of reality. It makes sense to pursue fusion, but we should understand that it cannot contribute to our needs "here and now."

Fuel Cells and the Hydrogen Economy

The **fuel cell** is the name given to the concept of producing electricity from the chemical reaction of hydrogen and oxygen. The aerospace program pushed this concept far enough along that fuel-cell power plants are now being built, though the first plants are small in size. At the site of electric generation, fuel cells produce very little in the way of air pollution, thermal pollution, or noise pollution. Hence, as urban neighbors, they are expected to be more welcome than the smoky, sooty, coal-fired plant or the nuclear power plant with its built-in hazard for populations.

Whereas the oxygen needed for a fuel cell can be supplied from the air, hydrogen gas must be obtained by chemical processing. Hydrogen may be obtained

Should Electric Companies Be Publicly or Privately Owned?

> [The] impetus toward municipal ownership now is the higher rates of the private utilities and the feeling that local ownership would be more responsive to public needs.
>
> *Alex Radin, General Manager,*
> *American Public Power Association*★

> Higher electric bills are simply a reflection of the higher cost of everything a utility must buy to provide service.
>
> *W. Donham Crawford, President,*
> *Edison Electric Institute*†

About 80% of the power generated in the U.S. comes from investor-owned utilities. However, a number of electric companies are not investor-owned but are owned by local governments. These publicly owned companies generally have lower electric rates than investor-owned utilities and provide a better reception for public opinion. The reasons for lower electric rates are that a municipally owned utility does not have to pay dividends to stockholders or federal taxes on its profits; nor does it pay local property taxes.

The interest on the bonds issued by an investor-owned utility are taxed as income for the owner of the bond. In contrast, the interest on the bonds issued by state and local governments is not liable for federal taxation. Because interest on the municipal bond is not subject to taxes, purchasers are willing to accept a lower interest rate. Thus, the municipal utility is able to raise money for new construction at a lower cost than the investor-owned utility. As a consequence of this and the lack of taxation on its property or income, the municipally owned utility is generally able to offer lower electric rates to its customers.

However, since a municipal utility escapes local taxes, the local government must obtain its required revenues by increasing taxes on its remaining property. The lower electric rates may, in part, then, be an illusory saving for customers because of the nonpayment of local taxes by the utility. Lower electric rates due to tax-free bonding power and lower electric rates because no stockholder dividends need be paid are very real, though.

In addition, public utilities should also be more willing to save the public money by using peak-load pricing; that is, charging higher prices for electricity generated at the hours of peak use. The resulting lower peak demands mean less investment required for new capacity and, thus, lower electric rates. Privately owned utilities, in contrast, like to expand their capacity and investment because they are allowed to earn a rate of return on their investment.

Finally, utilities owned by local governments are expected to be more accessible to public opinion on issues such as location of its facilities and whether or not to use nuclear power. On the negative side, one must face the issue of efficiency—whether public enterprises are as efficient as private ones. It is this issue that gives the argument for public power two sides.

Do you think power companies should be publicly owned? If yes, why? If not, why not? Do you think a publicly owned power company will be equally as efficient as an investor-owned company? More responsive?

★ *Wall Street Journal*, 23 October 1974.
† *Wall Street Journal*, 23 October 1974.

by processing a liquid hydrocarbon fuel such as naptha, or processing natural gas or petroleum gas. A first-generation fuel cell, with about 5% of the power capacity of large modern coal or nuclear plants, is being installed in New York City to demonstrate the feasibility of fuel cells. An inverter will be needed to convert the direct current electricity produced by the fuel cell to the alternating form that we use throughout the country.

In a second-generation concept of the fuel cell, two fuels combine with oxygen: hydrogen and carbon monoxide. Both of these fuels are produced in the initial stage of coal gasification (see the supplemental section of Chapter 17). At present, this cell is the subject of research; full-scale cells are at least a decade away. The fuel cell's advantage of almost pollution-free power generation would disappear if the coal gasifier were nearby because the coal gasifier is a generous producer of air pollutants.

The hydrogen-using fuel cell is one example of how hydrogen may be used in the future. Hydrogen is not a new energy resource, however. Instead, it is a new way to convey energy. Hydrogen from the electrolysis of sea water might be produced by first generating electrical energy from the wind; the wind is the primary energy source. Hydrogen might also be produced from sea water using electricity from conventional power plants; the fuel that fired the power plants is the base energy resource.

In addition to its use in fuel cells, hydrogen is being explored as a fuel for the internal combustion engines of automobiles. It is also being considered to power turbine engines such as those in jet planes, and since a similar turbine engine is used to produce extra electrical power at times of peak demand, hydrogen might be used here as well. Such future uses of hydrogen are not at all certain. Some see such uses in their crystal ball; others do not.

The Uses of Energy and the Fuels that Supply It

To save energy, we first need an idea of how we presently use energy. We also need to know what energy resources we rely on. This information will help us determine how we can conserve energy and the fuels that can be saved.

Figure 15.7 shows how we have historically consumed energy. In general, the pattern of consumption has not changed drastically over the past quarter century. Roughly the same proportions exist in each economic category from 1950 to the present. Actual energy consumption increased by more than 100% between 1950 and 1981. A population increase of about 50% accounts for a portion of this increase in consumption. However, the average annual energy con-

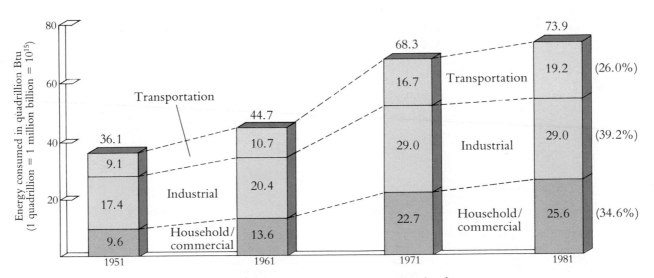

Figure 15.7 The pattern of growth in U.S. energy consumption by economic sector. (Source: Energy Information Administration.)

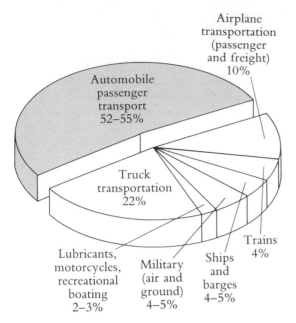

Figure 15.8 How transport energy in the United States is consumed.

of this transport energy in the United States is consumed as some form of petroleum. About 52–55% is burned in automobiles; another 22% is consumed in transportation of goods by truck. Airplane operations account for another 10% and railroads for only about 4%. Buses, pipelines, ships and barges, military air and ground vehicles, and other minor consumers use the remainder of U.S. transport energy (Figure 15.8). Although nearly all of the transportation energy in the U.S. comes from oil, in Europe this is not the case. In Europe, trains still play an important role, and most rail systems in Western Europe have been electrified, with the electricity being generated from coal and nuclear energy.

We can break down the household/commercial category of energy use as well. Nearly three-quarters of the energy consumed in this category goes into keeping people either warm or cool [Figure 15.9(a)]. When figures were last tabulated, heating used 63% of all energy in this category. Air conditioning consumed 8% of the energy in the household/commercial sector.[3] Water heating accounted for 14% of this energy;

sumption per person increased about 55%, from 210 million Btu per person annually in 1950 to 330 million Btu per person in 1979.

Transportation has steadily accounted for about 25% of the energy we consume each year. Nearly all

---

3 The figures are for 1970, but a hotter summer or colder winter could change the numbers, especially as air conditioning spreads to more of the population.

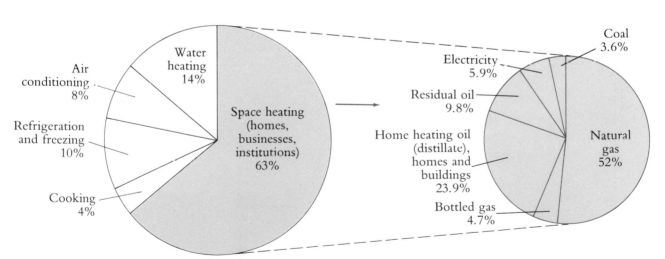

(a) Household/commercial energy use, 1970

(b) Space-heating energy use by fuel

Figure 15.9 (a) Pattern of household/commercial energy consumption by use in 1970. Of the energy consumed in this sector, 28% was electrical. (b) Pattern of space heating energy consumption by fuel type in 1970. The 9.8% residual oil was used principally to heat office buildings, schools, hospitals, etc. (Adapted from D. Kash et al., *Energy Alternatives,* A Report to the President's Council on Environmental Quality, 1975)

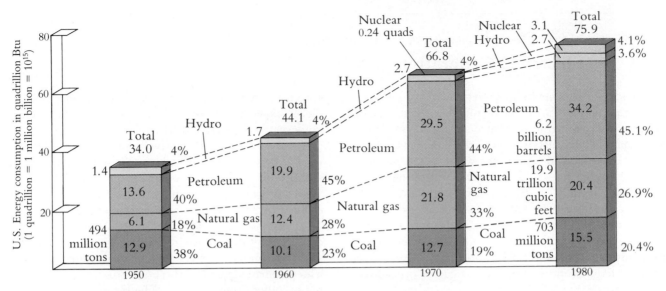

Figure 15.10 Pattern of growth in U.S. energy consumption by fuel type. (Source: Energy Information Administration)

refrigeration and freezing, 10%; and cooking, 4%. Air conditioning and refrigeration together make up 18% of the use of this category. Both of these uses draw on electricity as virtually their sole source of energy. If we include the electrical energy consumed in water heating, space heating, and cooking, another 10% of this category's energy derives from electricity, for a total of 28%.

In Figure 15.9(b), the space heating category is broken down by fuel type to show the contributions of the various fossil fuels.

As the energy uses in various sectors expand, so also must our use of fuels. Figure 15.10 displays our energy growth in the last quarter century in terms of the fuels we have consumed. From this chart, we can see that petroleum and natural gas have expanded their portion of the energy market since 1950, while the portion using coal has declined. From 58% of the energy used in 1950, petroleum and natural gas expanded to 72% of the energy market in 1980. While coal appears to have lost ground in percentage terms, it has not lost ground in actual amounts used. The 1980 production of this fossil fuel was 40% more than in 1950. Coal did lose growth in two important markets: the railroads, where coal gave way to petroleum, and home heating, where coal was displaced by oil and natural gas.

From the 1950s onward, coal as a means of heat-ing homes lost ground steadily to natural gas and oil. Coal was dirty and produced large quantities of ash, which the home owner added to the trash. Coal also required a large storage area in the home (the coal bin), and the coal furnace needed frequent shovel feeding. Natural-gas heating systems were compact, automatic, and needed no storage area in the home. Oil heat, although requiring a tank for storage, was also compact and automatic. These fuels, which were inexpensive when they began to become popular, displaced coal in most space heating applications and still hold most of this market today.

Since the oil embargo of 1973–74, with the following sharp increase in oil prices and the shortages and price increases in natural gas, the nation has begun to re-evaluate the importance of our coal resource. It is the fossil fuel in which we are richest. Perhaps 20% of the world's supply of coal lies within our borders. All plans to fill the growing U.S. demand for energy focus upon coal as a central element. Our coal, however, is a mixed blessing. It has serious effects upon the land and upon our health. Because coal is now seen as so important to the nation's economic health, and because it also has a very large negative impact, we need to explore fully the consequences of our plans for its use. As we enter onto a new path, we should be aware, to the largest extent possible, of its condition. It may be a rocky road indeed.

New Developments in Conventional Electrical Generation

Although improvements in the efficiency of conventional electric generation have come almost to a halt, novel and unconventional routes of improvement remain open. The thrust of most of this research is to put processes together to capture the waste heat from electric generation. We will discuss three promising ideas. The first is called "district heating" and uses spent steam for heating. The second, known as "combined cycle," puts together the turbine engine and the conventional boiler/turbine system. The last is **magneto-hydrodynamics** (MHD, for short) and is designed to extract energy from the hot exhaust gas of a coal-fired electric plant.

District Heating

One option to make electric generation more efficient is to make use of the waste heat, the heat from combustion or fission not converted to electrical energy. The spent steam from the power plant can be transported via underground pipes to homes, offices, and factories to provide space heating during winter months. District heating, as this process is called in Europe, has been achieved in Finland and elsewhere, but its prospects are diminished by the large quantity of pipes that must be placed to deliver the heat to customers. The concept also requires small power stations, with customers for the heating service located very nearby to decrease heat losses. Since smaller plants are less economical than larger ones and since most Americans do not like to live near power plants, the prospects for district heating are not bright at the present time.

Combined Cycle

To describe the combined cycle method requires a brief description of the turbine engine, a device used to power jet aircraft. These improved turbine engines have been adapted to power generation and have become common equipment in electric generating stations.

The turbine engine is turned by the combustion gases from the burning of kerosene or natural gas, instead of by steam. Only about 25% of the heat energy produced is converted to electric power, and the fuel is expensive compared to coal. The equipment, however, compared to the boiler/turbine system of conventional plants, is relatively inexpensive. Electric utilities buy these turbine generators and hold them on standby for times when power loads reach levels beyond the capacity of their base-load equipment. At those times, the turbine engines are pressed into service. For these peak loads, a slightly more expensive fuel does not drive costs too high.

The combined cycle idea is really quite simple. The turbine engine and the boiler/turbine system are put together (Figure 15.11). The burning gases in the turbine engine generate power from the rotation of the engine shaft. The very hot gases discharged from the turbine engine are then used to boil water to steam. The steam, in turn, is used in the ordinary way in electric power plants to turn another turbine for additional electric generation.

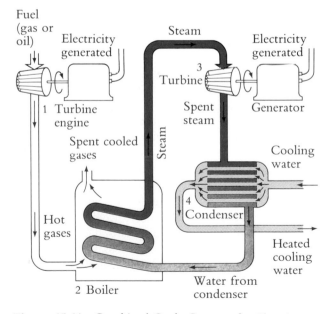

Figure 15.11 Combined Cycle Concept for Electric Power Generation. This concept puts together the conventional power generation scheme with "front-end" generation using a turbine engine. The hot gases from the turbine engine are used to boil water to steam for conventional electric generation.

Magnetohydrodynamics (MHD)

The last major extension of modern generating plants is known as magnetohydrodynamics; it makes use of the hot gases from combustion. The combustion gases are seeded with potassium, which ionizes into charged particles in the very-high-temperature flow. The hot gases with ionized particles are directed through a channel surrounded by coils to induce a magnetic field. The movement of charged particles through the magnetic field sets up a current flow that is drawn off by electrodes spaced along the channel. Upon exiting the channel, the hot gases are used to produce steam, which is then used to turn a turbine in the ordinary way.

The combined MHD–conventional system could reach efficiencies of up to 60%. That is, 60% of the heat in coal would be converted into electrical energy. If such plants are built, they will be far less polluting than the 40%-efficient conventional plants. For every kilowatt-hour of electrical energy produced, the 40%-efficient plant will consume 50% more fuel than the 60%-efficient plant. That means 50% more thermal pollution, particles, sulfur oxides, and nitrogen oxides from the conventional plant.

One of the problems still to be overcome is obtaining materials that will not corrode in the high temperatures (2000°C) in the MHD equipment. No large generators have yet been built, although a good-sized developmental unit has been operated in Russia. MHD research in the U.S. seems to be well funded at the experimental level.

Questions

1. From what you have read about electric power, do you think that the use of electric autos will help solve the pollution problems of the United States? Explain your position as fully as possible. Will it help one kind of geographical area and hurt another?
2. According to the U.S. Energy Information Administration, electric energy consumed in *end uses* for 1980 and projected for future years (all in quadrillion Btu) is

1980	7.80
1985	8.92
1990	10.23
1995	11.97

Yet in Figure 15.4, the consumption of energy in electric generation shows as higher numbers (again in quadrillion Btu):

1980	26.0
1985	29.7
1990	34.1
1995	39.9

What is the explanation for these two different figures on electrical energy consumption?

Further Reading

Phillips, Owen, *The Last Chance Energy Book*. Baltimore; Johns Hopkins University Press, 1979.
This short (140 pages) book is an eminently readable, entertaining and thoughtful treatment of our energy dilemma. Its discussion of alternative energy sources is very well done, as is the treatment of the theory of resource exploitation.

Energy: The Next Twenty Years, Report of a Study Group sponsored by The Ford Foundation and administered by Resources for the Future. Cambridge, Mass.: Ballinger Publ. Co., 1979.
Oil, coal, nuclear, solar, projections, economics—all are here, thoroughly documented and comprehensive; one of the most complete studies of the 70s.

Hayes, E. T., "Energy Resources Available to the United States," *Science,* **203** (19 January 1979), 233.
Although this work is a nontechnical treatment of conventional sources, the reader will find very little discussion of natural sources of power (see Part V of this book).

Kash, Don, et al., *Our Energy Future*. Norman, Okla.: University of Oklahoma Press, 1976. (Originally published as Energy Alternatives: A Report to the President's Council on Environmental Quality, 1975.)
Slightly dated, but still a comprehensive source of information on most conventional energy alternatives, resources, and uses.

Metz, W. D., "Magnetic Containment Fusion: What Are the Prospects?" *Science,* **178** (20 October 1972), 291.

Metz, W. D., "Laser Fusion: One Milepost Passed—Millions More to Go," *Science,* **186** (27 December 1974), 1193.

Rose, D. J. and M. Feirtag, "The Prospect for Fusion," *Technology Review* (December 1976), 21–43.

Holdren, J. P. "Fusion Energy in Context: Its Fitness for the Long Term," *Science,* **200** (14 April 1978), 168.

Fickett, A. P., "Fuel-Cell Power Plants," *Scientific American,* **219**(6) (December 1978), 70–76.

Waldrop, M., "Compact Fusion: Small Is Beautiful," *Science,* **219** (14 January 1983), 154.

CHAPTER SIXTEEN

Coal: A Mixed Blessing

Introduction

Coal and Its Uses
How Much Coal Is There?/How Is Coal Being Used?

Environmental and Social Impact of Coal
How Coal Is Mined/Surface (Strip) Mining/Underground Mining

Health and Safety of the Coal Miner
Mine Safety/Black Lung Disease

Impact of the Coal-Fired Electric Plant

Transportation of Coal

CONTROVERSIES:

16.1: *Strip Mining: What Are The Implications of Expansion on Jobs? on the Environment? on Human Health?*

16.2: *What Should Be Done with the Coal in the West?*

Introduction

Coal took us out of the era in which only human muscle, wind, and water power were available to manufacture goods. Coal brought us into the era of machinery powered by combustion. Watt's steam engine, powered by coal, made it possible for us to have machinery that was not linked to rivers. Coal was also the fossil fuel that first changed our methods of transportation—coal powered the steam locomotive.

Although the major use of coal in the U.S. today is as a fuel for steam electric power plants, it was once an important fuel for home heating and for transportation. After World War II, home owners began to switch from coal to oil and the newly available natural gas. Railroads also converted, from coal-fired locomotives to the more dependable and cleaner diesel engine.

There are four basic kinds of coal: peat, lignite, bituminous, and anthracite. All are derived from natural processes acting on wood, and all are used as fuels. These basic forms reflect differing levels of carbon content.

Peat, the least developed form of coal, is still highly woody in nature. Peat is found in bogs where wood, immersed in water, has decayed in the absence of oxygen. Historically, peat has been used only for heating.

Lignite, or brown coal, has a slightly greater heat and carbon content than peat, but still has less than the higher forms of coal. Most lignite resources of the country can be mined from the surface (strip-mined).

Bituminous coal is characterized by a higher heat and carbon content than lignite. Sub-bituminous coal is a grade of coal whose heat content is just below that of bituminous coal, but above that of lignite. Of the four types of coal, we rely most heavily on bituminous coal because of its relatively high carbon content and its abundance in the United States.

Anthracite coal has the highest carbon content of all coals. However, its current production and use in the United States is very small because its reserve base is so low.

Coal is not a pure substance. In addition to carbon, it contains inorganic material that remains after coal has been burned; we know the material simply as ash. Sulfur also occurs in coal, sometimes as iron sulfide, and sometimes combined with organic compounds. Arsenic is also present in coal, as are radioactive elements. In fact, coal is the dirtiest of the fossil fuels.

Dirty though it is, it is a magnificent source of heat energy. Burning one pound (0.454 kilograms) of bituminous coal releases 13,000 Btu (13,700 kilojoules) of heat energy. Furthermore, coal is the most plentiful fossil fuel we possess in the United States.

Coal and Its Uses

How Much Coal Is There?

As we have seen, coal has a range of heating values, depending on its form. Peat has the lowest heating value; lignite is next, followed by sub-bituminous and bituminous coal; and finally, by anthracite, which has the highest heating value. To give an accurate accounting of our coal reserves, we need to provide reserve estimates for each major kind of coal. We also need to know how much of the reserves can be surface mined and how much must be mined underground.

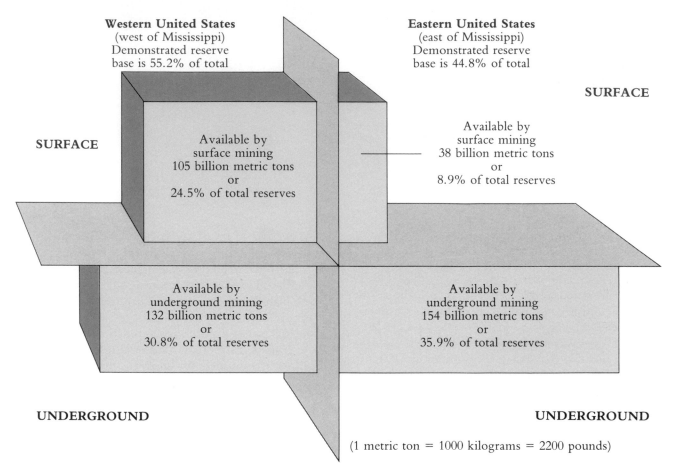

Western United States
(west of Mississippi)
Demonstrated reserve
base is 55.2% of total

Eastern United States
(east of Mississippi)
Demonstrated reserve
base is 44.8% of total

SURFACE

SURFACE

Available by
surface mining
105 billion metric tons
or
24.5% of total reserves

Available by
surface mining
38 billion metric tons
or
8.9% of total reserves

Available by
underground mining
132 billion metric tons
or
30.8% of total reserves

Available by
underground mining
154 billion metric tons
or
35.9% of total reserves

UNDERGROUND

UNDERGROUND

(1 metric ton = 1000 kilograms = 2200 pounds)

Figure 16.1   Demonstrated reserve base of coal in the U.S. as of January 1, 1980 (430 billion metric tons). Note that numbers may not add due to independent rounding. (Source: Energy Information Administration)

Of the 397 billion metric tons[1] of discovered coal reserves in the United States, 55% occurs in the western states (see Figure 16.1). Of these western coal reserves, 44% can be obtained by surface mining methods, while the remaining 56% requires underground mines. Eastern coal, in contrast, is concentrated as underground reserves, which constitute about 80% of total eastern coal. The total coal reserves in the U.S. amount to about 20% of the world's coal reserves.

These reserves are discovered resources; that is, resources we are reasonably confident exist. However, they are not fully recoverable by mining technology. The coal obtained from underground mining is only

about 50% of the actual coal in the deposits, since supporting structures or pillars must be left in the mine. In contrast, surface mining recovers 80–95% of a coal deposit, leaving only minor amounts of residue. Even if we apply these recovery factors, these reserves are sufficient for several centuries at the present rate of use.

Coal reserves must also be described by rank (type), since heating values per pound differ by rank. About 92% of the coal reserves in the United States is of either bituminous or sub-bituminous grade. Lignite is the other major coal resource, at about 6% of the total, and virtually all this low-heating-value coal is available by surface mining. Anthracite coal reserves, which all require underground mining, make up only about 2% of U.S. coal reserves. Because the resource

---

1   One metric ton = 1000 kilograms = 2200 pounds = 1.10 English ton.

Table 16.1 Coal Reserves in the U.S., in Energy Units and Tons, as of Jan. 1, 1980

	Billions of metric tons	Quadrillion Btu (Quads)[a]	Quadrillion kilojoules	Percent of coal resource, based on energy
Bituminous	217	6206	6547	59%
Sub-bituminous	165	3630	3830	35%
Lignite	40	616	650	6%

a 1 quad = 1 quadrillion Btu = 1000 trillion Btu.

base of anthracite is so small, it makes up only a tiny portion of the coal market, and is not considered further.

The heating value of bituminous coal is 13,000 Btu per pound (30,200 kilojoules per kilogram) and that of sub-bituminous coal 10,000 Btu per pound (23,200 kilojoules per kilogram). Lignite provides only about 7000 Btu per pound (16,250 kilojoules per kilogram). Using the heating values of the different coals, we can restate U.S. reserves of coal in Btu as well as tons (Table 16.1).

The coals of various ranks are not evenly distributed in the United States. Lignite and sub-bituminous coal, for instance, are found in large quantities only west of the Mississippi River. Of the bituminous coal reserves, in contrast, 85% is concentrated in the eastern portion of the United States; only about 15% of this resource lies west of the Mississippi (Figure 16.2).

Certain states have much more coal than others. In the East, coal resources are concentrated in Ohio, Illinois, Pennsylvania, West Virginia, and Kentucky. In the West, North Dakota, Montana, Wyoming, and Colorado have rich coal resources (Figure 16.3). About 45% of the land area of North Dakota and 41% of Wyoming contain substrata of coal-bearing rock. About 35% of Montana and 28% of Colorado sit atop coal resources. Taken together, Montana and Wyoming account for 40% of all U.S. coal reserves.

Nearly 85% of the coal that is less than 1% sulfur by weight is found west of the Mississippi. Thus, to obtain low-sulfur coal in the eastern United States, where a larger percentage of the coal is used, the East must bring its coal from the western United States. These imports may come in by train or barge, or if the newest technology takes hold, by coal-slurry pipeline.

However, the lower heating values of most western coals require greater quantities to be shipped.

How Is Coal Being Used?

The use of coal is growing, after a period in which it had fallen steadily (see Figure 16.4). The long decline in coal consumption came about as the railroads switched from coal to diesel fuel and home owners switched to gas and oil. The present upward trend is the result of the growing demand for electricity, a de-

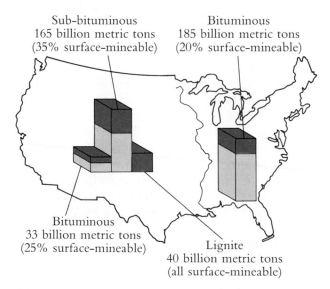

Sub-bituminous
165 billion metric tons
(35% surface-mineable)

Bituminous
185 billion metric tons
(20% surface-mineable)

Bituminous
33 billion metric tons
(25% surface-mineable)

Lignite
40 billion metric tons
(all surface-mineable)

Figure 16.2 Distribution of grades of coal as of January 1, 1980. Note that numbers in figure may not add to numbers in Table 16.1 due to independent rounding. (Source: Energy Information Administration)

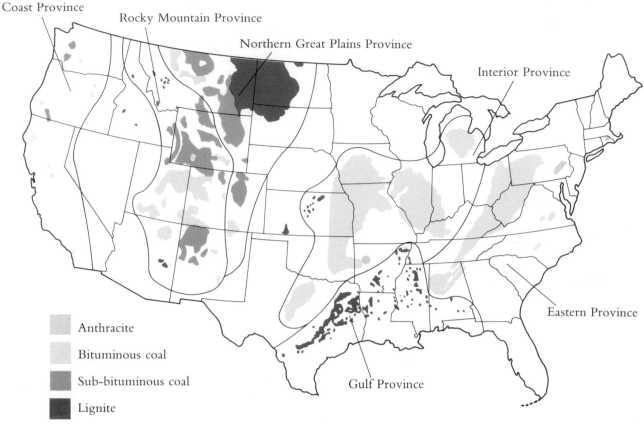

Figure 16.3　Distribution of United States coal resources. (Source: Bureau of Land Management, 1974)

mand that increasingly is being filled by coal-fired electric generating stations.

In 1980, about 754 million metric tons of coal were produced in the United States. With discovered reserves of about 400 billion metric tons, coal appears to be a secure energy source for the next several centuries. Of these 754 million metric tons of coal produced in 1980, about 60% was obtained by surface mining methods. The dramatic growth in surface mining is shown in Figure 16.4. In 1980, more coal was surface mined in the U.S. than was produced by all methods in 1960.

Coal is currently finding greatest use in the electric utility industry. In 1980, coal consumed in the generation of electricity amounted to about 70% of the annual coal production in the U.S. Other uses of coal include coking in the iron and steel industry (8%), general industrial use (7%), and export (11%) (Figure 16.5). A small amount of coal is still used in home

heating, although this use has fallen dramatically since the end of World War II. Foreign countries that purchase our coal generally use it in their iron and steel industries.

Environmental and Social Impact of Coal

With so much coal available, it appears to make good sense to get the maximum use from this fuel. It can replace oil and gas in electric generating stations that now use these costly or scarce fuels. Also, it can be converted from solid form to a liquid fuel or from solid form to a gaseous fuel, although there are costs to the conversion.

The National Coal Association, an industry organization, calls coal "America's Ace in the Hole." Although they urge us to let the mining industry go in

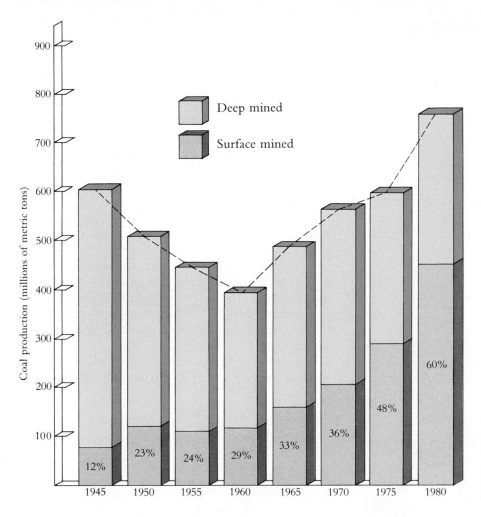

Figure 16.4 Coal production in the U.S. from 1945–1980. (Data from U.S. Department of Energy) Note that 1 metric ton = 1000 kilograms = 2200 pounds.

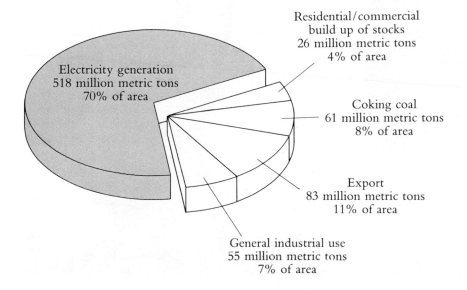

Figure 16.5 Consumption of coal by sector in 1980. (Source: *U.S. Coal, At a Glance*, 1980, U.S. Dept. of Energy)

and get it, coal does not rise from the earth like magic. Both the natural environment and humans may be sacrificed along the way. And at its point of use, the burning of coal pollutes the air, injures vegetation, and affects human health.

How Coal Is Mined

The impact of coal mining on the environment is enormous. Where coal is mined underground, human safety is threatened by mine explosions and collapse. Underground miners are subject to black lung disease, a crippling shortness of breath. Acid drainage from underground mines pollutes streams and rivers.

Where coal is surface mined without adequate reclamation, unchecked erosion chokes streams with particles of earth. Mountains of rubble degrade the landscape and highwalls ring the scalped hillsides. To understand the origins of these environmental insults, we need to explore the ways in which coal is mined.

Coal is mined in two principal fashions, although variations exist. Coal may be surface mined, a practice that has grown dramatically in the last decade. Or coal may be drawn from underground mines, a traditional method that is giving way to surface mining.

Why is surface mining capturing such a large share of the coal-mining market? It is a matter mainly of economics: the price of surface-mined coal at the mine is roughly about half the cost of coal from underground mines. The low price of this coal is due to the technology of surface mining, which has eliminated a great deal of human labor. A surface "miner" can haul each day about twice the tonnage hauled by an underground miner. In addition, the special safety precautions to avoid cave-ins and explosions in underground mines are not needed. Furthermore, health precautions relative to particles in the air in mines become unnecessary in surface mining.

The price of surface-mined coal has also been low because, until 1977, very few states required the recla-

(a)

(b)

Figure 16.6 The equipment used in strip mining is massive. (a) This is the bucket of a "dragline" machine at the Medicine Bow Coal Mine in Hanna, Wyoming. The bucket holds 78 cubic yards, a volume equal to that of a 16-foot square room with an 8-foot ceiling. (b) This dragline operating at a surface mine at Marrisa, Illinois, is so large (20 stories tall) that it had to be assembled at the mining site. (Photos courtesy of U.S. Department of Energy)

mation of land scarred by the surface mining process. The Federal Surface Mining Control and Reclamation Act of 1977 changed the rules for reclamation, requiring restoration of disturbed land. The difference in price due to a lack of adequate environmental protection has now diminished somewhat, but a gap in price remains. Surface-mined coal, where transport costs do not alter the cost structure, will still be cheaper than coal mined from underground mines.

Surface (Strip) Mining

How coal is removed. Whether a modern coal company decides on surface or underground mining of a coal seam depends on the depth of the seam below the surface of the earth. The more layers of soil and rock (called overburden) that must be removed, the greater the cost to mine from the surface. The decision also depends on the thickness of the seam, since the amount of coal recovered will influence profits.

Much modern surface mining depends on giant equipment (Figure 16.6). To justify bringing this costly equipment to a site, a very large deposit of coal must be present. Thus, a small number of huge surface mines account for a large fraction of the coal produced each year in the United States. In 1982, the thirteen surface mines operating in Campbell County, Wyo-

ming, supplied nearly 9% of the nation's annual coal production. Operating at full capacity, these mines could have supplied about 16% of 1982 production, or 126 million metric tons.

There are two basic kinds of strip mining; their difference stems from the different land forms on which they operate. **Contour mining** is used in hilly terrain where coal seams lie beneath hills. A power shovel cuts a groove into a hillside, exposing the coal seam. The overburden may be discarded to lower levels of the hill, but reclamation is more difficult if this is done (Figure 16.7). When the overburden becomes too thick to remove, giant auger drills, up to 7 feet in diameter, may bore into the coal seam to bring out more coal.

When no attempt is made to restore the area, a **highwall** remains (Figure 16.8). This vertical cut into the side of the hill may have vegetation at the top, followed by a layer of soil and rock. Then comes a dense black stripe—the coal left behind when further removal of the overburden became too costly. Such a

Figure 16.8 A highwall is left after a vertical cut into a hillside has been made to expose a coal seam for mining. The dense black stripe is coal that was too costly to recover. Highwalls, which may run for miles, isolate the hills from many animal species and so destroy valuable wildlife habitat. Leaving highwalls is now illegal, but the hills of Appalachia are scarred from pre-1977 operations.

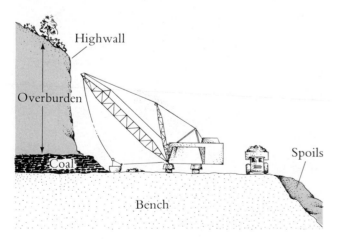

Figure 16.7 Contour mining using a power shovel. Disposal of the spoils down the hill slope makes later reclamation more difficult. (Adapted from D. Kash et al., *Energy Alternatives,* U.S. Environmental Protection Agency, EPA 430/9-73-011, 1973.)

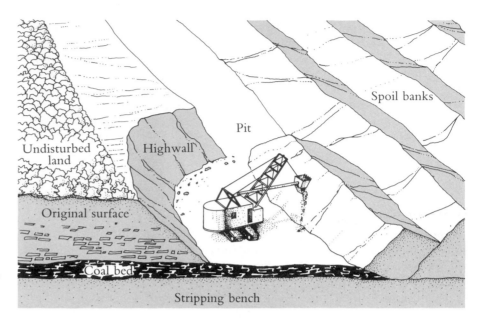

Figure 16.9 Area mining, one of the methods of surface mining for coal. It is used when the land above the coal seam is flat or gently rolling.

stretch of highwall may run for miles. Over 20,000 miles (32,000 km) of highwalls ring the hills of the Appalachian region, scars from an era when profit came before environment.

Area mining, in contrast to contour mining, operates on flat terrain (Figure 16.9). Area mining lays up the soil in rows, exposing the coal in a long path. When the coal is removed from that open path, the overburden on the coal next to the path is excavated and dumped into the mined area, exposing another long path of coal. The process is repeated for row after row, leaving hills of rubble resembling miniature mountain ranges (Figure 16.10).

In larger surface mine operations, a power shovel may be utilized, along with a dragline. The power shovel scoops overburden and loads coal into trucks. The dragline is used primarily to move overburden from place to place. The bucket of the largest dragline can carry up to 200 cubic yards (168 cubic meters) of earth. This is equivalent to a room 27 feet (8.1 meters) wide in both directions with an 8-foot (2.4-meter) ceiling. The weight of earth in the bucket of the dragline may run to 200 tons (182 metric tons).

Reclamation of surface-mined land. Even into the 1970s, strip miners were allowed to withdraw coal and depart. The result was a surrealist picture of huge mounds of rubble, sometimes arranged in rows, sometimes strewn haphazardly over the landscape.

The land from which the coal was torn was often left useless and without vegetation. Black-striped highwalls ringed the tops of hills, cutting off any plant and animal life from the nearby surroundings.

Figure 16.10 Area strip mining left unreclaimed. This land near Nucla, Colorado, was area mined for coal by the Peabody Coal Co. and left. Rapid erosion of soil occurs from such areas, clogging streams and making them subject to flooding. Acid conditions in the rubble and soil hinder the regrowth of trees or other vegetation. The land is unfit for human or wildlife habitation. (EPA-Documerica, Bill Gillette)

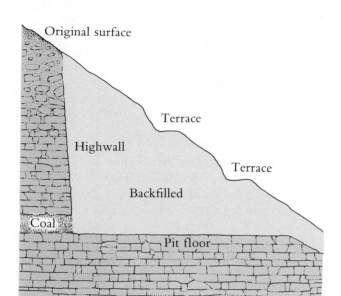

Figure 16.11 Full backfilling of contour-mined land. This procedure is used to reclaim hill slopes that have been contour mined. The terraces are used to channel runoff gradually down the hill. (Adapted from D. Kash et al., *Energy Alternatives,* U.S. Environmental Protection Agency, EPA 430/9-73-011, 1973.)

From unreclaimed land, soil particles can come washing off the barren slopes of the hills and can fill stream beds. Because the stream channels are decreased, the flows from large rainstorms cannot be carried in the channel and the stream floods more frequently. Furthermore, the barren hillsides with loose earth are unstable. A soaking rain can increase the weight of a soil mass and decrease its hold to the stable earth. The result is landslides occurring off the sides of strip-mined hills.

Eventually, vegetation will return to many of these areas, but the process is slow. The rubble left on the land is acidic because of the presence of impurities in coal (see Acid Mine Drainage, p. 308); this acid condition hinders the regrowth of grasses and trees. In addition, acid rivulets run off these lands and enter nearby streams. Acid waters do not typically support aquatic life, and humans cannot drink them.

Reclamation can prevent such scenes of devastation. The concept of reclamation is simple. Restore the original contour of the land, if possible with topsoil at the top, and re-establish vegetation to "anchor" the soil.

Where area mining is utilized on gently sloping or flat areas, the topsoil is saved so that it can be replaced on top of the regraded rubble from the mining operation. Once the topsoil has been replaced, fertilizer and grass seed are applied. With adequate rainfall and 6 inches (15 cm) or more of topsoil, the vegetation has a good chance of being established, preventing rapid runoff of water and checking the erosion process.

Where contour mining has occurred, special methods have been designed to repair the damage. In one reclamation procedure, *full backfilling,* the spoils are pushed back up the bank and fully cover the highwall (Figure 16.11). The restoration matches the surface of the filled area to the surface at the top of the highwall so that water from runoff will not fall rapidly over a sharp slope. To prevent erosion gullies from forming, terraces may also be cut around the restored hillside to catch and slow the water that runs off the slope.

Backfilling is an effort to repair damage after it has been done. In contrast, the *modified block cut* is designed to minimize damage during the process of mining (Figure 16.12). In the first step or cut, overburden is

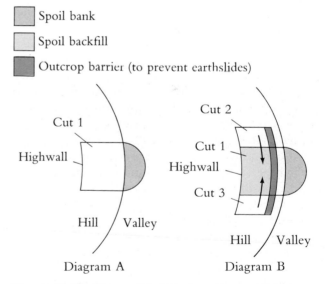

Figure 16.12 The modified block cut is one of the best methods to reclaim contour-mined land, and appears to cost no more than ordinary mining. In diagram A, an initial cut is made into the hill. Diagram B shows the second and third cuts being made; the spoils from these cuts are deposited in the void left by the first cut. The spoils from later cuts are deposited in the new voids created by cuts two and three. (Adapted from D. Kash et al., *Energy Alternatives,* U.S. Environmental Protection Agency, EPA 430/9-73-011, 1973.)

excavated from just above the coal seam and is dumped downhill. The second cut and subsequent cuts are smaller in size and are taken at the same elevation as the first, but to either side of the initial cut. The overburden removed from the second and third cuts is dumped into the adjacent space made by the first cut. More cuts are taken to either side of these initial cuts, and the spoils are discarded into the spaces made by the second and third cuts, and so on.

These are only a few of the many methods that have been used to restore contour-mined land. The modified block cut is reported to have been successful on slopes of 20° and above. In addition, the modified block cut appears to be no more expensive than ordinary contour mining, suggesting that it may become an important procedure in the future.

Replanting a strip-mined area. The best revegetation cannot soon bring back forested areas. Wildlife characteristic of forests cannot survive in grassland; their food supply and nesting places do not exist. Thus, the ecological balance is unavoidably altered by strip mining, even with the best of reclamation.

The whole effort of reclamation, be it of contour-mined land or area-mined land, appears to be very dependent on rainfall. In regions where rain is sparse, it may happen that the best of methods are inadequate to restore vegetation to the land. The concern here is mainly with the vast coal fields of the West, where huge strip mining operations are underway and increasing in scope; the consequences of such operations in low-rainfall areas are difficult to predict.

A committee of the National Academy of Sciences studied the issue of western surface mining from the standpoint of the potential for reclamation of the lands.[2] The panel indicated that areas with less than 10 inches of annual rainfall were probably not good candidates for reclamation in the sense that full restoration of native vegetation might not be possible within a reasonable time scale.

The committee felt that sandy soil supporting only desert shrubs probably cannot be restored to its original form because no topsoil is available to cover the earth after surface mining. Because so few seeds would be present in the sterile, exposed soil, the growth of new plants would be extremely slow. The natural succession of plant species, in which one diverse community of plants replaces another, would be long delayed in this difficult environment, so that 100 years or more might pass before the native vegetation returned.

Figure 16.13 summarizes the views of the committee on the probabilities of successful restoration of the four vegetative zones encountered in the western coal states.

Surface mining law established. In 1977, after years of acrimonious debate and disagreement, Congress finally passed a law establishing partial control over strip mining. The Surface Mining Control and Reclamation Act forbids leaving highwalls after mining. It also forbids dumping overburden on slopes of over 20°. The legislation demands that strip-mined lands be returned to approximately original contours. Prime farmlands, according to the legislation, are not to be mined unless they can be returned to their original productivity. Furthermore, a tax is now collected on coal currently being produced. This tax is to be used to pay for restoration of land that was marred in the era before reclamation. While enforcement of these standards is left to the states, the Department of the Interior may step in when the states fail to act.

Although reclamation costs could run from $1000–$5000/acre ($2500–$12,500/hectare), depending on the slopes involved, it will still be profitable to surface mine coal. Further, the relative position of strip-mined coal as cheaper than deep-mined coal generally will not change.

Surface mining: Is the issue settled? During the 1970s, federal laws to control strip mining were proposed again and again, only to be defeated either by Congress or by presidential veto. One group in Congress favored a complete ban on all strip mining of all minerals. On the opposite side were the coal companies and other mining companies. Arguments were fierce as each side had its truth to tell. Strip mining was ravaging counties in Kentucky, Virginia, and West Virginia, where irresponsible operators would simply devour a coal seam and depart, leaving rubble and destruction in their wake. At the same time, responsible companies were restoring the land they had mined. In general, however, state laws on surface mining were weak[3] and the conscience of the operator was all

---

2 National Academy of Sciences, *Rehabilitation Potential of Western Coal Lands*. Cambridge, MA: Ballinger Publ., 1974.

3 A notable exception was Pennsylvania, where a tough law was enforced by a colorful and tough administrator, William Guckert.

(a) *Ponderosa Pine and Mountain Brush*

"... where precipitation is favorable for plant growth and where rather deep, fertile soils have developed, [such sites] do not present a revegetation problem; it must be remembered, however, that many years are required to grow mature pines."

(b) *Mixed Grass Plains*

"... a rather high probability for satisfactory rehabilitation ... predicting such results assumes the best technology will be applied, including the addition of top soil, ..."

(c) *Sagebrush Foothills*

"... a delicate reconstruction process in the handling of substrata and top soil is demanded for favorable rehabilitation ... in some places disturbed ground may have to be repeatedly reseeded ...."

(d) *Desert*

"Disturbing such areas for the surface mining of coal amounts to sacrificing such values permanently for economic reward."

Figure 16.13 Reclaiming surface-mined lands in the West. (Quotes are from National Academy of Sciences, *Rehabilitation Potential of Western Coal Lands*. Cambridge, MA: Ballinger Publ., 1974.) (Photos: (a) Kathleen Sullivan; (b,c) U.S. Bureau of Land Management; (d) National Park Service, Richard Frear.)

people had to rely on. In such cases, pangs of conscience meant lower profits for operators.

Now, as strip miners move into the West under a new law, it seems wise to review the local impact of strip mining as it took place in the 1970s. Perhaps the most eloquent spokesman against strip mining was the Honorable Kenneth Heckler, representative to Congress from West Virginia. He gave a speech in the House of Representatives entitled "The Hills of Appalachia are Bleeding";[4] the following statement is from that speech:

> The human suffering of those who live near strip mining sites is pitiful. The blasting and the bulldozers have frequently sent boulders onto the property and even into the homes of those on the fringes of strip mining. I have called at homes where embarrassed owners have almost cried because they cannot draw me a glass of water, for their water supply has turned black or brackish. The entire Appalachian area is honeycombed with moonscapes. Foundations of homes have been destroyed. The giant mounds of earth which are pushed over the hillsides are unstable. They swell with the rains, get heavy, crack and slide. They erode away, washing rock and soil down the hillsides into the streams below . . . Do you want more areas of the country to become instant Appalachias? Do you want to allow these strip-and-run exploitation artists to rip up your land the way they have ripped up ours?

Surface mining is now increasing dramatically in the West. In 1975, a portion of Death Valley National Monument was being surface mined for coal by the Tenneco Company. Montana, Wyoming, North Dakota, and other western states are being surface mined extensively.

The law that now governs surface mining is the Surface Mining Control and Reclamation Act of 1977, a federal law described earlier. In terms of western mining, there are several flaws in that legislation. When a state fails to enforce the law, and illegal and destructive stripping takes place, much damage could occur before the federal government discovers the operation and is able to take action. The act dictated that

prime farm lands could not be stripped unless their productivity could be restored, but who is to judge what has not yet been determined even by scientists? In addition, there is no prohibition on surface mining the desert, where we know that restoration of the land is extremely unlikely. Finally, it appears that an administration unsympathetic to the law and to the environmental values it is meant to preserve, has the power to limit enforcement of the law by simply cutting the budget and staff of the enforcement agency.

Thus, it appears that the issue of coal surface mining is not yet laid to rest. As the 1980s proceed, you may wish to observe the effect of vastly increased surface mining on the states of the western plains.

Underground Mining

How coal is removed. In an underground coal mine, men and machines cut the mineral from the coal face and load it into cars or onto conveyors for removal to the surface. The **room and pillar** method of mining is the traditional means of extracting the coal (Figure 16.14). From a central passage, parallel rooms or tunnels up to 20 feet (6 meters) in width may penetrate the coal seam for 200–300 feet (60–90 meters). Another set of parallel tunnels driven perpendicular to the first results in a tic–tac–toe pattern of rooms. The pillars of coal between the rooms, along with large timbers and roof bolts, hold the earth up. In room and pillar mining, about 45–50% of the coal in the seam is actually brought out. Nearly three-quarters of the cutting and loading is now done by machine, while the remaining one-quarter is still accomplished by the more labor-intensive operations involving either hand cutting or hand loading.

Of growing interest is the highly efficient method imported from Europe known as **longwall mining.** Only in the last decade has this procedure been used in the United States. Longwall mining removes the coal without disturbing the layers of earth above the coal seam and without creating permanent deep mines (Figure 16.15). Movable jacks hold the roof of the mine in place while mining is going on, but when the coal has been removed, the jacks are moved forward to the new mining "room." When the jacks have been relocated, the earth collapses above the mined space because no columns support it. The surface subsides as the jacks are gradually moved forward, but otherwise

---

4 Honorable Kenneth Heckler, Member of Congress, *Regulation of Surface Mining,* Part I, Hearings before Subcommittees of the Committee on Interior and Insular Affairs, House of Representatives, Serial 93-11, 1973.

the surface remains untouched. In addition, most of the mining area has been tightly sealed off from water or oxygen. No tunnels into the interior of the mine are left behind, so water has no easy route to escape to the surface and oxygen has no easy route to enter. Thus the formation of acid mine drainage is largely prevented.

Coal waste piles or refuse banks. Once coal has been extracted from a mine, it may be sent directly to customers, such as the electric utilities. Two-thirds or more of current bituminous coal production moves directly to use in this way. The remainder is sent to a preparation plant, where it is crushed, washed, and graded in size.

The wastes from the preparation plant consist of low-grade coal, shale, slate, and coal dust. These wastes are stored in huge hills known as **refuse banks** or "gob" piles. Water contaminated with coal dust from the washing process may also be produced. This water may be settled in ponds to remove the coal particles, but in some cases the water is discharged directly into streams. One practice is to dam streams with the

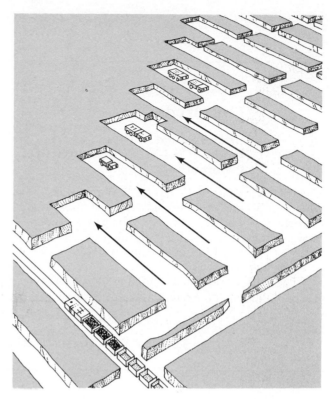

Figure 16.14 Looking down on a room-and-pillar mining operation. Arrows indicate the direction of further mining. The tunnels may be up to 20 feet (6 meters) in width.

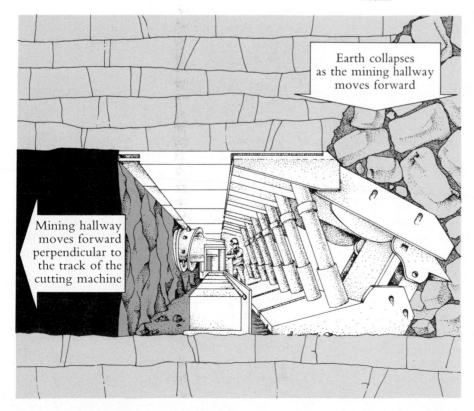

Earth collapses as the mining hallway moves forward

Mining hallway moves forward perpendicular to the track of the cutting machine

Figure 16.15 Longwall Mining. The cutting machine proceeds along the coal seam, enlarging the width of the hallway. As the hallway is enlarged, the roof supports are moved forward, closing the room and allowing the earth behind the supports to collapse. No underground tunnels are left when the longwall mining operation has been completed. (Adapted from *Coal,* Utah Power and Light, 1982.)

West Virginia Flood Toll at 60 with Hundreds Lost

Bodies Found in River 24 Miles Away After Water Behind Coal-Waste Pile Rushes Down Over 14 Communities

MAN, W. VA., FEB. 27—Up on Buffalo Creek, where people used to live, only the bulldozers and the helicopters belonged today. The people were gone, scattered by yesterday morning's flash flood that was unleashed by the collapse of a coal waste pile accumulated over the last 15 years. . . .

The tragedy occurred around 8 A.M. yesterday when a huge pile of waste material from the operations of the Buffalo Mining Company gave way on a mountain outside Man. The wall of water, estimated by some residents as ranging from 20 to 50 feet high at the beginning, rushed downhill through 14 mining communities, wrecking most of the houses in its path and displacing as many as 5,000 persons . . . .

George Vecsey,
Special to **The New York Times,** 28 February 1972

refuse bank, thus creating a settling pond behind the dam. In West Virginia in 1972, such a coal waste pile gave way, resulting in a devastating flood (see excerpted news story). A vice president of the Pittston Coal Co., the owner of Buffalo Mining, indicated that the company considered the flood to be "an act of God."

Refuse banks that are not used to dam streams may also be a source of acid water as rainfall drains through them. One infamous coal pile in Aberfam, Wales, collapsed in the 1960s, burying a school and all inside. These mounds of coal debris may often catch fire, producing air pollution, and can smolder for years. One bank in the northern Appalachian region, discovered burning in a 1963 study, had been burning since 1884, despite efforts at control. Controlling the fire in a refuse bank may require actual burial of the mound under an earth cover. There are known to be 800 refuse banks of varying composition in the anthracite region of Pennsylvania alone.

Subsidence. The removal of coal from underground mines leaves another legacy: **subsidence.** The practice of removing some of the supporting pillars for their coal at the end of a mining operation increases the risk of subsidence. As the roof supports in a mine give way, the mine fills with overburden; sink holes and cracks may appear at the surface; and tremors may shake the area.

Though only a nuisance when it occurs in farmland or forest, subsidence is a serious problem when human dwellings are nearby. Roads and foundations may crack, buckle, and crumble as the earth settles into firmer position below. Railroad tracks may be bent and lose their support; gas mains may break and threaten explosion; and sewer lines may crack. In some areas of Pennsylvania, homes have had to be abandoned for safety.

Acid mine drainage. Still another effect of underground coal mining is acid mine drainage. From underground mines and strip mines flows a substance called "yellow boy," which eventually settles and coats stream bottoms. Yellow boy, derived from iron pyrites, one of the impurities in coal, makes water acidic. Water that carries yellow boy is called **acid mine drainage.** Acid mine drainage results because the mining of a coal seam is rarely complete. Some coal is unavoidably left behind in both surface and deep mining operations.

Acid mine drainage arises from strip-mined areas in the following way. An area that has been strip

CONTROVERSY 16.1

Strip Mining: What are the Implications of Expansion on Jobs? on the Environment? on Human Health?

> . . . Superscale western mining means . . . between twenty-five thousand and forty thousand jobs will be lost in the East.
>
> *Arnold Miller**

> Black lung and silicosis are rare diseases among strip miners. Very few men are injured by rockfalls in surface mines. In an underground mine, if something goes wrong, there is nowhere to go; the miner is surrounded by rock . . .
>
> *David L. Kuck†*

As surface mining expands its share of the coal market, the number of mining injuries and fatalities declines. From the viewpoint of worker health, it appears that strip mining holds an edge over deep mining. Also, strip-mined coal costs less to produce than deep-mined coal. Yet Arnold Miller, former President of the United Mine Workers, sounds like an environmentalist when he commented on strip mining:

> Fifty years ago they promised to develop Appalachia; and they left it a wreckage. Now they promise to develop the Great Plains. They will leave it a ruins.

As environmentalists, we hear Miller warning one section of a country of what the coal industry did to another, warning of the destruction of the land and of vegetation. In counterpart to Miller's concern for the environment, David Kuck, in a letter to *Science,* asks

> . . . Which resources are we most concerned about conserving—human lives, the terrain, the vegetation, or the mineral?

Miller, however, has another motive for his warning, a motive he does not hide:

> A headlong commitment to superscale Western mining means that over the next five years between twenty-five thousand and forty thousand jobs will be lost in the East. Of course, that concerns us as a union of miners. It concerns us also because we have lived through an unending depression in Appalachia . . . Finally it concerns us because you cannot turn underground coal production on and off like a light switch. If we arrive at a rational fuels policy (finally) and decide to strengthen our emphasis on Eastern mining, the mines will not be there, and neither will the miners.

Do you agree with Miller now that you understand his motivation? Defend your position. Do you think coal mining in the East should be fostered? If you do, can you think of ways to keep coal mining in the East alive?

* *Center Magazine,* November–December 1973.
† *Science,* **183** (11 January 1974), 4120.

mined and has not been restored has numerous ditches covering the mined landscape. Rain leaves pools of surface water in these ditches, and this water comes in contact with coal still in the soil. The surface water from these ditches follows a winding route to a stream, coming in contact with more coal en route. The surface water leaches iron pyrites from that coal, and this substance gives rise to the acid waters.

The process that creates acid mine drainage in surface mining is much the same for an underground mine. Water may enter the mine directly through the mine shaft, or groundwater may seep through the layers of soil into the mine. Sometimes the earth above the mine has subsided (sunk into the mine). In this common situation, water may easily penetrate the mine through the "sink hole" that has been formed. It is even possible that entire surface streams may go underground through such sink holes. Iron pyrites from the coal remaining in the mine dissolve in these waters.

Iron pyrite is known chemically as ferrous sulfide, a compound composed of iron and sulfur. It is also well known as the shiny golden mineral "fools' gold." When ferrous sulfide dissolves in water, oxygen in the water begins to oxidize both sulfide and ferrous ions. The oxidation, in which bacteria play a role, converts the sulfide ion to sulfate ion and the ferrous ion to ferric ion. Oxygen from the water is used up in the conversion, resulting in such low levels of oxygen that the normal aquatic life of the stream is threatened. The effects of low levels of dissolved oxygen include destruction of the fish population, the normal insect population, and much of the normal microscopic plant and animal life. These effects are further discussed in Chapter 12.

Other reactions make matters worse. The ferric ion combines with hydroxyl ions in the water to form the yellow–brown precipitate ferric hydroxide, the substance we referred to as "yellow boy." Yellow boy may sink and coat the bottom of the stream. Creatures such as the freshwater crayfish, which feed on the stream bottom, will thus be deprived of their food source. Visitors to the coal mining areas of Appalachia often remark on the yellow–orange color of the streambeds in the region.

Although yellow boy coats the streambed and the low levels of dissolved oxygen harm aquatic life, the worst consequence is the acidity of the water. The precipitation of ferric hydroxide leaves an excess of hydrogen ions in solution. The water is now rich in both hydrogen ions and sulfate ions. This situation is equivalent to adding sulfuric acid to water. The highly acidic water with low levels of dissolved oxygen and a stream bottom coated with yellow boy together create an environment unfavorable to the survival of aquatic species.

Acid mine drainage has historically been a regional problem. Pennsylvania, West Virginia, and neighboring coal states have had the greatest share of the problem. An estimate in the 1970s by the Environmental Protection Agency placed the total length of streams affected by acid mine drainage at more than 11,000 miles, mostly east of the Mississippi River. However, the surface mining of coal in the western states was only beginning in this period. Because so much of the nation's coal now originates in the West, the regional aspect of the problem of acid mine drainage is disappearing. Its control is a problem of national scope.

Solutions to the problem of acid mine drainage. Even though coal mining operations will undoubtedly continue and expand, a number of ways exist to possibly prevent or reduce acid mine drainage. The methods are applied in different situations and are not equally successful.

Where an area is to be deep mined, it is helpful if the mining company uses "down-dip" as opposed to "up-dip" recovery (Figure 16.16). The shaft of the down-dip mine slopes downward to the coal seam; the shaft of the up-dip mine angles upward to the coal seam. The pooled waters in a down-dip mine have difficulty escaping to surface waters.

Unfortunately, traditional methods of coal mining favor up-dip mining. In years gone by, mine shafts would be carved into the hillside at an upward angle so that coal could be loaded onto carts at the coal face and allowed to roll down to the outside of the mine. Water that accumulated in the mines from groundwater seepage or that "wept" off the walls due to condensation would simply run out of the mine by gravity. Abandoned mines are responsible for a very high portion of the acid mine drainage that occurs.

How, then, can we control the pollution from long-abandoned mines? You might think one way would be to seal the entrance to the mine to prevent the escape of acid drainage. However, when the mine

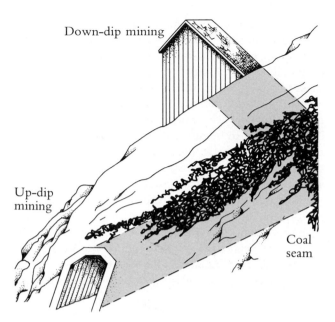

Figure 16.16 Two methods of reaching a coal seam. In down-dip mining, surface water may enter through the downward-sloping shaft or via groundwater flow. It may also condense on the walls of the mine because of humid conditions there. Such waters are likely to be safely trapped in an abandoned mine. In contrast, water that enters the up-dip mine via groundwater flow and through condensation flows out naturally down through the mine shaft itself. When such a mine is abandoned, it is extremely difficult to seal the entrance.

shaft rises upward at an angle from the entry, water accumulates behind the seal, and the tremendous pressure of a column of water can cause the seal to leak or give way. In contrast, where the mine shaft is level or where it falls away at an angle from the entrance, mine sealing has been effective.

Another option to control acid drainage from abandoned mines is to prevent surface water from entering the mines. This can be done by filling sink holes and by rerouting gulleys that may sometimes flow above an abandoned mine. If the stream channel can be routed around the mine site, surface water will be unable to seep into the mine. Another method of controlling acid mine drainage is to chemically treat acid waters with crushed limestone to neutralize the acidity. This last technique can be applied to acid drainage from both active or abandoned mines.

Health and Safety of the Coal Miner

Mine Safety

Coal mining, more than any other process of fuel resource extraction, has an impact on human life itself. We often read and hear of mine disasters where rescue operations have failed, of cave-ins or coal mine explosions. Several hundred coal mining deaths occur each year. It is a grim, grimy, and dangerous business. Although statistics tell us that mines are getting safer, coal mining is still one of the most hazardous jobs in modern times. Some dangers are immediate: a loading car may break loose and careen down a slope, striking a miner. Other dangers are delayed: the sooty air, laced with coal particles, may implant in a miner's lung and eventually destroy his ability to breathe.

About 60% of the injuries in underground coal mining are related to roof falls and face falls (the coal face landslides in miniature). About half of the deaths in underground coal mining stem from roof falls alone. Hauling coal underground is also extremely hazardous. Accidents occurring in haulage are often fatal. About 15–20% of underground accidents occur in hauling operations. Explosions and fires account for another 15–20%.

Explosions in coal mines result from the ignition of methane gas (though coal dust, itself, can explode). Pockets of methane commonly occur in underground coal deposits. If these pockets go undetected and gas builds up in the mine, a simple spark can detonate an explosion. While the explosion itself can injure and kill, the rock falls that result may be even more hazardous, even to the extent of sealing men inside a portion of a mine. Such explosions from methane gas can be prevented by adequate ventilation of the mine, but to be sure the mine air is safe, methane detection equipment must be used.

The high accident rate in the coal industry leads to a great number of disability cases. Unable to work, these former miners may also require extensive medical attention. Not only do accidents contribute to disability, but working conditions do as well. Performing heavy labor in a stooped position or crawling on hands and knees lead to inflammations of the joints, conditions known as "beat hand" and "beat knee."

Black Lung Disease

Accidents, fatalities, crippling deformities, and finally, **black lung:** a wheezing and shortness of breath, an inability to climb stairs or perform labor, a disease so disabling that some victims are only able to sit in chairs, their lungs destroyed by the particles they inhaled. The public and the coal miners call this condition black lung disease because the lungs of miners who die from it are black (Figure 16.17). The physician knows the disease as "coal workers' pneumoconiosis" (pronounced new-mō-cō-ni-ō'-sis) and abbreviates the name as CWP.

The disease is caused by the inhalation of dust composed of minute particles of carbon and rock. The carbon particles account for the black color of the lungs. A growth of fibers occurs within the lungs around the sites where particles have deposited. These fibers can, if they grow extensively within the lung, destroy the normal elasticity of the lung. The elasticity of the lung, its ability "to bounce back," is what makes breathing easy for most of us. A miner's lung, overgrown with fibers due to coal dust, lacks the ability to bounce back. For him, breathing is torture. There appears to be no treatment for black lung other than for the bacterial infections that are side effects of the disease. That is, the symptoms cannot be relieved and the disease cannot be cured.

The risk of black lung disease appears to increase with quantity of coal dust inhaled. Two factors influence this quantity significantly. First, the greater the number of years spent underground, the greater the quantity of dust inhaled. The specific mining job—whether it is dusty or relatively clean—also influences

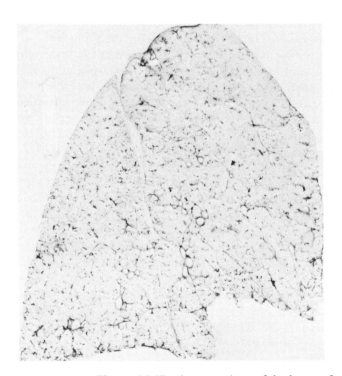
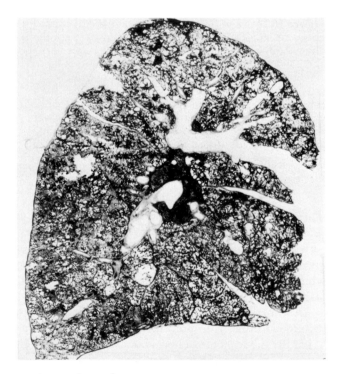

Figure 16.17 A comparison of the lungs of a non-miner and a coal miner. Photo left shows a section from the lung of a man who died at age 86 and was never a coal miner. Photo right shows a section from the lung of a man who died at age 78 and was an underground coal miner for 36 years. Although the man did not smoke, he suffered from severe emphysema, a lung disease in which breathing is badly impaired. (Photos courtesy of Dr. Frank Green, M.D., Chief, Pathology Section, Appalachian Laboratory for Occupational Safety and Health.)

the risk of contracting the disease. Cutting-machine operators show a very large risk compared to workers stationed outside the mine. To bring the number of new cases down, the levels of dust in the mine must be brought under control. Although a federal standard for dust in coal mines has been set, the extent of compliance with the standard is not clear because of the absence of checks on company data.

How many individuals are afflicted with black lung disease? The question requires answers at a number of levels. Although there were about 400,000 working coal miners in 1947, mechanization of the mines had reduced the number of miners to 255,000 by 1980. In fact, at one point in the 1920s, there were about 700,000 active coal miners. This means that many men have retired from the mines in the past 30 years or so. We need to know this fact to understand why over 250,000 claims of black lung disease have been approved by the Social Security Administration. These claims were submitted either by miners or their widows and more claims are continually being approved. More than a billion dollars per year were being paid in 1978 as benefits to coal miners or their widows. These are tax dollars, not dollars from the coal industry where the miners contracted their disease.

In 1978, a bill to amend the first Black Lung Law (passed in 1969) was enacted by Congress. The original 1969 bill called for the coal industry to pay black lung compensation, but the coal industry successfully fought this law in court, and it was never implemented. The 1978 Black Lung Benefits Reform Act established a Black Lung Disability Trust Fund, which comes from a tax on each ton of coal produced. Money from this fund would be used to pay any newly established claims of black lung disease.

Impact of the Coal-Fired Electric Plant

Coal, as we pointed out earlier, finds its main use in the generation of electricity.

The burning of coal produces carbon dioxide, but only a very small amount of carbon monoxide. Carbon dioxide is of concern because of its hypothesized effect on the earth's climate. (See Chapter 20 for a further discussion of carbon dioxide and climate.) The three most important pollutants in the stack gases of coal-fired power plants are sulfur oxides, particulate matter, and nitrogen oxides.

Sulfur oxides injure plants, materials, and people. (See Chapter 21.) Sulfur dioxide was present in the air in great quantities during the air pollution disasters at Donora, Pennsylvania, and the Meuse Valley in Belgium, in the infamous "London Fog" and in the episodes in New York City. A single 1000-megawatt coal-fired electric plant burns 4–5 million tons of coal a year. If the coal is 2.5% sulfur (not an unusual level), the plant will discharge 200–250 thousand *tons* of sulfur dioxide per year. As the nation uses more and more coal, the sulfur dioxide problem will multiply unless something is done. That something consists of two steps, which may be used in combination: ridding the coal of its sulfur by cleansing it, and "scrubbing" the stack gases. (See Chapter 21.)

Particles or particulate matter are "the partners in crime" of sulfur oxides. Particles are known to aggravate human respiratory problems. (See Chapter 21.)

Particles are discharged into the atmosphere when coal is burned at power plants; these particles are known as fly ash. If the emission of fly ash were not controlled by electrostatic precipitation, the annual production of particles from a 1000-megawatt plant could be as high as 200–250 thousand tons. Instead, because particles are so well controlled, only about 10% or less actually reaches the atmosphere.

Nitrogen oxides are also produced when coal is burned at electric power plants. Although nitrogen oxides have direct effects upon our health, they are noted primarily for their interaction with hydrocarbons to produce ozone, which aggravates diseases of the lungs and respiratory tract. Control of nitrogen oxides appears to be a very difficult technical problem.

Coal also contains arsenic in small quantities. Arsenic, which is a carcinogen, is released in fly ash or particulate matter when coal is burned. It is estimated that 1000–3000 tons of arsenic are released annually from coal burning. If an electric plant has a scrubber or an electrostatic precipitator, 70–90% of the arsenic is removed.

Selenium is also present in small amounts in the fly ash discharged by coal-fired power plants. Much of this selenium seems to appear in the smallest fly ash particles, which often elude capture by electrostatic precipitators. High doses of selenium ingested by cattle from the grasses they feed upon would be injurious.

The last category of emissions from coal-fired power plants is unexpected. These are radioactive ele-

What Should Be Done with the Coal in the West?

Assuming coal is to be mined, and there is every indication that it will continue to be, the options are as follows:

1. ship it east and west to population centers via train or coal-slurry pipeline;

2. burn the coal in generating stations in the West and transmit the electricity to the population centers;

3. gasify the coal, removing sulfur, and send the gas via pipeline to the population centers.

If the coal is shipped to population centers, it will be burned there. Even if scrubbers are installed, the total pollution burden on individuals living in the cities will be increased. Not only will pollution levels rise, but the number of individuals exposed will increase as well. The pollution impact of burning the coal is thus more serious. However, even when electric utilities build their plants in the remote areas of western states, they are being forced by federal government regulation to install such devices as scrubbers and precipitators.

On the other hand, a task force, not connected with the government, composed of industrial executives and environmentalists, appears to be recommending "that new coal-fired power plants should be built in the regions where the power will be consumed—which is to say, the people who get the electricity should also have to live with the environmental effects of generating it."★ The issue is whether to keep clean areas clean or allow clean areas to be polluted in order to prevent polluted areas from being degraded still further.

There are other issues as well. The panel of the National Academy of Sciences that studied strip mining in the West (see reference on p. 305) referred to the huge water demands that would be placed on the arid West if energy development takes place in that region (Table 16.2).

If water resources are to be committed to large electrical generation and coal gasification plants, huge supplies of water will be necessary. Based on current technologies and projected rates of consumption many power plants are planned for an operating life of less than fifty years. What will be done with the proposed aqueducts, dams, and water when the coal economy no longer needs them? Does society wish to commit water resources to an ephemeral industry at a particular place?

Water demands aside, it has also been pointed out that if coal is to be gasified (made into a synthetic natural gas) and sulfur removed in the process, there is no reason why gasification should not take place in the East. Not only could the coal be gasified in the East, where water is more abundant, eastern coal could be used rather than western coal. Thus, water is not taken from a place where it is scarce and neither gas nor coal needs to be transported long distances. Unfortunately, coal gasification plants also produce air pollution, again placing the pollution burden on the East.

These questions of policy are not easily settled; the debate will continue. Where do you stand on the issue of where energy development should go and why? Do you think the plants should go in remote areas of the West? Should they have pollution control equipment? Why did the Task Force feel that people who live in population centers should bear the pollution burden?

★ *Science,* **198** (21 October 1977), 276.

ments and their decay products, which are found naturally in fly ash. The quantity of radiation from a coal-fired electric plant is very low, less than 1% of natural background radiation, about the quantity of radiation from a nuclear plant operating under the stringent standards set in the mid–1970s.[5]

---

5 See J. McBride, R. Moore, J. Witherspoon, and R. Blanco, "Radiological Impacts of Airborne Effluents of Coal and Nuclear Plants," *Science,* **202** (8 December 1978), 1045.

Table 16.2 Comparison of Water Consumed in the West for Different Uses of Coal[a]

	Thousands of acre-feet consumed per year[b]
Rail transport	less than 1000
Coal-slurry pipeline	9,200
Coal gasification	28,000
On-site electric generation	44,000

a The figures indicate acre-feet of water consumed for each 12.5 million tons of coal that are either transported by rail or slurry or are consumed in coal gasification or in on-site electric generation.
b One acre-foot = 3050 cubic meters.
(Data are for Yampa River Basin. Source: U.S. Geological Survey, Open File Report 77-698, August, 1977)

Transportation of Coal

Coal does not reach its destination without having an impact on the environment. Coal may move from mine to point of use in several ways: by train, by barge, by slurry pipeline, as synthetic natural gas, or as electricity. In recent years, the "unit train," a train devoted exclusively to moving the huge quantities of coal needed by coal-fired electric plants, has come into use. Coal also may be moved by barge on a portion of the trip from mine to plant. The most recent development in coal transportation is the coal-slurry pipeline; in this system, a mixture of water and crushed coal is pumped from mine to point of use. Finally, the energy in coal may be transported by converting coal to a synthetic natural gas (SNG) or directly into electricity at a coal-fired generating station.

Train transportation of coal seems a harmless enough activity at first glance; after all, the trains move on existing lines along rail networks long established and accepted. In fact, however, the vastly increased movement of coal will have an impact. The coal car is an open car, so dust from coal is scattered along the right of way. Where trains pass through communities, the dust will affect people as it settles from the air. A 1000-megawatt coal-fired electric plant has a coal requirement equal to about 12,000 tons (10,900 metric tons) of coal per day. This is the quantity delivered daily by a single unit train of about 100 cars.

Coal trains may be moving enormous distances. Contracts have already been completed for the sale of Montana coal to a power plant in Paducah, Kentucky, on the Ohio River and to Detroit, Michigan, on Lake Huron. Coal from Wyoming has been sold to power companies in Illinois and Indiana, where coal production is already large. These sales were probably based on the low sulfur content of western coal. A geographical portion of these long-distance trips may be undertaken by barge. Coal trains reaching the Mississippi River may transfer their loads to barges, which complete the trip to market. Coal trains may also unload at Lake Superior ports, leaving barges to finish the journey to the midwestern cities on the Great Lakes.

In contrast to barging, the coal-slurry pipeline is viewed as a serious competitor to the train as a means of coal shipment. First attempted successfully near the turn of the century, coal-slurry pipelines are now attracting close and favorable attention from coal companies. Although a coal-slurry pipeline was delivering coal to London in 1914, not until 1957 was a 100-mile (160-kilometer) pipeline for coal established in Ohio because of the state's high rail rates for coal. Reaching from Cadiz, Ohio, to Cleveland, the 10-inch (25-cm) pipeline was capable of moving 1.3 million tons (1.2

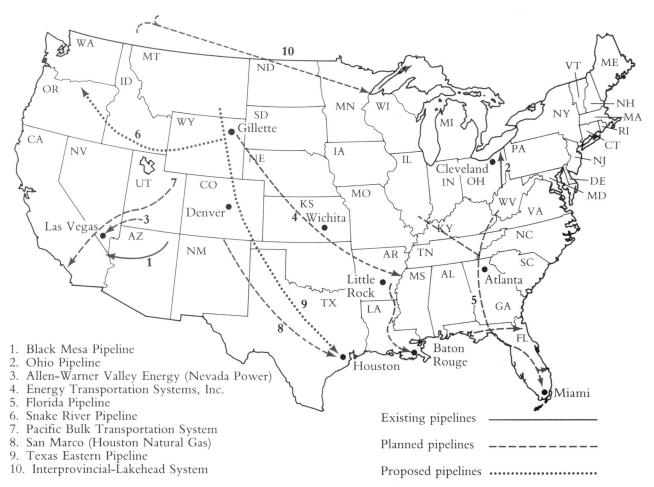

1. Black Mesa Pipeline
2. Ohio Pipeline
3. Allen-Warner Valley Energy (Nevada Power)
4. Energy Transportation Systems, Inc.
5. Florida Pipeline
6. Snake River Pipeline
7. Pacific Bulk Transportation System
8. San Marco (Houston Natural Gas)
9. Texas Eastern Pipeline
10. Interprovincial-Lakehead System

Existing pipelines ——————————

Planned pipelines — — — — — —

Proposed pipelines ·················

Figure 16.18 The status of coal-slurry pipelines. (Source: Coal Traffic Annual, 1979)

million metric tons) of coal each year. The pipeline closed in 1963 because local rail rates had finally become competitive.

The pipeline had been well publicized, however, and its success encouraged another installation in Arizona. There, the Peabody Coal Co. was to supply a power plant in the southern tip of Nevada, some 270 miles (432 km) distant from the mine. Because of the rugged terrain, direct rail haul was considered unfeasible. A roundabout routing by railroad, avoiding the mountains, would have been very costly. Completed in 1970 by the Black Mesa Pipeline Co., the pipeline now carries a river of water and finely ground coal through the deserts and mountains of the rugged

Southwest. The mixture is about 50% water and 50% coal.

With the Black Mesa experience in mind, companies have been planning more coal-slurry pipelines. The longest pipeline currently under consideration would reach 1036 miles (1658 km) from Gillette, Wyoming, to White Bluff, Arkansas. The 38-inch (1-meter) pipeline is capable of delivering 25 million tons (22.7 million metric tons) of coal each year to the power plants in Arkansas.

Unfortunately, surface water is scarce in the West. The most important uses for water are irrigation, community water supply, and recreation. Water destined for these purposes should not be diverted for coal-slurry pipelines. The Black Mesa pipeline, as well as the pipeline proposed from Wyoming to Arkansas, draw water from deep wells rather than consume precious surface water.

In the face of these problems, the slurry pipeline remains attractive because of its comparatively low cost. When the coal must move long distances where a railroad bed does not yet exist, rail haul will not compete. Neither will rail/barge shipment, nor the transmission of electricity (Figure 16.18).

Questions

1. What are the four types of coal, ranked by heating value? Which is most abundant in the United States?
2. Why are lignite and sub-bituminous coal shipped from western states to the Midwest and further east? Give two reasons. What portion of lignite is to be mined by underground mining?
3. Where are the majority of coal reserves that can be mined by surface mining: in the eastern U.S. or the western U.S.? Where are the reserves of low-sulfur coal concentrated: in the eastern U.S. or the western U.S.? What does this imply for the percentage of future coal production that might take place in the East or West?
4. What single end-use consumes two-thirds of all coal mined each year in the United States? Why was coal use falling until about 1960? What fuels were replacing it? Why has coal use increased since that time?
5. Why is surface-mined coal gradually replacing coal from deep mines? You may wish to give several reasons, but one is most important.
6. What are the two basic kinds of strip mining and how does a mine operator decide between them?
7. What is a highwall and how does it come about?
8. When strip-mined land is not "reclaimed," that is, left "as is" after mining, a number of problems occur. List and briefly explain three of the problems.
9. What is reclamation? What is the best way to reclaim a hillside that has already been strip mined? is about to be strip mined? Why were coal companies generally unwilling to reclaim the land they surface mined until laws were established?
10. Strip mining is on the increase in the West. What are the two vegetative areas with the least probability of being successfully reclaimed?
11. The Surface Mining Law passed in 1977 calls for reclamation of lands previously stripped. Where will the funds come from to do this?
12. Why are people concerned about coal waste piles (coal refuse banks) created from the wastes from underground mining? What is subsidence?
13. What is acid mine drainage?
14. What environmental problems does acid mine drainage cause?
15. Why could you say that acid mine drainage is one of the environmental costs of generating electric power?
16. Who should pay for controlling acid drainage from long-abandoned inactive mines and from mines only recently in operation? Should it be the coal company, local or state governments, or federal agencies? Why?
17. Coal miners are at risk from explosions in underground mines. What substance(s) is (are) exploding?
18. How does coal dust affect a miner's health? Why do the miner's lungs lose their elasticity?
19. What air pollutants, not including carbon dioxide, come from burning coal? Include three little-known pollutants among the six you list.

Further Reading

Acid Mine Drainage

Ackenhail, Alfred, "Pennsylvania Erases its Mining Scars," *Civil Engineering Magazine* (October 1970), p. 54.
This is a short but extremely well-written article that covers causes, effects, and simple control approaches.

Goldberg, Everett, and Garrett Power, "Legal Problems of Coal Mine Reclamation," University of Maryland School of Law, prepared for the Environmental Protection Agency, 1972. For sale by the Superintendent of Documents, U.S. Government Printing Office, Washington, D.C., 20402.
While aimed at legal problems, this publication provides clear, well-written, and extensive descriptions of the pollution problems that stem from coal mining.

Lackey, James, "Aquatic Life in Waters Polluted by Acid Mine Waste," *Public Health Reports,* **54** (1939), 740–746.
This is a classic article identifying the nature of the problem of acid waters. It is nicely written and well illustrated.

Coal-Slurry Pipelines

Gray, W. and P. Mason, "Slurry Pipelines: What the Coal Man Should Know in the Planning Stage," *Coal Age* (August 1975).

Clearly written, surprisingly environmental, description of the status, promise, and problems of these pipelines.

"Coal Slurry Pipelines," *Coal Traffic Annual,* 1979, National Coal Association, copyright 1980.

Status of plans for coal-slurry pipelines in the U.S.

General References

Kash, D., et al., "Energy Alternatives," prepared for the Council on Environmental Quality, 1975. Published as *Our Energy Future.* Norman, OK: University of Oklahoma Press, 1976.

Energy: The Next Twenty Years, Report by a Study Group Sponsored by the Ford Foundation and Administered by Resources for the Future. Cambridge, MA: Ballinger Publishing, 1979.

Oil, coal, nuclear, power, solar, projections, economics—all are here, thoroughly documented and comprehensible; one of the most complete studies of the 1970s.

"Demonstrated Reserve Base of Coal in the United States on January 1, 1980," Report of the Energy Information Administration, U.S. Department of Energy, DOE-EIA-0280(80).

Coal—Bridge to the Future, Report of the World Coal Study. Cambridge, MA: Ballinger Publishing, 1980.

Probably the most comprehensive study of coal on a worldwide basis that has been done.

Strip Mining

Alexander, Tom, "A Promising Try at Environmental Detente for Coal," *Fortune* (13 February 1978), p. 94.

This is the story of the National Coal Policy Project, a fascinating experiment in which environmentalists and coal industry people met head to head to resolve their differences. This is the group referred to and quoted in Controversy 16.2.

"Processes, Procedures, and Methods to Control Pollution from Mining Activities," U.S. Environmental Protection Agency, Washington, D.C., Document Number EPA-430/9-73-011, 1973.

This publication describes approaches to controlling mining pollution rather than the causes and effects of mine drainage. As such, it is probably most interesting to consult when studying strip mining.

Surface Mining: Soil, Coal and Society, National Research Council. Washington D.C.: National Academy Press, 1981.

This clearly written volume by a committee of the National Academy of Sciences updates and expands *Rehabilitation Potential of Western Coal Lands,* a previous effort of the National Academy of Sciences.

Surface Mining and Our Environment, U.S. Department of Interior, 1968.

Vivid color pictures of strip mining and resulting erosion and acid mine drainage.

The American Coal Miner, The President's Commission on Coal, 1980.

A report on community conditions and living conditions in the coal fields. Many photographs. Covers housing, health, safety, black lung disease, company towns. The meaning of coal as a resource is expanded by this document.

CHAPTER SEVENTEEN

Oil and Natural Gas

A Brief History of Oil in the United States

Finding and Producing Oil and Natural Gas

What Are Oil and Gas?

Oil and Gas Resources
*Oil Resources of the United States/United States and World
Use of Oil/Natural Gas Resources, Production, and
Use/Other Sources of Gas*

Synthetic Fuels
*Synthetic Fuels from Coal: Gasification and
Liquefaction/Synthetic Fuels from the Tar Sands of
Athabasca/Synthetic Fuels from Oil Shale in the Rockies*

**The Power of the Cartel and Its Impact on Oil
Companies**

**Oil Conservation and the Control of
Inflation—More than One Reason to Save**

CONTROVERSIES:

17.1: *Social Effects of Natural Gas Pricing*

17.2: *The Economics of Synthetic Fuels: Will
Synthetic Fuel Provide Energy Security or Become
a White-Elephant Investment?*

A Brief History of Oil in the United States

In the decade just preceding the Civil War, a black oily liquid was being collected from salt wells near Pittsburgh, Pennsylvania. The producers originally bottled the liquid as a medicine; but some of the liquid was processed by a small "refinery." The refinery could produce five barrels of "carbon oil" each day for use in oil lamps. The carbon oil competed with whale oil and coal oil for the lamp oil market, eventually capturing the lighting market.

From these modest beginnings, a little over a century ago, the oil industry has grown to be the largest industry in the world, dominated by giant multi-national companies. The principal instrument, or rather vehicle, propelling this growth was, of course, the automobile. This turn-of-the-century invention has accounted for a steadily growing demand for oil through most of this century.

Until the 1950s, the United States produced nearly all the oil it consumed. In that decade, the United States found itself importing cheap foreign oil in larger and larger quantities. Even into the early 1970s, oil on the world market was only about $2.00 per barrel. By the late 1970s, the proportion of U.S. demands met by imports was approaching 40%.

The 1968 discovery of oil on the North Slope of the Brooks Range in Alaska brought a dozen oil companies exploring for the riches buried in that frozen land. The nation decided to develop Alaskan oil as a means of establishing greater independence from foreign sources. Alaskan oil extended our proved reserves by nearly 25%, or 10 billion barrels. In the mid-1970s, the Trans-Alaska pipeline was cut across the Brooks Range to the Alaskan port of Valdez for shipment of oil via tanker to West Coast ports. When the first Alaskan oil finally arrived, the West Coast did not need it (as predicted by those who advocated a pipeline through Canada to the Midwest). A new pipeline to carry Alaskan oil inland may yet be built.

In 1978, drilling began offshore in the eastern United States. Though populated areas and ecologically important wetlands were not far away, national needs apparently dictated that the search for petroleum go on. By 1982, leasing of most U.S. offshore areas for oil exploration was being contemplated.

Finding and Producing Oil and Natural Gas

We link oil and natural gas together not only because they are often found together geologically, but because these two preferred fossil fuels are both in short supply in this country, and require imports to fill demands. We also link these two fuels because conservation efforts can help make these fuels last longer.

Liquid and gaseous hydrocarbons, known collectively as **petroleum,** are found in underground reservoirs within sedimentary rock; often both liquids and gases occur together. Whether they are present in any particular sedimentary rock is always unknown until drilling is undertaken. This uncertainty makes predictions of recoverable oil in the world a subject of sharp controversy.

Because oil reservoirs are deep underground, precise knowledge about the limits of a field and the volume of oil in place are not available. A large number of wells, many of them dry holes, would have to be drilled at considerable expense in order to define the field size precisely. Quite naturally, petroleum engineers tend to estimate volumes of oil on the low side so that their recommendation for the development of a field is sure to have a positive payoff.

The quantity of oil that flows out of the well by natural pressure alone is referred to as **primary recovery** and averages about 20% of the oil in place. Since about 1960, petroleum engineers have developed **secondary recovery** methods to force up more of the petroleum. These secondary recovery techniques involve either injection of water beneath the oil or injection of gas above the reservoir to place increased pressure on the oil in place. Secondary methods can increase the yield of oil reservoirs to 90% of the oil in place.

What Are Oil and Gas?

Scientists believe that petroleum was derived from the algae plankton in the ocean. In much the same fashion that coal was derived from woody land plants, bacteria break down buried plankton in the absence of oxygen. Chemical and physical processes, operating over many thousands of years, further convert the organic material until at last petroleum compounds become abundant.

Oil consists mainly of liquid hydrocarbons; hydrocarbons are compounds whose only elements are carbon and hydrogen. About 90–95% by weight of oil is usually hydrogen and carbon, about 80% or more being carbon alone. Sulfur and oxygen each may account for up to 5% of the oil's weight. Oil with a sulfur content of less that 1% is referred to as "sweet" crude oil. High-sulfur oil, because of the odor of hydrogen sulfide, is called "sour" crude oil.

To separate oil into its components, it is distilled or **refined.** A distillation column with trays is used to separate the components of oil (Figure 17.1). Petroleum gases exit the top of the tower; liquids collect at various levels in the tower. The components that boil first condense at the topmost plate. Those that boil last, the tars and pitches, collect at the bottom of the column. On the trays between are the various components, separated in an order reflecting the ease with which they boil.

Natural gas, in contrast to oil, may be as little as 65% carbon by weight; its hydrogen content is variable. Although sulfur is usually very low, nitrogen

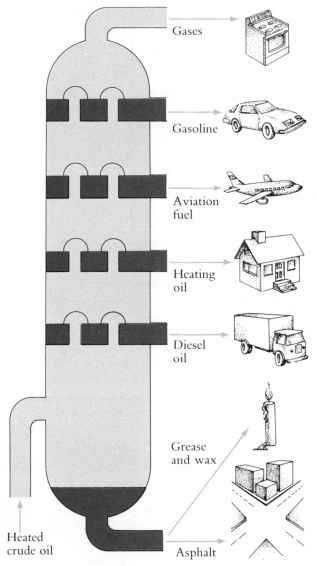

Figure 17.1 The refining of crude oil into its components. The components collect at levels that reflect how easily they boil. Gasoline and dissolved gases boil out first, followed by aviation fuel and heating oil. The most difficult components to boil, the tars and pitches, collect at the bottom of the refining tower. (Adapted from *The Open University, The Earth's Physical Resources, Block 2, Energy Resources,* The Open University Press, Walton Hall, Milton Keynes, United Kingdom, 1973.)

levels may be high, up to 15%. The same kinds of sediments and geological conditions that give rise to oil also give rise to natural gas. It is not unusual, therefore, to find oil and gas together. Often the gas is found dissolved in the liquid hydrocarbons. In this situation, the gas is separated from the oil when it is brought to the surface. Gas is also often found trapped above the liquid petroleum; the industry calls this "gas cap" gas. Gas is most often found alone, but about 25% of natural gas is found in the search for oil.

Oil and Gas Resources

Oil Resources of the United States

If we were to ask the question, "How much oil is there in the United States?" the reply would come in at least two parts. The first part of the reply would probably take this form: "The *Oil and Gas Journal* estimated **proved oil reserves** in the United States at 29.8 billion barrels as of January 1, 1982." The meaning of the term "proved" is that this oil is known to remain in the portions of oil fields upon which production drilling has already begun. Furthermore, we are assured that

this oil is profitable to recover. These reserve quantities, however, have been falling since the early 1960s in the U.S. as new discoveries have failed to keep pace with our rate of production.

The second part of the reply would be: "There is actually more oil than this, but for one reason or another we cannot count this oil at the present time." Some small portion we can estimate as quite likely to be recoverable on the basis of experience when new but proven methods of recovery are utilized. These are called **indicated reserves.** Another larger portion can be inferred on the basis of drilling to date; that is, we already have limited evidence of its existence. Another much larger quantity, for which we have much less assurance, is, as yet, **undiscovered resources.** One estimate of these quantities for the United States is displayed in Figure 17.2. Included in this figure are not only "undiscovered resources" and "indicated" and "inferred" resources, but also proved reserves of oil as well as oil extracted so far in history.

The quantity of undiscovered oil may be estimated on the basis of geological evidence; by the use of statistical methods; by extrapolation of past trends; or by combinations of these. Unfortunately, these procedures can lead to very different estimates of the oil

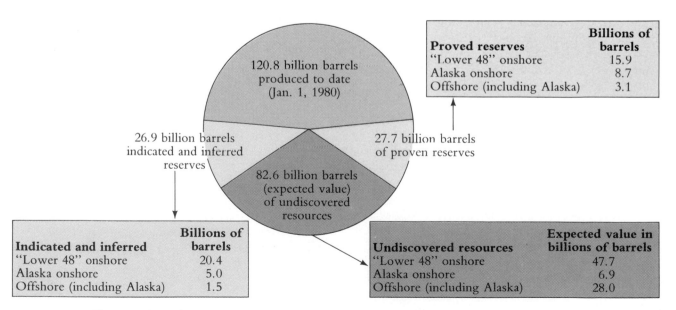

Figure 17.2 Oil resources of the United States by region and category. (Data source: U.S. Geological Survey Circular 860, 1981. Data as of January 1, 1980.)

remaining to be discovered. Some of the most recent, widely cited studies provide estimates of undiscovered liquid hydrocarbons ranging from 72 billion to 200 billion barrels.[1] The lowest estimate (72 billion barrels) of unproven U.S. oil reserves was provided by M. King Hubbert of the U.S. Geological Survey.

There is one further type of oil resource—oil not presently profitable to produce. Some is already discovered; the remainder is yet to be discovered. We mentioned earlier that North Sea oil became profitable after the price of oil jumped from $2.00 per barrel to $12.00 per barrel. In this regard, we do not yet count oil from shale in any of the above categories.

A remarkable search for oil is now going forward in the Arctic Ocean, north of Alaska and Canada. There the international oil companies are making tremendous investments to find and produce oil in one of the most inhospitable environments on earth. Ice breaks up and reforms there annually; subzero temperatures and winds gnaw at workers; and waves threaten structures built to explore for and withdraw the oil.

In 1982, the price of oil was falling as the OPEC cartel felt the pinch of a worldwide recession. The falling price must have sent waves of concern through the officers of those companies investing fortunes in Arctic oil. The oil might not be profitable to produce.

United States and World Use of Oil

In 1981, the United States was using about 29% of all oil consumed in the world each year. It was not always this way, however; we used to consume even a greater proportion. We are gradually losing our dominance as a consumer of oil, not because we are consuming less, but because other nations have increased their own uses of oil so dramatically.

In 1979, U.S. consumption of 18.51 million barrels of oil per day was equivalent to 6.76 billion barrels of oil per year. By 1981, in response to rapid increases in the world price of oil and a severe recession in the U.S., this figure had fallen to 16 million barrels per day.

Consumption of refined petroleum products in 1981 occurred in a number of end-use categories (Fig-

1 *Energy: The Next Twenty Years,* Report by a Study Group Sponsored by the Ford Foundation and Administered by Resources for the Future. Cambridge, MA: Ballinger Publ., 1979.

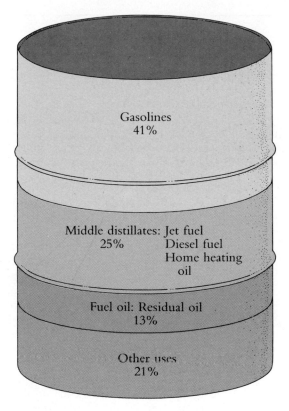

Figure 17.3 Oil consumption in the United States by use category, in 1981. (Source: Energy Information Administration)

ure 17.3). Gasolines accounted for 41% of consumption; use of middle distillate, which consists of home heating oil, diesel fuel, and kerosene (jet fuel), made up 25% of consumption; and residual fuel oil use accounted for 13% of oil consumption. (Residual oil is used for heating commercial and industrial buildings and as a fuel in a decreasing number of steam electric plants.) The total of all other uses, including asphalt, plastics, chemical manufacture, etc., gives us the remaining 21%. These shares have remained relatively constant over the last several decades. The portion of oil that goes for gasoline in Western Europe is only about 21%, but this figure is growing as car ownership continues to grow in Western European nations.

In 1981, world oil production was 55.9 million barrels per day. The U.S. used 16 million barrels of oil per day in 1981, about 29% of the world's oil production. Yet in 1979, 1980, and 1981, it produced only 10.2 million barrels of oil per day; the remainder of its consumption in those years had to be imported. The

A Fish Story

M. King Hubbert, who provided the lowest recent estimate of U.S. oil reserves remaining to be discovered, is one of the most renowned energy analysts in the world. In a chapter of a Congressional Committee Print, *Project Interdependence,* Hubbert explains the philosophical basis of his calculations in an amusing way.★

> The problem of estimating the ultimate amount of oil or gas that will be produced in any given region has been aptly likened to that of estimating the abundance of fish in a lake by the success in fishing. Although the fish in the lake are not seen by the fisherman, if he catches a fish by almost every cast the inference is justified that the lake is teeming with fish. If, after such a lake has been heavily fished by vacationers for a number of seasons, the same angler is able to catch only a fish or two per day, he is justified in the inference that the lake has been about fished out.
>
> Similarly, in a virgin petroleum-bearing region, if most of the initial exploratory wells succeed in discovering oil fields, the inference is justified that the given region is rich in petroleum resources. When, at a more mature stage of exploratory drilling, it is found that a steadily decreasing fraction of the exploratory wells drilled succeed in finding oil, the inference is unavoidable that most of the oil in the region has already been discovered, or that the lake is about fished out.
>
> However, in the case of the lake, if fishing is discontinued for a few years, the lake will become restocked, whereas in the oil-field analogy this cannot happen because oil fields do not breed. There was only a fixed and finite number of oil accumulations in the sediments of the region initially and every time one of these is discovered, the number remaining is reduced by one. It is inevitable that as the remaining fields become fewer in number, and probably also deeper and smaller in size, the difficulty of making a discovery must accordingly increase. Conversely, this record of fewer discoveries with increased exploratory drilling affords one of our more reliable means of estimating how far we have progressed toward the ultimate discoveries likely to be made in the region.

To many of us, a few fish per day sounds like a great deal, so Hubbert must have some very special fishing spots. Of course, Hubbert is right about how we might respond to a lake that once yielded fish in plenty and now provides only a few per day. Suppose, however, that we had a cabin on the lake and a canoe with which to explore. How quickly would you as a fisherman abandon your lake? Would you take other actions first in terms of tackle and bait? What actions? How about in terms of exploration? What advanced methods might you think of to locate the fish? Might you consider catching fish that were less desirable than those in your earlier catches? In the context of oil, a less desirable fish might be oil shale. Can you think of others? Perhaps a different way of cooking would make them better to eat. You begin to understand now how the oil companies have responded to their fewer successes on the U.S. mainland.

Hubbert's story gives us another insight on oil. Because oil exploration is likened to fishing, we are drawn to a comparison of fish to oil. Fish is a renewable resource; unless we remove nearly all the fish, the fish will breed and repopulate the lake. Oil is a nonrenewable resource; each removal is final and irrevocable. A much better energy resource on which to rely is one that is renewable, such as solar energy.

★ *Project Interdependence,* Chapter XIX, World Oil and Natural Gas Reserves and Resources, U.S. Government Printing Office, Washington, D.C., 1977.

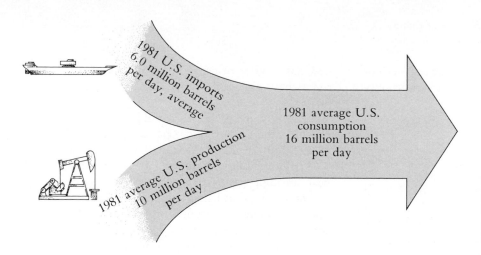

Figure 17.4 Production, imports, and consumption of oil in the United States in 1981 (figures are rounded). (Source: Energy Information Administration)

division between imports and U.S. production in 1981 is illustrated in Figure 17.4.

Proved reserves of oil are not distributed equally throughout the world. Figure 17.5 shows the inequity. As we can see, the Middle East has by far the greatest proved reserves in the world. It is not hard to see why our imports from Arab members of OPEC amounted to 31% of all our oil imports in 1981.

The future statistics on oil production and oil importation are uncertain for a number of reasons. If imported oil is taxed, if OPEC raises its prices steeply

once again, or if major oil discoveries occur in the offshore U.S., domestic production could climb. A shale oil industry is a distinct possibility.

If the economy is not healthy and many people are out of work, the use of the automobile will diminish as people try to save money. This was seen in the period following the 1973–74 embargo, when the country went through an economic slump due to combined inflation in the price of oil and other goods. This decrease in automobile use and hence in the use of gasoline was seen again in 1982 when a severe recession

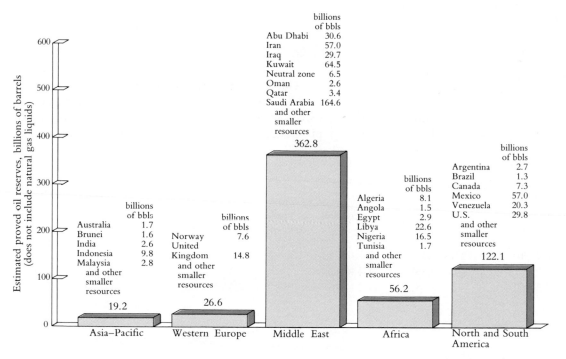

Figure 17.5 Proved oil reserves as of 1 January 1982. (Data source: *Oil and Gas Journal*, 28 December 1981.)

gripped the nation. Any combination of these various events could occur, leading to enormous uncertainty in predicting the U.S. oil future.

Natural Gas Resources, Production, and Use

We can describe natural gas resources in the United States in the same way in which we described oil resources. Proved reserves are in place and profitable to produce. There are indicated and inferred reserves for which we have some degree of evidence. Finally, undiscovered resources are yet to be found. Table 17.1 provides estimates of natural gas resources in these various categories and indicates the potential in Alaska and the offshore for natural gas.

Proved reserves of natural gas have been falling since 1970. The natural gas industry explained this phenomenon as being the result of government control of the price of natural gas sold in interstate commerce. The industry argued that the price they were allowed to charge did not allow them enough profit to justify looking for new sources of gas. Finally, in 1978, the industry convinced Congress that an end to control was necessary. In that year, Congress passed and the President approved a bill that would phase out all price controls on natural gas by 1985. Since then, the price of natural gas has risen, as all predicted. The average price to the consumer jumped from $2.56 per 1000 cubic ft in 1978 to $3.68 per 1000 cubic ft in 1980, and has continued to rise as price controls are gradually lifted. The gas purchased by most consumers, however, is a mixture of old gas (controlled in price) and new gas (higher priced) and its price will continue to rise as new gas gradually replaces old in the marketplace.

Although reserves of natural gas have been falling, consumption of natural gas has been fairly stable since about 1969. Between 20–22 trillion cubic feet (0.51–0.62 trillion cubic meters) of gas were consumed each year through the 1970s and into the early 1980s. Of this quantity, about one trillion cubic feet (0.028 trillion cubic meters) were imported annually from Canada by pipeline. In the early 1980s, the U.S. began to import natural gas by tanker, principally from Algeria. In order to save space in shipment, the gas was made liquid by cooling it to very low temperatures and putting it under pressure. At the coastal terminals, the liquid was to have been converted back to gas and injected into pipelines that already carried U.S.-produced natural gas.

Although liquefied natural gas (LNG) is still being received in small shipments at U.S. coastal terminals, two factors have limited the imports of LNG. First, U.S. gas supplies have improved, apparently as a result of higher prices being allowed for U.S.-produced gas. Second, Algeria has been asking exorbitant prices for its liquefied natural gas.

Other Sources of Gas

Geopressured methane. If there were ever a phantom resource, the natural gas known to be dissolved in deeply buried caverns of salt water is that phantom. Trapped subterranean waters are at very

Table 17.1 Natural Gas Resources of the United States[a]

Trillions of cubic feet[b]	Production to date	Proved reserves	Indicated and inferred	Undiscovered (expected value)
Lower 48 states	519.3 (14.69)	123.3 (3.49)	132.1 (3.74)	390.2 (11.04)
Alaska onshore	1.2 (0.03)	30.0 (0.85)	4.4 (0.12)	36.5 (1.03)
Offshore (including Alaska)	57.5 (1.63)	38.2 (1.08)	41.0 (1.16)	167.0 (4.73)
Total	578.0 (16.4)	191.5 (5.42)	177.5 (5.02)	593.7 (16.80)

a Source: *U.S. Geological Survey Circular 860,* 1981.
b Trillions of cubic meters in parentheses.

high temperatures and under enormous pressure. In this environment, natural gas may be formed from the breakdown of oil deposits and dissolved in the salt water.

If the salt water caverns themselves can be tapped, a very large resource of natural gas may await us. The theory is that if the pressure on the reservoirs can be released, perhaps by allowing some of the water to escape as steam, the gas will come out of solution forcibly and can be captured at a well. Controversy abounds over whether or not this gas can, in practice, be recovered, but experts agree it is there. A Chevron well driven 21,000 feet (6300 meters) into the earth in Louisiana is said to have produced geopressured gas in 1977, but the company asserts it was an ordinary gas deposit. There are potential environmental problems if we use this resource, such as disposal of hot brine and concern over subsidence above the cavern if geopressured gas is tapped. The resource may, however, be worth thinking about.

The U.S. Geological Survey estimates that there may be 24,000 trillion cubic feet (680 trillion cubic meters) of gas in the geopressured zone of the continental U.S. This estimate should be compared to current annual use in the U.S. of about 20 trillion cubic feet (0.57 trillion cubic meters) of gas; this is a 1200-year supply at the current rate of use. How much of the geopressured gas could be recovered is not known.

Methane associated with coal. There are other ways to produce a methane gas of the quality presently produced from drilling. One such method is to remove the methane produced in coal deposits before opening the mine. This procedure both produces gas and makes the mine safer to work. In fact, the Equitable Gas Company of Pittsburgh has been quietly producing natural gas from a coal field since 1949. Still another source of natural gas is the shale created in the geologic era known as the Devonian. There may be as much as 500 trillion cubic feet of gas in these deposits in the eastern and central U.S.

Methane in tight sands. One unusual plan for gas production from wells is quietly slipping into oblivion. Known as the Plowshare Program,[3] the undertaking began in the 1960s under the leadership of the

Atomic Energy Commission (now part of the Department of Energy). Underground nuclear explosions were to be used to free gas contained in "tight" formations by fracturing the rock walls between small deposits.

Although three experiments were conducted in New Mexico and Colorado, the program came to an end because residents of the states where the atomic explosions were to be set off were concerned about the effects. In addition, the gas itself had a low level of radioactivity.

Methane from organic wastes. Another way to obtain gas is to capture the methane produced by sanitary landfills, the areas where trash and garbage are buried. The decomposition of the buried garbage in the absence of oxygen is known to produce methane, sometimes contaminated with the foul-smelling gas, hydrogen sulfide. The methane level, however, is good and the gas can be burned.

In addition, methane has been produced for some years in sewage treatment plants in the device known as the anaerobic digester. The methane is burned in the plant to provide the heat needed for the digester and elsewhere in the plant.

Methane may also be produced from animal wastes that may build up on feedlots for beef production. The decomposition of these wastes in the absence of oxygen produces methane. Plans to use gas from cattle manure have been formulated by the Peoples Gas Company, a Midwestern firm.

Synthetic Fuels

The term "synthetic fuels" refers to liquid and gaseous fuels derived from coal or shale oil (or other sources), rather than from naturally occurring petroleum and natural gas. The name does not indicate a precise class of compounds; only fuels derived from other than their usual source.

Synthetic fuels have been around for a long time. Oil derived from shale rock and oil from coal may be the synthetic fuels of today, but a century ago, these fuels, along with whale oil, were the only liquid fuels available. In the first half of the nineteenth century, before we learned to exploit petroleum, the lamps of North America were lit by whale oil, coal oil, and shale oil.

---

3 A hopeful name derived from the biblical phrase ". . . They shall beat their swords into plowshares, . . ." (Isaiah, 2: 4).

Social Effects of Natural Gas Pricing

Let the Bastards Freeze in the Dark!

A popular bumper sticker in oil- and gas-rich states

"I stayed in bed to keep warm and to keep from being sick,"
she said. "My brother used to sit by the oven."

The New York Times, 7 February 1977,
*an unidentified elderly woman in Queens, New York,
describing how her brother had gotten frostbite and gangrene.*

The conflict on natural gas pricing probes deeply into one of the controversies in our free-enterprise society. On the one hand, we have a society that believes in free enterprise as a practical and efficient means of providing the goods and services people demand. On the other, we have chosen to provide social security and medical care to the elderly. We attempt to be at once a society that both supports a free market system and cares for those who have least.

The issue of natural gas pricing illustrates the conflict between these objectives. The notion of a free marketplace involves people who need a product and producers who can provide it. When many producers can provide a product, let us say bread, we assume the producers compete; that is, we assume they lower their prices to similar levels so that they can continue to capture a portion of the demand. Unfortunately, perfectly free markets do not exist for every product.

A free market for the sale of natural gas, which moves in interstate commerce, has not existed for some years. The market has been controlled by the government because it was thought to be open to abuse. The rationale was that monopolies could occur when a single pipeline supplied an urban area. A single gas-supply firm could charge very high prices for natural gas because it would be the only supplier and because gas is essential. To prevent excessive prices in this monopoly situation, the government controlled the price of natural gas.

Government control of the price of gas sold interstate, however, appears to have been too tight. The low price of this gas led many millions of families in urban areas of the eastern U.S. and the Midwest to heat their homes with natural gas. Meanwhile, gas sold within the state in which it was produced became a valued good. In addition, its price was not regulated. Gas was used in producer states for fertilizer manufacture and chemical manufacture and for electric power generation, as well as for heating homes and hot water. The price of gas sold within the state in which it was produced soared to four and five times the regulated price for interstate gas. With such a market in the producing states, it was no wonder that producers did not want to sell gas on the interstate market.

Supplies to the East and Midwest did not keep pace with demand; in the late 1970s, new customers for natural gas had to be turned away by the local utilities. Some industries had their supplies of natural gas diverted to homes and schools as severe winter weather set in. The producers told us that the government should not regulate prices. Mobil Oil Company put it this way:*

What is needed is decontrol of prices of new supplies of natural gas. . . . This approach would give producers the incentive to step up their already active search for gas . . .

On another occasion, Mobil added,†

The government's focus on low prices to the consumer has ignored his need for secure and adequate supplies.

In 1978, Congress ordered the regulation of natural gas prices to end by 1985, with a gradual phasing out of price controls. It would be a mistake, however, to suppose that the decision of 1978 will no longer be debated. When Congress originally ordered the regulation of natural gas in 1954, the gas industry lobbied and buttonholed Congress for deregulation the next year, the year following that, the year after, and so on until 1978, almost a quarter of a century later. And the vote to end regulation in 1978 was very close. In 1985, we are scheduled to have our first taste of a "free" market in natural gas. Some senators and representatives, as well as ordinary citizens, do not trust the oil and gas industry and can be expected to work to re-establish price regulation to prevent sky-rocketing gas prices. In a sense, the issue of regulation boils down to whether or not you trust the oil and gas industry.

It is clear that both sides have something of value in their argument. Low interstate prices have caused shortages as new supplies have been sold within their state of origin, but low prices have also protected the consumer. An unregulated natural gas industry has the potential to exploit the consumer. Furthermore, if prices rise sharply, reserves may come up, but the poorer people among us will be less able to afford heat for their homes. Do you have suggestions for what we should do? Do you favor the government's removing or restoring controls on natural gas? Why or why not?

---

* Mobil Oil Company advertisement, *The New York Times,* 18 January 1976 and 26 January 1977.
† Mobil Oil Company advertisement, *The New York Times,* 18 December 1976 and 26 January 1977.

Synthetic gas is not a new idea either, although the quantity and quality of the product now being considered is quite different from earlier versions. Gas was being manufactured from coal in the early 1800s on a wide scale, first in Britain, then in the United States. The product of the local "gasworks" was known as "town gas" or "illuminating gas"; this low-heating-value gas was used for community lighting, for heating homes, and for cooking. By the end of World War II in the United States, natural gas had displaced town gas. In Scotland, town gas is only now giving way to natural gas from the North Sea.

Today, synthetic fuels are planned from three sources: "tar sands," coal, and shale oil. All three sources have significant potential to extend fuel supplies in North America.

Synthetic Fuels From Coal: Gasification and Liquefaction

Coal gasification processes are of two types, based on the quality of the product. One process produces a gas of relatively low heating value, consisting primarily of carbon monoxide and hydrogen; the process is referred to as low-Btu gasification. The second process produces a gas with a heating value nearly that of natural gas, and the process is known as high-Btu gasification. The gas from this second process may be referred to as SNG (Synthetic Natural Gas), and consists primarily of methane.

This high-Btu gas is a reasonable substitute for the natural gas transported by pipeline to many sections of the country. The low-Btu gas, however, is likely to be used in industries only at the place where it is produced, or it may be used to generate electric power near the site of production.

The processes currently being studied are known by such trade names as Lurgi, HYGAS, BI-GAS, and SYNTHANE. The only commercially proven process for coal gasification, however, is the Lurgi process, which originated in Germany before World War II. Although these processes differ in their details, the broad picture is very nearly the same. In all processes, the undesirable constituents of coal are replaced with hydrogen, eventually producing methane in the high-Btu gas.

The typical coal gasification plant is expected to be large. A plant that is to produce 125 million cubic feet (3.34 million cubic meters) of high-Btu gas each day will require 12,500 tons (11,350 metric tons) of coal, or about one ton of coal for each 10,000 cubic feet of gas. The quantity of coal needed for a year's operation of such a plant (operating 90% of the time) is 4.25 million tons (3.35 million metric tons). In 1982, the cost projected for the Great Plains Coal Gasification plant, which was planned to produce pipeline-quality gas at this rate, was estimated at $2 billion.

Each coal gasification plant of this size is expected to consume 5000–7500 acre-feet of water per year in the process of producing the gas. An acre-foot of water equals a volume one acre in area and one foot in height, so that 5000 acre-feet is about 8 square miles, one foot deep with water.

All plants presently planned would use the Lurgi gasification process, which is presently being used in Sasol, South Africa. All plants are also planned to produce pipeline-quality gas. Nearly a dozen such plants were in the planning stages as the 1970s were drawing to a close.

Another method of coal gasification is not carried on in a factory. Designed to exploit coal that cannot be extracted at a profit, the process takes place underground and should not disrupt the land extensively. *Underground coal gasification* produces gas of low heating value, which may be upgraded to pipeline-quality gas by chemical processes in a plant on the surface. Although underground gasification has not been implemented commercially at this time, its potential for profitable gas production appears to be excellent.

The production of gas by underground gasification has the potential of multiplying our coal reserves, possibly by a factor of three, according to the Department of Energy. Since these reserves are uneconomic to exploit for mining, their use for gas production gives us an entirely new resource base.

A liquid hydrocarbon fuel can also be produced from coal, using one of three basic methods. The *catalytic method,* the oldest of these processes, uses various catalysts to convert gas from a coal gasification plant into liquids. A commercial coal-to-oil plant in Sasol, South Africa, using the catalytic method, has been in operation for over 25 years. The plant converts 3500 tons (3175 metric tons) of coal to about 30 million cubic feet (0.85 million cubic meters) of gas each day. The gas is then converted to a liquid. The principal products of the plant are gasoline and other motor fuels, along with a high-Btu gas suitable as a substitute for natural gas. The first step in the production of the oil is the step we referred to earlier as the Lurgi process for making synthetic natural gas from coal.

Unfortunately, making liquid hydrocarbons from coal is quite costly. It is likely to produce a synthetic oil that costs perhaps 50–100% more than the 1980 cost of crude oil. An industry that involves itself in the manufacture of oil from coal by one of these processes must be betting that OPEC will raise the price of oil quite steadily through the next decade.

In the *pyrolysis process* for coal liquefaction, coal is heated to drive off a hydrocarbon liquid, leaving behind a solid of nearly pure carbon, called char. Although several pyrolysis processes have been developed, none appears competitive with the coal-to-liquid process called *solvent refining of coal.*

Solvent refining of coal (abbreviated SRC) is an example of a process in which hydrogen is donated to coal to increase the hydrogen-to-carbon ratio. The product is either a liquid or a solid, depending on its temperature. It is solid at room temperature but melts at 350–375°F (177–191°C). Low in sulfur and ash, solvent-refined coal may find use as the boiler fuel for electric power plants. Although a pilot commercial facility was announced in 1978 in West Virginia, the $1.6 billion project was cancelled in 1981.

Synthetic Fuels from the Tar Sands of Athabasca

In the northern portion of the province of Alberta, Canada, is a region where deposits of a black sand coated with pitch or tar are being exploited to produce oil. These petroleum deposits, instead of being trapped in an underground reservoir, have migrated up into porous sands near the surface of the earth. The tar, a hydrocarbon known as bitumen, may account for up to 16% of the weight of the sandstone. Three tons of a rich tar sand, one that is 14% or more bitumen by weight, is sufficient to produce two barrels of hydrocarbon liquid. Geologists estimate there may be 300 billion barrels of oil in the tar sands deposits, about 10 times the proved oil reserves in the United States.

However, several efforts in Alberta to extract the oil have had a difficult history; the oil has proven costly to produce. A Sun Oil project known as Suncor began in 1963 and can produce nearly 60,000 barrels of oil per day, but it is plagued by mechanical breakdowns. The Syncrude project begun in 1978 by a consortium of oil companies has a production capacity of 120,000 barrels per day. Imperial Oil (an Exxon subsidiary) began a project at Cold Lake, Alberta, to produce 140,000 barrels per day, but has postponed completion indefinitely. The most recent project, known as Alsands, lost half its initial ownership in 1982 when five oil companies pulled up stakes from the $13-billion venture. One of the remaining partners was Petro-Canada, Canada's state-owned oil company. Alsands was planned to have a capacity of 137,000 barrels of oil per day.

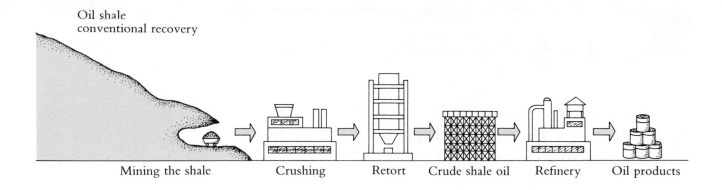

Oil shale
conventional recovery

Mining the shale   Crushing   Retort   Crude shale oil   Refinery   Oil products

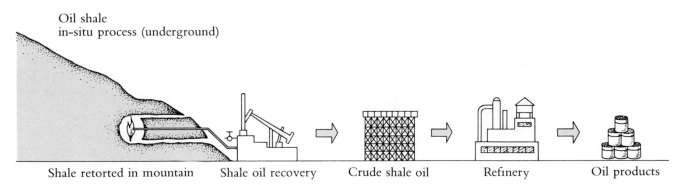

Oil shale
in-situ process (underground)

Shale retorted in mountain   Shale oil recovery   Crude shale oil   Refinery   Oil products

Figure 17.6 Shale oil can be produced in two basic ways. The mined material may be retorted ("cooked") above ground; or the rock may be cooked in place (underground). Of the two, the second is more acceptable from an environmental standpoint, as the surface is not modified and waste rock is not produced. The difficulty, or unknown, is how efficient the underground process can become. At the moment, it seems that surface mining and surface retorting are most economical. Oil produced from shale must, in any case, be cleansed of sulfur and nitrogen before it is comparable to oil used in refinery operations.

Tar sands are also found in the United States, but most deposits are small in comparison to Canada's. By far the largest U.S. deposits are on federal land in Utah; 19–29 million barrels of liquid hydrocarbons might be produced from these deposits. Venezuela has large deposits of tar sands that it was working vigorously to exploit until the price of crude oil began to fall in 1982.

Synthetic Fuels from Oil Shale in the Rockies

Another source of liquid hydrocarbons is the shale rock that contains the organic material known as kerogen. A liquid that is much like oil can be extracted from the kerogen by distillation. The organic matter locked in the shale is the result of geologic processes acting on the ancient sediments accumulated in inland lakes. The world's largest deposits of this oil-bearing rock are apparently in Wyoming and Colorado, where the equivalent of 600 billion barrels of oil may be present. Compared to U.S. proved reserves of 30–40 billion barrels of oil in ordinary reservoirs, this resource is enormous.

Why have we been so slow to develop this resource? Clearly, it could solve our oil shortage (see Controversy 17.2). There are several reasons for the long delay. First, removing the oil requires the mining of rock. About $\frac{1}{2}$–2 barrels of oil can be extracted from each ton of rock, leaving 1700 pounds of waste rock to dispose of. If we were to produce one billion barrels of oil from this shale, or about one-sixth of our annual (1978) consumption of oil, we would produce about 900 million tons of waste rock. This waste rock would have a huge impact on the land; the mining operation could be massive. Caution in pursuing this resource is appropriate (Figure 17.6).

The Economics of Synthetic Fuels: Will Synthetic Fuel Provide Energy Security or Become a White-Elephant Investment?

A concerted effort by government and industry to develop commercial-scale synthetic fuel technology is still vital to the nation's security.

Business Week Editorial*

Energy independence is a myth, a false idol that diverts our attention . . . from the real problem: preparation for oil supply disruptions.

William Bradley,†
U.S. Senator

The better way to solve the national security problem is to build up the strategic reserve (of petroleum).

John C. Sawhill,‡
Former Head of the Synthetic Fuels Corporation

. . . The American Government should . . . raise the question of what would be done for a corporation . . . that has invested heavily in a synthetic-oil process that turned out to be uneconomical.

Thomas Hughes,§
Professor, University of Pennsylvania

Synthetic fuels can be produced: the technology is available even now for coal gasification, coal liquefaction, and the retorting of shale rock. Yet again and again we have seen these industries begin to rise only to fall flat. After a brief flirtation with shale oil production in the 1950s by the Union Oil Company in Colorado, in the 1970s the energy industry once again expressed interest in synthetic fuels. Interest became serious; the giant oil companies leased and bought coal lands, and also acquired lands rich in oil shale. Driven by oil prices as high as $42 per barrel on the spot market (the official price went to $34 per barrel), plans moved forward rapidly for synthetic fuel production. At that selling price, industries could turn a profit on the sale of shale oil and gas and liquid fuel from coal. The government too was interested—an industry on the shores of North America that could make the continent self-sufficient in oil could help protect us from political blackmail by producing nations.

Nonetheless, in 1982, Exxon backed out of the massive $6-billion Colony shale oil project with the Tosco Corporation in Colorado, even though the project was about *halfway* to completion. And in 1982, WyCoal Gas Company suspended its $2.7-billion coal gasification project in Wyoming, citing high interest rates and low energy prices. Also in 1982, half of the oil-company partners withdrew from the $13-billion project in Canada to produce oil from the tar sands that cover portions of the province of Alberta. In addition, the $1.6-billion facility for solvent refining of coal in West Virginia was cancelled. Although several significant projects continued in progress, including a $2-billion effort by Union Oil in Colorado and the $2-billion Great Plains Coal Gasification plant in North Dakota, the cancellings of these keystone efforts for energy self-sufficiency were all related.

The high price that OPEC forced on the world for its oil in 1980 and 1981 ($34 per barrel) had made these synthetic fuel efforts look profitable, but the world slid and then fell into a deep recession in 1981 and 1982, partly as a result of these steep oil price increases. With 10–11% of the labor force out of work, travel by auto declined; thermostats were turned back; industries that used oil were shut down or cut back their production; and oil demand declined sharply. The recession and the decline in oil demand occurred worldwide. With the decline in demand, members of OPEC were unable to sell the quantities of oil they had expected, and upon which they had based their economic plans and activities. To sell more oil, OPEC members sold oil at a price substantially lower than their officially quoted price. This signal, the decrease in price as a result of the recession, was quickly received by the energy companies. These were the companies developing the tar sands, gasifying and liquefying coal, and preparing to produce oil from shale rock. Instead, they shut down: profits were no longer in the wind.

It is of interest that the pullout by Exxon took place even though loan guarantees of over a billion dollars had been provided by the Synthetic Fuels Corporation, an arm of the federal government; that is, financing of the investment was available. But the guarantee of financing was not sufficient; profit potential was still lacking.

One significant shale oil project that did survive the hard times was the Union Oil effort in Colorado. Its survival of the dip in oil prices apparently was the result of a guarantee by the U.S. Department of Defense to purchase Union's shale oil output at a fixed price. This price corresponded to the peak oil price that occurred on the spot market in 1981—about $42 per barrel. Thus, the economics of Union's shale project did not change when the world oil price dropped, but the government was committed to buy oil at an inflated price.

Many economists viewed the dip in world oil prices as momentary, corresponding to a low point in the economy, from which oil prices would not only recover, but rise to exceed their previous peak levels. Higher oil prices might again stimulate interest in producing synthetic fuels.

Synthetic fuels, as noted elsewhere in the text, have been around for a long time, and have always faced this economic problem. Farben, a German chemical concern, built a coal liquefaction plant in Germany in the 1920s. The plant was built in anticipation of rising oil prices, but the prices never rose. Eventually, Hitler bailed out Farben by buying its synthetic gasoline as fuel for the German war machine.

Another suggested approach to achieving national security in oil is to quickly build up the Strategic Petroleum Reserve, a government-owned-and-run storage facility. First conceived in the late 1960s, the concept was finally implemented in 1975 when Congress authorized storage of up to one billion barrels of oil (about two months supply in 1981). By December, 1981, about 225 million barrels had already been stored in 16 salt caverns in Texas and Louisiana, and in a salt mine in Louisiana. Each barrel stored by the government must be paid for by the government at the market price. During 1982, storage was being filled at about 200,000 barrels per day. At world oil prices of $34 per barrel, that is $6.8 million per day, or about $2.5 billion per year of government expenditures.

With all these facts at hand, can you recommend the best strategy for oil independence? Is it construction grants to the energy industry? or loans? or guaranteed prices? Or should we fill the Strategic Petroleum Reserve as quickly as possible? Remember that the synthetic fuels industry utilizes surface mining, generally, and demands water in the water-short West. Are a synthetic fuels industry and a reserve the only long-run options open to prevent disruption of our oil supplies? What other options might decrease dependence on foreign oil, and what are their environmental impacts?

* *Business Week,* May 24, 1982, p. 198.
† *New York Times,* January 18, 1982, p. A14
‡ *New York Times,* May 5, 1982, p. D1
§ *New York Times,* June 19, 1974, p. 30.

The Institute of Ecology provided a review of the government's Environmental Impact Statement on oil shale leasing. That report indicated other problems besides disposal of the waste rock, itself a considerable obstacle. Oil shale processing is expected to cause the release of mercury, cadmium, and lead to the air. Water used in the mining operation would become rich in dissolved salts; the brine must be disposed of in a way that will not contaminate groundwater. Revegetation of mined areas is probably difficult because of the saline content of the waste rock (Table 17.2).

The second reason we are not producing oil from shale has to do with costs. Estimates of the cost to produce a barrel of oil from shale are very uncertain. In 1977, some said shale oil could be produced for as little as $8 per barrel; others thought the cost of production (not the sale price) might be as much as $28 per barrel. More recently (1980), the Office of Technology Assessment (OTA), an arm of Congress, estimated that shale oil would have to sell for $48 per barrel to give the manufacturer a 12% return on investment, and

Table 17.2 Physical Impacts of Oil Shale Production

Production of one barrel of shale oil *requires*	Production of one barrel of shale oil *produces*
2–6 barrels of water; mining and retorting 1–2 tons of oil shale; 100 square feet of land	1/3 pound of airborne dust; 2½ pounds of polluting gases; 2–5 gallons of contaminated water; 1–1½ tons of spent shale

hence a reasonable profit. In the early 1980s, others have cited figures as low as $42 per barrel, a price that was being reached on "spot" purchases of oil in 1981, before a world-wide recession set in. If the government were to subsidize 50 percent of the manufacturer's construction cost, the price to earn a fair return could be reduced to $34/barrel, according to OTA. With such uncertainty, energy companies have been reluctant to invest in oil shale production.

The Power of the Cartel and Its Impact on Oil Companies (Economic and Political Aspects of the Global Oil Situation)

The oil cartel has had enormous influence on the quality of life in the United States. It could be argued that the cartel's pricing policy will have long-run benefits in stimulating our use of renewable energy resources. It could also be argued that the high prices will eventually lead us to develop shale oil processes, which could have adverse effects on the states in which the shale is located. In the short run, the pricing policy will have marked effects on our ability as a nation to afford other improvements to our environment. Constantly increasing oil prices increase inflation and cause citizens to ask if we can afford pollution control. To understand the urgent need for oil conservation, we need a closer look at the organization that sets oil prices.

To explain the power of the cartel, we must first discuss the notion of a monopoly. A single-firm monopoly is an enterprise that is alone in its field. No other firm exists that supplies the same product or service to customers. When one company produces such a very large portion of the total demand for the product, the other firms that supply the product are in a happy but risky situation. If the giant firm has the financial resources or its production costs are very low in comparison to competitors, it can decrease its price to a level at which competitors cannot make a profit. If the competitors are driven from the scene, the giant becomes a monopolist. On the other hand, if the giant firm does not care to increase its share of the market for some reason, competitors are in an excellent financial position. They can share the monopoly power of the giant firm to sell at an inflated price.

In the international oil market today, the giant firm is represented by OPEC (the Organization of Petroleum Exporting Countries). The competitors are the international oil companies, many of which are based in the United States. Even though OPEC consists of a number of countries, they are still able to dominate the marketplace as though they were a single firm; that is, they set the price at which they sell their oil. Such an organization, which attempts to dominate the international market, is referred to as a cartel. In order to ensure that each member of the cartel is able to obtain the price set by OPEC, each country participating in OPEC agrees to produce so many million barrels of oil per day and no more.

Since OPEC both sets a price and allocates production to the member nations, each member is, in effect, agreeing to a particular revenue from oil sales; each will earn the product of price and production. The stability of the cartel is anchored on the agreement that each member will continue to accept its stated share of production so that it can obtain the price set by the organization. The cartel, any cartel, may not be stable if one of its members needs increased revenues or if new suppliers appear on the scene.

We mentioned in our brief discussion of monopolies that the other producers in a near-monopoly situation would be in a "happy but risky" situation. The international oil market illustrates this point well. Just before the 1973–74 embargo, oil produced within the U.S. was selling on the U.S. market for $3.50 per barrel. Just after the oil embargo of 1973–74, OPEC began to demonstrate its strength; it set the price of its oil at $11 per barrel. By 1980, the price was set at $28 per barrel. Oil now being produced in the United States by the oil companies can be sold at the OPEC price.

This new high price for oil is the "happy" situation of producers who operate in the shadow of a monopolistic firm. The price of their goods rises to the giant firm's price. U.S. and international oil companies reap a handsome benefit from the cartel's control of the market.

Their situation is not without risk, however. In their search for greater profits, the firms do invest heavily in exploration, production, and in research for methods of enhanced recovery. To see the origin of the risk, we use the example of the North Sea, where the pace of exploration and development has quickened since OPEC's rise to power. In the middle 1970s, an offshore oil platform, built to withstand the lashing winds and towering waves of the North Sea, cost about 200 million dollars. This investment had to be made before the first drop of oil could be produced from a field (see Figure 17.7). It was estimated that

North Sea oil would cost on the order of $5–$6 per barrel to produce, but so long as the cartel keeps its price high, the investment is well spent. It is of interest that the oil in the North Sea was not even counted as a reserve in the early 1970s. The world price for oil was then about $2.00 per barrel, so that North Sea oil was not profitable to produce. If the cartel were to decide to lower its price to less than about $7 per barrel, the North Sea investments would again no longer be profitable.

Could the cartel lower its prices so drastically? The plentiful oil in the Middle East costs less than 25 cents per barrel to produce, as opposed to $5 in the North Sea, and transporting oil via tanker to market costs only about $1 per barrel. OPEC has awesome power to cut prices and destroy the competition's investments. This is the "risky" situation to which we referred earlier. Synthetic fuels derived from oil shale, coal, and tar sands face similar risks.

Figure 17.7 Production platform nearing completion, destined for the Brent Field in the North Sea. The investment to produce this oil is evident. (A Shell photograph.)

Oil Conservation and the Control of Inflation—More Than One Reason to Save

Conservation of oil is important to future generations. The basic goal of oil conservation is to postpone the days of oil scarcity until new sources of energy become available. It is an unarguable goal, worthy in and of itself. Nonetheless, there are two other important reasons to conserve oil. These other reasons make our interest and the interest of future generations one.

First is the hazard of dependence by the United States on foreign oil. In 1973 and 1974, we endured an oil embargo by the OPEC nations. That embargo, begun as punishment for the U.S. position in the 1973 Arab–Israeli war, proved to the OPEC nations their immense economic power, their ability to set prices for oil in the marketplace. Another embargo like the one undertaken in 1973–74 could stall the engine of our industrial society and influence the freedom we have in choosing our policies relative to other nations.

Second is the major influence of oil imports on the economic health of the nation. Our massive oil imports may go in part to heat our homes, but they also fuel the fires of inflation. How can this be? It is quite simple.

In 1981, the U.S. was importing oil at the rate of 6 million barrels per day at an average cost of $37 per barrel. More than 81 billion dollars left the country to pay for this oil; the money entered the coffers of foreign treasuries and foreign companies. When dollars leave the country, the companies, individuals, or nations that obtain them may be willing to sell their dollars, exchanging them for either their own currencies or the money of other nations. As an example, if Saudi Arabia were to purchase a plant from West Germany to desalt sea water, it would pay West Germany in German marks. If Saudi Arabia did not hold enough German marks at the time, but had many U.S. dollars from the sale of oil, it might offer to sell dollars in return for marks. It would seek a nation or buyer who was willing to accept dollars and give marks in return (Figure 17.8).

Of course, there is a competing process in which

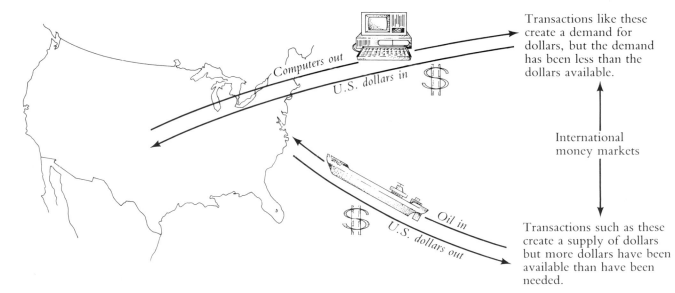

Figure 17.8 The flow of U.S. dollars into and out of the country. In recent years, more U.S. dollars left the country in foreign trade than entered it. This trade imbalance, known as a trade deficit, leads generally to a fall in the value of the dollar. A trade surplus, on the other hand, would be expected to lead to an increase in the value of the dollar.

U.S. goods such as computers are purchased by foreign buyers. Dollars are needed by the nations or companies that purchase these goods. To obtain the needed dollars, these firms or nations enter the international money market to purchase U.S. currency; they sell the currencies they hold in return for dollars. On any given day, some people are willing to sell dollars so they can have other currencies for their transactions, while others are seeking to buy dollars so they can purchase U.S. goods.

On a particular day, the average price at which the dollar is exchanged for the currency of some other nation may be viewed as the value of the dollar relative to that currency. Exchange rates exist for the dollar with the British pound, the dollar with the German mark, the Swiss franc, the Dutch guilder, the Japanese yen, and so on. All this has to do with oil, with energy conservation, and with inflation. We trace the effect in the following way.

Each year, economists add up the dollars paid for U.S. goods by foreign firms and nations and compare the total with the dollars leaving the country for the purchase and importation of foreign goods. Each year they find that massive flows of dollars leave the country. If more dollars leave the country than enter it, the balance of trade for the U.S. is said to be negative. That is, more dollars have fled the country than were needed by foreign interests to purchase U.S. goods; hence there must be a supply of U.S. dollars on the international money market that exceeds the demand. Whenever supply exceeds demand at a given price, the price decreases, causing demand to increase and the excess supply to be absorbed. More simply, when the supply of dollars exceeds the demand for dollars, the price falls. For most of the 1970s, the dollar fell in value, and at the root of the fall was the outflow of dollars for foreign goods and most particularly for oil.

In the early 1980s, the dollar actually climbed in value for several years, caused by high interest rates being paid by the U.S. government to finance its debt. The flow of dollars overseas for the importing of oil slowed the climb in value but could not stop it. When high interest rates have fallen, as they surely will, the trade deficit will again become negative because of the nation's enormous bill for oil imports. Then the value of the dollar will again decline.

The impact of this seemingly remote event, changes in value of the dollar, becomes clear when we consider the purchase of a foreign-made product. A consumer looking at a Japanese car or German camera will find that the price of these items has risen considerably since the early 1970s. The price increase resulted from a decline in the value of the dollar. Suppose a German camera sold in Germany for 1000 marks in the early 1970s and because of manufacturing efficiencies still sold for about that amount in the late 1970s. It took $330 to buy this camera in the early 1970s because each mark cost 33¢. In early 1980, however, the same camera cost $500 because each mark then cost about 50¢.

This kind of price increase on goods imported from the industrial nations was a contributing factor to the excessive inflation in the price of goods that the United States experienced in the late 1970s. Many of the goods and raw materials we utilize may have been produced abroad. Examples are autos, cameras, radios, tape recorders, watches, clothing, iron, steel, copper, cobalt, and so on. When the dollar falls in value, the cost in dollars of these imported goods and raw materials goes up in response. The impact of these price increases, caused by the fall in value of the dollar, is a rise in the cost of living, which contributes importantly to the annual inflation rate. The fires of inflation are fanned by this fall in value of the dollar, and the fall in value stems from the vast quantities of oil we import from foreign nations.

Questions

1. What is *primary recovery?* What are two *secondary recovery* methods? What level of recovery can be achieved by secondary recovery methods?
2. What are the two basic elements in oil? What other elements or substances may be present in the oil?
3. Why is crude oil distilled or refined? What are some of the substances into which it is separated?
4. In terms of oil, what are "proved reserves"? How do these differ from "undiscovered reserves"? To what extent do experts, agencies, and companies agree on the oil remaining to be discovered?
5. Hubbert likens a lake with intensive fishing to a region with petroleum deposits and intensive exploration. Finish the analogy.

6. What portion of the oil that we use is consumed in gasoline combustion? What are some other uses of oil?

7. What portion of the oil used worldwide is consumed in the United States? Why is this portion falling?

8. What part of our oil consumption was imported in 1981? What portion of this was from the Middle East?

9. Why was the price of natural gas controlled by federal law in the first place? What were the effects of price control as seen by the gas and oil companies?

10. The idea of making gas from coal is not new. How old is it? What was one of the names given to gas made from coal in those days?

11. How many tons of coal would be required *each year* by a coal gasification plant producing 125 million cubic feet of gas per day?

12. Underground coal gasification is not expected to decrease our estimated reserves of coal. Why?

13. Coal gasification and liquefaction processes have been operating in an overseas country for many years.

Where? Is this the first place these processes were used? If not, where were they used before?

14. What three sources of synthetic oil exist in North America and in what estimated quantities? Where are these deposits located? How will the price of OPEC oil affect whether they will be recovered?

15. What are some of the untapped resources of natural gas or synthetic natural gas? What two factors generally are responsible for the fact that we have not used these resources in quantity?

16. Cartels set prices. How do they manage to get their prices? Use the example of oil to explain.

17. How do competitors with the cartel benefit? What risk do they face when they develop a new and costly oil field?

18. Explain how the large quantity of oil we import each year can lead to a fall in the value of the dollar and hence to inflation.

Further Reading

General

Energy: The Next Twenty Years, Report by a Study Group Sponsored by the Ford Foundation and Administered by Resources for the Future. Cambridge, MA: Ballinger Publishing, 1979.

Oil, coal, nuclear power, solar, projections, economics—all are here. Thoroughly documented and comprehensive; one of the most complete studies of the 1970s.

International Petroleum Encyclopedia. See latest yearly edition.

For the facts as the industry sees them, on oil production, reserves, locations around the world, this is the document. Many tables and charts.

Kash, D., et al., *Energy Alternatives: A Comparative Analysis,* for the President's Council on Environmental Quality. Published as *Our Energy Future.* Norman, OK: University of Oklahoma Press, 1976.

As thorough a description of energy technology and resources as can be found. Language is largely nontechnical and new words applying to the technologies are defined as they are used.

Metz, W., "Mexico: The Premier Oil Discovery in the Western Hemisphere," *Science,* **202** (22 December 1978), 1262.

"The Natural Gas Shortage," *Business Week* (27 September 1976), p. 66.

A very readable and relatively unbiased article describing the background to the natural gas shortage.

Hefner, Robert, "The NGPA is Working Well," *Oil and Gas Journal* (December 28, 1981), p. 224.

This article indicates the impact of the gradual decontrol of natural gas prices on availability of gas.

Cook, James, "The Great Oil Swindle," *Forbes* (March 15, 1982).

The message is: the energy crisis isn't over.

"The Geopolitics of Oil," prepared for Senate Committee on Energy and Natural Resources (November 1980).

This book is summarized in an article in *Science,* **210** (December 19, 1980), 1324.

Oil Shale/Tar Sands/ Coal Gasification and Liquefaction

Fletcher, K. and M. Baldwin, eds., *A Scientific and Policy Review of the Final Environmental Impact Statement for the Prototype Oil Shale Leasing Program of the Department of the Interior,* The Institute of Ecology, 1973.

Understandable and worth reading for the environmentalist viewpoint.

Maugh, T., "Tar Sands: A New Fuels Industry Takes Shape," *Science,* **199** (17 February 1978), 756.

Oriented toward the science, technology, policy aspects of tar sands exploitation.

"Oil from Shale Is Still a Distant Hope," *Business Week* (23 April 1979), p. 126.

Nontechnical, oriented toward problems of oil companies interested in developing the shale oil resource.

"Oil Shale and The Environment," Office of Research and Development, U.S. EPA, EPA 600/9-77-033, October 1977. Available from National Technical Information Service, Springfield, Virginia.
Recommended: nontechnical and profusely illustrated.

Tippee, R., "Tar Sands, Heavy Oil Push Building Rapidly in Canada," *Oil and Gas Journal* (30 January 1978).
Oriented toward the commercial, economic aspects of the tar sands effort.

"An Assessment of Oil Shale Technology," Office of Technology Assessment, Congress of the United States, 1980.

"Uncertainty, Cost/Price Squeeze Hit Fledgling Synfuels Industry," *Oil and Gas Journal* (May 24, 1982), p. 21.
The drop in oil prices and its effect on synfuel projects.

"Gulf Successful in Gasifying Steeply Dipping Coal Beds," *Oil and Gas Journal* (December 28, 1981), p. 71.

Shohinpoor, Moshen, "Making Oil from Sand," *Technology Review* (February/March, 1982).
Nontechnical; indicates environmental impacts of tar sands exploitation.

CHAPTER EIGHTEEN

Nuclear Power

Introduction

Light-Water Nuclear Reactors
*How They Operate/Radioactive Wastes and Radiation
Releases from Light-Water Reactors/Safety of Light-Water
Reactors*

Problems in the Nuclear Fuel Cycle
*From the Mine to the Power Plant/Fuel Reprocessing and
Spent Fuel Rod Storage/Mixed Oxide Fuel and Plutonium
Transport/International Safeguards/Storage or Disposal of
Wastes from Fuel Reprocessing*

**Economic and Social Aspects of
Nuclear Power**

Alternatives to Reprocessing

Liquid-Metal Fast Breeder Reactors

CONTROVERSIES:

18.1: *Is Nuclear Energy an Inevitable Result of
Technical Progress?*

18.2: *Does Reprocessing Produce Weapons-
Grade Plutonium?*

18.3: *Does the Public Have the Right to Know
(How Easy It Is to Steal Nuclear Material and
Build an Atomic Weapon)?*

Introduction

It began under a cloud—a mushroom cloud rising over a Japanese city. The reputation acquired by that beginning has proved very difficult to change. Nuclear electric power remains suspect in the minds of many, despite continuing assurances of its safety.

The scientists who created the atom bomb also envisioned the peaceful use of atomic energy as a means of generating electric power. Under the program "Atoms for Peace" during the Eisenhower Administration, private corporations were given the privilege of owning nuclear reactors. Industries in joint projects with the Atomic Energy Commission then began the development of nuclear power reactors.

The first experimental production of electricity from a nuclear reactor occurred in 1956, when a boiling-water reactor began operating at Argonne National Laboratory. In the following year, a pressurized-water reactor at Shippingport, Pennsylvania, began delivering 60 megawatts of electrical power. The size of new plants increased quickly as operating experience grew. By 1963, several nuclear plants were delivering 200 megawatts of electric power and commitments had been made for the larger Oyster Creek plant in New Jersey and the Nine Mile Point plant in New York. When these new plants went on-line in 1969, they each delivered up to 600 megawatts of power.

A tide of commitments to nuclear plants began in 1965. Seven orders were placed that year, 20 the following year, and 30 in 1967. The rate then began to fluctuate, but continued strong into the 1970s. By mid-1974, almost 240 nuclear plants had been ordered, and most new plants on order were to provide 1000 megawatts or more of electric power. The new technology had taken hold with remarkable speed.

However, between 1974 and 1978, only 13 new orders were placed for reactors sited in the United States. Further, no electric utility in the United States has ordered a nuclear plant since the two ordered in 1978. Even further, 13 reactor orders were cancelled in 1978, eight in 1979, 16 in 1980, 6 in 1981, and 9 in 1982. In all, 78 reactor orders were cancelled between 1974 and 1981. The tide has begun to recede.

The course of nuclear power development has been slowed by debate in the scientific community and in Congress about the possible levels of radiation released from the plants. Scientists have also argued over the reliability of the containment structure in which the fission process is allowed to proceed. Several states have had referendums on the further development of nuclear electric power in their jurisdictions. Debate continues on the use of plutonium as a nuclear fuel to extend fuel supplies.

Is it the cloud at the beginning that has alone limited the acceptance of nuclear electric power? Or is it a problem of public misunderstanding of its actual safety and its potential for effectiveness? Is there more substance to the debate than simply poor public relations?

Understanding the controversy that swirls around nuclear power requires a discussion of the complete nuclear power system as it has evolved to this present time. Only with such a discussion can we knowledgeably examine the positions of the opponents and proponents of nuclear power and judge their validity. In no other way can the reader formulate sound opinions.

Light-Water Nuclear Reactors

How They Operate

To understand the operation of nuclear power plants, it is helpful first to examine the nuclear reaction known as **fission.** Fission occurs when a neutron strikes the nucleus of **uranium-235** or certain other heavy atoms. The reaction consists of the breaking apart (fissioning) of the atom, made unstable by the striking neutron. Two relatively large fragments and a large amount of heat energy are released in the breakup of the atom. The large fragments or fission products are themselves atoms, each consisting of a portion of the electrons and a portion of the nucleus of the "parent" atom. The fragments are often unstable and "decay" in gradual steps to stable atoms, at each step releasing radiation of various types.

Not only are fission products and energy released, but neutrons are also ejected in the fission reaction. If one of these neutrons strikes another uranium-235 atom, it can cause that atom to fission in turn, releasing still more neutrons and thereby sustaining the fission reaction. The sustained fission reaction is the basis of both the atomic bomb and the nuclear power plant.

Two fundamental reactor types are in common use in the United States and elsewhere around the world, although others do exist or are in the design stage. These two are the **boiling-water reactor (BWR)** and the **pressurized-water reactor (PWR).** Their technology is relatively proven and reliable, and they promise to be the workhorses of nuclear electric generation for the next several decades. Other reactors are being developed and investigated, but their contributions to our electric energy supply are likely to be small in the near future.

The boiling-water reactor (BWR) and pressurized-water reactor (PWR) are actually very similar. In both reactor types, uranium-235 undergoes controlled fission in a **core,** generating heat energy.

In the interior of the core, long cylindrical metal rods are arranged vertically; these rods contain the fuel (Figures 18.1 and 18.2). The fission that takes place within the fuel rods provides the energy to heat the water flowing through the core. Since neutrons and

Figure 18.1 Inspection of Fuel Rods. Westinghouse assembles nuclear fuel rods at a fabrication plant at Columbia, South Carolina. In this photo, the rods are undergoing final visual inspection. Each rod contains uranium dioxide pellets, the "fuel" for the nuclear reactor. At this stage, the rods are not yet emitting dangerous radiation. (Westinghouse Photo by Jack Merhaut)

Figure 18.2 Fuel-Loading Operation. A technician monitors the fuel-loading operation at Unit No. 1 of the Calvert Cliffs Nuclear Power Plant, owned by the Baltimore Gas & Electric Company. Each fuel bundle—of a total 217 bundles—is being positioned by an automatic fuel-handling machine, shown extending down to the upper core level. (Combustion Engineering, Inc.)

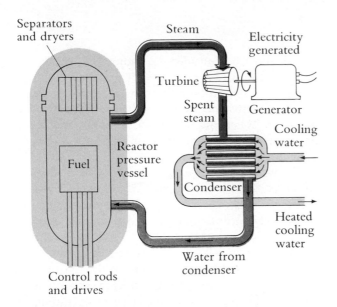

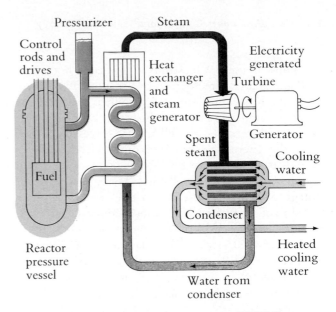

Figure 18.3 Boiling-Water Reactor (BWR) Power Plant. Water is boiled to steam in the reactor pressure vessel, just as in the boiler of a coal-fired electric plant. The steam turns a turbine to generate electric power and then is condensed to a liquid prior to return to the "boiler."

Figure 18.4 Pressurized-Water Reactor (PWR) Power Plant. In contrast to the BWR, where steam is produced in the reactor core, only liquid water flows through the primary loop of the PWR. The water, at a very high temperature and pressure, boils water to steam in the secondary loop. The operation of the secondary loop then is the same as that of the primary loop of the BWR.

gamma rays are produced by the fission process, the core must be heavily shielded to prevent the exposure of workers to these dangerous particles and rays. Layers of concrete and iron or steel are used to encase the core. Within the fuel rod, about 3% of the uranium fuel is in the form of the isotope uranium-235, which can undergo fission. The remainder of the uranium is uranium-238, which cannot undergo fission.

The core of the nuclear reactor is the "boiler" of the nuclear power plant. In the core of the nuclear reactor, uranium-235 atoms are struck by neutrons and break apart or fission into fragments, known as fission products. Heat energy and more neutrons are released. The heat energy from fission is taken up by the water that circulates through the core. The water used in the Light Water Reactors is ordinary water (light water) such as occurs naturally in the hydrologic cycle. Another type of reactor, which we discuss later (the CANDU), uses "heavy water" to take up the heat from the fission reaction. Heavy water has been enriched in atoms of deuterium (a heavier isotope of hydrogen) and has as a consequence more deuterium oxide molecules.

As the reactor operates over a period of time, the amount of uranium-235 within the rod decreases and fission product fragments build up. The fission process slows down because the fission products capture some of the neutrons that sustain the chain reaction. After

about three years, the rod is no longer an efficient source of heat because of a low rate of fission; it is then replaced by a fresh fuel rod. The spent fuel rod is now relatively low in uranium-235 and rich in fission product fragments.

This, in general, is how the core operates, but the BWR and PWR differ in how the heat energy is stored in the circulating water. The boiling water reactor functions in precisely the same way as the boiler in a fossil-fuel power plant; that is, the water in the core is converted directly to steam, which is used to turn a turbine (Figure 18.3).

However, the pressurized-water reactor does not convert the water in the core to steam. Instead, the reactor maintains enormous pressure on the water to prevent it from boiling. All the heat energy from fission goes into raising the temperature of the liquid water. This intensely hot water is cycled from the core to a steam generator, where its heat is transferred to water flowing in a secondary loop. The water from the core returns to the core to be heated again. This water, circulating between the core and the steam generator, flows in what is called the primary loop. The water in the secondary loop boils to steam as it leaves the steam generator under pressure. This steam in the secondary loop is used to turn a turbine and generate electricity (Figure 18.4).

Thus, in the BWR, steam is generated in the core itself and is used directly to turn a turbine. In contrast, in the PWR, steam to turn the turbine is produced in a steam generator and flows only in the secondary loop.

In both reactors, the steam that leaves the turbine lacks the energy to generate more electric power, though it is still quite hot. This steam condenses to liquid water in a heat exchanger. Here, heat from this "spent" steam is transferred to a flow of cold water, called cooling water. The condensed water circulates back to be converted to steam once again. In the PWR, the conversion to steam takes place in the steam generator. In the BWR, it occurs in the core itself.

The use of cooling water and its effects on aquatic life are discussed in the section on thermal pollution (Chapter 24).

Radioactive Wastes and Radiation Releases from Light-Water Reactors

Just as fossil-fuel power plants emit pollutants, so also do nuclear power plants. Whereas emissions from fossil-fuel power plants include the commonly known chemical air pollutants, emissions from nuclear power plants are radioactive elements. These radioactive elements stem almost entirely from the fission process.

Elements that are products of the fission process include radioactive forms of krypton, xenon, cesium, strontium, and iodine. These are among the fragments resulting from the fission of uranium-235. Although the fission process is supposed to be confined to the interior of the fuel rod, defects in the rod's metal shell allow fission products to escape into the water in the core. Fission products, then, appear in the water in the primary loop. The treatment of this water produces solid, liquid, and gaseous wastes.

Gaseous wastes from nuclear plants include radioactive forms of the gases krypton, xenon, and nitrogen. The plant discharges its gaseous wastes through tall stacks, which allow the substances to be widely dispersed. The total amount of radioactivity discharged from the stacks of a PWR is far less than that from a BWR.

The water that circulates through the core contains not only fission products, but also metals from the shell of the fuel rod. These metals became radioactive when struck by neutrons from the fission process. If this water should leak from a pipe or from a seal or

joint, it must be considered a radioactive waste, and must be cautiously handled and disposed of. Although the liquid wastes are, for all practical purposes, water, radioactive substances are dissolved or suspended in the water. When the liquid wastes are treated to remove these substances, solid and gaseous wastes are created.

As we mentioned, the gases are vented through tall stacks. Solid wastes, which are relatively low-level radioactive wastes, are encased in concrete within 55-gallon metal drums and shipped to special sites for burial.

Burial of low-level solid wastes is conducted at federally licensed burial grounds. Although they are commercially operated, the grounds are supervised by the states. Six such sites have been in operation, located in Illinois, Kentucky, Nevada, New York, South Carolina, and Washington. Political pressures have caused the Illinois, New York, and Kentucky sites to close, seemingly permanently. The other sites have been closed periodically as well, again largely because of political pressures.

How well do nuclear plants prevent routine radioactive releases to the environment? When one considers the potential for such releases, the answer is: "Exceptionally well." Most plants are meeting the rules set by the federal government on releases, and in the last decade these rules have become very stringent. The rules call for routine releases of less than 5% of the natural background radiation that most people receive. It should be added that these guidelines were not always so strict and that criticism by scientists helped to bring the guidelines down.

In addition to the spent fuel, one other type of waste arises from a nuclear power plant: the plant itself. No major plant has yet been fully "decommissioned," although the decommissioning of the Shippingport plant is about to begin. If a plant is to be dismantled and hauled away, the cost could run to $100 million for the modern 1000-megawatt plant, or about 10% of its initial cost, a cost not yet taken into account by utilities. The alternative is to seal and guard the site and reactor for 100 years or more.

Safety of Light-Water Reactors

Routine releases of radioactivity, then, are negligible. That is, on a day-to-day basis, a resident near a nuclear power plant is exposed to less radiation than natural background levels. Nonetheless, criticism is still directed at the safety of nuclear power plants. Day-to-day emissions, however, are not the target of this criticism; it is the radiation release that could occur if a serious accident or sabotage took place within the plant. One particular form of accident is of more concern than all others.

We recall that water circulates through the reactor core to withdraw the heat of fission. If the circulating water were suddenly lost, the heat in the core would accumulate rapidly, and the contents of the core could melt. In the jargon of the nuclear industry, this is called a "loss-of-coolant accident." This kind of accident might occur if a steam or water line carrying water into or out of the core should break. The high pressures in the line would probably cause most of the water or steam to be rapidly ejected from the core. The temperature in the core would rise rapidly, melting the fuel rods (a "meltdown"). Radioactive xenon and krypton gases, products of uranium fission, would be released from the melted rods and would enter the building through the break in the pipe. The molten mass of fuel and fuel rod shells would fall to the floor of the reactor vessel, perhaps melting through the floor and falling into the containment room that surrounds the reactor core. More radioactive gases would be released into the containment area, but the containment room is designed so that the leakage of radioactivity to the outside will be very small. A water spray system will condense the steam in the containment room and reduce the temperature there.

The first line of defense, however, is at the reactor core. The reactor core is also equipped with a spray system. The Emergency Core Cooling System is designed to inject water automatically into the core if a break in the coolant line should occur; its purpose is to cool the reactor core and prevent melting. The ability of the emergency core cooling system to flood the core with water was finally tested in 1979 at a test reactor, more than 20 years after the first commercial nuclear power plants went on line. The initial test did prove successful, but it was run at less than full-power operation.

A loss-of-coolant accident is regarded as the most serious accident that could occur at a light-water reactor. The severity of the accident, in terms of its impact on the population, would depend on how effectively the emergency core cooling system functions. If a meltdown were not prevented, the impact then would depend on how effectively the containment room prevented radioactivity from entering the environment. And if radiation escaped, the impact would depend upon the wind speed and direction and whether a populated area was downwind from the plant.

Because reactor safety is so prominent an issue in the nuclear power debate, the Atomic Energy Commission (now the Nuclear Regulatory Commission) directed that an extensive study (WASH-1400) of the problem be undertaken. In the late 1970s, Professor Norman Rasmussen of M.I.T. conducted the study, which cost 4 million dollars, involved more than 100 people, and took 3 years to complete. The study concluded that the odds against the worst-case accident occurring were astronomically large. The worst-case accident projected an estimate of about 3000 early deaths and 14 billion dollars in property damage due to contamination. Cancers occurring later due to the event might number 1500 per year. The odds against such an event, however, were estimated at 10 million to one, given that 100 reactors are in operation.

At the same time, the probability of a meltdown is not so small. With 100 light-water reactors in operation (a likely figure, by the time you read this; see Figure 18.5), the probability of a meltdown is 1 in 200 per year. Thus, although the worst-case accident is estimated only as a remote possibility, the meltdown alone is not such an unlikely event. The Rasmussen study attempts to evaluate the probability that a meltdown will turn into a serious, life-threatening accident. The study concludes that the safety features engineered into the plant are very likely to prevent such serious consequences.

The Rasmussen study has had serious criticisms, however. One of the most scathing was that of a panel of 21 scientists, economists, and political scientists assembled by the Ford Foundation for a study of nuclear power. The panel felt that the Rasmussen study seriously underestimated the uncertainties of reactor safety. Flaws in methodology, they asserted, could make some of the probability estimates of accidents low by a factor of 500.[1]

Another panel, this one composed only of nuclear scientists and created in 1977 by the Nuclear Regulatory Commission itself, also reviewed the Rasmussen

---

1 *Nuclear Power, Issues and Choices,* the Nuclear Energy Policy Study Group, S. Keeny, Jr., Chairman. Cambridge, MA: Ballinger Publishing, 1977.

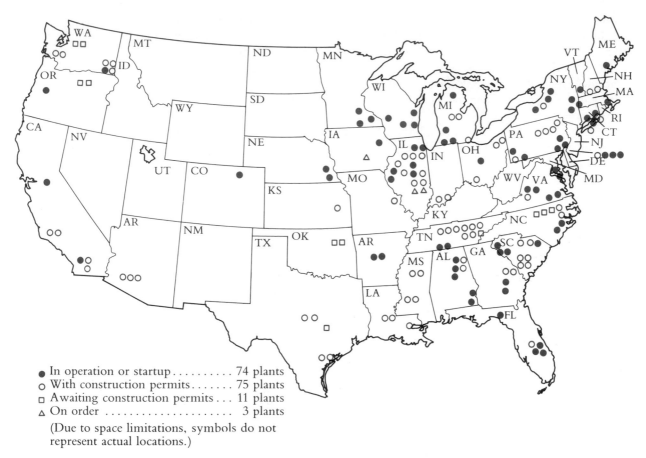

● In operation or startup . . . . 74 plants
○ With construction permits . . . . 75 plants
□ Awaiting construction permits . . . 11 plants
△ On order . . . . 3 plants
(Due to space limitations, symbols do not represent actual locations.)

Figure 18.5 Status of nuclear power plants, December 31, 1981. (Source: U.S. Department of Energy.)

report and found it lacking in a number of respects. In particular, the estimates of risk set forth in the report were criticized as being subject to much larger errors than the report implied, errors that could not be reduced because of a fundamental lack of data. That is, whereas the report provided estimates of the risk of various accidents, the panel felt the risks could not be stated nearly so precisely.[2] The Rasmussen report, once the cornerstone of arguments for the safety of nuclear plants, has been reduced in status to merely a piece of evidence, to be read very carefully for flaws.

When it appeared, the Rasmussen study silenced many doubters of the safety of nuclear power plants,

and until 1979, the issue of nuclear safety lay buried in the jumble of questions that had piled up about nuclear power. Accompanying the safety question were such concerns as thermal pollution, radioactive waste disposal, fuel reprocessing, the spread of atomic weapons, and the transport of nuclear materials. In 1979, however, after a nuclear reactor near Harrisburg, Pennsylvania, experienced an accident of a character not thought possible, nuclear safety came bubbling to the top of the nation's and the world's nuclear concerns.

The reactor incident at the Three Mile Island Nuclear Power Station in Middletown, Pennsylvania, began at 4:00 AM on March 28, 1979, with the failure of two pumps in the secondary cooling water loop of Reactor 2. This was the first in a series of events that eventually resulted in a partial meltdown of the reactor

---

2 *Risk Assessment Review Group Report to the Nuclear Regulatory Commission,* prepared for the U.S. Nuclear Regulatory Commission, NUREG/CR-0400, September 1978.

Is Nuclear Energy an Inevitable Result of Technical Progress?

> The anti-nuclear movement . . . seeks to undermine the scientific and technological revolution which has created the modern world.
>
> *Alex A. Vardamis*

> If man is to survive on this earth, . . . it will be through his wisdom in choosing between those innovations that he can control and those that he cannot.
>
> *George Dryfoos*

The excerpted letters that follow appeared on successive days in 1978, shortly after demonstrations at the nuclear plant site at Seabrook, New Hampshire.* They raise issues and questions that are worth exploring.

To the Editor:
   The recent anti-nuclear-energy demonstration in Seabrook, N.H., should be placed in perspective. The animating rationale of the protest movement is based upon a rejection of progress and technology . . . .
   The anti-nuclear movement is doomed to failure, for it seeks to undermine the scientific and technological revolution which has created the modern world . . . The Seabrook demonstrators . . . are temporary and inconsequential impedimenta to the tide of history and are destined to become a forgotten footnote to the constructive advancements of science in this century . . .

   Alex A. Vardamis

To the Editor:
   Questioned as to the possibility of another blackout in New York this summer, Con Ed Chairman Charles Luce sensibly pointed out that in a complex system of machines, operated by fallible human beings, there can be no 100 percent guarantee.
   His words should be carved in stone for the benefit of those who see nuclear power as the answer to our future energy needs . . .
   The catalogue of hazards ranges from an admitted inability to guarantee that storage of nuclear wastes will not fatally pollute our land, water and atmosphere, to the very real possibility that proliferation of weapons-grade fuel will make possible the production of explosive devices . . .
   If man is to survive on this earth, it will not be through his genius for technical innovation. It will be through his wisdom in choosing between those innovations that he can control and those that he cannot.

   George Dryfoos

Who was Ned Ludd? And who were the Luddites? Is Vardamis suggesting that anti-nuclear demonstrators are modern-day Luddites? Is the comparison a valid or invalid one? Defend your position. Has the United States ever rejected an otherwise useful technology for environmental reasons?

---

* The Vardamis letter appeared on July 6, 1978; the Dryfoos letter appeared on July 7, 1978, both in *The New York Times.*

core and an accident generally accepted to be the most serious at any commercial nuclear power reactor in the world. Although no one died as a direct result of the Three Mile Island (TMI) accident, it came very close to being much worse.

The accident at Three Mile Island resembled only slightly any of the many scenarios in the Rasmussen report. In that report, the accidents of interest were the major loss of coolant accidents, resulting from large equipment failures. The "accident" at Three Mile Island was really a collection of small events—small equipment failures and human errors, and it nearly led to a complete core meltdown. Releases of significant and potentially lethal amounts of radioactive material were fortunately avoided.

The role of human operators at Three Mile Island was terribly important; their actions made the accident far worse that it could have been. Poor control-room design and poor information display contributed to delayed and inadequate response. During the first eight minutes of the accident, the operators were subjected to 100 different alarms of bells, buzzers, and lights.

Further, the response of local, state, and federal authorities revealed a most disturbing lack of preparation for such accidents. The public was confused by contradictory information from many sources. No useable evacuation plan existed, although many people left voluntarily and children and pregnant women were advised to leave. We have also learned from the event that a federally maintained team, which could respond to a nuclear accident, may be required.

The question of clean-up and decontamination of the crippled plant still remains. In a referendum in 1982, the public in the surrounding counties demanded that the plant not reopen. Thus, we may see the first full-scale decommissioning take place at TMI.

Problems in the Nuclear Fuel Cycle

From the Mine to the Power Plant

A system for nuclear electric generation contains more than a single power plant. It encompasses a cycle that begins with the mining of uranium and proceeds through use of uranium at the plant to the recycling and/or disposal of spent fuel rods. We find serious safety questions focussed at a number of points in the nuclear fuel cycle. Figure 18.6 illustrates the operations in the nuclear fuel cycle, along with the transportation of products from one operation to another.

The first operation in the cycle is the actual mining and milling of uranium ore. This step is followed by the extraction of an oxide of uranium called "yellowcake." In the past, at older mining operations, the tailings from the processing of uranium ore were simply dumped on the ground (see Chapter 30); they are, however, a potential hazard.

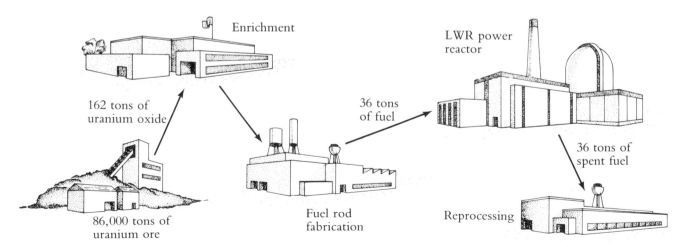

Figure 18.6 The Nuclear Fuel Cycle. The numbers on the diagram are approximate annual requirements for one 1000-megawatt (electrical) pressurized-water reactor.

Figure 18.7 Oak Ridge National Gaseous Diffusion Plant. The gaseous diffusion plant at Oak Ridge, Tennessee, is the first such facility built in the world for production of enriched uranium. The original gaseous diffusion process building—the U-shaped building—began operating in 1945, producing enriched uranium for the Manhattan Project of World War II. The plant, operated for the U.S. Department of Energy by Union Carbide Corporation, is one of three gaseous diffusion plants in the U.S. The other two are at Portsmouth, Ohio, and Paducah, Kentucky. (DOE Photo by Frank Hoffman)

Two types of uranium are found in the extracted uranium oxide: uranium-235 and uranium-238. Both are forms of uranium, except that an atom of uranium-238 has three more neutrons in its nucleus. In the ore, only about 0.7% of the uranium is present as uranium-235. The remainder is uranium-238. However, it is the fission (splitting) of uranium-235 that provides the heat energy in the core of the BWR and PWR. Since the concentration of uranium-235 is so low, a step is needed to increase the concentration of uranium-235. This step is called **enrichment,** and the process is known as "gaseous diffusion." In this step, the level of uranium-235 is increased to 3% of the total uranium present, although it can be increased further if, perhaps, a bomb were being made (Figure 18.7).

The enriched uranium is then formed into fuel rods at a fuel fabrication plant. The fuel rods are shipped, in turn, to the power plant where, as we discussed earlier, they produce heat in the core of the reactor. When fission products have built up in the rods to a level that slows the chain reaction, the rods are removed from the core of the nuclear power plant. Such spent rods were originally intended to go to a fuel reprocessing plant, where remaining fuel resources could be recovered. At the moment, however, the rods are simply piling up. The reason for this buildup is discussed shortly.

Fuel Reprocessing and Spent Fuel Rod Storage

The most serious problems in the fuel cycle begin after the spent fuel rods have been removed from the core of the power plant. Because quantities of reusable fuel are still in the rods, the advocates of nuclear power have viewed the step of fuel reprocessing as a crucial one. However, hazardous radioactive elements, the fission products, are also in the rods.

Of the reusable fuels, first is the uranium-235 that has not fissioned and can still be used if recovered. Another substance present in the spent rod can also be used as a nuclear fuel. **Plutonium,** which was not present in the rod initially, was formed, or "bred," from uranium-238 during the process of fission. Recall that uranium-238 does not participate in the fission process, even though it is the most abundant element in the rod. However, during the fission process, a small but significant portion of the uranium-238 is converted by nuclear rearrangement to plutonium. The fuel reprocessing plant is designed to separate the useful uranium-235 and the plutonium from the hazardous and unwanted fission products (Figure 18.8).

Through most of the 1970s, no reprocessing plant was operating in the United States. The plant at West Valley, New York, was closed for repairs at the begin-

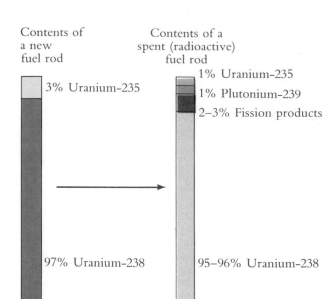

Contents of
a new
fuel rod

Contents of a
spent (radioactive)
fuel rod

3% Uranium-235

1% Uranium-235

1% Plutonium-239

2–3% Fission products

97% Uranium-238

95–96% Uranium-238

Figure 18.8 New and Spent Fuel Rods. Plutonium-239 is created when neutrons are captured in the nucleus of uranium-238. This capture process creates uranium-239, which decays in two steps to plutonium-239.

ning of the decade and never reopened. The G.E. reprocessing plant at Morris Plains, Illinois, never opened its doors because of design problems. And the Allied General Nuclear Services Plant at Barnwell, South Carolina, was unable to meet its licensing requirements.

Storage at the power plant. Although a fuel rod is not particularly hazardous when it enters the core, a spent fuel rod contains the products of uranium fission. We might refer to the intensely radioactive fission products from the "burning" of uranium as the "ashes" of the nuclear fuel cycle. Unlike the ashes from coal, though, these substances emit lethal quantities of radiation and are at high temperatures due to the radiation. The rods must therefore be unloaded from the reactor core through an underwater canal to minimize exposure of plant personnel to radiation. Each year about one-third of the rods in the core are removed and replaced.

The spent rods are removed through the underwater canal to storage in a basin also filled with water. In the first few months of storage in the basin, much of

the fast, initial radioactive decay within the rods has taken place. Most of the dangerous iodine-131 has disappeared in this interval. The rods can then be placed with greater safety into heavily shielded canisters for transport to a fuel reprocessing plant (Figure 18.9).

However, there was no fuel reprocessing plant operating in the United States through most of the 1970s. Thus, spent fuel rods have been accumulating at power plants for some time. During the administration of President Carter, the plan to deal with spent fuel rods relied on *Away-From-Reactor (AFR) storage* as the means of relieving the space squeeze on nuclear power plants. The concept was to transport the spent rods to several centrally located sites, where they would again be stored intact in swimming-pool-like vaults. There they would remain until the government could decide what to do next in the way of final disposal. With the arrival of the Reagan Administration in 1981, the plan for Away-from-Reactor storage was abandoned. As of 1983, no decision had yet been made nor plan formulated as to what to do with the spent fuel rods accumulating at the operating nuclear power plants.

By the late 1970s, a number of reactor operations had already begun reracking the spent fuel rods in their

Figure 18.9 Crash Test of a Spent Fuel Cask. A rocket-propelled rail car carrying a 74-ton spent fuel cask smashes into a concrete wall at a speed of 81.4 mph. The impact demolished the front of the rail car, turning it into a mass of twisted metal, but the cask was essentially undamaged. Such crashes are meant to duplicate the worst possible accident conditions in order to predict how well containers will survive. (Sandia Laboratories.)

storage basins to make more room, in order to proceed with refueling on schedule. There is, however, a limit to the number of spent fuel rods that can be stored at older power plants. The Nuclear Regulatory Commission has predicted that 27 of the 73 nuclear plants operating in 1982 will have filled their storage pools by 1990. Recognizing that the capacity to reprocess spent fuel is far in the future, if ever, utilities are now building their new plants with very large water-filled basins; these basins can store as much as 15 years' accumulation of spent fuel rods. In fact, we may soon find shipments of spent fuel rods moving on major highways as utilities move the rods from plants that are short of storage capacity to newer plants with remaining capacity for fuel rod storage.

At the fuel reprocessing plant (West Valley Days). In 1966, the first fuel reprocessing plant in the U.S. began operation at West Valley, New York, a town just south of Buffalo. Privately owned by Nuclear Fuel Services, the plant processed spent fuel rods from light-water nuclear power plants, separating the unwanted fission products from the reusable fuels, uranium and plutonium. When the plant closed in 1972 to modernize and expand, it had already processed 600 metric tons of spent fuel.

In 1976, Nuclear Fuel Services announced that it could not reopen its West Valley plant. It offered as reason for its action the additional expense of meeting new earthquake-protection requirements. In fact, the motive for the decision to remain closed may have been the presence of a carbon-steel tank holding about 600,000 gallons (2.25 million liters) of high-level radioactive wastes. The tank has been expected to leak. A tank of similar material in Hanford, Washington, eventually leaked its radioactive wastes because its steel corroded.

No one was certain how to prevent the tank at West Valley from leaking. An estimate in the mid-1970s for the cost to transfer the wastes safely to a stainless-steel tank was 600 million dollars. In contrast to the tank that leaked at Hanford, the West Valley tank does have double walls. Nevertheless, eventual corrosion and leakage were still seen as certain.

In 1980, Congress directed the U.S. Department of Energy to "clean up" the West Valley plant, specifically to remove the wastes from the tank and encase them in glass. The leak of high-level wastes "waiting to happen" at West Valley, followed by Federal action

to prevent a serious situation, are one result of the plant's six years of operation. Other activities at West Valley do not bolster our confidence in industry's ability to handle nuclear wastes. Surveys of the environment around the plant found radiation doses to people to be very small, but the control of radiation received by employees inside the plant was less successful.

Employees got high doses of radiation because the company did not build into the plant the capability for remote repair of broken equipment. The choice was made to do without remote repair, apparently to save money in plant construction. Employees repaired broken equipment by hand, even though the broken machines were probably contaminated by radioactivity. The technology for remote repair of broken machines was not used, even though the methods used earlier at Hanford could have been employed.

This resulted in high doses of radiation to the regular employees of the plant. The "dirty jobs," those with the worst exposures, were given to unskilled labor hired off the streets of Buffalo. The unskilled workers were sent in to do a job, allowed to continue until reaching a maximum level of radiation exposure, and then taken off the job. For their daring or ignorance, they received a half day's pay for as little as a few minutes of work.

Two other reprocessing plants have been built. The General Electric plant at Morris Plains, Illinois, never worked, and so never opened. The Allied General Nuclear Services[3] Plant at Barnwell, South Carolina, has never been licensed to operate by the Department of Energy, although it is substantially complete. Supplemental federal funding required for the Barnwell plant ceased in late 1983, and the companies that built Barnwell have filed suit against the federal government to recover their investments. Though the facility could conceivably be brought back to a state of readiness, most movable heavy equipment was being sold off to recover a portion of costs.

The closing at West Valley, the failure at Morris Plains, and the delays at Barnwell have contributed to an uncertainty in the utility industry about the future of nuclear power. In the late 1970s, it was apparent that the electric industry's enthusiasm for nuclear power was decreasing; the number of orders for new reactors fell dramatically. No reactors for the U.S.

---

3 A partnership of General Atomic and Allied Chemical.

have been ordered since 1978. This halt in orders will mean a declining rate of new nuclear units going into service in the mid-1980s.

Arguments for and against reprocessing. Three reasons have been put forward for the reprocessing of spent fuel. First, some experts have asserted that there "may be" a reduction of costs for nuclear fuel when the uranium and plutonium remaining in the spent fuel are purified and recycled. About 1% of the uranium in spent fuel is uranium-235; there is about 3% in fresh fuel. Further, plutonium "bred" from uranium-238 during the fission process may have increased from virtually nothing to a bit less than 1% of the contents of the spent fuel rod.

The reuse of the fissionable uranium-235 and plutonium can decrease the need for mining and enriching new fuel, but there is controversy over how much savings this makes possible. Those who favor reprocessing estimate up to a 20% savings in the cost of electric power if spent fuel is reprocessed. Other responsible scientists estimate savings in electric costs of only 1% or 2% if reprocessing of spent fuel is carried out.[4]

The second reason cited for fuel reprocessing in the U.S. is to decrease dependence on foreign sources of uranium. If foreign suppliers have a large share of a nation's market, they could raise prices without worry or could put an embargo into effect in order to influence the nation's foreign policy. France and Germany, two industrial nations that have decided to reprocess nuclear fuel, have little uranium resources of their own. They also have relatively low amounts of other energy resources. Reprocessing makes these nations less dependent on uranium fuel imports.

The third reason given for reprocessing is to reduce the difficulty of disposing of radioactive wastes. This reason, however, is vigorously debated since, if no reprocessing takes place, the high-level wastes remain neatly packaged inside the fuel rods.

In the United States, plans to proceed with reprocessing were brought to a standstill in the late 1970s. Probably one of the reasons for the delay was the need to ship plutonium by road or rail from the reprocessing plant to the plant that manufactured the fuel rods. If plutonium, the raw material for atomic bombs,

4 See *Nuclear Power, Issues and Choices.* Cambridge, MA: Ballinger Publishing, 1977.

were hijacked, the risks would be great. If spent fuel were not reprocessed, there would be no plutonium recovery nor shipment.

Concerned that plutonium could be hijacked for nuclear weapons, U.S. diplomats have asked Britain, France, and Germany to reconsider their decisions to reprocess nuclear fuel. Yet Britain is building a reprocessing plant at Windscale in Cumbria, with a capacity twice that of Barnwell. The British public is aroused, however, because of a nuclear accident at an earlier plant at Windscale in 1972. The British people have also protested a decision to use the plant at Windscale to reprocess the spent nuclear fuel of Japan. The nation's energy authority was accused of making Britain the world's nuclear "dustbin," a British term for trash can. A number of delays in reprocessing have occurred as the British ponder their decision.

In Marcoule, France, an existing plant offers a reprocessing capacity of 1000 metric tons (MTU) per year. At La Hague, France, an 800-MTU plant has been built, which is scheduled to be expanded several times over. In Germany, a plant with a capacity of 1500 MTU per year has been planned for operation in 1984. The plant is to be located at Gorleben, near the border with East Germany.

In addition, France and Germany are selling their reprocessing technology. France has contracted to sell reprocessing technology to Pakistan, and West Germany has offered to sell reprocessing technology to Brazil. France and Germany have agreed under pressure not to export reprocessing technology in the future, but their agreements with Pakistan and Brazil may still be honored.

Why did the United States oppose these sales and work against the ownership of reprocessing plants by other countries? The answer lies in what reprocessing—specifically, the PUREX process—does. It produces plutonium that can be used in bombs. PUREX stands for Plutonium and Uranium Extraction, and was developed during World War II.

The PUREX process begins with the mechanical chopping of fuel rods into fragments; the fuel is then dissolved in nitric acid (Figure 18.10). Uranium and plutonium are extracted from the solution by mixing it with a solvent that selectively dissolves these elements. The reprocessing plant produces three separate products: (1) uranium, a portion of which is uranium-235; (2) plutonium; and (3) wastes (the fission products). The amount of plutonium that could be recovered an-

Does Reprocessing Produce Weapons-Grade Plutonium?

Conflicting views exist on the potential for plutonium recovered from spent fuel rods to be used in bombs. W. P. Bebbington[*] tells us:

> It is not generally appreciated that during the long exposure of fuel in a power reactor, there is an accumulation of plutonium isotopes other than plutonium-239, particularly plutonium-240, which makes it much more difficult to assemble a supercritical mass of plutonium without an inefficient premature explosion. Weapons-grade plutonium is made with much shorter exposure in the reactor.

In contrast, an anonymous official in Washington is quoted in *Science*[†] as saying, "For proliferation, you can't find anything worse than Purex. It was developed for bombs."

A member of the Nuclear Regulatory Commission also addressed the issue of the suitability for bombs of plutonium recovered from the spent fuel of power reactors.[‡]

> There is an old notion, recently revived in certain quarters, that so-called "reactor-grade" plutonium is not suitable to the manufacture of nuclear weapons. . . .
>
> The obvious intention here is to create the impression there is nothing to fear from separated plutonium from commercial power plants. This not true . . .
>
> The fact is that reactor grade plutonium may

be used for nuclear warheads at all levels of technical sophistication. . . . we now know that even simple designs, albeit with some uncertainties in yield, can serve as effective, highly powerful weapons—reliably in the kiloton range.

In support of this latter view, we note that of the seven nations that had tested nuclear weapons (by 1983), six had obtained the materials for their bombs from reprocessing. It should be pointed out that the material reprocessed was not necessarily produced in nuclear power reactors.

From these excerpts, it is clear that experts disagree on the usefulness of reactor plutonium for atomic weapons. What would you tell people if you were asked whether plutonium recovered from the rods from nuclear power plants was useful for bombs? Given that the experts disagree, suppose the issue of plutonium recovery were on a national election ballot as a yes or no question—how would you vote and why? Why might you like more information on who these "experts" are in order to make your decision? Does where these statements were published influence your opinion?

---

[*] "The Reprocessing of Nuclear Fuels," *Scientific American*, **235(6)** (December 1976), 30.

[†] *Science*, **196** (1 April 1977).

[‡] Victor Gilinsky, "Plutonium, Proliferation and Policy," *Technology Review*, **79(4)** (February 1977).

Figure 18.10 A crane prepares to place a spent fuel cask inside a vertical decontamination structure. Note the receiving pool in the foreground, within which the washed-down cask will later be opened so that fuel bundles can be removed. (Photograph courtesy of Nuclear Fuel Services, Inc.)

nually from the spent fuel of a 1000-megawatt light-water reactor is about 550 pounds (about 250 kilograms), or enough to produce about 15 atom bombs per year.

The leverage that the United States used under the Carter Administration to induce Europe and Japan to withhold nuclear technology from non-nuclear nations was the fact that the U.S. is a principal supplier of enriched uranium to the European nations and Japan. That supply, Europe and Japan were told, could be interrupted if exports had gone forward. Hints have surfaced that the Reagan Administration may no longer care to deter the export of nuclear technology to non-nuclear nations. This could be a signal that U.S. suppliers may also be allowed to compete with their European counterparts in overseas sale of nuclear

technology. The impact of such policies on the spread of nuclear weapons is predictable.[5]

Mixed Oxide Fuel and Plutonium Transport

The reprocessing of spent fuel produces separate streams of uranium, plutonium, and waste material, the last consisting principally of fission products. The uranium is transported elsewhere to be enriched in uranium-235. The plutonium can be transported to a special fuel fabrication plant. There plutonium is mixed with uranium that has been enriched in uranium-235; the mixture is then encased in fuel rods (Figure 18.11). These "mixed oxide" fuel rods can be used in today's light-water reactors. They are called mixed oxide fuel because both uranium and plutonium are present in the oxide form.

If spent fuel is reprocessed, should plutonium be recovered for use as a supplemental fuel in light-water reactors? One issue is most important to us here: transportation of the plutonium. Plutonium is likely to move by truck along ordinary highways from the reprocessing plant to a plant where the mixed oxide fuel will be made. Although the truck's contents will be clearly labeled as "radioactive" to warn the public, there will occasionally be collisions, overturns, flat tires, etc.

The standards that the specially designed packing must meet are strict, so that the contents of the shipment cannot escape should an accident occur. Furthermore, the shielding that prevents radiation exposure must stay intact so that no radiation escapes. In addition, the component packages of the shipment cannot come in close enough contact to one another to allow a chain reaction (and explosion) to occur.

The older design package for plutonium transport was built up with three layers of steel and insulation, all surrounding about 10 pounds (4.5 kg) of plutonium. Improved designs call for even more insulation, and may allow transport of up to 15 pounds (6.4 kg) of plutonium in a package. At 40 drums per truck, about 600 pounds (0.26 metric tons) of plutonium can be moved in a given shipment. If the plutonium were extracted from the spent fuel of 100 operating plants,

---

5 "Reagan Changes Course on Non-Proliferation," *Science,* **126** (25 June 1982), 1388.

each of 1000–megawatt capacity (not far from the current status of the nuclear power system), the total mass of plutonium produced *annually* would be 25 metric tons. At 0.26 metric tons per shipment, this would involve 100 shipments of plutonium annually. Centers of population could be avoided by the trucks and the individual shipments could be made smaller, but the added mileage of the extra trips would increase the probability of an accident.

We are concerned about the movement of plutonium on the highways for two reasons. First, it is a dangerously poisonous substance; inhaling the most minute quantities can cause cancer. Second, forceful redirection of the shipment (hijacking) could occur. The plutonium could be used by terrorists for either atomic bombs or dispersal weapons. Escorts for the trucks carrying plutonium were temporarily rejected in the mid-1970s as too costly, but armed guards may eventually be needed to escort armored convoys of plutonium; such convoys could be equipped with special radio links to police. Knockout gases also may be carried to foil sabotage attempts. Finally, the plutonium might be "spiked" with a radioactive element that emits high levels of radiation, exposing hijackers to additional risk.

Perhaps the most effective way to reduce the possibility of hijacking plutonium is to place the reprocessing plant and mixed oxide fuel fabrication plant at the same site. The plutonium produced at the reprocessing plant would need to move only to an adjacent building to be combined into fuel rods. Nuclear electric power plants might be located at this same site.

International Safeguards

The term "safeguards" refers to the procedures and methods of preventing both the misuse of nuclear materials and the spread of nuclear weapons. A nation receiving a nuclear power plant may be subject to international treaties that force it to accept safeguards.

When the United States first began to export nuclear technology, it insisted on treaty guarantees from the buying nations that neither the material nor the technology would be used for atomic weapons. The United States reserved the right to inspect all records and reports from the buyers, as well as the nuclear power plants themselves. Over a thousand inspections were conducted by the United States through 1974, but an international agency has now picked up this

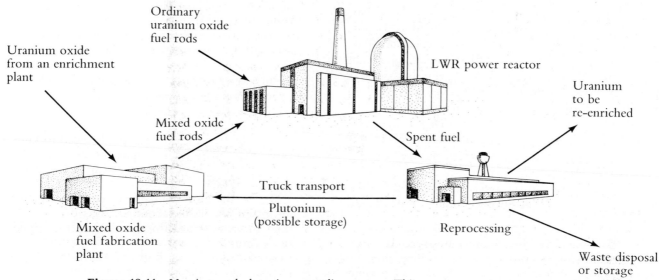

Figure 18.11 Uranium and plutonium recycling system. This diagram shows the nuclear fuel cycle as it was envisioned to operate in the early days of nuclear power in the United States. At present, no recycling system exists in the U.S., although such systems do exist in Britain and France.

Does the Public Have the Right to Know (How Easy It Is to Steal Nuclear Material and Build an Atomic Weapon)?

> The choice [on nuclear safeguards] is ours to make as a nation, and we believe it should be made on broad economic and social grounds after full public discussion.
>
> **M. Willrich and T. Taylor**
> **Nuclear Theft: Risks and Safeguards.**
> **Ballinger Publishing, 1974**

> As seasoned professionals know only too well, truly constructive contributions to an effective, balanced safeguards system . . . require competent, dedicated and sustained hard work—nearly always at low profile.
>
> **G. R. Keepin, in**
> **"Nuclear Materials Safeguards: A Professional Speaks Out,"**
> **Nuclear News (September 1974), p. 76**

A book published in 1974 by Ballinger Publishing Company set off a controversy in the nuclear community. The book, *Nuclear Theft: Risks and Safeguards,* was written by Mason Willrich and Theodore Taylor. Willrich is a professor of law who has been involved with the Arms Control and Disarmament Agency of the United States Government. Taylor is a nuclear physicist, now a corporation executive, who has been involved with the design of atomic weapons.

The second chapter of *Nuclear Theft* contained a rather general discussion of bomb building, with extensive quotations from an encyclopedia entry by J. S. Foster, a nuclear weapons expert. The authors chose to omit material, though the references were not classified, on the specifics of bomb building. Willrich and Taylor conclude that:

> Under conceivable circumstances, a few persons, possibly even one person working alone who possessed about ten kilograms of plutonium oxide and a substantial amount of chemical high explosive (the implication is that the material will be squeezed by the explosive to decrease the needed critical mass) could, within several weeks, design and build a crude fission bomb . . . This could be done using materials and equipment that could be purchased at a hardware store and from commercial suppliers of scientific equipment for student laboratories.

In their opening chapter, Willrich and Taylor explain their reasons for writing the book and for the level of detail they provide:

> . . . how much does the public need to know about these matters? This question haunts us, and we believe it merits discussion before proceeding further.
>
> To us, the most compelling argument against informing the public about the risks of nuclear theft is that such an effort might inspire warped or evil minds. . . . However, a large amount of information in much greater detail than we present here is already in the public domain. . . .
>
> . . . when security risks are inherent in a long-term activity, which is clearly the case with

nuclear theft, the public in a democratic society has a right to know, and those with knowledge have a duty to inform. . . .

. . . The years just ahead provide the last chance to develop long-term safeguards that will deal effectively with the risks of nuclear theft. Once the material flows in the nuclear power industry are as enormous as expected a few years from now, it will be too late.*

Willrich and Taylor closed their introductory chapter with an appeal to readers to arrive at their own conclusions on risks and safeguards, saying that the choice belonged to the people and (by implication) not to the technical experts.

A number of nuclear professionals, however, felt that Willrich and Taylor had provided aid to would-be terrorists by collecting so much information in one source on ways to steal and use nuclear materials. G. R. Keepin, who reviewed the book in a magazine read by nuclear professionals, disagreed with Willrich and Taylor. His position is that the public would be better off if the book had not been printed, that the nuclear industry was moving along satisfactorily with safeguard methods.

Without elucidation, this reviewer would simply record here the opinion that it is both unseemly and counterproductive for a former professional in the weapons field to speculate—however hypothetically—on how a would-be diverter might proceed with design and fabrication of an illicit atomic bomb. Genuine concern

with safeguarding . . . nuclear materials could surely take more constructive forms than indulging in what could turn out to be self-fulfilling prophecy.

The dictates of reason and prudence would appear to reject Taylor's assertion that "it seems necessary to be quite specific" in order to make the risks of nuclear terrorism credible, and to convince the public of the gravity and urgency of the nuclear materials diversion problem. . . .

The important over-all point to be made here is that—notwithstanding certain glaring shortcomings and sins of the past—the AEC and much of the nuclear industry are in fact making great strides toward effective, stringent control of the nuclear materials which are the lifeblood of that industry.†

Does the public have the right to know that a technology is unsafe if publishing this information could increase the risks? Would you rather leave the management of nuclear material to the experts, as Keepin suggests, or did Willrich and Taylor make the right decision? Is the discussion here in itself a disservice because it brings the controversy to light?

References to assist your understanding of this controversy are provided at the end of the chapter.

---

* M. Willrich and T. Taylor, *Nuclear Theft: Risks and Safeguards,* Cambridge, MA: Ballinger Publishing, 1974, pp. 2–4.
‡ G. R. Keepin, in *Nuclear News* **17(12)** (September 1974), 76.

---

function. This agency is the International Atomic Energy Agency (IAEA). The IAEA's purpose is to:

ensure that special fissionable and other materials . . . equipment . . . and information made available by the Agency . . . are not used in such a way as to further any military purpose.

The IAEA does not itself protect fissionable material; this function is left to the nation that owns the material. Nor can the IAEA prevent theft or sabotage, even if it is aware of them. All "police-type" responsibilities are reserved to the owning nation. If a prohibited act is detected, the IAEA may notify the supplier

nation so that it can end its cooperation agreement with the violating country, but that is virtually all the agency can do.

The Nuclear Non-Proliferation Treaty (NPT), which gave the IAEA its power to monitor and inspect, is designed to prevent new nations from obtaining nuclear weapons or from obtaining nuclear explosives for "peaceful uses." However, a party to the Treaty can legally make all the preparations needed to manufacture a nuclear weapon so long as it does not actually assemble the warhead. A summary of the treaty, which was established in 1970, is given in Table 18.1.

Table 18.1 The Non-Profileration Treaty[a]

Article I prohibits the transfer of nuclear weapons or other nuclear explosive devices (including devices for peaceful nuclear explosions) to any state. . . . Nuclear-weapon states are also forbidden to assist non-nuclear-weapon states to acquire nuclear weapons or explosive devices.

Article II prohibits non-nuclear-weapon signatories from manufacturing or otherwise acquiring nuclear weapons or devices, including peaceful nuclear explosives.

Article III obligates the non-nuclear-weapon parties to accept international safeguards . . . to ensure that there is no diversion of nuclear material to the manufacture of nuclear explosives.

a Stockholm International Peace Research Institute, "Preventing Nuclear Weapons Proliferation," Stockholm, Sweden, January 1975.

Since 1970, more than 100 nations have signed the treaty and most have approved it internally as well. Nevertheless, many nations have not signed, among them Argentina, Brazil, Chile, China, France, India, Israel, Pakistan, South Africa, and Spain.

Of the nations that have not signed the NPT, India was able to explode an atomic bomb in 1974, violating a treaty with Canada, which was supplying it with nuclear technology. By diverting nuclear material from a power plant to the manufacture of explosives, India was able to manufacture a nuclear explosive, illustrating the hazard of giving nuclear technology to a nation that has not signed and ratified the NPT. Sales of full nuclear technology, including reprocessing, have also been made to Brazil (by West Germany) and to Pakistan (by France). Although these nations deny intentions to obtain nuclear weapons, neither has signed the Non-Proliferation Treaty, and hence cannot be inspected by IAEA. There is a great deal of insecurity in providing such nations with nuclear technology.

Besides the Non-Proliferation Treaty, a number of European nations belong to the European Atomic Energy Community, known as EURATOM.[6] In the 1950s, this organization created a safeguards system including inspection and audit. The IAEA accepts the safeguard findings of EURATOM, but the IAEA re-serves the right to conduct independent studies. The IAEA cannot, however, investigate procedures in France, since France has not signed the NPT.

Unfortunately from the safeguards standpoint, it appears from recently published information that a reprocessing plant could be secretly built in as little as 4–6 months. Once in production, the plant could process spent fuel rods from light-water reactors and produce enough plutonium for half a dozen bombs per month. The time for the discovery of violations and sanctions against the offending nation is terribly small.

Storage or Disposal of Wastes from Fuel Reprocessing

The safe storage of high-level wastes from a fuel reprocessing plant is an issue of great importance to the nuclear industry. The plans developed for storing these wastes are drawn from experience in nuclear-weapon manufacture, which produces liquid wastes similar to those from fuel reprocessing plants.

At the Hanford site in Washington, at the Savannah River site in South Carolina, and at the National Reactor Testing Station in Idaho, experience has accumulated in handling these high-level liquid wastes. At Hanford, a concrete tank with an exterior carbon-steel liner was used initially. At Savannah River, a carbon-steel tank was placed inside a concrete tank, and steel was used to line the outside. Leaks have occurred from both these tanks because of corrosion of the steel. Exposure of the public has been negligible, but now the underground storages sites of the tanks must be controlled for several hundreds of years.

The corroding or rusting evidently took place at points of stress in the metal tanks. New tanks are now being heat-treated to relieve such stresses. Although no leaks have yet occurred from tanks so treated, a long time passed before the leaks were discovered in the original types of tanks. The tanks used at the Idaho site, on the other hand, are of stainless steel, and leaks from these tanks have not been reported.

The era of liquid wastes from reprocessing is ending, however. At the Idaho site, as well as at Hanford, methods to solidify the liquid wastes have been developed. Solidification makes it much more difficult for these wastes to contaminate the environment. Concrete and clay are viewed as useful substances into

6 France, Belgium, Luxembourg, West Germany, the Netherlands, and Italy were initial members. The United Kingdom, Ireland, and Denmark joined later.

which the wastes can be physically mixed and fixed in solid form. Fixing the wastes into glass may provide even greater safety because of the continuous character of the solid form of glass. Studies of this method are under way. New regulations for high-level wastes, if reprocessing should ever resume in the U.S., call for the wastes to be solidified before shipment from the fuel reprocessing plant.

A number of proposals have been studied for safe storage alternatives or for disposal of high-level wastes. One of the most unusual suggestions is to rocket the wastes into deep space, but the costs to dispose of many tons of high-level wastes in this way are very high. Furthermore, one cannot rule out an accident at the launching pad or a failure to exit the earth's gravity due to a failure of the boosting rockets.

Other unusual suggestions have been put forward, all involving burial in the deep earth. One concept would see the wastes placed in the geologic trench in the ocean where the ocean's tectonic plates are gradually sliding beneath those of the continent. Another idea is to dispose of the wastes in a deep shaft drilled into the earth's core or in a cavity carved out by a nuclear explosion. In none of these processes can the wastes be retrieved at some future time. Nor can the results of these plans be carefully evaluated with present knowledge.

As early as 1955, however, nuclear scientists and engineers were considering the possibility of storing high-level wastes in deep geologic formations, such as salt beds. Natural underground salt beds exist in a number of places in the United States and apparently remain stable in both their physical and chemical characters for thousands of years. The judgment that salt beds are a safe place to store nuclear wastes has been reaffirmed many times, but recent criticism suggests that the heat generated by the wastes could alter the salt formations.

In 1972, the Atomic Energy Commission selected a site near Lyons, Kansas, for Project Salt Vault. The project at Lyons was designed to show that storage in salt beds would work, but the project was abandoned when it was discovered that salt mining (by dissolving salt in water) within the formation was being conducted not far away. The unknown effect of mining on the stability of the formation was the key factor in ending the project at Lyons. The idea of storing high-level wastes in bedded salt, however, was not aban-

doned because of this setback. A site in New Mexico is still being considered.

A plan to store the wastes on a temporary basis was still needed, however, and the concept of a retrievable surface storage facility (RSSF) has been studied to solve the temporary storage problems for high-level wastes. Retrievable storage is designed to hold high-level solid wastes accumulating from reprocessing while a permanent solution is sought.

Two concepts are being considered for retrievable surface storage. In one, the wastes would be placed in canisters stored under water. In the other, the canisters would be open to the air, either indoors or outside on concrete pads (Figure 18.12).

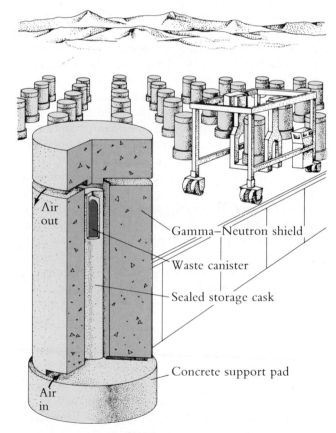

Figure 18.12 Retrievable Surface Storage Facility for High-Level Radioactive Wastes from Fuel Reprocessing. In this artist's concept, the storage cask, stored out of doors, is shown in cross section, mounted on a concrete pad. Air cooling is used to keep the cask from overheating. The site is carefully guarded from trespassers.

Economic and Social Aspects of Nuclear Power

The terror and destruction of Hiroshima and Nagasaki hung on the consciences of the nuclear scientists who built the first atomic bombs. Many saw in the awesome new technology hope for supplying electric power, for turning the power of the atom to peaceful uses; these scientists pushed their dream. It was a dream that would in part release them from the responsibility they bore for the birth of atomic power.

These scientists then did not know or foresee all that would eventually be known about radiation effects. Nor could they foresee all the issues involved in the safe disposal of radioactive wastes. But they had confidence that all the problems in the nuclear fuel cycle could be resolved—and so they persevered. They were very successful at pushing their dream, and they were very successful at resolving many problems of the nuclear fuel cycle. They were the architects of the current age of nuclear electric power. To some, however, the dream of the nuclear scientists was ill-considered, was too hastily pushed, and could lead too many nations into possession of nuclear weapons.

Almost three decades have now elapsed since President Eisenhower opened nuclear technology to private use. It is logical to ask, "What is the future of nuclear electric power in the United States? in other parts of the world?" On the one hand, more and more nuclear electric plants are coming on line, and the fraction of our electric power supplied by nuclear energy is steadily increasing (see Chapter 15, Figure 15.6). On the other hand, the nuclear industry—the collection of companies that build the generators and the major components that are assembled at the sites—is in trouble. No order has been received for a new nuclear reactor to be constructed in the United States since 1978. Only orders for reactors in overseas countries have kept the U.S. nuclear industry afloat. What are the barriers that have stalled and halted, at least for the moment, the spread of nuclear technology in the United States? What are the impacts of the sinking status of the U.S. nuclear industry?

The factors that have slowed construction and halted new orders for nuclear reactors are several-fold and interrelated. Difficulties in obtaining full licensing for the operation of new plants have so delayed these plants coming on line that their costs have soared. Of course, the licensing procedures were designed to en-sure safe operation of the plants. From about a billion dollars for a 1000-megawatt (electrical) nuclear plant in the mid-1970s, the 1985 cost is now expected to be on the order of $1.5 billion. Construction costs, driven up by delays and by historically high interest rates, are one aspect of added costs. The price of uranium fuel has also risen; it quadrupled in the late 1970s from $10 per pound to $40 per pound when an international cartel of nations and companies successfully imitated the tactics of OPEC, the oil cartel. The economics that pointed utilities toward nuclear energy and away from fossil fuel have thus been confounded.

The delays in licensing are part of an effort to ensure the safety of each nuclear plant and to ensure that the impacts of each plant on its local environment are acceptable. The fact that nuclear plants are less efficient than coal-fired plants means that more heat is discharged to the cooling water and that thermal pollution effects must be more closely examined. Safety has become an even greater concern since the reactor incident at Three Mile Island. In large part, the concern with safety must reflect the point of view of the public at large, and not simply be the opinion of the engineers, scientists, and others who regulate the nuclear industry. Thus, the slowing of the industry growth represents the effect of public opinion. In addition, the search for disposal sites for radioactive wastes has been slowed by hostile receptions in many states. The concern is that storage will not be as safe as promised.

The nuclear industry's orders for new plants have ground to a halt, and cancellations of previous orders continue. Yet plants that have cleared the hurdles are still being built. Nuclear engineers are needed to staff them. What of the training of new nuclear engineers in the U.S.? With declining enrollments, future staffing of plants by nuclear engineers may be difficult.

We are witnessing a very unusual event in the history of a nation. We are gradually rejecting nuclear electric power. The current plants and plants in the building stages will be with us for a long time, but the trend is clear. For the moment, coal appears to be the route chosen for the next generation of power plants; other clear choices are not on the horizon.

Overseas, the trends are not so clear; smaller nations want nuclear power, want to be part of the nuclear club. France, Germany, and Britain, though public opposition exists, are building the entire fuel cycle, including reprocessing. We are either being left behind or are choosing another route to the future.

✒ **Alternatives to Reprocessing**

In the original PUREX process, uranium and plutonium are recovered as separate products. Nonetheless, it is possible to modify the PUREX process so that uranium and plutonium are recovered together as a mixture. The mixture of uranium and plutonium would be mainly uranium-238, a substance that cannot be used for bombs. Only about 1% would be uranium-235 and 1% would be plutonium. Since this material could not be used for bomb building, it would be much less valuable to would-be hijackers. The uranium–plutonium mixture can be used in fuel rods if it is first mixed with uranium extra-enriched in uranium-235. The final mixture would then be about 3% fissionable material. The modification of the PUREX process is called "coprocessing" and lately has been called CIVEX (for civilian extraction) (Figure 18.13).

The CIVEX process would, in addition, leave some small quantity of fission products in the recovered uranium–plutonium mixture. This radioactivity from the fission products would make the mixture physically unapproachable for humans. Only remote handling would be possible. This mixture of uranium and plutonium, made dangerously radioactive by the fission products, will seem much less attractive to

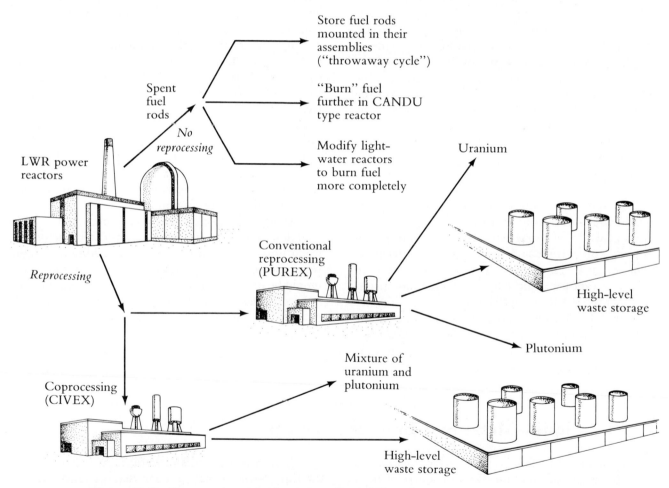

Figure 18.13 Reprocessing and Alternatives. Spent fuel can be reprocessed by either the PUREX or CIVEX methods, the latter process providing a less useful product to would-be hijackers. Alternatively, spent fuel can be simply discarded (although very carefully), or "burned" further in a CANDU reactor or in a modified light-water reactor.

would-be hijackers when it moves from the reprocessing plant to the fuel fabrication plant. Advocates of CIVEX claim that a nation given only this reprocessing technology would not have the pure plutonium needed for bombs. The separation of plutonium from the mixture would, however, require only one additional chemical step. Coprocessing would also continue to produce fission products as high-level wastes from reprocessing, just as the PUREX process does.

We pointed out earlier, but it bears repeating, that reprocessing is based on the concept of recovering "unburned" uranium-235 as well as plutonium, substances that are viewed as usable energy resources. It is also possible that light-water reactors could operate in a "throw-away" cycle. In such a cycle, spent fuel rods would never be opened. They would simply be stored until the levels of radiation and radioactive elements have decreased enough to allow ordinary disposal.

The throw-away cycle, while simpler, may be wasteful of resources. If a way existed to recover the energy in the spent fuel without reprocessing, the methodology would be very attractive, because plutonium would not become available. In this way, the spread of nuclear weapons would be delayed. There are, in fact, several ways to recover the energy remaining in spent fuel without reprocessing.

The light-water reactors, which dominate the United States and European markets, are not the only way to generate nuclear electric power. Another economically competitive and relatively proven technology has originated in Canada. This technology is more efficient at capturing the energy of uranium fission than either the BWR or the PWR. It is the heavy-water reactor built by Atomic Energy of Canada, Ltd., and it is known as the CANDU reactor. The CANDU reactor circulates "heavy water" through the core to capture the heat from the fission process. More than a dozen commercial-scale units of this type are operating already or are under construction. By 1983, CANDU reactors were capable of delivering 12,500 megawatts of electric power.

The CANDU reactor does not require enriched uranium fuel, as light-water reactors do. While light-water reactors use a fuel enriched to about 3% uranium-235 (the rest of the fuel is uranium-238), the CANDU reactor can operate with unenriched uranium fuel, which may be only about 0.7% uranium-235. When the fuel rods are finally withdrawn from the CANDU reactor for replacement, the level of uranium-235 has fallen to 0.25% or slightly less. This level is far below the natural concentration of uranium-235 in uranium ore.

In contrast, the level of uranium-235 in the spent fuel rods withdrawn from light-water reactors is nearly 1%. This level in the fuel rods that are discarded from the light-water reactors is actually higher than the level in fresh fuel rods used by the CANDU. It follows that what is "spent" to the light-water reactors can potentially still serve as fuel for the CANDU. Thus, it has been suggested that the discarded rods from light-water reactors be transferred to CANDU reactors, where more energy could be derived from the rod without reprocessing.

Another suggestion that makes use of the Canadian design is to modify pressurized-water reactors by using heavy water in the coolant in addition to light water, so that more complete fission of uranium-235 would be possible. However, heavy water would have to be manufactured in the U.S. in order to accomplish this change. These alternatives are summarized in Figure 18.13.

Liquid–Metal Fast Breeder Reactors

We have discussed so far the light-water reactors that utilize the fission of uranium-235 as a source of heat, but surprisingly, light-water reactors have been viewed by nuclear advocates as a short-term source of energy. These people see light-water reactors as a means to supplement and replace fossil fuels until the time when more secure sources of nuclear power, such as the breeder reactor, become available.

Viewed as the successor to the light-water reactor, the **liquid-metal fast breeder reactor (LMFBR)** is under development in the U.S.S.R., Japan, France, the United States, and possibly other countries as well. The light-water reactor uses uranium-235 for fission, but uranium-235 is not as plentiful in nature as uranium-238, the dominant component of uranium ore. The breeder reactor utilizes the abundant uranium-238

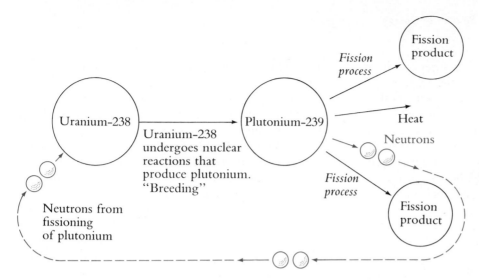

Figure 18.14 Nuclear Reactions in the Breeder Reactor. Plutonium is "bred" by bombarding uranium-238 with neutrons. These neutrons are furnished by the fission of plutonium-239. The fission process produces heat, which eventually is used to make steam, and releases the neutrons needed to breed more plutonium-239 from uranium-238.

to produce power. The idea of using uranium-238, even though it does not naturally undergo fission, is ingenious.

In our discussion of light-water reactors, we pointed out that the fission of uranium-235 "bred" plutonium-239 from uranium-238. Plutonium-239 undergoes fission just as uranium-235 does when properly struck by a neutron. In much the same fashion as uranium-235, plutonium breaks apart into two fragments, with a release of neutrons and energy. The neutrons released sustain the fission of more plutonium-239, producing heat. The neutrons also "breed" more plutonium from uranium-238 atoms. The process, illustrated in Figure 18.14, both produces heat from fission and "breeds" more fuel to undergo fission. This explains the word "breeder" in the name of

the reactor.

Instead of using water to withdraw heat energy from the core, this reactor utilizes a liquid metal (pure, melted sodium) to withdraw the heat of fission from the reactor core. Liquid sodium has some advantages. Sodium melts at 210°F (99°C), but does not boil until it reaches 1640°F (893°C). Because its boiling point is so distant from its melting point, it is unlikely to "flash" to a vapor if an accidental rupture of a line occurs. The loss-of-coolant accident, which is a concern in light-water reactors, is therefore thought to be much less probable in a breeder reactor. In addition, the sodium in the core will not slow down the high-energy neutrons released in the fission process, which are essential for breeding more plutonium.

The design of the LMFBR, shown in Figure 18.15,

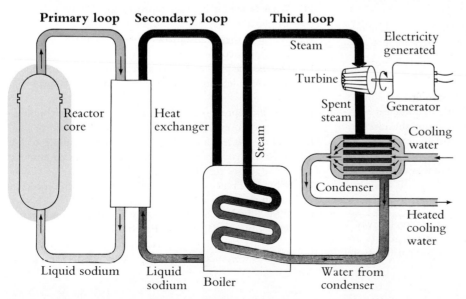

Figure 18.15 Schematic diagram of the liquid-metal fast breeder reactor (LMFBR).

calls for a primary loop containing liquid sodium, which circulates through the core of the reactor. This sodium, which will become radioactive as it circulates through the core, picks up the heat produced by the fission process. A second loop also contains circulating liquid sodium. This sodium picks up the heat from the first loop in a heat exchanger. A third loop contains water, which boils to steam by the heat from the second loop. The steam turns the turbine, is condensed, and returns to the boiler.

The fuel rods in the LMFBR contain a mixture of plutonium oxide and uranium oxide (called "mixed oxide"). In time, as fission proceeds, fission products build up in the rods and have the effect of "poisoning" the reaction by capturing a portion of the neutrons, just as in light-water reactors. When neutron capture slows heat production too greatly, the rods are withdrawn and sent to a reprocessing facility. There, plutonium and uranium are separated from the fission prod-

ucts. The plutonium content will have increased during the breeding cycle in the reactor. The fission products are wastes and require disposal. New mixed-oxide fuel rods are fabricated from the plutonium and from uranium.

The familiar problems of reprocessing, plutonium movement, and disposal or storage of fission-product wastes are all present in the breeder reactor concept. Furthermore, the quantities of plutonium will be enormous, even in comparison to the quantities potentially produced by light-water reactors. A core meltdown after a loss of the sodium coolant would pose an even greater problem than a meltdown in a light-water reactor, because the core contains so much more of the highly toxic plutonium. Thus, although the breeder reactor is an idea of cunning ingenuity for extending fuel supplies, the problems and issues that surround its development must be critically examined.

Questions

Light-Water Nuclear Reactors
How they operate

1. What are the two basic types of light-water reactors?
2. Describe the basic differences between the two reactor types.
3. Which process is more like that of a fossil-fuel power plant?
4. As the reactor operates, what eventually happens to the level of uranium-235 in the fuel rods? to the level of fission products?
5. What is the function of the heat exchanger, as shown in Figures 18.3 and 18.4, in the electric generation process?

Radioactive wastes and radiation releases from light-water reactors

6. What accounts for the presence of fission products in the water that circulates through the core of a nuclear reactor? Why should we be concerned about the ultimate disposal of water in which these fission products are dissolved?
7. What is done with the solid radioactive wastes generated by nuclear power plants?

Safety of light-water reactors

8. Explain what is meant by core melt, or "meltdown." What would cause a meltdown? What safeguards are designed into nuclear power plants to prevent this from happening?

Problems in the Nuclear Fuel Cycle
From the mine to the power plant

9. What is accomplished by "enrichment" in the processing of nuclear fuels? Is uranium-235 or uranium-238 used for the fission process?

Fuel reprocessing and spent fuel rod storage

10. What is initially done with "spent" fuel rods from nuclear power plants?
11. What is meant by fuel reprocessing? What substances are separated out?
12. Which of these substances can be used again? Where did plutonium come from?
13. Discuss the pros and cons of reprocessing as an alternative to indefinite underground storage of spent fuel rods.
14. Was the reprocessing plant at West Valley a technological success, even though it went out of business?
15. Did the West Valley Plant use remote handling of fuel rods? Did it use remote repair of broken equipment? What abuses resulted?
16. Why did France and Germany proceed with reprocessing in the 1970s?
17. Why did the U.S. work against the sale of reprocessing technology to Brazil and Pakistan?

International safeguards

18. What rights does the Nuclear Non-Proliferation Treaty give to the International Atomic Energy Agency (IAEA)?
19. Does the IAEA have the right to seize fissionable material to prevent manufacture of nuclear weapons?

Storage or disposal of wastes from fuel reprocessing

20. Are carbon-steel tanks adequate for storing reprocessing wastes?
21. List and discuss briefly at least three methods that have been proposed for the storage and/or disposal of high-level radioactive wastes.

Further Reading

General References

Bupp, I., and J. Derian, *Light Water: How the Nuclear Dream Dissolved.* New York: Basic Books, 1978.

Hohenemser, D., R. Kasperson, and R. Kates, "The Distrust of Nuclear Power," *Science,* **196** (1 April 1977), 25–34.

"Nuclear Dilemma: The Atom's Fizzle in an Energy-Short World," *Business Week* (25 December 1978), p. 54.

"Nuclear Energy: How Bright a Future," *Environmental Science and Technology* **11(2)** (February 1977), 128–130.

"Nuclear Power: Can We Live With It?", a panel discussion including D. Kleitman, N. Rasmussen, R. Stewart, and J. Yellin, *Technology Review,* **81(7)** (June/July 1979), 32.

Nuclear Power, Issues and Choices, The Nuclear Energy Policy Study Group, S. Keeny, Jr., Chairman. Cambridge, MA: Ballinger Publishing Co., 1977.

Weinberg, A., "Social Institutions and Nuclear Energy," *Science,* **177** (7 July 1972), 27–34.

Weinberg, A., "Salvaging the Atomic Age," *The Wilson Quarterly* (Summer 1979), pp. 88–112.

Nuclear Plant Reactor Safety

"NRC Panel Renders Mixed Verdict on Rasmussen Reactor Safety Study," *Science,* **201** (29 September 1978), 1196.

"Reactor Safety Study: An Assessment of Accident Risks in U.S. Commercial Nuclear Power Plants," Nuclear Regulatory Commission, October 1975, Wash-1400 (the Rasmussen Report).

"Risk Assessment Review Group Report to the U.S. Nuclear Regulatory Commission," prepared for the U.S. Nuclear Regulatory Commission, NUREG/CR-0400, September 1978 (review of the Rasmussen Report).

Shapely, D. "Reactor Safety: Independence of Rasmussen Study Doubted," *Science,* **197** (1 July 1977), 29–31.

"The Accident at Three Mile Island (The Need for Change: The Legacy of Three Mile Island)," Report of the President's Commission, J. Kemeny, Chairman, October 1979.

Reprocessing, Wastes, and Nuclear Weapons Spread

Lester, R. K., and D. J. Rose, "The Nuclear Wastes at West Valley, New York," *Technology Review* (May 1977), pp. 20–29.

Rose, D., and R. Lester, "Nuclear Power, Nuclear Weapons, and International Stability," *Scientific American,* **238(4)** (April 1978), 45–57.

Bebbington, W. P., "The Reprocessing of Nuclear Fuels," *Scientific American* (December 1976), 30–41.

References to "Controversy: Does the Public Have the Right to Know?"

Keepin, G. R., "Nuclear Materials Safeguards: A Professional Speaks Out," *Nuclear News* (September 1974), p. 76.

Willrich, M., and T. Taylor, *Nuclear Theft: Risks and Safeguards* Cambridge, MA: Ballinger Publishing, 1974.

Other publications that stir the pot of controversy are:

Gillette, R., "Nuclear Safeguards: Holes in the Fence," *Science,* **182** (14 December 1973), 1112.

Lapp, R. E., "The Ultimate Blackmail," *The New York Times Magazine* (4 February 1973), p. 13.

McPhee, J., "The Curve of Binding Energy," *The New Yorker Magazine* (3, 10, and 17 December 1973).

McPhee, J., *The Curve of Binding Energy: A Journey into the Awesome and Alarming World of Theodore B. Taylor.* New York: Farrar, Straus, & Giroux, 1974.

Shapely, D., "Plutonium: Reactor Proliferation Threatens a Nuclear Black Market," *Science,* **172** (9 April 1971), 143–146.

(*National Archives/Documerica; Paul Sequeira*)

PART FIVE

Air Pollution and Energy-Related Water Pollution

When we burn the fossil fuels—coal, oil, and gas—we furnish heat for homes, energy for mobility, and power for the production of goods. Fossil fuels have become our servants, and our skill at capturing their energy has transformed civilization. This mastery of energy has brought abundance and comfort. But energy has turned out to be a dangerous servant.

Our misuse of these fuels has resulted in air pollution episodes so severe that people have died. Ships carrying oil to fuel automobiles have foundered, spilling cargoes of black oil that foul beaches and bring death to countless sea birds. And while we fear such events, they are little more than the "tip of the iceberg," visible to us because of their nightmare drama. Beneath these episodes is a history of chronic and accumulated effects on humans and other animal species and on plants. Taken together, these effects may be found to be even more dangerous. Some of the environmental effects of power generation were mentioned in Part IV: acid mine drainage, strip mining

damage, and nuclear waste contamination of the environment.

Part V begins a chapter on atmosphere and climate (Chapter 19). The concepts in this chapter are essential to understanding how air pollutants, produced when energy is generated, affect climate and the natural cycles of materials such as carbon, sulfur, and nitrogen. Such disturbances eventually threaten the health and survival of human and wildlife populations.

Numerous pollutants foul the air we breathe. Solid particles such as lead, asbestos, and soot are dispersed in the air. Liquid droplets of hydrocarbons and sulfuric acid float in the atmosphere. Gases such as carbon monoxide, nitrogen oxides, and sulfur dioxide are dissolved in the atmosphere that surrounds us. The contaminants in the air have direct biological effects upon us. Our breathing may be impaired; diseases of the heart and lung are complicated and made worse. In addition, the landscape is influenced because vegetation is also sensitive to air pollutants. Construction materials such as mortar and metal are corroded by certain pollutants, and the fibers of our fabrics are damaged by other pollutants.

Near Los Angeles, the most sprawling city in the country and the most dependent upon the automobile, smog is destroying the pine trees in the hills that surround the city. Almost 200 miles away, about the distance between New York and Boston, the desert holly, a plant that grows in the isolation of Death Valley, is threatened by the creeping smog from Los Angeles. The smog results from reactions between hydrocarbons and nitrogen oxides. The predominant source of these substances in the Los Angeles basin is the automobile. Sunlight stimulates the reactions between hydrocarbons and nitrogen oxides.

In the northeastern United States, as well as in Scandinavia, an "acid rain" is turning inland lakes increasingly acid, destroying fish life and other aquatic species. The acidity is caused by the solution in rain of sulfur oxides and nitrogen oxides in the atmosphere. Sulfur oxides are pollutants produced primarily from the burning of coal and residual oil, two sulfur-containing fuels. Coal is burned to generate electric power and to manufacture steel. Residual oil is burned to produce heat and to generate electric power. Oxides of nitrogen stem principally from combustion occurring at fossil-fuel power plants and from the combustion of gasoline in automobiles.

Chapters 20, 21, and 22 examine the air pollutants produced by the combustion of fossil fuels and possible means of controlling these pollutants. Chapter 20 treats carbon dioxide and carbon monoxide. Chapter 21 discusses sulfur oxides and particulate matter. Chapter 22 focuses on photochemical air pollution, which stems from the reactions of hydrocarbons and the oxides of nitrogen.

However, the impact of energy on the environment extends far beyond its effect on the air we breathe. Energy activities reach out as well to influence oceans and rivers around the globe. The effects begin the moment energy is extracted from the earth. Where oil is drawn from offshore wells, blowouts may occur that threaten both marine species and beaches. The transport of oil in tankers from the producing nations of the world to the industrial consumers is also hazardous. Tankers may be shipwrecked or may spill oil during normal operations. Chapter 23 describes this water pollution problem, which is a direct result of our use of oil as a source of energy.

Other major problems occur when coal and oil are burned in the steam boilers of electric power plants. About $2\frac{1}{2}$ times more heat is produced than can be converted into electricity. At nuclear power plants, three times more heat is produced than can be converted into electrical energy. Heat produced but not converted to electricity is wasted; it goes into raising the temperature of the cooling water of the power plant or into hot stack gases.

Environmental problems can arise in three areas as a consequence of the steam power plant cycle. First, drawing cooling water at high velocities into the narrow tubes of a condenser subjects organisms living in the water to an extraordinary shock. We might compare it to the trauma we would suffer if we attempted to ride on the outside of an airplane. The second area of concern is the effect of elevated temperature on aquatic life in a body of water into which the heated condenser water is discharged. The third area of concern is the effect of the numerous chemicals added to the cooling water to prevent the cooling apparatus from clogging. If most future power plants were designed to use the method known as once-through cooling, then, by the year 2000, over two-thirds of the daily runoff in the U.S. could be required to cool power plants. Although such an extreme situation is unlikely, the potential certainly exists for waste heat to have a serious effect on the water environment. Thus, it is necessary to examine the effects hot water can have on aquatic organisms and also to consider what alternative cooling methods are available.

In Chapter 24 we look not only at the problem of thermal pollution, but also at a related energy pollutant: noise. Noise and thermal pollution are related in that both are forms of waste energy; the waste energy is heat energy in the case of thermal pollution and sound energy in the case of noise. Energy pollutants differ from other air and water pollutants, such as sulfur oxides or oil, because the pollutant itself is relatively short-lived in the environment. If you ignore excess heat or excessive sound, it will disappear, unlike oil, which can often be found in sediments years after an oil spill. Both thermal energy and noise can, however, cause lasting harm in the environment. Such effects are the focus on Chapter 24. One other form of energy as a pollutant is discussed in Chapter 24. This is the electric field generated under high-voltage power lines.

CHAPTER NINETEEN

Lessons from Ecology:
Atmosphere and Climate on Earth

The Donora Story, October, 1948

The Earth's Atmosphere
Components/Origin of the Oxygen in Air/Distribution of the Atmosphere

Climate on Earth
Weather versus Climate/World Climates

Changing Climates
Natural Changes/Human Influences on Climate

Inversions

The Donora Story, October, 1948

During the last week of October 1948, a heavy smog settled down over the area surrounding Donora, Pennsylvania. Weathermen described it as a temperature inversion. . . . Smogs of short duration are not unusual and except for discomfort due to irritation and nuisance of the dirt and poor visibility, no unusual significance is attached to such occurrences.

This particular smog, encompassed the Donora area on the morning of Wednesday, October 27. It was even then of sufficient density to evoke comments by the residents. It was reported that streamers of carbon appeared to hang motionless in the air and that visibility was so poor that even natives of the area became lost.

The smog continued through Thursday, but still no more attention was attracted than that of conversational comment.

On Friday, however, a marked increase in illness began to take place in the area. By Friday evening the physicians' telephone exchange was flooded with calls for medical aid, and the doctors were making calls unceasingly to care for their patients. Many persons were sent to nearby hospitals, and the Donora Fire Department, the local chapter of the American Red Cross, and other organizations were asked to help with the many ill persons.

There was, nevertheless, no general alarm about the smog's effects even then. On Friday evening the annual Donora Halloween parade was well attended, and on Saturday afternoon a football game between Donora and Monongahela high schools was played on the gridiron of Donora High School before a large crowd.

The first death during the smog had already occurred, however, early Saturday morning—at 2 A.M., to be precise. More followed in quick succession during the day and by nightfall word of these deaths was racing through the town. By 11:30 that night 17 persons were dead. Two more were to follow on Sunday, and still another who fell ill during the smog was to die a week later on November 8.

On Sunday afternoon rain came to clear away the smog. But hundreds were still ill, and the rest of the residents were still stunned by the number of deaths that had taken place during the preceding 36 hours. That night the town council held a meeting to consider action, and followed with another on Monday night. By this time emergency aid was on its way to do whatever possible for the stricken town . . .

From "Air Pollution in Donora, PA,"
Public Health Bulletin No. 306,
U.S. Public Health Service, 1949

What are inversions, those weather conditions that can combine with pollution to cause such havoc to human health? In this chapter we explore that subject and other aspects of the atmosphere and climate on earth. In Chapters 5 and 8 we examined land habitats and water habitats before discussing their environmental problems. In a similar way, we now discuss clean air before looking at air pollution.

The Earth's Atmosphere

Components

Although we usually speak of the material we breathe simply as "air," it is actually a mixture of substances. Clean, dry air is mostly (99%) nitrogen gas and oxygen gas. However, clean air contains several other substances, which, although present in only small quantities, are still very important. **Carbon dioxide,** which has a strong effect on the earth's atmospheric temperature, is an example (Table 19.1), and **ozone** gas is another. Water vapor is also an important component of the atmosphere. The amount of water vapor varies from 0% by volume in dry air to about 4% in very humid air. Dust particles from both human and natural sources (e.g., volcanoes) may also be a significant component of air, even though present in a relatively small percentage.

Origin of the Oxygen in Air

The oxygen in air is essential for respiration by plants and animals. Ozone, which is discussed more fully in Chapter 30, provides protection from the sun's ultra-

Table 19.1 Components of Clean Dry Air[a]

	Percent by volume
Nitrogen (N_2)	78.08
Oxygen (O_2)	20.94
Argon (Ar)	0.93
Carbon dioxide (CO_2)	0.03
Ozone (O_3)	less than 0.00005

a Small amounts (less than 0.002%) of neon, helium, methane, krypton, and hydrogen are also present. Nondry air, of course, also has water vapor in it.

violet rays, which can be harmful to life. However, neither oxygen nor ozone has always been part of the earth's atmosphere. Scientists believe that 4.5–5 billion years ago, the earth's atmosphere was similar to the mixture of gases released when volcanoes erupt; primarily water vapor, carbon dioxide, and nitrogen. (Heavy rains during the period when the earth was cooling and becoming more solid washed out most of the carbon dioxide.) Oxygen, however, came from a completely different source: green plants.

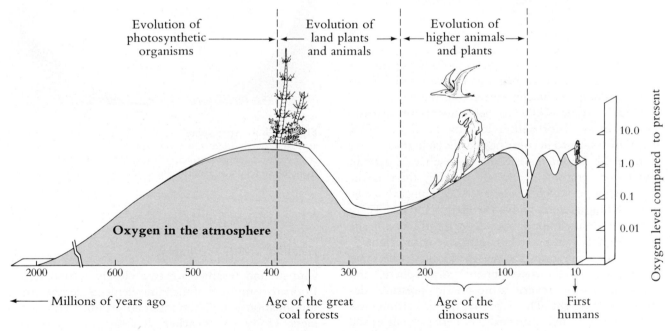

Figure 19.1 Origin of Oxygen in the Atmosphere. About 2 billion years ago, the level of free oxygen in the earth's atmosphere began to rise. After a protective layer of ozone was formed from part of the oxygen, land plants and animals evolved. Oxygen levels have varied a great deal as levels of production and use have changed. (Source: E. P. Odum, *Fundamentals of Ecology*, 3rd ed. Philadelphia: Saunders, 1971)

At some point, more than three billion years ago, simple cells, which were nourished by chemicals floating in the water in which they lived, evolved into organisms capable of photosynthesis. While using the sun's energy and carbon dioxide to make their own organic compounds, they also produced oxygen. This oxygen began to build up in the atmosphere, and a new era began (Figure 19.1). Part of the oxygen (O_2) was changed by sunlight into ozone (O_3).

Distribution of the Atmosphere

Like the term "air," the word "atmosphere" is often used loosely to describe the whole envelope of gases surrounding the earth. However, the components of the atmosphere are not distributed uniformly throughout this envelope. Scientists who study the atmosphere recognize several zones at different heights above the earth, depending on their temperature characteristics (Figure 19.2).

The layer closest to the earth's surface is called the troposphere. In this layer, 5–10 miles (9–16 kilometers) high, most of what we call weather occurs. All rainfall, almost all clouds, and most violent storms occur in this layer. The temperature in the troposphere usually decreases as we go up. Above the troposphere is the stratosphere. Temperatures in this layer first remain constant and then begin to increase with height. Most of the ozone is concentrated in the stratosphere and this is responsible for the temperature increase. That is, ozone absorbs ultraviolet light and this causes the stratosphere to become heated. Even higher, above 30 miles (50 kilometers), is the mesosphere, a zone in which temperatures again fall. Finally, there is the thermosphere, which is the layer beginning about 50 miles (80 kilometers) above the earth's surface. There is no well-defined upper boundary to this layer. In the thermosphere, temperatures again rise with height.

Most of the gases are present in about the same percentages in the troposphere, stratosphere, and mesosphere. However, atmospheric pressure decreases with height. That is to say, the air becomes less and less dense going upwards. Ninety percent of the atmosphere is concentrated within 16 kilometers of the earth's surface. You may already be familiar with this fact if you've ever travelled into high mountain areas. Difficulty in breathing results from the fact that oxygen concentrations are lower at higher altitudes.

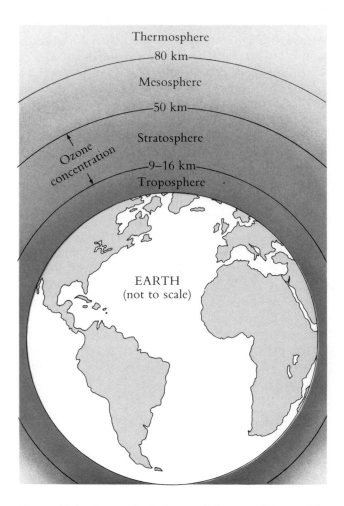

Figure 19.2 Atmospheric Layers. Diagram of the earth's atmosphere, divided into layers on the basis of temperature. Heights given are approximate, as they vary from place to place around the globe.

Climate on Earth

Weather versus Climate

Climate and weather are two closely related terms. Both are composed of the same elements: temperature, precipitation, wind, humidity, air pressure, and cloudiness. However, **weather** describes the day-to-day changes in these elements, while **climate** is a more long-term description of these factors for a given region. Climate is not simply average weather, but also includes the kind of extremes of heat, cold, wind, etc. that occur in the region.

World Climates

Various ways exist of describing the world in terms of climate, but the most useful appears to be a system using temperature and precipitation. These two elements have perhaps the greatest effect on the distribution of soil types, vegetation, animal life, and human activities. In Chapter 1 we noted that the severity and constancy of temperature and rainfall determine the kinds of communities that grow up in a region. In Chapter 5, we discussed how climate (principally temperature and precipitation) determines the soil type that develops in a region. Because of these interactions, it is not surprising that a map of world climate corresponds in many major details to a map of world biomes or to a map of soil types around the world (Figures 19.3, 19.4, and 19.5).

Several factors control the climate for a given region. Most important is the amount of sunlight or solar radiation. Figure 19.6 summarizes some of the effects of solar radiation on climate. If this were the only controlling factor, however, the earth's climate would be a series of bands running horizontally around the globe, corresponding to the amount of solar radiation at different latitudes.

A glance at Figure 19.3 shows this is not the case. A major factor modifying the influence of sunlight is the distribution of land and water. Land heats and cools more quickly than water; thus, continental climates tend to be more extreme than marine climates. Similarly, the presence of mountains affects the amount of precipitation in a region, while cool or warm ocean currents cool or warm nearby coastal areas.

A more detailed description of the factors controlling climate is unfortunately beyond the scope of this text. More on this fascinating subject can be found in the references listed at the end of the chapter.

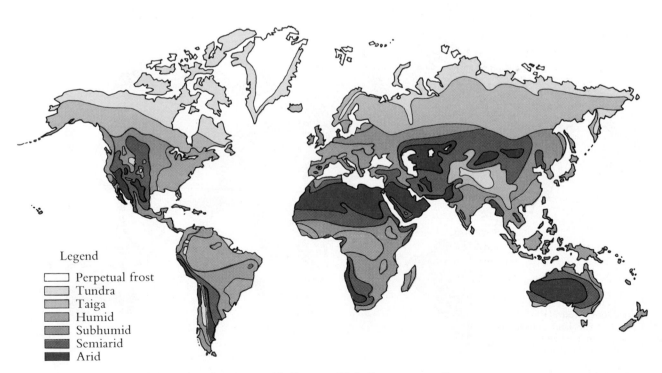

Legend
- Perpetual frost
- Tundra
- Taiga
- Humid
- Subhumid
- Semiarid
- Arid

Figure 19.3 Map of major world climates. (U.S. Department of Agriculture Yearbook, 1941)

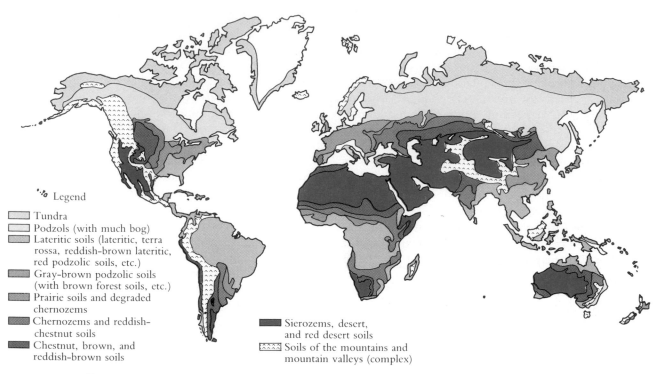

Figure 19.4 Map of major world soil types. (U.S. Department of Agriculture Yearbook, 1941)

Legend
- Tundra
- Podzols (with much bog)
- Lateritic soils (lateritic, terra rossa, reddish-brown lateritic, red podzolic soils, etc.)
- Gray-brown podzolic soils (with brown forest soils, etc.)
- Prairie soils and degraded chernozems
- Chernozems and reddish-chestnut soils
- Chestnut, brown, and reddish-brown soils
- Sierozems, desert, and red desert soils
- Soils of the mountains and mountain valleys (complex)

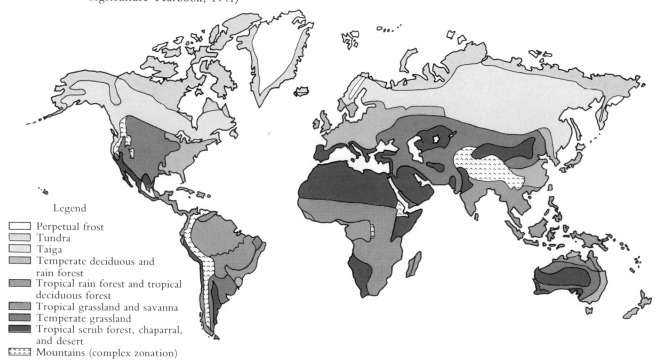

Figure 19.5 Map of major world biomes. (From E.P. Odum, *Fundamentals of Ecology,* 3rd ed. Philadelphia: Saunders, 1971)

Legend
- Perpetual frost
- Tundra
- Taiga
- Temperate deciduous and rain forest
- Tropical rain forest and tropical deciduous forest
- Tropical grassland and savanna
- Temperate grassland
- Tropical scrub forest, chaparral, and desert
- Mountains (complex zonation)

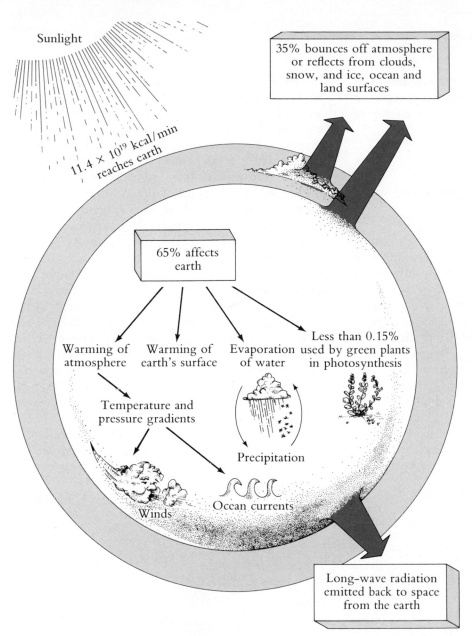

Sunlight

11.4×10^{19} kcal/min reaches earth

35% bounces off atmosphere or reflects from clouds, snow, and ice, ocean and land surfaces

65% affects earth

Warming of atmosphere

Warming of earth's surface

Evaporation of water

Less than 0.15% used by green plants in photosynthesis

Temperature and pressure gradients

Precipitation

Winds

Ocean currents

Long-wave radiation emitted back to space from the earth

Figure 19.6 Solar Radiation and the Earth's Energy Budget. Of the solar radiation reaching the earth and its atmosphere, about 1/3 is reflected away immediately by clouds, snow and ice, and water or land surfaces. The remaining 2/3 is absorbed by the earth and its atmosphere. Of this amount, less than 0.15% is used by green plants in photosynthesis. The rest causes water to evaporate and also warms the earth's surface and atmosphere. This warming is uneven over the globe, leading to temperature and pressure gradients. Along with the rotation of the earth, these gradients cause winds and ocean currents. All of the radiation absorbed by the earth and its atmosphere (except for a minute amount stored as fossil fuels) is eventually reradiated into space, mainly as long-wave radiation in the infrared range.

Changing Climates

Natural Changes

Although it is possible to identify different climatic regions in the world, these climates should not be regarded as unchangeable. Over periods of time, climatic conditions vary. You are probably familiar with the fact that over the past 2 million years, ice sheets have periodically advanced toward the equator and then receded again. But world climate varies on a much shorter time scale than the one on which ice ages

are measured. For instance, it appears that global temperatures have been dropping over the past 30 years (0.5 degrees Centigrade since the 1940s), as part of a cooling trend that some scientists think may continue and others think will soon end.

Various theories have been proposed on the causes of natural changes in the earth's climate. These theories involve such phenomena as drifting continents, variations in the earth's orbit, variations in solar energy (e.g., sunspots), or volcanic activity. Although it is not possible to choose among these various theories at the present time, it is important to understand that climate does change naturally on both long and short time scales. Thus, the effects of human influences on climate may either add or subtract from the severity of natural changes.

Human Influences on Climate

A number of human activities are influencing world climates in several possible directions.

Carbon dioxide. Carbon dioxide in the atmosphere acts to absorb radiation from the earth that would otherwise escape into space. By absorbing this radiation and reemitting it again, carbon dioxide keeps the earth's atmosphere warmer than it would otherwise be. Carbon dioxide in the atmosphere has been increasing since the late 1860s, paralleling the growth of industry. However, world temperatures increased only until the 1940s, when they began to drop. This could be the result of a natural cooling trend, which would be even more severe without the CO_2 effect. The addition of carbon dioxide to the atmosphere from the burning of fossil fuels is discussed in Chapter 20.

Particulates. The cooling trend just mentioned could also be due partly to the effects of particles in the atmosphere. Natural sources of airborne particles occur, notably volcanoes. When volcanoes erupt, they may spew huge quantities of dust high into the stratosphere, where it can remain for several years. These particles can reflect solar radiation away from the earth, causing a cooling of the atmosphere. In fact, notable cooling trends have been experienced several times after volcanic eruptions. Not all scientists agree that man-made particles (discussed in Chapter 21) contribute to a cooling trend. In fact, some feel that they may have a net effect of warming the atmosphere. (See the references at the end of this chapter for a more complete discussion of the many factors involved.)

Chlorofluorocarbons, too, may have effects on climate. These compounds, once used extensively as propellants in spray cans, appear to destroy ozone in the stratosphere (see Chapter 30) and also seem to absorb infrared radiation. This latter effect could cause atmospheric warming. What will the final outcome be? We don't know.

Whatever the outcome, it seems fairly obvious that humans are influencing climatic processes that are not

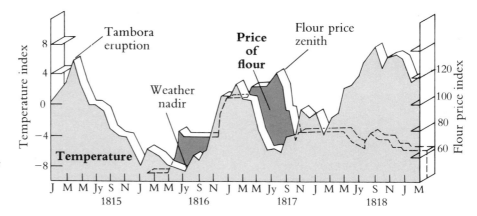

Figure 19.7 Effect of Tambora Eruption on Temperature and possible effect on the asking price for a sack of flour in the London, England, commodities market. The temperature indices are 5-month running sums of departures from the long-term mean for the corresponding month, using data from historical reconstruction of the central English climate by Manley (1959). Flour prices are from *Duffy's Farmers Journal*. This is not a fortuitous relationship, in all likelihood, because newspapers of the time describe the effect of inclement weather conditions on harvesting operations and crop growth following the eruption. Also, we expect a lag in the effect of weather on crop prices, because most countries have about a one-year carryover of grain in storage. (Source: K.E. Watt, *Principles of Environmental Science.* New York: McGraw-Hill, 1973.)

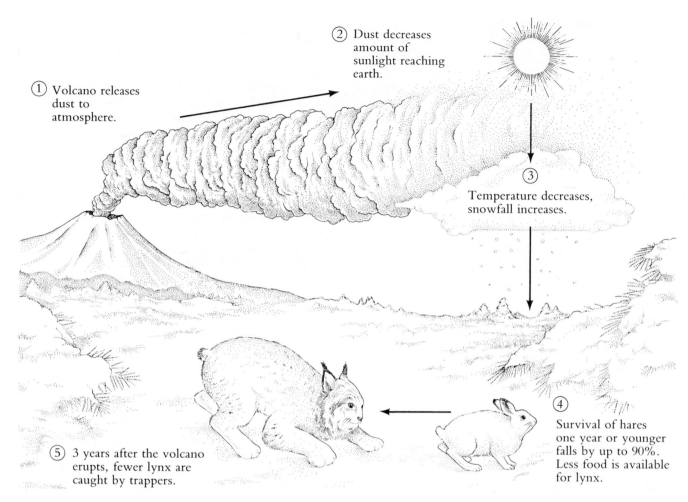

① Volcano releases dust to atmosphere.

② Dust decreases amount of sunlight reaching earth.

③ Temperature decreases, snowfall increases.

④ Survival of hares one year or younger falls by up to 90%. Less food is available for lynx.

⑤ 3 years after the volcano erupts, fewer lynx are caught by trappers.

Figure 19.8 The volcano–temperature–snow–hare–lynx–trapper effect. Note that the effect of the volcano was not seen in the lynx population until 3 years after the eruption. In a similar way, the effects of human-caused climate changes may not be seen immediately.

well understood in ways that are not completely clear. We are concerned about these possible changes because by a sort of domino effect, humans feel economic and social effects, and possibly even political effects, from changes in climate. For instance, Figure 19.7 shows how a volcanic eruption in 1815, by cooling and worsening weather conditions, probably caused the price of flour in London to double eighteen months later. Think of the social and political events that might follow this sort of climatic event.

Changes in climate affect agriculture because of the possibility of crop failures. Temperatures may be too low, more or less rainfall than normal may occur, or more severe storms may arise. Wildlife is also affected. For example, the survival of fish eggs or of young hares depends on certain temperatures. Furthermore, effects of climate on one species may, in turn, affect other organisms in food webs. An interesting example of this sort of delayed effect was pointed out by K. E. Watt (Figure 19.8).

Cities and climate changes. Infrared photos taken by satellites show another characteristic change in climate due to human activities: the urban heat island. That is to say, cities are warmer than the surrounding countryside. Partly this is due to changes in the surface of the land when countryside is changed to city. The concrete and asphalt areas of the cities absorb and hold heat better than soil and vegetation, and release the heat slowly at night. Another factor is waste heat from air conditioning and space heating. The pollution over cities, too, may contribute, as carbon dioxide, water vapor, and particulates absorb and reradiate some of the long-wave radiation from the city. This radiation would otherwise escape into space.

Cities also appear to increase the amount of rainfall over the city itself and even downwind of the city.

Acid rain. Another way that human activities are affecting climate involves a change in the quality of rainfall. The production of sulfur oxides (Chapter 21) and nitrogen oxides (Chapter 22) is causing rainfall to become more acid in many parts of the world. This acid rain then interferes with the growth and development of plants and animals.

Inversions

Finally, we return to the weather phenomenon with which we opened this chapter, because it clearly illustrates how human activities can interact with natural phenomena of weather and climate to cause serious ecological or health disturbances.

As noted in Chapter 2, we depend on the second law of thermodynamics to help dispose of air pollutants. That is, air pollutants emitted from smokestacks tend naturally to disperse and become diluted in air to nonharmful levels. Winds help speed this mixing, as do upward air currents, which carry pollutants aloft to mix into the upper atmosphere. A condition can arise, however, in which the air layers are very stable. The pollutants, instead of mixing into the upper atmospheric layers, are restricted to the layer of air close to the ground. There the pollutants accumulate to unhealthy levels. The term for this weather condition is an **inversion.** The term refers to the fact that the normal temperature situation in which air temperatures decrease with height does not exist. Rather, colder air sits beneath a warmer layer of air. In this situation, the colder air, which is heavier, does not rise up through the lighter air layer above it. Pollutants accumulate below this "lid" of warmer air.

Inversions commonly form during the cold clear nights of fall. On a clear day in the fall, the sun's rays heat the surface of the earth, which, in turn, heats the layer of air next to the surface. However, earth is a better radiator of heat than the atmosphere. Thus, during the night, the earth radiates heat into space. As the earth cools, it chills the layer of air next to the surface. By morning, a temperature inversion exists: cold air lies close to the earth, while the air above is still fairly warm. Once the sun rises, the earth's surface is warmed again, and so is the layer of air next to the surface. The inversion then disappears as the day advances. These so-called surface inversions tend to be shallow except where there are valleys. Here the cold air drains off the uplands during the night, forming deeper inversions that may be slow to disappear.

A more lasting inversion may occur as a result of the sinking of a high-pressure air mass (anticyclone). As such air sinks to lower altitudes, it becomes compressed and its temperature rises. This phenomenon often sandwiches a lower layer of cool air between the warm subsiding air and the earth's surface, giving rise to an inversion aloft. Sometimes both kinds of inversions can occur together.

Once the inversion is formed, pollutants begin to accumulate in the cooler lower layer. Autos carrying people to work and trucks carrying goods discharge carbon monoxide, nitrogen oxide, and hydrocarbons to the atmosphere. Industrial activities discharge these contaminants plus sulfur oxides and particulate matter. Concentrations in the cold lower air begin to increase even further. The sun attempts to work. Vertical thermal currents are set in motion, which rise until their temperature has reached that of the surrounding air. However, if the cold layer is thick and extends far above the surface of the earth, the currents may be unable to escape into the upper air. The concentrations of contaminants grow through the day as little pollution escapes. Sometimes, these conditions are repeated for several days. The level of contaminants becomes increasingly dangerous. Strong winds are needed desperately to break up the cool lower layer. Sometimes they arrive in time; sometimes, as in Donora, Pennsylvania, October 1948, they do not.

The following chapters, 20–22, explain the effects various air pollutants have on living organisms. Also explained are the ways in which pollutants are generated and how they can be controlled.

Questions

1. What are the major components and the important minor components of the earth's atmosphere?
2. What is the source of oxygen in the earth's atmosphere?
3. Differentiate between weather and climate.
4. Briefly describe the factors that influence the major climates on earth.
5. Are climates fixed and unchanging? Explain.
6. What is an inversion? How can inversions increase the effects of air-pollution emissions? Outline the process that results in the formation of an air-pollution inversion. Begin with a description of the "normal" temperature stratification of the earth's atmosphere.
7. Briefly describe human influences on climate. What do you think the final effect will be?

References

Lutgens, F. K., and E. J. Tarbuck, *The Atmosphere,* 2nd ed. Englewood Cliffs, N.J.: Prentice-Hall, 1982.

Kerr, R. A., "Climate Control: How Large a Role for Orbital Variations," *Science, 201* (14 July 1978), 144.

Levin, H. L., *The Earth Through Time.* Philadelphia: Saunders, 1978.

Watt, K. E., *Principles of Environmental Science.* New York: McGraw-Hill, 1973.

CHAPTER TWENTY

Air-Pollution Episodes, Carbon Dioxide, and Carbon Monoxide

Air-Pollution Episodes: The Awakening

Carbon Dioxide
The Carbon Cycle/Carbon Dioxide from Fossil Fuels/The Greenhouse Effect

Carbon Monoxide
Motor Vehicles: The Major Source/Carbon Monoxide and Hemoglobin/What Emissions Controls on Autos Will Bring

The Effects of Carbon Monoxide on the Body

CONTROVERSY:

20.1: *Should Society Protect the Unhealthy from Air Pollution?*

Air-Pollution Episodes: The Awakening

Strangely, our recognition on a national scale of the permanent threat of air pollution comes long after serious events in the past momentarily riveted our attention on these hazards. These events serve today as constant reminders of the consequences of doing nothing. As such, they are the touchstone of the clean air movement.

The earliest recorded incident of air pollution in the United States occurred in October, 1948. A blanket of cold air enveloped the town of Donora, Pennsylvania, and refused to rise. Pollutants were trapped in the stable air mass and began to accumulate. No air pollution sensors were present to signal the dramatic rise in pollution in the small valley town. Most of the residents carried on their normal occupations, but as evening approached, the evidence could no longer be ignored; people were dying. Twenty deaths that day and afterward were attributed to the fog, and America suddenly entered an era of awareness of air pollution. (For a fuller story of the air-pollution disaster at Donora, see Chapter 19.)

Four years later, in December, 1952, an air-pollution episode gripped the city of London for five successive days. More than 4000 deaths occurred in that interval as a result of the pollutants that hung in the air. These deaths, attributed to the "London Fog," were the deaths in excess of those normally expected to occur during that time of year. The elderly and those with chronic respiratory diseases were often the victims and suffered most. In the years following, episodes occurred again in London, but fortunately the tragic extent of the 1952 fog was never repeated.

Episodes have occurred in New York City, and alerts have been announced in other major cities in recent years. The mathematical science of statistics and the monitoring of air pollution have, however, provided new information. When our senses have failed to tell us of the decline in the quality of our air, chemical sensors have detected pollutants. When the effects on health have not been visible to the untrained eye, statisticians have shown deterioration of health at times of high pollution levels.

These episodes or incidents are not the result of sudden and vast discharges of pollution from many sources. That is, we cannot contend that they are caused by certain individuals or industries that have taken a momentary irresponsible action. In fact, industries have probably operated in a "business as usual" fashion. The pollutants present during an episode are typically only a portion of the normal output from domestic, industrial, and transportation activities.

Instead, an air-pollution episode represents a massive accumulation, a gathering together, of these ordinary pollutant discharges on a huge scale. The massive accumulation is the result of weather conditions—specifically, inversions that hinder the natural mixing of the atmosphere and prevent the natural dispersion and scattering of pollutants. (Inversions are described more fully in Chapter 19, page 378.) Because the emissions leading to and continuing during an episode are occurring at no more than a normal rate, if an episode happens once in a particular city, the potential exists for an episode to happen again. It will occur again when the weather conditions that trap pollutants occur again. The potential for an episode will only decrease if some positive action is taken to reduce the output of pollutants.

Sadly, the weather conditions that now threaten to bring on air-pollution episodes were regarded in the past as among the finest weather of the year (Figure 20.1). In rural areas, the fine clear days of fall that follow cool cloudless nights are still enjoyable. In polluted urban areas, however, such weather is a danger signal. Such weather extended over many days may lead to the inversions that act to trap air pollutants in the lower layers of the air. Weather conditions such as these subjected the town of Donora to a deadly concentration of pollutants more than three decades ago. Only the control or removal of air pollution sources can prevent episodes from happening again.

What are the air pollutants that so injure our health? Where do they come from? How do they harm us? How can we control them? In the sections that follow we shall describe the various air pollutants, their origins, their effects, and methods to decrease their discharges.

Carbon Dioxide

The Carbon Cycle

Burning of fossil fuels produces two oxides of carbon. One, carbon dioxide, is not poisonous, although it could change the earth's climate. The other, carbon monoxide, has known harmful effects on humans.

As we have said before (see Chapter 1), plants use carbon dioxide, and animals and plants produce it. Plants use carbon dioxide and produce oxygen during **photosynthesis.** Most plant and animal species consume oxygen in respiration. They also produce carbon dioxide as a waste product of that respiration. Hence, carbon dioxide is naturally present in the atmosphere in reasonable quantities. A normal sample of air contains about 0.03% carbon dioxide by weight.

Not only is carbon dioxide consumed in photosynthesis, but it also dissolves in the oceans. When carbon dioxide dissolves in the ocean, carbonic acid is produced at a *very low concentration.* The carbonic acid dissociates in part to bicarbonate and carbonate ions. These ions combine with calcium and magnesium (from the natural weathering of rocks), which have

Figure 20.1 The sunny days of fall are beautiful to behold in rural areas. In urban areas, however, these same days can threaten air-pollution episodes. If the air over a city remains windless for too many days, the pollutants from autos, industries, homes, and ships may become dangerously concentrated, to the point where human health can be harmed. (USDA—Fish and Wildlife Services; photo by Wilbrecht)

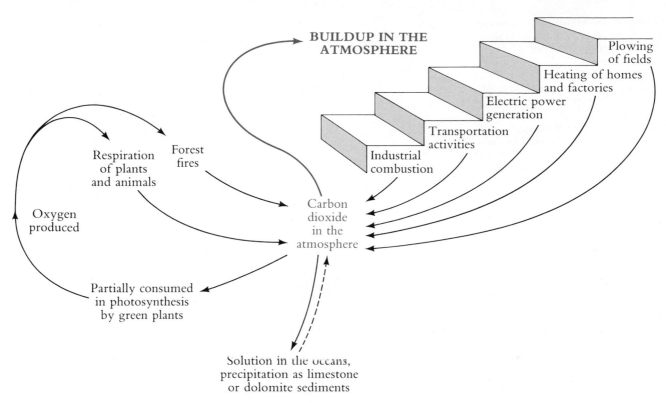

Figure 20.2 The Carbon Dioxide Cycle as Modified by Human Activity. If human activities are excluded, carbon dioxide is only added to the atmosphere by the respiration of plants and animals and by forest fires. It is removed from the atmosphere by photosynthesis and by solution in the oceans. The removal rate and the restoration rate were in rough balance before humans began burning fossil fuels. The combustion of fossil fuels has upset the balance and carbon dioxide is now accumulating in the atmosphere.

been carried into the ocean. The reaction of the calcium ion with carbonate ion produces calcium carbonate. We know this relatively insoluble substance as limestone. Magnesium and calcium may jointly react with the carbonate ion to produce dolomite. These precipitation reactions remove carbonate from the water, making room for more carbon dioxide to dissolve.

Thus, there are two natural mechanisms for carbon dioxide removal from the atmosphere:

1. solution in the ocean, followed by precipitation; and

2. the utilization of carbon dioxide by green plants in photosynthesis.

(See Figure 20.2.)

Carbon Dioxide from Fossil Fuels

Carbon dioxide is now being produced in massive quantities from the burning of fossil fuels. The complete oxidation of carbon produces the gas carbon dioxide:

$$C + O_2 \longrightarrow CO_2$$

In the early 1980s, the United States alone was burning annually about 600 million tons of coal in electric power plants. We also consumed annually a comparable tonnage of gasoline in that period. And the rest of the world consumes about twice the quantity of petroleum products that we do. This worldwide increase in combustion has increased the rate at which carbon dioxide enters the atmosphere. Carbon

Figure 20.3 Atmospheric concentration of carbon dioxide at Mauna Loa Observatory, Hawaii. Combustion of fossil fuels is producing carbon dioxide faster than green plants can use it and faster than the ocean can dissolve it. As a consequence, levels of carbon dioxide in the atmosphere are building up and will build up faster as combustion rates increase. Scientists suggest that carbon dioxide may prevent the earth from radiating heat back to space and so cause a warming trend. Whether the trend will occur and the ice caps melt are still large unknowns. (Source: Charles D. Keeling, private communication, April, 1983)

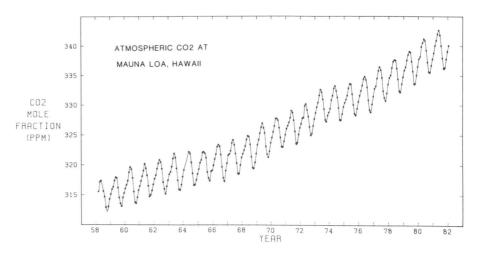

dioxide from fossil fuel combustion now adds about 5–7% annually to the amount of carbon dioxide production from green plants. The amount of carbon dioxide released annually to the atmosphere due to fossil fuel combustion has been estimated at about 18 *billion* tons. About half this additional quantity is accumulating. That is, it is not being consumed in photosynthesis, nor is it being dissolved in the ocean; it is remaining in the atmosphere. (See Figure 20.3.)

As time goes on, the amount of carbon dioxide produced annually from combustion is expected to grow enormously. As an example, one projection of the use of coal for electric power generation in the United States shows consumption doubling from 1980 to the year 2000.

The extent of the increase in fossil fuel combustion depends on energy conservation measures taken. It also depends on the extent to which nuclear and solar energy replace coal, oil, and natural gas. Total fossil fuel combustion might be expected to increase by 3–4% annually through the year 2000. On that basis, the amount of carbon dioxide in the atmosphere could potentially increase by as much as 25% from today's level by the year 2000. Another factor that could hasten the increase in carbon dioxide levels is the destruction of forests for agriculture, as is happening in countries like Brazil. Carbon dioxide intake by the plants in these forests would then decrease.

The Greenhouse Effect

For about three-quarters of a century, scientists have been aware of the buildup of carbon dioxide. It is now being carefully watched. The concern of thoughtful scientists is that carbon dioxide will trap heat in the earth's atmosphere. The effect of carbon dioxide has been compared to the effect of the glass panes of greenhouses. The panes let sunlight into the greenhouse, and the greenhouse is warmed by the solar radiation. At night, heat radiates away from the building. However, the glass panes decrease the rate of heat radiation out of the structure.

Carbon dioxide is expected to act in a similar way. Solar energy would continue to reach the earth without being affected by the gas, and the earth would be warmed. But the radiation of heat away from the earth would be slowed by the carbon dioxide. The earth would then heat up (Figure 20.4). How quickly would the earth be expected to heat up? How much? Could the polar ice caps melt?

From the turn of the century until World War II, a small warming trend was noted. Since about 1940, however, the trend has reversed, and a modest cooling has occurred that scientists feel is the definite trend of the earth's temperature at this time. A study that seemed to show a warming trend in the Southern Hemisphere was apparently incorrect in its mathematical approach. Even scientists who agree that there will be a greenhouse effect note that the warming trend suggested in the Southern Hemisphere was incorrectly calculated. Thus, we have as yet not seen the global increase in temperature that has been predicted. From appearances alone, the theory has not stood up.

Other factors may be operating, though. In the first place, the global cooling trend is regarded by some as being part of totally natural changes in the earth's climate and temperature, perhaps due to the gradual and cyclic changes in the earth's orbit. The trend may also be related to volcanic activity putting small particles high up in the earth's atmosphere.

In 1980, Mount St. Helens in the State of Washington spewed forth enormous quantities of volcanic dust. In 1982, El-Chichón, a volcano in Mexico, rocketed even more massive amounts of dust into the stratosphere. A volcano in Indonesia was also releasing volcanic dust into the atmosphere during this period. Such particles, suspended in the atmosphere, reflect sunlight away from the earth, robbing the earth of the heat that the sun's radiation would have provided. Years with extremely cold winters and cold summers followed the massive eruption of Mt. Tambora in Indonesia in 1815.

In 1982, meteorologists were almost unified in predicting a series of three or four very cold winters for the United States, beginning in that year. Their reasoning differed, but their conclusions were the same.

As with many long-range predictions, time did not immediately prove the forecasts correct. The winter of 1982 was colder than usual in some parts of the United States and warmer in other parts. Proof of the quality of the predictions would have to wait for future winters.

Particles stem not only from volcanic activity; particles are also discharged in abundance by our industrial society. Combustion produces not just carbon dioxide, but also produces atmospheric particles in enormous quantities. The fraction of sunlight reaching the earth has been decreasing, and presumably this has been due to particles suspended in the atmosphere.

Scientists who have studied the carbon dioxide question suggest that the cooling trend is, in fact, fortunate, for it affords us temporary protection from the earth beginning to heat up. Many scientists now agree on the ultimate effect of carbon dioxide in the atmo-

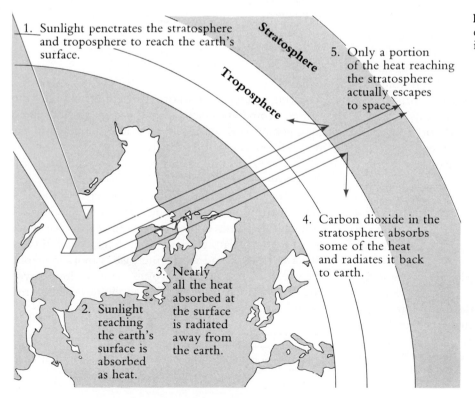

1. Sunlight penetrates the stratosphere and troposphere to reach the earth's surface.

Stratosphere

Troposphere

5. Only a portion of the heat reaching the stratosphere actually escapes to space.

Figure 20.4 The greenhouse effect of high levels of carbon dioxide on the heat balance of the earth.

4. Carbon dioxide in the stratosphere absorbs some of the heat and radiates it back to earth.

3. Nearly all the heat absorbed at the surface is radiated away from the earth.

2. Sunlight reaching the earth's surface is absorbed as heat.

sphere. A National Academy of Sciences committee of experts has estimated the impact of a doubling of atmospheric carbon dioxide levels. They see a global average surface warming of 3°C, with a 50% error to either side, as the result of such a doubling of carbon dioxide levels from 300 parts per million to 600 parts per million. From 1880 to 1980, the carbon dioxide level went from between 280–300 parts per million up to 335–340 parts per million, a 10–12% increase.

A conceivable figure for the percentage increase in the carbon dioxide level by the year 2000 is 25% from its current value. Such an increase could produce a change in the average worldwide temperature of about 1°C. This temperature change might be even larger in the polar regions because the haze of particles that surrounds our industrial civilization is not present. Furthermore, the melting of ice in the polar regions would reduce the reflectivity and increase the absorbance of light and heat in those regions. Thus, the polar regions are expected to react to changing carbon dioxide levels more quickly than other areas of the earth.

In fact, one pair of investigators compared satellite photos of the Antarctic ice pack with earlier maps and suggest that the summer ice pack in Antarctica had decreased by about 35%, or 0.9 million square miles (2.5 million square kilometers), from 1973 to 1980. No decrease in the Arctic ice pack was noted in the same period. Another pair of scientists estimates that 50,000 cubic kilometers of ice has melted since 1940, raising the global mean sea level an average of 3 millimeters per year during the period. Mean global temperature, however, decreased by 0.2°C during this interval. Such a quantity of polar ice melting requires the input of a vast amount of heat. Thus the ice sheets absorbing heat and melting could be delaying the onset of surface warming.

If the average Antarctic temperature were to increase 5°C, enough ice could melt in Antarctica to raise the sea level by 15–25 feet worldwide. That would put the city of New Orleans and large portions of Florida under water. Drought-prone regions in North America and Asia could also develop as global climate shifts.

When will the carbon dioxide effect be seen as a clear increase in temperature, distinct from other trends? Scientist J. M. Mitchell suggests that by the turn of the century, the upward trend of surface temperature will become apparent (see Figure 20.5). He also notes that "by the time the CO_2 disturbance will

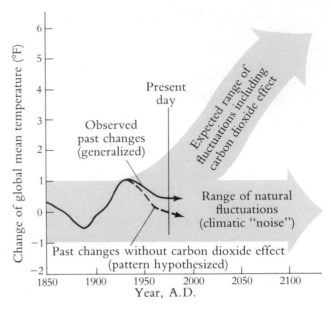

Figure 20.5 Time Trends of Global Mean Temperature with and without the Carbon Dioxide Effect. By about the year 2000, scientists expect to see the trend in mean global temperature begin to deviate significantly from the pattern it would normally have been expected to follow. (Source: J. Mitchell, "Workshop on the Global Effects of Carbon Dioxide from Fossil Fuels," published by the U.S. Department of Energy, May 1979, p. 98)

have run its full course, more than a millenium into the future, temperature levels in the global atmosphere may possibly exceed the highest levels attained in the past million years of the earth's history."[1]

To be fair, some scientists, lead by S. B. Idso, see the predictions as grossly overstated. Their physical measurements produce results in conflict with the predictions from the global models of carbon dioxide and temperature. Built-in mechanisms may also take effect to soften the heating trend. One likely possibility is that a warming trend would increase evaporation from the oceans. This, in turn, would increase cloud cover and decrease the quantity of sunlight reaching the earth.

---

1 J. Mitchell, "Workshop on the Global Effects of Carbon Dioxide from Fossil Fuels," published by the U.S. Department of Energy, May 1979, p. 98.

Carbon Monoxide

Motor Vehicles: The Major Source

When carbon is not oxidized completely, the colorless and odorless gas **carbon monoxide** results. Carbon monoxide surrounds the city dweller in concentrations greater than any other air pollutant. Because the gas is colorless and odorless, however, we cannot detect it with our senses. It cannot be seen; it cannot be smelled; but it is present nevertheless.

The greatest source of carbon monoxide in our cities is the motor vehicle. Over 120 million vehicles are on the road in the United States, discharging the gas into the atmosphere. In most cities, over 90% of the carbon monoxide in the air comes from the incomplete combustion of carbon in motor fuels. The reaction is:

$$C \underset{\substack{\text{carbon in} \\ \text{motor fuel}}}{} + \tfrac{1}{2}O_2 \longrightarrow CO$$

Whereas incomplete combustion of carbon produces carbon monoxide, complete combustion of carbon produces carbon dioxide (CO_2) as the end product.

There is another source of carbon monoxide to which people are exposed; but only smokers and their neighbors have this special privilege. We can compare the individual who smokes moderately and lives in a clean environment with an individual who does not smoke and lives in a highly polluted environment. The smoker absorbs daily twice as much carbon monoxide as his nonsmoking counterpart.

Carbon Monoxide and Hemoglobin

In the bloodstream, carbon monoxide competes with oxygen for the hemoglobin molecule that carries oxygen to the body's cells. For this reason, at high enough concentrations, it is a deadly poison. Hemoglobin, a complex protein present in blood, transports oxygen from the lungs to the cells and carries carbon dioxide from the cells back to the lungs. Carbon monoxide, however, attaches more strongly to hemoglobin than oxygen does. The more carbon monoxide present in air, the more hemoglobin is "tied up" and the less oxygen can reach the cells. (What happens to you when your hemoglobin is tied up? See page 391.)

In workplaces such as tunnels and loading platforms, carbon monoxide may reach concentrations of 70 milligrams per cubic meter. After 8–12 hours at such a concentration, a worker may lose the service of 10% of his hemoglobin. If an individual were exposed to an average of 16 milligrams per cubic meter for an 8-hour period, about 3.0% of his hemoglobin would be unavailable for oxygen transport. This level is not uncommon on city streets. At this point, the impact of smoking on the availability of hemoglobin can be seen. Even the *light* smoker ties up 3.0% of his hemoglobin *by smoking alone*. This is equivalent to being in a room with a carbon monoxide concentration of 16 milligrams per cubic meter. The moderate smoker achieves nearly a 6% loss of useful hemoglobin.

The standard for carbon monoxide in the atmosphere is 10 milligrams per cubic meter, averaged over 8 hours. This level is not to be exceeded more than once a year. We can see that smoking alone produces a personal environment worse than that required by the air-quality standards. In 1977, a Los Angeles suburban resident or a resident of Phoenix, Arizona, would have experienced about 50 days per year on which carbon monoxide levels exceeded the standard. A resident of Portland, Oregon, would have experienced nearly 60 days in which the standard was exceeded, and a Seattle, Washington, resident nearly 90 days. In 1976, 120 days exhibited levels above the standard in Los Angeles. New York City, specifically Manhattan, was even worse.

The effects of carbon monoxide at low levels are subtle and difficult to determine. When 3% of the hemoglobin is bound by carbon monoxide, tests have shown that the ability of individuals to judge differences in light intensities decreases. In experimental situations that produced levels of 10% unusable hemoglobin, the skills needed to drive a car were impaired; responses to brake lights and to the speed of the auto ahead were poorer. The possible influence on safety is obvious.

The effects of low levels of carbon monoxide on health have been surmised from experimental data rather than from observing the health status of people and relating their health to levels of carbon monoxide in the atmosphere. The use of experimental data is necessary because when outdoor carbon monoxide concentrations are high, the concentrations of other pollutants in the air are typically also high, and the effects cannot be separated.

Should Society Protect the Unhealthy from Air Pollution?

Would it not be better to close the EPA and buy each person sensitive to carbon monoxide a condominium in Key West?

Paul MacAvoy,*
Professor of Economics, Yale University

Should the entire populace assume the burden of preventing aggravation of a disease in a relatively small group of people who unfortunately live in large cities?

R. Jeffrey Smith*

In setting standards . . ., we must be concerned with the health effects on the most vulnerable in our population rather than upon the healthy groups.

Edmund Muskie, U.S. Senator from Maine†

For over a decade, we have sought to protect particularly sensitive citizens, such as children, the aged and asthmatics from polluted air. I don't think the American people would stand for abandoning these sensitive populations by misguided use of cost/benefit analyses. . . . It is useless to pretend that some kind of utilitarian calculus can give us the answers to what are essentially moral and political questions.

Henry Waxman, U.S. House of Representatives‡

The [briefing] paper [from President Carter's Council of Economic Advisors] says that EPA should not base its policy [on smog] on the problems of marginal people but should try to make its decision "consistent with the preferences and behavior of the general population."§

As originally written, the Clean Air Act required the Environmental Protection Agency to set air-quality standards at levels that would protect the public health "with an adequate margin of safety." The "public health" meant not only the health of the general population, but the health of those most susceptible to damage from air pollutants. Those with chronic respiratory disease, such as emphysema and asthma, and those with heart conditions, such as angina, are among the people most susceptible to ill effects from air pollutants. Children, because their rate of breathing is more rapid than that of adults, in some cases show greater susceptibility to air pollutants. The original Clean Air Act assumed that protection of the most sensitive segment of the population was a reasonable proposition, but this assumption has been challenged by industry and economists, who claim some substances are too expensive to control. The National Commission on Air Quality does not agree with industry about costs, but notes that: "The costs of meeting primary air quality standards are best taken into account in determining what control programs should be implemented in specific areas of the country, not in establishing a national air quality standard to protect public health."[||]

MacAvoy is suggesting that carbon monoxide control is so costly that the solution is to relocate people sensitive to carbon monoxide. It is now clear that carbon monoxide can aggravate *angina pectoris,* an extremely common form of heart disease, at levels in the air not far above the air-quality standard. MacAvoy is suggesting that it costs too much to protect people with angina. His solution is a restatement of the sentiment of conservative economists: "Let them vote with their feet," which translated means: "If they can't handle poor air quality from a health standpoint, they should move."

There are two issues here in suggesting that people move if they don't like air quality. The first is the cost of moving. Do all people have the personal wealth to move? Are there social costs as well as economic costs involved in moving? What are the social costs? Are the social costs higher to the elderly? The second issue is the choice of where to move. An example of the risks in choosing a destination occurred in a family known to the authors. Because of his heart condition, the father moved to an area noted for clean air. But Tucson, Arizona, went in 15 years from one of the cities with the cleanest air to one of the most polluted cities in the nation (see Table 21.2, page 408). If pollution is not controlled everywhere—if the most susceptible are not protected everywhere—there may well not be a safe place to move.

Does anyone in your family have angina? Do you think they need to be protected by safe air quality? Would they consider moving? If they would consider moving, is their destination safe today and in the future?

Two other common diseases that make people susceptible to poor air quality are emphysema and asthma. Sulfur dioxide particularly has an impact on people with these lung conditions. Asthma is a disease that often begins in childhood. What bearing does this have on MacAvoy's suggestion that afflicted people move?

---

* *Science,* **212** (June 12, 1981), p. 1251.
† "Air Pollution, 1970", Part I, Hearings Before the Subcommittee on Air and Water Pollution of the Committee on Public Works, U.S. Senate, March 16, 17, 18, 1970, p. 74.
‡ Joint Hearing Before the U.S. Senate Committee on Environment and Public Works and the U.S. House of Representatives Committee on Energy and Commerce, March 2, 1981, p. 59.
§ Quoted from *Science,* **202** (December 1, 1978), p. 949. The air-quality standard was loosened in 1979 (see Chapter 22, p. 430).
|| *To Breathe Clean Air,* Report of the National Commission on Air Quality, 1981, p. 8.

What Emission Controls on Autos Will Bring

The initial control of carbon monoxide emissions was achieved by increasing the ratio of air to fuel in the gasoline engine. Additional air is provided to burn the gasoline more completely, resulting in less carbon monoxide in the exhaust. Some vehicles also utilized "air injection." Air was mixed with the exhaust stream for a final combustion of the carbon monoxide to carbon dioxide.

$$CO + \tfrac{1}{2}O_2 \longrightarrow CO_2$$

The typical motor vehicle of the mid-1960s exhausted an average of 73 grams of carbon monoxide in every mile of travel. A standard of 23 grams of carbon monoxide per mile was set for 1971 vehicles, and tests by the Environmental Protection Agency established that new passenger cars met the standard easily. Carbon monoxide emissions of 3.4 grams per mile were achieved by new autos in 1981.

To achieve this standard, the exhaust gases are mixed with a stream of air in the presence of a catalyst. Further oxidation of the remaining carbon monoxide occurs in this **catalytic converter.** The catalyst system appears for the present to be the chosen method of reducing carbon monoxide emissions. Nevertheless, auto manufacturers are probably studying the potential of the stratified-charge, dual-carburetor engine system. This system has been used in the Honda, a vehicle of Japanese manufacture, and has achieved the required reduction in carbon monoxide.

To assess progress in carbon monoxide control more precisely, we need to compare in a number of areas the actually occurring levels of the gas against the atmospheric standard that has been set for it. The standard is 10 milligrams per cubic meter averaged over eight hours, not to be exceeded more than once per year.[2]

---

2 In 1982, the Environmental Protection Agency proposed allowing violations up to five days per year, a significant loosening of the standard.

Twenty-eight areas of the country were not expected to meet this standard in 1982. Of these 28 areas, 13 heavily populated areas were expected to have average carbon monoxide levels more than 50% above the standard in 1981. These areas were Portland, Oregon; Washington, D.C.; Los Angeles; Denver; Seattle; Phoenix; San Jose; the New York Metropolitan Area; Cleveland; Pittsburgh; Atlanta; Chicago; and Louisville. These are all urban areas, but growing urbanizing areas may also experience problems meeting the standard. Experts hope that compulsory inspection of automobile emission control equipment, followed by maintenance as necessary, will help prevent such problems, as well as assist the larger urban areas in meeting the standard.

By 1987, the only violations of the carbon monoxide standard should be occurring in Los Angeles and Denver, but this assumes that inspection and maintenance of air pollution control equipment will be re-

quired. If inspection and maintenance are not required, all thirteen of the areas listed above could still be exceeding the carbon monoxide standard in 1990. (See Figure 20.6.)

Progress in carbon monoxide control must be considered mixed. Even though the allowable emissions per mile for new cars were being steadily pushed down during the decade, total tonnage of carbon monoxide emissions fell only a modest 5% between 1970 and 1979. By 1979, the average, taken over all cars, of carbon monoxide emissions in grams per mile had fallen by about one-third from the 1970 level. A 35% increase in vehicle miles travelled, however, prevented any substantial reduction of the total tons emitted.

Nonetheless, the decrease in emissions per mile did show up at sites in a number of cities. At 223 urban monitoring sites that were already saturated with traffic in 1972, average levels of carbon monoxide in the air dropped from about 3.4 to 2.2 milligrams per cubic

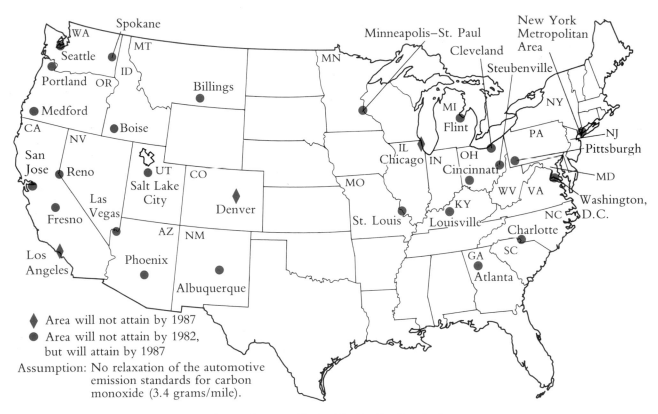

Figure 20.6 Status of Compliance with the Air Quality Standard for Carbon Monoxide. Assumption: No relaxation of the automotive emission standards for carbon monoxide (3.4 grams/mile). (Source: National Commission on Air Quality, 1981)

meter of air from 1972 to 1979. That is, traffic volumes did not increase appreciably at these sites during this time, and the reduction in emissions was seen directly in a decline in atmospheric levels of carbon monoxide. (See Figure 20.7.)

It is disturbing that total emissions are not declining, even though a reasonable control of vehicle emissions is being achieved. If emissions are going down at sites where traffic is remaining at the same level as earlier, emissions are likely to not be declining so clearly at sites where traffic levels are increasing. And carbon monoxide levels are likely to be increasing *somewhere* if vehicle miles are rising so quickly. Thus, substantial across-the-board progress is still elusive as long as vehicle miles are increasing. The failure of emissions control to bring down total carbon monoxide tonnage suggests strongly that other means may be necessary to decrease total emissions. One such method is to build and operate superior public transportation systems.

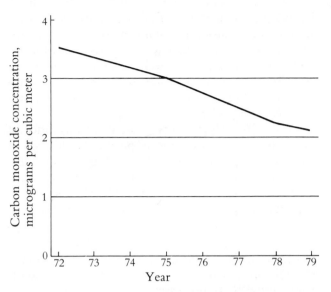

Figure 20.7 Carbon monoxide levels through time, as averaged over 223 urban monitoring sites. (Source: "Trends in the Nation's Air Quality," EPA, October 1980)

The Effects of Carbon Monoxide on the Body

In an atmosphere rich in carbon monoxide, death results from asphyxiation. This is another way of saying that the body tissues become starved for oxygen. Scientists have suspected for more than a decade that the concentrations of carbon monoxide found in our cities are harmful. Nevertheless, it has only been in the last few years that the needed data have been obtained to brand carbon monoxide in the air as a serious health hazard.

Carbon Monoxide Disrupts Oxygen Transport in the Body

To understand the danger of small concentrations of carbon monoxide, we need to review the process that supplies oxygen to the tissues of the body. Oxygen, a component of the air, is brought into the lungs with each breath. (An illustration of the lungs is provided in Figure 31.1.) At the alveoli, the small sacs at the end of the tree-like branches of the lungs, the oxygen gas is transferred into the bloodstream. In the blood, the oxygen attaches to hemoglobin, a complex protein mole-

cule carried in red blood cells (erythrocytes). The red cells transport the oxygenated hemoglobin through the arteries of the body and finally to the capillaries, the narrowest tubes of the arterial circulatory system. Here, oxygen is transferred across the walls of the capillaries into the cells.

Carbon dioxide, one of the waste products of cell activity, flows in the opposite direction, from the cell into the bloodstream. Some of the carbon dioxide takes the place of the oxygen that was attached to hemoglobin, and some of the gas dissolves in the blood fluid as bicarbonate ion. The blood, now rich in carbon dioxide, returns via the veins to the lungs. There, carbon dioxide diffuses from the blood into the alveoli, while oxygen from the air in the sacs moves into the blood. The carbon dioxide is then exhaled in the breath.

This normal pattern of transportation is disturbed when carbon monoxide is present in the air we breathe. Even minute quantities can disrupt oxygen transport, because carbon monoxide is about 200 times more attractive to hemoglobin than is oxygen. The carbon monoxide attaches tightly to the hemoglobin, depriving oxygen of its carrier to the cells. The greater the amount of carbon monoxide present in the air, the more hemoglobin is "tied up" and unavailable for oxygen transport. When hemoglobin binds carbon monoxide to itself, it is called *carboxyhemoglobin*. In contrast, when hemoglobin has oxygen attached to it, it is called *oxyhemoglobin*.

Table 20.1 shows how even small quantities of carbon monoxide gas in the air produce high levels of carboxyhemoglobin in the blood. Note that the table lists carboxyhemoglobin percentages after 8–10 hours of breathing contaminated air. This level is termed the "equilibrium value." A longer exposure at that concentration will not increase the percent of carboxyhemoglobin any further. Also note that even when the air is free of carbon monoxide, some small amount of hemoglobin is tied up. This carbon monoxide stems from natural body processes.

Effect of Low Carbon Monoxide Levels

Individuals with increased levels of carbon monoxide-bound hemoglobin are subject to two important effects. One effect is a decreased ability to perceive one's personal environment. This perception has been measured by a number of tests. For instance, individuals have been asked to report on sound signals. At levels of carboxyhemoglobin in the range of 2–5% of total hemoglobin, the signals were often missed. The ability to tell which of two tones was longer in duration decreased when carboxyhemoglobin was in the range of 2.5–4%. The processes of the mind are also interfered with. Simple tests, such as adding columns of numbers, took longer to complete as carboxyhemoglobin levels increased. The ability to distinguish when a light becomes brighter also decreased. Tests of brightness perception showed fewer correct answers even when carboxyhemoglobin levels were as low as 3%.

No formal link has yet been established between elevated carbon monoxide in the outdoor atmosphere and driving performance. Nevertheless, carbon monoxide levels may rise to 60 milligrams or more per cubic meter of air in freeway traffic, producing carboxyhemoglobin levels close to those at which skills are impaired. One scientist found higher carboxyhemoglobin levels in individuals involved in auto accidents, but we are still unable to blame carbon monoxide as a cause of auto accidents.

The past decade has brought us much new information on carbon monoxide's relation to heart at-

Table 20.1 Small Amounts of Carbon Monoxide in the Air Can Remove a Great Deal of Hemoglobin from Service

Level of carbon monoxide per cubic meter of air (milligrams)	Estimated percentage of hemoglobin tied up as carboxyhemoglobin after 8–10 hours[a]
0	0.4
5	1.0
10	1.6 (1972 air-quality standard and resulting carboxyhemoglobin)
20	2.9
30	4.1
40	5.4
50	6.6
60	7.8

a This is the percentage of hemoglobin to which carbon monoxide is attached. This hemoglobin is not available to carry oxygen to the cells.

tacks. Earlier, medical scientists suspected that carbon monoxide might be a cause of heart attacks because they found that more heart attacks occurred during periods of high carbon monoxide concentrations. Because other pollutants were also at high concentrations at the same time, no firm conclusions could be drawn. Now the suspicion is being substantiated in new data from individuals with angina pectoris.

Angina pectoris is a chronic form of heart disease characterized by chest pain; it is less severe than an acute heart attack, which is life-threatening when it occurs. Individuals who suffer from angina pain have been tested for their susceptibility to carbon monoxide. They first were asked to breathe air with carbon monoxide at levels sufficient to raise their carboxyhemoglobin to 3%; then they were asked to exercise. At this blood level of carboxyhemoglobin, when the subjects were stressed, the onset of angina pain arrived sooner than expected under normal conditions; furthermore, the pain continued longer than expected. Data reported in 1981 now indicate that even a 2% level of carboxyhemoglobin will aggravate angina.

Angina is only one form of heart disease. In total, 35% of the annual deaths in the United States are attributed to some form of heart disease. Carbon monoxide is known to decrease the supply of oxygen to the tissues. One tissue that is very frail if deprived of oxygen is the myocardium (muscle tissue of the heart). Experiments with angina patients tend to support the contention that carbon monoxide may be an agent in causing heart attacks. (We do not say that carbon monoxide causes heart disease itself, which is most frequently defined as a narrowing of coronary blood vessels.)

Another factor supports the idea that carbon monoxide is one of the villains in bringing on heart attacks. Smoke inhaled from a cigarette may contain as much as 4% carbon monoxide. Heavy smokers may have

carboxyhemoglobin levels as high as 10–15%. One study showed an average carboxyhemoglobin level of 4.4% in smokers of all types who were not exposed to carbon monoxide in their job. We know from statistical studies that when people quit smoking, their risk of heart attack is quickly reduced. It appears that a causative agent has been removed from their personal environment. That causative agent may be carbon monoxide, but other substances are absent as well when they stop smoking, so that we are still uncertain of the relationship.

Questions

1. Regions A and B experience air-temperature inversions with the same frequency. Describe at least three differences between the two regions that could cause region A to experience air-pollution episodes much more often than region B.
2. Name and describe the two natural processes by which carbon dioxide is removed from the atmosphere. What factors account for the fact that the total quantity of carbon dioxide in the atmosphere is increasing? Discuss the possible effects of this accumulation.
3. What is the major source of carbon monoxide in urban areas?
4. What process in the human body is hampered by carbon monoxide?
5. What health effects does carbon monoxide have?
6. What measures have been taken to reduce carbon monoxide emissions by motor vehicles? Are these measures working?

Further Reading

General References

Controlling Air Pollution, American Lung Association, 1974.
A very well-done short book on methods to control air pollution from stationary and mobile sources. Drawings and text are excellent. Writing level is for the layperson.

Waldbott, George, *Health Effects of Environmental Pollutants,* 2nd ed. St. Louis: C. V. Mosby Co., 1978.
This well-illustrated book (136 pages of illustrations) has a wealth of information and documentation on all of the major air pollutants. It is a technical work, but clearly written. The book could serve as an excellent reference for the health aspects of air pollutants.

Carbon Monoxide References

Air Quality Criteria for Carbon Monoxide, U.S. Environmental Protection Agency AP-62, 1970.

Control Techniques for Carbon Monoxide, Nitrogen Oxide and Hydrocarbon Emissions from Mobile Sources, U.S. Environmental Protection Agency AP-66, 1970.
While technical, these documents are very clearly written and very complete.

Carbon Dioxide References

Woodwell, G., "The Carbon Dioxide Question," *Scientific American,* **238(1)** (January 1978), 34–43.
A clear exposition by a research scientist in the field of the global carbon cycle.

Lovelock, J. E., *GAIA: A New Look at Life on Earth.* New York: Oxford University Press, 1982 (available in paperback).
This is a fascinating book for a mature student with good biology and chemistry training. Lovelock argues for the evolution of the earth's atmosphere to a remarkably stable and self-correcting medium that has evolved a special relationship with life on earth.

Kuklas, G., and J. Gavin, "Summer Ice and Carbon Dioxide," *Science,* **214** (October 30, 1981), 497–503.
This is the technical article dealing with the discovery of the melting of the Antarctic icepack.

Kellog, W., and R. Schware, *Climate Change and Society: Consequences of Increasing Atmospheric Carbon Dioxide.* Boulder, CO: Westview Publ. Co., 1981. 178 pages (available in paperback).
An atmospheric scientist and political scientist teamed to produce this small but well-done volume with an international perspective.

Breuer, George, *Air in Danger—Ecological Perspectives on the Atmosphere.* New York: Cambridge University Press, 1980. 189 pages (available in paperback).
While carbon dioxide is a principal focus, other air pollutants with global consequences are discussed. Recommended for the nonscientist.

CHAPTER TWENTY-ONE

Sulfur Oxides, Acid Rain, and Particulate Air Pollutants

Sulfur Oxides
Sulfur Comes from Coal and Oil/Oxidation of Sulfur to a Corrosive Mist/Effects on Materials, Plants, and People/Ways to Reduce Sulfur Dioxide Emissions/Progress in Control of Sulfur Oxides

Acid Rain
Description and Sources/Biological Impacts of Acid Rain/Other Effects of Acid Rain/What Makes Some Lakes and Soils Sensitive to Acid Rain?/Control of Acid Rain

Particulate Matter
Particles Stem from Combustion and Industrial Activities/Effects of Particulate Air Pollutants/Ways to Control Particle Emissions/Progress in Control of Particulate Matter: An Uphill Battle

Lead Compounds in the Air
Sources of Lead Air Pollutants/A Special Hazard to Children

Trends in Air Quality—Challenge in Display

The Health Effects of Sulfur Oxides and Particles

Children and Lead Poisoning

CONTROVERSY:

21.1: *Costs of Air Pollution Control*

Sulfur Oxides

Sulfur Comes from Coal and Oil

Sulfur compounds in the air derive mainly from the burning of sulfur-rich fuels, such as coal and heating oils. These **sulfur oxide** compounds pollute the air in many sections of the country (see Table 21.1), but the fuels from which the sulfur came are needed to produce heat, generate electricity, and power machinery.

Not all fuels contain significant quantities of sulfur. Some coals may have as little as 0.5% sulfur, while other coals may contain as much as 6% sulfur. Coal is used extensively in steel manufacture, but its primary use is to generate steam to produce electricity. While the average sulfur content of the coals used in electric power generation is on the order of 2.5%, the coal used in making steel must be much lower in sulfur. In fact, long-term contracts by the steel industry tie up a large proportion of the low-sulfur coal in the United States.

For every million metric tons of coal burned in electric power generation, about 25,000 metric tons of sulfur are released. Of course, the sulfur is not released in the elemental form, but mainly as sulfur dioxide gas. In 1981, about 597 million metric tons of coal were burned for the sole purpose of generating steam to produce electricity. How many tons of sulfur were released?

Sulfur is also present in crude petroleum (raw or unrefined), but probably at less than 1% sulfur. Refining coaxes much of the sulfur out of the common petroleum products such as gasoline and kerosene. The waste sulfur compounds are "flamed" at the refinery;

Table 21.1 Monitoring Sites with Very High Mean Annual Sulfur Dioxide Concentration and Large Populations, 1978

	Annual mean concentration (micrograms[a] per cubic meter)	Maximum 1978 averages over each 24-hour day (micrograms[a] per cubic meter)
Pittsburgh, Pa. (2 sites)	140	602
Magna (suburb of Salt Lake City), Utah	93	811
Toledo, Ohio	84	915
Buffalo, N.Y.	78	267
New York City, N.Y.	77	296

a A microgram is one one-millionth of a gram.
(Source: U.S. Environmental Protection Agency)

they are burned to sulfur oxides at the top of a tall metal stack. Gasoline and kerosene, then, are responsible for only a small portion of the sulfur compounds in the atmosphere. In the United States, probably less than 5% of the yearly release of sulfur compounds comes from the burning of gasoline. Home heating oil is also relatively low in sulfur, having an average sulfur content of about 0.25%.

In the refining process, much of the sulfur is shifted to residual oil, one of the most dense of the refining products. Anywhere from 0.5–5% of residual oil may be sulfur, although special refining steps may be used to reduce the sulfur content further. Alternatively, low-sulfur crude oil may be used at the outset so that a low-sulfur residual oil is produced. Residual oil is used to heat apartments and institutions such as schools and hospitals. In the past, residual oil was once widely used in boilers to produce steam for electric generation; utilities chose residual oil then because of its relatively low cost and its low sulfur content compared to coal.

Natural gas, in contrast to coal and oil, is almost free of sulfur. From this standpoint, it is an environmentally useful fuel. In 1978, burning coal and oil in electric generating plants produced 19.4 million tons or 67% of the total sulfur oxides (28.8 million tons) emitted that year. Other coal and oil burning accounted for half of the remainder. Sulfur compounds also enter the air from the smelting and refining industries, as well as from other sources.

Oxidation of Sulfur to a Corrosive Mist

When coal or oil is burned, the sulfur in the fuels is oxidized. Two compounds are formed: sulfur dioxide and sulfur trioxide. Less than 3% of the sulfur oxidizes to the trioxide form during the initial burning. The remainder, the primary form in which sulfur enters the atmosphere, is sulfur dioxide. The sulfur dioxide in the air is oxidized gradually by oxygen in the air to the trioxide. Any sulfur trioxide formed will immediately react with water vapor to yield sulfuric acid. The sulfuric acid is present in the air as a fine mist of liquid droplets. This very corrosive mist eats away many materials, including building materials such as marble and mortar.

Sulfur dioxide produced in combustion also reacts with water vapor in the atmosphere to produce sulfurous acid. Sulfurous acid is a weak acid that also reacts with oxygen to produce sulfuric acid. Thus, sulfuric acid is produced by two routes. The route that includes sulfur trioxide is the major route when air is dry. The route that includes sulfurous acid is the major route when humidity is high.

Oxides of calcium and iron are also formed in the process of combustion. These oxides enter the air in great quantities when coal is burned, but particles from the burning of oil are far less numerous. The oxides often react with sulfuric acid to yield calcium and iron sulfate in particle form. That is, calcium oxide may react with sulfuric acid to produce calcium sulfate and water. Iron oxides may react in a similar fashion.

The reactions are summarized in Figure 21.1.

Such sulfate particles plus sulfuric acid droplets may account for 5–20% of the particulate matter found in urban air. It is thought that much of the sulfur dioxide is converted to the sulfate and sulfuric acid forms within a few days from the time of its emission. During this time, winds can transport the pollutants hundreds of miles.

Effects on Materials, Plants, and People

Many materials may be attacked by the mists of sulfuric acid, among them carbonate building materials such as marble and mortar. Calcium sulfate, water, and carbon dioxide are produced in the reaction. Metals such as steel, copper, and aluminum are also corroded, and common cloth fabrics are damaged as well.

High concentrations of sulfur dioxide and its derivatives may cause acute injury to plants. After such acute injury, where sulfur dioxide concentrations may reach nearly 3000 micrograms per cubic meter, leaves, needles, and other plant tissues may appear as though bleached. Eventually, leaves or needles may take on a scorched, reddish-brown look and the damaged foliage falls from the plant. Smelters that refine copper ores and lead ores containing sulfur may pollute extensive areas with their gaseous emissions. A near absence of vegetation may be observed in the amazing scene of devastation round the International Nickel Company smelting operation at Sudbury, Ontario, for example.

Even where sulfur dioxide levels average only on the order of 100 micrograms per cubic meter, such as occur in cities, plants may have a yellow discoloration. Fruit trees such as apple and pear and forest trees such as ponderosa pine and tamarack (larch) are susceptible to damage from sulfur oxides. The cotton plant is also susceptible, as are alfalfa and barley.

Elevated levels of sulfur oxide have been linked to human illnesses and even to deaths. In air-pollution episodes in New York, Osaka, and London, investigators have noted an increase in the normal death rate following periods of high concentrations of sulfur oxides. Effects on human health of sulfur oxides and particulates are difficult to separate because the two kinds of pollutants tend to interact. The effects of the two together are discussed in the supplementary material on p. 414.

Respiratory illnesses such as bronchitis are seen to

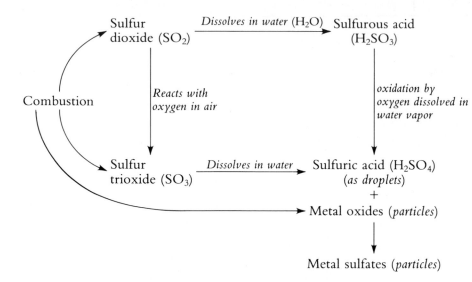

Figure 21.1 The reactions of sulfur in air.

increase with sulfur oxide levels. One study found that the illness rate increased in an area where the average annual concentration was only 100 micrograms per cubic meter, a level not far from observed concentrations in a number of areas in the United States (see Table 21.1).

To understand how severe sulfur oxide air pollution has been in the past and where we stand today in its control, we need to describe the air-quality standards for these substances. The daily standard for sulfur oxides set in 1971, based on the maintenance of human health, is 365 micrograms per cubic meter (an average over 24 hours), and this level is not to be exceeded more than once per year. An air pollution "alert" is to be called when the 24-hour average exceeds 800 micrograms per cubic meter. The annual standard, the average over one year, is 80 micrograms per cubic meter.

Rural areas have background concentrations of about 0.5 micrograms per cubic meter, but urban areas have levels 50–1000 times greater. The people of Chicago, for example, lived at an *average* concentration of 470 micrograms per cubic meter in 1964. Levels in Chicago have since decreased considerably under the stimulus of the air-quality standards issued in 1971.

In the famous London Fog of 1952, the concentration of sulfur oxides reached 4000 micrograms per cubic meter. In a 1962 episode in New York, the average concentration during one day was 2500 micrograms per cubic meter. Such concentrations are still occurring in some places in the world. In Ankara, Turkey,

in January, 1982, sulfur dioxide levels reached 2835 micrograms per cubic meter during a period of several days. All industries and business, except hospitals and bakeries, were ordered closed in an attempt to reduce emissions. Ankara's air pollution problem is chronic because it uses much high-sulfur lignite coal and high-sulfur fuel oil.

Sulfur oxides have clearly been shown to aggravate and cause respiratory problems by increasing resistance to air flow in the pulmonary tract. That alone is reason enough to control sulfur oxides emissions. Sulfur dioxide may, however, have an additional uncounted effect. The gas has been shown to be a cocarcinogen in rats. In the presence of benzo(a)pyrene, sulfur dioxide increases the frequency of cancer occurrence above the natural rate. At the moment, we can say that sulfur dioxide has not been proven to be either a carcinogen or a cocarcinogen in humans, but it is suspect.

Ways to Reduce Sulfur Dioxide Emissions

Cleaning up coal itself. In addition to the possibility of using expensive low-sulfur coal, emissions of sulfur dioxide from power plants can be reduced by cleansing the coal of its sulfur prior to use. Two forms of sulfur occur in coal: inorganic sulfur and organic sulfur. Inorganic sulfur is the sulfur present as pyrites; these are the metal sulfides, such as iron sulfide (also

Costs of Air Pollution Control

The good and sweeping intentions of many environmentalists are now an obstacle blocking those less-fortunate Americans who desire economic justice.

Bayard Rustin,
The New York Times, 11 August 1975

. . . Using that resource [air] as an inducement to promote development is as short-sighted as would be the reckless use of any other resource.

Cubia L. Clayton
Science, 193 (10 September 1976), 953

Faced with the high costs of pollution control, it is natural that one of an industrial executive's first impulses may be to relocate where the standards are less strict. This kind of reaction is, of course, viewed as a serious problem by labor and local political leaders, who are understandably concerned with a loss of jobs and with a lack of growth in their areas.

Bayard Rustin, a labor leader, writes:

Many in the environmental lobby, in a hysterical effort to reverse the effects of the industrial revolution in a few years, have supported sweeping and uninformed legislation which has had unnecessarily harmful economic consequences. The Clean Air Act is but one example. It has resulted in the closing down of hundreds of plants, from drop forges to specialty organic chemical plants. The environmental damage of the plants forced to close had been minimal and could have been easily corrected. The economic damage of the legislation has been severe.

The poor black in the ghetto or the unemployed white worker would not find very much with which to identify in [Senator Stuart] Udall's semi-revivalist call that we must no longer "over-indulge ourselves" or seek "to satisfy unlimited greed and desire for luxury." They would agree that "we must change our way of life," but they would take this to mean rather more luxury than they have been accus-

tomed to in the past. The good and sweeping intentions of many environmentalists are now an obstacle blocking those less-fortunate Americans who desire economic justice.★

But this is not the whole story. An official of the New Mexico Environmental Improvement Agency has noted that:

. . . environmental quality is thus too often viewed as the tool with which to bargain.

The difficulty, apart from environmental degradation, is that this approach fails to consider the reality of air quality as a natural resource which is as depletable as any other. Whether one agrees philosophically with national air quality standards, they do exist, and their existense means the end of the age-old concept of an unlimited air resource. Hence using that resource as an inducement to promote development is as short-sighted as would be the reckless use of any other resource.

The imposition of a uniform, nationally designated Class II ceiling . . . actually ensures more development than would otherwise occur. This is because (i) those states that desire to use air quality as an inducement to development will not be allowed to develop at the expense of neighbors who are interested in maintaining as much of a quality environment as possible, and (ii) a tighter ceiling than that imposed by national standards will help impress on everyone that air is a depletable resource and that new industry must be required to utilize the best control technology in developing new energy supplies.

The question of available technology is the crux of the problem. Existing industry faced with the problems of the retrofit of control devices is finding the job difficult and expensive. In many cases, the result has been an unwillingness to accept the fact that the job of control is even possible. But a difficult job is not the same as an impossible one, particularly in the case of new industry where controls can be made an integral part of plant design.

The end result is that, rather than having an air shed used up by three or four inadequately controlled industries, more industry can be accommodated.†

Who is right? How would you balance air cleanup and jobs? Can you think of ways the poor and unemployed could benefit from air pollution control?

---

* Bayard Rustin, President, A. Philip Randolph Institute, *The New York Times,* 11 August 1975.
† Cubia L. Clayton, *Science,* **193** (10 September 1976), 953.

---

called iron pyrite). Organic sulfur is sulfur that is chemically bonded to the carbon in the coal. Special washing steps are sufficient to remove inorganic sulfur from coal. The organic sulfur, in contrast, requires chemical treatment for removal. Sulfur as pyrites (metal sulfides) may range from 30–70% of the total sulfur present in coal, but, on average, equal quantities of organic and inorganic sulfur are present.

To cleanse coal of the inorganic or mineral sulfur, the coal is first crushed to expose the mineral veins. The ground coal is then mixed with water in a large tank. Pyrite has a higher density than coal and hence sinks faster; the cleaned coal is skimmed from the top of the tank. Up to 500–1000 tons of coal can be processed per hour. Washing is most effective on coals with higher proportions of pyrites.

"Chemical cleaning" of coal is a general name for a variety of processes, all of which remove the sulfur bound to the carbon in coal. Several dozen processes are competing for research and development funds in this arena. Some of the chemical methods remove both the pyrite sulfur and a portion of the organic sulfur.

One promising cleaning method developed by the Batelle Memorial Institute mixes finely ground coal into water containing sodium and calcium hydroxide. Treated under pressure and high temperature, the coal is cleansed of most of the pyrites and half of the organic sulfur; these substances remain in the liquid phase. The coal is washed, then dried, and is ready for use. The method has the potential to make the high-sulfur coal from the eastern United States an acceptable fuel in coal-fired electric plants. The coal could potentially be used without the need to "scrub" sulfur dioxide from the stack gases, but a commercial-scale application is needed to assess the cost.

One process that appears to remove up to 60% of the organic sulfur is known as **solvent refining of coal (SRC).** Ground-up coal is mixed with the organic solvent anthracene; nearly 95% of the carbon in the coal dissolves in the anthracene. Exposed to high pressures and temperatures, the solution is then treated with hydrogen. The coal can be recovered either as a solid or a liquid. The solid material, black and brittle, has less than 1% ash, though the original coal may have had anywhere from 8–20% ash. Sulfur is reduced to less than 1% as well, and the heating value has increased by 30% on a per-ton basis. Solvent refining of coal, if it turns out to be commercially feasible, has bright prospects. With sulfur lowered sufficiently and ash nearly eliminated, it may be possible not only to avoid scrubbing the stack gases, but also to avoid electrostatic precipitation for particle removal.

Scrubbers. A number of methods are being and have been developed to "scrub" sulfur dioxide from the gas that exits the smokestack. The term "scrub," unfortunately, is not scientifically descriptive; it simply means that the stack gases are cleansed of sulfur dioxide. The operations of the various "scrubbers" are based on chemical reactions with the sulfur dioxide in the stack gas of coal-fired power plants. The chemicals formed in these reactions may be waste products or marketable items.

Of the more than 50 control concepts, the single most thoroughly tested and reliable process is the lime–limestone wet scrubber. About 85% of the scrubbers in place in 1982 were of the lime–limestone type. In this device, the flue gas is passed through a slurry mixture of limestone and lime in water. Sulfur dioxide is absorbed into the slurry and reacts to form calcium sulfite and calcium sulfate (gypsum). The flue gas is not only cleansed of about 80% of its sulfur dioxide, but also 99% of the fly ash it carried. The flue gas has been cooled so much, however, that it must be reheated to make it buoyant enough to rise up the chimney for discharge to the atmosphere.

The resulting sludge must be disposed of in some way. Because it is a gloppy mass, it is not by itself a good landfill material. However, it appears that a hardener such as fly ash can be added to react with the slurry, producing a clay-like substance.

Because sludge disposal is seen as a serious problem, scrubbers that produce a marketable product instead appear very attractive. One such scrubber produces sulfur of high purity; another produces a dilute sulfuric acid. The sulfuric acid is not economical to transfer across large distances; but the sulfur, which is used in such products as pharmaceuticals, industrial chemicals, and fertilizers, is quite compact. There is some concern that so many plants will be producing sulfur that the price will fall. The scrubber design known as the Wellman–Lord System is one that produces sulfur; it was tested in Gary, Indiana, and removed 91% of the sulfur dioxide while producing high-purity sulfur.

A recent concept combines both coal cleaning and scrubbing. Cleaning mineral sulfur from coal is relatively inexpensive, but is limited to the fraction of sulfur contained in pyrite (metal sulfide) form. The U.S. Environmental Protection Agency has suggested removing this fraction of the sulfur, then scrubbing the stack gas produced when the coal is burned. Since the stack gas will have a lower level of sulfur dioxide, it need not be as fully treated to meet emission standards.

During most of the 1970s, portions of the utility industry kept up a steady criticism of scrubbers. Financed principally by the American Electric Power Company, the largest electric power company in the country, a massive (3.6-million-dollar) advertising campaign derided the scrubbers. They were called unreliable and expensive; the amounts of sludge were said to be enormous. However, as experience with scrubbers has grown, their reliability has proven to be very good. The mountains of sludge predicted have not materialized either. A typical 1000-megawatt coal-fired plant is expected to require 400–700 acres for disposal of *both* fly ash and scrubber sludge over a lifetime of 30 years. EPA estimated a total acreage requirement for scrubber sludge and fly ash of 18,000 acres by 1985 and 63,000 acres by the year 2000.

The opponents of scrubbers advocated control of sulfur oxides by the use of very tall stacks that would disperse the gas more widely. When levels at the ground became hazardous, they suggested that the boiler burn a low-sulfur fuel until the episode was past. Alternatively, it was suggested that power generation be cut back during such times. The opponents of scrubbers also suggested that the industry should wait to see how well coal cleaning turns out before installing scrubbers.

Now the pattern has reversed; scrubbers are an accepted technology, although acceptance is still grudging. By 1982, some 225 scrubbers (or flue gas desulfurization units) were either in place, being built, or on order. In all, 100,000 megawatts of coal-fired electric capacity would ultimately be covered by these units, about 35% of coal-fired capacity. This is the rough equivalent of 100 full-size modern coal-fired plants. Further, as new generating plants are planned, the number of units will increase still further. As one publication put it,

> To scrub or not to scrub, that is the question.
> Whether 'tis more economical and efficient to remove particulates and SO₂ from the end of the stack after the burning of coal
> Or to clean coal before it is consigned to the flames. . . .

Environmental Science and Technology, June 1974, p. 510

Fluidized bed. Coal washing (physical cleaning) and scrubbers are existing technologies to reduce sulfur dioxide fumes. In contrast, the fluidized bed combustion system is an experimental design to remove sulfur dioxide. The fluidized bed system burns coal as a conventional boiler does, but the coal is mixed with granular limestone and is layered on metal plates. Further, the air for combustion comes from below, passing through holes in the metal plates up past the coal. The air flow is so large that the particles of coal and limestone are lifted or floated (fluidized, as the name implies) above their bed on the plate.

The limestone reacts with the sulfur dioxide from the burning coal to form fine particles of calcium sulfate, which are carried off in the stack gases and removed with the fly ash, probably by an electrostatic precipitator. The temperature of the burning coal in the fluidized bed is lower than the temperature of burning coal in a boiler; as a consequence, fewer nitrogen oxides are formed. Thus both sulfur dioxide and nitrogen oxides are lower for the fluidized bed combustion system. We should have results on the effectiveness of this experimental design by the mid-1980s.

Progress in Control of Sulfur Oxides

What progress has been made since 1970 and where do problems remain? In 1977, the sulfur dioxide standard was violated in portions of Pittsburgh on over 20% of the days. In portions of Salt Lake City in 1977, the standard was violated on over 50% of the days and the alert level was exceeded on nearly 10% of the days.

The National Commission on Air Quality saw four urban areas as not achieving the health-based standard for sulfur dioxide in 1982: Pittsburgh, Indianapolis, Gary, and Chicago. In addition, a number of areas in the western U.S. near nonferrous smelters were also not expected to achieve the standard by 1982 (see Figure 21.2). The possibility for not achieving the sulfur dioxide standard also existed in an additional 24 areas, including both urban areas and areas near smelters. Whether an area attained the standard depended not only on cleanup achievements but also on the weather that occurred in these areas.

Progress through the 1970s appears to have been almost nil in reducing sulfur dioxide emissions, and the progress may even be reversed as more coal-fired electric capacity comes on line in the 1980s (Figure 21.3). At the same time that emissions were *not* coming under control, concentration of sulfur dioxide in the air in urban areas was decreasing fairly steadily. (See Figures 21.9 and 21.10.) What was happening?

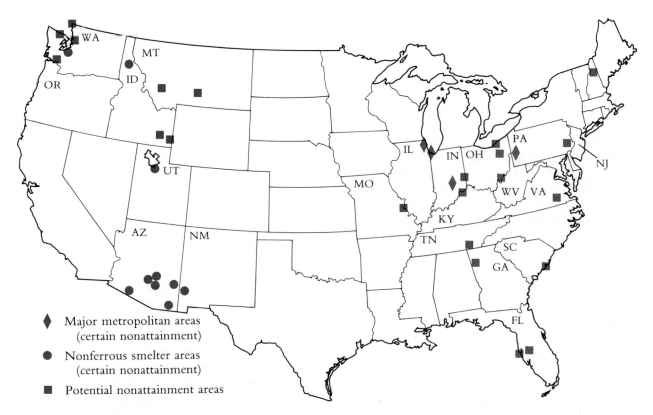

◆ Major metropolitan areas
(certain nonattainment)

● Nonferrous smelter areas
(certain nonattainment)

■ Potential nonattainment areas

Figure 21.2 Areas that Were Not Expected to Achieve the Health-Based Sulfur Dioxide Standard in 1982. Actual 1982 air quality data were not available at time of printing. (Source: National Commission on Air Quality, 1981.)

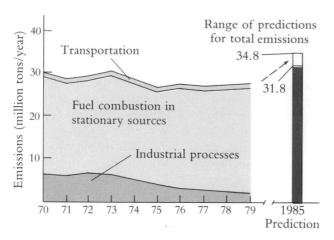

Figure 21.3 Sulfur Oxides Emission Trends. Even with wider application of control technology, increased coal burning causes higher emissions. "Fuel combustion in stationary sources" refers to the burning of coal and oil for energy in electric power plants and factories, as well as the burning of these fuels in commercial and industrial buildings and residences for space heating. Stationary sources are distinguished from mobile sources such as cars, trucks, trains, planes, etc. "Sulfur oxides from industrial processes" refers to the production of sulfur oxides from such activities as the smelting of ores. The sulfur referred to here comes from the ore in these cases and not from the burning of the fuel. (Sources: U.S. EPA, "Trends in the Quality of the Nation's Air," October 1980, and SEAS Model and REPS Model.)

The emissions were simply being shifted. Air pollution control requirements forced polluters in urban areas to change fuels or otherwise control emissions and new sources were forced to more remote locations. The atmospheric burden of sulfur was simply being more widely dispersed. In terms of controlling acid rain, then, which depends on total emissions, little progress if any was made over the decade of the 1970s.

Acid Rain

Description and Sources

First in the Scandinavian countries, then in the Northeast United States and Southeastern Canada, then in Northern Europe, Taiwan, and Japan, scientists have been finding that rainwater—water thought of as amongst the more pure available in nature—has become highly acidic.

This **acid rain** is the result of the presence of both sulfur oxides and nitrogen oxides in the atmosphere. Sulfur oxides, as we have pointed out, are the result of burning the fossil fuels that contain sulfur: coal is the prime source of sulfur among the fuels; oil is second; and natural gas, a distant third. Although nitrogen oxides result from the burning of fossil fuels as well, the source of the nitrogen is likely to be the air itself, since most fossil fuels are relatively low in nitrogen. The high-temperature combustion of all three fossil fuels results in significant emissions of nitrogen oxides.

Oxides of sulfur originate largely from power plants and smelters. Oxides of nitrogen also stem to a very large degree from power plants, but automobiles are another significant source (see Chapter 22). Power plant emissions are often from tall stacks; nitrogen oxides from such sources are more likely than ground-level emissions to be caught up in the long-distance transport so characteristic of acid rain. In these combustion processes, sulfur is generally oxidized to sulfur dioxide and nitrogen to nitric oxide.

These gases, sulfur dioxide and nitric oxide, undergo chemical reactions in the atmosphere. The sulfur dioxide is oxidized to sulfur trioxide, which then dissolves in water droplets to form sulfuric acid. Nitric oxide is oxidized to nitrogen dioxide, which dissolves in water droplets to form nitric acid. These two acids, as well as salts of these acids, are responsible for "acid rain." The more of these acids present in the atmosphere, the more acidic that rainwater becomes. The two acids do not contribute equally to acid rain. In the Northeastern United States, about 15–30% of the acidity is due to nitrates (nitric acid); the remainder is attributed to sulfates (sulfuric acid). In California, in contrast, nitrates are the most common component of acid rain. Measurements of acid rain in Pasadena, California, showed 57% of the acidity due to nitric acid and the remainder (43%) due to sulfuric acid. The large portion of acidity due to nitric acid is the result of the high tonnage of nitrogen oxides emissions from automobiles in Southern California.

The measure of acidity is the number of hydrogen ions per liter of water. Water molecules (H_2O) are normally dissociated into hydrogen ions (H^+) and hydroxyl ions (OH^-). In a sample of pure water, we would expect to find about 0.0000001 (one ten-millionth) of the water molecules dissociated into hydrogen ions and hydroxyl ions; in pure water the two ions are present in approximately equal numbers. A solution with equal concentrations of hydrogen and hyroxyl ions is called "neutral." It is neither acidic (having more hydrogen ions) nor basic (having fewer hydrogen ions).

Rainwater is not pure; it comes in contact with and dissolves carbon dioxide, a natural component of the atmosphere. The solution of carbon dioxide in water produces carbonic acid, a weak acid. The concentration of hydrogen ions relative to the number of water molecules in unpolluted rainwater containing dissolved carbon dioxide would be about 0.000001, or one hydrogen ion per million molecules of water. If you count the zeros after the decimal you will see that the hydrogen ion concentration has increased by a factor of 10 due to the solution of carbon dioxide. This level of hydrogen ions is presumed to be the natural condition for rainwater.

The rain falling in New England, on the other hand, has a ratio of about 0.0001 hydrogen ions per molecule of water (1 hydrogen ion per 10,000 water molecules). This is about a 100-fold increase above the expected concentration of hydrogen ions in rainwater. New England has acid rain because the area is downwind from the major industrial centers of the Northeast, where sulfur-bearing fossil fuels are burned in enormous quantities.

Acidity is not commonly measured by the ratio of hydrogen ions to water molecules. Instead, it is measured by the negative logarithm of the hydrogen ion concentration, which is called the **pH.** Thus, $-\log_{10}(.0000001) = 7$, and a pH of 7 indicates water that is neither acid nor basic, but "neutral." The rain in New England, with about 0.0001 hydrogen ions per molecule of water, has a pH of $-\log_{10}(.0001)$, or 4. In general, the lower the value of pH, the more acidic the water (see Figure 21.4). A pH of 4 for rainwater used to be very unusual. No pH values even less than 4 are commonly observed in some areas (Figure 21.5).

Biological Impacts of Acid Rain

Of all the biological impacts of acid rain, the most obvious is the reduction and even elimination of fish populations in lakes that have been made acidic. In Sweden, Norway, and Eastern North America, commercial and sport fishing have suffered as fish populations have declined or disappeared.

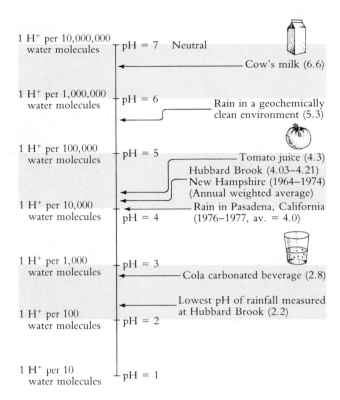

Figure 21.4 Relation between pH and acidity, including the pH levels of some common substances and of rain in different environments. The lower the pH, the more acidic the substance.

In the numerous lakes of the Adirondack Mountains at high elevations (610 meters and above), acid waters are now the norm. In 1975, 51% of these lakes had pH values less than 5 and 90% of these low-pH lakes had no fish. In total, 45% of the high-altitude lakes were without fish populations; only 4% of these lakes were without fish in the 1930s. It has also been found that one-third of the more than 2000 lakes in Southern Norway are devoid of fish, a change that has taken place since the 1940s.

It is clear that acid waters have something to do with declining fish populations, but it is generally not the case that mature fish are killed in massive numbers by the acidity in these waters. Instead it appears that acid waters are preventing fish from reproducing. Female fish may not be able to spawn (release their eggs) in acid waters, and if they are able to do so, eggs and larvae may die at an increased rate. Thus, it is common in lakes that are in the process of becoming more acidic to see no young fish, but only mature adults.

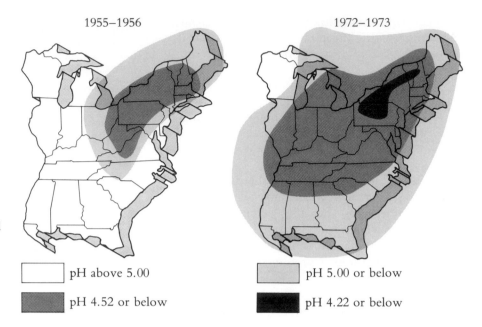

1955–1956   1972–1973

Figure 21.5 The Weighted Annual Average of pH of Precipitation in the Eastern United States in 1955–1956 and 1972–1973. (Source: *Nitrates: An Environmental Assessment,* National Academy of Sciences, 1976)

pH above 5.00

pH 4.52 or below

pH 5.00 or below

pH 4.22 or below

Many areas where fish populations have declined due to acid rain experience cold winters, with the accumulation of snow. When the snow pack melts, a rush of acid water may enter the lake, turning the lake sharply acid. The melting of the snow pack and the sharp increase in acidity coincide roughly in time with the spawning activity of fish. Thus, it appears that spawning is subjected to the maximum acid conditions that may occur through the year. As fish populations decline, it would be expected that those species that feed on fish, such as the bald eagle, the loon, and the osprey, mink, and otter, would also see some decline.

Scientists are predicting that populations of frogs, toads, and salamanders may all be reduced by acid rain. Many of these species breed in the temporary pools of water found during spring rains; the pools are likely to be more acid than the lakes because they are supplied only by the acid rainwater. Again, these species are likely to decline quietly rather than in massive numbers; new young individuals will simply not survive to join the population.

Acid rain is thought to have an effect on plants as well as on animals. Changes in plant species where acid rain falls are difficult to detect; long periods of time are thought to be necessary to pick up changes such as forest growth. To fill the gap, experiments have been performed in the laboratory in which plants were irrigated by water that had been made acid to the same extent as rainwater in the northeastern United States. The water was sprayed on pine trees and tomato plants to mimic the way rain arrives. Pine needles grew to only half their normal length in these experiments and fewer tomatoes were produced than normally. In the field, however, more factors operate than in the laboratory, so that these observations give us concern but not necessarily the power to predict.

Other Effects of Acid Rain

At the moment, scientists have not detected any direct impact of acid rain on human health, but there are clearly possibilities for such effects. All the possibilities revolve around the increased ability of acid water to dissolve or otherwise act on minerals. Mercury in natural waters may be converted to monomethyl mercury under acid conditions. Fish will accumulate monomethyl mercury in their flesh. High mercury concentrations in fish have been found in areas where acid rain is falling and lakes are acidifying. Mercury is a known human poison, which has contaminated fish in the past under other circumstances (see Chapter 11).

If reservoirs used for drinking water were to become acidic, toxic metals from the watershed may dissolve in water being used for human consumption. In addition, acidic drinking water could dissolve lead from household plumbing systems. Such a case has apparently already occurred in New York State. There, the highly acid water from the Hinckley Reservoir near Amsterdam, New York, leached lead from household plumbing, causing concentrations of lead above the drinking water standard. The water had stood in pipes overnight; once this water was run out of the system, lead concentrations returned to their typically low levels.

Acid rain also damages mortar and stone by chemically reacting with the calcium and magnesium in them. Irreplaceable statuary is particularly at risk from such damage. In addition, iron products and other metals are very susceptible to corrosion from acid rain. Swedish investigators found a high correlation between acid rain and the corrosion rate of steel.

What Makes Some Lakes and Soils Sensitive to Acid Rain?

Some lakes seem not to be subject to becoming acidic; these lakes or their watersheds seem to have the capability of neutralizing acid additions. Other lakes, based on their present conditions, are obviously sensitive to acid rain. Vulnerable lakes typically are fed by a watershed in which the bedrock contains igneous rock (such as granite) or metamorphic rock (such as gneiss). These rock types are resistant to solution; water flowing over them dissolves little in the way of minerals and hence the lakes into which they flow have very "soft" waters.

The minerals in sedimentary rock, on the other hand, are more easily dissolved by water flowing over them. Lakes whose watersheds are composed of sedimentary bedrock tend to have "hard" water, water rich in dissolved minerals. If limestone bedrock (a sedimentary rock composed of calcium carbonate) is present in the watershed, runoff tends to be a "hard" water. Lakes whose watersheds are composed of sedimentary bedrock tend to resist becoming acidic because the carbonate minerals neutralize acidity.

The Adirondack lakes have watersheds whose bedrock is largely granite. The waters in the lakes are typically soft, low in dissolved minerals. Acid rain makes a big impact on these lakes. Regions of North America whose lakes are thought to be sensitive to the impact of acid rain are shown in Figure 21.6.

The composition of the bedrock is not the only factor in determining whether lakes are sensitive to acid additions. The soil in the region also plays a role. Watershed soils that are rich in soluble minerals, such as calcium or magnesium carbonate, tend to protect lakes because these minerals neutralize acidity. As an example, although granite bedrock is common in areas of Maine, the lakes in these areas do not seem to have acidified, apparently because of the neutralization of acid rain in the lime-bearing soil—this in the face of rain with an average pH of 4.3. Some lakes in Florida, in contrast, have acidified despite having a watershed in which there was sedimentary bedrock. These lakes apparently did not get much groundwater inflow and the surface flows drained over soil in which little calcium carbonate was present.

The acidity in some lakes has been controlled in the past by the addition of lime. The practice of "liming," while it can help individual lakes, is not a long-term solution. The quantities of lime required to maintain all lakes at reasonable levels of hydrogen balance would be enormous.

Not only are lakes and their populations sensitive to acid rain, but soil and land-based ecosystems are sensitive as well. Soils have differing sensitivity to acid rain, just as lakes do. Again, soils derived from sedimentary materials, such as carbonate minerals, or rich in organic material tend to neutralize acid rain. Those soils that stem from rocks such as granite and gneiss, which resist solution, tend to be easily acidified. About 70–80% of the land in the Eastern United States possesses soils sensitive to acidification. Acid rain may leach the already small mineral content and plant nutrients from these soils, decreasing their productivity. That such productivity effects have occurred has been hard to demonstrate except under laboratory conditions. Such possibilities, however, should be enough, coupled with the other known effects of acid rain, to stimulate action.

Acid rain has no respect for boundaries of states or nations. Britain and Northern Europe export acid rain to Sweden and Norway. Emissions in the United States contribute to acid rain in Canada, and Canada donates emissions that produce acid rain in the United States. In fact, the largest source of sulfur oxides in all of North America is the smelter complex near Sudbury, Ontario. Canada and the United States negotiated for much of the early 1980s on how each nation would handle its sulfur oxides emissions and which nation bore the most responsibility for acid rain. The discussions were not always pleasant. It was clear that the U.S. was the greater producer of sulfur oxides and that prevailing winds blow toward Canada from the Eastern U.S. Nonetheless, the Reagan Administration contended that it could not be determined how much of each nation's production crossed the boundary.

Control of Acid Rain

The means of controlling acid rain are clear; they are the same means that should be used to control sulfur oxides and nitrogen oxides emissions. At power plants, this means scrubbers to remove sulfur oxides and combustion modifications to decrease the forma-tion of nitrogen oxides. (Scrubbers were discussed more extensively earlier in this chapter.) Automobiles are only a minor source of sulfur emissions, but a major source of nitrogen oxides. To control nitrogen oxides from autos, catalytic converters and engine modifications are used. Control of automobile emissions is discussed further in Chapter 22.

Particulate Matter

Particulate matter is another serious air pollutant. Unlike the pollutants discussed earlier, particles are not a single chemical type. Instead, numerous solid and liquid compounds are dispersed in the air from many sources.

Figure 21.6 Areas Containing Lakes Sensitive to Acid Precipita-tion.(Source: *To Breathe Clean Air,* National Commission on Air Quality, March 1981, published by the U.S. Government Printing Office. Washington, D.C.)

Particles Stem from Combustion and Industrial Activities

Although we mention them briefly here, two types of particles are treated separately from the main discussion. Lead and asbestos have special properties and sources that require specific discussion. Lead is treated later in this chapter; asbestos is discussed in Chapter 30.

The burning of coal produces not only ash particles (calcium silicates) and carbon particles, but also metal oxides such as calcium and ferric oxide. The metal oxide particles may react with the mist of sulfuric acid droplets. The reaction produces still other particles, the metal sulfates. The sulfuric acid droplets themselves are particles derived from the reaction of sulfur trioxide with water vapor. Both the acid droplets and the sulfate particles then, are derived in part from coal burning (see Figure 21.1).

The quantity of particles derived from coal burning is enormous. Fortunately, however, a large proportion of the particles is removed from the stack gases. In 1978, about 2.4 million metric tons of particles were released to the atmosphere from coal-burning electric plants in the United States. Probably seven times that quantity was generated, but most of it was removed. The 2.4 million metric tons from electric plants, however, made up approximately one-fifth of the 12.5 million metric tons of particles reaching the air from all sources in 1978. A small amount of ash does come from the burning of oil. Whereas 80–120 pounds of ash are emitted per 1000 pounds of coal, only about 2 pounds of particles stem from burning an equivalent amount of oil.

Liquid hydrocarbons (compounds of carbon and hydrogen) and liquid derivatives' of hydrocarbons come from the incomplete combustion of gasoline and of diesel fuel. Still another kind of particle results from the photochemical reactions of nitrogen oxides and hydrocarbons in the air. These sunlight-stimulated reactions produce liquid organic substances that are scattered as tiny droplets in the air. The term **smog** has been coined to describe the resulting fog-like condition. Photochemical smog is particularly evident in cities such as Los Angeles, where automobile use is excessive. In the early 1970s, up to 40% of the airborne particles in Los Angeles stemmed from use of the automobile. Particles of lead are also emitted from automobile exhaust. These derive from the lead compounds put into gasoline to increase its octane.

Surface mining of coal and of other substances also produces large quantities of particles. The refining of ores and manufacture of metals are among the industrial processes releasing particles to the air. The grinding and spraying that accompany construction are also sources. Taken together, these industrial activities may account for more particle emissions than occur from coal combustion.

Asbestos escapes to the atmosphere from new building construction, where it is being sprayed into place as insulation; it also escapes in the demolition of older buildings. Finally, the incineration of solid wastes may, in some cities, be a significant source of particles when incinerators are located centrally.

Effects of Particulate Air Pollutants

Particles may settle on surfaces, leading to a dirty gray appearance. In addition to soiling, particles may cause corrosion, acting as centers from which corrosion spreads. The annual repair of these surfaces has a significant cost.

The sulfuric acid mist may damage plant tissue. Compounds from photochemical reactions produce a burn on the leaves of many vegetables. Beets, celery, lettuce, and pepper are among the many species susceptible to such damage. Inert, unreactive particles may simply soil plant surfaces.

The health of human beings is also affected by particles. The frequency of respiratory infections such as colds and bronchitis is seen to increase with particle levels. Furthermore, at high concentrations of particles, deaths in excess of the number expected for the time of year have been seen to increase. The health-based standard for particle concentrations has been set at 260 micrograms per cubic meter averaged over a 24-hour period, and this level is not to be exceeded at any one point more than once per year. At a particle concentration of 375 micrograms per cubic meter, an air pollution "alert" is to be declared. Under an alert, industries might be requested to curtail or postpone their activities. The annual standard for particulate matter is 75 micrograms per cubic meter, averaged over one year. (See Table 21.2.)

Particles have a subtle effect on weather. They may act as nuclei upon which water vapor condenses.

Table 21.2 The Nine Worst Cities in Terms of Particles in the Air, 1977

City	Number of monitoring sites	Numbers of sites with an annual average greater than annual standard	Lowest average annual concen. over all sites	Highest average annual concen. over all sites	Maximum 24-hour value at any site
Tucson, Ariz.	7	3	67	156	591
Pocatello, Ind.	4	3	65	218	1371
Chicago, Ill.	25	12	50	170	1106
Granite City, Ill.	8	8	85	185	485
Taos County, NM	1	1	— 168	—	577
Middletown, Ohio	3	2	64	192	707
Cleveland, Ohio	23	13	48	152	705
Youngstown, Ohio	5	4	66	172	602
El Paso, Texas	14	10	60	158	691

Notes: 1. Standard for average over one year = 75 micrograms per cubic meter.
2. Standard for 24-hour average = 260 micrograms per cubic meter.
(Source: U.S. Environmental Protection Agency)

Lengthened periods of fog may result from high levels of particles in the air. Increased rainfall has also been seen where particle levels have risen. Furthermore, particles may be reflecting solar energy away from the earth. As such, they may have been responsible in part for the small but noticeable cooling in the Northern Hemisphere during the last quarter century. The recent eruptions of Mount St. Helens and El Chichón in Mexico introduced such quantities of particles to the global atmosphere that emissions due to human activity were dwarfed. These volcanic emissions, however, do not contribute to the particle levels in our cities, except on rare occasions.

One particular group of particles that has been implicated with relation to human health are the sulfates we mentioned earlier. These particles are often among the smallest in size in urban air and so have easy entry to the lungs. Because sulfate particles are small in size, they remain in the air longer than larger particles and are subject to long-distance transport over hundreds of miles. The sulfates are the sulfuric acid droplets and ammonium, calcium, magnesium, and iron sulfates, among others. These particles apparently are formed by chemical reactions of emitted substances in the air, rather than being themselves emitted from a combustion source. The sulfates are implicated not only in human health, but also in acid rain. Much of this sulfate in the air undoubtedly stems from power plant emissions.

No air-quality standard has yet been stated for sulfates. Nonetheless, a 1975 position paper from the U.S. Environmental Protection Agency suggested that health effects begin to show up at sulfate levels of 10 micrograms per cubic meter, averaged over a 24-hour day. Data on sulfate have not been as extensive as on other pollutants in the past, but measurements in August 1977 showed almost all of the Northeast U.S. having 10 or more days with sulfate levels at or above 10 micrograms per cubic meter.

High sulfate concentrations occur in areas where sulfur dioxide emissions are high, but they are higher still at times when ozone, another air pollutant, is present in high concentrations in the air. The relation is explained by noting that ozone hastens the oxidation of sulfur dioxide to the trioxide form and hence fosters the formation of sulfuric acid and metal sulfates.

Rainfall appears to bring sulfate levels down rapidly. Sulfate concentrations do not appear to peak in a geographic sense, but tend to be spread widely across regions. The wide spreading is the result of the small size of sulfate particles. Winds can transport sulfate particles further than larger particles before they are deposited.

Ways to Control Particle Emissions

Numerous devices are available to reduce particle emissions. These include the settling chamber, the after-burner to ignite and burn particles, and the **electrostatic precipitator.** First applied to the collection of fly ash in 1923, the precipitator is now in use at nearly 1000 plants in the U.S. The operation of the precipitator consists first of attaching electric charges to fly ash particles. As the flue gas moves between the plates of the device, ions or electrons formed by a high-voltage discharge bombard the particles of fly ash and provide them with charges. The charged particles are attracted to and deposited on the grounded metal plates or collection electrodes. The plates must be cleaned periodically by vibrating them or rapping them with a mechanical device (Figure 21.7). The devices can attain efficiencies of particle collection up to 99%, and can operate with little maintenance while drawing very little electric power.

An alternative to the electrostatic precipitator is the "baghouse," composed of fabric bags to capture the fly ash. The fabric filters have a long history of use in removing particles in industries such as grain mills, asbestos factories, and cement plants. Electric utilities have been purchasing these units on a trial basis in the last few years in order to evaluate them for use in power plants. The principal of the baghouse is simple; it is a very large vacuum cleaner. Air is drawn up through the bags; the particles cannot pass out through the fine weave of the bags and are trapped inside it. Only clean air exits. Despite impressive efficiencies (over 99%), utilities have not adopted this technology to any significant degree, citing lack of experience with the technology.

Particulate control methods have been widely applied and are having some impact on air quality. Although particulate emissions have been steadily declining since 1970, the achievement of particle standards is proceeding with painful slowness.

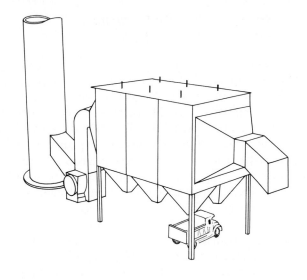

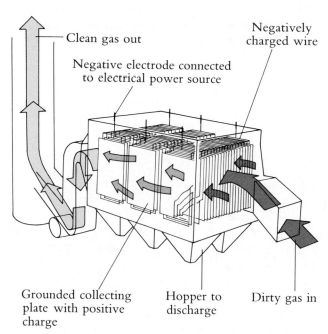

Figure 21.7 Electrostatic Precipitator. Dirty air flows between negatively charged wires and grounded metal collecting plates. The particles in the flowing air become charged and are then attracted to the plates, which hold the accumulated dust until it is periodically knocked into hoppers. The clean air is then pumped out through the stack. (Adapted from *Controlling Air Pollution,* American Lung Association, 1974.)

Progress in Control of Particulate Matter: An Uphill Battle

Progress in bringing down particle levels in the air has been fairly steady, but now may be slowing down. In the 20-year period from 1960 to 1979, the average concentration of particles in the air, over a number of areas and monitoring sites, fell by 32%. By the end of the 1970s, however, improvements in air quality, as measured by particle levels, seem to have come to an end (see Figure 21.12). In 1980, about 21% of the population still lived in areas in which the annual standard (75 micrograms per cubic meter) was exceeded.

According to the National Commission on Air Quality, particulate levels actually increased in 16 major areas of the nation from 1975 to 1979. These areas included Kansas City, Missouri; Buffalo; Baltimore; Seattle; Houston; St. Louis; Denver; and Portland, Oregon. In addition, five areas of the nation exceeded the daily particle standards on over 140 out of 365 days in 1979. These were Phoenix (285 days); Los Angeles Basin (202 days); Las Vegas (175 days); Tucson (156 days); and Spokane (146 days).

In their 1981 report, the National Commission on Air Quality foresaw that 27 major urban areas would *definitely* be unable to achieve the standard for particles in the air by 1982. Another 40 areas, smaller in size, some rural, some urban, were likewise *definitely* expected to be unable to achieve the standard by 1982. An additional 25 areas were *likely* to be unable to meet the standard in 1982 (Figure 21.8).

Not only does much remain to accomplish in the control of particles, but a new concern is emerging as well—namely, that the air-quality standard for particles, as presently stated, is inadequate to protect people's health. The present standard is stated as a maximum weight per cubic meter. The removal of large particles from the air might achieve the standard for

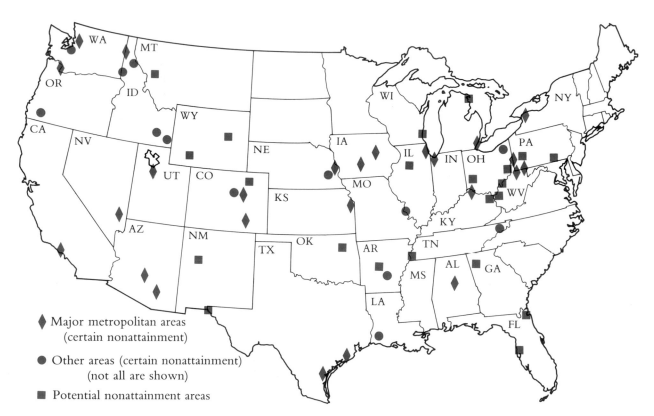

♦ Major metropolitan areas
(certain nonattainment)

● Other areas (certain nonattainment)
(not all are shown)

■ Potential nonattainment areas

Figure 21.8 Areas Not Expected to Achieve the Health-Based Standard for Total Suspended Particulates by 1982. (Source: National Commission on Air Quality, 1981.)

some city, but many small particles could remain. Larger particles are less likely to reach the lungs because of respiratory defense mechanisms, so their removal is not so important in the first place. Fine particles, on the other hand, those with sizes less than two or three microns, are more likely to reach the lungs and have a serious effect, yet they add little to the mass of pollutants in the air. A new air-quality standard for fine particles in the air is needed.

Lead Compounds in the Air

Sources of Lead Air Pollutants

Lead has been added to gasoline for many years as either the compound lead tetraethyl or lead tetramethyl. These compounds are used to improve the antiknock quality of gasoline; they eliminate the "ping" during engine operation. These substances are not absolutely necessary, however. A slightly higher-priced refining process produces a fuel with the required antiknock properties. Lead need not be present in gasoline. In the past, lead levels have ranged from 2.5–4.23 grams of lead per gallon of gasoline. About 75% of this amount is released to the atmosphere in the exhaust stream of cars. In 1975, over 142,000 metric tons of lead were added to the atmosphere from gasoline combustion.

A Special Hazard to Children

Lead, a substance present in food, water, and air, is a cumulative poison; that is, it gradually builds up in the human system because of a slow rate of natural removal. Its presence in the atmosphere adds to the burden of lead we carry in our body. Lead slows the rate at which the bone marrow produces red blood cells. Lead also blocks the body's manufacture of hemoglobin, the molecule that carries oxygen from the lungs to the tissues and carbon dioxide from the tissues to the lungs.

In addition to breathing lead in the air, small children in poorer and older urban areas are exposed to another and more dangerous source of lead: peeling paints, which an unaware child may eat. Since regulations were established in the early 1970s, most paints have become almost lead-free, although absolute control has not yet been achieved. Paint manufactured before the controls were set had from 5% to as much as 40% lead in the dried film.

The dangers of lead poisoning due to eating paint chips increase if there is lead in the atmosphere. Where lead concentrations in the air are highest, people carry a burden of lead in their blood and tissues. The reason is that the body absorbs and retains a portion of the inhaled lead. City residents, then, are closer to the threshold levels of lead at which the symptoms of poisoning appear. Children have a threshold level about one-half that of adults, so that they are even more susceptible to lead poisoning. Lead in the air is pushing lead tissue levels in children too close to the poisoning threshold. (This point is discussed in more detail on p. 415.)

Although lead poisoning may lead to death, in milder cases it causes mental retardation. Even levels below the poison threshold appear to cause subtle inadequacies in learning. Although the fact is little mentioned because the immediate health effects of lead are so great, lead has been shown to cause cancer in rats, suggesting that a similar effect in humans might be found. The hazard to health from acute and chronic lead poisoning is one of the reasons for the 1974 decision to eventually make gasoline lead-free. The other reason is that lead poisons the catalytic converter, the device that further burns the hydrocarbons and carbon monoxide remaining in exhaust gases.

In 1977, the Environmental Protection Agency established a standard for lead in the atmosphere of 1.5 micrograms per cubic meter, averaged over one month. Lead emissions from automobiles will be declining each year as fewer and fewer cars on the road use leaded fuel.

Trends in Air Quality—Challenge in Display

We've all been warned: there are "lies, damn lies, and statistics." We present two sets of "statistics" on nationwide sulfur dioxide concentrations and one on particles for you to see. Figure 21.9 appeared in a widely distributed bulletin from the U.S. Environmental Protection Agency. The bulletin, "Trends in the Quality of the Nation's Air," was intended for the general public. Figure 21.10 appeared in a draft technical report from the U.S. Environmental Protection Agency; this report was intended largely for scientists and engineers. Public viewing was unlikely.

In Figure 21.9, we have added two black lines to indicate the same time period used in Figure 21.10.

Figure 21.9 shows dramatic progress between 1964 and 1971. Note, though, that the bottom portion of the graph (shown with dotted lines) was missing as it originally appeared. A quick glance at Figure 21.9 gives the impression that levels have been driven nearly to zero, when that is not at all the case. The impression of dramatic progress is achieved by compressing the time scale in which the decreasing trend is shown. The axis of time is 25 years in length in Figure 21.9; the time axis is only nine years in length in Figure 21.10. Figure 21.10 shows only steady, undramatic progress, and the bottom portion of the graph is shown.

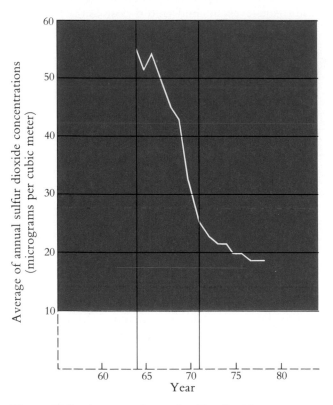

Figure 21.9 Average of annual sulfur dioxide concentrations over many monitoring stations. This graph was intended for a wide audience, primarily of lay people. (Adapted from: "Trends in the Quality of the Nation's Air," U.S. Environmental Protection Agency, October 1980)

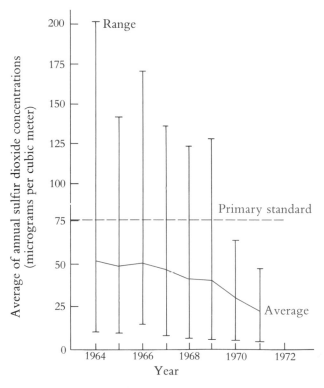

Figure 21.10 Average sulfur dioxide concentrations are shown for the same 32 monitoring stations as in Figure 21.9. This graph was intended primarily for an audience of scientists and engineers. (Source: Draft Technical Report, U.S. Environmental Protection Agency)

Furthermore, Figure 21.10 shows the range of data used to calculate the average. Both the highest and lowest numbers that entered the calculation of the average are shown. Some very low numbers, representing very good air quality, were used to calculate the trend. Neither the highest numbers, representing the worst air quality, nor the lowest numbers, are shown in Figure 21.9. By not showing the worst air quality, an impression that all is well is created.

The same bulletin reported particles in the air in a similar way; the graph that appeared in "Trends in the Quality of the Nation's Air" is reproduced here as Figure 21.11. We have added and shown with dotted lines the portion of the graph that was missing in the bulletin. We have also added the health-based standard to the graph. With the bottom portion of the graph missing, a quick look at Figure 21.11 suggests there is little left to accomplish in controlling particles in the atmosphere. How wrong that conclusion is can be seen by looking back at Figure 21.8 and at Figure 21.12.

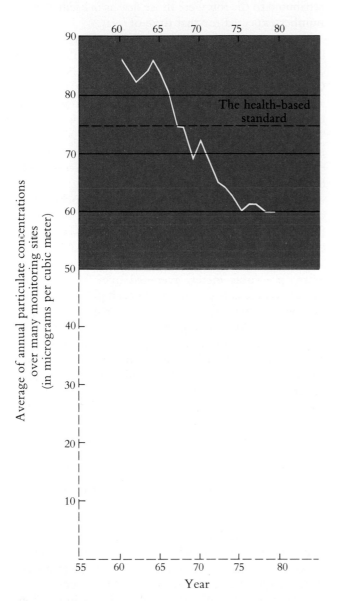

Figure 21.11 Particle levels in the atmosphere have been declining since about 1960. (Adapted from: "Trends in the Quality of the Nation's Air," U.S. Environmental Protection Agency, October 1980)

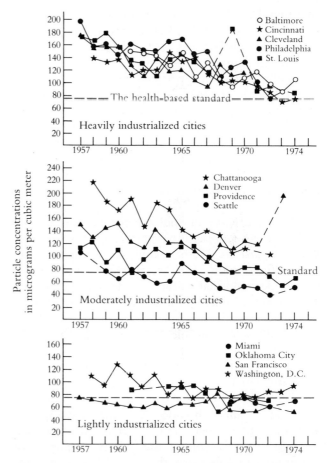

Figure 21.12 Particle levels have not been declining so rapidly in some cities. (Source: External Review Draft Report No. 2, Air Quality Criteria for Sulfur Oxides and Particles, U.S. Environmental Protection Agency, February 1981)

The Health Effects of Sulfur Oxides and Particles

The two categories of pollutants, sulfur oxides and particles, are discussed together for two reasons. High levels of both these pollutants have been measured during some of the worst air-pollution episodes recorded. By worst, we mean those episodes with the largest numbers of deaths and illnesses. Sulfur oxides and particles in the air also potentiate one another. That is, when both substances are present at high levels, the effects on health are worse than if only one of the substances were present.

Why have sulfur oxides and particles appeared together in the past at high concentrations? The answer is that they have had a common source: the burning of coal. Coal is burned for electric power generation; before World War II it was used widely to heat homes in the United States.

Sulfur oxides and particles aggravate respiratory disease. When we first recognized that contaminants that fouled the air could cause death, sulfur oxides and particles were present at the scene. In the Meuse Valley of Belgium in 1930, in the tragedy at Donora, Pennsylvania, in 1948, and in the London Fog of 1952, a killer struck. Today, there is no longer any doubt that sulfur oxide pollution in the presence of particles was the killer.

For a time, we did not understand why in these pollution episodes the problem was always sulfur oxides *plus* particles. We have long known that particles act as a focus upon which water vapor condenses. We now understand that sulfur dioxide quickly dissolves in the water droplets, producing a highly acid, highly corrosive mist. It is this sulfurous acid mist that is thought to be responsible for so much damage to life and health.

In the London Fog of 1952, more than 4000 deaths were attributed to the excessive pollution levels. The number of individuals who became ill was obviously far greater. Sulfur dioxide levels may have reached 4000 micrograms per cubic meter of air. The death toll was heaviest among the elderly, a population group already afflicted with lung and heart diseases. This appears to be the way sulfur oxides and particles act, by making worse existing respiratory and cardiac diseases. Although many deaths during and after the fog were due to such respiratory diseases as pneumo-

nia and bronchitis, investigators were able to blame pollution levels because more deaths occurred than had been expected for that time of year. The 4000 deaths attributed to the fog were those deaths *in addition to* the number expected for that time of year.

Since those early episodes, scientists have accumulated masses of data, some indicating health influences even at very low levels of air pollutants. In New York City, for instance, when sulfur dioxide levels (averaged over 24 hours) exceeded 500 micrograms per cubic meter of air, a small increase in deaths was noted. The daily standard set for sulfur dioxide is 365 micrograms per cubic meter, so the margin of safety between the standard and the level at which excess deaths have occurred is not very large.

Individuals already suffering from bronchitis (excessive phlegm and difficulty in breathing) have been found to experience greater illness when particle concentrations exceed 250 micrograms per cubic meter and sulfur dioxide concentrations exceed 500 micrograms per cubic meter, averaged over 24 hours. At *annual* average concentrations for each pollutant of 100 micrograms per cubic meter of air or greater, higher rates of bronchitis among adults and respiratory infections among children have been observed. The standard set for the annual average concentration of sulfur dioxide is 80 and for particles 75 micrograms per cubic meter, so again, there is not a wide margin of safety. In the period 1960–1965, many major cities exceeded these standards. By 1975, only 5% of the monitoring stations were showing violations of the daily standard and 2.5% were showing violations of the annual standard. However, many violations were in highly populated areas, meaning that population risks were still high.

In London, since the devastating fog, particle levels have been cut from an annual average of 300 micrograms to 50 micrograms per cubic meter, and sulfur dioxide has been cut from 300 to 200 micrograms per cubic meter, averaged over a year. Investigators now find much less air pollution-related disease in London.

Particles in the air have been implicated as a cause of cancer. A **carcinogenic substance** is one that has been shown in the laboratory to produce cancer among experimental animals. Alternatively, the evidence that it produces cancer may come from observation of occupational groups inadvertently exposed to the substance. An example is the increased risk of lung cancer among asbestos workers.

A number of carcinogenic substances have been found in polluted city air, but no direct evidence exists that these compounds cause cancers in humans at the concentrations observed. That is, we can say that carcinogens are present in city air, but we cannot accuse them. In the language of the detective, the substances are "suspicious." They are suspicious because, in a number of countries, investigators have found more cases of lung cancer among urban dwellers than among people living in rural areas. Remember, however, that individuals who live in cities may have industrial jobs and different habits, such as smoking and so on. Since it is now clear that smoking is a principal cause of lung cancer, scientists have had to eliminate or correct for smoking in their studies. Even after this is done, the conclusion remains that urban dwellers suffer more lung cancer than rural residents.

In addition to these surveys within countries, studies have been made on British migrants to the United States. Of the study participants, those born in Britain who came to reside in the United States had two-thirds the frequency of lung cancer as the permanent residents of Britain. However, the incidence of their lung cancer still exceeded that of the general population of the United States. It bears mentioning that up to the present time, British life styles have differed from our own. Coal has been the main fuel for heating homes and buildings and for generating electric power, as well as for the production of "town gas," a gas with properties resembling natural gas but manufactured from coal. Furthermore, the number of autos per 1000 population has been about half that in the United States.

The airborne substances under greatest suspicion as potential carcinogens are polycyclic aromatic hydrocarbons, an example of which is benzpyrene. Known to cause cancer in experimental animals, these compounds have been found free or associated with particles of soot in urban air in the United States and Britain.

Children and Lead Poisoning

Lead is found in water and in food in trace amounts. It is also present in layers of old house paint. And lead is in the air because gasoline contains lead additives to make it burn more smoothly. These sources of lead, taken together, put urban children at great risk of developing lead poisoning. Children may begin to experience symptoms when the level of lead in their blood reaches 50 micrograms per 100 grams of whole blood, although the threshold for adults is about 60% higher. Surprisingly, many urban children have been found to have blood lead levels that are not far from the toxic limit. Evidence suggests that such levels just short of actual lead poisoning may be associated with learning difficulties.

Lead poisoning in a child is first seen as a variety of symptoms, including appetite loss, problems in discipline, and a lack of interest in play. The disease progresses to constipation, vomiting, seizures, and, finally, coma. A permanent loss of mental ability may result from a case of lead poisoning.

The risk of developing lead poisoning comes about in the following way. The food that we eat has

trace amounts of lead. It comes from soil, from pesticides, or from settling of airborne lead particles on the leaves of vegetables. Lead is also in water, arising from the weathering of rocks that contain lead substances and from lead particles that settled from the air. Scientists estimate that a one-year-old child ingests 130 micrograms of lead each day in food and drink.

City air is fouled with lead particles from the combustion of gasoline. The act of breathing moves these particles into the lungs. Lead has been measured in the air of cities at monthly averages up to five micrograms per cubic meter. At this concentration, a one-year-old inhales and absorbs almost as much lead from the air as he is absorbing from food and drink. The child's exposure to this toxic metal, then, can be nearly doubled because of the presence of lead in gasoline. The inhalation of street dust, which has been found to have high levels of lead, is still another route by which children are exposed to lead.

If lead in the air and in the diet were the only sources of lead, the threat to the urban child would be less severe. There is yet another source of lead, however, that exposes large numbers of children to the danger of lead poisoning: this is the lead in paint. Lead has long been used as a pigment in house paints, but since World War II, titanium dioxide has been gradually replacing lead oxide in paint. Since the end of 1973, lead has been banned from most paints in the United States. Nevertheless, houses from the pre-war era remain. Today many of these houses have fallen into disrepair, and the paint is crumbling from their walls. Mostly poor people inhabit these structures; their children may eat the poisonous paint chips.

Eating paint chips is tragically common. Small children of all social and economic classes are often observed eating nonfood items. It is the small child's way of sampling his environment. In this case, the sample is deadly indeed. A one-gram paint chip may contain up to 50,000 micrograms of lead. One estimate suggests that up to 2% of the children living in decaying pre-war dwellings may have levels of lead in the blood exceeding the toxic limit.

Lead poisoning is a public health problem of major importance. To guard against new cases, parents and children must be informed of the risk involved in eating paint chips. Detection programs are necessary to find those children already near the toxic limit. Where cases are found, the surface from which the paint peeled should be covered up or the paint should be completely removed.

Finally, lead should be removed from gasoline so that this extra source no longer contributes to the burden of lead that a child must carry. At present, lead is being removed from gasoline because it harms the catalytic converter. This is the device that has been added to new cars to burn hydrocarbons and carbon monoxide more fully. We may eventually see the day when nearly all gasoline is free of lead.

Questions

1. Where do sulfur oxides come from? Where do particles come from? What industrial activity using what fuel produces the most sulfur oxides?
2. Why does the gas sulfur dioxide cause mist to be acid?
3. What kinds of particles are produced by an auto?
4. Why do some individuals in the electric power industry say that tall smokestacks are the correct response to air pollution from sulfur oxides? What does the Environmental Protection Agency suggest instead? What kinds of pollution will be prevented if sulfur dioxide is removed from stack gases?
5. Salmon no longer run in certain streams in Norway where high acidity has prevented eggs from developing. That is, they no longer breed in, and hence run in, such streams. The salmon does not, however, "make a decision." Instead, the life cycle of the salmon plays a role in this process of adjustment. From study of salmon life cycles, biologists have determined that the salmon, an ocean fish hatched in fresh water, somehow nearly always returns to spawn in the same freshwater stream in which it began its life. Explain how this remarkable homing instinct on the part of the salmon would eventually lead, in the presence of a high acid content, to a stream without salmon. Would you expect the salmon to be eliminated quickly or over several years?
6. Why are particles regarded as an "accomplice" of sulfur dioxide?
7. What is the meaning of "excess deaths" and how does this concept fit into a study of the health effects of air pollution?
8. List five ways in which urban children are exposed to lead that may enter their systems.
9. Explain how the lead from gasoline comes to contaminate air, water, and food.
10. What is pica (you will need to look this up), and how is it related to the occurrence of lead poisoning?

Further Reading

Air Pollution Engineering Manual, 2nd ed., United States Environmental Protection Agency, U.S. Government Printing Office, Washington, D.C., 1978.

Although this is a highly detailed technical work, a number of portions are general descriptions that can be understood by the layperson; there are even discussions of the history of the technology. Whether all the details given here are of general interest, however, depends on the intent and training of the reader.

Air Quality Criteria for Particulate Matter, AP-49, U.S. Environmental Protection Agency, 1969.

Control Techniques for Particulate Matter, AP-51, U.S. Environmental Protection Agency, 1969.

Air Quality Criteria for Sulfur Oxides, AP-50, U.S. Environmental Protection Agency, 1969.

Control Techniques for Sulfur Oxide Air Pollutants, AP-52, U.S. Environmental Protection Agency, 1969.
While technical, these documents are very clearly written and very complete.

Controlling Air Pollution, American Lung Association, 1974.

"Landfill Made by Scrubbers," *Business Week,* 16 January 1978.
An easy-to-read description of the problems associated with the sludge from scrubbers.

ReVelle, C., and P. ReVelle, "Lead," in *Sourcebook on the Environment.* Boston: Houghton-Mifflin, 1974, p. 134.
A view of all sources of lead exposure and their relative contributions to lead levels in humans.

Waldbott, G., *Health Effects of Environmental Pollutants,* 2nd ed. St. Louis: C. V. Mosby Co., 1978.
This well-illustrated book (136 pages of illustrations) has a wealth of information and documentation on all major air pollutants. It is a technical work, but clearly written. The book could serve as an excellent reference for the health aspects of air pollutants.

To Breathe Clean Air, Report of the National Commission on Air Quality, March 1981, available from the Superintendent of Documents, Washington, D.C., 20402, 356 pp.
A most thorough compilation of regulations and recommendations. An excellent source of data.

Sulfur Emission: Control Technology and Waste Management, U.S. Environmental Protection Agency Decision Series, May 1979, EPA 600/9-9-79-019.
Description of the scrubbers and their impacts.

Hileman, B., "Particulate Matter: The Inhalable Variety," *Environmental Science and Technology,* **15,** No. 9 (September 1981), p. 983.
The new standards proposed by the U.S. EPA for particles with diameters less than 10 microns is discussed here. The conclusion is that EPA is leaving no "margin of safety" as dictated by the Clean Air Act.

Goldsmith, B., and J. Mahoney, "Implications of the 1977 Clean Air Act Amendments for Stationary Sources," *Environmental Science and Technology,* **12,** No. 2 (February 1978), pp. 144–149.
A discussion of the Prevention of Significant Deterioration (PSD) section of the Clean Air Act Amendments of 1977.

Patterson, Walt, "A New Way to Burn," *Science 83,* April 1983, p. 67.

Further Reading on Acid Rain: all articles are readable and nontechnical.

Luoma, J., "Troubled Skies, Troubled Waters," *Audubon,* November 1980.

"How Many More Lakes Have to Die," *Canada Today,* **12,** No. 2, February 1981, available from Canadian Embassy, 1771 N Street, N.W., Room 300, Washington, D.C., 20036.

Likens, G., "The Not So Gentle Rain," in *Yearbook of Science and the Future,* Encyclopedia Brittanica, 1981, pp. 212–227.

Gorham, E., "What to Do About Acid Rain," *Technology Review,* October 1982, p. 59.

Likens, G., "Acid Precipitation," *Chemical and Engineering News,* **54** (November 22, 1976), p. 29.

Cowling, E., "Acid Precipitation in Historical Perspective," *Environmental Science and Technology,* **16,** No. 2, 1982, p. 110A.

Burton, P., "Acid Rain: The Water That Kills," *National Parks,* July/August 1982, p. 9.

"Acid Rain," United States Environmental Protection Agency, Office of Research and Development, EPA-600/9-79-036, July 1980.

CHAPTER TWENTY-TWO

Photochemical Air Pollution

What Is Photochemical Pollution?
Nitrogen Oxides and Hydrocarbons: By-Products of Combustion/A Complex of Reactions/Effects on Plants and Animals

Control of Photochemical Air Pollution

Progress in Limiting Photochemical Air Pollution

Controlling Pollutants from Automobiles

Pollutant Standards Index and Summary of Standards

The Effects of Nitrogen Dioxide on Our Health

Chemistry of Photochemical Air Pollution

CONTROVERSY:

22.1: *Air Pollution Control: A "Policy beyond Capability"?*

What Is Photochemical Pollution?

A photochemical reaction requires light energy in order to take place. Certain pollutants in the atmosphere—nitrogen oxides and hydrocarbons—undergo photochemical reactions. These reactions produce new pollutants, including ozone, aldehydes, and exotic organic compounds. The new pollutants are referred to, in sum, as **photochemical air pollution** because they arise from photochemical reactions.

Thus, the emission sources responsible for photochemical air pollution are the sources of nitrogen oxides and of hydrocarbons. By the same reasoning, the control of photochemical air pollution requires controlling emissions of nitrogen oxides and hydrocarbons.

Nitrogen Oxides and Hydrocarbons: By-Products of Combustion

While nitrogen oxide is produced naturally from such occurrences as forest fires, the activities of human society are responsible for the high concentrations that occur in cities and near industries. During the high-temperature combustion of fossil fuels, two types of combustion occur that produce nitrogen oxides. In the first type, oxygen from the air and nitrogen *from the fuel* react to produce nitrogen oxides. Coal may typically have about 1% nitrogen content. Oil and gas may have 0.2–0.3% nitrogen content; it is this nitrogen that is likely to be oxidized.

In the second type of reaction, oxygen from the air reacts with nitrogen *from the air* to produce nitrogen oxides. Thus, a fuel with no nitrogen in it can produce nitrogen oxides during combustion. Nitrogen oxides

are produced whether the fuel is natural gas, coal, gasoline, or home heating oil. Approximately 95% of the annual emissions of nitrogen oxides stem from the combustion of fossil fuels. About 40% of the total emissions are from motor vehicle operation and other transportation modes. Another 30% result from the burning of natural gas, oil, and coal in electric power plants. Industrial combustion of fossil fuels produces an additional 20% of the total. The manufacture of explosives and nitric acid, two noncombustion sources, also result in emissions of nitrogen oxides. (See Figure 22.1.) Of the three basic fossil fuels, the

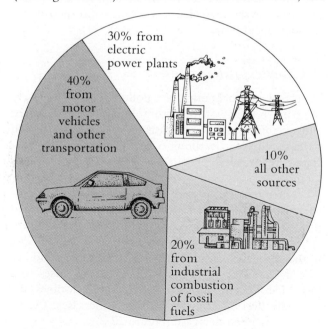

Figure 22.1 Sources of Nitrogen Oxides Emissions. (Data from: *To Breathe Clean Air,* National Commission on Air Quality, 1981)

burning of natural gas in all uses accounts for about 20%, the burning of coal about 25%, and the burning of oil about 47% of all nitrogen oxide emissions.

Hydrocarbons are released from many sources. Methane gas is present naturally in the atmosphere due to emissions from coal fields, gas, and petroleum fields. Methane emissions also stem from fire, and emanations of methane arise from swamps as well.[1] Methane, however, does not typically react in the atmosphere. Thus, even though methane background levels may reach a milligram per cubic meter, its presence does not concern us, since it appears to be harmless at these concentrations.

The hydrocarbons that do concern us are mainly by-products of the activities of civilization. Currently, one-third of the annual emissions of hydrocarbons stem from motor vehicle operation. In the cars of the 1960s, hydrocarbons evaporated from the carburetor and the fuel tank. They also escaped around the piston during compression. Most important, however, incomplete combustion left unburned hydrocarbons as droplets and gases in the exhaust stream. Although these sources of hydrocarbon emissions have now come under some control on new vehicles, significant quantities of hydrocarbons still enter the atmosphere from transportation activities. Smaller sources of hydrocarbons are petroleum refining and petroleum transfer operations.

A Complex of Reactions

Levels of photochemical air pollution follow closely the pattern of use of the automobile. During morning and evening rush hours, peak emissions of nitrogen oxides and hydrocarbons occur. These are the substances that react to produce the photochemical air pollutants.

Nitrogen and oxygen unite during the high-temperature combustion of fuel in an automobile engine to produce the gas nitric oxide, which enters the atmosphere. In several hours' time, the level of nitric oxide in the air decreases substantially. During the period of that decline, the level of nitrogen dioxide rises to a peak. Later, as the nitrogen dioxide level declines, the concentration of a third gas, ozone, increases. Ozone levels, too, then decrease (Figure 22.2).

---

1 The will-o-the-wisp seen in swamps is actually burning methane gas.

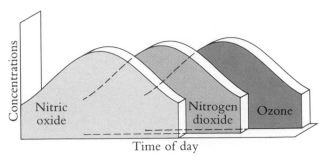

Figure 22.2 Levels of nitric oxide, nitrogen dioxide, and ozone during and after the morning "rush hour."

The reactions that produce high concentrations of nitrogen dioxide and ozone are not perfectly understood. Nevertheless, certain key relationships have been identified. The presence of hydrocarbons and the photochemical properties of nitrogen dioxide make these complex reactions possible. These reactions produce a typical pattern of pollutant levels over a period of time. Nitric oxide rises during the period of high emissions from motor vehicles. Nitrogen dioxide rises and nitric oxide falls due to the reaction of hydrocarbons with nitric oxide. After nitric oxide levels have fallen, ozone levels rise due to the photodissociation of nitrogen dioxide. (These complex reactions are explained in more detail on p. 431.)

Still other pollutants result from the consequences of these reactions. Hydrocarbons react with nitrogen dioxide to give peroxyacyl nitrate (PAN) compounds. Ozone reacts with hydrocarbons to produce aldehydes. Eventually, winds disperse all of these contaminants (Figure 22.3).

The term **oxidant** is applied to compounds capable of oxidizing substances that oxygen in the air cannot oxidize. Nitrogen dioxide, PAN compounds, ozone, and aldehydes are all oxidants. A concentration of oxidants is reported in terms of the weight of all such substances per cubic meter of air. The air-quality standard for photochemical air pollutants is given in terms of an allowable oxidant concentration. The concentration does not indicate how much of any specific compound is present, although ozone is usually the largest component. The Environmental Protection Agency under the Reagan Administration is now proposing to state the standard in terms of only ozone, since this is the only contaminant actually being measured. Such a dilution of the standard seems ill-advised.

Figure 22.3 The buildings in this city fade into a haze of pollution. In the city itself, respiratory problems are likely to be increasing. (National Archives, Documerica Collection)

Effects on Plants and Animals

Hydrocarbons appear to have little direct effect on plants or animals, although one hydrocarbon, ethylene, seems to stimulate plant growth. Elevated nitrogen dioxide levels, in one study, were seen to produce a higher incidence of human respiratory illness. (Health and biological effects are discussed in detail on p. 428.) The gas also causes some dyes used on synthetic fabrics to fade.

The photochemical oxidants, on the other hand, appear to have a wide variety of effects on plants and humans. Ozone and PAN compounds damage the leaves of common vegetables. Ozone also causes citrus trees to lose their fruit early. In 1982, the Office of Technology Assessment released figures on the value of crop damages from air pollution. The report studied the damages due to only one air pollutant, ozone. In addition, the study limited its analysis to just four crops: soybeans, corn, wheat, and peanuts. The conservative estimate of damages to these four crops from ozone was between $1.9 billion and $4.5 billion per year, as reflected in lower crop yields.

In humans, ozone and aldehydes irritate the respiratory tract. The eyes may be irritated by aldehydes and by PAN compounds. Individuals with asthma have found their disease aggravated by these oxidants, often severely.

The health-based standard for photochemical oxidants measured as ozone is 240 micrograms per cubic meter, averaged over one hour. It is not to be exceeded more than once per year. The standard was relaxed to this level in 1979 from a level one-third lower under pressure from industry on the claimed basis of more health effects data.[2] Recall that hydrocarbons and nitrogen oxides levels are controlled in order to control the level of photochemical oxidants. Thus, we can expect a push from auto industry lobbyists to loosen the standards on hydrocarbons and nitrogen oxides. It happened once before in just this way.

In the interval 1973–1976, Los Angeles experienced one-hour oxidant levels exceeding 300 micrograms per cubic meter on over 15% of the days; that is, during about 60 days per year, on average, oxidant levels were 25% higher than the standard. (See discussion of the new standard on p. 430.)

Control of Photochemical Air Pollution

The control of oxidants is accomplished by controlling hydrocarbons and nitrogen oxides. Because the automobile is responsible to such a large degree for these pollutants, efforts are under way principally to decrease the amounts from this source (Figure 22.4).

The hydrocarbons that blow past the pistons in an automobile engine are now being recycled to the combustion chamber. Those that evaporate from the carburetor and fuel tank are also recycled. Emissions of hydrocarbons from the exhaust system are decreased by mixing more air with the gasoline when it is burned. Beginning with the 1975 models, many new cars were equipped with a catalyst system. The cata-

2 See *Science,* **202** (December 1, 1978), p. 949; **203** (February 9, 1979), pp. 500, 529.

Figure 22.4 The automobile is responsible for much of the air pollution in cities. It is the source of most of the carbon monoxide, and the auto generously contributes nitrogen oxides and hydrocarbons to urban air. Nitrogen oxides and hydrocarbons are the substances out of which photochemical smog is formed. (EPA Documerica)

lyst system oxidizes unburned hydrocarbons in the exhaust to carbon dioxide and water vapor. Unleaded gasoline must be used by vehicles with the catalyst system to prevent "poisoning" the catalyst and making it ineffective.

Another catalyst system is used to control emissions of nitrogen oxides. This device reduces the nitrogen oxides back to molecular nitrogen.

Power plants that burn coal, gas, or oil are another large source of nitrogen oxides. A shift to nuclear electric power would reduce nitrogen oxides from this source, but there are many problems with nuclear power. (See Chapter 18.) Decreasing the air intake in gas and oil boilers may also decrease emissions of nitrogen oxides from these sources. In general, changes in the combustion process in industrial burners seems to offer a great deal of promise for controlling nitrogen oxides.

Progress in Limiting Photochemical Air Pollution

Recall that photochemical air pollution results from the reactions of nitrogen oxides and hydrocarbons. Progress is measured, then, not only in terms of ozone and oxidant levels but also in terms of emissions and levels of nitrogen oxides and hydrocarbons. In 1981, the National Commission on Air Quality concluded that additional controls on hydrocarbons and nitrogen oxides would be required for a number of areas to meet the ozone standard.

On a nationwide basis, nitrogen oxides emissions increased by 18% between 1970 and 1979. The increase results from a 35% increase in vehicle miles during the 10-year period and from an increase in the use of coal for electric generation. EPA has projected that nitrogen oxide emissions could increase by the year 2000 by as much as 50% from levels in the late 1970s. (See Figure 22.5.)

Emissions of hydrocarbons over the decade have remained relatively steady. The 35% increase in annual vehicle miles travelled did not result in a significant increase in hydrocarbon emissions, most likely because of the successful control of hydrocarbon emissions from automobiles. (See Figure 22.6.)

Because emissions of nitrogen oxides are rising and emissions of hydrocarbons are not decreasing, it is not surprising that many sections of the country are having difficulty meeting the ozone standards. They would have even greater difficulty if the allowable level had not been increased in 1979 from 160 to 240 micrograms per cubic meter. That 50% increase in allowable level by itself may have removed 10–20 cities from the "not meeting the standard" list. Figure 22.7 indicates areas of the country that are very unlikely to be able to meet the ozone standard by 1982, but can be expected to meet it by 1987. Figure 22.7 also shows the areas that are unlikely to meet the standard even by 1987.

Ozone has been a problem in Southern California for many years and continues to blight that area because of an extremely high rate of automobile use. The Eastern United States is not taking a back seat, however; severe ozone pollution now afflicts the Northeast Corridor stretching from Boston down to Washington. In 1979, about 30% of the population in this densely inhabited area was exposed to ten or more days in which the ozone standard was exceeded. In fact, the entire northeastern United States from Wash-

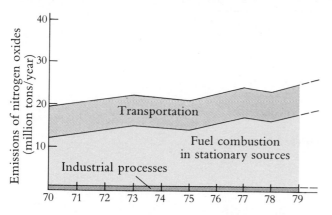

Figure 22.5 Emissions of Nitrogen Oxides Are Rising. Nitrogen oxides from "fuel combustion in stationary sources" refers to the burning of coal, oil, and gas for energy in electric power plants and factories, as well as the burning of these fuels in commercial and industrial buildings and in residences for space heating. "Emission from industrial sources" refers primarily to the manufacture of nitric acid. (Source: Trends in the Quality of the Nation's Air, U.S. EPA, October, 1980)

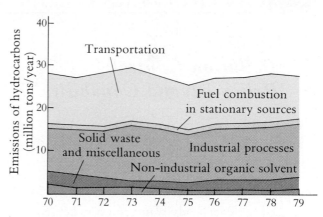

Figure 22.6 Emissions of Hydrocarbons Are Not Decreasing. "Fuel combustion in stationary sources" has the same meaning as in Figure 22.5. "Industrial processes" refers to refineries and the organic chemical industry. "Nonindustrial Organic Solvent" refers primarily to dry-cleaning processes. (Source: Trends in the Quality of the Nation's Air, U.S. EPA, October, 1980.)

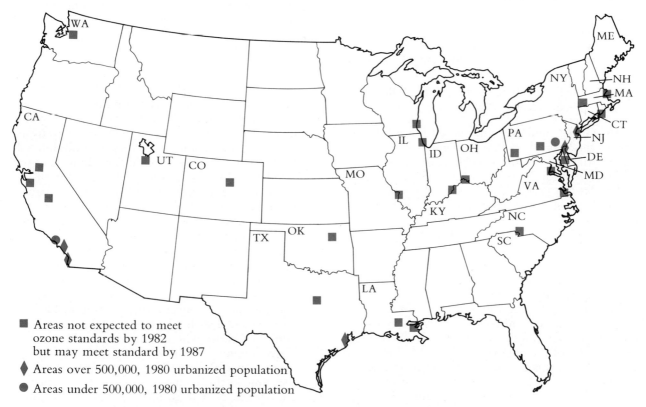

Figure 22.7 Areas Not Expected to Meet Ozone Standards by 1982 or by 1987. (Source: *To Breathe Clean Air,* National Commission on Air Quality, 1981)

Air Pollution Control:
A "Policy beyond Capability"?

> We think this is a necessary and reasonable standard to impose
> upon the [auto] industry.
>
> *Senator Muskie*

> Everybody wants to go to heaven, Miss Morgan, but nobody
> wants to die.
>
> *Bobby Gentry, "Casket Vignette"*

The 1970 Clean Air Act, called the "toughest air pollution law," was designed to promote the air quality we would like to have, not the quality of air we were sure we had the technology to achieve. Ever since then, scientists, industrial leaders, and politicians have been arguing over whether this was a good idea. Is it reasonable to base a law on what has been called "policy beyond capability"; that is, what you hope to achieve rather than what you are fairly sure can be accomplished?

In the Senate floor debate on the bill, the following exchange occurred:

> *Mr. Griffin.* Did the committee have any hearings in this session on this problem as to the state of the art—on the likelihood or possibility that this goal can be reached by 1975?
>
> *Mr. Muskie.* Yes, we had testimony jointly before the Commerce Committee and before our committee from the automobile companies on the state of the art. With respect to this specific deadline, no.
>
> *Mr. Griffin.* On this particular bill?
>
> *Mr. Muskie.* No.
>
> *Mr. Griffin.* No hearings?
>
> *Mr. Muskie.* The deadline is based not, I repeat, on economic and technological feasibility, but on considerations of public health. We think, on the basis of the exposure we have had to this problem, that this is a necessary and reasonable standard to impose upon the industry. If the industry cannot meet it, they can come back. . . .

> *Mr. Griffin.* . . . without adequate expertise, without the kind of scientific knowledge that is needed—without the hearings that are necessary and expected, this bill would write into legislation concrete requirements that can be impossible—and that will literally force an industry out of existence. . . .★

One scientist has commented:

> It goes without saying that "policy beyond capability" has many risks. . . . Reaching accommodations with the regulated industries . . . carries with it the risks of re-emergence of public concern, subsequent congressional investigation, and, perhaps most important, only limited progress in reducing air pollution. Enforcing the standards as written, on the other hand, will, in all likelihood, greatly expand those to be regulated. As one example, meeting the present federal air quality standards may well require controlling the number of automobiles in our cities. Since automobiles are such an important part of American life, regulating their use obviously can affect where we live, how we get to work, how we spend money, how we entertain ourselves. Whether public concern for pollution in 1969–1970 included such a broad mandate is uncertain—and those who assume that it did are clearly taking a risk in drawing those conclusions. Super-planning has never been super-welcome in this country. . . . In environmental matters, as with so many public policy issues, we are well advised to heed the wisdom of Bob-

bie Gentry in her song *"Casket Vignette"*: "Everybody wants to go to Heaven, Miss Morgan, but nobody wants to die."†

What do you think? Does setting standards that we are not positive we can meet act as a spur to the development of pollution control? Are we asking for trouble with this type of policy? What about the air we have to breathe? How much do you think Americans are willing to give up, if that is necessary, to assure themselves of clean air? See the section in this chapter, "Controlling Pollutants from Automobiles."

---

* *Congressional Record,* 21 September 1970, pp. 5160–5195.
† Charles O. Jones, annual meeting of the American Political Science Association, Washington Hilton Hotel, Washington, D.C., September 5–9, 1972.

---

ington, D.C. north to the Canadian border, bounded by the Atlantic Ocean on the east and Ohio on the west, has poor air quality in terms of ozone.

As bad as the northeastern U.S. is, Southern California is still worse. In 1979, the ozone standard was exceeded in the Los Angeles basin on 194 days; in San Diego, 66 days; and in the Ventura–Oxnard area, 73 days. Houston, Texas, logged 48 days above the standard in that same year, and St. Louis County, 50 days. The sunbelt cities of Reno, Las Vegas, Phoenix, Tucson, and Albuquerque likewise exhibit poor air quality in terms of ozone. In total, as the 1980s opened, nearly 144 million people lived in areas that did not meet the ozone standard. With emissions of nitrogen oxides rising and emissions of hydrocarbons fairly steady, progress toward achieving the ozone standard can be expected to be slow.

Controlling Pollutants from Automobiles

The preceding chapters, as well as this one, have listed the sources of the air pollutants that enshroud our cities. Two sources have predominated: motor vehicles and electric power plants, and of the two, the greater source of pollutants are motor vehicles.

In terms of carbon monoxide, 75% of the annual emissions in 1978 were from highway vehicles. In most major cities, 90% or more of the carbon monoxide burden is due to motor vehicles. In terms of particles in the air, motor vehicles are responsible for 7% of the annual emissions; the particles include both lead

oxides and droplets of photochemical pollutants.

Although motor vehicles account for only about 1% of the annual emissions of sulfur oxides, the auto is responsible for 55% of the yearly emissions of nitrogen oxides. Highway vehicles also produce 33% of the annual burden of hydrocarbons. With a record like this, it is no wonder that many control efforts have focused on the automobile.

The record of success overall has been very positive, largely as a result of the willingness of Congress to become involved on a regular basis, instead of delegating responsibility for regulation to a commission. With control of auto emissions began a new concept in environmental regulation—namely, "technology forcing." (See Controversy 22.1.)

In the Clean Air Act of 1970, Congress directed a 90% reduction in automotive emissions of carbon monoxide, nitrogen oxides, and hydrocarbons. The reduction was measured from the emission levels then occurring in 1970-model automobiles. When the auto industry was unable to meet the emission standards, Congress and the Administrator of the Environmental Protection Agency were forced to grant delays. To a very large degree, however, only delays were granted. The basic goals remained, to a considerable extent, firm. Only the standard for nitrogen oxides emissions was loosened.

Table 22.1 shows the time pattern of emission control that resulted from the Clean Air Act of 1970. Except for nitrogen oxides, the original goals set for 1975 and 1976 were finally met in 1981. The law stated that a 90% reduction was to be achieved from 1970 emissions levels. The figures in column (3) of Table

Table 22.1 Auto Emissions and Emission Standards in grams/mile

	(1) Typical auto of the mid-1960s	(2) 1970 Standards	(3) Originally required for 1975	(4) Originally required for 1976	(5) Achieved in 1980	(6) Achieved in 1981
Hydrocarbons	11	2.2	0.41	—	0.41	0.41
Carbon monoxide	80	23	3.40	—	7.0	3.4
Nitrogen oxides	4	—	—	0.40	2.0	1.0

22.1 do not correspond to a 90% reduction from column (2) because the testing procedure has changed several times since the original goal was set. If the 1970 emissions had been determined by the current method of measurement, the figures in column (3) would correspond to a 90% reduction from the 1970 emissions so measured.

Why has it taken so long to bring auto emissions almost under control? Was the means of regulation wrong? Should Congress not have set emission standards for automobiles, but left the task to the Environmental Protection Agency? Should Congress have paid more attention to the costs of achieving the emission standards? In fact, they commissioned a study on costs and feasibility by the National Academy of Sciences as part of the Clean Air Act of 1970. The Academy reported in 1972, two years after the act, that control was feasible and costs were reasonable. As Mills and White described it:[3]

> Congress had earlier simply authorized the administrative bureaucracy to set standards; now Congress itself was setting standards—and in the process showing contempt and vindictiveness for both the automakers and the federal agencies.

On the other hand, if the automakers had not delayed implementing then-feasible exhaust control technology in California in the 1960s, Congress might have been more trusting. Congress was suspicious that the auto companies were simply foot-dragging again. (See Controversy 22.1.)

The auto industry lobbies to this day for the relaxation of emission standards and will probably continue to do so until it decides to take credit for the cleanup.

A different problem confronts us today, however.

It is how to make sure that pollution control devices on autos are adequately maintained and are removing the quantities of pollutants they are supposed to remove. Vehicle inspection in order to be sure the devices are working has been strongly resisted by motorists. Tampering with pollution control devices has been found to occur on 12–19% of cars sold with these devices. No wonder vehicle inspections have been resisted. Now the enemy is no longer the auto companies; in the words of Pogo Possum, "it is us."

Pollutant Standards Index and Summary of Standards

The Pollutant Standards Index is a means by which the general public is informed of air quality. A radio or television announcer may say that the state health department has declared air quality is "moderate," or that air quality is "good," or perhaps "unhealthful." "Moderate" air quality means that one or more of the pollutants exceed 50% of its air-quality standard, but no pollutant exceeds 100% of its air-quality standard. "Good" air quality, on the other hand, means that none of the pollutants exceeds 50% of its air-quality standard. There is one exception to measuring each pollutant against its standard: ozone is measured against a different yardstick. Ozone's benchmarks are indicated in the accompanying chart (Table 22.2).

When one or more pollutants exceed 100% of their air-quality standards, but none exceeds the respective alert level, the air is regarded as "unhealthful." During intervals when the air is "unhealthful," persons with existing heart or lung disease should reduce exertion and outdoor activity. During such a period, even healthy people show evidence of pollutant irritation.

---

3 E. S. Mills and L. F. White, "Auto Emissions: Why Regulation Hasn't Worked," in *Approaches to Controlling Air Pollution*, A. F. Friedlander, Ed. Cambridge, Mass.: MIT Press, 1978.

An "Alert" is declared when one or more of the five major pollutants exceeds the value listed in Table 22.2 in the row labeled "Alert," but when none exceed the values in the row labeled "Warning." During an "Alert," the air quality is said to be "very unhealthful." Health scientists advise that those with heart and lung disease *and* those who are elderly should stay indoors and reduce exertion.

A "Warning" is declared if the concentration of one or more of the major pollutants exceeds the appropriate value in the row labeled "Warning," but none exceed the values in the row labeled "Emergency." Now, in addition to the cautions offered during the "Alert," *all* the population should reduce outdoor activity.

An "Emergency" occurs when the concentration of one or more of the pollutants exceeds the appropriate value in the row labeled "Emergency," but none yet exceed the values in the row labeled "Significant Harm." In addition to the previous advice, *all* people should now remain indoors and avoid physical exertion.

Table 22.2 Pollutant Standards Index and Air-Quality Categories

Index value	Air quality classification	Pollutant levels					Air quality regarded as
		Particulate matter (24-hour), micrograms per cubic meter	Sulfur dioxide (24-hour), micrograms per cubic meter	Carbon monoxide (8-hour), milligrams per cubic meter	Ozone (1-hour), micrograms per cubic meter	Nitrogen dioxide (1-hour), micrograms per cubic meter	
500	Significant Harm	1000	2620	57.5	1200	3750	
400	Emergency	875	2100	46.0	1000	3000	Hazardous
300	Warning	625	1600	34.0	800	2260	
							Very unhealthful
200	Alert	375	800	17.0	400	1130	
							Unhealthful
100	NAAQS	250	365	10.0	235		
						No short-term standard for nitrogen dioxide	Moderate
50	50% of NAAQS	75[a]	80[a]	5.0	180		
							Good
0		0	0	0	0		

a Annual primary NAAQS (National Ambient Air Quality Standard).
Note: An Alert, Warning, or Emergency is declared when any one of the pollutants exceeds the appropriate value, but none exceeds the value listed in the next more serious category.
(Reference: Pilot National Environmental Profile, 1977, U.S. Environmental Protection Agency, October 1980)

Table 22.3 Primary (Public Health) Standards for Air Quality (As of January, 1983)

Pollutant	Time period of standard[a]	Maximum permissible concentration
Suspended particles	annual geometric mean[b]	75 micrograms per cubic meter
	maximum 24-hour concentration	260 micrograms per cubic meter
Sulfur oxides	average annual concentration	80 micrograms per cubic meter
	maximum 24-hour concentration	365 micrograms per cubic meter
Carbon monoxide	maximum 8-hour concentration	10 milligrams per cubic meter
	maximum 1-hour concentration	40 milligrams per cubic meter
Oxidants/Ozone	maximum 1-hour concentration	240 micrograms per cubic meter
Nitrogen dioxide	average annual concentration	100 micrograms per cubic meter
Hydrocarbons	maximum 3-hour concentration	160 micrograms per cubic meter

a All standards based on concentrations over 24 hours or less are not to be exceeded more than once per year.

b The geometric mean is the antilog of x, where x is the sum of the logarithms of the data divided by the number of data points.

Finally, when one or more pollutants exceed the level associated with significant harm, air quality is at its lowest ebb. Vehicle use may be curtailed by closing all but the most vital government offices, closing banks, schools, food stores, and the like.

The standards for the major air pollutants are summarized in Table 22.3.

Effects of Nitrogen Dioxide on Our Health

The combustion of fossil fuels in engines and furnaces produces such intense heat that pollutants are formed from two natural substances in the air: nitrogen and oxygen. These pollutants are the nitrogen oxides. Nitrogen and oxygen combine at the extremely high temperatures that exist in internal combustion engines such as that of the automobile. The combination also takes place in boilers such as those of the coal furnaces of electric power plants. About 90% of the nitrogen oxides are produced in the form of nitric oxide (one atom of nitrogen plus one atom of oxygen). The remaining 10% are in the form of nitrogen dioxide (one atom of nitrogen and two atoms of oxygen).

Most of our information on health effects concerns nitrogen dioxide. Nitrogen dioxide initially composes only 10% of nitrogen oxides emissions. However, a complex series of chemical reactions in the air converts much of the nitric oxide to the more hazardous nitrogen dioxide form. These reactions, which are not, strictly speaking, an oxidation by molecular oxygen, are stimulated by the presence of sunlight. Reactions induced by the presence of sunlight are termed "photochemical reactions." At this time, the details of the conversion of nitric oxide to nitrogen dioxide in sunlit air have not been fully understood. Whatever the process, we know that nitrogen dioxide is a product, and scientists have recently learned some of the hazards of this substance.

Nitrogen dioxide is a foul-smelling gas. Even at levels as low as 120 grams in a *billion* grams of air, about one-third of a group of volunteers were able to detect its presence. This is equivalent of about 0.23 milligrams of nitrogen dioxide in one cubic meter of air (at sea level). The ability to detect the gas disappeared, however, after only 10 minutes of exposure, although people reported a dryness and roughness of the throat. Even these irritations vanished after prolonged exposure to levels 15 times the odor threshold discussed here.

Not only does nitrogen dioxide affect the sense of smell, but night vision is altered by this gas. The ability to adapt one's eyes to see in the dark is decreased. This phenomenon has been observed at levels as low as 75 grams of nitrogen dioxide per billion grams of air (or 0.14 milligrams per cubic meter). The visual and olfactory responses to nitrogen dioxide may be called sensory effects. More important are the effects of nitrogen dioxide on disease and on the functioning of the human body.

Two effects of nitrogen dioxide on body functions have been noted. One is the greater effort required in breathing; medical investigators refer to this as increased airway resistance. The response has been observed in normal healthy individuals at gas concentrations as low as 30 grams per billion grams of air (0.056 milligrams per cubic meter). Individuals with chronic lung diseases experienced breathing difficulty at only 20 grams per billion grams of air (0.038 milligrams per cubic meter).

Another body function response to nitrogen dioxide has been suggested but remains to be verified. Measurements by a team of Czech scientists indicate that nitrogen dioxide can form methemoglobin in the blood. The investigators feel nitrogen dioxide gas can attach to hemoglobin, just as carbon monoxide does, thus keeping hemoglobin from carrying oxygen to the tissues. It is accepted that nitrites in food can attach to hemoglobin, forming methemoglobin, but this is the first suggestion that nitrogen dioxide gas can also.

A number of studies have noted increased respiratory disease in areas polluted by nitrogen dioxide, but the presence at the same time of other pollutants at relatively high levels make the results less useful. Firm conclusions on the causative agent become impossible. On the other hand, the study by Shy and others (see "Further Reading," second item) of over 4000 schoolchildren and parents in several portions of Chattanooga, Tennessee, does point toward nitrogen dioxide alone as causing increased respiratory disease. Although we say "causing," a more accurate statement would be that nitrogen dioxide made people more susceptible to the *pathogens* that cause respiratory disease. Nitrogen dioxide from a nearby TNT plant was thought to be the most concentrated by far of air pollutants in the study area. Researchers observed more colds, bronchitis, croup, and pneumonia among the population group exposed to these higher nitrogen dioxide levels than among a less exposed but otherwise similar population nearby. It was not clear, however, whether the increased respiratory illnesses were a response to sustained levels of nitrogen dioxide (as indicated, for example, by the average annual concentrations) or to short-term concentrations (as indicated by average daily levels). Studies of the effect of nitrogen dioxide on the susceptibility of animals to respiratory disease have given similar results. These animal studies offer very useful support to an hypothesis about the response of humans to a pollutant.

Investigators have also sought to link nitrogen dioxide to increased mortality (death) in addition to increased morbidity (disease) rates. Statistical analysis indicates a positive relation between nitrogen dioxide concentrations and a higher mortality rate from heart disease and cancer. That is, areas with higher nitrogen dioxide levels do have a greater number of deaths from these causes. In the two notable studies, though, the presence of other pollutants at the scene make it difficult for scientists to draw conclusions. In addition, a limited number of sampling stations in each study area leaves the pollutant profile in each area incompletely described.

Exposure and Susceptibility

Individuals with chronic (continuing) respiratory diseases, such as emphysema and asthma, and individuals with heart disease may be more sensitive to direct effects of nitrogen dioxide; we do not yet know whether this is true. What we do know, however, is that nitrogen dioxide is associated with increased cases of short-term respiratory disease. Individuals with chronic heart and respiratory disease are more likely to develop complications from these short infections—dangerous complications, such as pneumonia. On the order of 10–15% of the population in the United States are thought to have some form of chronic respiratory disease.

This line of reasoning leads us to conclude that the standard for nitrogen dioxide should be set at a level that protects the population from increased respiratory infections. The standard set for nitrogen dioxide is an average annual concentration that should not be exceeded. Its value is 50 grams per billion grams of air or 100 micrograms per cubic meter of air.

Long-term exposure to nitrogen dioxide at levels of 500 grams per billion grams has been shown to decrease resistance to respiratory disease. Thus, the allowable concentration is set at one-tenth the level known to have effects on health. This is a fairly common margin of safety employed in setting standards. On the other hand, one of the statistical investigations—not fully accepted for reasons just cited—linked an increase in cancers to levels near the annual air-quality standard.

At present, no short-term standard (for instance, an average daily concentration) has been set, although a value was briefly proposed in 1971 and then abandoned. In the past, peak (instantaneous) levels of nitrogen dioxide in excess of 400 micrograms per cubic meter of air have been commonly observed in major cities.

Biological Effects of Photochemical Air Pollution (Smog)

The standard for oxidant concentration in the air, as pointed out earlier, is 240 micrograms per cubic meter, averaged over one hour. This concentration is not to be exceeded more than once each year. The new standard is for ozone alone, rather than all oxidants as the previous standard was, even though other oxidants can account for as much as one-fourth of the total oxidants present in the air. The new standard is also 20% higher than the level at which "alerts" were called previously.

The standard was revised to this level in 1979 from 160 micrograms per cubic meter, a level set in the early 1970s. During 1974–75, the old standard was violated on 38% of the days at one or more of the stations in Los Angeles and Philadelphia. Denver and Washington reported such violations on 27% of the days in 1974–75, and Cincinnati and Houston noted violations on 23% of the days in the same interval. Thus, the relaxation of the standard from 160 to 240 micrograms per cubic meter gives the appearance of significantly better air quality.

The new standard appears to be a compromise

between air quality and the expense of controlling pollution. At the level of the new standard, performance of a high-school cross-country team was observed to deteriorate. Asthmatics have been noted to have an increased frequency of attacks in the range of 300–500 micrograms per cubic meter, the lower level a mere 25% above the current standard. The fraction of the population afflicted with bronchial asthma has been estimated at 3–5%. Individuals with chronic lung disease also have been found to be affected by photochemical oxidants. Eye irritation is a significant effect of photochemical air pollution. At hourly averages as low as 200 micrograms per cubic meter, such eye irritation could be expected to begin. The standard, or rather the goal to be attained, would not exclude irritation. An unanswered question is "Is eye irritation a health effect?" Does it indicate that other effects are beginning?

A number of studies of laboratory animals were also used in setting the old standard. In the presence of oxidants in the air, mice became more susceptible to bacterial infections; red blood cells and the cells of heart muscles became misshapen. A three-month exposure to ozone levels at about two-thirds of the old standard produced a decrease in the weight of rats and disrupted the physiological chemistry of the animals. Hamsters were found to have damage to the chromosomes of white blood cells on short-term, low-level exposures.

Even before effects of oxidants on health were documented, investigators found damage to plants. Tiny dark spots called "stipples" were discovered on the leaves of plants in the 1940s. The condition was first observed in Los Angeles, but has since been seen across most of the United States. The spots are the result of elevated ozone levels. Such levels result when excessive automobile use leads to high concentration of hydrocarbons and nitrogen oxides. Laboratory studies show that many common vegetables are damaged by ozone and also by PAN compounds. The gas also affects citrus trees by causing the fruit to ripen and fall at a smaller size. Many plants show damage at oxidant concentrations between 80–160 micrograms per cubic meter, levels that are below the new standard.

Chemistry of Photochemical Air Pollution

Nitrogen oxides and hydrocarbons are the substances that react to produce photochemical air pollutants. Nitrogen and oxygen unite during the high-temperature combustion of fuel in an automobile engine. The product is principally nitric oxide; less than a hundredth of the product is in the form of nitrogen dioxide.

$$N_2(g) + O_2(g) \longrightarrow 2\,NO(g)$$

$$\text{nitrogen} \quad \text{oxygen} \qquad \text{nitric oxide}$$

As the exhaust mixes with air, about one-tenth of the nitric oxide is oxidized to nitrogen dioxide.

$$2\,NO(g) + O_2(g) \longrightarrow 2\,NO_2(g)$$

$$\text{nitric oxide} \quad \text{oxygen} \qquad \text{nitrogen dioxide}$$

However, once nitric oxide is well diluted in air, further conversion to nitrogen dioxide by this direct oxidation step no longer takes place. In most American cities, nitric oxide levels reach their peak during the morning rush hours. Even though further direct oxidation of nitric oxide to nitrogen dioxide does not take place, we note that, in several hours' time, the level of nitric oxide has declined considerably. Coincident with this decline, nitrogen dioxide levels are rising. As the concentration of nitrogen dioxide diminishes, a third gas, ozone, increases. Ozone levels, too, then wane. The reactions producing high levels first of nitrogen dioxide and then of ozone are not yet fully known or explained. What is certain, however, is that the presence of hydrocarbons and the photochemical properties of nitrogen dioxide make the reactions possible.

Hydrocarbons, which are simply compounds of carbon and hydrogen, are of two types. Saturated hydrocarbons have no room for additional hydrogen atoms. The carbon atoms are, in essence, completely reacted (or saturated). Unsaturated hydrocarbons, on the other hand, react readily with many substances. The reactions eventually exhaust the capacity of the carbon atoms to react. Methane, the background hydrocarbon mentioned earlier, is saturated and generally not capable of entering further chemical reactions in the atmosphere.

There is a series of reactions in which the primary pollutants, nitric oxide and hydrocarbons, are converted to the photochemical pollutants nitrogen dioxide, ozone, aldehydes, and PAN compounds. The energy of light makes these reactions possible. Light falls on molecules of nitrogen dioxide, produced as exhaust gases are emitted. In the presence of light energy, the nitrogen dioxide dissociates (comes apart) into a molecule of nitric oxide and a single free atom of oxygen.

$$NO_2(g) \longrightarrow NO(g) + O(g)$$

Free oxygen atoms are not normally present in the atmosphere because of their enormous reactivity. These free oxygen atoms react in two ways. One series of reactions is circular and leads back to the initial substances. The series, known as the "photolytic cycle," begins with the dissociation step. Atomic oxygen then reacts with oxygen molecules in the air to produce ozone. Ozone, in turn, reacts with nitric oxide to return nitrogen dioxide and oxygen. The cycle is shown as follows:

$$
\begin{array}{ccc}
NO_2 & \xrightarrow{\text{Dissociation}} & NO + O \\
+ & & + \\
O_2 & & O_2 \\
 & \nwarrow & \downarrow \\
\text{Cycle} & & NO + O_3 \\
\text{completed} & &
\end{array}
$$

While there is only a small decrease in nitrogen dioxide, the presence of ozone in the air is established by this continuously operating cycle.

The second series of reactions that follows the dissociation step is more important, however. The free atomic oxygen also reacts with unsaturated hydrocarbons to produce extremely reactive organic molecules called *free radicals*. These free radicals react with the nitric oxide initially present, producing nitrogen dioxide. This is a crucial reaction because more nitrogen dioxide is produced in this step than dissociated to begin the reactions. The result is a build-up of nitrogen dioxide levels. We cannot attempt to write balanced equations here because scientists still do not thoroughly understand the reaction. In words, the reactions are:

atomic oxygen + hydrocarbons ⟶

free-radical hydrocarbons

free-radical hydrocarbons + x NO →

x NO$_2$ + hydrocarbons

where x is the number of nitric oxide molecules that react with each free radical. The value of x is greater than 1.

Because of this sequence of reactions, nitric oxide levels fall and nitrogen dioxide levels rise. You may have noticed on smog-ridden days that the air takes on a brown tinge. This is due to the presence of brown nitrogen dioxide gas.

Another mechanism for nitrogen dioxide build-up and nitric oxide depletion is through the reaction of nitric oxide with carbon monoxide and oxygen:

$$NO + CO + O_2 \longrightarrow NO_2 + CO_2$$

Remember that atomic oxygen is being produced all the while by the dissociation of nitrogen dioxide in sunlight. Further, ozone is being produced by the reaction of atomic oxygen with the molecules of oxygen that are normally present. When nitric oxide concentrations become very low, the reaction of ozone with nitric oxides (see the photolytic cycle) cannot proceed. Hence, ozone levels build up.

In summary, a typical pattern of pollutant levels through the day is observed. Nitric oxide levels rise initially due to motor vehicle emissions. Then nitrogen dioxide concentrations rise and nitric oxide levels fall as the reaction of free-radical hydrocarbons consume the nitric oxide and produce the dioxide. Ozone from the photodissociation of nitrogen dioxide then builds up because of the absence of nitric oxide.

Questions

1. What are the chemical categories of air pollutants produced by the automobile? List three categories. Which categories are known to affect our health?
2. Why did Congress require that hydrocarbon emissions from the automobile be decreased?
3. After 10 minutes of exposure to an atmosphere with 0.23 milligrams of nitrogen dioxide per cubic meter of air, volunteers lost the ability to notice the smell of the gas. People who visit Los Angeles often note the brown haze over the city and smell pollution. Residents notice the smell less. Why?
4. Nitrogen dioxide decreases the ability of the eyes to adapt to seeing in the dark. Can you think of a common activity where personal safety could be threatened by the decreased ability to adapt to darkness after brightness? Explain. Is the activity likely to be undertaken in air with elevated levels of nitrogen dioxide? Why?
5. Animal studies done in the laboratory indicate an increased susceptibility to respiratory disease in the presence of nitrogen dioxide. These studies support epidemiological evidence, that is, evidence obtained by observing the disease response of human population. Why are scientists careful about applying animal studies to setting pollution standards for humans? What animal might be advantageous to use from this standpoint? What is their principal disadvantage?
6. What are the names of the photochemical air pollutants? Why are they called "photochemical" air pollutants? How are they controlled?
7. The air-quality standards for photochemical pollutants were set in such a way as to protect the health of a portion of the population that was particularly susceptible to the substances. Who are these people?

Further Reading

Air Quality Criteria for Hydrocarbons, U.S. Environmental Protection Agency AP-64, 1970.

Air Quality Criteria for Nitrogen Oxides, U.S. Environmental Protection Agency AP-84, 1971.

Air Quality Criteria for Photochemical Oxidants, U.S. Environmental Protection Agency AP-68, 1970.

Control Techniques for Carbon Monoxide, Nitrogen Oxide, and Hydrocarbon Emissions from Mobile Sources, U.S. Environmental Protection Agency, AP-66, 1970.

Although technical, these documents are very clearly written and very complete.

Waldbott, George, *Health Effects of Environmental Pollutants,* 2nd ed. St. Louis: C.V. Mosby Co., 1978.

This well-illustrated book (136 pages of illustrations) has a wealth of information and documentation on all of the major air pollutants. It is a technical work, but clearly written. The book could serve as an excellent reference for the health aspects of air pollutants.

CHAPTER TWENTY-THREE

Energy Use and Oil Pollution

How Oil Pollution Starts
Used Motor and Industrial Oil/Normal Shipping Operations/Tanker Accidents/Offshore Drilling

Biological Effects of Oil
Oil and Birds/Toxicity of Various Kinds of Oil/Detergents and Marine Life/Effects on Ecosystems/Oil in Food Chains/Biological Recovery after an Oil Spill

Economic and Social Effects of Mixing Oil and Water
Offshore Drilling/Deep-Water Ports

Lessening Oil Pollution and Its Effects
Reusing Waste Oil/Moderating the Effects of Offshore Drilling/Improving Tanker Safety/Cleaning Up Oil Spills

CONTROVERSIES:

23.1: *Save Palau: For What? For Whom? (Economic and Social Effects of Deep-Water Ports)*

23.2: *Are National Interests More Important than Local Concerns? (Offshore Drilling)*

How Oil Pollution Starts

"Oil pollution." The words bring to mind pictures of wrecked tankers grinding over submerged rocks or of geysers of flaming oil shooting from well blowouts. Yet, historically, such dramatic happenings have accounted for only a small portion of the 2–5 million tons of oil added each year to the world's waters.

Most of the oil spilled into natural waters has been the result, not of accidents, but of normal operations. Even in 1979, the worst year on record for accidents, twice as much used motor and industrial oil entered the oceans as was spilled in oil-well or tanker accidents (Figure 23.1).

Used Motor and Industrial Oil

In most years, half of the oil entering natural waters is used motor and industrial oil from sewer outflows and from runoff. Although anyone found dumping oil into a waterway in the U.S. is libel to heavy fines, there are still few convenient depots for the collection of such wastes as used automobile oil. Thus, used oil most often finds its way into sewers or garbage dumps. From there it runs into nearby waterways.

Normal Shipping Operations

All ships collect both oil and water in their bilges. The simplest method of disposing of this oily water is to pump it into the ocean. An international treaty, written in 1973 by the United Nations Intergovernmental Maritime Consultive Organization (IMCO), forbids the discharge of oily wastes near shores and limits the amount that can be dumped on the high seas. However, several important shipping nations did not even sign the treaty. In areas where dumping is prohibited by the treaty, illegal dumping still goes on, especially at night when it is hard to detect.

A further portion of oil pollution from shipping operations is due specifically to oil tankers. Many tankers pump sea water into empty oil tanks as **ballast** and use sea water to wash out tanks between cargoes. When such a tanker nears an oil-loading port, it has tanks full of oily water to dispose of. For many years, it was simply pumped back into the ocean. Newer ships are fitted with separate ballast and oil tanks or with slop tanks where oil can be recovered from ballast water. Further, some tankers now wash out tanks with oil rather than with water. However, all tankers and all ports do not have the special equipment needed for these procedures.

Tanker Accidents

Although accidents to tankers and offshore oil-well blowouts normally account for only a small percentage of the total amount of oil spilled, they result in most of the visible damage. The reason is that a grounded tanker or an offshore blowout releases an enormous quantity of oil at one time. Because it is cheaper to ship oil in large quantities, the size of the tankers traveling the oceans is increasing rapidly. The average-size supertanker now in use carries about 250,000 tons; it is as long as three football fields. Even larger ships have been built or are being considered, up to 1,250,000 tons. The possibilities for ecological damage if one of these tankers is wrecked are immense.

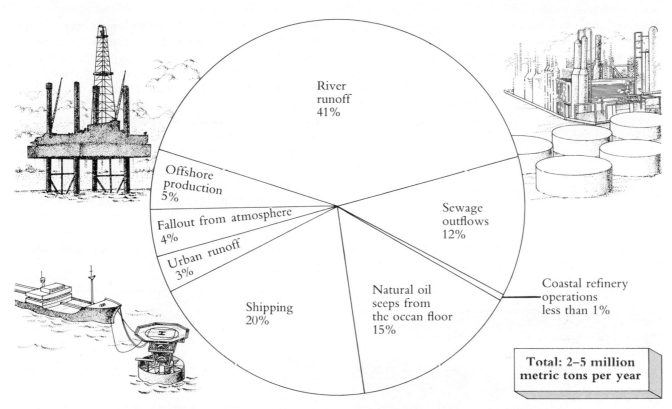

Figure 23.1 Sources of Oil in the World's Waters in the Early 1980s. Most of the oil is used motor and industrial oil carried in runoff and sewage inflows. Of the portion from shipping operations, the largest part is due to normal procedures. In most years, only a small portion comes from accidents. (Adapted from: Farrington, J. W., in *Petroleum in the Marine Environment,* Petrakis, L., and F. Weiss, eds., American Chemical Society, Washington, D.C., 1980)

Damage from the wreck of the 233,690-ton *Amoco Cadiz* in 1978 included total loss of the oyster beds along the Brittany coast, as well as effects on the fishing and resort industries. In tidal flats and marshes, these effects will last up to 10 years. Total costs could eventually exceed 30 million dollars.

Offshore Drilling

The **blowout** of the Mexican oil well *Ixtoc* in the Gulf of Mexico in 1979 caused the spill of more than 3 million gallons (about 10,000 tons) of oil before it was brought under control 9 months later. Oil and dead sea birds washed up on the beaches of barrier islands in South Texas, more than 3600 miles (6000 kilometers) away. The possibility of accidents such as this is a major source of concern about the speed-up in offshore oil development being undertaken by the U.S. government.

It is estimated that some 82 billion barrels of oil are yet to be found in the United States, and about one-third of this probably lies offshore under the continental shelf. Government officials want to accelerate the development of this resource, both to decrease reliance on foreign oil and to increase government revenues from the royalty payments oil companies make on the oil they obtain. Oil-well blowouts are not the only problem connected with development of this outer continental shelf oil resource, however.

During normal well-drilling operations, drilling muds and cuttings are released and chronic oil spills occur. The drilling muds contain a variety of chemical additives, including heavy metals as well as diesel oil. Later, while oil is pumped from the well, water comes along with it. This oil-and-water mixture is usually separated on the oil rig. The water, which still contains an average of 25 parts per million of oil, is discharged back into the sea. In addition, transporting the oil to shore involves small spills due to the loading and unloading of barges or pipeline leaks.

Biological Effects of Oil

Oil and Birds

Birds suffer during an oil spill because the oil soaks into their feathers, ruining the waterproofing and insu-lating qualities. The birds can no longer keep warm or float. Estimates of the number of birds killed during an oil spill are often low because many birds simply sink out of sight. As birds try to preen away the oil, they swallow it and are blinded and poisoned (Figure 23.2).

Oil contaminates or destroys sea birds' natural foods. Diving birds are especially affected, as they dive through the oil slick again and again in search of food.

Rescue attempts, in which oil is washed off the birds, also remove much of the natural waterproofing from their feathers (Figure 23.3). The birds must then be kept in captivity until they molt and grow a new set of feathers. It is extremely difficult to keep and feed large numbers of wild birds for any length of time. Thus, bird salvage rates have averaged only a disappointing 1–15%.

Figure 23.2 Santa Barbara Oil-Well Blowout. Much of what we know about the effects of oil on marine life is a result of offshore oil-well blowouts or tanker accidents. A case in point is the blowout that occurred in 1968 in the Santa Barbara Channel off the coast of California. An explosion at Union Oil Company's Platform A caused oil and gas to spurt up in a 100-foot column. Amidst a mist of oil droplets, the crew worked frantically to shut off the well. But then a second explosion occurred and poisonous gases began bubbling up. Faced with this further danger, the crew abandoned the platform. Oil began to pour out of a rupture in the sea floor. The oil slick grew to 200,000 gallons per day (about 700 tons/day). Seven days later, it came ashore on the beautiful Santa Barbara beaches (left) during a series of storms. Workers, including many volunteers, began to spread straw on the beaches and rake it up again, to absorb the oil (right). Others organized to try saving the sea birds, which staggered along the beaches or were washed into shore. At least 3600 birds perished as a result of the Santa Barbara blowout. (Left: Standard Oil Company, R. Isear; Right: Massachusetts Audubon Society, Saunders)

Figure 23.3 Rescue workers attempt to clean oil from birds. Diving birds are especially affected by oil slicks because they dive into the slick again and again in search of food. There is some evidence that the birds are attracted to slicks, perhaps by sick fish near the oil. (New York Zoological Society Photo)

Toxicity of Various Kinds of Oil

Although the slick of crude oil at Santa Barbara (Figure 23.2) washed in and out several times, which meant that the beach had to be cleaned and recleaned, marine life, aside from the birds, was not severely affected. Crude, or unrefined, oil is not as poisonous as certain refined oils. The portions of oil that boil off at very low heat are the most toxic. Gasoline and home heating oil are rich in these phenols and aromatic hydrocarbons. Since these substances also evaporate readily, the longer an oil slick can be kept away from a beach, the better. Part of these substances probably evaporated from the Santa Barbara slick before it reached the shore. In contrast, a spill of Number 2 home heating oil at West Falmouth, Massachusetts, in 1969, resulted in the death of numerous fish, bottom dwellers, and creatures in the intertidal zone (such as mussels, crabs, and starfish).

Detergents and Marine Life

When the tanker *Torrey Canyon* went aground in 1967, spilling 30 million gallons (about 100,000 tons) of oil, it was the first major oil spill ever to occur. Not much was known about the eventual effects of oil on marine life. The main concern of almost everyone involved was that the oil would foul the vacation resort beaches along the coasts of England and France. Since the summer holiday season was due to begin, local hotel keepers and the government directed almost all their concern toward keeping the oil off the beaches. When this proved impossible, the next thought was to get rid of the oil as quickly as possible. Since oil and water do not mix well, detergents were used to break up the oil into small droplets and enable it to mix into the water and wash away. Although this method was fairly successful in cleaning up the oily beaches, it proved disastrous to sea creatures.

We now know that detergents make oil more toxic to marine life. The oil–detergent mixture sticks to wet surfaces, like fish gills, where oil alone would not stick. The detergent enables oil to penetrate deeply into the sand and kill even those creatures who might otherwise burrow to safety. Further, those early detergents used low-boiling petroleum hydrocarbons as solvents, exactly those fractions of crude oil most toxic to marine life! Because of the enormous amount of detergent used (two tons per ton of oil), a massive kill of marine life resulted. Except for a few anemones, almost everything within a quarter of a mile from shore and in waters up to 7 fathoms deep perished. In a few rocky coves that were too isolated or difficult to spray with detergent, the crude oil disappeared from view by itself in about two years.

New detergent formulas no longer contain toxic solvents. Nevertheless, they are still hazardous to marine creatures. The Environmental Protection Agency limits their use in the U.S. to oil fires or other cases where human life or the major portion of a bird population would be endangered by an oil spill.

Effects on Ecosystems

In addition to its effects on individual aquatic organisms, oil affects whole ecosystems. There is a salt marsh in Southampton, England, where an oil refinery discharges, each day, 1500 gallons (5800 liters) of water contaminated with very low levels of oil (10–20 parts per million parts water). This chronic pollution has killed off over 90 acres (36 hectares) of marsh grasses in the area around the refinery. Once the grasses were gone, the sandy soil began to erode so that the affected area is now lower than its surroundings. Birds and other aquatic creatures that once found food in the area have been forced to move elsewhere. Thus, very small amounts of oil can, over a long period of time, have serious effects on an aquatic community.

In areas where many oil spills occur, such as harbors or the Main Pass oil field in the Gulf of Mexico, changes in the *kinds* of organisms are becoming apparent. For example, one study showed that organisms growing in bottom sediments in Timbalier Bay in the Gulf of Mexico were mainly two hardy species known to take over in polluted areas. The Gulf of Mexico has been contaminated with oil for so many years that it is no longer possible to find an unaffected area there to see what the natural communities were like.

In the North Sea, on the other hand, drilling operations began in 1973, and studies have been carried out since then. The studies found gradual increases in the amount of oil in sediments around the drill sites. In addition, definite decreases were seen in the number of species found and in the total number of organisms found. The area in which these decreases were seen became larger as time went on.

Oil in Food Chains

Many reports have been published about the tar lumps found floating far from shore in the oceans (Figure 23.4). These lumps are what is left of oil spills after the low-boiling fractions have evaporated. The lumps range in size from tiny crumbs to pieces as big as a man's fist. A bacterial film covers the lumps and even barnacles grow on them. Lumps have been found in the stomachs of saury, surface-feeding fish that are eaten by tuna. Thus, the lumps are finding their way into food chains that lead ultimately to humans.

Both oil and oil tars contain some cancer-causing substances. Several studies of shellfish grown in polluted waters have shown that they have an abnormally high number of growths similar to human cancer tumors. Oil, which is concentrated by shellfish such as oysters and clams, may be at least a partial cause of these tumors.

Figure 23.4 On Nauset Beach, Cape Cod, Massachusetts, workers use rakes to clean up oil pellets. Thirty miles of beach were littered with the tar balls, which ranged in size from raisins to footballs and washed ashore for a period of several weeks. The clean-up cost over $100,000. It is not known where the spilled oil, which led to the formation of the tar balls, came from. (NYT Pictures)

Biological Recovery after an Oil Spill

After oil is spilled, recovery time would certainly include the length of time it takes for all traces of oil to disappear. However, it also means the time necessary for the polluted area to be repopulated with the kinds and sizes of organisms that lived there before the spill. If a spill does not kill all the resident organisms, those left begin to repopulate the area, once poisonous parts of the oil have disappeared. Organisms from other areas also begin to move in, either swimming and floating (e.g., larvae) or creeping in from nearby colonies (sea grasses). Competition among species and predation begin to establish a balance between the different groups. But how long does oil persist in the environment after it has been spilled?

When the coastal barge *Florida* ran aground in Buzzards Bay at West Falmouth, Massachusetts, in 1969, there was one positive aspect. Scientists at nearby Woods Hole Oceanographic Institute were well equipped to investigate closely the effects of the spill. They discovered that even though oil may have disappeared from the surface of the water, its effects can still be far-reaching and serious. In Buzzards Bay, heavy seas and onshore winds ensured that the spill of home heating oil was well mixed with the water. In a few days, no oil could be seen floating on the surface. Five years later, however, oil could still be found in the sediments. Where the oil was found, there were fewer organisms than in neighboring regions. The oiled sediments had also eroded away to some degree. This seemed to allow the oil to spread further. The area in which oil was found was much larger than the original spill. Certain of the sea creatures studied, such as fiddler crabs, were still absorbing oil into their bodies. This oil caused them to behave strangely and prevented the population of crabs from increasing to the levels that existed before the spill. Scientists therefore warn that although oil may have only short-term effects in some areas, in others, such as marshes, the oil becomes mixed into the sediments and lasts for many years (Figure 23.5).

Economic and Social Effects of Mixing Oil and Water

Offshore Drilling

The highly populated states along the Atlantic coast are understandably uneasy about offshore drilling. The Northeast needs oil because it is located far from areas presently producing oil. But, while the area as a whole could benefit from the oil, the people living in the coastal states would bear all the environmental costs of producing the oil. These include possible oil spills from drilling and transporting the oil, as well as water and air pollution from onshore development of oil-refining facilities. Localized overcrowding may occur as workers stream in to develop the oil fields. In some economically depressed areas, people may welcome the possibility of oil-related jobs. However, unless there is careful planning by local officials and agreements are drawn up between the oil companies and the towns, few local people may be hired. In such a case, the native population will suffer inflation; housing shortages; overcrowding of schools, roads, and recreational facilities; and higher taxes, which accompany an oil boom, without sharing in any of the money it brings.

Deep-Water Ports

The development of deep-water ports presents some of the same economic and social problems onshore as does drilling for oil on the continental shelf. A deep-water port is a complex of buoys and a platform. It is located in deep water, some distance from the coast. Such distance is necessary to provide deep enough water to float the supertankers bringing the oil (Figure

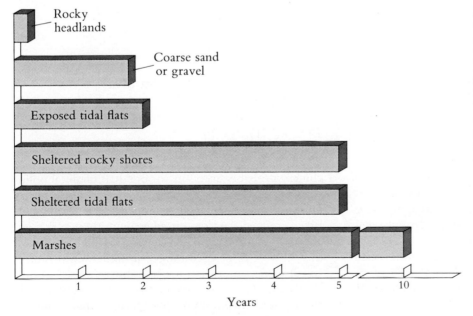

Figure 23.5 Persistence Time for Oil. On exposed rocky headlands, oil may disappear in days or weeks, while it can still be found five years later (as asphalt) in sheltered rocky shores. In sheltered tidal flats or marshes, the oil is still found 5–10 years later or even more. (Data collected after the *Amoco Cadiz* oil spill, from Gundelach, E. R., et al., in *1981 Oil Spill Conference,* American Petroleum Institute, Washington, D.C., 1981, p. 525)

Figure 23.6 A single-point mooring (SPM) facility can load tankers of any size in any weather condition. Crude oil is fed from land through a submarine line to this structure, standing in deep water several miles offshore. (An Exxon photo)

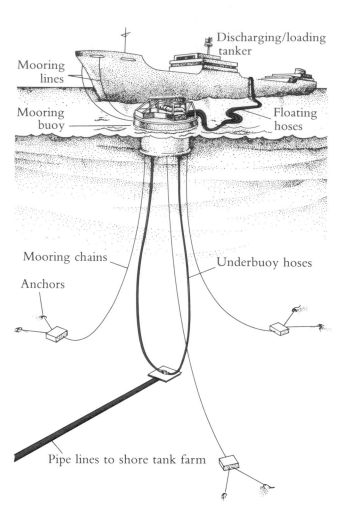

23.6). Currently, U.S. ports are served by tankers in the 50,000-ton range because the harbors are not deep enough to accommodate supertankers. From an environmental point of view, one benefit of the deep-water ports is that they can be located out of the mainstream of normal harbor traffic, thus reducing the possibility of tanker collisions. However, oil spills during loading and unloading, as well as pipeline leaks, pose a threat to ocean communities. In addition, dredging ship channels and other construction activities produce material that can cause disposal problems.

One of the best sites for a superport is the Palau Islands in Micronesia. The sociological effects of building a port there, however, are expected to be enormous. (See Controversy 23.1.)

Lessening Oil Pollution and Its Effects

Reusing Waste Oil

Used oil from automobiles, trucks, and industrial machinery probably accounts for one-half of all oil pollution. Schemes for reducing the pollution of water by oil must take this into account. Waste oil can be collected, refined, and then used again, although this is a complicated process. Many substances are added to oil that allow it to be used longer at higher temperatures and at increased bearing loads. Some examples of materials added to oil include compounds of barium, calcium, zinc, magnesium, phosphorus, and chlorine.

Save Palau: For What? For Whom?

(Effects of Deep-Water Ports)

Two hundred islands in a chain almost 100 miles (161 km) long make up the Palau Island district of the U.S. Pacific Trust Territory.* A few are inhabited; some are only a few yards across. But all are scattered among the coral reefs located in deep, clear, blue Pacific waters. The location of the islands and the depth of the surrounding waters make the area an ideal port for supertankers carrying oil from the Middle East to Japan. In Palau, the oil could be loaded off the supertankers and into small tankers, which are able to enter Japan's relatively shallow harbors. However, the dredging and blasting needed for construction of the superport would destroy parts of the coral reef and the very delicately balanced community of marine life it supports. Further, the construction workers, oil-port personnel, and all of the accompanying hustle and bustle might overwhelm the predominantly farming and fishing culture of the native islanders. The following quotes are from an article by Andrew Malcolm in *The New York Times,* 7 February 1977:

> An outside project of this magnitude is too big for this little place. It would control our politics, our economy, our institutions, our lives. It is far beyond us. And after centuries under foreigners, we want to be free, man, free.

> **Moses Uludong,**
> **a leader of the Save Palau Organization**

Speaking as an individual, the superport would be an environmental disaster of the worst order. The people would trade a few years of money for their whole fragile environment and culture.

The potential for environmental disaster here is much greater than even on the Alaskan pipeline.

> **Mr. Owen,**
> **Palau Conservation Officer**

I can't accept all this so-called environmental concern. The world has always had pollution and overcome it. These people want to keep Palau a human zoo so they can come and swim and take pretty pictures and then go home to their own lives while the people here starve without work. Save Palau, they say. For what? For whom?

> **Roman Tmetuchl,**
> **Businessman**

It is never easy to draw conclusions when environmental protection appears to be pitted against the human desire to upgrade living standards. What are your feelings about these questions: What environmental damage do you think the superport might do in Palau? What sort of disruption might the construction crews cause? (Consider, for instance, what services they will need, where they will live, what they will do for recreation.) How likely is it that native Palau Islanders will either be hired by or provide services to the construction and operating crews of the port? Would you side with Mr. Tmetuchl or Mr. Owen? Supposing you were an islander, with whom would you side?

---

* In January, 1980, the islands became the Republic of Belau.

Occasionally even vegetable and animal fats are added. Along with these additives, used oil picks up contaminants from the air and the machine itself.

However, new processes have been worked out that appear to remove all contaminants and produce an oil that works as well as virgin oil. Because the cost of virgin oil is rising, it is becoming economically attractive to re-refine waste oil. One of the main problems involves collecting this waste oil so it can be reused. In some areas, citizen groups have organized waste oil collection campaigns. This involves setting up depots where car owners can drop off used oil after they change the oil in their cars; it also involves educating people about the need to dispose of oil properly. The Federal Energy Administration has prepared a "Waste Oil Recycling Kit" to show citizen groups how to begin recycling centers for used oil in their communities.

Moderating the Effects of Offshore Drilling

One result of offshore drilling accidents such as the Santa Barbara disaster has been an improvement in the safety devices designed to prevent blowouts. Producing oil wells have safety valves designed to close off the wells if storms or earthquakes destroy the platform. Experts believe that these safety devices will work about 96% of the time (Figure 23.7).

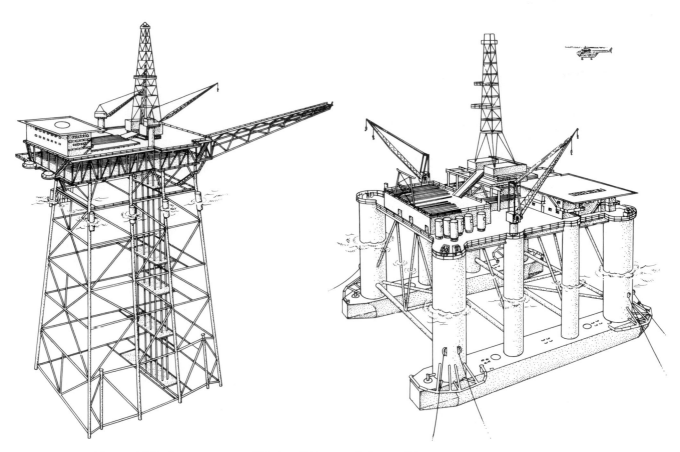

Figure 23.7 In waters up to 900 feet (300 meters) deep, oil is produced from permanent platforms fixed to the sea bottom. Several safety valves are attached to the system to prevent well blowouts. In deeper waters, or where storms are more frequent, floating platforms, or even platforms located almost entirely under water, may be used for drilling and production.

Are National Interests More Important than Local Concerns?

(Offshore Drilling)

> In the past few years we have written so many editorials
> supporting the Trans-Alaska Pipeline project, that now that
> the project is beginning we feel a bit hypocritical in . . .
> complaining about its impact. . . . By and large the average
> citizen is being hit, and hit hard, [by a] whirlwind of inflation,
> . . . skyrocketing real estate prices, . . . unconscionable rent
> increases . . . or eviction, . . . and traffic jams.

Fairbanks All-Alaska Weekly, *10 May 1974*

> This is a classic example of the broad public interest versus
> narrower local interests, and in this case local concerns do not
> always reflect national or even regional needs.

James Watt, Former Secretary of the Interior ★

Offshore oil drilling in several areas has been vigorously opposed by many environmental groups and sometimes by the states themselves. Clearly, such development could cause oil pollution due to catastrophic accidents or even just normal operations. For instance, oil is believed to lie beneath the sediments along 2/3 of Northern California's coastline. Wildlife also abounds along the coast, including all of the nation's endangered southern sea otters.

Coastal areas directly opposite the proposed drilling sites fear the social and economic upheavals a large influx of workers bring. Similar upheavals occurred during construction of the Alaskan pipeline. But such concerns are not limited to offshore drilling proposals. Communities and environmental interest groups also protest the siting of nuclear power plants, hazardous waste dumps, and oil refineries. James Watt (see above quote) appears to feel that national interests must take priority over local concerns.

Do you agree that national priorities (such as a need for oil to reduce dependence on foreign sources or to increase federal revenues) come before the interests of local residents? What efforts can the federal government make to reduce harmful impacts on local residents? Do you feel that the federal government should provide compensation for unavoidable harmful effects? Would this change your opinion about the first question?

What about potential harm to local wildlife or plant populations? Should local feelings about this be taken into account?

★ *The New York Times,* 3 June 1981, p. A20.

Figure 23.8 Corralling Spilled Oil. Workmen off a Perth Amboy, N.J., marina maneuver part of a 13,800-foot (4200-m) boom in an effort to contain a massive oil spill in the Arthur Kill and sweep it toward the shore and vacuum cleanup trucks. (William E. Sauro/NYT Pictures)

Still, because safety devices are not 100% effective and because there is always room for human error, a high probability exists that some spills and blowouts will accompany any offshore drilling. According to a report by the Council on Environmental Quality, Alaskan areas are unusually hazardous with respect to offshore drilling. An earthquake of a magnitude of at least 7 on the Richter scale can be expected in the Gulf of Alaska every 3–5 years. An earthquake in 1964 wiped out the original town of Valdez. (The new town is across the harbor from the terminal end of the Alaska pipeline.)

Environmentalists are rightly concerned about possible effects of offshore drilling on the fishing industries. Bristol Bay, one of the areas slated for development, has the richest fishing grounds in Alaska. The Outer Continental Shelf Bill, passed in 1978, includes strict safety procedures and also specifies that oil companies must set up a compensation fund for fishermen and landowners hurt by offshore spills.

Federal laws can also be used to control the effects offshore oil development has on nearby shore areas. The Coastal Zone Management Act provides funds for the states to examine their coastal resources and decide which areas must be protected as wilderness, which are suitable for development, and so forth. The law also requires that once a state has developed a coastal zone management plan that is accepted by the federal government, the federal government cannot take actions that are not in line with the state plans.

Improving Tanker Safety

The U.S. Department of Commerce estimates that as many as 85% of shipping accidents are due to human error. Part of the problem may be due to the existence of "flags of convenience." Certain countries, notably Panama, Liberia, Singapore, and Cyprus, have much lower standards for the ships and crews to which they grant registration. Because it is cheaper to register in such countries (both because standards are lower and because taxes are lower than in countries like the U.S.), a large proportion of the world's shipping fleet is registered in them. The Law of the Sea Treaty deals with this problem by confirming the right of any state to set standards for vessels entering its ports. (However, the treaty is yet to be signed by all the countries concerned and may founder on the issue of rights to seabed mining of minerals.) Safety features such as double hulls and bottoms on tankers, special tanker lanes, and traffic controls should certainly be considered to prevent disastrous spills. Size limitation should also be explored for tankers serving the U.S.

Figure 23.9 Fishing boats tow an oil skimmer through an oil slick to demonstrate how the booms (V-shaped arms) funnel oil into the skimmer (at the point of the V), where it is skimmed off the surface of the water and collected. Note the darker area of clean water following in the wake of the skimmer. Such systems do not work in choppy waters, where the oil slops over the top of the booms. (Photo courtesy of Clean Seas Inc., Santa Barbara, California)

Cleaning Up Oil Spills

If an oil spill does occur, what can be done about it? The oil slick must first be kept from spreading with barriers such as floating booms (Figure 23.8). The oil can then be soaked up with adsorbent materials or skimmed off the surface of the water. Various devices are available to skim oil from water (Figure 23.9). However, booms and skimmers only perform well in calm waters, on small to moderate spills. Ocean currents can carry oil under the booms, and in rough seas, oil does not form a slick at all, but forms droplets that mix into the ocean. In a storm, little can be done to prevent spilled oil from coming ashore on beaches. When oil does reach beaches, it is still best cleaned up by broadcasting straw and raking it up again. According to a report by the Maine Department of Environmental Protection:[1]

> Once the volume spilled exceeds the capabilities of present state of the art technology, man can then only realistically chip away at the oil with available recovery machines, walk behind the spill on the beaches physically recovering oil with rather primitive tools, attend to claims for various oil spill related losses, and measure the effects by counting nature's victims as they wash up on the beach.

---

1 *Statistical Report—Oil Spills 1976,* Maine Department of Environmental Protection, Bureau of Oil Conveyance Services. Quoted in *Maine EnvironNEWS,* **3**(13) (11 February 1977), 7.

Questions

1. What is the major source of oil pollution in the world's waters? Why are tanker accidents and oil-well blowouts significant, even though they contribute only a small percentage of all the oil spilled in water?
2. Suppose your class decided to start a waste oil recycling project in your community. How would you begin? How would you explain the need for such a project to citizens in your community?
3. What are the major effects of spilled oil on the water environment?
4. What are the problems associated with drilling for and producing oil on the outer continental shelf?
5. From the point of view of environmental safety, how would you rank obtaining oil from overseas by tanker versus by offshore drilling? Is environmental safety the only important consideration?

Further Reading

Wilson, S., and K. Hayden, "Where Oil and Water Mix," *National Geographic,* February 1981, p. 145.
An exploration of the wildlife refuges along the Texas coast where the protection of whooping cranes and oil production are in a fragile balance.

"Offshore Oil Drilling," *Environmental Science and Technology,* **15,** No. 11, November 1981, p. 1259.

Richardson, Eliot, "Law of the Sea," *Environmental News Region I,* March/April 1981, p.11.

A good summary of the provisions and rationale of this controversial treaty.

"The Language of Oil," Mobil Oil Corporation, 150 E. 42nd Street, New York, NY 10017 (1974).
This small booklet, produced by Mobil Oil, clearly defines 100 commonly used terms relating to oil production and oil pollution. It is a handy reference to have while reading on the subject of oil pollution. Write to the above address for copies.

References

Kerr, R. A., "Oil in the Ocean: Circumstances Control Its Impact," *Science,* **198** (16 December 1977), 1134.
This article provides a good summary of several studies done to determine the long-term effects of spilled oil.

Sanders, H., et al., "Long-Term Effects of the Barge *Florida* Spill," EPA PB 81 144-792, January 1981.

Petrakis, S. L., and F. T. Weiss, eds., *Petroleum in the Marine Environment,* American Chemical Society, Washington, D.C., 1980.

"Safety and Offshore Oil," National Research Council, National Research Press, Washington, D.C. (1981).

1983 Oil Spill Conference, American Petroleum Institute, Washington, D.C. 20037.

CHAPTER TWENTY-FOUR

Energy as a Pollutant: Thermal Pollution; Noise Pollution; High-Voltage Power Lines

Thermal Pollution

Review of the Power Generation Cycle/Efficiency and the Second Law of Thermodynamics/Cooling Cycles/Biological Effects of Thermal Pollution/Factors that Act in Combination with Temperature/Better Ways to Dispose of Waste Heat/ Closed-Cycle Cooling Schemes/Cooling Water and the Future

Noise Pollution

Measuring Sound/Noise versus Sound/Health Effects of Noise/Noise and Wildlife/Noise Exposure/Psychology of Noise versus Sound/Laws Governing Noise Pollution

High-Voltage Power Lines

Corona Effects/Electric Shocks/Electric Field Effects/Magnetic Field Effects/The Outlook

CONTROVERSY:

24.1: *Economics and Thermal Pollution Control*

In this chapter we examine three pollutants that are forms of waste energy: thermal pollution, noise pollution, and electromagnetic fields under high-voltage power lines. Unlike the air and water pollutants discussed in the previous chapters, which were pollutant substances resulting from society's generation or use of energy, these pollutants are forms of energy itself. Thermal pollution is waste heat energy from the generation of electric power. Noise pollution is sound energy that comes from a variety of sources. Noise pollution is also related to energy use, however, in that the major source of noise to which most people are exposed is transportation noise from airplanes, cars, trucks, and buses. Although both heat and sound have the somewhat unusual property (for a pollutant) of disappearing if they are ignored, both have effects on humans and on the environment that cannot be ignored.

Thermal Pollution

The generation of electricity produces waste heat that must be disposed of. When this waste heat is released into the environment, it can have harmful effects. For this reason it is classed as a pollutant: **thermal pollution.**

Review of the Power Generation Cycle

Whether a power plant operates by burning fossil fuels (coal, oil and gas) or by nuclear fission, the plant first generates heat, which is used to boil highly purified water to steam. The steam, under very high pressure, turns a turbine. The pressure of the steam when it leaves the turbine is markedly reduced. Such "spent" steam is no longer capable of turning a turbine. However, it is still very hot; that is, not all of the heat energy provided by combustion was transferred to the turbine. The energy that was not transferred is the origin of the "waste heat" that is discarded to the environment.

The water originally turned into steam is highly purified, to prevent solid deposits from building up. It is reused to avoid having to continually purify more water. Thus, the low-pressure steam is passed through a condenser after exiting from the turbine. In the condenser, the heat in the spent steam is transferred to cool water drawn from a lake, river, or the ocean. The steam condenses, and, at the same time, the cool water is warmed. The steam that condensed is now liquid water and cycles back to the boiler to be converted again to steam. The process is continuous in the sense that water is entering the boiler, being converted to steam, turning a turbine, and being condensed continuously (Figure 24.1).

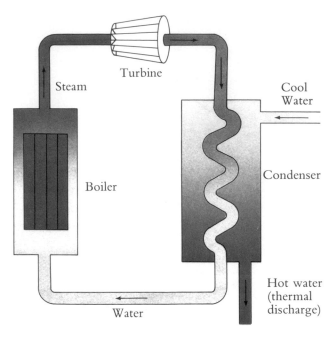

Figure 24.1 Summary of how thermal pollution occurs from an electric power generating station. Whether the power plant uses fossil fuels (coal, oil, gas) or nuclear fuel to boil water to steam, the basic cycle is similar to the one shown here.

Efficiency and the Second Law of Thermodynamics

A fossil-fuel power plant, burning coal or oil, is said to be, at best, 40% efficient. That is, for every kilowatt-hour of electricity produced (which is equivalent to 3400 British Thermal Units), the plant produces 5100 Btu of waste heat. Thus, almost two units of waste heat are produced for every usable unit of electricity. Nuclear power plants are even less efficient: only about 32% of the energy they produce goes to generate electricity. A small portion of the waste heat from fossil-fuel plants goes up the stacks. The rest of the waste heat from a fossil-fuel plant and all the waste heat from a nuclear plant goes into the cooling water.

Electric power generation provides an example of the **second law of thermodynamics.** That is to say, when energy is transformed from one type into another, some is always lost as heat. Here, the energy in fuels such as coal or uranium cannot all be used to produce electricity. Some is lost as heat. (This point is explained more fully in Chapter 15.)

Cooling Cycles

There are two basic types of cooling arrangements in electric generating plants. In the one we have been discussing, cool water is drawn from a nearby body of water, is used to condense steam, and then is discharged back to the body of water. This is called "once-through" or "open-cycle" cooling. Thermal pollution of water results from open-cycle cooling. In the second basic arrangement, called "closed-cycle" cooling, the heat that cooling water takes up to condense the steam is released to the atmosphere through a cooling tower. Since the cooling water circulates continuously through the plant in closed-cycle cooling, it would seem that no further water need be withdrawn or discharged after an initial quantity of cooling water is taken into the plant. Unfortunately, this is not the case; the cooling tower loses some water to the atmosphere and another small fraction of the water is drawn off and discharged. (This is discussed further on p. 454.) Water loss is made up by drawing new (makeup) water into the cooling system. The two cycles are shown in Figures 24.2 and 24.3.

In summary, then, there are three, and only three, media into which the heat in condenser water may be directed. The heat can be disposed of, with no further steps, in the body of water from which the condenser water was drawn. This is how heat was disposed of at most electric generating stations until the 1960s. Heated condenser water also can be applied without

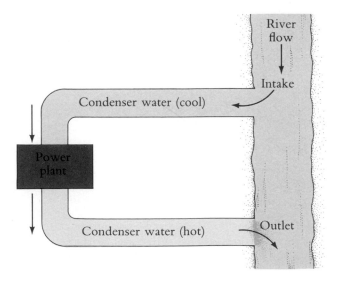

Figure 24.2 Open-cycle (once-through) cooling.

further steps to farmland to furnish both moisture and warmth to crops. Such use is uncommon, but has been experimented with on a small scale. Finally, the heat can be discharged to the atmosphere, through the use of cooling towers. This technique has been in favor since the first attempts to control thermal effluents.

In the next section we discuss the effects of thermal pollution on aquatic life when once-through cooling is used. Following that, we will look at the advantages and disadvantages of other ways of disposing of waste heat.

Biological Effects of Thermal Pollution

Off Turkey Point on Biscayne Bay, Florida, in the early 1970s, biologists measured an area of almost 75 acres (30 hectare) that was barren of life. Surrounding this was another 100 acres that was only sparsely populated. Yet Biscayne Bay is known to yield over 600,000 pounds per year of marketable seafoods such as spiny lobsters and stone crabs. Sport fish abound in the bay and a valuable bait-shrimp fishery is located there. Many fish and shellfish use the bay as a breeding grounds, while a variety of wading birds feed in the shallows. What caused the apparent ecological disaster

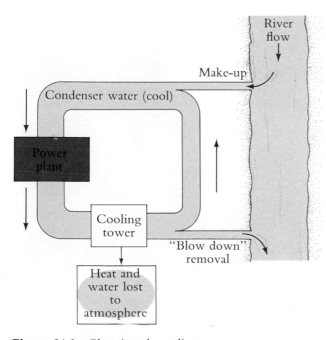

Figure 24.3 Closed-cycle cooling.

at Turkey Point in the middle of an area ordinarily teeming with aquatic life? The barren area was centered around an effluent canal from Florida Power and Light's Turkey Point Power Plant. At that time, the plant took cool water from the bay and returned heated water through the outflow.

Temperature is an important factor in the well-being of all organisms. For each species, there is a particular range of temperatures that supports life. Some organisms have adapted to living in the hot springs at Yellowstone National Park, where temperatures may reach 70°C. There are even fish in arctic waters that survive being frozen in ice. But for any particular species, the temperature range necessary for survival is relatively narrow; in some cases, very narrow indeed. For instance, some organisms that build coral reefs in the Caribbean can withstand no more than a few degrees change in temperature.

Warm-blooded animals like humans have evolved a variety of mechanisms to keep their body temperatures in the proper range. Digestion of food produces heat. Sweating increases heat loss when the body is hot and needs to cool off. Most aquatic creatures, however, are not able to maintain a particular body temperature. These creatures must stay the same temperature as the water in which they live. Those that can swim, such as fish, move about to find suitable temperatures. This is called behavioral regulation of temperature. Those that cannot move, such as adult oysters or rooted plants, are at the mercy of water temperatures. Outside certain limits, they simply cannot survive.

Acclimation. Both laboratory and field studies have shown that the temperatures that are lethal to an organism depend in part on the temperature at which the organism has been kept. Thus a fish held at 27°C for a few days or weeks may be able to survive longer when moved to 31°C than a fish that had been kept at 20°C. This improved ability to withstand higher temperatures is called **acclimation**. Nonetheless, in a series of increasing temperatures, there is always a temperature at which no length of acclimation time at a lower temperature can help an organism survive. When raised to this "ultimate upper lethal temperature," the organism dies within a short time. Organisms can acclimate in a downward as well as in an upward direction. There are lower as well as upper temperature limits for survival.

Furthermore, if given a choice, creatures like fish, which can move about, spend most of their time at a particular temperature, one they seem to "prefer." Thus, in the winter fish may congregate in warm-water outflows from power plants, not because they can't live in the naturally cold waters, but because they prefer their environment to be somewhat warmer. If the power plant must shut down during the winter for repairs or refueling, a fish kill can result. The fish will not be able to withstand the normal, cold water temperatures because they have acclimated to the warmer effluent temperature.

Survival and well-being of organisms. Thermal fish kills are relatively rare events. And, although fish kills are dramatic evidence of the effects of thermal pollution, less obvious effects can be even more serious. Temperature can affect the ability of organisms to reproduce without actually killing them directly. For instance, trout require cool summer temperatures for the formation of eggs and sperm. Although adult fish might be able to survive in warm summer water, they will not reproduce. For certain insects, hatching is triggered by increasing temperature. If waters are artificially warmed, the hatching temperature may be reached earlier in the year than is normal. At the Hunterston Generating Station in Ayrshire, Scotland, a species of intertidal sand-dwelling copepod hatches earlier than usual due to a heated effluent. It then dies off in greater than usual numbers because not enough food is available early in the year for the larvae.

Fish that are not killed by high temperatures still may be unable to catch food. Even more subtle changes have been shown to occur that cause thermally shocked fish to be "picked out" by predators. Heat-stressed fish may also be more susceptible to disease. In the long run, these sorts of effects can be just as devastating to the population as a direct thermal kill.

The effect of excess heat on ecosystems. Temperature can affect the whole community structure of an aquatic environment. For instance, different species of freshwater algae compete for light, space, and nutrients. Temperature changes can alter the competitive position of different species, even though the changes are not severe enough to be lethal for any species. At low temperatures (around 21°C), yellow-green algae may predominate in a lake community. As the temperature is raised to 26 or 32°C, green algae become

more abundant, and finally blue-green algae begin to dominate at very high temperatures (Figure 24.4). In this manner, heat can seriously affect aquatic food chains, since blue–green algae tend to be more resistant to grazing than other kinds of algae. In addition, blue-green algae characteristically have a greater mass than the species they replace. (See also Figure 24.5.)

As mentioned previously, heat can change the types of insects hatching at various seasons. Thus the available kinds of food can be changed by heat, leading to changes in the kinds of fish or other creatures a lake or river can support.

In fresh water, fish appear to be the most sensitive species. Protection of fish from excess heat would thus appear to protect freshwater ecosystems in general. In salt water, however, plant species may be more sensitive than fish to heated discharges.

Overall, the effect of heat is to simplify aquatic communities. That is, fewer species are found, although there may be many individuals of each species. One study found that fewer than half as many species were found at 31°C than at 26°C. Another 24% disappeared at a temperature of 34°C. Such a simple ecosystem may be less stable than the original more complex one.

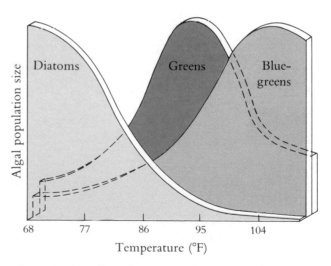

Figure 24.4 Effect of temperature on types of phytoplankton. [Reprinted from J. Cairns, Jr., "Effects of Increased Temperature on Aquatic Organisms," *Industrial Wastes,* **1**(4), 150 (1956). Used with permission.]

Figure 24.5 Wood Piling Riddled by Shipworms. Warm-water effluents from a New Jersey Central Power and Light Company nuclear plant on Oyster Creek appears to be allowing the spread of shipworms into Oyster Creek. The worms require warmer temperatures than are normally found in Oyster Creek. The larvae of shipworms eat their way into wooden structures like ship hulls or dock pilings, growing from pinhead size to more than two feet (0.6 m) long and one-half inch (1.25 cm) in diameter. (Carl Gosset/NYT Pictures)

Many natural waters are subject to seasonal variations in temperature. This variation allows different species to be dominant at different times, and therefore a greater number of species are able to compete for space and nourishment in a given area. A partial explanation for the effect of thermal pollution on populations may be that thermal additions can even out this temperature fluctuation. Thus, fewer species are able to coexist in a given area.

Many thermal-effect studies have been carried out with respect to the location of power plants. The studies seem to show that obvious harmful effects on ecosystems are more likely to occur from power plants located on naturally warm waters because organisms living in warm regions are often already near their upper thermal limits. The additional heat from the power plant pushes the organisms over their thermal limits.

Factors that Act in Combination with Temperature

Many factors act in combination with temperature to affect the survival and well-being of organisms.

Oxygen content of water. Higher temperatures increase the amount of oxygen fish need by increasing their metabolic rate. At the same time, the warmth decreases the amount of oxygen the water can carry. As a result, lethal temperatures are lower and acclimation may not occur. Hot water may also stimulate the growth of bacteria in polluted water. The bacteria then use up even more of the available oxygen. When measured in terms of BOD (Biochemical Oxygen Demand, see p. 243), heat can have the same harmful effect on water quality as a sewage outflow.

Chemicals in water. Increasing the salinity, or salt content, of water changes the effects of temperature, making some organisms more resistant to heat and cold and others less so. Other chemicals, such as pesticides, copper, and mercury, can also make aquatic creatures less able to withstand high temperatures.

Chlorine, used to prevent the fouling of cooling-water pipes by organisms such as slime-forming bacteria, creates problems because aquatic organisms are very sensitive to chlorine. Studies have shown that oysters are sensitive to less than 0.05 parts per million chlorine. Levels as low as 0.12 parts per million have caused almost three-quarters of the photosynthetic activity of algae to stop. In some cases, power plant damage to nearby waters is probably due more to the chlorine used than to the heat given off. Other techniques can be used instead of chlorine to prevent fouling. Examples are mechanical scouring and short periods of high-velocity water or high temperature in the condenser pipes. (As little as 30 minutes per week of water at 38°C can be effective.) All have been used alone or in combination with chlorine. In any case, it has been recommended that the amount of total residual chlorine[1] not exceed 0.02 parts per million.

Entrainment. Microscopic floating plants, called **phytoplankton,** as well as fish, insects, and larval stages of various creatures, can be sucked into the con-

---

1 The measurement of residual chlorine is mentioned on page 205.

denser of a power plant along with cooling water. This is called **entrainment.** Although most of the harm to entrained organisms is probably due to water velocity and pressure, entrained organisms are exposed to higher temperatures than those in the receiving waters of the estuary, lake, or river. This can be a very serious problem in cases where the power plant takes a major portion of a river's flow during dry weather. Various intake and screen designs can reduce the number and size of organisms entrained in the condenser flow. One study has estimated that the flounder population in Long Island Sound has decreased 6% as a result of 35 years of entrainment of the larval fish in local power plants. In another study, a power plant on Lake Erie was estimated to entrain and kill the equivalent of $30,000 worth of smelt (as larvae) per year.

Better Ways to Dispose of Waste Heat

The thermal pollution problem can be approached in two ways. Ideally, the waste heat, which is, after all, a form of energy, could be made use of, rather than simply dumped into nearby bodies of water (Figure 24.6). If this is not possible, technical solutions such as closed-cycle cooling can be used.

Figure 24.6 Peach blossoms protected by ice formed when warm water was sprayed on them. Orchards protected this way yielded full crops in years when nearby unprotected crops were damaged by frost. (Research done under auspices of Corvallis Environmental Research Laboratory.)

Space heating. One of the first proposals for using heated water from power plants was as a source of home heating and hot water. Because the power plant and the homes that are to use the water must be designed with this in mind, little use has so far been made of the concept. In West Germany, studies are being done to determine whether communities of over 40,000 people can be switched onto such a system. Four German cities and four towns will be involved in a pilot project because they already have suitable regional heating systems. In other cases, expensive pipes would have to be laid to carry the hot water.

Aquaculture. Hot water can help food organisms grow faster or farther north than normal. A number of projects using power-plant effluents in aquaculture, or underwater farming, have been devised. For instance, a hatchery at Vineyard Haven, Massachusetts, has used hot water to speed the growth of lobsters to marketable size in two years, one-quarter of the time usually required. Farms are in operation off the coast of England that produce clams in water normally too cold for clams. Other farms produce oysters in the water of Long Island Sound, which is too far north for oyster farms. Both the clam and oyster farms use hot water from power-plant effluents to maintain water at the necessary warm temperatures.

Problems are involved in using thermal effluents for aquaculture, however. Aquatic creatures often concentrate minute amounts of contaminants in water. Oysters and clams, which filter enormous quantities of water in order to sieve out their food, are especially likely to magnify toxic substances. Thus, if effluents from power plants are to be used in aquaculture, they must be free of such pollutants as copper from pipes, small amounts of radioactive materials, and pesticides.

In addition, the large numbers of organisms grouped together on a "fish farm" produce wastes. It may be necessary to install some sort of sewage treatment facility to prevent polluting the fish-farm water and also nearby areas. The fish farm itself is subject to disaster if the power plant must shut down for emergency repairs. Placing farms in areas where two or more plants operate near each other could solve this problem.

Other uses. In short, hot water should be used in ways that enhance the quality of life rather than cause it to deteriorate. Table 24.1 lists some of the ways that

Table 24.1 Possible Beneficial Uses of Waste Heat from Power Plants[a]

Water desalinization
Agricultural uses: irrigation; frost protection
Waste disposal: desalinization or demineralization of sewage water; sterilization and drying of sewage
Sterilization of drinking water: using heat instead of chemicals
Refrigeration: using gas-absorption refrigeration
Climate control: district heating and cooling; greenhouse heating; melting arctic ice and snow into irrigation water for arid regions
Heating intake waters at power plants: prevent fouling of pipes
Transportation: keep shipping lanes and harbors ice-free
Re-refining of waste oil
Aquaculture: lure fish for catching
Power from new energy technologies: thermoelectric elements, etc.
Wildlife protection: warm-water ponds for water fowl
Airport safety: defogging and de-icing runways
Mining: hot water and steam for hydraulic mining techniques

a Adapted from S. R. Fields, "Morphological Analysis of Beneficial Uses of Waste Heat from Power Plants," U.S. A.E.C. HEDL-TME 71-97 (1971), p. 8.

have been suggested for using power plant thermal discharges.

Closed-Cycle Cooling Schemes

Evaporation and cooling. The problems discussed so far have all been caused by the use of water to receive waste heat. Technologies have developed that can transfer much of the waste heat from cooling water to the atmosphere. These technologies provide two important benefits to the environment. If heat is transferred from the cooling water to the atmosphere, there is no need to discharge heated cooling water to a lake or river. In this way, damage to aquatic life is avoided. Not only is thermal pollution avoided, but withdrawal of cooling water from the lake or river can be cut to several percent of the amount normally withdrawn for once-through cooling. The reason is that the water that has been made cool can be reused to condense steam again and again. Thus, withdrawal of water from the main body can be cut dramatically, allowing use of the water in other ways.

The technologies currently in use that cool condenser water are cooling towers and cooling ponds. Both make use of the concept that when water evaporates, a great deal of heat energy is absorbed by the evaporating water molecules, allowing them to shift from the liquid to the vapor state. In evaporative cooling, evaporation of a small amount of cooling water withdraws a large amount of heat from the water that remains behind; this is the source of the cooling effect. The temperature drop of water cooled by a cooling tower is on the order of 14°C.

The use of the evaporative cooling principle is widespread. Bedouins keep their waterskins wet on the outside; the evaporation cools the water within the skin. In the American Southwest, "evaporative coolers" are often used in place of air conditioners to chill homes.

Cooling ponds. In cooling towers and ponds, the evaporative cooling principle is put to good use. Cooling ponds are technically simple: they are shallow ponds designed with very large surface areas to make evaporation take place more easily. Large areas of land are required if a plant cools its condenser water with such a pond. A design factor commonly quoted is 1–2 acres per megawatt of installed capacity. Thus, a modern 1000-megawatt plant could require 2–3 square miles of cooling-pond area. For many plants, such land areas are either not available or too expensive (Figure 24.7).

Wet cooling towers. Where land is too expensive for a cooling pond, towers are virtually the only cooling option available. **Cooling towers** are more complicated than ponds, and come in several types. The natural-draft evaporative cooling tower is a massive concrete structure (Figure 24.8). The natural-draft tower is referred to as a "wet" cooling tower because a portion (up to 2.5%) of the recirculating water is lost (evaporated) to the atmosphere. Addition of a wet natural-draft cooling tower to an electric plant may add about 5% to the cost of constructing the plant and 1% to the residential electric rate (see Controversy 24.1).

Another type of wet cooling tower utilizes fans to move air from outside the tower into and through the structure. This mechanical-draft tower requires electrical energy to operate, so that its operating cost is higher than the natural-draft tower. In addition, the fans may be noisy and often need repair. Nonetheless,

Figure 24.7 The Turkey Point Power Plant no longer discharges water into Biscayne Bay. Instead, a 6000-acre (2400-ha) canal system cools condenser water. In this photograph of the canal system, the power plant is in the upper right. (Photograph courtesy of Florida Power and Light Company)

Figure 24.8 A typical evaporative cooling tower. Note its enormous size relative to its surroundings. Such structures can be seen for miles. This is a picture of the Trojan Nuclear Power Plant on the Columbia River near Prescott, Washington. (EPA Documerica)

since the towers are much smaller [about 50 feet (15 m) wide and high and 300 feet (92 m) long], their construction costs are much lower. The evaporation process for both towers involves only water; salts in solution in the water remain behind. Hence, continual evaporation of water results in an increased concentration of salts in the water. To slow the rate of salt build-up, a portion (about 1–3%) of the circulating stream is continuously removed and replaced by fresh water with lower salt content. The removed water is called "blowdown" in the cooling tower trade. In addition, a mist of fine droplets may be carried by the air currents into the surrounding atmosphere. The sum of the losses due to evaporation, blowdown, and mist may account for 4% of the circulating water, and this quantity must be constantly replaced by a freshwater stream drawn from a lake, river, or other body of water. Thus, even with closed-cycle cooling, there is some demand for water by the plant.

Dry cooling towers. Different from "wet" towers are the "dry" cooling towers. The dry cooling tower does not use the process of evaporation to cool condenser water. Instead, waste heat is transferred directly to air passed upward through giant heat exchangers. Dry cooling towers have been used rarely, and only on power plants of 330 megawatts or smaller, due to their higher costs. However, increasing short-

Economics and Thermal Pollution Control

One is somewhat appalled at the relative emphasis given to thermal pollution.

Frank L. Parker

A more appropriate determination of the costs of pollution control would balance those costs with the social and economic benefits of a healthier environment.

Douglas M. Costle

In a letter to the editor of *Science,* Frank L. Parker wrote:[*]

> . . . industry will have to invest an additional $8 billion to meet 1977 requirements of "best practicable" water pollution control technology at existing plants. However, industry will have to spend an additional $9.5 billion to meet the 1977 standards for thermal discharges. The costs for thermal plant discharge elimination then exceed those from all other industrial wastes put together. When one compares the known incidence of disease and environmental destruction from the discharge of heavy metals, carcinogens, mutagens, pesticides, and so forth, to the known incidence of disease from thermal discharges (zero), one is somewhat appalled at the relative emphasis given to the thermal pollution abatement.

Environmental Protection Agency administrator Douglas M. Costle gave a somewhat different argument in another article:[†]

> . . . a more appropriate determination of the costs of pollution control would balance those costs with the social and economic benefits of a healthier environment. For example, when natural systems become so contaminated that higher local and state taxes must be spent to clean them up, people have less to spend as consumers . . . more energy is needed to make water safe for drinking, community income from recreational activities is lost—these are just a few of the considerations overlooked when the plea is made that environmental controls are unproductive corporate costs.

Frank Parker seems to suggest that the ability to cause disease is the main criterion on which to judge the seriousness of a pollutant. On the other hand, Costle mentions some problems that thermal pollution can cause; for example, overgrowth of problem algae leading to increased costs for drinking water treatment or loss of recreational waters. Should we differentiate between health effects and other effects when it comes to paying for pollution cleanup?[‡]

Thermal pollution can have lethal effects on fish populations or the population of other aquatic species. Should our concern for these organisms be equivalent to our concern for human health and well-being?

---

[*] F. L. Parker, *Science,* **185** (1974), 568.
[†] D. M. Costle, *EPA Journal,* **4**(1) (January 1978), 2.
[‡] The article by Reynolds (Further Reading) notes in its bibliography several references dealing with the costs of thermal pollution.

ages of available water may force many utilities to turn to dry cooling, especially in the West and Southwest. At the Wyodak Power Plant located at the mouth of a coal mine near Gillette, Wyoming, dry cooling was used because it was cheaper than shipping the coal to the nearest available water source.

Are cooling towers an environmentally sound solution? For several reasons, cooling towers are not perfect controls for thermal pollution. The towers are enormous in size and visual impact. For a wet cooling tower, a height and base diameter of 400 feet (121 m) is not uncommon. Such a structure, located in rural areas next to a generating station, is the equivalent of a 40-story building and dominates the landscape for many miles (Figure 24.9).

Wet towers may also present other problems to the area in which they are located. To illustrate, a pair of wet natural-draft towers on the Monongahela River at Fort Martin, West Virginia, are used to cool water from two 540-megawatt coal-fired power plants. The towers decrease the temperature of the condenser water from 45°C to 32°C. Of the 250,000 gallons/minute

cooled by each tower, about 6000 gallons/minute evaporate. That quantity of evaporation is approximately equivalent to a room 10 feet (3 m) on a side with a ceiling height of 10 feet (3 m); this quantity of water is evaporating every minute. The towers must be taller than the depth of the river valley, which is 300 feet (92 m) at this site. If they were not, mists could settle in the valley, enshrouding the area in a permanent fog.

Fog is a potential problem in cold weather, when it could produce icy highways and low visibility. The taller natural-draft tower has a lesser tendency to fog because its moisture is released at such a high elevation. The dry cooling tower, which we discussed earlier, has no fog associated with it since water is not evaporated to provide the cooling.

Drift is like fog, an atmospheric problem that must be controlled. Drift, in contrast to fog, which is evaporated moisture in the air, consists of particles of liquid water captured by the upward flow of air through the tower. These water droplets are dispersed throughout the surrounding area by winds. The much shorter mechanical-draft towers with their high-veloc-

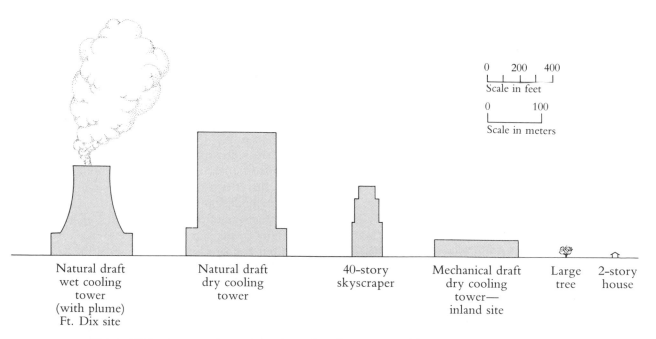

Figure 24.9 A comparison of the sizes of cooling towers with some familiar objects. (Adapted from P. Meier and D. Morrell, "Issues in Clustered Nuclear Siting," Brookhaven National Laboratory, 1976)

ity airflows tend to produce more drift than the natural-draft towers, where airflow is much slower. Drift could cause corrosion of the power plant's electrical equipment, and would be unacceptable over densely populated areas. Where the cooling water was drawn from a saltwater source such as an estuary or the ocean, the drift from a tower would be salt water and could increase the salt content of agricultural land. Because of its potential for damage, the designs of new cooling towers have been altered to decrease drift losses substantially.

One further potential problem of cooling towers and waste heat disposal ought to be mentioned. Because of the growing scarcity of sites, and for security reasons as well, active consideration has been given to "power parks." At such power parks, 10–15 nuclear plants, each capable of producing more than 1000 megawatts, would be constructed. Such parks would be capable of supplying the electrical demands of a state or several states. Currently, the total generating power of plants at any one site equals no more than several thousand megawatts.

Recall that nuclear plants are 30–32% efficient at converting heat energy to electrical energy. Thus, for every megawatt of electrical output, two megawatts or more of heat must be disposed of. A park producing power at the rate of 40,000 megawatts would have to dissipate waste heat at a rate of more than 80,000 megawatts. The waste heat could have an effect on weather patterns, since this rate of heat output rivals the rate at which heat is dissipated by thunderstorms and volcanoes. The creation of new weather patterns by the disposal of this quantity of heat is not unthinkable. Rain and thunderstorm activity are possible. Since whirlwinds can result from the release of energy in a thunderstorm, tornadoes could occur as well. The mathematics for the prediction of weather cannot yet tell us whether these potential problems could become reality.

Cooling Water and the Future

Closed-cycle cooling via towers or ponds has become the main way in which new steam electric plants dispose of waste heat. The use of these cooling methods is required on new plants (with some exceptions) because of the almost total ban on heated discharges imposed in 1974 by the Environmental Protection Agency (EPA). Many of the plants built in the 1960s and early 1970s also utilize closed-cycle cooling. An EPA ruling forced this action on many of the larger plants built between 1970 and 1974. The need to conserve water led still other plants in water-short areas to adopt closed-cycle cooling.

Although new steam electric stations built on inland waters are likely to employ closed-cycle cooling methods, not all new plants will require it. Power plants on the cold ocean or on cold estuaries may escape the requirement. Power companies can win exemptions for their plants if they can demonstrate that the heated discharges will not alter the ecological balance in the receiving waters. That is, they must show that the shellfish, fish, and wildlife normally native to the area will be able to survive and reproduce. Such a demonstration is most likely on cold and abundant natural waters, such as the estuaries of the Atlantic Northeast and the Pacific Northwest, as well as the oceans and bays of these areas.

Although we have discussed dry cooling as an alternative that avoids the problems of fog and drift, it may become important as a means of condensing steam for power plants when virtually no water is available. Wet cooling requires water to replace losses due to evaporation, drift, and blowdown. Even this make-up quantity may, in some water-short areas, become impossible to obtain. Under such circumstances, dry towers become the only alternative.

The need for dry cooling is expected to appear in the 1990s in parts of the United States. In many states west of the Mississippi, particularly the Dakotas, Colorado, Wyoming, Utah, New Mexico, Arizona, and Texas, dry cooling is likely to become necessary in the 1990s. California is a special case. Although placing plants near the ocean would relieve the need for dry cooling, most ocean sites are too near the San Andreas Fault for nuclear power plants. Water is largely imported to the lower portion of the state for municipal and agricultural use and may not be available for steam power plants. However, municipal wastewater may be "reclaimed" by treatment and used for cooling. The concept is not new. Baltimore has long been selling its treated wastewater to the Bethlehem Steel Works at Sparrows Point for cooling. Even with reclaimed wastewater being used for cooling, however, the rapidly growing power demands of California will probably make dry cooling necessary in the 1990s for a portion of its power-generating capacity.

Noise Pollution

The range of facts and opinions on whether or not environmental noise is a serious problem is very wide. Consider the following examples.

Studies show that children living under the flight path of jet planes at Los Angeles International Airport have higher blood pressures than children from similar social and economic (but quiet) neighborhoods. Laboratory experiments show that animals exposed to loud noise have increased cholesterol levels and develop atherosclerosis. The California Supreme Court upheld a judgment allowing a man to recover damages for mental and emotional distress caused by constant jet noise over his Los Angeles home.

Yet manufacturers attempting to market quiet vacuum cleaners, blenders, and motorcycles have by and large failed to gain public acceptance for them. And the EPA Office of Noise Control was phased out as part of budgetary cuts in the early 1980s.

Almost everyone agrees that too much noise can cause hearing loss, but what about other possible health effects? Is noise a serious environmental problem? How is noise defined and measured? These questions are the focus of the following sections.

Measuring Sound

Sound is produced when something vibrates: someone's vocal cords, a jackhammer, or the tiny diaphragm in a telephone receiver. The vibrating body alternately compresses the adjacent air and retreats from it, allowing it to expand. This alternate compression and expansion of air molecules moves through the air and is called a sound wave. If the sound wave reaches a surface that can vibrate, such as an eardrum, it sets that surface vibrating (Figure 24.10).

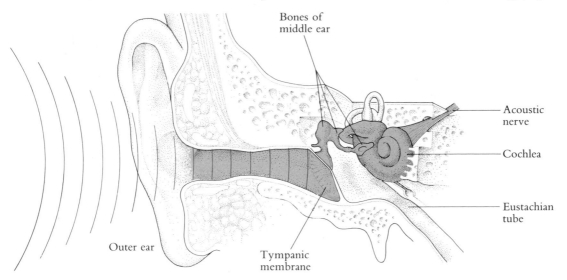

Figure 24.10 Schematic Representation of the Ear. Sound waves move down the ear canal to the eardrum, or tympanic membrane. Vibrations from this membrane are amplified by the bones of the middle ear and passed on to the fluid in a structure of the inner ear called the cochlea. These vibrations cause pressure changes in the cochlear fluid. Tiny cells called hair cells embedded in a membrane of the cochlea sense the changes in pressure caused by the transmitted sound waves. Messages from the hair cells are carried to the brain by the auditory nerve and interpreted as sound. Very loud sounds (over 140 dB) actually destroy hair cells, and may rupture membranes of the inner ear, leading to deafness. However, even lower dB levels over long enough periods can selectively destroy hair cells, especially those sensitive to high frequencies. (Adapted from *Life: The Science of Biology,* copyright 1983 by William K. Purves and Gordon H. Orians. Used with permission of Sinauer Associates and Willard Grant Press.)

Sound has both a frequency and an intensity. The *frequency* of a sound, measured in hertz, is the number of compression–expansions per second. As the frequency of sound increases, we hear it as a higher and higher pitch. Middle C on a piano is 264 hertz; high C is 1056 hertz.

Sound *intensity* is related to loudness. The human ear is sensitive to a wide range of sound intensities, from a falling leaf to the lift-off of a Saturn rocket. In order to compress this wide range into an easily used tool, sound is measured on a logarithmic scale called the decibel scale. On this scale, a doubling of sound intensity shows as a 3-decibel (3-dB) increase. Thus, if one plane gives a reading of 76 dB on a sound meter, two identical planes, generating twice the sound, will register 79 dB. Ten planes would generate ten times the sound and produce a 10-dB increase on the meter. Decibel levels of some common sounds are shown in Table 24.2.

Humans, however, do not register sound in the simple, unbiased way that sound meters do. Humans judge a 10-dB increase on a sound meter to be about a doubling of sound. Furthermore, high-frequency sounds are usually judged louder than low-frequency sounds of the same intensity. Various weighting scales have been devised to correct sound meter measurements for this human bias. One common scale is the A scale, denoted db(A).

Noise versus Sound

Noise is more difficult to define than sound because noise requires a value judgment. Noise has been called unwanted sound. However, some people are annoyed by any sound not of their own making, while others are not inclined to complain no matter how loud a sound is.

Most of us would probably agree that sounds loud enough to cause harmful physical or emotional effects are properly called noise. Thus, one way to differentiate noise from sound is to look for these kinds of effects.

Health Effects of Noise

Hearing loss. Sounds of 60–80 decibels can cause a threshold shift. That is, a particular sound must be of higher intensity to be heard again. In some cases, the ear can recover from these shifts, which are thus tem-

Table 24.2   Noise Level of Common Sounds

Decibels[a]	Sound
140	Physically painful to hear
130	
120	Maximum in a boiler shop
110	Unmuffled motorcycle / Jet engine test control room
100	Power mower / Rock and roll band
90	Inside a subway car / Electric blender
80	10-horsepower outboard motor / 20–50 feet from heavy traffic
70	Tabulating machines in an office / 20 feet from a passing automobile
60	Dishwasher / Accounting office
50	Conversation from a distance of three feet / Private business office
40	Average home
30	
20	Broadcasting studio
10	
0	

a Measured on the C or unweighted scale, re 0.0002 microbar. (From "Noise—Sound Without Value," Committee on Environmental Quality, Federal Council for Science and Technology, September 1968.)

porary shifts. But as sound levels increase and as the time of exposure increases, recovery can take longer. At high-enough intensity or duration, the shift is permanent: hearing loss results. Above 80 dB, and possibly even from 70–80 dB, frequent exposure will probably lead to hearing loss, although individuals vary in their susceptibility to hearing damage.

Between 6 and 16 million Americans suffer from partial to severe hearing loss from noise on their jobs, from military service, or from other sources. Accidents, absenteeism, and compensation claims (mainly for hearing loss) due to noise cost U.S. industry some four billion dollars per year.

Several studies have shown that hearing loss, particularly at higher frequencies, is increasing among young people. A number of observers believe this is caused by listening to amplified rock music.

Cardiovascular problems. Noise can cause those physiological responses seen in general stress reactions. That is, an increase in blood pressure and increased secretion of certain hormones. Many studies have shown that high blood pressure occurs 60% more frequently among workers with prolonged exposure to over 85 dB. Arrhythmias and vascular disorders are also found.

Residents of communities near airports have been shown to be more likely to have high blood pressure and other cardiac problems than similar groups in quiet neighborhoods. Even children exposed to airport or traffic noise are subject to blood pressure rises.

Mental effects. Studies show that noise can cause a narrowing of attention span. That is, the ability to keep track of more than one thing at a time decreases. The performance of complex tasks becomes more difficult, while simple tasks are not affected. Noise, especially uncontrollable and unpredictable noise, has been shown in the laboratory to cause feelings of loss of control over one's environment. This can lead to depression and loss of motivation to try again at problem solving. Other studies show that for some people, noise produces anxiety symptoms such as headache, nausea, and irritability.

Studies of children living or going to school in noisy versus quiet environments have shown a variety of noise-related effects. Children from noisy environments were not as good at puzzle solving as those from quiet environments. In several studies, children from noisy environments had significantly lower reading levels than those from quiet ones.

Effects on sleep. Sounds above 38–48 decibels interfere with sleep for most people. Even between 33 and 38 decibels, many people are bothered. Steady noises are not as annoying as intermittant ones. However, contrary to general belief, people do not become accustomed to repeated noise during sleep. Laboratory studies of sleeping subjects show that heart rate and electroencephalograph responses to a repeated noise are the same as the initial response to the noise.

Noise and Wildlife

Wild creatures show an aversion to loud sounds, which is the basis for commercial devices that scare birds and mammals away from crops. Other behavioral effects of noise include huddling, piling up, decreased exploration, or avoidance of certain areas.

Besides hearing loss, experiments show that noise causes sexual and reproductive problems in animals. There are few documented examples of the effects of noise on wildlife. The best-known involved sonic booms that scared a colony of sooty terns off their nests in 1969, causing a massive hatching failure.

Noise Exposure

Another way of differentiating between sound and noise involves looking at environmental noise exposure—the noise level to which people are exposed during most of their day.

A general definition of a quiet environment would probably include the requirement that normal, relaxed conversation be possible. For this, background noise must be below 66 dB(A) for two people sitting close together and below 50 dB(A) for opposite ends of a normal classroom. However, small children and older people need lower background levels than these to understand speech at the same distances.

Depending on where one lives, the background noise level varies a great deal. The major background noise to which most people are subjected is traffic noise. Downtown in a large city, levels can easily reach 76 dB(A). Normal conversation is not possible at this level. In contrast, rural noise levels are customarily 30–50 dB(A) (Figure 24.11 and Table 24.3).

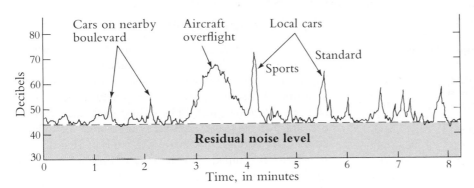

Figure 24.11. Noise Profile of a Suburban Neighborhood. The noise profile of an area shows a steady background level of noise (here traffic noise), on which are super-imposed a variety of noise events. (From: "Community Noise," NTID 300.3, U.S. EPA 1971)

Table 24.3 Noise Profiles

Location	L_{90}[a]	L_{50}[a]	L_1[a]
Third-floor apartment next to freeway	73	77	87
Second-floor tenement in New York City	63	68	81
Urban residential near airport	51	55	92
Urban residential	46	51	66
Suburban residential near city	39	43	62
Small-town residential cul-de-sac	38	42	55
Farm valley	33	36	48
North Rim of Grand Canyon	19	23	43

A noise profile can also consist of the dB levels that are exceeded 90% of the time (L_{90}), 50% of the time (L_{50}), and 1% of the time (L_1). The L_{90} represents the background noise level, while the L_{50} gives a measure of the average noise level. Note that on the rim of the Grand Canyon, noise levels are more than 40 dB lower than in a New York City tenement.

a Arithmetic averages of the 24-hour values in the entire day.

Psychology of Noise versus Sound

Simple intensity, however, is not the only component of noise. Studies have shown that the annoyance value of noise depends on the sound frequencies in the noise, as well as how long the noise goes on. Also involved is the time it takes for the noise to reach its maximum intensity. Further, annoyance is governed by people's feelings about the need for a noise; for example, noise from a military airport necessary for national defense is usually less annoying than noise from a civilian airport.

Another factor is whether the hearer has control over the noise: the user's own vacuum cleaner is generally less annoying than a neighbor's power lawn mower.

Laws Governing Noise Pollution

According to EPA data, over two million Americans are consistently exposed to noise levels over 75 db(LDN)[3] (Table 24.4). Long-term exposure to noise above these levels is generally agreed to be unhealthful. In 1972, Congress passed the Noise Control Act, which directed the Environmental Protection Agency to set national noise-control standards and promote the development of "an environment for all Americans, free from noise that jeopardizes their health and welfare." EPA set about designing standards for the worst noise sources. Eventually, maximum noise standards were issued for a variety of sources, including new medium and heavy trucks, portable air compressors, interstate rail carriers, buses, and motorcycles. A time table was also worked out for reduction of aircraft noise. New aircraft were required to meet standards for quiet operation immediately, while airlines were required to retrofit noise-control equipment to aircraft already in service on a gradual schedule. Although the control of aircraft noise was to be accomplished by 1983, the deadline was later extended to 1986. The Occupational Safety and Health Administration issued regulations in 1982 for noise control in the workplace.

---

3 This dB level is a weighted average of the day and night noise level in which an additional 10 dB is added to night noises to reflect the fact that they are more disturbing.

Table 24.4 Noise Exposure in the U.S.

| Kind of noise | Proportion of population exposed to: | | |
	Severe adverse effects	Major adverse effects	Moderate adverse effects
Road traffic	1%	10%	36%
Aircraft	0.4%	3%	19%
Railroads	0.2%	—	1%
Industrial	—	—	3%

The most serious noise hazard in the U.S. is transportation noise from road traffic, aircraft, and railroad equipment. The next largest problem is industrial noise, especially for those working in noisy industrial sections.

(Adapted from: Pilot National Environmental Profile, U.S. EPA, 1980)

EPA had planned to issue standards for many other sources, as well as labeling requirements to help consumers choose among various brands of noisy small appliances such as blenders and vacuum cleaners. However, budgetary cuts in the early 1980s felled most of the projected programs.

Is noise a serious problem? Yes, if we take into account the number of Americans living in noisy environments that can promote physical and mental problems, or the amount of money paid in workmen's compensation claims for hearing loss or noise-related injuries. Perhaps not, if programs to control noise can run afoul of budgetary considerations. However, it should be noted that even if the national political recognizance of noise seems at a low point, legally the case is otherwise. Several recent court decisions have affirmed that people should be awarded compensation not only for noise-related hearing loss but also for loss of property values due to noise (near airports) and for mental and emotional stress caused by noise. In addition, the courts have upheld the right of local governments to set noise-control standards, even where such rules infringe on the regulation of interstate traffic (i.e., at airports). As a result, many localities are setting restrictions on night traffic at airports in order to reduce noise.

High-Voltage Power Lines

The third example of energy itself as an environmental pollutant involves high-voltage power lines. These are electric power lines designed to carry large quantities of electric power from generating plants to large population centers. The largest such lines now in use are 765,000-volt (765-kV) lines, one of which could carry enough power for the cities of Boston and Baltimore combined. For the future, lines carrying up to 2200 kV are planned. Power lines this size have both electric and magnetic fields surrounding them (Figure 24.12).

An ordinary kitchen has an electric field of about 3 volt/meter due to the electric appliances in it. Directly under a 765-kV transmission line, the field at ground level is about 10 kV/meter (10 kV/m). However, if one moves 500 feet (152 m) away from the line, the field decreases to 0.1 kV/m. Thus the possible problems center on effects that might occur directly around or under the lines. These effects include electric shocks, biological effects induced by electric and magnetic fields, and the effects of corona.

Corona Effects

Corona, which occurs mainly in bad weather, is the breakdown of air directly surrounding a power line. It is most notable for the noise it generates—a crackling or frying sound. Although well below levels that cause hearing damage, this noise can be annoying. Corona can also cause interference with radio and T.V. signals, which could be a severe problem in fringe areas of reception. In addition, ozone and nitrogen oxides may be formed. Levels, however, seem too low, compared to other sources, to be of concern.

Electric Shocks

High-voltage power lines cause electric shocks to people or animals moving below them. For a distance of several feet around the line itself, there can be a "flashover" or breakdown of the air between the line and a conducting object, allowing a dangerous current to flow. Power lines must be set high enough so that no objects passing underneath (such as a boat with a high mast) come into range to allow such flashover.

However, the electric field surrounding the power

Figure 24.12 High-Voltage Power Lines. Lines such as these carry up to 765,000 volts. Possibly harmful environmental effects may be caused by electric shocks or by electric and magnetic fields generated directly around or under the lines. (U.S. Department of Energy)

line can also cause a shock hazard. The reason is that objects in an electric field collect an electric current. For instance, a large tractor under a 765-kV power line can collect up to 4–5 milliamps (5 mA). Such a current is still not a hazard unless someone touches the tractor while he is grounded (for instance, standing on wet ground), thus allowing current to flow from the tractor through the person and into the ground. The shock hazard in this example is probably just at the limit of one that would be very painful but not otherwise harmful to a child. Higher-voltage power lines than this, however, could lead to more serious results.

Electric Field Effects

Internal electric fields and cell membranes. Besides the possibility of shocks, electric fields can have other effects on living matter. The external power-line field causes an internal electric field to form in living tissue. For humans, the internal current density caused by an external electric field of 10 kV/m is still 10–100 times less than the current density that acts on the membrane of a muscle or nerve cell and causes it to

react, or "fire." Whether electric current densities of this magnitude can cause other, more subtle, effects on cells is still hotly debated. None are yet proven, but several experiments are in progress.

Surface effects on skin and leaf tips. At the surface of a body or tip of a pointed leaf, the local field can be much higher than the internal field, however. This leads to a tingling sensation in humans caused by vibration of hairs on the skin. In addition, pointed leaves may show burned leaf tips (round leaves are not affected). Neither of these effects seem to have any harmful results for the organism as a whole, although some people find the tingling sensation unpleasant.

Mental effects. Other effects, such as excess fatigue, have been reported from Russian experiments, but could not be repeated by U.S. investigators.

Pacemakers. The electric field under 765-kV power lines can definitely affect certain types of pacemakers. Although such pacemakers have a fail-safe mechanism that takes over, farmers or other workers with pacemakers who have to spend time under high-voltage power lines should discuss the problem with their doctors. People with pacemakers driving under such power lines are not at risk because the metal car body shields them from external electric fields.

Magnetic Field Effects

At ground level under a 765-kV power line, the magnetic field is about 0.56 Gauss (0.56 G), but it decreases rapidly to 0.016 G at 500 feet (152 m) from the line. Migrating birds appear able to detect the 0.4 G magnetic fields generated by large-scale antennas such as the Navy's Project Sanguine. However, the birds appeared able to compensate by using other cues (i.e., sun and star positions). No actual disturbances of migration patterns were seen.

Other harmful biological effects of magnetic fields have not been confirmed at the levels found under currently operating power lines.

The Outlook

In summary, the electric and magnetic fields generated under high-voltage power lines to date, do not appear to cause serious biological effects. If transmission voltages are increased, however, problems, especially with electric shocks, could result. Power companies will need to introduce safety devices to shield people, plants, and animals from the higher electric fields that might be generated.

Questions

1. Most of the thermal pollution in this country is waste heat from power plants using once-through cooling. Describe, from the point of view of the operation of power plants, why thermal pollution occurs.
2. Contrast winter and summer thermal fish kills.
3. Why is an actual "fish kill" not necessary in order to destroy a population of fish or other aquatic creatures?
4. Briefly outline the harmful effects of thermal pollution. Contrast the possible beneficial uses of waste heat.
5. What are the advantages of closed-cycle cooling over open-cycle cooling?
6. What are the disadvantages of closed-cycle cooling?
7. Will thermal pollution become a problem if new energy technologies such as fusion, solar, wind, geothermal, and tidal power are used? Explain.
8. Table 24.1 lists a variety of suggested ways for using waste heat from power plants. Analyze the suggestions, noting where you feel other environmental problems could result from these uses.
9. How would you define noise compared to sound?
10. What is the major source of background noise exposure for people in the U.S.?
11. What are the major sources of noise pollution?
12. What physical and mental effects can result from excessive noise exposure?
13. Do you think noise is a serious environmental problem? Explain.
14. What are the major causes of concern with respect to transmission of electric power by high-voltage power lines (765 kV and greater)?

Further Reading

Thermal Pollution

Barnett, P. O., "Effects of Warm Water Effluents from Power Stations on Marine Life," *Proceedings Royal Society of London B* **180**, 497 (1972).
A brief resume of the effects of thermal pollution and one example of the harmful effects an operating power plant can cause.

Bush, R. M., et al., "Potential Effects of Thermal Discharges on Aquatic Systems," *Environmental Science and Technology,* **8** (6) (1974), 561.

The effects of heat on different species of organisms are compared in this paper, which is intended to help in the choice of suitable sites for new power plants.

Gentele, G. H., et al., *Power Plants, Chlorine and Estuaries,* U.S. Environmental Protection Agency, Ecological Research Series, EPA 600/3-76-055 (1976).
Contains a thorough, well-written review of chlorination, its biological effects, and its effects in receiving waters.

Reynolds, John Z., "Power Plant Cooling Systems: Policy Alternatives," *Science*, **207** (25 January 1980), p. 367.
The author (director of environmental services for Consumers Power Company) argues that once-through cooling is really not such a bad policy. See if you agree with him or if you can refute his ideas.

Noise

Summary, Conclusions and Recommendations from Report to the President and Congress on Noise, NRC 500.1, U.S. Environmental Protection Agency, December 31, 1971.
This summary and the other 8 volumes that go with it are an exhaustive treatment of the economic, social, and health effects of noise on people and animals. In addition, the major sources are examined, along with possible control measures.

Toward a National Strategy for Noise Control, U.S. Environmental Protection Agency (1977).
EPA's goals in noise abatement and the means to achieve them are discussed.

References

Thermal Pollution

"Industrial Waste Guide on Thermal Pollution," Federal Water Pollution Control Administration (now part of EPA), Pacific Northwest Water Laboratory, Corvallis, Oregon, 1968.

Crawshaw, L., et al., "The Evolutionary Development of Vertebrate Thermoregulation," *American Scientist*, **69**, September–October 1981.

Lihach, N., "Water Water Everywhere But . . .", *EPRI Journal*, October 1979, p. 6.

Noise

Smith, R. Jeffrey, "Government Weakens Airport Noise Standards," *Science*, **207** (14 March 1980), p. 1189.
Some of the political and legal implications of noise are highlighted in this article.

Environmental Noise Control Act of 1972, Report of the Committee on Public Works, U.S. Senate, September 19, 1972.

Identification of Products as Major Noise Sources, Federal Register, June 21, 1974, p. 22297.

Bibliography of Noise Publications (1972–1982), United States Environmental Protection Agency.

Cohen, S., D. Krantz, G. Evans, and D. Stokols, "Cardiovascular and Behavioral Effects of Community Noise," *American Scientist*, **69** (1981), p. 528.

High-Voltage Power Lines

Miller, M. W., and G. E. Kaufman, "High Voltage Overhead," *Environment*, January/February 1978, p. 6.

(U.S. Department of Energy)

Natural Sources of Power and Energy Conservation

We found in Parts IV and V that many pollutants stem from the use of energy and the production of electric power. The list of such pollutants was surprisingly long and the quantities of pollution remarkably large.

It is logical to ask the question: "Do the methods of energy use and power generation have to be accompanied by degradation of the environment?" While it is true that human activities have some unavoidable effect upon the environment, technologies do exist whose environmental impacts are far less than others. It is practically impossible for us to have no effect on the environment in which we live because living itself consumes and utilizes energy. (This point was explained in terms of thermodynamic laws in Chapter 2.)

Though human life must have some impact, there are natural balances, or righting mechanisms, that tend to keep environments and the natural communities living in them in an equilibrium, where changes come only slowly. Nonetheless, in many cases, human activities overbalance these mechanisms, causing rapid changes, with which neither humans nor wildlife tend

to cope well. Conventional power generation and the direct use of fossil energy, which produces quantities of air and water pollutants, are such activities.

In this sixth part we shall look at natural sources of power: wind, sun, tides, geothermal energy, and hydropower. These methods of power generation seem to promise softer impacts on the environment than combustion of fossil fuels or the fissioning of uranium. In addition, most of these energy sources are renewable, in the sense that nature makes them available virtually forever.

It is remarkable to think that only two centuries ago, society had, besides human and animal energy, only three forms of energy at its disposal. All three of these forms could be traced to the sun.

Hydropower was used to operate mills, which ground grain or wove cloth. Hydropower requires water running downward to the sea from the uplands, where water fell as rain. Of course, the sun caused the water to evaporate in the first place. Wood was another source of energy; it was used for cooking, heating, and smelting iron. Again, the use of the sun's energy in the process of photosynthesis is responsible for the existence of wood. Also, wind was used to pump water for irrigation and to fill the sails of the great wooden ships. Wind is the result of temperature differences in the atmosphere caused by uneven heating by the sun. The giant windmills of Holland are vivid reminders of an early era when only such ingenuity and muscle were available to do our work.

In the past one hundred years, our industrial society has relied on the heat from fossil fuels, the heat from fission, and the energy from flowing water to power the enormous development that has taken place in our civilization. To only a very minor extent has wind been utilized, or the sun's energy tapped, or heat withdrawn from the earth. Yet these sources of energy, which stem from the earth's natural processes, are all around us.

In the last decade, we have increasingly been turning to these natural sources of energy because, in many respects, they provide energy without limit (Figure A). Furthermore, the pollution that comes from them is often minor in comparison with that from fossil fuels and fission. We have been turning to these sources for another fundamental reason, though. As fuel supplies become less secure and more expensive, these sources become more attractive and more economical. The rising prices of oil and gas are, in a large measure, responsible for our renewed interest in sun, wind, and water. It seems strange to be saying "Thank you" to the oil cartel, but it has made the point of dwindling energy resources more vividly than the arguments of any resource economist.

Thus, we are standing today on the edge of an exciting era: a frontier in much the same way that the western United States was a frontier (Figure B). People will be laying claim to economic territory as a solar industry is built. Challenges of exploration and discovery in energy generation and energy conservation lie before us. Once again, we have the opportunity of

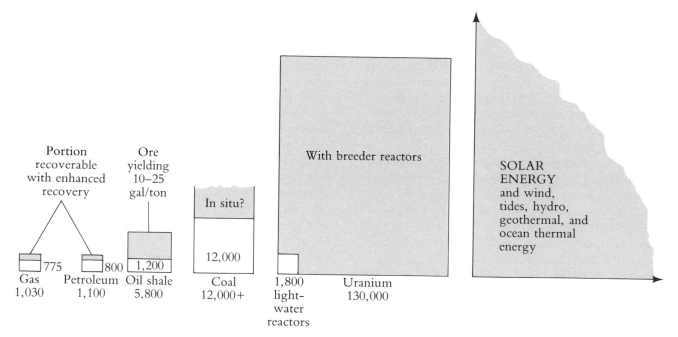

Figure A Available energy resources in the United States measured in quads. (1 quad = 1 quadrillion Btu = 1 million billion Btu.) The total U.S. energy comsumption in 1974 was 73 quads. [Adapted from *A National Plan for Energy Research, Development and Demonstration: Creating Energy Choices for the Future,* Energy Research and Development Administration (now part of the Department of Energy), 1975.]

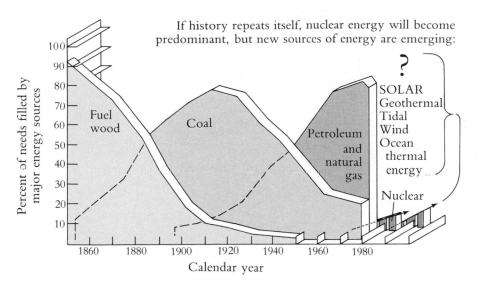

If history repeats itself, nuclear energy will become predominant, but new sources of energy are emerging:

Figure B U.S. energy consumption pattern through time. [Adapted from *A National Plan for Energy Research, Development and Demonstration: Creating Energy Choices for the Future,* Energy Research and Development Administration (now part of U.S. Department of Energy), 1975.]

becoming self-sufficient in energy resources, this time by learning to use the natural energy sources that surround us. It is a period that will give new meaning to existence as we struggle to secure a stable, fruitful, and peaceful life in harmony with nature for generations to come.

In Chapter 25 we discuss how electrical energy is generated from falling water. We describe not only conventional hydroelectric generation but also the potential for small-scale hydropower installations and the development of power from the tides.

In Chapter 26 we discuss the generation of electric power from the wind. We also show how heat stored in the interior of the earth can be used for both space heating and the generation of electric power; environmental impacts are indicated as well.

In Chapter 27 the many useful forms of solar energy are discussed. We include not only water and space heating, but also solar electric generation, of which there are several important forms. The conver-

sion of solar energy to biomass fuels and the use of temperature differences in the layers of the ocean are also described.

Chapter 28 details methods of energy conservation. When energy was inexpensive, hot water from electricity was reasonable, and large heavy automobiles could still be operated at low cost. Home owners could afford not to have storm windows or adequate insulation. Electric heating of homes was even a possibility. Today we know that electric units for heating homes and water are expensive luxuries. People choose lighter, more efficient autos as gasoline grows steadily more expensive. Public transportation is being installed or upgraded in many parts of the country and car pooling is being encouraged. Insulation of homes and other energy-saving improvements are rewarded by tax incentives, as we reach out to save energy in many ways. The savings are turning out to be large, but to change old patterns of behavior takes time and requires a national commitment.

Power from Falling Water

Conventional Hydroelectric Generation

Small-Scale and Low-Head Hydropower

The Redevelopment of Hydropower

Power from the Tides

Pumped Storage

Methods to Increase and Steady Tidal Energy

Conventional Hydroelectric Generation

Water, an ancient source of power, remains today a good option for supplying electrical energy to our industrial civilization. The energy from falling water, captured by a waterwheel, has been used directly to grind grain, cut lumber, and weave fabrics. But the gristmills and sawmills on our rivers began to fade when, in the 1880s, the generation of electric power began at waterfalls. In 1895, the power of the mighty Niagara River was harnessed and used to produce electricity. This was the first large-scale effort for the production of hydropower in the U.S., although earlier hydroelectric plants had been built in Europe. No one will say that hydroelectric power is without problems. (See Chapter 9 for a discussion of the environmental impacts of reservoirs.) Nonetheless, when we consider the pollution and potential hazards and dangers of other forms of power generation, hydropower begins to look much less menacing.

Hydropower now provides about 12% of the nation's electrical energy needs. A generating capacity of 73,000 megawatts is available at some 1500 hydropower plants across the country. In the 1930s, hydropower was furnishing 30% of the nation's generating capacity, but decreasing costs of fossil-fuel power stations gradually made hydropower generation at smaller sites (up to 25 megawatts) uneconomical. Many of the smaller hydropower plants in the Eastern U.S. closed, and newer, larger hydropower plants were constructed in the West—at Hoover Dam, Grand Coulee Dam, and at other sites.

The principle of hydropower generation is simple.

The kinetic energy[1] of falling water is used to turn a turbine, which is linked to an electrical generator. Early hydroelectric plants were of the "run of river" type, in which water flowing in the river was not dammed, but merely directed through a turbine. Large changes in river elevation were needed for these plants; the Niagara Falls development was of this type. Most modern hydropower installations, however, use dams to increase the volume of water that can be steadily discharged through the turbine (Figure 25.1). Dams do more than provide a reservoir of water upon which to draw; they increase the height of the water surface. The greater pressure provided by this higher water surface gives the falling water a higher velocity and hence more kinetic energy. The energy derived from this more powerful flow of water is, as a consequence, much larger.

In practice, water is drawn from the reservoir downward through a long smooth channel called a penstock, and directed across the blades of a turbine, which rotate horizontally (Figure 25.2). The turbine shaft is directed upward into the generator unit. Many turbine/generator units are needed at a typical installation. Efficiency factors on the order of 60–70% are common. That is, 60–70% of the energy in the falling water is converted to electricity.

Hydropower units are costly to install and require maintenance, but they use a fuel that is free and not subject to inflation. The fuel is supplied by the sun, which evaporates water into the atmosphere from

---

1 Kinetic energy is the energy associated with the movement of matter. The energy is proportional to both the mass being moved and the velocity squared.

Figure 25.1 Shasta Dam and Reservoir on the Sacramento River, north of Redding, California. The dam is a curved concrete gravity structure, 602 feet (183 meters) high, and the reservoir stores over 4.5 million acre-feet (556,000 hectare-meter). Shasta Power Plant has five main generating units, with total capacity of 442,310 kW. (Bureau of Reclamation.)

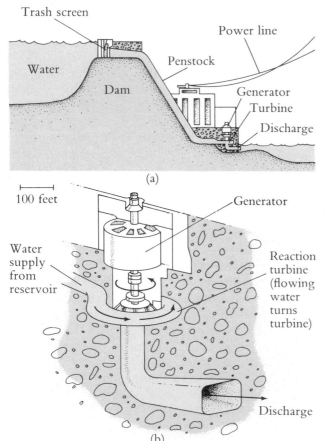

Figure 25.2 (a) Components of a hydropower system. (b) Turbine-generator unit. (Adapted from D. Kash, et al., Energy Alternatives, *Report to the President's Council on Environmental Quality,* 1975)

low-lying oceans, lakes, and rivers. The water vapor condenses as rain, which falls in the highlands and flows downward to the sea. Hydropower units interrupt that flow to the sea and capture the energy in the moving water, energy which would otherwise be used up in carrying sediments to the sea.

Hydropower, however, is not free of environmental impacts. Briefly, we mention here that dams and reservoirs modify not only the land that is flooded, but also the quality of the water that is stored and released. In addition, the stream channel downstream from the reservoir is affected. The loss of land habitat is one clear effect. The decrease in the quality of the water is surprising, though. Water released from a reservoir, depending on the season, can be very low in dissolved oxygen, and hence an unfavorable environment for fish and other normal aquatic species. Finally, the water released from a reservoir erodes and scours the stream channel to a greater extent than the undammed stream would have. All these effects are discussed more completely in Chapter 9.

It is not actually necessary to dam a free-flowing stream to create the benefits of an elevated water surface or a steady flow. Partial diversions of upstream water can be used to create a man-made lake off to the side of the river. Such a lake has the benefits of both elevation and steady availability. As you read about pumped storage at the end of this chapter, you will see that off-river storage lakes are actually quite common, but have not been thought of as replacements for reservoirs.

Table 25.1 Regional Distribution of Hydropower Resources[a]

	Potential (1000 MW)	Percent of world total	Developed (1000 MW)
North America	313	11	59
South America	577	20	5
Western Europe	158	6	47
Africa	780	27	2
Middle East	21	1	—
Southeast Asia	455	16	2
Far East	42	1	19
Australia	45	2	2
USSR, China, etc.	466	16	16
	2857	100	152

a Reproduced from M. King Hubbert, *Resources and Man*, copyright 1969. Used with the permission of the National Academy of Sciences, Washington, D.C.

The hydropower resources of North America relative to other continental areas are extensive (see Table 25.1), although most sites in the U.S. with a large potential have already been put to use. Nonetheless, much capacity at low-head sites remain to be exploited, as we shall discuss in a moment.

In Canada, on the other hand, huge hydropower resources are just now being tapped. Hydro-Quebec is now building hydropower projects on a massive scale, including an installation on James Bay that will be the largest hydroelectric plant in North America. Power from these projects has already been sold to New York State and is likely to be available in other eastern states of the U.S. as well. High-voltage transmission lines will stretch down from Quebec to New York to carry the power.

Small-Scale and Low-Head Hydropower

It is surprising to note, as a U.S. Corps of Engineers 1977 study did, that hundreds of hydropower sites, with dams already in existence, are no longer in use for electric generation. Many of these sites are small in capacity, 5 megawatts or less, but their potential is real. Abandoned during an era of cheap fuel for central electric power stations, the sites, if only put into service once again, could nearly double our capacity for hydroelectric power generation.

In the middle 1970s, about 57,000 megawatts of hydroelectric capacity were available in the United States. By 1980, this figure had expanded to 73,000 megawatts. The Corps of Engineers estimated that by restoring the abandoned dams and adding generating equipment to those dam sites never equipped for hydropower, 55,000 megawatts of capacity could be added. About half this quantity would be drawn from the small dams, 5 megawatts or less. Since a new thermal power plant is usually rated at about 1000 megawatts, adding these 55,000 megawatts is equivalent to building 55 major plants with very little more disruption than has already occurred because of previous development of the site (Figure 25.3).

We mentioned earlier that the first hydroelectric developments were run-of-river plants; the sites chosen for these plants typically were waterfalls with sufficiently large changes in elevation to turn the turbines. When all such sites were exhausted, dams were built to create the needed changes in elevation. The types of turbine, the Francis, Kaplan, and Peyton wheels, all required large "heads" of water, 100–2000 feet (30–610 meters), depending on the type.

In the early 1950s, a French firm, Neyrpic, invented a new kind of turbine—the **bulb turbine**—that is so versatile it can generate power just from rapidly flowing water. Not even a dam is required. This development has dramatically changed the potential contribution of hydropower to electric power needs. The bulb turbine has made tidal power, with its low heads of 30–50 feet (9–15 meters), technically feasible as well. The tidal power plant on the Rance River Estuary in France uses the Neyrpic bulb turbine. (See

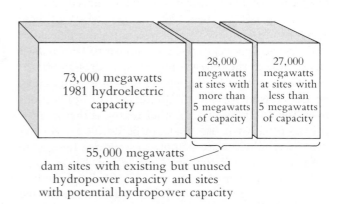

Figure 25.3 Used and unused hydropower capacity in the United States (not including bulb turbine run-of-river possibilities).

Figure 25.7 for a diagram of a bulb turbine.) The bulb turbine can be installed at sites previously too small to use and it can also be used to convert existing dams to hydropower installations. On existing dams, it could be installed just downstream of the outlet works, in such a way that reservoir releases could be channeled through the turbine.

Another development in small-scale hydro has also taken place in France. The Leroy-Somer Company, a manufacturer of motors, has recently produced a miniature hydro plant they call Hydrolec. Marketed for $7800 in 1978, the smallest plant can produce power at the rate of 5 kilowatts. A change in the water level of a stream of as little as 3 feet (0.9 meter) is sufficient to generate power, as long as the flow rate exceeds 70 gallons (266 liters) per second. Operating full-time, the smallest Hydrolec could produce power worth over $2000 in a single year (at 5 cents per kilowatt-hour).

The American giant in turbine manufacture, Allis Chalmers, is now producing a standardized low-head turbine with a maximum power output of 6 megawatts. This low-cost unit was used in the redevelopment of power capacity at the Barker Mill Dam in Auburn, Maine, a project that once again began producing power in 1980, after a 33-year interruption in power production at the site.

The Redevelopment of Hydropower

Applications to the Federal Energy Regulatory Commission for permits for exclusive preliminary evaluation of hydropower sites have surged rapidly, from 25 in 1978 to over 1200 in 1981. Many of these applications have been submitted from municipalities and electrical cooperatives, who appear to be playing a major role in the redevelopment of hydropower.

The tax-free bonding power of municipalities may be one reason for the predominance of applications from the public sector. Another reason is that municipalities often own sites that can be developed for hydropower: their water-supply reservoirs. Indeed, in 1981, the city of Patterson, New Jersey, began refurbishing the power plant on the Great Falls of the Passaic River, originally built in 1914. The city expected to begin selling the power for about a million dollars per year in 1983, the projected completion date of the plant. Their investment cost a total of about 10 million dollars. In 1981, in Litchfield, Connecticut, Northeast Utilities reopened a 320-kilowatt generator originally built in 1905; their cost was only $30,000.

The New England River Basin Commission identified some 1511 dam sites in the State of Connecticut; of these, 200 were felt to hold potential for hydropower generation. In New York State, the Commission counted 5300 dam sites, 1672 of which were suitable for electric generation. The "Energy Master Plan" of New York State calls for development of 1050 megawatts of power capacity at these sites by 1994, the equivalent rating of a new coal or nuclear generating station. These dams are in the right place at the right time to relieve the high cost of power generation in New York State, where oil is often burned as a power-plant fuel to relieve extensive air pollution with sulfur compounds.

Even irrigation canals are now being trapped for electric power. In California, hydropower has been generated since 1982 by water flowing in the Richvale Irrigation Canal. Using a Schneider generator, the power plant generates a steady 75 kilowatts (0.075 megawatt) from water falling only 9.5 feet (2.8 meters). A similar power plant is opening on the Turlock Irrigation Canal in California, as well.

Also in California, the Metropolitan Water District, which supplies water to Los Angeles and San Diego, is installing generating turbines in their water-supply system. Their basic water supply from Northern California and the Colorado River in Arizona has to be pumped over mountains to reach the city. It arrives with heads of up to 1500 feet (450 meters), ripe for conversion to electricity. The Metropolitan Water District is investing about $100 million in 15 projects. Power will be sold to the California Department of Water Resources.

The redevelopment of hydropower is not without problems, however. Breaking through tangles of red tape at both the state and federal level is one such problem. Cost is another. A new kilowatt of installed hydropower capacity at existing sites can cost between $300 and $2000, depending on how much refurbishing and site development is necessary. With 1985 construction costs of nuclear plants and coal-fired plants projected at $2000 and $1500 per kilowatt, respectively, the economics of installing hydro do not always appear favorable—unless the fuel cost enters the equation. Free fuel almost over the life of the structure makes hydro quite competitive when the longer view is taken.

Finding a market for this power is a potential problem as well. In the past, utilities discouraged private development, either by refusing to purchase surplus capacity or by paying discouragingly low prices for the electrical output. A federal law (1978) now requires utilities to purchase such power at fair prices. In addition, utilities have been levying a stand-by charge on hydro producers. The stand-by charge is a fee paid by a hydro developer who may occasionally have to purchase power from the utility rather than sell his surplus power. Nonetheless, several manufacturers in New England have redeveloped existing hydropower installations for their own use.

On the positive side, one factor favoring the redevelopment of small-scale hydropower is the relatively short time needed to bring new capacity on line. It takes from one to perhaps five years to be producing electricity once activity is begun. Compare this to the 10–12 years of lead time needed for nuclear plants.

Power from the Tides

Franklin Roosevelt, President of the United States, 1932–1945, made his summer home on Campobello Island, which lies along the western edge of the Bay of Fundy. From his home, he could see the rise and fall of the largest tides in the world, and he recognized and spoke of the potential of this region for power generation. A tidal power project was actually begun on Cobscook Bay in 1935 because of his urging, but Congress suspended funding of the project in 1936. Though nearly half a century has passed since that era, Roosevelt's vision has not been fulfilled. Yet the promise in those towering tides remains.

Tides are the result of the gravitational pulls of the moon and, to a lesser extent, the sun on the great oceans. As the earth rotates, a portion of the ocean waters are lifted and held in position for a time by these gravitational pulls. When the swell of water in the grip of the moon reaches the land, as it must because of the rotation of the earth, it appears as a high tide. Further rotation of the earth releases the grip of the moon on that portion of the ocean and the tide falls away. Tides rise and fall twice each day, although the times shift with the season and the moon's position (Figure 25.4).

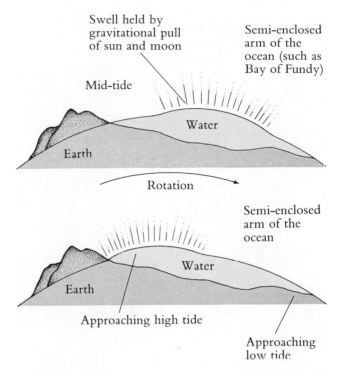

Figure 25.4 Tides. The swell of water that we call a tide is the result of the gravitational forces of the moon and, to a lesser extent, the sun. When the earth rotates, the swell seems to travel toward the shore. In reality, the swell is simply remaining in the same position relative to the moon and sun, and the shore is approaching the swell.

The average height of the tidal swell is only a few feet (half a meter), *except* when the ocean tides move within relatively narrow bodies of water. In this case, an oscillation wave is set up that may be up to 10–20 times the normal height of the tidal swell. Bay of Fundy tides, the largest in the world, run up to 52 feet (16 meters). Between England and the European coast (France, Belgium, and Holland), such large tides are also created. The highest tides of the year occur when the moon and sun are most nearly in line and their separate gravitational pulls come together to increase the volume of water dragged across the sea.

The possibility of using the energy in the tides has been turned to reality in France, where in 1968 Electricité de France finished a tidal power station on the Rance River, which flows into the Atlantic Ocean. To understand how tidal power works, we can discuss the operation of the station on the Rance (Figure 25.5).

Figure 25.5 La Rance Power Station lies across the estuary of the Rance River, which empties into the Atlantic Ocean between St. Malo and Dinard on the coast of Brittany. (Photo from the French Engineering Bureau, 1825 Jefferson Place, Washington, D.C. 20036)

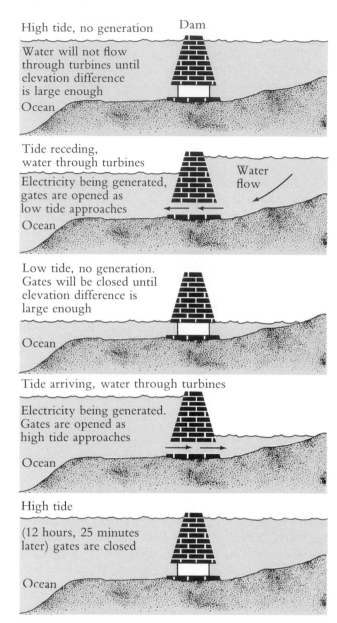

High tide, no generation   Dam

Water will not flow through turbines until elevation difference is large enough
Ocean

Tide receding, water through turbines

Electricity being generated, gates are opened as low tide approaches
Ocean

Water flow

Low tide, no generation. Gates will be closed until elevation difference is large enough

Ocean

Tide arriving, water through turbines

Electricity being generated. Gates are opened as high tide approaches
Ocean

High tide

(12 hours, 25 minutes later) gates are closed

Ocean

Figure 25.6 Schematic diagram of the operation of a tidal power station.

The Rance River experiences tides nearly as high as those in the Bay of Fundy, up to 44 feet (13.5 meters). Although nearby coastal tides are lower, the Rance is a relatively narrow river, and when the tide moves up and down within its banks, the water moves at a very high speed. A half-mile long dam has been built across the Rance and is used to store the waters of the arriving high tide. When tidal waters are receding, the stored water is released to the ocean through bulb turbines below the dam, and electricity is generated. To many people, this is the way tidal power is expected to operate. At La Rance Power Station, there is more. Energy can be generated on both the falling tide and the rising tide.

The rising tidal waters are captured inland by opening a set of sluice gates, which allow the arriving tide to flow upstream in the direction of the river source. The gates are closed when the tide reaches its highest stage; then the water that was barricaded inside is allowed to flow seaward through turbines as the tide recedes. At low tide, most of the water has been released. (See Figure 25.6.) As the tide builds again, it does so against the closed gates, so the water levels on the seaward side exceed those on the land side of the dam. When a sufficient head is built up, the water is allowed to flow upstream through the turbines, again generating electricity. Thus, electricity is generated both on the receding tide and on the arriving tide. The generation of electrical energy from these low heads of water is made possible by the use of the bulb turbine, the device we described in our discussion of small-scale hydro. (See Figure 25.7.)

La Rance Station has 24 separate bulb turbine units, capable of generating a total of 320 megawatts of electricity. Together, the units capture 25% of the energy possible to capture. Because the tide is moving up a river rather than along a broad coastline, the turbine and dam rest about three miles inland and are thus protected from the ravages of the open sea. Along the Bay of Fundy, the Passamaquoddy Bay on the easternmost coast of Maine could provide similar shelter for the dam and turbine. The potential for annual power production on the Passamaquoddy is more than five times that of the Rance project.

In March, 1980, the U.S. Corps of Engineers completed a study of the potential for tidal power in Maine. They concluded that tidal power on Passamaquoddy Bay was still not economical. Nevertheless, on the Canadian side of the Bay of Fundy, the Tidal Power Corporation of Canada has contracted for a small-scale tidal demonstration project. The project, which was to have been completed in 1983, was for a hydroelectric turbine in the mouth of the Annapolis River, which empties into the Bay of Fundy. Tidal flows from 15–20 feet (5–6 meters) would be used to produce 20 megawatts of power. The Tidal Power Review Board is also studying the possibilities for a 1085-megawatt installation in the Cumberland Basin in the northern portion of the Bay of Fundy. If the Canadians captured all of the available energy in the Fundy tides through multiple projects, they could harness more than 4000 megawatts of electric capacity.

Because of the enormous cost of these structures, governments are reluctant to invest in tidal power. The Rance project cost two-and-one-half times the estimated cost of a river hydroelectric station of the same average power output, primarily because of the added cost of coffer-dams both ahead of and behind the project. Yet, once the initial investment is made, the production of power requires no fuel. Only maintenance

(a)

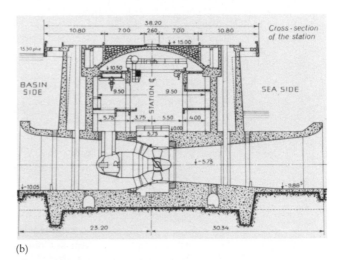

(b)

Figure 25.7 (a) The bulb turbine at La Rance Power Station, as viewed from the basin side. Note the size of the workman for scale. (b) A cutaway view of the power station, showing the bulb turbine. There are 24 bulb sets at the La Rance Power Station. (Photographs courtesy of the French Engineering Bureau, 1825 Jefferson Place, Washington, D.C. 20036)

of the structure is required. Thus, the costs of power are held down. The Rance station is proof that the concept of tidal power can work on a large scale and it is already influencing the public's views on tidal power.

A number of sites are located around the world where tides could be used to generate electricity. Some of the most attractive are listed in Table 25.2.

Other drawbacks to tidal power arise besides cost. If the tidal project is a great distance from the nearest large load center, long and costly transmission lines will need to be built to provide the tide-generated electricity an entry into the transmission network. On the other hand, such long-distance transmission is becoming more common as the new and more efficient 765-kilovolt lines are installed. New York City may soon be getting power from Hydro-Quebec via a 765-kilovolt line that runs several hundred miles. Tidal power from the Passamaquoddy or Cobscook in Maine reaching down to Boston no longer seems so unthinkable. Furthermore, the distance of load centers may not be so important. Industries that are heavy users of electricity may find it desirable to locate near tidal power stations. Examples of such industries are aluminum companies, steel companies using the electric arc furnace, and chemical companies producing chlorine and caustic soda via electrolysis.

Distance from the load center aside, there is another drawback to tidal power: the production of electricity is not steady. A moment's reflection on the nature of the tides should explain this. In the ordinary production of tidal power, electricity is only generated when the tide is receding; that is, when the height of water stored in the reservoir exceeds that of the receding water by a sufficient amount. As the tide drops and low tide approaches, the rate of power generation falls to zero, since there is no height difference. If the tidal project has reversible turbines, as La Rance does, power can be generated on the incoming tide as well, but still, power cannot be generated until the height of the rising tide exceeds the height behind the dam by a sufficient amount. As high tide is reached, the rate of power production again falls toward zero. Thus the curve of power production rises and falls twice each day, once for each of the two tidal cycles.

This cyclic production of electricity is unlikely to match the daily cycles of demand at load centers. Peak demands and peak production will occasionally occur together because the two tides shift in time as the seasons progress, but more often, the peaks of demand

Table 25.2 Selected Tidal Resources[a]

	Average difference between high and low tide		Average rate at which power could be produced (megawatts)
	Feet	Meters	
Severn, England	32.7	9.8	1,680
Mont St. Michel, France	28.0	8.4	9,700
White Sea, USSR	19.0	5.7	14,400
Mozen Estuary, USSR	22.0	6.6	1,370
Passamaquoddy, U.S., Canada[b]	18.3	5.5	1,800
Cobscook, U.S.[b]	18.3	5.5	722
Annapolis, Canada[b]	21.3	6.4	765
Minas–Cobequist, Canada[b]	35.7	10.7	19,900
Cumberland, Canada[b]	33.7	10.1	1,680
Petiteodiac, Canada[b]	35.7	10.7	794

a Source: J. McMullan, R. Morgan, and R. Murray, *Energy Resources.* New York: Halstead Press, John Wiley and Sons, 1977.
b All of the U.S. and Canada sites listed are along the Bay of Fundy.

and production will occur at different times of the day. Somehow, the tidal power must be fed into the transmission network at the proper rate. This means that other central power stations must usually be phased down in power output as the rate of tidal production reaches its maximum and phased up as the tidal production rate falls. A computer performs this task for Electricité de France at La Rance Station. The electricity from the tidal power station is, in effect, replacing on a fairly regular basis the electricity generated by other means. If the electricity replaced is from a coal-fired power plant, it is coal that is being saved.

A final word of caution is in order. As we look longingly at the tides and the awesome energy they carry, we must reflect as well on the environmental impacts of tidal reservoirs. The tidal reservoir is very likely to have an effect on the important biological area that extends along the shore of the ocean. The area, which is known as the intertidal zone, reaches from the point of the highest tide (or spray from the tides) to the lowest point exposed when the tides recede. (Both of these points are not fixed, but vary, within limits, with the seasons.)

Communities in this region consist first of those organisms that spend all or most of their time there. On sandy beaches are burrowing creatures such as crabs, shrimp, worms, and clams. Rocky shores support organisms attached to the rocks, such as mussels, oysters, barnacles, and the larger algae. In the waters of the intertidal zone, another set of organisms occurs: the phytoplankton. These microscopic floating green plants include the diatoms, the dinoflagellates, and the microflagellates, organisms that are swept in and out with the tides. One of the species of dinoflagellates is responsible for the infamous "red tides," which kill fish and sometimes make the flesh of shellfish poisonous to humans.

The phytoplankton, along with the attached larger red, brown, and green algae or "seaweeds," are the "producers" of the intertidal region. A variety of consumers also live in the intertidal zone. Some come in with the tide, among them zooplankton consumers, which spend their entire lives as plankton, and others that are the larval stages of crabs, jellyfish, sea urchins, snails, starfish, and other creatures. The intertidal zone is thus an important part of the sea's "nursery grounds." Still other consumers include the mature stages of crabs, clams, barnacles, snails, starfish, and other creatures that are present even at low tide.

Tidal power schemes, especially those maintaining a constant high pool (see the following section) change the natural periodicity of the tides. Such schemes also affect water temperatures in the area. Furthermore, we are not at all certain how the larval stages of marine species will survive passage through the turbine. Tidal power thus has the potential to change the relative balance among species that make up the communities of the intertidal zone. It is conceivable that nuisance species, such as those responsible for red tide, could be favored. Conversely, spawning of desirable species such as crabs or oysters could be harmed. In addition, we are not sure whether erosion could be accelerated or the deposit of sediments hastened by such projects.

Pumped Storage

The conventional method of hydropower generation on rivers is one way that falling water can be used in an electrical generating system. However, water can be used in another way in the electrical generating system. It can provide supplemental power during times of peak electrical demand. Ordinary reservoirs can be used in this way; that is, water can be saved and released through turbines at times of peak demand. However, water can also be stored off-river for this purpose. **Pumped storage** is the name given to an off-river reservoir that is drawn on at the time of peak loads.

Pumped-storage reservoirs are specially created by pumping water to high elevations during times when power demands are low. At these times, when the full capacity of an electric station, perhaps a coal-fired plant, is not being used, energy can be generated to pump water uphill into the pumped-storage reservoir (Figure 25.8). There the water remains until peak power loads occur. Then, the water is allowed to fall through the penstocks and turn the turbine pump to generate electricity. The same pump used to force water uphill is now turned in the opposite direction by the falling water to generate electric power.

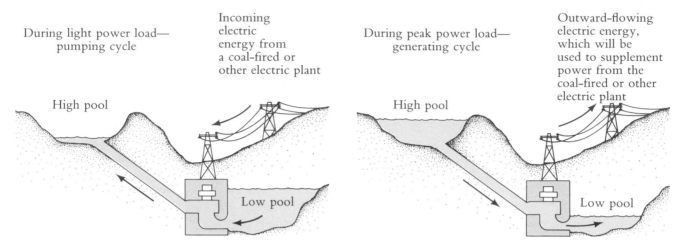

During light power load— pumping cycle

Incoming electric energy from a coal-fired or other electric plant

High pool

Low pool

During peak power load— generating cycle

Outward-flowing electric energy, which will be used to supplement power from the coal-fired or other electric plant

High pool

Low pool

Figure 25.8 Pumped-storage operation. (Adapted from D. Kash, et al., Energy Alternatives, *Report to the President's Council on Environmental Quality,* 1975.)

Essentially, a pumped-storage system uses spare generating capacity during periods of low demand. Electrical energy is converted into potential energy during periods of low electrical demands by pumping water to a higher elevation. During peak demand times, the potential energy is converted back to electricity to supplement power from the base load plant, which is now taxed to its capacity (Figure 25.8).

There are two principal disadvantages of pumped storage. The first is that the high pool used for storage has daily changes in elevation, going from nearly full to nearly empty (Figure 25.9). The earth sides of the reservoir exposed by this daily fluctuation are quite unattractive, resembling dark desert mounds with an undulating shape. In addition, the pumped-storage system is not fully efficient. Of the quantity of electric

Figure 25.9 The Blenheim–Gilboa Pumped-Storage Project. The high pool is atop Brown Mountain and water falls 1000 feet (300 meters) when electricity is produced from the unit. The turbine-generators can produce power at a rate of 1 million kilowatts (1000 megawatts). (Photo courtesy of Power Authority of the State of New York)

energy used to raise the water to the storage pool, only two-thirds is recovered. If coal is used to generate the electricity operating the pumps, the efficiency of converting the combustion heat to electricity is decreased by the two-thirds factor.

Pumped storage can be costly to build. The unit in Ludington, Michigan, which pumps water from Lake Michigan 358 feet (110 meters) up to a 1.6-square-mile (4.5-square-kilometer) storage pool, cost 340 million dollars to construct. An alternative being investigated for pumped storage is the feasibility of underground water storage in a tunnel or other subsurface chamber. Cost and safety of such underground structures are issues that need to be resolved, but the underground reservoir would obviously not have the usual environmental impacts of a pumped-storage project.

Methods to Increase and Steady Tidal Energy

Engineers have devised some clever, even remarkable, ways to assist the tides in giving us energy. On the clever side, they first had to deal with the issue of steadying the production of electricity through the entire tidal cycle. To accomplish this objective, two side-by-side basins can be operated jointly (Figure 25.10). The high basin is allowed to fill from the ocean on high tide. This high basin empties continually through a turbine to a second and lower basin. The second or lower basin empties on the low tide to the ocean. Operation always proceeds in the same direction, with continuous power generated by the water flowing through a turbine from the high to the low basin. The idea remains a concept, to the best of our knowledge. Because of the increased cost of the development, no one has yet attempted a double basin.

A remarkable technique has already been put to use at the single basin at La Rance. During the final portion of the arriving tide and beginning portion of the receding tide, the difference in elevation between the water in the reservoir and that in the ocean may be only a few feet. During this time, electricity from some other source can be used to pump ocean water (using the turbines) up into the tidal basin. The water is pumped up against a difference of only a few feet, so not much energy is required. When the tide has receded, that extra water falls through a distance of 20–30 feet, generating far more electric power than was used.

The same idea works at low tide except that the

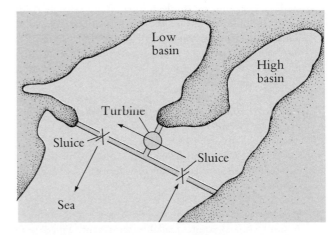

Figure 25.10 Linked Basins Concept for Tidal Generation. The high basin fills from the ocean on the high tide, but empties continuously into a second (low) basin. A turbine is turned by the water that empties from the high to the low basin. The low basin empties to the ocean on low tide. The Linked Basins concept has not, apparently, been implemented anywhere.

water is pumped out of the tidal basin into the ocean. The water level in the basin is thus reduced below sea level, and the incoming tide falls a greater distance. This scheme of pumping at high and low tides and recapturing later in the tidal cycle more energy than was put in, is sheer ingenuity.

Questions

1. What do we mean by "hydropower"? Where does the energy actually come from?
2. What are the environmental advantages and disadvantages of hydropower schemes? Briefly compare these to the environmental disadvantages of power from the burning of coal.
3. Explain how tides could be used to generate electric power. Where does the energy come from?
4. Discuss the possible economic and environmental advantages and disadvantages of tidal power. Compare to the disadvantages of generating power by burning oil.
5. Is there a net gain of electric energy when a pumped-storage system is used to generate electricity? Explain. What is the advantage of a pumped-storage generating system?

Further Reading

Hydropower

Carter, L., "Con Edison: Endless Storm King Dispute Adds to Its Troubles," *Science,* **184** (28 June 1974), 1353.
  This is the story through 1974 of the 14-year fight to block the pumped-storage unit called Storm King.
Erskine, G., "A Future for Hydropower," *Environment,* **20**(2), p. 33.
  An engineer converts his expert knowledge of hydroelectric generation and the bulb turbine into public property. A rare article.
Kirschten, R., "Hydropower: Turning to Water To Turn the Wheels," *National Journal* (29 April 1978), p. 672.
  A political, newsy article on low-head hydro developments.
Seltz-Petrash, A., "The New Energy Boom: Small-Scale Hydropower," *Civil Engineering* (April 1980), p. 66.
  A nontechnical, quite readable treatment. Filled with facts on developments in small-scale hydro.

Tidal Power

Bernstein, Lev, "Russian Tidal Power Station," *Civil Engineering Magazine* (April 1974), p. 46.
  A well-written article by the engineer in charge of a tidal power station built in Russia.
McMullan, J., R. Morgan, and R. Murray, *Energy Resources.* New York: Halstead Press, John Wiley and Sons, 1977.
  Tidal power is but one element of this paperback, which is thorough if a bit technical/scientific.
Project Interdependence, Committee Print 95-33, Congressional Research Service, U.S. Senate, 1977.
  This is a clearly written article on tidal energy without technical or engineering jargon. Recommended.
Ryan, Paul, "Harnessing Power from the Tides," *Oceanus,* **22,** No. 4 (Winter 1979–80), p. 64.
  A review of current activity in tidal power and description of a new concept in tidal power generation, in which a huge reinforced plastic sheet replaces the conventional dam.

CHAPTER TWENTY-SIX

Power from the Wind and
Power from the Heat in the Earth

Power from the Wind

*A Brief History/Windmill Design/Wind Resources of the
United States/Environmental Problems and Costs of Wind
Power/The Development of Wind Power*

Geothermal Energy

*Origin of Geothermal Heat/Two Important Uses for
Geothermal Heat/Resources of Geothermal Heat and
Projections of Use/Cautions and Environmental Impacts*

CONTROVERSY:

26.1: *Geothermal Energy Development—Should It
Take Precedence Over Risk to a National Park?*

Power from the Wind

A Brief History

Wind has been in the service of humankind since primitive people first raised a sail above a fragile log canoe. The prevailing westerlies were the winds that powered the voyages of discovery to the New World, and wind carried the Spanish Armada to victory after victory. The Trade Winds caught the sails of the great clipper ships and opened India and China to commerce with the West.

Long before the wind was harnessed by the ingenious windmills of the durable Dutch people, the ancient Persians had captured the wind to grind their grain. The Persian windmill turned on a vertical shaft, a design reinvented in the modern age. The windmills in Holland were used not only to grind grain but also to pump water out of the low-lying lands (polders) so that the fertile delta of the Rhine River could be farmed.

The Dutch windmills, in contrast to the Persian windmills, had the familiar horizontal axis (Figure 26.1). The blades of these wooden windmills reached

Figure 26.1 The two basic Dutch windmill designs. The design on the left had a "cap," which rotated so that the blades could be oriented to the wind. In the older design on the right, the entire building, except for the base, was rotated to set the blades in proper position relative to the wind.

80 feet (24 meters) in diameter, and the windmill could operate in winds of 25 miles per hour (42 kilometers per hour). Operators in the base of the mill oriented the blades to the direction and force of the wind. Many of the Dutch windmills, now 500 or more years old, are still in operating order, and courses are offered in Holland on windmill operation!

In the 1850s, a new type of windmill was invented in the United States. This American innovation, the multivane windmill, became common in rural America in the following years, finding its use first in raising water from wells. The multivaned mill pumped water for the steam locomotives that began crossing the American continent in the 1870s. The multivaned windmill is still in production today, but steel blades are now used in place of the hand-made wooden blades of the early models. With blades up to 30 feet (9 meters) in diameter, multivaned fans could produce up to 4 horsepower (3 kilowatts) in a 15-mile-an-hour wind (25 kilometers per hour) (Figure 26.2). In the 1930s, about six million multivaned windmills were in use, pumping water across the U.S.

(a)

(b)

(c)

Figure 26.2 The Multivane Windmill, invented in the U.S. in the 1850s, is still used on farms today to lift water from wells for livestock. Here an early model and two modern adaptations of the multivane concept are shown. (a) The self-regulating Perkins windmill as advertised in the *American Agriculturist,* December, 1892 (Vol. 51, No. 12). (b) The windmill supported by the pole with guy wires is from Chalk Wind Systems. (c) The windmill atop the tower is the product of Dempster Industries, Inc.

A new invention from Denmark gave windmills another chore on the American farm. In 1890, the Danes became the first people to generate electricity from a windmill. This windmill used a propeller with two or three thin blades instead of the multivaned design, and it could capture the wind's energy more efficiently and at higher wind speeds. Windmills brought electricity to the farms of America long before power lines were extended from central power stations. By the 1930s, these one-kilowatt windmill generators were helping rural America "tune in" to an exciting medium, the radio.

The windmill generators were used to charge automobile-type batteries, which could then be used for lights or radio, even when the wind was calm. As central station electric power reached more and more of rural America through public power programs like TVA, the use of such wind generators decreased. Until the present era, only relatively remote settings relied on small-scale wind power. Such sites may be rarely visited yet require small amounts of power on a steady basis. It is uneconomical to run power lines to such sites, so small-scale windmills are natural choices. Navigational beacons on islands at sea may often be powered by windmills.

Just for a moment, though, about 40 years ago, it looked as though electricity from the wind could compete with electricity from fossil-fuel generating stations. It was 1941, and a Smith–Putnam Generator rated at 1250 kilowatts in a 35-mile-an-hour wind (60 kilometers/hour) was installed at Grandpa's Knob, near Rutland, Vermont. Hooked into a power grid, the windmill delivered commercial power for $3\frac{1}{2}$ years. The tower was 110 feet (33 meters) tall and its two blades weighed eight tons apiece and were 175 feet (54 meters) in diameter. But in 1945 a sudden strong wind broke one of the blades, hurling it 750 feet (283 meters). A wartime shortage of materials prevented the windmill from ever being rebuilt, but the windmill at Grandpa's Knob proved that wind could be used to deliver commercial power.

In the present era of high fuel prices, it appears that such windmills can become cost competitive and contribute to the electricity needs of the nation. In this chapter we focus on electrical energy from the wind, but the reader should note also that sail power, abandoned on commercial ocean vessels around 1910, is once again a live option. Combining motor and sail on a single vessel and using durable new fabrics for sail-cloth may return us to the era of the square-rigger, with none of the delays and uncertainties of arrival.

Windmill Design

Windmills produce power when the wind pushes on the blades of the mill. The greater the "reach" of the blade, the more wind energy it can capture. Also, the greater the velocity of the wind, the greater the force on the blades and the greater the amount of energy captured.

The response to the diameter of the blade and the speed of the wind is not one-to-one. The power produced goes up with the square of the diameter of the blade and with the cube of the velocity of the wind. Table 26.1 indicates how power output changes for a typical horizontal-axis windmill at different wind speeds and blade diameters.

Note that at a wind speed of 20 miles/hour (33 kilometers/hour), multiplying the blade diameter by 4 (from 50 feet to 200 feet) multiplies the power output by 16. Observe also that at a blade diameter of 100 feet (30 meters), a wind of 30 miles per hour (50 kilometers/hour) generates 26 times more power than a wind speed of 10 miles/hour (17 kilometers/hour). This is why engineers lean toward big windmills and why they try to capture the higher winds.

Most large windmills now being built or in operation are designed to operate at wind speeds between 10–35 miles/hour (17–58 kilometers/hour). Winds less

Table 26.1 Power Output at Various Wind Speeds[a]

Wind speed		Power output in kilowatts		
		Diameter 50 feet (15 meters)	Diameter 100 feet (30 meters)	Diameter 200 feet (60 meters)
mi/hr	(km/hr)			
10	(17)	3	14	54
15	(25)	11	46	182
20	(33)	27	108	432
25	(41)	53	211	844
30	(50)	91	365	1458

a Source: *Alternative Long Range Energy Strategies,* Joint Hearing before the Select Committee on Small Business and the Committee on Interior and Insular Affairs, U.S. Senate, 9 December 1976.

Table 26.2 Observations and Events at Increasing Wind Speeds[a]

	Miles/hour	Kilometers/hour
Calm; smoke rises vertically.	1–3	2–5
Wind can now be felt on the face; leaves rustle.	4–7	6–11
Leaves and small twigs move constantly.	8–12	12–20
Dust is raised and loose paper blows; small branches move.	13–18	21–29
Small trees sway; the waves on water show crests.	19–24	30–39
Large branches are in motion. It is difficult to use an umbrella.	25–31	40–50
Entire trees are set to swaying in the wind, and walking against the wind becomes difficult.	32–38	51–61
At these speeds, the wind snaps twigs off trees.	39–46	62–74
Structural damage starts to occur.	47–54	75–87
Trees may be uprooted; severe structural damage is possible.	55–63	88–101
Such velocities are unusual inland; they cause widespread damage.	64–72	102–115
Hurricane conditions.	73–132	116–212

a Adapted from Industrial Instruments, Ltd., Stanley Road, Bromley, Kent, United Kingdom.

than 10 miles/hour produce little useful energy; winds higher than 35 miles/hour could wreck the windmill. To appreciate more fully the wind speeds at which the windmills operate, Table 26.2 lists events that occur at different wind speeds.

In a sense, the accident of Grandpa's Knob was an early indication of design problems in windmill technology, especially in durability and safety. The serious accident indicated that the huge propeller blades could "fatigue." "Fatigue" is a term from metallurgy, describing a metal that has undergone strong forces and whose structural strength is weakened as a result. Metal fatigue caused the accident at Grandpa's Knob.

Windmills should not be designed to capture gale winds. Even though such winds deliver far more power than do low-speed winds, they exert such a strong force on the blades that the machine itself may be destroyed. Furthermore, the proportion of time that gale winds blow is so small, that the contribution of gale winds to total power output is extremely small, making such risks not worthwhile. To combat the problem of gale winds, windmill blades are curved in such a way that they turn slightly to one side, out of the direct force of the wind, so that the full impact of large gusts does not damage the propeller. This old practice is known as "feathering." Newer materials that can withstand higher forces are also being used to prevent the breaking of blades.

Other problems in windmill design occur, by and large, simply because of the nature of the system needed to capture the power of the wind. Windmills typically stand on tall towers that enable the blades to reach the stronger winds that occur at higher elevations. Near the ground, houses, trees, small hills, and the like interrupt and obstruct the wind. Thus, tall masts are needed, but the heavy equipment, consisting of a propeller, a gearbox, and a generator, must sit atop the tall mast and this requires a very strong structure (Figure 26.3).

Another problem in using the power from windmills is the nature of wind itself. From little freshets to great gusts, the speed of the wind varies over a wide

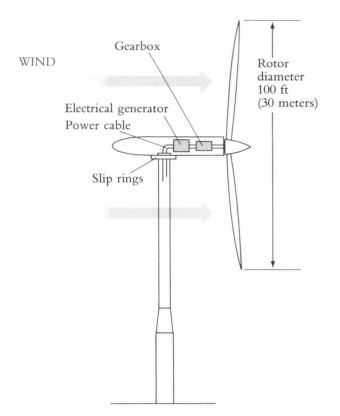

WIND

Gearbox

Electrical generator
Power cable

Slip rings

Rotor
diameter
100 ft
(30 meters)

Figure 26.3 Typical wind rotor system. (Adapted from
D. Kash, et al., *Energy Alternatives,* Report to the
President's Council on Environmental Quality, 1975)

range. As a consequence, the cycles per second of the
electrical output from a windmill varies. To correct
this, the alternating current generated by the turning
shaft is "rectified"; that is, it is converted into a steady,
one-directional flow. For large windmills, this steady
flow of current is fed to an "electronic inverter,"
which produces a stable alternating current that can be
fed into a power grid. Small windmills, such as those
used on isolated farmsteads or on islands at sea, feed
the "rectified" output to a very large storage battery
instead of to an "inverter." The batteries are essential
to store electrical energy for periods when the wind is
too calm to produce any energy.

The problem of cycles per second can be corrected, but the adjustment of power output is more
difficult. Just as with tidal power, there are times, with
a windmill, when little or no power is being produced.
At such times, conventional electric power will have
to be increased elsewhere to meet power demands.

Wind Resources of the United States

Wind varies with the features of the natural landscape
or urban terrain; it varies with the nearness to bodies of
water, with the weather and the season, and with the
height above ground. Winds over the land tend to be
slower, on average, than those on the coast or over the
ocean. Ocean winds are not only stronger, they are
steadier as well.

Recognizing that some of our best wind resources
are at sea, engineer William Heronemus of the University of Massachusetts proposed several banks of steel
windmills off the coast of New England. He envisioned that the electricity developed by the mills
would be converted to a steady, direct current that
could be used to electrolyze water to hydrogen. The
pure water to be electrolyzed would be provided by
distillation of sea water, also using the electrical energy developed. The hydrogen would come ashore
through a pipeline and be recombined with oxygen in
a **fuel cell.** Operation of the fuel cell would produce
electricity and water. Heronemus's vision is, however,
a long way off. For now, we need to look at the wind
energy we can reach easily, the energy on the land.

Across the U.S., the average annual wind speed is
about 10 miles/hour (17 kilometers/hour). It varies
from a lowest average of 6.5 miles/hour up to 37
miles/hour, the average wind speed at the top of
Mount Washington, New Hampshire. In fact, the
winds on Mount Washington have been known to
reach an incredible 150 miles/hour (250 kilometers/
hour). Figure 26.4 indicates the wind resources available across the United States. Interestingly, there is a
steady wind across the Great Plains, though the Plains
are relatively free of obstructions.

The President's Council on Environmental Quality has estimated that wind resources, if rapidly developed, could be used to displace 4–8 quads of thermal
energy by the year 2000. This is energy that would be
consumed to generate electricity. Depending on how
well we conserve energy, this could amount to 5–7%
of annual U.S. energy needs. The Council quotes one
study that estimates the wind potential of the U.S. at
1–2 trillion kilowatt-hours per year.[1] U.S. consumption of electricity in 1976 was on the order of 2 trillion
kilowatt-hours, indicating excellent potential for wind
power.

---

1 *Solar Energy: Progress and Promise,* Council on Environmental
Quality, April 1978.

Environmental Problems and Costs of Wind Power

Does wind power cause air pollution? No. Does it require water for cooling and cause thermal pollution? No. Does it consume fuels? No. It does cause noise, it does use land, and it takes materials to construct. It also has a visual impact, but the towers of long-distance electric lines have a height near that of the tallest windmill presently being considered. And cooling towers are even taller.

There is one other impact of wind power. Large windmills rotate at about 30 cycles per second. This is also the synchronization speed of television in the United States. For a distance of up to one mile, these large windmills may interfere with television reception. With the use of fiberglass blades, which are proving to be lower in cost than metal ones, the distance falls to about $\frac{1}{2}$ mile. Again, this is for large windmills; it is not expected to be a problem for small windmills.

Birds may be hurt by windmill blades, but it is difficult to predict to what extent this would occur.

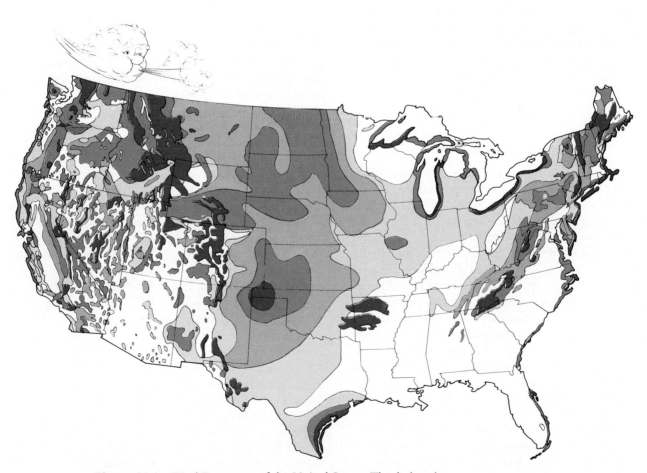

Figure 26.4 Wind Resources of the United States. The darker the shading on the map, the more wind power is available (in watts per square meter, averaged over a year). The most wind energy is available along the ocean coasts, the lower Gulf coast, the coasts of the Great Lakes, the Great Plains, and the Eastern and Western Mountain Regions. (From U.S. Department of Energy)

Certainly there are other environmental costs in the mining of iron ore, the manufacture of storage batteries, and the many more transmission wires and lines needed to collect electrical energy from multiple sources. But basically, when we count all the *environmental* costs, wind power comes out with a very low cost. What, however, is the *economic* cost of wind power?

The first experimental windmills produced for the federal government were definitely not cost-competitive with conventional power plants, but later test units are approaching a competitive status. The wind turbines at Goodnoe Hills, Washington, were to produce power at about 8 cents per kilowatt-hour. If the rotors can be mass-produced at lower costs, then the cost drops to about 4 cents per kilowatt-hour. Power produced at this cost would be competitive where high-priced oil is being used to generate electricity. When new blade designs have made the capture of wind energy even more efficient, costs on the order of 2–3 cents per kilowatt-hour will make wind energy fully competitive with conventional power.

The Development of Wind Power

Since the "energy awakening" began in 1973 and 1974, a growing number of large experimental wind generators have been built—principally by the U.S. Department of Energy, but more recently by private groups. A 100-kilowatt generator was first tested by the federal government at Sandusky, Ohio, in 1975. Although the station was beset by a number of difficulties, the Department of Energy went ahead with five additional wind turbines, each at a different site. Four of the wind turbines were rated at capacities of 200 kilowatts each. A fifth windmill was substantially larger. The four 200-kilowatt windmills are at Clayton, New Mexico; Block Island, off the coast of Rhode Island; the island of Culebra, a part of Puerto Rico; and Oahu, the largest and most populous of the Hawaiian Islands. The generator at Clayton has been delivering power to the town's municipally owned utility since early 1978. The Culebra wind turbine began operation in July 1978 and is delivering electricity to about 150 island homes. The Block Island wind station began operation in 1979 (Figure 26.5) and the Oahu wind turbine began to generate power in July, 1980.

A significantly larger windmill began operation in July, 1979, near Boone, North Carolina. The mill stands atop a mountain, and a steel tower reaches 140 feet (42 meters) into the air. The 200-foot (61-meter) diameter twin blades were able to produce 2000 kilowatts of electric power from winds of 25–35 miles/hour (40–50 kilometers/hour). The Boone wind station was shut down in 1981 after 20 months of successful testing. During this time, the wind turbine delivered power to a utility grid, the first time this has occurred since the ill-fated generator at Grandpa's Knob was operating in 1945.

In May, 1981, a cluster of three still larger windmills (they are being called wind turbines) began operation at Goodnoe Hills, Washington. Built by Boeing Engineering for the U.S. Department of Energy, these machines are rated at 2500 kilowatts (2.5 megawatts) in a 27.5-mph (44-kph) wind. The 200-foot (61-meter) towers are built to survive 125-mph (200-kph) winds. The rotor blades of each of the turbines are 300 feet (92 meters) in diameter and weigh 80 tons apiece. The power from the "wind farm" at Goodnoe Hills, which is sufficient for 2000–3000 average homes, is fed into the Northwest power grid by the Bonneville Power Administration. A still larger multiple-unit "wind farm" is planned by the U.S. Department of Interior for Medicine Bow, Wyoming. The Hamilton Standard Corporation has already sold the Department of Interior a 4-megawatt wind turbine for the installation. On the drawing boards is one wind turbine with a 400-foot (120-meter) blade span from the General Electric Company and another with a 420-foot span from Boeing Engineering. The 400-foot-blade machine will be rated at 6.2 megawatts; the 420-foot-blade machine will be able to generate electricity at the rate of 7.2 megawatts.

In the Netherlands, where wind power got its start, the U.S. tests have had an impact. In 1981, the Dutch committed to an experimental wind farm of 10–20 giant modern windmills. The experimental program is envisioned as a forerunner to "test the waters" for a $5.5-billion investment in a network of 1,100 large windmills, each with 262-foot (80-meter) rotors. If the system is completed as envisioned, the system would be supplying 12% of the Netherland's annual electric needs by the year 2000. To utilize the wind as fully as possible, the system would, when surplus power is available, pump water to an elevated lake for later release through turbines when the windmills are not operating. This is the well-known concept of pumped storage (see Chapter 25).

Figure 26.5 A new landmark now dominates the Block Island, Rhode Island, skyline. Sitting atop a 100-foot tower, this experimental wind turbine with its graceful 125-foot blades is capable of meeting nearly half the island's power demands at peak operation during the windy winter months. Experience with this machine will be used by the Department of Energy and NASA to design wind turbines suitable for commercial use in the 1980s and 1990s. (Photo courtesy of U.S. Department of Energy; R. R. Peabody, photographer.)

All of these developments have been projects of governments, but the data and experience collected has now interested private corporations in wind-power development. In 1981, Windfarms Limited, a San Francisco-based firm, began a project to produce and sell electricity on the Island of Oahu, Hawaii. The Hawaiian Electric Company will be purchasing up to 80,000 kilowatts (80 megawatts) of power from the cluster of about 30 windmills that the firm is building there.

Windfarms Limited has also contracted to deliver power to Pacific Gas and Electric in California. At a site in Altamount Hills, 50 miles east of San Francisco, the firm will be building some 600 relatively small windmills, each with a 50-kilowatt capacity under peak conditions, for a total power capacity of 300,000 kilowatts (300 megawatts). Hamilton Standard will be building its windmills.

U.S. Windpower, a Massachusetts company also planning to sell power, will be building its own machines, in contrast to Windfarms. It already has 100 windmills, each rated at 50 kilowatts, operating near Livermore, California. Pacific Gas and Electric is buying the power generated.

In Southern California, near Palm Springs, a 3-megawatt windmill is being erected for the local utility itself. Southern California Edison purchased the 200-foot (60-meter) diameter machine from the Schachle Company in Seattle. Its power rating will be higher than any wind turbine yet erected. Such private investments are exactly the response the government had been hoping to create.

Not to be forgotten, however, are the small windmills that have served the countryside since the early 1900s. These modernized private wind systems include 8–40-kilowatt machines that could be used on farms, at rural and suburban residences, and even at city residences. According to one study, if these wind systems can be made reliable, given a 20–25 year life, and are made cost competitive because of quantity manufacture, a U.S. market for over 17,000,000 machines awaits the industry. These systems have a long way to go, however, to become affordable. One manufacturer of 25-kilowatt windmills is charging $25,000 including installation for his machine, and that doesn't include delivery charges, which are on a mileage basis.

How fast could wind power grow? The 1981 Annual Report to Congress by the Energy Information Administration contained a projection made jointly by the Department of Energy and the Solar Energy Research Institute. Wind power produced in commercial quantity for utilities, as opposed to individual residences, should be delivering 264 megawatts of electricity by 1985, a remarkable spurt from virtually no contribution just ten years earlier in 1975. Hold onto your hats though, because the projection calls for a production rate of 1,993 megawatts by 1990 and 18,397 megawatts by 1995. Here comes the wind!

Geothermal Energy

"Healthful and Refreshing Warm Baths," the advertisements read for the resorts of Lake County, California. It was the late 19th century, and residents of the bustling city of San Francisco, some 75 miles to the south on the California coast, would travel to these "geyser" baths to be "restored." The odor of hydrogen sulfide gas was so noticeable in the area of the hot springs that the local stream became known as Big Sulphur Creek. Restorative though the baths must have been, no one saw any other commercial potential in the hot springs until the 1920s, when, perhaps inspired by success in Italy in the early 1900s, drilling for steam began in the area. But the drillers could not interest the electric companies in their steam.

Though U.S. technology first sought to exploit geothermal energy in the 1920s, Italy was able to produce electricity from steam from the earth in 1904. In the U.S., not until the 1950s could two companies interest Pacific Gas and Electric in using the steam that came bursting forth in Northern California. By 1960, Pacific Gas and Electric was generating electric power using geothermal steam at the rate of 11 megawatts. In all, nearly 30 generating units with a capacity of almost 2000 megawatts will be in place at the Geysers by 1986. The capacity of the sprawling plant had already reached 1400 megawatts in 1983 (Figure 26.6). One thousand megawatts is about the capacity of a modern coal-fired or nuclear power plant.

Wells are going ever deeper at the Geysers. Although the first steam was found at about 1000 feet, recent wells have been drilled to more than 9000 feet. In Italy, the Larderello plant had grown to 380 megawatts by the mid-1970s.

What is geothermal heat? How can it be used and where can we find it? Is it a clean and renewable source of power?

Origin of Geothermal Heat

Simply put, geothermal heat is the energy from the earth's interior. Of course, the eruption of a volcano, such as Mount St. Helens, is visible evidence of the enormous heat inside the earth. Scientists estimate the temperature in the core of the earth at thousands of degrees Centigrade. From the intensely hot interior of the earth, where molten metal and molten rock are thought to be the only form possible, up to the surface of the earth, a steady decrease in temperature occurs.

Only a few miles beneath the surface, one can occasionally find molten rock at 1000°C or more. The more likely find, however, is hot solid rock at a tem-

Figure 26.6 A Geothermal Power Plant at the Geysers in Northern California. This is one of nearly 30 plants in the steam fields of Northern California. Plans calls for the completion of a total of nearly 2000 megawatts (electrical) of capacity by 1986. Of this quantity, about 1400 megawatts of capacity were already in place in 1983. A bank of mechanical-draft cooling towers in the foreground of the plant are a reminder that thermal discharges are a part of geothermal electric power. (Photo courtesy of The Marley Cooling Tower Company.)

perature of perhaps 300°C. At a number of points on the earth's surface, usually in areas of volcanic and earthquake activity, this immense heat comes bubbling to the surface in the form of water and steam at temperatures as high as 300°C. This water and steam comes from groundwater that has wound its way from the surface down through porous rock and through rock fissures into a region of very hot rock. Heated by the rock, even boiled, this water comes bursting to the surface under pressure as steam and hot water. We call this erupting column of water and steam a "geyser." Old Faithful in Yellowstone National Park is our best-known geyser because of its towering spray and its insistent punctuality.

Two Important Uses for Geothermal Heat

Geothermal heat has the potential to be used in two basic ways: in the production of electricity, and in the heating of homes, offices, and factories. Whether the heat is used for electricity or for heating homes and factories depends on the form in which the resource makes its appearance. Sometimes the water comes billowing up from the earth as pure "dry" steam—that is, entirely vapor; no water droplets are mixed in the vapor. This dry steam can be used directly to turn a turbine and generate electricity. Condensed water may be reinjected into the earth or, if the quality is good enough, disposed of in a nearby body of water.

The Geysers, the field in California where Pacific Gas and Electric has its plant, and the Larderello field in Italy are both examples of fields that produce dry steam. Japan has developed one small generating unit using geothermal steam in Matsukawa. Finding dry steam is still by chance, however; geysers, discharging steam at the surface, led us to the California fields and to others as well.

In some fields, the geysers spew forth a mixture of steam and water droplets. Underground, these "wet steam" fields really have only water in them, but the water is at unusually high temperatures $(180–370°C)^2$ and is under very high pressure from the overlying rock. The release of pressure when the water comes to the surface causes some of the water to "flash" to steam. About 10–20% of the flow may flash to steam; the rest remains as hot water.

This mixture of steam and droplets cannot be used directly for the generation of electricity; the impact of the droplets would damage the turbine. In addition, geothermal water contains corrosive salts that make the use of wet steam even less advisable. The mixture must be separated into "dry" steam and water by a centrifugal separator, a device that whirls the heavier water droplets to the outer edge of the separator for collection. The steam is then directed to the turbine for the generation of electricity, while the hot water is disposed of in various ways. At Roosevelt Hot Springs, Utah, a device is being tested by Utah Power and Light that generates electricity with the water as well as the steam by directing the water droplets through a separate liquid turbine.

The steam that turns the turbine is condensed and must be gotten rid of. Although some novel ideas have been suggested for using the hot water that remains, reinjection into the earth appears to be favored because the water is often high in dissolved salts. The salts could damage a body of water into which the water was allowed to flow. The largest electric power plant in the world that uses wet steam is in Wairakei, New Zealand, where a plant with a power rating of 192 megawatts is in operation. The plant began operation in 1958. Whereas the Geysers plant and the plant at Larderello, Italy, are the models for power production with dry steam, the New Zealand plant is the model for power production with wet steam (Figure 26.7).

The first plant in the United States to use wet steam began operation in 1980 at Brawley, California. Operated by Southern California Edison and the Union Oil Company, the plant will be producing electricity at the rate of 10 megawatts. Another demonstration using water flashed to steam has been built in Valles Caldera, New Mexico, about 60 miles north of Albuquerque. The Union Oil Company teamed with the Public Service Company of New Mexico to build the 50-megawatt electric plant there, which will tap geothermal water at 280°C.

There is a third type of water-containing geothermal field in addition to the dry-steam and wet-steam fields. This is a field that produces hot water only. Such fields are even more common than wet- and dry-steam fields. While at first glance, hot water alone, especially if it has a high salt content, does not seem terribly desirable, it does, in fact, find good uses. Reykjavik (Ra′k-ya-vik), the capital of Iceland, with a population of 85,000, is heated almost entirely by hot

---

2 Under ordinary atmospheric pressure, water boils at 100°C.

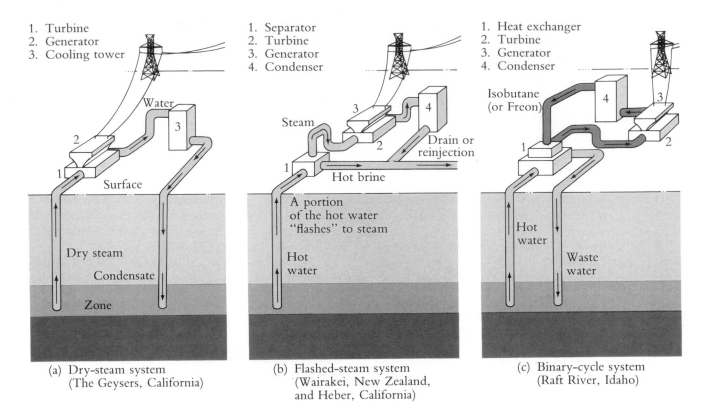

1. Turbine
2. Generator
3. Cooling tower

1. Separator
2. Turbine
3. Generator
4. Condenser

1. Heat exchanger
2. Turbine
3. Generator
4. Condenser

(a) Dry-steam system
(The Geysers, California)

(b) Flashed-steam system
(Wairakei, New Zealand,
and Heber, California)

(c) Binary-cycle system
(Raft River, Idaho)

Figure 26.7 Electricity can be produced in three ways from geothermal resources. (a) Dry steam, when it is available, can be used to turn a turbine directly for electric generation. (b) When only hot water is available, a portion of the flow is allowed to "flash" to steam. The steam, after being separated from the water, is then used to turn the turbine. (c) Another alternative for electric generation is to use the hot water to boil a fluid like isobutane to a vapor. The isobutane "steam" is used to turn a turbine to generate electricity. (Adapted from "Western Energy Resources and the Environment: Geothermal Energy," Report to the Office of Energy Materials and Industry of the U.S. Environmental Protection Agency, April 1977.)

water drawn from deep geothermal wells under the city (Figure 26.8). The use of geothermal water in Iceland began only in 1943. After it has been used to heat the city, the water is sent to greenhouses that produce fresh vegetables for Icelanders. In the U.S., geothermal heat has been used for several decades to heat homes in Boise, Idaho, and Klamath Falls, Oregon.

Hot water may be used for more than heating homes and offices. Geothermal water can be used to produce a "steam" or vapor from a "working fluid" that boils at a lower temperature than does water. Isobutane, one such fluid, might be boiled to an isobutane "steam" in a heat exchanger by using hot geothermal

water as the heat source; this "steam" is then used to turn a turbine. The spent isobutane vapor is condensed and returned to the heat exchanger to be boiled again. [See Figure 26.7(c).] The U.S., Japan, and the USSR are experimenting with this approach, which is called "binary cycle." It is called "binary cycle" because it uses two loops of fluid: the water loop, which heats the working fluid, and the working fluid loop. In the working fluid loop, the fluid is boiled, turns the turbine, condenses, and is reboiled.

A significant demonstration of electric power production from a binary cycle plant is the 45-megawatt facility at Heber, California. The Heber Project in the

Imperial Valley is a joint effort of numerous agencies and organizations, among which is Southern California Edison. The binary cycle plant is scheduled to go "on stream" in 1984; its power will be added to the power from a "wet-steam" geothermal plant that flashes hot water to steam and which was finished at the same site in 1982, another project of Southern California Edison.

The three kinds of geothermal electric plants—dry steam, flashed steam, and binary cycle—are all represented by facilities in the U.S. Plants of these types, and district heating projects, are shown in Figure 26.9.

There are still two additional ways to extract energy from the earth. One is not highly developed but shows promise. The other is only a concept at the moment. Geologists have noted that buried dry rock at temperatures as high as 300°C is about 10 times more abundant than water-bearing hot rock. Experiments indicate that this dry rock can be fractured by pumping down water under very high pressure. This technique, known as hydrofracturing, originated in the petroleum industry to get at gas deposits that were cut off from one another. Once the rock is fractured, water can be forced through the cracks and returned to

the surface at a much higher temperature. The water produced in this way may be as hot as ordinary geothermal hot water, depending on the temperature of the rock from which it was produced. There are, however, significant energy costs for pumping, depending on the depth at which the hot rock is found.

There are also deposits not far below the surface where molten rock (known as magma) resides. At least one such deposit lies in the United States. The deposit of molten rock in Marysville, Montana, may be a little more than a mile below the surface. Apparently, it is the remainder of a magma flow that was pressing toward the surface to become a volcano, but never made it. The temperature in the molten mass may be on the order of 1000°C. Can the heat in the molten rock be tapped by drilling? We do not know. In many ways, the molten rock, or magma, resembles the interior of an active volcano. Thus the question is even broader: can we tap volcanic heat? Will it be possible to bring the heat to the surface in the form of steam? Sandia Labs, New Mexico, are charged by the Department of Energy with developing the answers to these questions, but at the moment, there is no known method of drilling into magma.

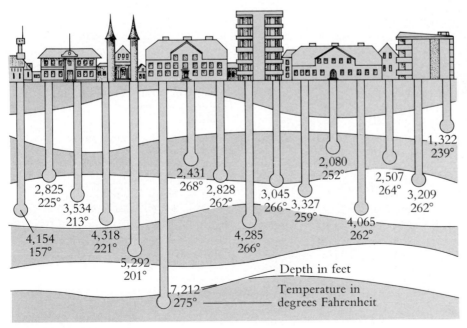

Figure 26.8 The city of Reykjavik, Iceland, has been using geothermal water to heat its homes, offices, shops, and factories since 1943. To provide the city with pollution-free heat, 32 holes have been drilled through the underlying lava beds to tap reservoirs of extremely hot water. A number of the holes have been productive, as shown in this diagram. Nine holes are currently in use. (© 1974 by The New York Times Company. Reprinted by permission.)

Representative U.S. Geothermal Projects

Location	Purpose	Technology	Capacity (MW)	Starting date	Sponsors
The Geysers, Calif.	Electricity, commercial	Natural steam cycle	800	1960–1980	Pacific Gas and Electric Co.; Union Oil Co. of California
Heber, Calif.	Electricity, demonstration	Binary cycle	45	1984	DOE; EPRI; San Diego Gas & Electric Co.; Chevron Resources Co.
East Mesa, Calif.	Electricity, pilot	Binary cycle	11	1979	Magma Power Co.
Raft River, Idaho	Electricity, experiment	Binary cycle	5	1980	DOE
Valles Caldera, NM	Electricity, demonstration	Direct-flash steam cycle	50	1982	DOE; Public Service Co. of New Mexico; Union Oil Co. of California
Northern Nevada (site to be selected)	Electricity, commercial	Direct-flash steam cycle	50	1984	Sierra Pacific Power Co. and other utilities
Heber, Calif.	Electricity, commercial	Direct-flash steam cycle	41	1982	Southern California Edison Co.; Chevron Resources Co.
Roosevelt Hot Springs, Utah	Electricity, commercial	Direct-flash steam cycle	20	1981 in stages	Utah Power & Light Co.; Phillips Petroleum Co.
Brawley, Calif.	Electricity, pilot	Direct-flash steam cycle	10	1980	Southern California Edison Co.; Union Oil Co. of California
Boise, Idaho	District heat, commercial	NA	NA	1981	DOE; State of Idaho; City of Boise
Crisfield, Md.	Hydrothermal, exploration	NA	NA	1979 (reached 60°C water)	DOE
Brazoria County, Texas	Geopressure, exploration	NA	NA	1979 (well complete)	DOE

Figure 26.9 **(Opposite page)** Geothermal resources and projects are concentrated in the Western States plus Louisiana and Texas. The resource areas shaded on the map represent deposits of geothermal water and steam. The resource areas shown shaded and surrounded by dots are geopressured areas. The water in these deposits is extremely hot and under an intense pressure from overlying sediments that prevents it from boiling. Methane may be dissolved in the geopressured water; the methane and hot water together represent a potent resource of energy. (Adapted from *EPRI Journal,* May 1980.)

Resources of Geothermal Heat and Projections of Use

The U.S. Geological Survey estimates our reserves of high-temperature water and steam useful for generation of electricity at a quantity capable of producing 11,700 megawatts each year for the next 30 years.[3] A plant of 1000 megawatts, the size of a modern nuclear or coal plant, is needed to supply an average city of one million people. These reserves, then, are capable of supplying the electricity needs of about 12 million people, or about 5–6% of current demands.

In contrast to these *known* reserves, the as yet undiscovered resources of high-temperature water and steam are estimated to be capable of producing 127,000 megawatts each year for the next 30 years (if they were immediately discovered and put to use). Some of these resources, however, might not yet be profitable to recover.

The resource base we have described so far is only that portion useful for the generation of electricity. In addition, lower-temperature water at 90–150°C is abundant.

Intermediate-temperature water systems are useful for heating, and the U.S. Geological Survey estimates the heating value of identified resources as equivalent to that of 14.3 billion barrels of oil. Given that the U.S. uses about seven billion barrels of oil each year *for all purposes* and has only about 30 billion barrels of reserves, this amount of heat is staggering. Furthermore, the value does not include as yet unidentified resources, which the Survey suggests may be three times the value of the identified resource base. We must be cautious, however, in interpreting these numbers, since population centers are not often on top of geothermal resources, as is Reykjavik, Iceland.

Two areas of the U.S. have unique geothermal resources. In the Gulf Coast region are huge deposits of extremely hot water under very high pressures, due to sediment layers above them. We refer to these deposits as "geopressured systems." The Geological Survey places these resources (in regions which they have assessed so far) at the equivalent of 11.2–43.3 billion barrels of oil. Perhaps three times these energy resources are thought to exist in areas as yet unassessed. In our discussion of natural gas, we pointed out that methane is in solution in the geopressured water and that some geologists consider the quantity of recoverable methane to be enormous, far more by factors than our current natural gas reserves. Thus, the geothermal deposits of the Gulf Coast are very attractive real estate for exploration. (See Figure 26.8 for the location of geopressured systems.)

In California extensive deposits of very hot water are found in the Imperial Valley. An intriguing idea for this water-short region is to flash the hot water to steam, generate power, condense the steam, and use it for irrigation. Though the water is initially high in salts, the condensed steam has, in fact, been distilled; that is, it has been made first into vapor and then condensed. Whereas the briny water would have laced the soil with salt if used as irrigation water, the condensate will not.

The estimates we provided here are only of underground water and steam systems. There is also the hot rock beneath the surface, which may be useful for heating water that is pumped into it, and these resources are likely to be far more abundant than the water and steam systems. Finally, geothermal energy in many cases may be renewable. Where the rock contains the heat and not merely the water, the heat in the rock can continue to be extracted by circulating water through the formation. The water may be water originally drawn from the formation or water brought to the site for that specific purpose.

---

3 These resources and reserves exclude those in national parks.

Geothermal Energy Development—Should It Take Precedence over Risk to a National Park?

The grizzly bear can't survive if Yellowstone Park is its only refuge. It also needs portions of the five adjacent national forests.

John Townsley,
Superintendent of Yellowstone National Park*

Any activity that would tend to damage them (the geothermal features of Yellowstone Park) . . . would be an insult of the worst kind—insensitivity typical of nineteenth century America.

Ralph Maughan*

One must simply accept the conclusion that our geothermal resources should be developed promptly, regardless of their location.

Alan Buck,
Stewart Capital Corporation*

Union Oil has pioneered the development of geothermal resources in the United States. The Union Oil Company operates the Geysers geothermal plant for Pacific Gas and Electric and is a partner in the geothermal development at Brawley, California, and at Valles Caldera, New Mexico. The company is now seeking to develop the Island Park Geothermal Area (IPGA) in Wyoming. The potential of this area is thought to rival that of the Geysers in California.

Unfortunately, the IPGA lies directly alongside Yellowstone National Park, the first (1872) National Park to be established in the United States. Yellowstone, as you know, is the home of Old Faithful, the world's most famous and unique geyser. Yellowstone is also home to the grizzly bear, a threatened species in the "lower 48," as well as home to bison, elk, and western waterfowl such as the trumpeter swan and Canada goose. The IPGA is outside of the park, but many are concerned that the proposed development will affect not only the geothermal features of the park but its wildlife as well.

The grizzly bear is the species thought to be most at risk from development of the IPGA. About 20% of the IPGA, which is in the Targhee National Forest, has been classified as land so critical to the grizzly that decisions about its use must first take account of the need to protect the bear. In fact, much more of the IPGA is crucial habitat for the bear, whose range includes both Yellowstone and parts of six adjoining National Forests. The presence of the additional people that would come to the area with geothermal development is thought to be the largest threat to the bear from geothermal-related activities.

The unique features of the 2.5-million-acre park may also be at risk from development at the IPGA. Even though Old Faithful is interior to the park and is some 13.5 miles from the IPGA, the possibility certainly exists of a deep underground connection between the thermal area and the geyser. Geologists have been unable to rule out the existence of such a connection, though they think it is unlikely. Could Old Faithful fall silent if development proceeds?

The opening of the Wairakei geothermal plant in New Zealand silenced the Great Geyser of Geyser Thermal Valley in that country. Geyser Thermal Valley, which had been the fifth most active geyser area in the world, last had a geyser eruption in 1965. Similar declines in activity have occurred with development and drilling in Italy and Iceland, as well as at Beowaive and Steamboat, Nevada. Geysers in the Boundary Creek area of Yellowstone are within a few miles of the proposed development; an effect on this geothermal area would seem likely.

Can the U.S. take a chance on geothermal development at the IPGA? What will be saved or replaced by this development? Given that power from the development will still be almost as costly as conventional power, what would be the benefits of development? To whom would they go? What would be the risks? Who would bear the risks? Should we preserve areas that have economic value to the nation if they are developed?

★ All quotes are from "The Incredible Shrinking Wilderness," *National Parks*, January/February 1982, p. 21.

Projections for geothermal electric power are built up in part from a survey of expansion plans of companies. One company, Pacific Gas and Electric, is planning on significant expansion of geothermal power generation. By 1990, PG&E hopes to have expanded its facility at the Geysers to 2000 megawatts from its current level of 910 megawatts. The 2000 megawatts should be able to be produced for 30 years, based on the steam resources existing there, and since most of the heat is in the rock, water injection may extend the resource still further. PG&E is lucky enough to have dry steam and plans to make the most of their resources.

Most other sites will be using wet steam and binary cycle systems, and these systems are just coming out of the development phase. In all, if present growth rates continue, geothermal electric generation could expand to 16,000 megawatts by the year 2000. The facilities for this generation would probably be in the western states, where most geothermal resources exist.

In summary, the geothermal potential of the continental U.S. appears to be very large and worth developing.

Cautions and Environmental Impacts

When geothermal energy is used, hot water must be disposed of. This is true even if "dry steam" is used to generate electricity. When this water is wasted from "wet steam" plants, it often has considerable amounts of salts dissolved in it.

Brines from the Imperial Valley may be as much as 20% salt, or about six times the salt concentration in seawater. At a geothermal deposit in El Salvador, where salt content is about half that of seawater, the brine has been reinjected into a nearby well almost a kilometer deep. Up to 800 tons of water per hour are injected there, making it unnecessary to carry the water via pipeline to the sea, an earlier alternative. Nonetheless, we still are not certain that waste geothermal water can be reinjected for the long periods that may be necessary. The quantities of salt are astounding.

Reinjection may prove to be more than a way to dispose of brine. It may also prevent the surface above geothermal wells from subsiding as water is drained from the deposit. The water, reinjected into the earth via a well not far from the drawing well, will help hold the earth intact. Not only may it prevent subsidence, it may also provide a "recharge" to the geothermal deposit. That is, if the rate of withdrawal exceeds the rate at which surface water descends through the earth to the deposit, the well could go dry. Reinjection may help to prevent such an occurrence. There is a possible drawback to reinjection: the water reinjected is typically more salty than that withdrawn, and this could lead to increasing levels of salt in the geothermal water.

Geothermal electric plants require a source of cool or cold water, as well as a source of heat for steam. The cool water is needed to condense the steam prior to disposal. In this regard, then, a geothermal electric plant is no different from a coal-fired or nuclear plant. The water used to condense the steam may come from a nearby freshwater source, such as a lake or river, or it may come from the ocean. If it comes from a freshwater source, cooling towers will be needed so that the heated fresh water from the condenser can be cooled and used again for condensing steam.

The volume of cooling water needed by a geothermal power plant is larger than that for a nuclear or coal plant of the same electric capacity. This is because the geothermal electric plant is less efficient than its cousins, a defect often cited by critics of geothermal energy. Even the Geysers plant in Lake County, California, which uses dry steam, has this problem. The problem stems from the nature of the geothermal resource; the steam is under lower pressure and at a lower temperature than steam used in most electric plants. Because of this, the transfer of energy to the turbine blades is less efficient. At the Geysers, the thermal efficiency is about 22%. This compares with an efficiency of 30% for most new nuclear plants and nearly 40% for most new coal-fired plants. Accordingly, cooling water consumption is much larger for geothermal electric plants.

Geothermal water is also likely to contain hydrogen sulfide. About 25% of geothermal water sources have hydrogen sulfide present as a contaminant. Hydrogen sulfide not only smells bad (it is the smell of "rotten eggs"), it may affect human health at high enough levels. Emissions from the Wairakei plant in New Zealand have caused silverware to blacken in a nearby village. The hydrogen sulfide content at the Geysers is high, about 200 ppm, but experiments at the Geysers show that more than 90% of the hydrogen sulfide can be removed.

To summarize briefly: geothermal energy may have a real, even dramatic contribution to make to our energy resource base. It is worth exploring further, even with its drawbacks.

Questions

1. How can the wind be used to generate power? Where does the energy that is captured originate?
2. Why do wind and hydropower, together, make a good system?
3. What major environmental and economic disadvantages and advantages does wind power have? Compare these to the disadvantages of generating power from nuclear fuels.
4. Briefly describe how the energy in geothermal heat can be used as a substitute for other energy sources.
5. What environmental impacts are associated with the use of geothermal heat? Compare to the impacts of fossil-fuel (coal, oil) power plants.

Further Reading

Wind Power

Merriam, M., "Wind Energy for Human Needs," *Technology Review,* **79** (3) (January 1977), 28.

The author, an engineer, treats the subject of wind energy at the level of *Scientific American.* Although there are a few formulas and engineering graphs, most of the paper can be understood without these elements.

Putnam, Palmer C., *Power from the Wind.* New York: Van Nostrand, 1948; reprinted, Van Nostrand Reinhold, 1974.

This book is listed for historical as well as practical interest. Note the author's name and recall the generator at Grandpa's Knob.

Perlman, Eric, "Kilowatts in Paradise," *Science 82,* **3,** No. 1 (January–February 1982), p. 78.

Describes efforts in Hawaii to use biomass, the wind, geothermal heat, and ocean thermal energy conversion as sources of power. Easy reading.

Marx, Wesley, "Seafarers Rethink Traditional Ways of Harnessing the Wind for Commerce," *Smithsonian,* **12,** No. 9 (December 1981), p. 51.

Discusses the possibilities of combining sail and motor as the moving force of ships and boats. Easy reading.

Sorensen, Brent, "Turning to the Wind," *American Scientist,* **69** (September–October 1981), p. 500.

A reasonably nontechnical treatment of wind power,

with cost comparisons of electricity generated from the wind with electricity generated by coal-fired and nuclear power plants.

"Wind: Prototypes on the Landscape," *EPRI Journal,* December 1981, p. 27.

A clearly written nontechnical review of past developments and future plans.

Geothermal Energy

Keifer, Irene, "Earth Boils Below While We Scratch the Surface for Fuel," *Smithsonian,* **5** (8) (November 1974), 82–88.

Wheeler, Romney, "The Geothermal Option," *Panhandle Magazine* No. 4 (1978), pp. 2–10.

"Geothermal: New Potential Underground," *EPRI Journal,* December 1981, p. 19.

"Tapping the Mainstream of Geothermal Energy," *EPRI Journal,* May 1980, p. 6.

These articles are clearly written without excessive technical jargon; they should be accessible to a wide audience.

Rinehart, John S., *Geysers and Geothermal Energy.* New York: Springer-Verlag, 1980, 256 pp.

A textbook on geothermal energy covering scientific aspects of geysers, such as geology, chemistry, and physics. Includes a section on uses and environmental effects. Recommended only for the scientifically trained student with deep interest in the subject.

Heiken, G., H. Murphy, G. Nunz, R. Potter and C. Grigsby, "Hot Dry Rock Geothermal Energy," *American Scientist,* **69,** July–August 1981.

Explores the option of injecting water into otherwise dry rock that is heated by underlying magma. Fairly technical and scientific.

Sourcebook on the Production of Electricity from Geothermal Energy, U.S. Department of Energy, 1980, DOE/RA/4051-1. Available from the U.S. Government Printing Office, Washington, D.C.

Comprehensive on its chosen subject.

Johnson, T., "Hot Water Power from the Earth," *Popular Science,* January 1983, p. 70.

Color illustrations and no mathematics make this article relatively easy to read. Probably the best "concept" pictures of geothermal power you will find.

Solar Energy

Introduction and History

Applications of Solar Energy

Solar Heat and Hot Water/Direct Conversion of Sunlight to Electricity: Photovoltaic Electricity/Interior Lighting from Sunlight/Central Station Electricity from Solar Energy/Electric Power from Space/Biomass: Biological Conversion of the Sun's Energy/Ocean Thermal Energy Conversion (OTEC)

Prospects for Solar Energy

The Sun as a Source of Heat for Water and Buildings

CONTROVERSY:

27.1: *Should Energy Sources Be Centralized or Decentralized?*

Introduction and History

Legend tells us that Archimedes saved his home city of Syracuse in Greece with solar energy. Ordering a thousand soldiers to turn their shields to the sun and lining them up in the shape of a parabola, Archimedes focused the sun's rays on the sails of the ships of an invading navy and burned them. No further practical applications of solar energy are recorded until the 19th century, when inventors began to experiment with the sun's ability to heat and even boil water.

In the early decades of the 20th century, it was still an open question as to what fuels would power society. Although fossil fuels jumped ahead because they were then inexpensive, the pattern is now beginning to reverse.

Early and exciting efforts to harness solar energy took place in the late 19th and early 20th centuries. In France, a solar steam engine was invented by Mouchot; his collector and engine were displayed in the 1878 World's Fair (see also Figure 27.1). By the early 1900s, farmers in California and Arizona had constructed solar irrigation pumps. By focusing the sun's rays on a boiler, they were able to produce steam, which was used to provide the turning power of a pump. Such a steam engine was even introduced in Meadi, Egypt, in 1912. Built by Shuman and Boys of Philadelphia, this engine was a long parabolic collector that could be turned to track the sun. The collector provided enough heat to boil steam for a 100-horsepower piston engine.

Such pumps and engines were only one aspect of a budding solar industry. Another dealt with solar hot

Figure 27.1 Solar collector exhibited at the 1889 Paris Exposition. The collector is driving a steam engine, which is furnishing the power to drive a printing press. (Courtesy of the Bettmann Archive)

water.[1] Around the turn of the century, residents of Southern California found themselves paying dearly for the coal needed to make hot water. To reduce these costs, they painted tanks black and placed them so that

1 The discussion of the history of solar hot water is drawn from two articles: K. Butti and J. Perlin, "Solar Water Heaters in California, 1891–1930," *Co-Evolution Quarterly* (Fall 1978), p. 4; K. Butti and J. Perlin, "Solar Water Heaters in Florida, 1923–1978," *Co-Evolution Quarterly* (Spring 1978), p. 74.

the tanks would absorb heat from the sun. On most days, the water would not be warm enough for showering until afternoon and then would cool quickly in the evenings. The Climax solar water heater, invented and patented in 1891 by Clarence Kemp of Baltimore, gave better results. With tanks packed in an insulated box with a glass window to let the sun in, the Climax Water Heater was often roof-mounted or attached to a wall. The water in the black tanks warmed more quickly and retained its heat longer because of the decreased heat losses from the box. The Climax water heater sold for $25 in Pasadena around 1900.

An even more efficient design was invented by Frank Walker of Los Angeles in 1898. Although double the cost of the Climax, Walker's heater was nonetheless popular because it was hooked into a conventional water-heating system, so that hot water could be drawn at all times, not just in late afternoon. Walker's design was mounted in the building so that the glass-window cover was flush with the roof; water was drawn from the top of the tank, where the hottest water naturally collected. Even though hot water could be drawn at all times of the day, the need to use coal or gas for early-morning hot water made the system expensive to operate. The sun's energy had somehow to be stored if one wanted hot water early in the day without using coal or gas.

To accomplish this, William Bailey created the basic elements of the modern solar hot-water system—in 1909! Bailey's design, called the Day-and-Night water heater, separated the component used for water heating from that used for heat storage (Figure 27.2). While a collector with copper tubes, much like a modern collector, was mounted on the roof, an insulated tank within the attic of the building stored the heated water. Water rose from the collector to the tank by the thermosyphon principle, which we shall discuss later. Bailey's system could hold heated water through the night and into the next day. Although the Day-and-Night sold for $100, it could cut gas heat bills by 75%, or about $25 per year. In four years, the water heater paid for itself.

In 1913, however, a record deep-freeze hit some parts of Southern California. The water froze and cracked the metal tubes, destroying the plumbing of the system; water leaked into houses. Bailey was called upon to invent a system that would not freeze. His new design simply mixed alcohol with water, creating an antifreeze, which was used as the collecting liquid,

Figure 27.2 Day-and-Night Brochure of 1923. In the first two decades of the century, solar water heaters were popular items in Southern California. In fact, the basic design still in use today was invented in California in that era. The arrival of inexpensive natural gas brought an end to the industry, however. Now, with fuel costs rising rapidly, solar hot water seems economical once again— this time throughout the country. (Brochure redrawn from *Co-Evolution Quarterly,* Fall, 1977.)

and the collector loop was separated from the hot-water loop. The collector liquid, now resistant to freezing, passed through a coil in the storage tank and heated the water in this way. By the end of World War I, more than 4000 units of Day-and-Night had been sold.

When natural gas discoveries in California made solar hot water less attractive in the 1920s, Day-and-

Night built and sold gas water heaters. In 1923, Bailey sold the rights to manufacture the solar heater to a Florida businessman, H. M. Carruthers. In 1932, an employee of Carruthers, Charles Ewald, added many new design features to the basic Day-and-Night model. One important change was to use soft copper tubing in the collector. The soft copper tubing, which proved resistant to cracking even in Florida freezes, allowed Ewald to avoid the alcohol-filled loop that Bailey invented and to heat the water directly for use.

Since the basic patents on the Day-and-Night had expired by the 1930s, other companies entered the business as well. In 1941, when America entered World War II, sales of solar water heaters in Miami were twice that of electric and gas water heaters. More than 15,000 solar hot water units had been placed in service in Miami by that time. The war brought the business to a quick halt, though; civilian uses of copper were forbidden.

The industry did not recover after the war because electric rates were falling while costs in both labor and materials were shooting up. Although the industry has yet to recover, thousands of solar hot water units are still in service in Miami. Today, as electric rates rise, solar hot water is once again looking attractive in Miami. (See "Prospects for Solar Energy" and Table 27.1, page 520.) From coast to coast in south Florida, new homes are once again being offered with solar hot water. In a very real sense, it is something of a "homecoming."

Applications of Solar Energy

Solar Heat and Hot Water

When people talk about the promise of solar energy, they may be talking about any of a number of solar "options." If they are pressed for the solar option that can deliver energy today, however, they will tell you that one option is ready right now. This option does not require further research and development to be installed in homes today. It is solar heat and hot water, produced by the flat-plate collector that has been used for so many years.

The production of solar heat and hot water begins with a water–antifreeze mixture flowing through the tubes of a flat-plate collector. The tubes, bathed in sunlight inside a glass-covered box, are heated by that sunlight and the heat is transferred to the liquid that flows through the tubes. The heated liquid is pumped to and through a heat exchanger, which is usually a coil of pipe inside a large water-filled storage tank. The heat in the flowing liquid is transferred to the water in the tank, which stores the heat for later use and for rainy days. (See Figure 27.3.)

The water in the storage tank serves as a reservoir, from which heat can be extracted for hot water for bathing, washing dishes, washing clothes, and the like. Hot water for these purposes is drawn by passing a portion of the cold water supply through another heat exchanger within the storage tank. That flow of water

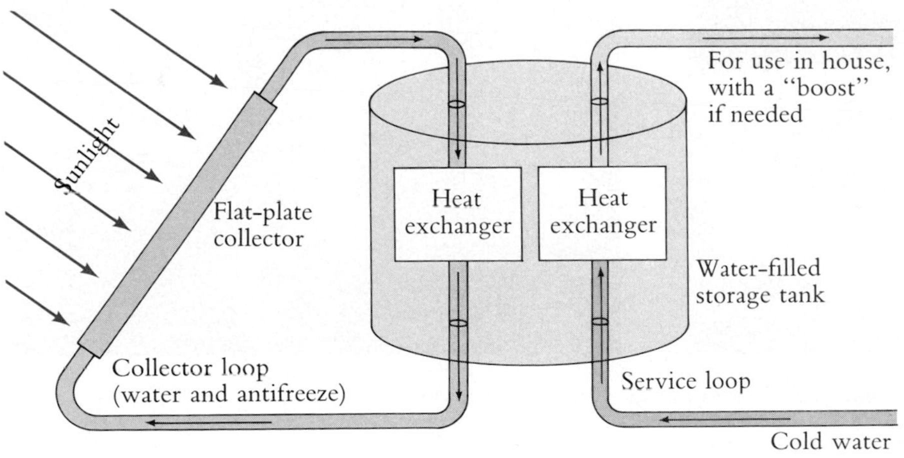

Figure 27.3 A Typical Arrangement of Collector and Storage Tank for Solar Hot Water. The water and antifreeze mixture in the collector loop circulates through the collector where the sun's energy is captured. The fluid then passes through a heat exchanger within a water-filled tank, transferring its heat to the water in the tank. The heat in the water in the tank is extracted by cold water, which flows through a second heat exchanger in the tank. The heated water in this service loop is now available for use in the house, either directly if it is hot enough, or with a boost from a conventional hot-water heater. The large water-filled tank is used to store heat from sunny days for use on cloudy days.

is heated by the hot water in the tank; it then passes to the conventional electric or gas water heater, where it is given a boost up to service temperature if a boost is needed.

To use the heat in the storage tank to heat a home, several options exist, depending on what form of conventional heat the home uses. If the heat for the home is supplied by forced air, as is so common now, water from the tank can be used in a water-to-air heat exchanger to heat the air flow to the house. If the air flow is not heated sufficiently, the conventional furnace turns on to provide a boost in temperature. The use of solar energy for heating requires a far larger investment in collectors than does solar hot water. Nonetheless, structures using solar heat are now being built. (See Figure 27.4.)

These arrangements for supplying heat and hot water are termed "active" because pumps using electrical energy are necessary to move liquids from one place to another. A second arrangement of solar equipment uses little or no outside energy to assist in providing heat for the home. These systems are termed "passive."

One design for passive solar heating avoids collecting heat from the sun during the summer and captures the sun's energy in the winter, when heat is needed. Large-paned windows are installed on the south wall of a building. These windows face the sun and the rays of the sun would be expected to enter the window both summer and winter. To block the rays of the summer sun, a roof overhang extends out over the windows (Figure 27.5). The overhang shades the window in summer, when the sun would otherwise contribute unwanted heat, but does not block the rays of the winter sun. The winter sun hangs low in the southern sky throughout the day and can furnish welcome heat to the dwelling.

The overhang concept can be extended to multiple-floor office buildings. Here an overhang extends over the outer rim of every floor, blocking the sun in summer but allowing its entrance in winter. Interestingly, for the individual dwelling, a deciduous tree can serve much the same purpose as an overhang. Again, assume we have large windows on a south-facing wall. The leaves of a deciduous tree growing in front of the window will block out much of the searing summer sun, but in winter the sun's rays can penetrate through the bare branches to enter the window. Pine trees elsewhere around the house help to cut the winter wind and thus decrease the loss of heat from the house.

Although a large south-facing window admits and captures more sunlight than a smaller one, it also pro-

Figure 27.4 Solar Heating. The left photo is the Town Elementary School in Atlanta, Georgia (Westinghouse Photo). The right photo is of a house in District Heights, Washington, D.C. (U.S. Department of Energy/Photo by Jack Schneider). On both structures, solar panels collect the sun's energy for use within the buildings. Provision is made in both buildings to store heat energy for cold, cloudy periods when sunlight is not available.

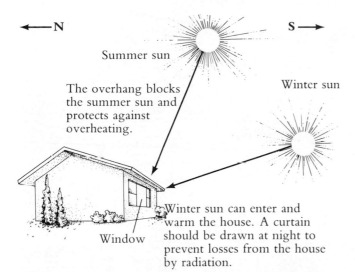

N ←

S →

Summer sun

The overhang blocks the summer sun and protects against overheating.

Winter sun

Window

Winter sun can enter and warm the house. A curtain should be drawn at night to prevent losses from the house by radiation.

Figure 27.5 Proper orientation and design of a house can decrease heating and air-conditioning requirements. The south-facing house has an overhang that blocks the sun's rays in summer but does not block them in winter. (The house illustrated is in the Northern Hemisphere.)

vides a wide surface that radiates heat into the night sky, an undesirable loss of energy. Opaque insulating curtains are one solution to the problem of back radiation. Another novel solution to the problem is to reduce the size of the window and arrange mirrors to capture and reflect additional sunlight into the smaller window. Reasonable quantities of sunlight are captured in this way and the potential for escape of back radiation is reduced.

Another architectural feature used for passive collection of the sun's energy is the solar greenhouse. The device is simply a greenhouse attached to a larger structure. Solar energy passes through and into the greenhouse. Drums painted black, which are filled with water, may be stored in the greenhouse. These drums and their contents are heated and become the means by which solar heat is stored for periods of no sunshine. (See Figure 27.6.)

These ideas for passive solar heating have been tried and improvements have been suggested to correct some problems. The first problem was one of condensation. Moisture tended to accumulate in passive homes because of their tight construction. Ceiling fans to keep air moving should cure such problems. The second difficulty encountered was the occurrence of local "hot spots" in the passive solar structures.

Again, fans can be used to distribute heat from these "hot spots" through the rest of the house. Basically, passive seems to need a slight assist to make it fulfill its promise.

Those portions of the country where solar energy has a distinct advantage have a high frequency of sunny days. The rates of solar heat reaching the United States are shown in Figure 27.7.

Direct Conversion of Sunlight to Electricity: Photovoltaic Electricity

The process of converting sunlight directly to electricity is known as **photovoltaic conversion**. The devices called semiconductors, made from silicon, which have so transformed the computer industry, are the basis of solar cells.

The cells are constructed by layering two wafers of silicon crystal next to each other with junctions between them. Light falling on one of the wafers "boils" electrons out of the crystal and across the junction into the other wafer. This current continues as long as the cell is bathed in light. The current is a directed (or direct) current (dc), and so requires conversion to alternating current (ac) before its use by a consumer.

The efficiency of conversion to electricity is low; only about 12–14% of arriving solar energy is converted to electricity by the current generation of commercially available silicon-based cells. Although silicon is the only material available commercially, cadmium sulfide and gallium arsenide also can be used in solar cells. Gallium arsenide, in particular, holds promise for higher conversion efficiencies, perhaps up to 30% of arriving solar energy.

Solar cells were first used in dramatic applications where cost was not a matter of concern. For instance, solar cells furnished the power for the Apollo moon rocket and the Viking space stations. In addition, the cells have been used for mountaintop radio, ocean signal buoys, highway call boxes, pipeline corrosion signal systems, and other uses. Typically, these uses require only modest amounts of power and are extremely remote relative to an electric transmission grid. Although there is a fledgling U.S. photovoltaics industry that supplies these needs, there is also an active federal program for solar cells, which supported more than 200 research and development projects in 1981.

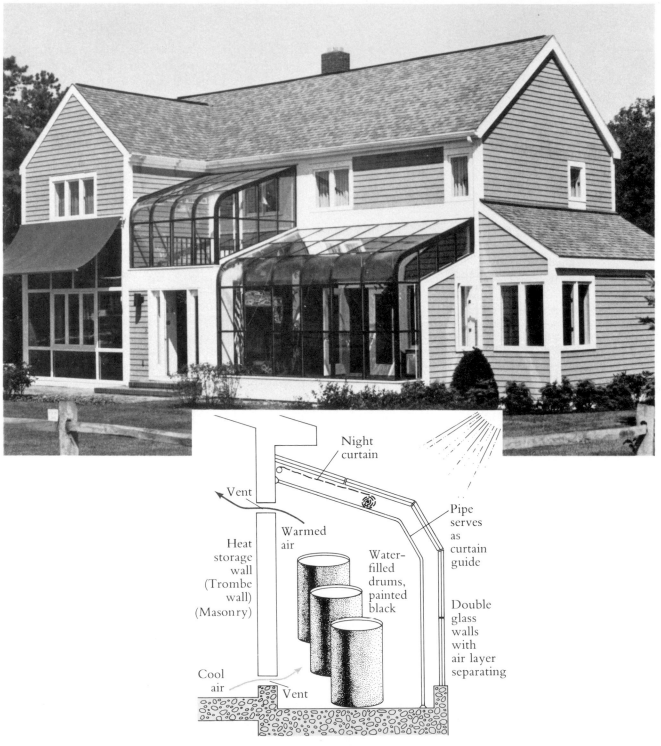

Figure 27.6 The Solar Greenhouse or Solarium, A Passive Solar System. This old idea from the Victorian era uses a greenhouse attached to and open to a residence. The furniture, floor, and perhaps drums of water are heated by the sun's rays. These fixtures radiate heat to the solar room and to the rest of the house when sunshine has ceased. (Photo courtesy of Four Seasons Solar Products Corp.)

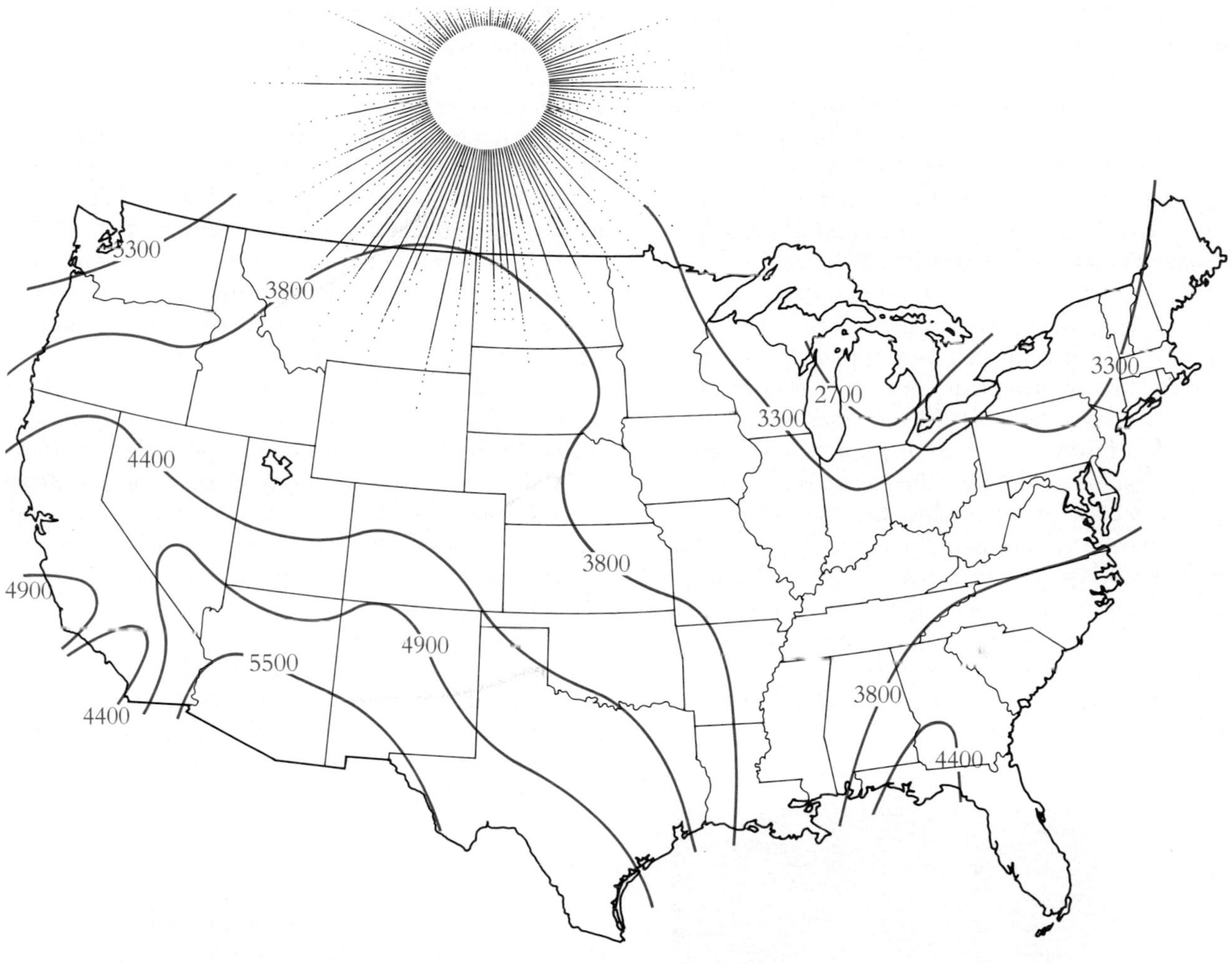

Figure 27.7 Distribution of solar energy over the United States. The figures give solar heat in Btu/square ft per day. Although some sections of the country have more sunshine than others, the attractiveness of solar energy depends also on the cost of alternate energy sources, such as electricity, oil, etc. (Source: D. Kash, et al., *Energy Alternatives,* Report to the President's Council on Environmental Quality, 1975.)

In 1981, completed photovoltaic systems (as opposed to unconnected modules) cost between $15 and $60 per peak watt.[2] Prices of $10 per peak kilowatt were anticipated in the not-too-distant future. Since the 1981 cost for conventional central station power is on the order of $1.20 per watt, the cost of power from solar cell arrays is still about 12 times more expensive than conventional power.

The U.S. Department of Energy hopes to stimulate the capability of producing photovoltaic systems at $1.60 per peak watt by 1986 (in 1980 dollars). The goal for 1990 is $1.10–$1.30 per peak watt, prices that would make photovoltaic attractive against the competition. The Department of Energy also set a goal for 1986 for the sale of 500 megawatts of photovoltaic generating capacity per year. Only about 4 megawatts of photovoltaic capacity were being sold each year in

---

2 The cost per peak watt is calculated by dividing the total cost by the number of watts produced during the portion of the day when the maximum solar energy is being received.

the early 1980s, though. (See Figure 27.8.) The most significant application of solar cells to date is a 1-megawatt photovoltaic power plant near Hespena, California, which uses tracking collectors; the plant, owned by Southern California Edison, was built by Arco Solar. Both optimism and pessimism have been voiced on the feasibility of these cost and sales goals, but as 1986 approaches, it seems that true commercialization of photovoltaic electricity is still some years away.

Thus, the photovoltaic system has a long way to go before it becomes useful for generating central station power or for turning on our lights at home. This is not to say that photovoltaic conversion is not promising or attractive. In a world in which technology advances at a rapid pace, there is hope that the semiconductors can be markedly reduced in cost, especially when they become mass-produced.

If cost barriers can be overcome, then photovoltaic electricity becomes very attractive indeed. No steam is generated; nothing is burned, heated, or fissioned; no fuel is mined and transported. No parts move and wear out, and the cells are expected to last a long time when enclosed in glass or plastic. In addition, the supply of fuel is inexhaustible.

Figure 27.8 The Solar Breeder. This unique plant in Frederick, Maryland, manufactures photovoltaic cells and panels using only energy from the sun. The 200-kW array of solar cells that forms the roof of the facility provides the entire electrical needs of the plant's production lines. A large battery storage system puts away energy on sunny days for continued production on cloudy days. Heating requirements are furnished by solar thermal energy, again with energy storage for sunless days. (Photo courtesy of the Solarex Corporation)

A radically different approach to the use of solar cells comes from Texas A & M University. Instead of using a sandwich of two thin silicon wafers, as the conventional cell does, the new system uses tiny (100-micron) spheres of silicon attached to a thin glass plate. The thin glass plate is immersed in an electrolyte solution, probably some kind of salt in water. Electric current is generated when sunlight falls on the spheres. Instead of being sent out to use or to storage in a battery, however, the electric current passes through and electrolyzes the electrolyte solution, producing hydrogen and another substance.

The substances are stored in the system and are intended for use as fuels at some later time. A likely use for the hydrogen is in a fuel cell (see Chapter 15, p. 287), a device that generates electric current from chemical reactions. Thus, the new "solar cell" does not have need of a storage battery, even though peak generation does not coincide with peak use. The energy is already stored in the system as hydrogen.

A pilot production operation for the new system got underway in 1982; commercial production was optimistically predicted for 1985. No cost predictions have been generally available. In 1982, a similar system was reported in the experimental stage at the University of California at Berkeley.

Interior Lighting from Sunlight

We always knew that the sun gave us light, and homes are often built with numerous windows to take advantage of this gift from nature. Although no electricity is generated in this application of sunlight, we discuss interior lighting with these other methods to generate electricity because in most situations, interior lighting is provided only by electricity. In industrial buildings and high-rise office buildings, interior rooms are often built without access to sunlight because windows are simply impossible to provide. Instead, incandescent and fluorescent bulbs furnish the light required for work. Artificial lighting in residential and commercial buildings has been estimated to account for 15% of U.S. electrical consumption. This fact motivates us to explore how we can turn the sun's rays to use in lighting our way.

The idea of "piping" sunlight through a building was given currency in 1974 when the Hyatt Regency Hotel in Chicago brought in sunlight to decorate the glass ceiling in its lobby. In 1976, two physicists at

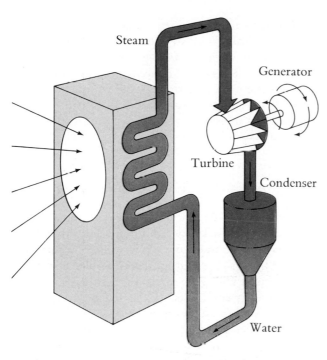

Figure 27.9 The Power Tower Concept. Water is boiled to steam by the concentrated rays of the sun. The steam is used to turn a turbine. (Adapted from D. Kash, et al., *Energy Alternatives,* Report to the President's Council on Environmental Quality, 1975)

Central-Station Electricity from Solar Energy

Picture again the invasion of ancient Syracuse and envision Archimedes directing his solar assault on the approaching warships. This is the principle of the **power tower**; the sun's rays are concentrated on a single point by properly adjusted mirrors. The mirrors, known as heliostats, rotate through the day to follow the sun across the sky. They reflect and focus the sun's rays on the power tower, where the enormous concentration of energy boils water to steam. The steam, piped to a turbine on the ground, turns the turbine and generates electricity (Figure 27.9).

A decade ago, the power tower was only a concept on a drawing board. Today, it has been turned to reality in the United States and overseas, and utilities see this new type of power plant as cost competitive with oil-fired and gas-fired electric plants.

The largest operating power tower is Solar One, a 10-megawatt (at peak) power plant occupying 130 acres in the Mojave Desert at Dagget, California, near Barstow (Figure 27.10). Begun in 1978, the unit,

Figure 27.10 The Power Tower in Operation. The Solar One facility, located near Barstow, California, can deliver up to 10 megawatts of power. A field of heliostats track the sun and focus the sun's rays on a boiler at the top of the power tower. Here water is boiled to steam, which is piped to a turbine–generator unit at the base of the tower for the generation of electricity. (Photo courtesy of the Electric Power Research Institute.)

Sandia Laboratory in New Mexico "piped" in sunlight to light their windowless office. To their pleasure, they found the lighting scheme not only economical, but pleasing as well in the quality of the light they obtained. The faint image of the sun was reflected on the walls, and clouds could be seen passing over the image.

The system to light interior rooms with sunlight has only one moving part, a flat mirror, which is controlled by a small computer to track the sun through the daytime hours.

Relative economics for solar versus electric lighting appear favorable because incandescent lights are only 10% efficient in converting electrical energy to light. Fluorescent bulbs, while more efficient, still convert only 20% of electrical energy into light. Sunlight directed into a building provides heat as well as light. Since lights furnish a portion of a building's heat requirement in winter, the use of sunlight for interior lighting can potentially save heating fuel as well.

which was built by McDonnel-Douglas, started delivering power to the grid in 1982. The $140-million facility is a joint effort of the Southern California Edison Company ($20 million) and the U.S. Department of Energy ($120 million). Although the construction cost was $14/watt, about ten times the cost of conventional power, there is promise that future plants will be far less expensive because of the economies that come from mass production of the heliostats. Costs in the range of $1.50 to $4.00 per watt seem feasible if heliostat costs can be reduced to $250 per square meter.

How does Solar One work? Its 1,818 heliostats, which were furnished by Martin Marietta Corp., each have 12 facets; each heliostat is 23 feet (7 meters) on a side. The heliostats track the sun through the sky, reflecting the rays of the sun to the top of the power tower. There, at the top of the 310-foot (95-meter) tower, the sun's rays focus on the boiler and produce steam at 950°F. The steam turns a turbine on the ground, generating enough power for a community of 7000–10,000 people.

The power tower has provision for heat storage— a one-million-gallon tank filled with rocks. The steam produced during periods of sunlight is used to heat oil, which circulates through the tank and heats the rocks. Using this stored heat to produce steam during hours of darkness, the facility can produce seven megawatts for up to four hours. But heat storage for off-sun hours may not be necessary if the power tower is viewed only as a device to provide power at peak times of demand.

Solar One *works*. It was expensive to build, but this is because components were specially built. The information provided by its operation, such as frequency of breakdown of heliostats and the failure of construction materials, will prove valuable in the next generation of power towers. Southern California Edison is already planning for that next generation: a 100-megawatt unit to be producing power by 1988. In the meantime, it has also contracted for another power tower in San Bernadino County; the 12-megawatt unit should be in service in 1984.

The power tower concept is being explored elsewhere in the United States, Europe, and Japan. There are units at Sandia National Labs (Albuquerque, New Mexico), and solar power towers have been built at Almeria, Spain; Adrano, Sicily (Italy); Themis, France; and Nio Town, Japan; though none are as large as Solar One.

In addition to planning for larger power towers, the electric utility industry sees value in joining solar power to an already existing conventional thermal electric plant. Known as *repowering,* the concept is that the sun's energy would replace some of the oil and gas presently used to boil water to steam. As much as 20% of fossil fuel use might thus be replaced by solar energy.

Electric Power from Space

A debate has been bubbling through the scientific community for the last decade about centralized versus decentralized sources of power. The debate asks whether numerous small sources of power matched to their end uses will serve society better than larger centralized power sources (see Controversy 27.1). In the debate, the larger centralized sources are the central-station electric plants. Nonetheless, even larger centralized sources are being envisioned. Perhaps none is more ambitious in scope than the Solar Power Satellite idea, which appears technically feasible, but whose economic costs could be staggering and whose environmental impacts are unknown. The Solar Power Satellite, even if research and development began now in seriousness, would be unlikely to contribute toward energy needs before 2015 or later.

The Satellite Solar Power Station would use solar cells to convert sunlight to electricity. The satellite's orbit could be adjusted so that it received light 24 hours a day. The electricity would be converted to microwaves and an antenna would direct a microwave beam back to a receiving antenna on earth. The microwave beam would be converted to direct current and then fed to a direct-current network.

Two other possibilities exist for operation of the Solar Power Satellite. In one concept, the solar radiation collected in space would be converted to infrared laser radiation, which would then be beamed to an earth receiver. In the other concept, mirrors attached to the satellite would continuously reflect and focus sunlight to a receiver on earth. The 24-hour continuous radiation at the receiver would then be converted to electricity.

Costs of building the satellite station, including the solar cells and other equipment, plus the cost of launching the payload into orbit, have been estimated by NASA at about 100 billion dollars for the first

5000-megawatt satellite, but there is much uncertainty attached to the estimate. Because of such costs and its environmental and political uncertainties, the Solar Power Satellite remains only a distant idea.

Biomass: Biological Conversion of the Sun's Energy (alias Photosynthesis)

Of all the conventional and widely used energy sources we have discussed, only nuclear power was not linked to the sun. Oil and gas are thought to be the buried remains of ocean plankton, species which derived their energy directly or indirectly from the sun. Coal was originally merely woody plants buried beneath the sediments in geological history. Hydropower can be traced to the sun's evaporation of water in the hydrologic cycle. Thus, our most common energy resources are linked closely to the sun.

Now, in an era of awakening to the finiteness of our energy resource base, we are turning to the sun to convert its rays directly to heat, hot water, electricity, and mechanical energy. That is, we are attempting to use our mechanical and engineering skills to gather the sun's rays directly to our purposes without any intermediate processing by nature.

It looks as though we will be reasonably successful at this conversion, but the fact that oil, gas, and coal all owe their energy to the sun reminds us strongly that nature already does an excellent job at converting the sun's rays to energy-laden molecules. We know the process as photosynthesis, the conversion of carbon dioxide and water to energy-rich organic molecules through use of the sun's energy. Why not take advantage of the energy being captured by green plants here and now? It is an idea that has already been used in earlier times.

Until the 20th century, most American families cooked and heated their homes with wood-burning stoves. Some people never stopped heating their homes with wood. They are rural people, typically, largely concentrated in Maine, Vermont, New Hampshire, and parts of the rural South. Wood was always available to them cheaply and it made sense to be economical. Now we are emulating them: about 11.5 million new wood-burning stoves were sold between 1974 and the end of 1981. By 1981, some 4.5 million households were using wood as their primary source of heat, and another 10 million households were burning wood as a secondary source of heat. It is estimated that 48 million tons of wood were burned to heat residences in 1981 in the U.S., double the tonnage used in 1969. Wood is also the fuel used by traditional societies throughout the developing world.

The burning of wood in industry has grown almost as fast. From 59 million tons of wood in 1969, industry increased its use to 81 million tons in 1981. This wood is chiefly scrap material from the pulp and paper industry and from the wood products industry, and it is burned in the boilers of these industries.

The saying goes that wood heats twice: once when you cut it and again when you burn it. Wood stoves, as opposed to drafty fireplaces that can cost additional heat, are also a valuable backup to conventional heat. If an ice storm were to snap transmission lines, as one did in New England in the winter of 1973, the loss of electricity would mean little or no heat in forced-air heating systems because no fan would be available to blow heat through the house. A wood stove with fuel would be a welcome feature at such a time.

Wood is a fairly clean fuel; apparently sulfur does not become incorporated in wood until it has reached the stage of advanced peat formation. Its heat content is also high: a cord of wood (a stack 4 ft × 4 ft × 8 ft) has about 20,000,000 Btu on average; a ton of coal has 24,000,000 Btu. The price of a ton of coal and a cord of wood vary by region and season, but are not far apart.

The fuel for wood stoves, if derived from scrap or from downed trees that would otherwise be wasted or would rot, does not diminish our resource base. Indeed, it takes pressure off the oil and natural gas markets. However, if the wood is drawn from woodlots that are not being replenished by the planting of new trees, there is an environmental effect. Natural regeneration of woodlots, depending on the locale, may take as little as 25 years or as long as 100 years. It is also possible to get wood from downed timber in State Forests and National Forests, where removal is often allowed. Two positive effects of this removal are that fire hazard may be reduced and sunlight will reach seedlings more easily. There is a negative effect as well. The downed wood would have decayed gradually, recycling its nutrients and minerals into the soil. Or fire would have consumed the downed timber, recycling the minerals quickly to the soil. When downed timber is removed, the cycle of growth and decay is interrupted.

If wood biomass were to become a major fuel, thousands upon thousands of new acres of trees would

Should Energy Sources Be Centralized or Decentralized?

Amory Lovins, a noted independent energy policy analyst, has suggested that there are

> . . . two energy paths that the U.S. (and, by analogy, other countries) might follow over the next 50 years. The first, or "hard" path is . . . high technology, centralized, increasingly electrified, and reliant chiefly on depletable resources (coal and uranium). The second or "soft" path is . . . relatively low technology and decentralized, electrified only where essential, based on renewable resources, and fission free. . . .*

Alvin Weinberg, director of the Institute for Policy Analysis of the Oak Ridge Associated Universities, takes issue with Lovins' suggestion that other forms of energy can be employed as effectively as electrical energy. He feels that

> . . . Lovins ignores electricity's advantages . . . The question of whether the advantages of electricity—its convenience, its cleanliness (compared with decentralized fossil fuel systems) and its costs are worth sacrificing . . . [because of] its admittedly poorer thermodynamic match with its end use and its possible social consequences if it is generated by fission.†

In the same article, Weinberg cites a speech by Franklin Roosevelt in 1936 to support his views on central station electric power:

> Sheer inertia has caused us to neglect formulating a public policy that would promote opportunity to take advantage of the flexibility of electricity; that would send it out wherever and whenever wanted at the lowest possible cost. We are continuing the forms of over-centralization of industry caused by characteristics of the steam engine, long after we have had technically available a form of energy which should promote decentralization of industry . . .

But David Lilienthal, former Chairman of the Tennessee Valley Authority, one of the largest suppliers of electricity in the nation, disputes what was once accepted wisdom:

> We were persuaded to accept the fashionable idea that great new generating stations and huge, regionalized transmission systems would deliver electric energy more efficiently at lower cost to the public than small, local decentralized ones. Nobody foresaw what would happen to the cost of oil and gas and coal, to the cost of transporting these fuels, or to the cost of constructing huge generating plants and mighty transmission lines. No one foresaw the rapid diminution of the world's reserves of oil. We placed major reliance on nuclear energy without being fully aware of its costs or its hazards.‡

Finally, Wilson Clark, commenting on the risks to our energy supplies in time of war, points out:

> . . . the increasing centralization of . . . our energy systems is increasing our nation's vulnerability, in the name of decreasing it.§

Alvin Weinberg favors central-station electric power; Amory Lovins advocates dispersed siting of power stations and individual energy generation. Their debate in many ways is "the" debate on centralized vs dispersed power sources; their exchanges are "the" exchanges of two communities. Both brilliant, both articulate, they have squared off, combating each other's views. To call one a conservative and the other a liberal, or one a traditionalist and the other a modernist, would probably be incorrect. Weinberg, however, has referred to Lovins and others as energy radicals. It is a most inappropriate label because of its political implications.

Both points of view have a long history. Individual energy generation is perhaps the oldest tradition, but the last half century has been an era of growth in centralized power generation. Thus, the debate is not one of tradition with new ideas, but one between two points of view that were both valid at different times in history. Each point of view says that current problems can be solved its way. Whose way do you think is most efficient and why? Or is it possibly a new time or a transition time when both points of view are valid? The issue of whose way is best should be approached by attempting to construct a list of advantages and disadvantages in each concept.

To answer a question such as this probably requires focusing on a single use of energy and answering specific questions. Since nearly half of America's homes heat hot water with electric energy, and since solar hot water is feasible and comparable in cost to electric hot water, it would be useful to compare these two energy uses. In making your comparison, recall that solar hot water may require electric energy to boost its temperature when the sun's rays are hidden for too long. Of course, burning oil or gas could be used to boost the temperature of the water as well.

The questions you might use to make your comparison include:

Which is more reliable: hot water from electricity from a huge central power station, or hot water from solar energy with an occasional electric assist? Which is

safer? Which is least subject to disruption? Which is least costly? Which is more convenient? Which is cleaner? Or, more accurately, where is the mess? If you add up the environmental problems, which method has the least impact? Which uses more resources?

Are there intangibles in your comparison between central and decentralized power? That is, are there elements in the one that you favor that you cannot qualify but that you simply prefer? What are those elements, if any? Can you identify a tradition that has the same point of view on what is preferred?

* *Alternate Long-Range Energy Strategies,* Joint Hearings Before the Select Committee on Small Business and the Committee on Interior and Insular Affairs, U.S. Senate, Interior Committee Serial No. (94-47) (92-137), 9 December 1976, for sale by the Superintendent of Documents, U.S. Government Printing Office.
† Review of *Soft Energy Paths: Toward a Durable Peace,* by Amory Lovins (Ballinger, 1977), appearing in *Energy Policy* (March 1978), p. 85.
‡ Quoted in *National Journal* (29 April 1978), p. 674.
§ Quoted in *Science,* **21** (February 13, 1981), p. 683.

have to be planted, because our present tree resources are largely devoted already to timbering and recreation. One proposal for producing wood in quantity for fuel is to grow sycamore trees for only five years to immature size, harvest them mechanically, and replant. Poplar is another fast-growing variety that may produce high yields. Although the burning of wood adds carbon dioxide to the atmosphere, just as the burning of oil, gas, or coal does, the growing trees would be withdrawing carbon dioxide from the air for photosynthesis. A rough balance of carbon dioxide addition and removal seems possible.

Sugar cane is another crop that gathers up the sun's energy efficiently. In fact, sugar cane may be the best biological converter we have. Its energy can be made available for nonfood use by fermenting it to ethyl alcohol. Sugar cane does not grow as well in this country as do other crops, particularly corn. Brazil, on the other hand, is well suited for sugar cane and is seriously pursuing the commercial production of fuel alcohol from sugar cane. By 1981, Brazil had 400,000 cars that ran on ethyl alcohol as fuel and 5000 alcohol-based service stations. Brazil was also planning for a

major increase in fuel alcohol production from sugar cane.

One unusual species of plant, the Hevea rubber plant, first found wild in Brazil, produces hydrocarbons that are used to make natural rubber. Grown almost exclusively in Indonesia and Malaya, the plant yields almost a ton per acre. Though this is half the yield of sugar cane, the growers think Hevea yields can be tripled. Although the Hevea plant has been our source of natural rubber for many years, enthusiasts see the hydrocarbons it produces as substitutes for oil. Hevea has competition, however, from another plant that produces hydrocarbons.

Native to the deserts of Mexico and the American Southwest, the wild desert shrub known as guayule (pronounced gwy-oo'-lee) has been cultivated in the U.S. in the past. In the early 1900s, guayule was one of our principal sources of rubber, but rubber from the Hevea plant replaced it in the marketplace. During World War II, when Southeast Asia was dominated by the Japanese, we again established a rubber industry based on guayule. That industry collapsed, however, when the rubber-producing nations of Malaya and Indonesia were liberated. Now, a number of scientists want to give guayule another try after first selecting the best strains for cultivation. Concern has been voiced, though, over the hybridized species of guayule that are being developed. The plant may cause severe skin rashes on the workers that harvest them, making mechanical harvesting a necessity. Still other desert plants have potential for cultivation as oilseed crops; these are jojoba, buffalo gourd, the gopher plant, guar, and devils claw.

Corn and corn wastes have been suggested as a source of biomass fuel. The burning of husks and stalks may be used in grain-drying operations, which presently call on propane and liquefied petroleum gas for heat. Corn can also be used to produce one component of the blended fuel popularly known as gasohol. Gasohol is a mixture of 90% unleaded gasoline and 10% grain alcohol. Corn carbohydrates can be fermented to produce the grain alcohol, which is then distilled to a high-purity alcohol. The technology of fermentation and distillation is well known and is the basis of the manufacture of whiskey from corn. The process is so simple that it has been done at home in illegal stills for many years.

The economics of gasohol are most uncertain. Gasohol is nearly competitive with gasoline, but the price of corn and the price of oil interact in a complex fashion to determine whether gasohol is profitable. A low price for corn and high price for oil is definitely most favorable for gasohol, but year-to-year variations in the prices of corn and oil make investment in the fuel alcohol industry an uncertain proposition.

The potential for biomass fuels is very large, but the impact of this development is uncertain. When use is made of otherwise discarded materials, producing biomass fuels is only good conservation practice, reflecting the ethic "waste not, want not." When new crops, new processes, or new acreages are proposed, though, we tread on less certain ground.

We do not know if we can afford the acreage for energy crops at the expense of food crops. Turning cropland devoted to food plants into cropland devoted to energy plants could cause a rise in the price of basic foodstuffs. U.S. consumers aside, for a nation that serves as a breadbasket to the world, the choice of energy crops over food crops could have wide effects. The use of residues left after harvesting conventional crops, rather than going to an agriculture that grows plants for energy, might be a more sound course of action while we learn about biomass energy. We also do not know if we can devote enough land to wood for burning because of the large demands for timber and paper. Also, many biomass processes do not have the advantage of some solar technologies because they require central processing. Solar house heating, in contrast to biomass energy plans, is a decentralized activity, carried out in each dwelling.

Of perhaps greatest concern, however, is the soundness of biomass proposals from a biological point of view. Planting only a single crop has a tendency to deplete soil of particular nutrients, attract insect pests, and reduce the survival of a diversity of animals. Caution is needed lest tropical forests be destroyed to produce a source of energy. Field crop research on these ideas is needed in order to evaluate biomass further. And careful economic analysis of the effects on food prices is needed as well.

Ocean Thermal Energy Conversion (OTEC)

Another idea to exploit the sun's energy draws on the temperature differences that exist between the surface and the depths in tropical oceans. The temperature differences are very large in some places, especially in areas in which the warm Gulf Stream flows. Between the surface and the water at a depth of 2000 feet (600 meters), the temperature difference may be as great as 22°C (40°F).

The OTEC idea is not at all new. First proposed by the French physicist d'Arsonval in 1881, the concept was actually tested more than half a century ago in the waters of the Bay of Matanzas, Cuba. D'Arsonval's pupil, Georges Claude, in 1929 built and tested the forerunner of current OTEC concepts. Waves and currents destroyed his experiment before he could make the idea practical.

The principle of OTEC is to use these waters at different temperatures alternately to boil and condense a "working fluid." In between, the vapor at high pressure is used to turn a turbine. The working fluids most studied and discussed are ammonia and propane. OTEC works this way: warm surface waters are used to boil the liquid working fluid to a vapor. The vapor is used to turn a turbine and generate electricity. Cold waters are brought up from the ocean's depths to condense the vapor in another heat exchanger (Figure

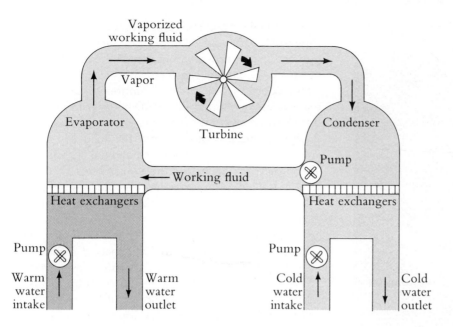

Figure 27.11 Schematic diagram of a closed-cycle ocean thermal power plant. (Adapted from John R. Justus, "Renewable Sources of Energy from the Ocean," in *Project Interdependence,* Committee Print 95-3, U.S. Congress, November 1977.)

27.11). The condensed fluid is returned to be vaporized again. Electricity would be generated continuously.

Of course, all of this is to be carried out at sea, so the electricity would require transport to land. Underwater cables would be used for this purpose. On the other hand, the electricity could be used at sea as it is generated, for such electric-intensive industries as aluminum or electric arc steel.

Both research and development on OTEC are being carried forward. The design of the huge heat exchangers, as well as of turbines, pumps, and other support structures, is being investigated. In addition, the pipe for a 100-megawatt plant must reach several thousand feet into the ocean depths; it would have a staggering 50-foot (15-meter) diameter. Can such a pipe be made to withstand buffeting by waves and currents? Or will multiple cold-water pipes be needed?

Two pilot projects, both in the waters of Hawaii, are coming to grips with OTEC problems. A DOE facility designated as OTEC-1 began operation in 1982 near Keahole Point, Hawaii; the facility, mounted on a former U.S. Navy tanker, is drawing cold water from 2100 feet (640 meters) below the surface. Ammonia is being vaporized in a heat exchanger by the warm water at the surface, and it is being condensed in a second heat exchanger by the cold water from the ocean depths. The ammonia "steam," however, is not used to generate electricity; the test is only of the heat ex-

changers—in particular, how to keep them effective and functioning in a marine environment.

The second major test facility, known as Mini-OTEC, was mounted on a barge off the Kona coast of Hawaii. In 1980, this 50-kilowatt plant, also using ammonia as a working fluid, successfully was able to generate 12 kilowatts of net power after electricity needs for such operations as pumping were deducted. The plant was a joint venture of the State of Hawaii and three private corporations. Although a milestone in OTEC testing, the facility resolved few of the engineering problems the OTEC concept faces.

On the drawing boards, DOE is planning a demonstration facility consisting of a cluster of four 10-megawatt OTEC units. Four separate units are used in order to ensure reliability; one unit going down will leave three others operating.

Many problems still must be solved. Corrosion by sea water is one problem area. Ocean organisms and slime may be so foul as to clog the heat exchanger, reducing its effectiveness. Large quantities of antifouling chemicals may be required. A novel idea is to bounce hard rubber balls back and forth through the heat exchanger pipes to keep them clear of scale.

Environmental questions such as impact on climate by diverting Gulf Stream waters still need to be answered. Nonetheless, vast quantities of electricity may be generated in this way. One estimate by the

Figure 27.12 Artist's conception of an OTEC factory ship. Ocean thermal energy conversion (OTEC) would exploit the temperature differences between the layers of tropical oceans. Here, an artist provides a vision of what an OTEC factory moored at sea might look like. (The Johns Hopkins University Applied Physics Laboratory.)

Department of Energy suggests that power equivalent to that from 200 electrical plants of a 1000-megawatt rating could be produced from OTEC plants moored in U.S. waters (Figure 27.12).

Prospects for Solar Energy

Solar energy can benefit people in a number of basic ways. First, by replacing fossil fuels, air and water pollution are decreased. Second, the replacement of fossil fuels means a decrease in fuel imports, especially of oil, and this will help secure the value of the dollar. Third, by replacing nuclear fuels here and abroad, the threat of the spread of nuclear weapons is decreased. Finally, solar sources can provide us some protection by making our fuel supply less subject to interruption.

What portion of our energy needs can solar energy contribute? There are certainly optimistic views of its potential. Both the President's Council on Environ-

mental Quality[3] and the Project Independence Report[4] suggested that under conditions of "accelerated development," solar power could contribute nearly 25% of the nation's energy requirements by the year 2000. Both reports had a wider definition of solar energy than we use here; solar included energy from wind and from hydropower. We have to remove the potential of these two energy sources from the solar category to arrive at our estimate of solar's potential contribution. The projection, then, for the year 2000 is that about 13% of the nation's energy needs could be supplied by solar energy in the form of heating, cooling, electric generation at central stations, photovoltaic electric energy, and biomass. Thirteen percent is a very large number for an economy that is expected to be using 80–120 quadrillion Btu of energy in the year 2000.

It is good that such a major contribution is in sight, but what are the conditions of "accelerated development"? "Accelerated development" has no precise definition; it means many things. Basically, however, it means some form of government intervention in the market for solar technology. Without such involvement, the paths that the solar market can take will differ dramatically.

Government intervention in markets is nothing new and nothing radical for Democrats or Republicans. Before the OPEC era, oil from overseas was brought in on a quota system for many years. The quota system prevented the domestic oil market from being flooded with cheap imported oil and thus protected the American oil industry. Strategic metals have been stockpiled for many years, both to have a store of the metals for emergencies and to shore up the markets for these materials. Farmers have received price supports from the Department of Agriculture on basic grains for many years to keep their incomes steady, and they have been paid to keep acreage out of production in order to keep crop prices up. In addition, a system of agricultural extension agents provide free advice to farmers on all aspects of farm operations. Why not have a system of energy extension agents providing free advice to home owners on the most appropriate ways to heat homes, laying out choices that include solar options and energy conservation?

---

3 President's Council on Environmental Quality, *Solar Energy: Progress and Promise,* April 1978, Government Printing Office, Washington, D.C.
4 Federal Energy Administration, *Project Independence Blueprint, Final Task Force Report: Solar Energy,* Government Printing Office, Washington, D.C., 1974.

The federal government, through the Atomic Energy Commission, has done most of the research and development on nuclear power and still provides the insurance against a nuclear catastrophe. The first nuclear fuel reprocessing plant, as well as one under construction, were heavily supported by government money. In fact, the entire nuclear industry has been backed by government involvement from the start. If nuclear power research and development merited government support, why not solar? The suggestion that the government stimulate the solar market by incentives and by bearing some development costs is not at all a radical proposal.

Incentives are necessary because active solar heating systems cost a lot. A house with an area of 1500 square feet (135 square meters) and 2–3 days of heat storage may need 750 square feet (68 square meters) of collector area. At a cost of $22 per square foot ($242 per square meter), the cost of the collectors alone is $16,500. The volume of the tank in gallons would be about twice the collector area. The 1500-square-foot house will then need a tank of 1500 gallons storage.[5] Installation and other equipment may push the cost to $20,000 and more for the entire solar house heating system.

Basically, two kinds of cost need to be brought down to help people buy solar technology. The cost of collectors can be brought down by mass production; that is, by expansion of the market to many thousands, even millions, of units. The cost of installation, pipes, plumbing, and the instruments for solar heating can be brought down by having a number of experienced companies willing to compete with one another for jobs. Again, a mass market is the key. These two costs, collectors and installation, make up the bulk of the cost of a solar heating system.

One mechanism the government is using to influence people to buy solar is tax credits. Congress has established a tax credit procedure that works in the following way. A credit is given against taxes owed of 40% of the investment in solar energy up to a total credit of $4000. Thus, a family that invests $4000 in solar hot water is able to deduct from taxes owed 40% of $4000, or $1600. This amount is subtracted from taxes owed, effectively reducing the investment in solar equipment by $1600, from $4000 to $2400.

There are many more variations on these basic

devices to stimulate use of solar energy, but this is basically what the CEQ means when they say "accelerated development"; they mean a market for solar equipment that is given government support through low-interest loans, tax credits, deductions, or similar concepts. Since the precise effect of any of these programs is unknown, the future is difficult to predict. Nevertheless, such a set of programs is sure to provide welcome extra growth to our already expanding solar market.

A number of states now offer tax credits for solar installations as well. A 1981 survey conducted by the National Solar Heating and Cooling Institute found 28 of the 50 states offering some form of tax incentive tied to the state income tax for people who install solar equipment. Also, 11 of the 50 states offer either an exemption from state sales tax or a refund of any sales tax applied to residential solar equipment. Finally, 29 of the 50 states offer some form of exemption from the annual property tax that would otherwise be paid on the value of the solar installation.

All these incentives and the removal of barriers has had a stimulating effect on the manufacture and sales of solar equipment for heat and hot water. In 1980, 5.31 million square feet (0.47 million square meters) of solar collectors for residential heat and hot water were manufactured. From federal tax returns comes information on the number of households that used the federal tax credit to help defray the installation of solar heat and hot water.

In 1978, 58,000 tax returns indicated solar installations; total value of these installations was $120 million. In 1979, reported investments grew to $166 million by about 61,000 tax payers. And the Energy Information Administration reports that 94,000 "active" solar systems were installed in 1980. As a consequence of this increase in solar installations, the number of manufacturers of solar collectors jumped from 45 in 1974 to 364 in 1980.

Nonetheless, solar heat and hot water still has a long way to go. One of the reasons solar has not taken hold even faster is that up-front cash is required to purchase solar equipment, and the interest rates on borrowed money have been at historic highs. To further stimulate solar purchases, President Carter proposed a Solar Development Bank, a repository of federal money that could be loaned at low interest rates for the purchase and installation of solar equipment. Although President Carter wanted to open the bank with $450 million to loan, President Reagan asked

---

5 A cylindrical tank with a 3-foot diameter and 7-foot height would provide roughly this volume.

Congress to rescind its appropriations to the bank in two successive years. By 1983, however, Congress had managed to allocate $21 million to the bank and hoped to allocate more.

Another suggested incentive to solar would be used to increase the number of solar homes being built. Home builders do not like to install costly equipment that may make a home too expensive to sell, so they have shied away from building solar homes. To increase the number of solar homes being built, solar advocates proposed a $2000-per-home tax credit to builders for each solar home erected. In a Congressional mood of budget cutting in 1981, the proposal went down to defeat.

As important as government incentives are, communication to the home owner on available solar technology is basic. Here, popular magazines are playing a significant role. One solar engineer has remarked that the regular feature articles on solar heating in *Popular Science* have done more to stimulate solar technology for the home than the government programs. It may be an overstatement, but it reflects the remarkably active interest of how-to magazines in solar technology. Not only are how-to magazines encouraging solar growth, but so are kit manufacturers. Heath-kit, probably the premier kit company in electronics and computing, has introduced a do-it-yourself Solar Water Heater kit. Heath offers use of a computer program to size the unit for each home and the use of its technical consultation service for installation and operation.

One of the greatest potential areas for solar energy, of course, is in the heating of water and buildings. All of the preceding discussion on progress in solar energy has focused on solar heating of buildings and on solar hot water. The reason is that these uses of solar are ready now; their application is not held up by lack of fundamental research. The only barriers are economic ones, which can be overcome.

Where are solar heating and solar hot water economical? The answer depends on the package of tax incentives being offered by the federal government and by the state governments at the moment the question is answered. Several studies have attempted to answer the question of where solar is economical, but each is built on now slightly out-of-date information.

A 1976 Study at the National Science Foundation[6]

Table 27.1 Where Solar Heat with Electric Backup Beats Electric Heat Alone

Albuquerque (0.96)	Miami (1.00)
Boise (0.80)	New York City (0.80)
Boston (0.50)	Oklahoma City (0.90)
Charleston (0.99)	Phoenix (0.96)
Grand Junction, Colo. (0.80)	Rapid City, S. Dak. (0.85)
Indianapolis (0.70)	Santa Maria, Calif. (0.90)
Los Angeles (0.80)	Washington, D.C. (0.80)
Madison (0.74)	

Numbers in parentheses are solar fractions: the fractions of annual heating loads met by solar energy. This chart is prepared for a 1500-sq-ft house with heating loads specific for the city.

(Source: National Science Foundation, 1976)

examined a number of U.S. cities to see where solar heating with an electric backup was less costly than electric heat alone. The study assumed that the costs of solar panels were reduced by 20% from their current levels due to mass production and that installation costs were reduced by 30%. Using costs current in 1975 and a real rate of growth of electric prices of 5% per year, the study found a wide list of cities in which solar heat was more cost effective than electric heat. (See Table 27.1.) The study assumed no federal tax incentives, as none were then in existence.

A 1978 study at the Department of Energy[7] assessed both solar heat and solar hot water for single-family residences. The study assumed arbitrarily that the solar installation would provide 70% of the hot water and 50% of the heating needs over an average year. The study considered only four cities; these were also cities investigated in the NSF study: Los Angeles, Calif.; Grand Junction, Colo.; Boston, Mass.; and Washington, D.C. Solar equipment was assumed to be exempt from property taxes and federal tax credits were counted in the analysis. (The tax credit figure, though, was at the level set by the 1978 National Energy Act; the tax credit has since increased by one-quarter to a full 40% of the investment in solar equipment.)

The analysis compared solar plus electric backup to electric alone, just as the 1976 NSF study did. For

6 Arthur McGarity, *Solar Heating and Cooling: An Economic Assessment*, National Science Foundation, Washington, D.C., 1976.

7 Bezdeck, R., A. Hirshberg, and W. Babcock, *Science,* **203** (March 23, 1979), p. 1214.

solar hot water, the time for accumulated savings to repay the full cost of the system was less than or equal to 13 years (see Table 27.2). For combined solar heating and solar hot water systems, the time for accumulated savings to repay fully the initial cost was less than or equal to 17 years in all four cities (see Table 27.2). Under all these assumptions, the study concluded that solar water heating is economical in most U.S. locations and that solar house heating is economical in many locations. Recall that these conclusions are drawn for the earlier and lower levels of federal tax credit and did not include state tax incentives.

Table 27.2 Time Required to Repay Solar Investments with Accumulated Savings

	Solar hot water	Solar heating plus solar hot water
Boston	13 years	17 years
Washington, D.C.	13	17
Grand Junction	9	12
Los Angeles	9	11

(Source: Department of Energy, 1978)

The Sun as a Source of Heat for Water and Buildings

To collect the sun's rays efficiently for the production of heat, special solar collectors have been designed, much like the one that Bailey invented (see p. 504). The designs are meant to decrease losses of heat energy. Typically, the collector is flat and stationary. This "flat-plate collector" is set on a roof at the angle that most nearly allows the rays of the sun to fall perpendicular to the surface of the collector.

Many different designs are possible for a flat-plate collector. The principles behind the designs, however, are much the same. We will discuss a design in which parallel metal tubes cross the collector; the metal tubes are welded to a metal plate. A liquid flows through the tubes. Figure 27.13 illustrates the design of the collector.

When the rays of the sun fall on the metal sheet, the sheet is warmed. The metal pipes, which are welded to the sheet, are warmed in turn. The liquid pumped through the pipes receives this heat and carries it off. The pipes and backing sheet are of metals such as aluminum, copper, and iron, chosen because they are good conductors of heat. Both pipes and sheet may be blackened by a paint containing carbon black in order to decrease reflection of the sun's rays and increase their capacity to absorb heat.

Above the pipes are two panes of glass. The glass absorbs and passes through much of the light, but reflects back a small portion (perhaps 10%) of the sun's rays. The primary purpose of the glass is to prevent back radiation from the metal plate to the sky. Just as the glass panes of a greenhouse prevent the warmth

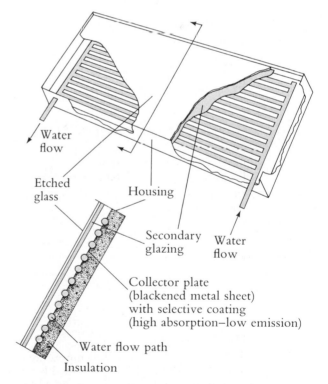

Figure 27.13 Flat-plate Solar Collector. Cold water is pumped in at the bottom of a row of parallel metal pipes. Warm water exits from the top pipe. The pipes are bonded to a blackened metal sheet. Below the sheet and around the outside of the box in which the pipes are housed may be glass-wool insulation. Two glass plates cover the top of the box. (Adapted from: *Solar Energy Handbook*, Honeywell, Form 74-5436, p. 9.)

inside from being radiated out of the structure, the glass cover of a solar collector traps heat within the device. The panes of glass do more besides. Because glass is a poor conductor of heat, it provides a barrier between the warm interior and colder air outside the collector. Thus, losses of heat by conduction to the outside air are decreased. Convection losses, as winds sweep by the collector, are also decreased because the air that is next to the panel and which is carried off by the wind, has a smaller heat content. Plastics may also be used over the heat collecting surface in place of glass.

The box in which the metal collector is housed should be made of a low-conducting material such as wood or plastic. The low thermal conductivity of these construction materials helps prevent heat losses to the surrounding air and structures. Insulation should be placed in the bottom portion of the box. The box itself is usually mounted on the roof in a stationary position, but can be mounted on the ground.

The flat-plate collector we have just described can be incorporated into the heating systems of new structures, including private homes, by replacing portions of the roofing that would otherwise be required. Also, the collector can easily be mounted on roofs of homes

already built. The cost of these collectors has been coming down as assembly-line production provides economies in manufacture.

The collector itself is the exterior symbol of a home or building that uses solar energy. Inside the building, attached to the collector, is a system of heat exchangers and storage tanks, which store and transfer the sun's heat. Diagrams of the system look slightly complex because the liquid in the collecting loop is not water but a mixture of water and antifreeze. We will describe a solar heating system that uses a loop containing a water and antifreeze mixture and another loop containing water alone for the heat transfer and storage functions. (Other systems use air and hot rocks, but we will mention these only briefly.) Ethylene glycol is the antifreeze now used in cars, but ethanol (ethyl alcohol or grain alcohol) has been used in the past as well.

A mixture of 70% water and 30% antifreeze (ethylene glycol) remains liquid to −15°C (5°F). This liquid, however, obviously cannot serve as a water supply as the circulating liquid of earlier systems did. Further, we would not wish to store great quantities of this heated liquid because antifreeze is very expensive. The liquid we store in great quantity must be cheap,

Figure 27.14 Schematic of a solar hot-water heater.

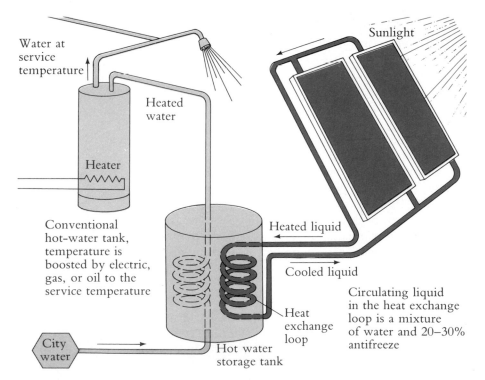

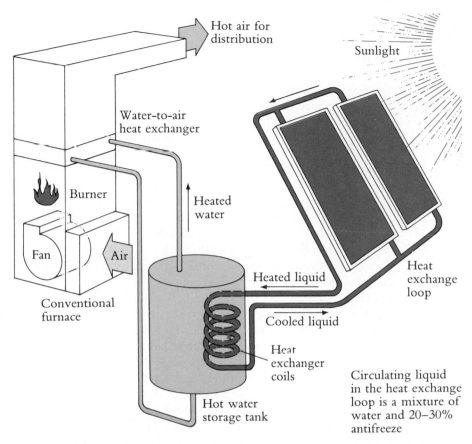

Figure 27.15 Schematic of solar heat for the home.

Labels within figure: Hot air for distribution; Sunlight; Water-to-air heat exchanger; Burner; Heated water; Fan; Air; Conventional furnace; Heated liquid; Heat exchange loop; Cooled liquid; Heat exchanger coils; Hot water storage tank; Circulating liquid in the heat exchange loop is a mixture of water and 20–30% antifreeze

and water is the logical choice. To heat the water for storage, the conventional method is to send the heated mixture of water and antifreeze from the collector through a heat exchanger that is within a large water storage tank. It is the water heated in the tank that is stored, rather than the water/antifreeze mixture.

The water in the storage tank heated by the water/antifreeze mixture can now be used in several ways. It can be used to provide hot water for such uses as dish washing, showers, etc. Or it can be used to provide hot water for heating the home.

Taking the case of domestic hot water first, we can see in Figure 27.14 that city water enters the dwelling and is passed through a set of heat-exchange coils in the storage tank. The flowing city water extracts heat from the hot-water storage tank. This heated water is then fed to the conventional hot-water tank where it can get a "boost," if necessary, to reach the temperature desired by the household—say 65°C (150°F). The "boosting" can be by burning gas or oil or by using electric resistance heating. The heat exchanger through which the city water passes may be a finned tube or

simply a small tank inside the larger tank.

Note that the sun's heat has been transferred from the collecting liquid to the water in the storage tank and then from the water in the storage tank to the city water that makes up the domestic water supply. The water in the storage tank has both accepted and given off heat. There is a reason for using the stored water between two loops of moving liquid. If the collecting loop should develop a leak in its heat exchanger, the water/antifreeze mixture will not leak into the home's hot water supply, but into the stored hot water. Unless both heat exchangers were to leak, a relatively unlikely event, the antifreeze will not reach the hot water supply. Even if both were to have minor leaks, the antifreeze level in the hot stored water would be very small and that reaching the hot water supply much smaller yet.

The other use for the hot water in the storage tank is for home heating (Figure 27.15). In contrast to the domestic hot-water supply system, the home heating system uses a closed loop of water that continually passes through the storage tank. This closed loop takes heated water from the storage tank and delivers it to a heat exchanger attached to a hot-air furnace. Except for the water-to-air heat exchanger, this furnace is the same as a conventional hot-air furnace. The heat exchanger on the furnace uses the hot water to heat the air that the blower fan forces through the heating system. An automatic thermostat in the house senses the temperature in the dwelling. If the air heated by the hot water is not warm enough, the thermostat controls turn the burner on so that an "assist" is given to the solar heating system.

As you can see, the solar hot-water supply system and the solar home heating system both rely on the storage tank. Both systems draw their heat from the storage tank. It makes sense, then, that if a family is installing a solar home heating system, they should install a solar hot-water system as well, since only a heat exchanger and some additional collector area are needed in addition to the equipment and piping already in place. Instead of going directly to heating a home with solar energy, a more likely first step toward solar is for a home to replace an electric hot-water system with a solar hot-water system. The collector area needed is far less for only solar hot water and savings add up more quickly.

If one is going to heat a home or produce hot water with solar energy, the storage tank is an essential feature. The heat stored in the water can carry over the sun's energy from today until tomorrow or the day after or the day after that, depending on the volume of water that is stored. The storage tank frees the home from variability in sunshine. It is much like a reservoir that stores water from periods of high flow until periods of need when the flow is low. The storage tank puts away solar energy from sunny days to days of need when there is little or no sunshine. When the stored heat runs out and there is still no sunshine, the conventional furnace goes on.

One of the obstacles to installing solar hot water or heating systems is the need for this large-volume tank; finding space for the tank may be difficult. In the late 1970s, a technology for heat storage that was long thought possible began to make its appearance on the solar components market. The technology consisted of a "phase-change salt" stored in a plastic or metal tray or tube. Glaubers salt (sodium sulfate decahydrate) melts in the 80°F-range when air or water heated by radiant solar energy is passed over the tubes or trays. When the salt changes to a liquid, which it does

without a change in its temperature, a great deal of heat is stored in a very small volume. This heat can be recovered gradually as it is needed for heating the house, simply by allowing the salt to solidify once again. The phase-change salt will provide heat storage, then, with a substantially reduced volume of heat-storage medium.

A more distinctly different system for solar heating relies on air as the collector medium. Air passes through the collector and absorbs the sun's energy. Of course, this air cannot be stored the way water can be; the volume of hot air that would have to be stored would be enormous! An easy substitute for hot-air storage is a pebble or rock bed. The hot air from the collector is passed through a tank, or a concrete or even wooden structure, filled with rocks and pebbles. The rocks and pebbles are heated by this flow of warm air and act to store the heat until the dwelling requires it. When heat is required by the house, a blower forces house air through the warm pebble bed. The air that exits from the bed is heated and is distributed through the house. If the temperature in the house is not kept high enough by the solar heat, a conventional furnace will then turn on and assist in heating the air.

The systems described are what might be termed "active" systems in which pumps, heat exchangers, valves, and the like must operate to ensure that the sun's heat is captured and distributed to the home. In contrast to these "active" solar heating systems are designs that utilize solar energy with no operating equipment. That is, the designs are able to capture and distribute the heat without any motors or mechanical devices. Such systems are termed "passive."

There are several good ideas for the passive collection of solar energy. One idea uses ordinary collector panels with tubes filled with the water/antifreeze mixture. Ordinarily, a pump is needed to circulate the liquid through the tubes to the heat exchanger in the storage tank. In the passive system, however, the collector is placed below the storage tank. When the liquid in the collector tubes is heated, its density decreases. Being lighter, the liquid rises naturally up into the storage tank where its heat is withdrawn. The cooled liquid falls through the return piping to the collector panels and there is heated again. The system is known as a "thermosiphon." While this does eliminate one pump, the remainder of the system requires active assistance.

A clever design of excellent potential using no

moving parts is the Trombe Wall, a solar collector that is a part of the house itself. In a sense, it, too, is a large panel of windows facing south to catch the sun, but it is also more. The large windows allow ready entrance of the sun's rays, but a few inches behind the windows is a concrete wall. Behind the concrete wall is the living space of the home. The side of the wall that faces south is painted black so that the maximum amount of sunlight is absorbed as heat. The wall heats up and the air space between the wall and the window heats up as well. Openings at the bottom of the wall allow cool air from the floor level of the living area to enter the space. Openings near the top of the wall allow the warm, less-dense air in the air space to rise up and out to the living area. That is, natural convection carries warm air into the room and cool air out. The dense concrete structure acts as a heat storage device as well. Clever as the Trombe Wall is, you the reader can probably improve the concept by devising a way to prevent the wall from radiating heat to the outside in the cold nighttime hours.

Questions

1. Briefly, sketch how the sun's energy can be collected and used to heat water for an average house. Why might you need a conventional "backup" heating system?
2. Contrast active and passive solar energy systems.
3. Describe how the sun's energy might be used to generate electricity, using solar cells. What environmental advantages does such a system have over generating electricity by burning coal or oil?
4. Suppose, instead, that electricity was generated by focusing the sun's rays to produce steam. Now compare the environmental advantages of solar versus conventional electric generation.
5. What environmental problems can you see arising if large tracts of land are devoted to growing trees for wood to be used as a fuel source? (Part of the answer is in Chapter 2, p. 30, but think about it yourself, first.) What about problems associated with burning the wood?
6. Discuss the economics of solar energy from the point of view of a home owner. What are the major costs? What are the savings? How can the government help home owners?

Further Reading

General References on Solar Energy

Butti, K., and J. Perlin, *A Golden Thread: 2500 Years of Solar Architecture and Technology*. New York: Cheshire Books/Van Nostrand Reinhold, 1980, 289 pp.

Metz, W., and A. Hammond, *Solar Energy in America*, American Association for the Advancement of Science, 1978, 218 pp.

"Solar Energy: Progress and Promise," Council on Environmental Quality, U.S. Government Printing Office, Washington, D.C. 20402, Stock No. 041-011-00036-0, April 1977, 52 pp.

Electric Power from Space

Office of Technology Assessment, "Solar Power Satellites," available from the U.S. Government Printing Office, Washington, D.C., 1981.

Glaser, P., "Solar Power from Satellites," *Physics Today* (February 1977), p. 373.

Herendeen, R. A., et al., "Energy Analysis of the Solar Power Satellite," *Science,* **205** (August 3, 1979), p. 451.

Ocean Thermal Energy Conversion

Metz, W. D., "Ocean Thermal Energy: The Biggest Gamble in Solar Power," Research News, *Science,* **198** (14 October 1977), 178–180.

Cohen, R., "Energy from Ocean Thermal Gradients," *Oceanus,* **22,** No. 4 (Winter 1979–80), p. 12.

Whitmore, W., "OTEC: Electricity from the Ocean," *Technology Review* (October 1978), pp. 58–63.

Office of Technology Assessment, "Renewable Ocean Energy Resources," Part I, Ocean Thermal Energy Conversion, 1978.

Photosynthetic Conversion of Solar Energy

Broad, W. J., "Boon or Boondoggle: Bygone U.S. Rubber Shrub Is Bouncing Back," *Science,* **202** (27 October 1978), 410–411.

Burwell, C. C., "Solar Biomass Energy: An Overview of U.S. Potential," *Science,* **199** (10 March 1978), 1041–1048.

Griffin, D., "Natural Rubber has a Future After All," *Fortune* (27 April 1978), p. 78.

Hammond, A., "Alcohol: A Brazilian Answer to the Energy Crisis," Research News, *Science,* **195** (11 February 1977), 564–566.

Plotkin, S., "Energy from Biomass: The Environmental Effects," *Environment,* **22,** No. 9 (November 1980).

Edelson, E., "The Great Gasohol Debate," *Popular Science* (July 1981), p. 53.

"Energy from Biological Processes, Summary," Office of Technology Assessment, Congress of the United States, July 1980.

Maugh II, T. H., "Guayule and Jojoba: Agriculture in Semi-arid Regions," *Science,* **196** (10 June 1977), 1189–1190.

Photovoltaic Conversion and Interior Lighting

Hammond, A. L., "Photovoltaic Cells: Direct Conversion of Solar Energy," Research News, *Science,* **178** (17 November 1972), 732–733.

Kelly, H., "Photovoltaic Power Systems: A Tour Through the Alternatives," *Science,* **199** (10 February 1978), 634–643.

Metz, W. D., "An Illuminating New Use for Solar Energy," *Science,* **194** (24 December 1976), 1404.

Smith, J., "Photovoltaics," *Science,* **212** (26 June 1981), 1472–1478.

Solar Heating

Dallaire, G., "Will This Low-Cost Solar Collector Revolutionize Home Heating?" *Civil Engineering* (February 1980), p. 38.

Hammond, A. L., and W. D. Metz, "Capturing Sunlight: A Revolution in Collector Design," Research News, *Science,* **201** (7 July 1978), 36–39.

McGarity, A., "Solar Heating and Cooling: An Economic Assessment," National Science Foundation, Directorate for Scientific, Technological and International Affairs, Division of Policy Research and Analysis, Washington, D.C. 20550, 1977.

Lunde, P., *Solar Thermal Engineering.* New York: John Wiley and Sons, 1980, 612 pp.

1980 Active Solar Installations Survey, U.S. Dept. of Energy, October, 1982, DOE/EIA-0360(80).

Solar Thermal Electric Power

Metz, W., "Solar Thermal Electricity: Power Tower Dominates Research," Research News, *Science,* **197** (22 July 1977), 353.

"Spinning a Turbine with Sunlight," *EPRI Journal* (March 1978), p. 14.

Van Atta, D., "Solar-Thermal Electric: Focal Point for the Desert Sun," *EPRI Journal* (December 1981), p. 37.

CHAPTER TWENTY-EIGHT

Energy Conservation

Looking at Two Possible Energy Futures

How We Heat Homes and Make Hot Water

The Manner of Transport of People and Goods
Passenger Transport/Freight Transport

Peak-Load Pricing—Decreasing the Need for New Power Plants

CONTROVERSY:

28.1: *How Do We Get Fuel-Efficient Cars on the Road?*

Looking at Two Possible Energy Futures

"We must extend our resource base to replace the fuels now becoming scarce." That is the theme of the energy companies, the businesses that profit from selling fuel and electricity. There is, nonetheless, another way to extend our energy resource base and to reduce fuel imports that are both costly and risky. This other way does not make profits for the energy companies, however. In fact, they earn less and consumers save money by finding ways to save energy.

Energy conservation ranges from simple and inexpensive to costly and complex. It can be as simple as turning down the thermostat, turning out unnecessary lights, wearing a sweater, showering instead of bathing, using a fan instead of an air conditioner, keeping a car in tune, or riding the bus instead of driving. Conservation can be accomplished by such relatively simple purchases as insulation for an attic, storm windows, weatherstripping for a door, or a flow-restricting device for a shower. Investments can go still larger: high-efficiency appliances or fuel-efficient automobiles may be purchased.

Energy conservation saves money; the dollars we do not spend for energy remain in our pockets. Energy conservation can decrease the need to import oil and gas from overseas. As we discussed earlier, this increases the value of the dollar in the international money market and helps prevent inflation. Energy conservation postpones the days of scarcity and keeps fuel prices from rising too fast. Finally, energy conservation means less of the air pollutants that are by-products of combustion. It may also mean fewer hazardous products of nuclear fission to dispose of or to guard for centuries.

Energy conservation also becomes a focus for attention when we observe the average quantity of energy consumed per capita in the industrial world and the less-developed countries. In 1950, the United States had a population of 152 million, which in that year consumed on average about 210 million Btu *per person*. (Btu stands for British thermal unit and is the quantity of heat required to raise the temperature of 1 pound of water by 1°F.) By 1979, the U.S. population had grown to 230 million people, roughly a 50% increase. Energy consumed in that year reached 330 million Btu *per person,* roughly a 55% increase in per person consumption alone (Figure 28.1). Similar per person increases were noted in Canada. The combination of a population rise and an increase in the per person consumption caused the total annual energy consumed in the United States to jump from 31 quadrillion Btu in 1950 to 79 quadrillion Btu in 1979. (One quadrillion is a million billion, or 1,000,000,000,000,000.)

These two aspects of energy use—the growth in per person consumption and the growth in population—reveal the two basic control points to dampen the growth in energy use in the United States. Conservation can decrease the energy consumption per person, and a stable population can limit the total energy consumption.

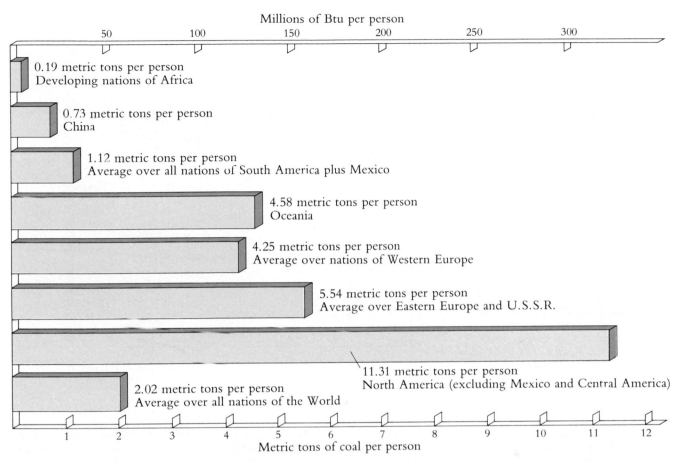

Millions of Btu per person

0.19 metric tons per person
Developing nations of Africa

0.73 metric tons per person
China

1.12 metric tons per person
Average over all nations of South America plus Mexico

4.58 metric tons per person
Oceania

4.25 metric tons per person
Average over nations of Western Europe

5.54 metric tons per person
Average over Eastern Europe and U.S.S.R.

11.31 metric tons per person
North America (excluding Mexico and Central America)

2.02 metric tons per person
Average over all nations of the World

Metric tons of coal per person

Figure 28.1 Energy consumption per capita by world regions, 1979. The people of North America are, far and away, the world's largest energy users on a per-person basis. The 330 million Btu consumed by the average person in 1979 in North America is equivalent to the energy in 11.3 metric tons of coal per person per year. (Source: 1979 Yearbook of World Energy Statistics, United Nations)

What specifically can energy conservation do? The Council on Environmental Quality[1] has described two alternate futures. Although both futures include a maximum commitment to solar energy, the first future assumes that conservation will be given sustained attention and emphasis. Future I, "The Conservation Route," is contrasted to Future II, "Business as Usual," in Figure 28.2. "The Conservation Route" will make it possible to decrease our reliance on oil and gas due to the combination of conservation and the development of solar energy. Our reliance on coal and nuclear power need increase to only a small degree. In stark contrast, the "Business as Usual" future, which projects energy demands increasing at 1.9% per year, would require a sixfold expansion of nuclear power and more than a doubling of our reliance on coal, even with an expansion of solar energy to its maximum possible contribution.

The relative impacts of these two energy futures are shown in Figure 28.3. The annual requirements for coal mined in the "Business as Usual" future are about double those in "The Conservation Route," as are

---

1 *The Good News About Energy*, 1979, available from the Superintendent of Documents, U.S. Government Printing Office, Washington, D.C.

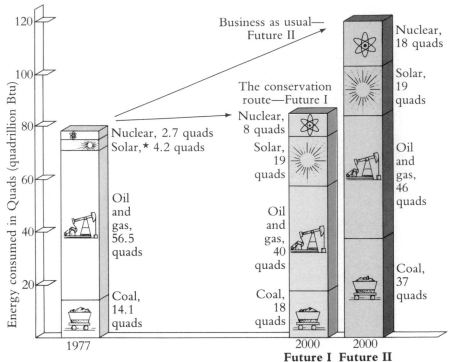

Figure 28.2 Two possible energy futures: "The Conservation Route" and "Business as Usual." Note that both futures include a maximum commitment to solar energy development. (Data Source: *The Good News About Energy,* Council on Environmental Quality, 1979.)

* Includes 1.8 quads of biomass; remainder is hydropower.

strip-mining requirements. And in the "Business as Usual" future, the number of coal-fired plants is double that in "The Conservation Route." The role of nuclear power in "The Conservation Route," measured in terms of the number of nuclear electric plants, is less than half that in the "Business as Usual" future. Spent fuel generated is likewise about halved.

How did the Council on Environmental Quality envision "The Conservation Route" being achieved? The specific steps to achieve the conservation route include more insulation in homes, apartments, public buildings, and factories; more efficient appliances and machines; better-insulated refrigerators; gas stoves without pilot lights; fluorescent lamps instead of incandescent bulbs. Total energy systems in industry may both generate electricity and produce process heat, as well as space heating. Automobiles can be made (and are becoming) far more fuel efficient. In short, energy conservation, if we can achieve it, holds as great an opportunity as discovering new sources of energy.

We have selected two areas on which we shall focus attention in our discussion of energy conservation.

These are (1) the way we heat buildings and make hot water, and (2) the way we transport people and freight. Building heating and water heating together account for about 20% of our annual use of energy. Transportation consumes about 25% of the energy we use each year.

There are choices in these two areas that can be made now by the consumer or business person that can influence energy consumption. New technology is not necessary in these areas; methods presently available are sufficient to make dramatic differences both in energy consumption and in pollution. For instance, if the cars on the road in the U.S. in the mid-1970s matched the average efficiency of the cars on the road in Europe at the same time, our annual travel nationwide could have been accomplished with 42% less gasoline. Because choices are now available in these areas that can make a big difference in our energy use, we discuss these two areas in detail.

Each of these two areas is what is called a "point-of-use." Other improvements at points-of-use are possible. Some we have already mentioned, such as natural and fluorescent lighting in place of incandescent lighting. A very effective and even charming method for saving energy is to use old-fashioned ceiling fans for cooling instead of air conditioning. The breeze from such a fan, whose power usage is about that of a standard light bulb, produces a cooling effect that is about equal to a 6°F-drop in temperature.

More efficient appliances are possible. Spark-ignition gas stoves can replace pilot-light gas stoves; better-insulated refrigerators can replace current models; better-insulated hot water heaters are now available; and high-efficiency oil and gas furnaces are now on the market. The labeling of appliances with their energy consumption per unit time will be a valuable aid to

consumers, not only helping them save energy but helping them save money as well.

In addition to improved point-of-use efficiencies in energy consumption, the penalty of using the "electric middleman" can be softened as well; that is, improvements in the efficiency of electric power generation are achievable. Cogeneration uses a jet engine and a conventional boiler one after another to increase the efficiency of electric generation. Magnetohydrodynamic generation in conjunction with a conventional steam boiler can also, as discussed in an earlier chapter, increase the efficiency of electric generation. "District heating," in which waste heat from electric generation

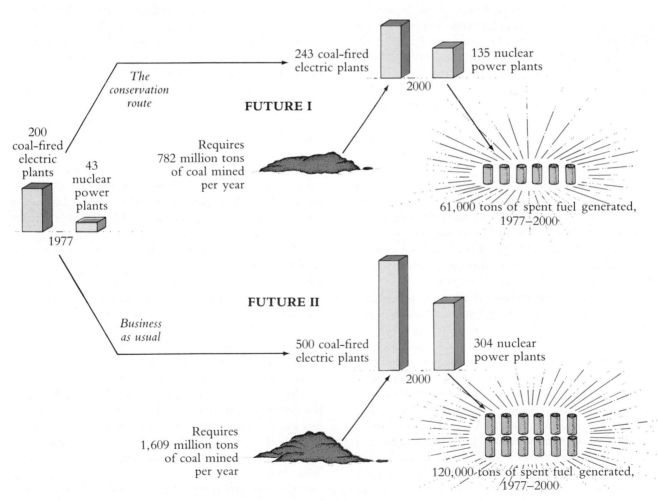

Figure 28.3 Some of the impacts of two possible energy futures. (Data Source: *The Good News About Energy,* Council on Environmental Quality, 1979.)

is put to use in heating buildings, makes the same fuel do two jobs. All three of these energy conservation methods have been treated fully earlier, although we did not then point to these methods as being ways to save energy. Thus, efficiency of conversion, in addition to efficiency at the point-of-use, offers opportunities for energy conservation.

Nonetheless, we return to the two areas mentioned earlier because of their importance in energy consumption and because of their potential for change.

How We Heat Homes and Make Hot Water

The choice about how to heat a home or heat hot water is simple to make in most cases, at least from the standpoint of what to reject if possible. In the recent past, there have only been three options, but now solar energy has entered the picture as a real alternative as well. These three options are natural gas, oil, and electricity. Let us look at these choices carefully.

Suppose you are in the market to purchase a home and you have seen three houses, all of them affordable and similar in other respects, except in the fuel used for heating. One heats with natural gas, another with distillate oil (home heating oil), and the third with electricity. In each house, hot water is produced in the same way that the home is heated; that is, the gas-heated home heats its water with natural gas, etc. Which home is preferable from the point of view of energy conservation, and which is preferable from the point of view of monthly costs of energy?

Assuming that natural gas is allowed to continue to rise in price, heating a home with natural gas will become, in short order, about as costly as it will be with oil. Gas had been much cheaper for some years, but the pattern is changing as the government allows the price of natural gas to rise in order to correct shortages. The energy efficiencies of a hot-air heating system that is oil-fired and one that is gas-fired are about the same. Either new system is 80% efficient. That is, about 80% of the energy derived from burning the fuel goes into raising the temperature of the house.

Electric heating, on the other hand, even when the house receives extra insulation as the electric company suggests, still costs more than heating by gas or oil. How much more depends on the local electric rates. Only in the Pacific Northwest have electric rates been low enough for electric heating to be an economical choice, and rates there are on the rise. In most cases, the electric home is considerably more expensive to heat than the home heated by gas or oil. The cost of creating hot water using electricity is large as well. A home heated with oil or gas but using an electric hot-water heater has an electric bill about double that of a home that heats water by gas or oil. Solar energy is rapidly becoming an available alternative for heating homes and hot water in many sections of the country. Its economics relative to oil, gas, and electricity are discussed under solar energy (Chapter 27), where we show that the sun has quickly become a serious rival for the other methods of heating water and homes.

Not only is the electric home more costly to heat, but in several instances it is extremely wasteful of resources. In an earlier chapter we described the result of the second law of thermodynamics. We noted that as a consequence of the second law, there is an upper limit on the fraction of heat energy that can be converted to work or into some other form of energy. For our purposes, the other form of energy is electrical energy, and the impact of the second law is that little more than about 40% of the heat liberated by burning coal, oil, or natural gas can be converted to electrical energy. The remainder of the heat from burning the fuel, the portion not converted to electrical energy, is wasted by the electrical plant into the environment, either to the air or to the water. The latter is the phenomenon we know as thermal pollution.

Let us suppose that the electric generating station is oil- or gas-fired. If we assume that the plant is 33% efficient, a reasonable efficiency figure, then two-thirds of the heat energy released by burning the fuel is lost into the environment. The remaining one-third in the form of electrical energy can be used to heat a home. If there is a 5% loss in transmission and in heating at the home, we have about 31% or 32% of the original energy in the fuel left to heat the home.

In contrast, if oil or gas is burned in the home to provide heat, the efficiency is much greater. New oil and gas furnaces attain 80% efficiency; but after some use, the heat exchanger efficiency falls and the conversion of combustion heat to useful heat decreases to 60–70%. It is reasonably clear in such circumstances that electric resistance heat, which is produced by a generating plant that burns oil or gas, is wasteful of that fuel. The same conclusion follows immediately for hot water. If the electric plant is coal-fired or nuclear, similar comparisons are not meaningful, since these fuels are not used to heat homes. (Coal use for home heating is very limited, though increasing.)

In the near future, however, the **heat pump** could change both the economics and energy cost of electric heating (Figure 28.4). This electrical device, a sort of refrigerator or air conditioner in reverse, is capable of heating a home during many, though not all, winter days; it exhausts cold air to the outside and warm air to the inside of a house. Measurements of the performance of these units have indicated that two units of heat energy can be produced for every one unit of energy (as electricity) delivered to the house. This cuts the cost of electric heating in half. Even better performances of heat pumps may be "around the corner" as more people try them as alternatives to conventional heating. One reason for their growing popularity is that the same unit that heats the home in winter can be used to cool the home in summer. Thus, one investment, not much more costly than central air conditioning alone, provides a unit that can both heat and cool.

If their economics are so unfavorable, why are electric heat and electric hot-water heaters seen so frequently in new homes? Why do builders seem to favor these types of water heaters? It is a matter of what are known as "first costs." An electric water heater may cost $150–200 installed; an oil-fired heater, $750 installed; a solar water heater with collectors runs several thousand dollars. If gas is available, gas water heaters will probably be installed, since their purchase price for the builder is about that of electric water heaters. If gas is not available, a builder will almost always give customers electric water heaters because they are cheaper for him to buy. A smaller investment is needed by the builder, and he can sell the house at a lower price, increasing the number of people who can afford his product.

Water heating can also be accomplished via heat pump, and two manufacturers are now offering heat-pump water heaters. Their electric requirements are expected to be about half that of conventional electric water heaters.

The means of heating and cooling a building and the means of heating water have a large influence on energy consumption and on cost. Nonetheless, there are numerous additional steps that can be taken to reduce the energy requirements of buildings. Of course, many of them come under the heading of

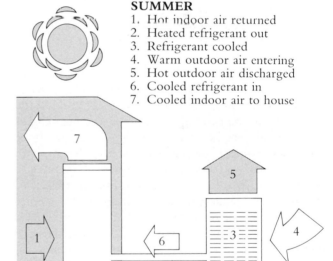

WINTER
1. Cold indoor air returned
2. Cooled refrigerant out
3. Refrigerant heated
4. Cold outdoor air entering
5. Colder outdoor air discharged
6. Heated refrigerant in
7. Heated indoor air to house

SUMMER
1. Hot indoor air returned
2. Heated refrigerant out
3. Refrigerant cooled
4. Warm outdoor air entering
5. Hot outdoor air discharged
6. Cooled refrigerant in
7. Cooled indoor air to house

Figure 28.4 How a heat pump works. (Adapted from GE Weathertron Heat Pump Brochure.)

insulation, the addition of materials that prevent the exit of heat from the building.

Two statements of caution about insulation are in order, however. The first caution has to do with careful selection and application of insulating material. Urea–formaldehyde insulation foam is now banned;

the manufacturers are out of business. The foam released irritant gases into a number of homes. Styrofoam panels, despite manufacturers' claims, are highly flammable, as is cellulose. To show that styrofoam is flammable, put a match to a styrofoam cup, out of doors; avoid breathing the black fumes. Traditional fiberglass insulation seems a reasonable alternative. Even with this material, caution is needed in application. Insulation applied to homes with poorly insulated old electric wiring should be placed in a way that allows free exit of heat from the wires. If this is not done, heat buildup around the wiring could lead to fires. Thus, careful selection and application of insulating material is in order.

The second caution relative to insulation is indoor air pollution. Gases and particles from cooking and smoking may accumulate in a house made too tight. Fresh filters on the furnace air flow, though needed to allow free circulation, become even more important to maintain. In addition, radon gas from the walls of homes built of stone can accumulate. Though uncommon in U.S. homes, air-to-air heat exchangers can be used to bring in fresh outside air with little heat loss.

These cautions on insulation are not meant to dampen our enthusiasm for saving energy. Most houses can use added insulation, but the type and the extent should be carefully chosen. Other possibilities for decreasing the heating and cooling requirements of buildings range from earth-sheltered housing to passive solar to the surrounding of homes with windbreaks. Windbreaks of trees can be especially effective in reducing heat requirements. Other energy-saving devices for the home include flow restrictors on faucets, ceiling fans, automatic setback thermostats, interior airtight shades on windows, awnings on the outside of windows, and jacket insulation on hot-water heaters. To find out about such developments, all that is necessary is to open the pages of an issue of *Popular Science.*

The Manner of Transport of People and Goods

Passenger Transport

No nation has more automobiles per person than the U.S., nor do other people drive as far as we do. We have a 55-mile-an-hour speed limit, however, and there are countries where people drive faster—much faster. Still, no society consumes as much energy per person in auto travel as our own. With so many cars and with so much space and sprawling development, we cover many more miles than do the peoples of other nations, and hence consume far more energy in travel.

Americans are often accused of wastefulness in this regard, compared to Europeans who consume far less energy in travel. Is the accusation true or false? It is both. It is true that we consume far more energy, and part of the reason for this has been our tradition of larger cars. That tradition results from a very low price of gasoline in the United States over many years. The taxes we placed on gasoline were always small in comparison to the taxes Europeans placed on their gasoline. So we hung on to our tradition of large cars, while the Europeans typically manufactured smaller cars that used less gasoline. In that sense, because we were never willing to tax gasoline sufficiently to discourage larger cars, the accusation of wastefulness is true. In another sense, however, the accusation of wastefulness compared to Europeans is false.

If we compare Europe to the United States, we see a very different society. Population centers are much closer together in European countries; families are, in general, less widely separated by distance. People in Europe tend to stay in the nations of their birth, and the nations are small, on the order of the size of our states. The huge distances between cities and towns in the American countryside are not characteristic of Europe. Nor is car ownership anywhere near as extensive. In the United States, there is slightly more than one car for every two people; in the Western European countries (the wealthier countries), there is about one car for every four people. In Europe, auto purchases are heavily taxed; road taxes levied on car owners increase the burden still further. Furthermore, the price of gasoline in European countries is two and one-half to three times that in the United States.

Public transportation developed in Europe before the era of the auto. Because of this, the networks of the bus, train, and streetcar systems are excellent; service, moreover, is frequent. Thus, Europeans generally have available to them a public transport system which, by appropriate transfers, can link almost any place in Western Europe with any other, nearly door-to-door. Such a journey would begin by walking to

Daylight Savings Time versus 64 Million Pounds of Candles Every Year

Benjamin Franklin was astonished. "An accidental sudden noise waked me about six in the morning, when I was surprised to find my room filled with light. I imagined at first that a number of lamps had been brought into the room; but rubbing my eyes I perceived the light came in at the windows," he said in a letter to the *Journal of Paris*. "I looked at my watch, which goes very well, and found that it was but six o'clock; and still thinking it was something extraordinary that the sun should rise so early, I looked into the almanac where I found it to be the hour given for the sun's rising on that day. Those who with me have never seen any signs of sunshine before noon, and seldom regard the astronomical part of the almanac, will be as much astonished as I was, when they hear of its rising so early; and especially when I assure them *that it gives light as soon as it rises*. . . . And, having repeated this observation, the three following mornings, I found always precisely the same result."[*][†]

The discovery gave rise to "several serious and important reflections." Dr. Franklin considered that had he slept to his usual rising hour, he would have slept six hours by the light of the sun, and in exchange have been up six hours the following evening by candlelight. Since the latter was a much more expensive light, he was induced by his "love of economy" to estimate how much could be saved by using sunshine instead of candles. For Paris, where he was residing, he calculated a saving of over 64 million pounds of candles every year. "It is impossible that so sensible a people, under such circumstances, should have lived so long by the smokey, unwholesome, and enormously expensive light of candles, if they had really known that they might have had as much pure light of the sun for nothing."[*][†]

Daylight Savings Time was first proposed in a serious way as a method by which to save energy in England in 1907. By 1918, the U.S. had adopted the procedure of setting the clocks ahead one hour during the spring, summer, and fall. The procedure was dropped, however, after the end of World War I. It was not until 1942, the year after the U.S.

entered World War II, that the nation returned to Daylight Savings Time, except that this time it was called "wartime" and it was in effect year-round. Turning the clock ahead brought a decrease in the peak demand for electric power that occurred in early evening. The war over, the nation again abandoned its manipulation of the clock, but now Daylight Savings Time had become popular enough for a number of states to adopt the plan on their own.

By 1966, to correct the lack of uniformity in times across the country, Congress enacted a bill that put all states (with several exceptions) on Daylight Savings Time from the last Sunday in April to the last Sunday in October. In 1974, year-round Daylight Savings Times went into effect again, this time for about a year and a half as an energy-saving response to the oil embargo. Public opinion on the potential hazard to school children in the dark morning hours was apparently responsible for the abandonment of year-round Daylight Savings Time.

Daylight Savings Time can be improved. We do need standard time during the winter months of November to February to make dark mornings brighter, but we can expand the period during which we use daylight time to include all of April and all of March. We can do this without having any darker mornings in those months than we do in September and October. An error was apparently made when daylight time was first adopted. Congress made daylight time from one month after the Spring Equinox, March 21, to one month after the Fall Equinox, September 21. To begin and end daylight time on days of the same length, the law should have placed the beginning of daylight time from one month *before the Spring Equinox* to one month *after the Fall Equinox*. Congress people do make mistakes.

---

[*] Quoted from B. Franklin, "An Economical Project," Letter to *The Journal of Paris*, 1784.
[†] David Prerau, "Changing Times: National Time Management Policy," *Technology Review*, **79**(5) (March/April 1977), 54.

the end of the block to catch a subway, bus, or street-car. From there, one proceeds to the central station, where long-distance buses and intercity trains are available. The need to use the automobile, therefore, is less frequent.

Even though such excellent systems are available, the importance of the auto to European travel is growing at a phenomenal rate. It is growing as fast as the growth of personal wealth allows it. In the Netherlands, a typical European country with regard to auto use, virtually all the growth in passenger kilometers is going into auto travel. Similar patterns are seen for other Western European nations. The Europeans enjoy their cars every bit as much as the Americans.

Having defended Americans, we now must also admit the truth. We have come to "need" our automobiles. We need them not simply because we have built a dispersed society in which the auto is a necessary means of communication. We need them for personal reasons as well. Kenneth Boulding, an economist, explained our relation with the auto in the following way:[2]

> The automobile, especially, is remarkably addictive. I have described it as a suit of armor with 200 horses inside, big enough to make love in. It is not surprising that it is popular. It turns its driver into a knight with the mobility of the aristocrat and perhaps some of his other vices. The pedestrian and the person who rides public transportation are, by comparison, peasants looking up with almost inevitable envy at the knights riding by in their mechanical steeds. Once having tasted the delights of a society in which almost everyone can be a knight, it is hard to go back to being peasants. I suspect, therefore, that there will be very strong technological pressures to preserve the automobile in some form, even if we have to go to nuclear fusion for the ultimate source of power and to liquid hydrogen for the gasoline substitute. The alternative would seem to be a society of contented peasants, each cultivating his own little garden and riding to work on the bus, or even on an electric streetcar. Somehow this outcome seems less plausible than a desperate attempt to find new sources of energy to sustain our knightly mobility.

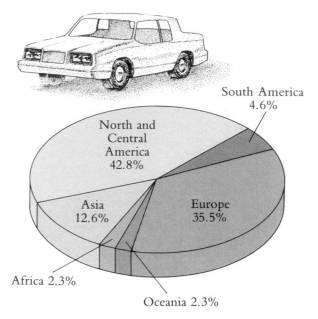

Figure 28.5 World distribution of motor vehicles in 1980. (Source: Motor Vehicles Manufacturers' Association, Facts and Figures, 1982)

Boulding may be suggesting that our "habit" is incurable, that the best we can hope for is a substitute fuel, such as a diesel "fix" or an electric "fix." He may be correct that the habit is beyond breaking. The growing use of the auto in Europe under such costly conditions gives us an idea of how enticing is the siren call of the automobile (Figure 28.5).

If the habit cannot be broken, what can we do? Try to imagine a society with 120 million private automobiles traveling as many miles as we do and using one-third less gasoline. This is where the Europeans have learned a trick we are only starting to learn. European cars are, on average, much more fuel efficient than American cars. An accepted figure for the efficiency of European autos, averaged across all cars on the road, is 10 kilometers per liter, or 22.5 miles per gallon. These high efficiencies are the response of motorists to the high prices of gasoline, since more efficient cars are less expensive to operate. The comparable figure, the average mileage for all cars in the U.S., was 15.2 miles per gallon in 1980. If our fleet of cars had the average efficiency of European autos, each million miles our fleet logged would require one-third less fuel.

2 Kenneth Boulding, "The Social System and the Energy Crisis," *Science*, **184** (19 April 1974), 255.

Put another way, a car that travels 15.2 miles on a gallon of gas requires 658 gallons to go 10,000 miles (an average yearly figure). The car that travels 22.5 miles per gallon needs only 444 gallons to go the same distance. At the current price of gasoline, the 214 gallons saved over a year amounts to a healthy sum. It seems that more fuel-efficient cars have a significant impact on our budgets as well as on energy conservation.

How do we arrive at an era of fuel-efficient automobiles? It is reasonably clear that high gasoline prices have encouraged the production of efficient cars in Europe. These high prices include, in general, taxes that amount to half or more of the total price of a liter of gas. Gasoline prices in the United States have risen rather sharply since the oil embargo and the rise of the cartel, from 30–35 cents per gallon in 1973 to their present levels. Gasoline prices in 1983 were nearly four times what they were before the embargo. Since the federal tax on gasoline has not increased from its original level of 6 cents/gallon during this period, most of the price increase is going into paying for more expensive crude oil. This rise in prices, which has channeled so many dollars to OPEC members, has had a desired effect in terms of new automobile manufacture. More energy-saving cars are coming down the production line and are entering the fleet of cars on the road, as consumers see the value of saving gasoline and dollars.

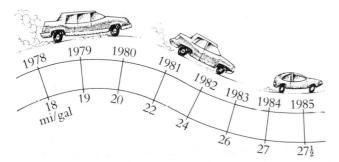

Figure 28.6 Fuel economy standards as set by Congress and the U.S. Department of Transportation. These standards are to be achieved "on average" by all manufacturers. That is, the average fuel economy of all vehicles sold by a particular manufacturer in a given year must meet the standard for that year. The standards for 1978, 1979, 1980, and 1985 were set by Congress as part of the Energy Policy and Conservation Act of 1975, while those for 1981, 1982, 1983, and 1984 were established in June, 1977, under the authority of the Act.

Table 28.1 Average Passenger Car Efficiencies in the U.S.

Year	U.S. passenger car registrations (millions)[a]	Average miles traveled per gallon of fuel consumed
1973	101.1	13.10
1974	104.8	13.43
1975	106.7	13.53
1976	110.1	13.72
1977	112.2	13.94
1978	116.5	14.06
1979	118.2	14.29
1980	121.7	15.15

a Federal Highway Administration data.
(Source: U.S. Department of Energy, Energy Information Administration)

The American public's recognition of the importance of fuel economy is mirrored in legislation passed by Congress in 1975. Congress set standards for the average fuel economy of autos manufactured after 1977. From an average value for 1974 models of 12.9 miles per gallon, auto manufacturers were able to achieve an average mileage for all 1977-model cars of 17.8 miles per gallon, in anticipation of meeting the first actual mileage standard of 18 miles per gallon, on average, in 1978 (Figure 28.6).

Auto manufacturers complained of the costs of retooling to meet the later standards and warned consumers of car price increases to come. They also argued that actual redesign of drive trains and new advanced engines would be necessary. In the first few years, they were able to increase the average mileage of their cars by selling smaller and hence lower-weight cars. They also achieved lower weights by substituting plastic and aluminum for steel in larger cars. General Motors introduced diesel engines on a number of its models in 1978. It is of interest that the *new* cars manufactured in 1981 only had to achieve a mileage standard that approaches the *average of all cars,* of all model years, on the road in Europe. Obviously the technology already existed for that achievement.

Even though more efficient new cars will be sold as time goes on, it will take time for energy-efficient cars to replace inefficient ones. The state of the nation's economy in terms of unemployment and inflation will influence how fast new cars are purchased and how fast older cars are retired. Table 28.1 shows the agonizing

How Do We Get Fuel-Efficient Cars on the Road?

To stimulate oil conservation, we need a generation of high-mileage cars, but how do we get them?

> We should adopt a surtax on all cars weighing more than 3,000 pounds escalated upward to as high as $1,000 tax on the big luxury cars weighing over 5,500.
>
> **John Quarles,**
> **in a Speech before the Coal and the Environment Conference,**
> **Louisville, Kentucky, October 1974**

> I would like to describe a proposal for a stiff gasoline tax, specifically, an increase in the tax to $2 a gallon (with rebates on a per adult basis) . . . A $2 tax is not much higher than present taxes in countries like France and Italy.
>
> **Robert Williams,**
> **appeared in The Dependence Dilemma,**
> **edited by Daniel Yergin, published by Harvard University Press, 1980**

> We must be cautious to see that a national policy of energy conservation does not unfairly burden the poor and those living in rural areas.
>
> **C. ReVelle,**
> **"Public Transport and the Netherlands: Implications for Transport Policy in the U.S.,"**
> **Center for Metropolitan Planning and Design, The Johns Hopkins University, 1977**

We can tax new car purchases by their weight or by their horsepower; we can tax gasoline; we can give income-tax rebates to individuals who purchase efficient new cars; we can ration gasoline, allowing one or two gallons per car per day. The many possible ways to direct people's attention to fuel-efficient cars are not all equal in their market effect; nor are they equal in their impact on upper-income, middle-income, and lower-income families.

There are four basic options listed above: (1) tax the purchases of inefficient new cars; (2) tax gasoline; (3) give rewards for purchasing efficient new cars; (4) ration gasoline. We can assemble many plans from these basic options, but for each plan there is a set of questions to be answered:

1. How will it work? That is, what is the effect on the rate of sales of different kinds of autos? On the sale of gasoline today and in the future?

2. (a) If the government pays, where should the money come from? From general tax revenues? From a special fund?
   (b) If a tax raises money that goes to the government, what should be done with the money? Should it go into general tax revenues

with no special earmarks? Or should it be set aside in a special-purpose fund? To give you an idea of what amounts of money are involved, each cent of federal gasoline tax raises one billion dollars each year. That means that the five-cent-per-gallon tax that began in April 1983 will produce revenues of about 5 billion dollars each year.★

3. Who is affected? Are the poor hurt more than the well-to-do? That is, is the policy "progressive," taking dollars from those better able to afford the expense? Can you think of ways that the impact on the poor can be softened?

4. What are the political chances?

5. Is there an impact on the national economy? On local economies?

To focus the way you answer these questions, here are some hypothetical individuals, all of whom are very much concerned about such policies. Explain how the various conservation policies could affect the following people:

Juan Lopez and his family live in Chicago; he maintains a set of apartments in the inner city in return for his own apartment. He just bought a gas-guzzling

1971 Zonker Grand Marquis.

Arthur Williams and Sally Adam-Williams have moved back to their home town of Wayside, Nebraska, but the nearest job is in Omaha, some 70 miles away. They do not want to leave their rural environment for a city, but they need to make a living. They are about to buy their first car.

Calvin and Nancy Beale own and run a diner in Ellsworth, Maine. Their diner, the Ellsworth Eatery, is well-patronized and locally famous for homemade pies and the "Maineburger." The bulk of their business comes in the summer, when tourists flock to nearby Acadia National Park. Tourists with tents and motor homes come from all up and down the Eastern Seaboard, even from as far south as Virginia. No trains serve Ellsworth.

Can you think of other scenarios showing how people will be hurt by the various gas-conservation policies? Can you think of people who will hardly be affected at all by the listed methods?

★ This small new tax is designated for road repair and jobs. It was not publicized as an energy-conservation measure prior to its passage.

---

slowness of improvements in the average fuel efficiency of automobiles in the United States. The near-depression level of the economy in 1981, 1982, and 1983 slowed even further the purchase of new automobiles and slowed severely the movement toward a fleet of fuel-efficient cars.

Even though energy-efficient cars save energy day by day, they cost more to buy initially, and because they are efficient, they depreciate in value more slowly. That means that lower-income individuals will hold their older cars longer before they can afford an efficient used car. Thus, both the economy and the higher cost of fuel-efficient cars could delay a transition to a fleet of cars with higher mileage.

What can we do in the meantime? Individual action and responsibility will help a great deal. First, and most simply, there is a speed at which automobiles and trucks are most efficient in terms of fuel consumption. For automobiles that speed is 35–40 miles per hour (56–64 kilometers per hour); for trucks it is a bit

higher. During World War II, a nationwide speed limit was imposed of 35 miles per hour. That speed limit was designed to conserve fuel for the war effort. The 55-mile-per-hour (89 kilometer-per-hour) limit currently mandated by the federal government represents a compromise between the speed that conserves the most energy and the speed people apparently prefer to go. The 55-mile-per-hour limit, while initially planned to save energy, turned out to be a life-saving device as well. Motor vehicle fatality and accident rates have fallen significantly since the speed limit was reduced (Figure 28.7). This additional reason to maintain the nationwide speed limit helps to support energy conservation.

There are other features in the way we drive that influence mileage. Underinflated tires can decrease mileage by a mile per gallon, but radial tires can increase mileage by a similar amount. Idling wastes gasoline. If a car idles for a half hour, about a quart of gasoline is needlessly burned. Racing starts, uneven

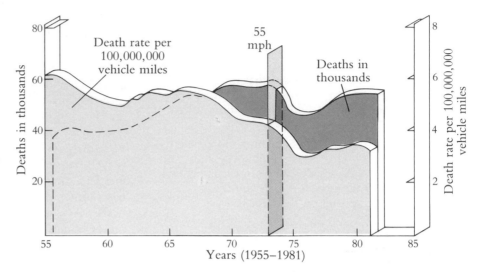

Figure 28.7 Traffic Deaths and Death Rates. Between 1973 and 1974, the year the 55 mile/hour limit was first enforced, the annual death toll dropped by 9000 and it has remained lower than the 1970–1972 levels every year since. In addition, the number of deaths per 100,000,000 vehicle miles fell from 4.24 to 3.59, or by 15%, between 1973 and 1974, and has not gone back up. (Source: National Safety Council, Accident Facts, 1979)

acceleration, and braking in city traffic can reduce the mileage a car obtains by up to 30%. Driving in lower gears takes more gasoline, as does an untuned engine. Short trips in cool and cold weather use more gallons per mile while the engine is warming up. Longer trips take advantage of the time already spent warming the engine.

When talk turns to saving gasoline, someone inevitably raises the possibility of the electric car. In the early days of the automotive industry, the electric vehicle (EV) was a competitor with the auto that was powered by an internal combustion engine, but cheap gasoline for the internal combustion engine car and a limited range for the EV determined that the EV would not survive. In recent years, research on new and longer-lasting batteries has again raised people's hope for an electric vehicle.

A new concept has also emerged, the "hybrid car." The hybrid vehicle, being developed by GE for the U.S. Department of Energy, would use an internal combustion engine for long trips and a battery for around-town trips. The battery would be charged by household current. The vehicle would definitely save petroleum, not a bad idea in a nation that is so dependent on oil imports. The electric energy it used would be likely to come from coal-fired and nuclear electric generating plants, so oil is conserved. Nonetheless, the energy savings of the hybrid vehicle would not be great, if indeed there are any savings. The reason is that electricity production is already so inefficient: 40% for the modern coal plant, 30% for the nuclear plant.

After electrical transmission losses have been deducted, the hybrid vehicle and conventional auto are not far different in their efficiency of energy use. Suppose an electric vehicle goes two miles per kilowatt-hour and uses energy from a 40%-efficient coal plant, which energy suffered a 5% loss in transmission. This electric vehicle would require the consumption of 4860 Btu at the coal plant per mile of travel. A conventional auto getting 27.5 miles to the gallon (the 1985 new car standard) uses 4727 Btu to go one mile. We use a fuel-efficient small car for this calculation because electric vehicles are expected to be small. Different assumptions about miles per gallon and generating efficiencies do alter these calculations, but not so significantly as to alter the general conclusion: electric vehicles are not significant energy savers, but energy shifters.

Not only is energy efficiency not much improved, but the pollution burden is shifted to other substances and to other sites—the sites of the electric power plants. The nation should think long and hard about a switch to electric vehicles.

So far, we have argued principally for changes in the automobile itself, not in the way our society transports itself. Public transportation (buses, train systems, and the like), if properly implemented, can have an important role in reducing energy consumption. Figure 28.8 illustrates the relative efficiency of various means of transportation. As can be seen, long-distance travel is typically most efficient because vehicles are more fully loaded and because fewer starts and stops save energy.

A vehicle not listed is the "auto in city traffic with

a load equal to 75% of capacity." Such a vehicle is known as a car pool. People in car pools most often live near one another or near the line of travel. They also work near one another. The four-person car pool with an average load equal to 75% of capacity consumes only about 3000 Btu per passenger mile, an efficiency level about that of a train traveling between cities.

Clearly, buses and trains are energy-efficient modes of travel, but how do we promote and encourage their use? To obtain a large proportion of people who will ride the bus, train, or subway, a number of factors must be positive. Routes must cover a large portion of the geographic area, so that access to public transport is not difficult. Stops must be frequent enough and spaced closely enough on the route that the time to walk to a stop plus the time to wait for a bus is not too great. Equipment must be clean; conditions must be safe and comfortable; costs must be reasonable when compared to use of the automobile. Routes cannot be too winding or the time from origin to destination will put off potential riders. In this regard, the use of express bus lanes on super highways has a positive effect in reducing travel time.

Even when all conditions are favorable, ridership on public transport may still be so low that the costs of providing the service in equipment, personnel, gasoline, and maintenance cannot be paid by the revenues generated from fares. Even in Western Europe, where public transport runs quite full, governments make up the difference between revenues earned and the costs of actually operating and maintaining the system.

If the costs of public transport must be supported in part by the government, we need to justify this expenditure. It can be justified on the basis of energy savings, but, of equal importance, air pollution will be reduced by diminishing the number of cars on the road. Ozone, nitrogen oxides, carbon monoxide, and lead will all decrease as use of the auto decreases. These are positive benefits of public transportation. In addition, those who continue to drive have some benefits in terms of less-crowded streets and fewer accidents.

Where should the government find the money to make up the costs of public transport that are in excess of the system's revenues? Should income taxes be raised to pay this bill? Or should some other form of taxation be used to raise the funds to support public transportation? Should gasoline be taxed more heavily? The tax revenues could be used not only to help pay for public transport, but would simultaneously make public transport economically attractive to more people.

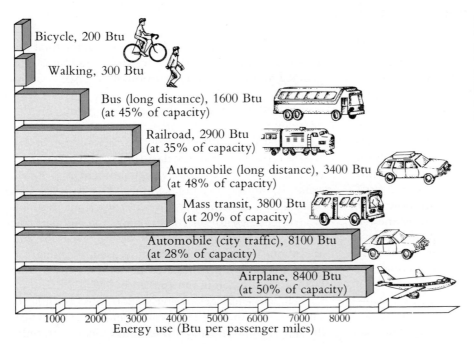

Bicycle, 200 Btu
Walking, 300 Btu
Bus (long distance), 1600 Btu (at 45% of capacity)
Railroad, 2900 Btu (at 35% of capacity)
Automobile (long distance), 3400 Btu (at 48% of capacity)
Mass transit, 3800 Btu (at 20% of capacity)
Automobile (city traffic), 8100 Btu (at 28% of capacity)
Airplane, 8400 Btu (at 50% of capacity)

1000 2000 3000 4000 5000 6000 7000 8000
Energy use (Btu per passenger miles)

Figure 28.8 Relative efficiency of various modes of travel. [Data Source: E. Hirst and J. Moyers, "Efficiency of Energy Use in the United States," *Science,* **179** (30 March 1973), 1299.] Capacity assumptions are those of Hirst and Moyers.

Freight Transport

About 30% of the energy we spend each year on transportation goes into the movement of goods; the remainder is expended on moving people. Of the various modes of freight transport, railroads, pipelines, and waterway transport all have comparable levels of efficiency. Trucks consume 4–5 times more energy per ton-mile than these other modes (Table 28.2).

The transport of goods, especially between cities, has evolved in much the same way as the transport of people. Where initially waterways and trains carried the bulk of goods moving between cities, trucks have been increasing their share of the load quite steadily. Thus, the energy-efficient modes of transport—barge and rail—are losing ground to an inefficient mode, the truck. Certain bulk goods such as coal, however, are unlikely to ever give way to truck transport. Instead, coal-slurry pipelines appear a possibility to capture a portion of the coal transportation market. (See page 315 for a discussion of coal-slurry pipelines.)

The reasons for the growth in trucking are rooted in the door-to-door nature of truck service. Once goods are loaded, they often need not be reloaded be-

Table 28.2 Energy Cost of Freight Transport Modes[a]

Mode	Btu/ton-mile
Pipeline	450
Railroad	670
Waterway	680
Truck	2,800
Airplane	42,000

a Source: E. Hirst and J. Moyers, "Efficiency of Energy Use in the United States," *Science*, **179** (30 March 1973), 1299.

fore arriving at the final destination. If the particular goods are not moving in sufficient quantity to justify a rail track from door-to-door, truck transport, though costly, becomes a very attractive alternative. A routing that involves first a truck, then a train, and then a truck may be low in transport cost because of the rail savings, but the two transfers consume time and money and also increase the likelihood of damage to the shipment or of theft.

Peak-Load Pricing—Decreasing the Need for New Power Plants

Utilities will be adding about 257,000 megawatts of new electrical generating capacity in the 15-year period from 1980 to 1995, according to a projection of the Energy Information Administration. This is the capacity projected as necessary to meet the growing demands for electrical energy. The consumer will pay dearly for this new construction. The 1985 construction cost of a single new 1000-megawatt electrical plant is estimated at about 1.5 billion dollars for a nuclear plant and 1.25 billion dollars for a coal-fired plant. Yet not all of these plants are truly necessary; most electric utilities are well aware of an economic method to decrease their need for new power plants. The method, known as peak-load pricing, has already been adopted in much of Europe, but most utilities in the United States have ignored its existence.

The numbers and capacity of the power plants scheduled for addition is a response to the future demands foreseen by the electric utilities. The additions,

however, do not respond to the growth in the total use of electric power. Instead, capacity additions are planned to meet a high proportion of the peak or highest power demands.

At certain times of the year, depending on the geographic area, electric demands, averaged over the day, reach their highest levels. Further, within a season, at certain times of the day, the rate of power drain from the central station may peak momentarily. As an example, during the hot summer, Los Angeles and surrounding areas may require enormous quantities of electrical energy for air conditioning. On any particular hot day, loads reach their highest levels in the late afternoon, when air conditioning is used to counteract the heat that has accumulated in homes, offices, and factories and when home appliances are in active use.

The electrical generating station must be able to supply even the highest peak demand levels, even though such momentary levels may occur for only several hours out of the year. To meet these demands, special equipment may be utilized to generate supplemental power. The special equipment may be used because the boiler/turbine system of the electric plant is not large enough to supply power at the desired rate. Turbine engines that resemble the engines in jet airplanes may be used to generate the additional electric power needed at such times, as may pumped-storage hydropower (see Chapters 15 and 25). Natural gas, when it is available, or home heating oil may be used to power the turbine engines.

If the power demand exceeds the capacity of the boiler/turbine system, the daily peaks must be met by this turbine-engine generation or from pumped storage. At such times, the boiler/turbine system will be operating at full capacity. On the other hand, in slack seasons when daily power demands are much less, the boiler/turbine system will not be operating anywhere near full capacity. Since the investment in an electric plant is considerable, it makes sense that the plant should not have idle capacity at some times and be taxed to its limits at others. It makes sense also that the plant should not be built with more capacity than is absolutely necessary, because of the enormous cost.

What we are saying is that maximum cost-effectiveness occurs when the load is as level as possible. Telephone companies decrease long-distance calling rates at night and in the evenings to increase the use of equipment that would otherwise be idle at those times. They are trying to level the load on their facilities and make use of slack capacity. Utilities can level their loads as well with **peak-load pricing.**

You already know how peak-load pricing works; the telephone company has even taught you to use it. Long-distance calls in the daytime are usually of a business nature, and the volume of such business calls falls in the evening and at night. On a daily basis, the phone company decreases the rates for long-distance, direct-dialed calls at night to encourage use of the phone at hours when it otherwise might not occur. The lower rates also shift some calls that would have been made in daytime into the evening and nighttime hours. This decrease of the peak calling rate during the day means that fewer long-distance trunk lines will need to be installed to carry the flow of calls across country. The telephone company also lowers long-distance rates on weekends and holidays when the decreased volume of business calls leaves much of the company's network idle. The lower night rate system

used by the phone company is a benefit to consumer and company alike.

Electric power companies should be encouraged to price electric energy in the same way; that is, to lower the price of electricity at times when the rate of use is low, and, as necessary, raise the price during times of peak demand.

At night and in the late evenings, when electric power plants are usually operating at only partial capacity, the rate per kilowatt-hour of electricity would be lowered (Figure 28.9). People using dishwashers, electric driers, washing machines, etc. would probably postpone these activities from peak hours of use into those relatively slack times with lower rates. Electric hot-water heaters would be put on timers so that water heating would occur during those slack hours. Industries might be encouraged to schedule some of their operations for the evenings.

During the day, electric rates would be somewhat higher to recover revenues from uses that have shifted to the evening hours. All in all, capacity would be used more efficiently, peak rates of use would be clipped off, and the need for investment in new generating capacity to meet the peaks will be diminished.

Peak-load pricing may also be applied on a seasonal basis. That is, higher prices may be used in the peak air-conditioning season so that loads will be "clipped" in the summer and so that investment in new capacity can be decreased.

With such advantages, why are power companies reluctant to go to peak-load pricing? The fault probably lies in the nature of the beast.

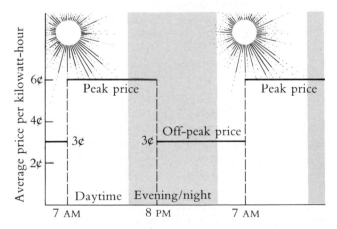

Figure 28.9 Peak-Load Pricing: A Simplified Example.

Utilities are regulated by governments because the utilities are the sole suppliers of electric power for the regions they serve. If they were not regulated in the prices they charge, they could charge very high prices without losing customers because they are the only suppliers in the region. Thus, Public Service Commissions limit the prices utilities can charge. The commissions allow the utilities to charge prices high enough to earn a "fair rate of return" on the investments they make in generating and transmission facilities.

By the "rate of return," we mean the ratio of profit to invested capital. If you put $100 in a savings bank, the $5 or $6 earned each year would be the return on the $100 investment. Your rate of return is 5% or 6%. Since utilities are generally allowed to earn about 12% on their invested capital, they have an incentive to spend as much money on their facilities as is reasonably possible. This is one reason that nuclear plants have probably been so popular with utilities: the greater cost to build a nuclear power plant means that the fair rate of return applies to a bigger investment base. Since peak-load pricing would decrease the number of new power plants that are needed, the utility will make a smaller investment. This means, in turn, fewer dollars of income for the utility, since the rate of return will apply to a smaller investment base. Thus, utilities have been reluctant to use peak-load pricing.

Peak-load pricing, however, does more than benefit the consumer's pocketbook. By decreasing the number of power plants required, the area of land as well as the water flows that must be committed to these enterprises are decreased. Hence the earth is spared unnecessary disruption and despoiling. Fewer transmission lines will crisscross the landscape, and the number of new pollution sources will decrease. Furthermore, the massive energy commitments required to build power plants are delayed. Since peak loads are clipped off, less oil and natural gas are used to generate the inefficient turbine engine power used at such times. Peak-load pricing is a simple but powerful concept that can benefit the environment considerably.

Questions

1. Compare heating water for the home by gas, electricity, and solar energy in terms of: initial cost of the equipment; cost to run the system; efficiency of energy use. Which system would you choose if you expect to live in your home three more years? Ten more years? The rest of your life?
2. What is (are) the purpose(s) of "Daylight Savings Time"?
3. How is it that people in some European countries manage to travel almost as many miles by private auto as we do in the U.S., but they use one-third less gasoline?
4. How can individuals, all by themselves, reduce the amount of gasoline used?
5. What energy savings exist in mass transit?
6. If you do not ride a bus to work or school, what would it take for you to do so? If you do ride one, why do you choose this method of transportation?
7. Are electric vehicles energy savers? Explain.

Further Reading

General

Boffey, P. M., "How the Swedes Live Well While Consuming Less Energy," *Science,* **196** (20 May 1977), 856.

Hirst, E., and J. C. Moyers, "Efficiency of Energy Use in the United States," *Science,* **179** (30 March 1973), 1299–1304.

Landsberg, H. H., "Low-Cost Abundant Energy: Paradise Lost?" *Science,* **184** (19 April 1974), 247–253.

Mazur, A., and E. Rosa, "Energy and Life-Style," *Science,* **186** (15 November 1974), 607–610.

Schipper, L., and A. J. Lichtenberg, "Efficient Energy Use and Well Being: The Swedish Example," *Science,* **194** (3 December 1976), 1001–1013.

Starr, C., "Is Sweden More Energy-Efficient?" *Science,* **196** (8 April 1977), 121–124.

Energy in Transition, 1985–2010, Committee on Nuclear and Alternative Energy Systems, National Academy of Sciences, 1980.

"Energy Conservation: Spawning a Billion Dollar Business," *Business Week* (April 6, 1981), pp. 58–69.

Ross, M., and R. Williams, "The Potential for Fuel Conservation," *Technology Review* (February 1977), pp. 49–57.

Automobiles and Energy Conservation

Lave, L., "Conflicting Objectives in Regulating the Automobile," *Science,* **212** (22 May 1981), 893–899.

"The Big Shortfall in Auto Fuel Economy," News and Comments, *Science,* **205** (21 September 1979), 1232–1235.

Gray, C., and F. von Hippel, "The Fuel Economy of Light Vehicles," *Scientific American,* **244,** No. 5 (May 1981), pp. 48–60.

Dallaire, G., "Transportation Innovations That Would Banish America's Energy Crisis," *Civil Engineering,* November 1981, pp. 47–50.

"Flywheels: Energy-Saving Way to Go," *Environmental Science and Technology,* **10,** No. 7 (July 1976), pp. 636–639.

"Energy Storage (II): Developing Technologies," Research News, *Science,* **184** (24 May 1974), 884–887.

Residential Energy Conservation

Dallaire, G., "Zero-Energy House: Bold Low-Cost Breakthrough That May Revolutionize Housing," *Civil Engineering,* **52,** No. 5 (May 1980), 47–59.

"Saving Energy Dollars, Special Guide," *Consumers Reports,* October 1981, pp. 563–594.

Bartos, Jr., M., "Underground Buildings: Energy Savers?" *Civil Engineering,* May 1979, p. 85.

For an understandable and directly useful discussion of methods you can apply to save energy, we recommend:

"Tips for Energy Savers," U.S. Department of Energy, Assistant Secretary for Conservation and Solar Applications, DOE/CS-0020, March 1978.

(National Archives/Documerica; Gene Daniels)

PART SEVEN

Human Health and
the Environment

A startling and disturbing fact, emerging from the mass of scientific data on cancers, is that 60–90% of human cancers are probably caused by environmental factors. "Environmental" is used, in this case, in the broadest sense. The term covers such things as eating and smoking habits, as well as exposure to possible carcinogens (a carcinogen is something that causes cancer) in air, water, and manufactured goods.

About 400,000 Americans die each year from cancer and, at the same time, 700,000 new cases are detected (Figure A). A large portion of these cancers must be attributed to smoking, and another smaller

Cancer deaths (1969)	🚶🚶🚶🚶🚶🚶🚶🚶🚶🚶🚶🚶🚶🚶🚶🚶🚶🚶🚶🚶🚶🚶🚶🚶🚶🚶🚶🚶🚶🚶🚶🚶🚶🚶🚶🚶🚶🚶🚶
World War II battle deaths	🚶🚶🚶🚶🚶🚶🚶🚶🚶🚶🚶🚶🚶🚶🚶🚶🚶🚶🚶🚶🚶🚶🚶🚶🚶🚶🚶🚶🚶🚶🚶🚶🚶🚶🚶🚶🚶🚶🚶
Automobile accident deaths (1969)	————————————————————————————————🚶🚶🚶🚶🚶🚶
Vietnam war deaths (6 years)	————————————————————————————————🚶🚶🚶🚶
Korean war deaths (3 years)	————————————————————————————————🚶🚶🚶
Polio deaths (1952; worst year)	————————————————————————————————————

Figure A U.S. Deaths from Various Causes. (Each figure represents 10,000 deaths.) [Source: *The Implications of Cancer-Causing Substances in Mississippi River Water,* The Environmental Defense Fund, 1525 18th Street N.W., Washington, D.C. 20036 (1974), p. 15]

Table A What Causes Cancer?[a]

Cause	Estimated percent due to cause	Possible range
Smoking	30	25–40
Diet (not including food additives)	35	10–70
Workplace hazards	5	2–8
Alcohol	3	2–4
Environmental radiation (ultraviolet, natural x-rays, and cosmic rays)	3[b]	1–4
General air and water pollution	2	1–5
Medical treatment (drugs, x-rays)	1	$\frac{1}{2}$–3
Food additives	1	2–minus 5 (protective effect)
Consumer products (for instance, Tris; urea–formaldehyde foam insulation; asbestos in patching compounds; trichloroethylene in aerosol cans; some hair dyes)	1	1–2
Nonenvironmental causes (infection, pregnancy, childbirth, sexual development)	17	2–?
Unknown	?	

The only real agreement is that about 30% of U.S. cancer deaths are due to smoking. A smaller percentage (4–5%) result from exposure to carcinogens in the workplace. Choice of diet probably has a large influence, but we don't yet know how large. General air and water pollutants, food additives, and industrial products probably account for another 4–5% of cancers.

a Data adapted from Doll, R., and R. Petro, *Journal of the U.S. National Cancer Institute,* June 1981.
b Does not include skin cancers other than melanomas.

portion to working in certain industries. There is also evidence that the general public may be endangered by such factors as their chosen diet or by toxic substances that are widely distributed in the environment (Table A).

In Parts III, IV, and V, we discussed a number of water and air pollutants, emphasizing where they come from and why they are found in our environment. These pollutants can have a variety of effects on human health. The health hazards of bacterial or viral water pollution and the special case of trihalomethanes (carcinogens that may be formed from water purification procedures) were noted in Chapter 10. The health effects of chemicals such as mercury, cadmium, or nitrates, which sometimes contaminate drinking water, were detailed in Chapter 11. In Chapters 20, 21, and 22 we examined the harmful effects on human health of those air pollutants that result from power generation.

Now we take a closer look at certain other environmental factors and pollutants. Of particular concern are those suspected of contributing to, or causing, cancer deaths. Although there are a number of important health problems in the U.S., cancer remains overall the greatest cause of death from disease in people under 55.

In Chapter 29, we discuss how the risks of health hazards such as carcinogens are evaluated. An interesting feature of much of the information used is that it comes from studies of the occurrence of diseases (such as cancer) in groups of people, or populations. This is the science of **epidemiology.** In this chapter we also explore the difficulties in balancing risks against possible benefits of hazardous substances.

In the following chapter, a number of pollutants

suspected of being contributors to the burden of carcinogens or otherwise harmful to human health are discussed. Most of these pollutants could be classified as "environmental" in the sense that individuals have little control over their exposure. Examples are asbestos, radiation, and workplace pollutants.

Chapter 31 covers three major contributors to the class of environmentally caused cancer: diet, drugs, and smoking. Although these factors are also called environmental, individuals can exercise a great deal of control over exposure to them. (There are exceptions; for instance, pesticide contamination of foods or smoke from a neighbor's cigarette.)

The last chapter in this part addresses the problem of protecting people from toxic substances in the environment. There has been relatively little attempt at government control of some of the voluntary environmental factors that may be harmful to health, such as choice of diet or smoking habits. However, many laws have been passed in an attempt to control toxic substances in air, water, and foods, and to see that hazardous wastes are disposed of properly. These laws are the focus of the concluding chapter.

CHAPTER TWENTY-NINE

Science and Public Protection

Risks and Benefits

Lifestyle and Risks/A Basis for Judgments

The Epidemiologist as a Detective

An Atlas of Cancer Deaths/A Mystery Case/Latency Period of Cancers/Can Drinking Water Cause Cancer?

Experiments to Determine Potency of Carcinogens

Animal Experiments/The Strengths of Carcinogens/New Tests

Risk–Benefit Analysis: Policy or Panacea?

Quantitative Risk Assessment/Risk Assessment and Limited Resources/Adding the Benefits to the Risk Decision

Carcinogenesis

CONTROVERSIES:

29.1: ***The Murder of the Statistical Person: Risk–Benefit Analysis Applied to Carcinogens***

29.2: ***Economics and Cancer Protection: Removing Chloroform from Drinking Water***

Risks and Benefits

Lifestyle and Risks

In 1958, Congress passed a law that has become known as the Delaney Clause. This law specifies that no additive can be called safe (and thus used in food) if it has been shown to cause cancer in animals or humans. At the time this law was passed, it seemed a reasonable precaution. To many scientists, it still seems a good policy. To others, however, an outright ban on carcinogens appears naive. It is argued that consideration must be given to the possible benefits of a substance as well as to the possible risks. Basic to such a risk–benefit analysis is the idea that we can, in some way, measure both risks and benefits so that they can be compared.

What kind of data do we have on cancer-causing substances and how good is it? Who should make the final comparisons of risks versus benefits? These questions are important not only in legislating carcinogens but also in developing laws for many other environmental pollutants. Radiation standards, for instance, involve the same kinds of questions about public safety and how large a risk is tolerable. In fact, we all make risk–benefit decisions every day. Even such simple decisions as whether to cross the street, go for a ride in a car, or fly in an airplane involve this type of decision, although most of us hardly think about it. We are generally willing to accept a certain amount of risk as the price of the lifestyle we choose. But is such a procedure suitable for regulating carcinogens?

A Basis for Judgments

With respect to cancer-causing substances, we need to know how potent a carcinogen is, who is likely to be exposed to it, and how great that exposure is. Then this risk could be compared to the possible benefits gained from allowing the use of the substance. Although this may sound a perfectly reasonable and relatively straightforward procedure, in fact, a number of pitfalls exist.

Probably the greatest problem is that we do not have reliable scientific data on how carcinogenic most substances are to humans. Since we cannot experiment directly on humans, we must rely on three other kinds of data. The first is **epidemiological** data: the study of disease in large groups of people. The second kind of data come from experiments in which animals are dosed with a test substance and then examined for cancer. A third type of evidence can be gained from tests on bacteria, or animal or human cells grown in test tubes. All of these methods have limitations, however, and none can be used directly to judge human cancer risk. The following sections examine more closely the various methods of testing for carcinogens.

The Epidemiologist as a Detective

An Atlas of Cancer Deaths

In 1977, the U.S. Public Health Service published an atlas of cancer mortality for the 20-year period 1950–1969. The atlas consists of a series of maps of the United States, showing the rates of death from various forms of cancer, county by county. Some striking variations can be found in the occurrence of various forms of cancer. For instance, in the Northeast (New Jersey, southern New York, Connecticut, Rhode Island, and Massachusetts), there is a very high rate of mouth, throat, esophageal, laryngeal, and bladder cancers. The rates are high for males but not for females in this area, which suggests that the cancers may be related to the jobs at which the men work. In fact, this area is highly industrialized, with a large concentration of chemical industries. Several manufactured chemicals (2-napthylamine is one) are known to cause bladder cancer. In Salem County, New Jersey, one-fourth of the work force is employed in the chemical industry or related industries. The rate for bladder cancer in Salem County is the highest for any U.S. county with a reasonably large (10,000 or over) population. On the other hand, colon and rectal cancer is high for both males and females in this area. Although working in the chemical industries may eventually be shown to increase a person's chance of developing bladder cancer, some other explanation must be sought for the high incidence of colon cancer.

Examples are found all over the world of the occurrence of specific cancers in certain groups of people. Linxian County, in the People's Republic of China, has a very high incidence of cancer of the esophagus. Cancer of the stomach is very common in Japan, while in the U.S. it has steadily decreased to very low levels since the beginning of this century. Liver cancer is a serious problem in Africa, Southeast Asia, and coastal China, but is relatively rare elsewhere in the world. More examples could be given; however, it is clear that there are concentrations of various types of cancers.

Thus, there is a great deal of support for the idea that cancers are caused by some condition or some combination of conditions in the environment of different areas. This method of matching disease in populations with factors that might have caused them is part of the science of **epidemiology.** Such figures as those presented in the atlas cannot prove what causes a cancer, but they can provide clues as to where to look further.

A Mystery Case

For example, on the Baltimore census is an area where the cancer mortality rate is $4\frac{1}{2}$ times that of the city as a whole. The tract surrounds a former Allied Chemical plant that manufactured arsenic compounds for 100 years. Arsenic is a known carcinogen, and former plant employees have a lung cancer rate 14 times higher than residents in other areas of the city. When investigators tried to determine whether emissions from the plant could have been responsible for the high neighborhood rate of lung cancer as well, they were at first puzzled by the pattern of where cancer victims lived with respect to the plant. Most victims seemed to live east and north of the plant, while prevailing winds blow north and northwest. If plant emissions contributed to the occurrence of lung cancer, one might have expected the pattern of cancers to follow the wind direction, the way pollutants would have blown. Eventually, former plant employees told researchers that there was once a railroad spur that ran to the plant from the north. Arsenic compounds were shipped to the plant along this line. If the cars were not sealed after unloading at the plant, arsenic could have blown out and settled along the rail tracks. In fact, a high concentration of arsenic was found in the soil along the old rail bed, the concentration decreasing as samplers moved away from the plant.

Actually, some 150–200 chemicals were manufactured at the plant at one time or another. Thus, it may never be possible to know for sure what specific substances caused the high cancer rate at the plant and in the neighborhood.

Latency Period of Cancers

This case points out another problem involved in determining the causes of cancers: a long time usually goes by between exposure to a carcinogen and diagnosis of cancer. For instance, it is now known that lung cancer caused by breathing asbestos dust does not appear for 20 years or more after exposure to the dust. In one study, an average of 39 years elapsed between the time workers breathed the asbestos and the time they developed lung cancer (Table 29.1).

Table 29.1 Latency Period of Some Cancers

Type of cancer	Approximate number of years between exposure to carcinogen and diagnosis of cancer
Lymphomas	2–5
Bladder cancers (due to aromatic amines)	18
Mesotheliomas (asbestos-caused)	30
Lung cancers (asbestos-caused, bronchogenic)	20–40

For moral reasons, epidemiologists cannot experiment on humans to obtain data on causes of cancers. (Except, perhaps, when such information surfaces as a result of using chemicals to treat illnesses. Certain drugs have been found to be human carcinogens.) Further, human cancers often take 20–30 years to develop; thus, even methods that involve comparing groups of people who were unintentionally exposed to chemicals (pesticides workers, industrial employees, construction workers) and groups of people who were not exposed, are of limited value. Information can be gained this way, but it comes too late.

Because there is such a long time period between exposure to a carcinogen and the proof that harm has been done, many scientists and legislators feel that we must control substances released into the environment. Otherwise we may find out in 20 or 40 years, when it is too late, that some commonly used chemical is a carcinogen.

Can Drinking Water Cause Cancer?

A good example of the strengths and shortcomings of epidemiology is the controversy over carcinogens in the New Orleans water supply. In 1974, the Environmental Defense Fund (EDF) published a report noting that the number of cancer deaths in New Orleans was higher than expected. The report suggested this was due to carcinogens, both chlorinated and nonchlorinated organics, in the drinking water. (See Chapter 10 for a discussion of how these chemicals come to be in drinking water.)

Although no one argues with the fact that carcinogenic materials were found in New Orleans drinking water, it is not an easy matter to decide how much of a hazard this represents. According to a 1972 EPA study, various organic chemicals were present in the parts-per-billion range. In fact, except for a few chemicals,[1] most were present at less than 1 part per billion. We do not know the effects of these very low concentrations of toxic chemicals, day after day, over a person's whole lifetime. Scientists from EDF attempted to solve this problem by looking at statistics on people drinking less-polluted water. Ten parishes in Louisiana and one-third of another draw their drinking water primarily from the Mississippi; the rest use groundwater (generally believed to be less polluted than surface water) or other surface water supplies. When these two populations are compared, it can be shown that there are more cancer deaths among those drinking Mississippi River water. Unfortunately, this is not proof that the water is causing excess cancer deaths. Some other factor might be the true cause. The association with drinking water might be chance or associated in some way with the true cause (for instance, some food in the diet of people living in the affected area could be the cause). Researchers thus checked a variety of other possibilities.

Cancer deaths were divided into groups by race and sex. The deaths were still found to be associated with the source of drinking water for white and nonwhite males and for nonwhite females. Dividing people into groups by income made no difference, nor did dividing them by occupation. Cancer deaths were not correlated with living in southern versus northern parishes. (The southern parishes are different socioeconomically from the northern parishes and thus might be expected to have differences in diet, etc.) Even elevation was checked, since air pollution could be expected to be worse at low elevations than at high ones. No correlation was found, however.

Cancer rates in the city itself might be falsely high if people have moved there for better medical care after the disease is diagnosed. However, subtracting the New Orleans data from that of all other Louisiana people drinking Mississippi River water did not change the positive relationship with cancer deaths.

The epidemiologist must be even more of a detective, however. Why are cancer deaths not correlated with the source of drinking water for white females?

---

1 Chloroform was found at 113 ppb, four others were found at 1–8 ppb.

What about the role of smoking or alcohol consumption? Are the excess cancers of a kind that could reasonably be expected to develop from contaminated drinking water (liver or digestive or urinary cancers), or are they more likely to have developed from air pollutants (lung cancers)? Can you think of other factors to check?

More detailed studies, comparing cancer cases with the actual concentrations of carcinogens in the victims' drinking water, would come closer to proving a link between drinking water and cancer. Studies such as this are now under way. At the moment, most epidemiologists feel that the evidence linking chlorinated organics in drinking water to excess cancers is becoming stronger.

Epidemiological evidence, however, can probably never be as strong as evidence obtained from animal experiments. The reasons are that it is so difficult to be sure that all possible interfering causes for a problem have been thought of, and because groups of humans vary widely in their susceptibility to different diseases. Even so, eventually the weight of epidemiological evidence becomes so great that most people consider the case proven beyond reasonable doubt. This is the case with evidence about cancer and smoking (Chapter 31).

Experiments to Determine Potency of Carcinogens

Animal Experiments

Many different animals have been used to test for carcinogens. Experiments on large animals such as monkeys or dogs are extremely expensive and also take years to produce useful results. Such experiments may be justified in cases where many people are being exposed to suspected carcinogens. Some of the information we have on smoking and lung cancer or chemicals and bladder cancer comes from this type of research.

The accepted method of testing for carcinogens is generally to use two species of rodents—rats and mice or hamsters. The animals are given the test chemical in the highest doses possible without causing them to become sick or die (from some cause other than cancer). Periodically, the animals are examined to see whether more than the normal number of cancers can be found in the treated group of animals.

Very large numbers of animals may be needed to detect a cancer-causing substance. For instance, suppose a chemical caused cancer in 1 out of 10,000 humans exposed to it. If the entire U.S. population were exposed, 20,000 people would develop cancer, a fairly large number of cases. Yet, to detect this hazard in an animal experiment, 10,000 rats would be needed to find one cancer. At least 30,000 would be needed to call the results significant. Further, humans may be more sensitive to some chemicals than rodents. For instance, humans are 60 times more sensitive to the effects of thalidomide than mice and 100 times more sensitive than rats.

The Strengths of Carcinogens

We know that, in general, some substances are able to cause cancer at lower concentrations than others. However, we do not yet have the information that would make it possible to state the relative potency of most carcinogens for humans. In part, this is due to the fact that most of the information comes from animal experiments and cannot be directly applied to humans. The effects of carcinogenic materials on humans have been studied well enough in only six cases[2] to rank their relative strengths. In these six cases, the carcinogenic effect in humans is roughly comparable to that obtained in rat experiments. It is on this basis that results in rats are often applied to humans (Figure 29.1). Animal data, however, may not always be a valid basis to use in assigning human potency to carcinogens, because carcinogenic effects differ among animal species. To give two examples: aflatoxin is not carcinogenic in adult mice, although it is in rats; 2-napthylamine is not a rat carcinogen, although it is carcinogenic to humans.

Another problem is that some materials that are not carcinogenic themselves seem to cause cancer in combination with other substances. In addition, there is evidence that small amounts of carcinogens add together to increase the total carcinogenic effect (see the supplemental material at the end of this chapter).

Of the almost 2 million chemicals known, only about 6000 have been tested in animal experiments to determine whether they were carcinogenic. Of this

---

2 Benzidine, chlornaploazin, diethylstibesterol (DES), aflatoxin B$_1$, vinyl chloride, and cigarette smoke.

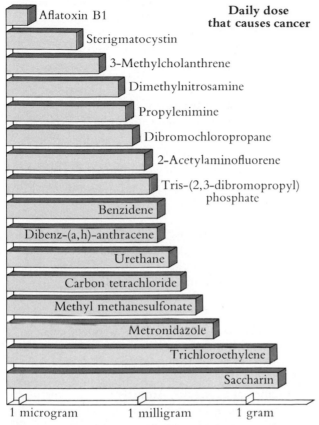

Daily dose that causes cancer

Aflatoxin B1
Sterigmatocystin
3-Methylcholanthrene
Dimethylnitrosamine
Propylenimine
Dibromochloropropane
2-Acetylaminofluorene
Tris-(2,3-dibromopropyl) phosphate
Benzidene
Dibenz-(a,h)-anthracene
Urethane
Carbon tetrachloride
Methyl methanesulfonate
Metronidazole
Trichloroethylene
Saccharin

1 microgram   1 milligram   1 gram

Figure 29.1 Relative Strength of Carcinogens in Rats and Mice. The doses shown are those that cause cancer in half of a group of test rats or mice when they receive it daily over their lifetime. As you can see, aflatoxin is over a million times more potent than saccharin. [Source: T. H. Maugh II, *Science,* **202** (6 October 1978), p. 38. © 1978 by the American Association for the Advancement of Science]

6000, 1000 have given some evidence that they do cause cancer. Of the 1000, a few hundred are considered proven carcinogens.[3]

New Tests

Animal welfare groups are pressing for new methods of testing. They want to see the use of tests that do not require animals (especially dogs, cats, and monkeys) be killed or injured in order to determine the safety of chemicals for humans.

---

3 See Further Readings, *Annual Report on Carcinogens.*

Several new methods are being developed to provide faster, cheaper screening for carcinogenicity. One group of methods uses animal cells grown in test tubes. Changes in the **DNA** of cells exposed to chemicals can be correlated to the development of cancers. Results are promising, but much work remains to be done on this group of methods. Ideally, the animal cells used should be human cells, but this is not yet possible.

Another method, called the Ames Assay, is already in use. It involves adding the chemical to cultures of the bacterium *Salmonella typhimurium.* The bacteria are then examined to see if the chemical has caused a mutation. (Mutations are heritable changes in the composition or arrangement of genes, which are composed of DNA. See page 559 for more on this subject.) In this particular system, researchers look to see if bacteria that require the nutrient histidine to grow have changed, or mutated, so they can grow without it. (This is called a back-mutation.) Although the system detects mutagens, not carcinogens, it has been found that there is a good correlation between the two. That is to say, 80% of the carcinogens tested have been found to be mutagens. Furthermore, most known human carcinogens give a positive result in the Ames Assay. Of the noncarcinogenic substances tested, 87% were negative in the assay, while 13% gave a false positive result. These percentages may improve even more as the method is refined. For instance, it has been shown that certain substances undergo chemical changes in the body that turn them into carcinogens. Thus, benzo(a)pyrene is a carcinogen after conversion to an active form by microsomes. If benzo(a)pyrene is used in the Ames Assay, it is not mutagenic. However, if a preparation of liver microsomal enzymes is also added to the test system, mutations occur (Figure 29.2). Further refinements such as this will undoubtedly increase the accuracy of the Ames Assay and of the other "quick tests," such as those using cell cultures. The Ames Assay is quick (about three days to run a test) and inexpensive (about $200 per chemical, compared to $100,000 per chemical in the conventional rat and mouse assays). It is the possibility of false negatives, however, that limits the usefulness of the test. That is, up to 20% of carcinogenic compounds may not be mutagenic in the Ames Assay. Thus, no chemical could be given a clean bill of health on the basis of the Ames Assay alone.

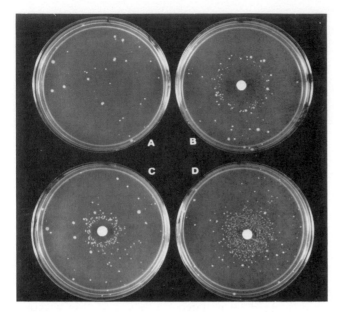

Figure 29.2 The "spot test" for mutagen-induced revertants. Each petri plate contains, in a thin overlay of top agar, the tester strain TA98 and, in the cases of plates C and D, a liver microsomal activation system (S-9 Mix). (Plate B did not require the liver system.) Mutagens were applied to 6-mm filter-paper discs, which were then placed in the center of each plate. (A) Control plate. Spontaneous revertants: (B) plate showing revertant colonies produced by the Japanese food additive furylfuramide (AF-2) (1 microgram); (C) by the mold carcinogen aflatoxin B_2 (1 microgram); (D) by 2-aminofluorene (10 micrograms). Mutagen-induced revertants appear as a circle of revertant colonies around each disc. [From Bruce N. Ames, Joyce McCann, and Edith Yamasaki, "Methods for Detecting Carcinogens and Mutagens with the Salmonella/ Mammalian-Microsome Mutagenicity Test," *Mutation Research,* **31:** 347–364 (1975). Reprinted by permission. Copyright © 1975 by the Elsevier Scientific Publishing Company, Amsterdam. Permission: Elsevier/North Holland Biomedical Press.]

A final group of methods uses mathematical modeling and computers to determine, by looking at the structure of a chemical, whether it is likely to be carcinogenic or perhaps a good anticancer drug.

Some combination of testing methods will eventually have to be worked out in order to catch all potential carcinogens before they can be distributed in the environment.

Risk–Benefit Analysis: Policy or Panacea?

Quantitative Risk Assessment

Animals as substitutes for people. There are many uncertainties in determining the carcinogenic risk to humans when using data from animal or test-tube experiments. Nonetheless, many reputable scientists are pressing for a policy of quantitative risk assessment. This means assigning actual numbers to the risks of contracting cancer from particular substances or treatments.

We have already discussed some of the problems involved in using animals or bacterial cells to measure the potency of carcinogens to humans. This part of risk assessment is sometimes called biological extrapolation.

Thresholds. Another part of quantitative risk assessment involves numerical extrapolation. This means extending the results of animal experiments at high doses to the lower doses involved in workplace or environmental exposure. Animals in experiments receive much larger doses of test compounds or of radiation than those to which workers or the general public are at all likely to be exposed. This is done on the assumption of some positive relationship between the dose of a harmful substance and its effect: the larger the dose, the more likely the effect will show up in an experiment. By using these high doses, scientists can find evidence of carcinogenicity even for weak carcinogens, using reasonable numbers of experimental animals. The results are then extrapolated down to the probable exposure levels of workers or the general public. The question is often posed, however, whether this is reasonable; whether there is not, in fact, a level below which no effect occurs (the threshold level) (Figure 29.3). Thresholds could be the result of such processes as repair mechanisms by which the body repairs damage caused by carcinogens. There is evidence for such repair (see the supplemental section at the end of this chapter).

At present, the more conservative policy is to assume that there is no threshold. In other words, we assume that smaller and smaller doses result in fewer, but still measurable, ill effects.

Risk Assessment and Limited Resources

Those who advocate quantitative risk assessment point out that we actually do not have the resources to test properly the 9000 synthetic chemicals manufactured on a large scale today, much less the 500–1000 new ones introduced each year. They argue that, flawed as

it may be, such analysis would at least give a basis for deciding how to spend limited resources. Substances would be ranked both on how many people would be exposed and how toxic or carcinogenic they appear in a combination of animal and other tests. Substances judged to present higher risks would be tested first and most extensively.

Adding the Benefits to the Risk Decision

Where opinion divides most widely is on whether to use quantitative risk assessment to determine the cost to society of a hazard and then compare this cost to the possible benefits obtained. Some scientists, such as Arthur Upton, Director of the National Cancer Institute, feel that the risk evidence is too flimsy to make such comparisons the basis for regulating carcinogens (for setting exposure standards in the workplace, for instance).

In addition to uncertainties in risk assessment, critics of cost–benefit analysis point out several drawbacks. Such analysis is intended to reduce all factors to a common denominator for purposes of comparison. Dollar values must therefore be put on such things as a human life or a pleasant environment. Many people feel such things are not quantifiable.

The analysis also cannot consider intergenerational effects. That is, many substances are both **mutagens** and **carcinogens.** People exposed may develop cancer and their germ cells may undergo mutations. The effects of these mutations must be borne by later generations.

Also, there may well be inequities in the distribution of costs versus the benefits. The benefits of a regulation go to the people protected by it, while the costs may be borne by another group (i.e., industry). On the other hand, without regulation, the health risk is borne by people who may not benefit at all from the process causing the risk. Cost–benefit analysis does not deal with this sort of problem. Even determining the cost of regulating a substance, which may seem straightforward, involves extrapolating the probable cost of regulation into the future. The discount rate chosen is important to such analyses but not easy to agree upon, especially if inflation is a likely complication.

Once risk assessment is an accepted technique, however, the temptation will certainly exist for law-

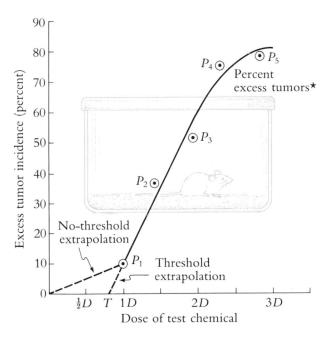

* Excess tumor incidence (percent) is defined as:

$$\frac{\dfrac{\text{tumors in}}{\text{exposed population}}}{\dfrac{\text{number of}}{\text{exposed population}}} - \frac{\dfrac{\text{tumors in}}{\text{control population}}}{\dfrac{\text{number of}}{\text{exposed population}}} \times 100$$

Figure 29.3 Determining the Effect of Carcinogens at Low Doses. When the results of animal experiments (which look at the number of excess cancers versus several doses of a test chemical) are graphed, a curve like that shown from P_1 to P_5 is obtained. But the first point, P_1, is at 10% incidence; that is, 1 animal in 10 developed cancer. Regulatory agencies are worried about much lower levels of cancer incidence than this. (For instance, a dose causing a 1% incidence of cancer in the U.S. population would cause 2 million excess cancers—obviously an unacceptable risk.) So, the graph must be extrapolated down to these lower levels. Some scientists argue that a threshold exists below which no effect occurs. This would be shown by the dotted line P_1 to T. Most regulators, however, would agree that a no-threshold extrapolation such as the dashed line between P_1 and 0 is a safer policy. (Source: Office of Technology Assessment.)

The Murder of the Statistical Person: Risk–Benefit Analysis Applied to Carcinogens

> Public policy is fundamentally an economic exercise. It cannot evade the balancing of risks and benefits . . .★
>
> **Gio Batta Gori, Deputy Director,**
> **Division of Cancer Cause and Prevention,**
> **National Cancer Institute**

> Although the occurrence of very large errors [in risk assessment] should be rare, each such error could be a catastrophe.†
>
> **Arthur Upton, Director,**
> **National Cancer Institute**

The use of risk–benefit analysis to regulate carcinogens is shot-through with controversy. One of the most serious limitations is that the data for estimating risk must come from animal or test-tube experiments, and we are not yet sure how well this data predicts human risk. Risk–benefit analysis also involves such arguable features as assigning value to human life or a pleasant environment and the probable cost of regulation at some date in the distant future. Further, it does not take into account who bears the costs of regulation or nonregulation, compared to who derives the benefits.

Another problem arises when costs and risks have been determined as well as possible and are ready to be compared. If risks are not to be reduced as close as possible to zero (because of the high cost), some judgment must then be made as to what is an acceptable risk. For instance, one might decide that since 20,000 out of every 100,000 Americans die of cancer, increasing the number to 20,001 out of 100,000 would be an acceptable risk. This represents 2200 excess cancer cases (assuming a U.S. population of 220 million people), or 30 more deaths per year due to cancer in the U.S. (2200 deaths spread over a 70-year life span). Now, if these 30 people each year were actually known to us, the risk might not seem so negligible. This is called the "murder of the statistical person."

Do you favor regulation of carcinogenic substances by risk–benefit analysis, or by a policy of reducing the risk to as low a value as possible, regardless of cost? Does the kind of substance to be regulated make a difference? Suppose it were a food additive? Suppose it were a cosmetic? Suppose it were an air pollutant? Does it make a difference whether exposure is voluntary (food additive) or nonvoluntary (air pollutant)? Does it make a difference who is exposed (children, pregnant women, workers)? Who should make these kinds of decisions? Congress, the judiciary, or regulatory agencies? (All of these bodies have actually made such decisions—can you think of examples?) Or perhaps a special court that includes members of the general public and/or special-interest groups?

★ *Science,* **208** (18 April 1980), p. 256.
† *Science,* **204** (25 May 1979), p. 813.

makers and regulators to use cost–benefit analysis to regulate hazardous substances. It will be all too easy for regulators and the general public to lose sight of the uncertainties and limitations of this apparently simple but possibly treacherous tool. (See Controversy 29.1.) In Chapter 32 the use of risk assessment and risk–benefit analysis in current U.S. laws is discussed further.

Carcinogenesis

We are just beginning to understand how a carcinogen affects cells. There are many different carcinogens and there is probably more than one way that cancers arise. However, a large group of chemicals appear to cause some permanent change in a cell by reacting with its DNA. The DNA in a cell is in the form of one or more large molecules, composed of nucleic acids, phosphate groups, and the sugar deoxyribose linked in a long chain. Sections of the DNA chain, containing specific sequences of nucleic acids, are now recognized as genes, those units of cellular material that control heredity and the normal workings of a cell. A permanent change in one of these genes is called a **mutation.** Most, although not all, carcinogens can be shown to cause mutations in specially designed experiments. Of course, not all mutations cause cancer (Figure 29.4). A mutation involving genes that control a cell's reproductive mechanisms is an example of the kind of change believed to lead to cancers. This first stage of carcinogenesis is called initiation.

Second is a stage called preneoplasia, or the latent period. This stage is long, often lasting 20 years or more in humans. Little is known about what happens during this long, apparently quiet period, but there is evidence that many cells are repaired, or revert to normal during this time. Vitamin A may play a role in repairing cell damage, as do a variety of enzymes in cells. Cells that are not repaired go on to stage three, or transformation. Now they show the characteristics of cancer cells and begin to proliferate. Much research is being directed toward finding substances that can help cells repair themselves during preneoplasia in order to undo the effects of exposure to carcinogens.

Chemicals in another group are called **promoters.** Promoters are substances that do not cause cancer by

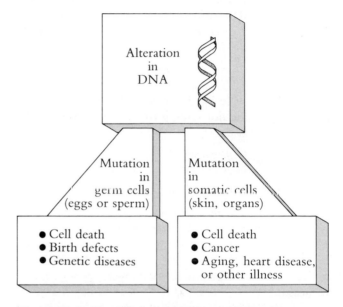

Figure 29.4 The Effects of Changes in DNA. (From: *Short-Term Tests for Carcinogens, Mutagens and Other Genotoxic Agents,* Technology Transfer Report # EPA-625/9-79-003, U.S. Environmental Protection Agency, July 1979, p. 5.)

themselves. They seem to "turn on" cells that have been initiated by a carcinogen but are in a latent period. Cocarcinogens are substances that cause cancer in combination but not by themselves. The artificial sweeteners saccharin and sodium cyclamate are both examples of compounds that are either very weak carcinogens or not carcinogens at all, but promoters. In animal experiments, sulfur dioxide in combination with benzo(a)pyrene causes cancer, while neither of the two chemicals alone does so.

Economics and Cancer Protection: Removing Chloroform from Drinking Water

> Before we are forced to spend tens of billions of dollars for [chloroform] removal, someone should ask whether these estimates of risk are based on reason or an Environmentalist theory.
>
> *Charles A. Stewart, Jr.**

> Who has informed public utilities that average householders are not willing to pay an additional $7 per year for clean water?
>
> *D. J. Baumgartner,*
> *EPA Corvallis Environmental Research Laboratory†*

Chloroform is considered a carcinogen. It has also been found in trace concentrations (parts per billion) in most public water supplies (Chapter 10). Scientists, however, are not in agreement over whether the amount of chloroform in drinking water poses a threat to health.

Several studies have shown that chloroform (at doses much higher than are found in water supplies) causes cancers in animals such as rats and mice. The National Academy of Sciences, using these experimental results, calculated that:

> At a concentration of 10 pbb (10 micrograms per liter) during a lifetime of exposure this compound would be expected to produce one excess case of cancer for every 50,000 persons exposed. If the population of the United States is taken to be 220 million people, this translates into 4,400 excess lifetime deaths from cancer, or 62.8 per year.
>
> In view of [this potential in humans], and taking the risk estimates into account, it is suggested that very strict criteria be applied when limits for chloroform in drinking water are established.‡

Not everyone agrees that you can use this type of animal data to calculate human risks, however. Her-

bert Stokinger, Chief of Experimental Toxicology at the National Institute for Occupational Safety and Health, says:

> Why is EPA concerned about the presence of such minute amounts of a substance shown to be carcinogenic only in animals and at levels that would correspond to a daily intake of $CHCl_3$ of from 17,000 to 34,000 mg/day?
>
> Extrapolating animal data to man for the purpose of estimating the incidence of cancer in human populations can prove a fallacy.§

The Environmental Protection Agency, directed by Congress to set standards that will protect the American public, has decided to err on the side of caution. It is requiring the removal of chloroform from drinking water by activated carbon filters (GAC).

Although the EPA held hearings at which interested people could state their views, the public was not asked directly—for instance, by a ballot referendum—whether they were in fact willing to pay extra for the added protection. In contrast, when fluoridation of drinking water (to prevent tooth decay) was an issue, in the 1960s and early 1970s, the Safe Drinking Water Act was not yet in effect. At that time, the EPA did not set drinking water standards. Each state set its own standards. The fluoridation issue was decided, in general, by a separate vote in each community. For this reason, we now have a patchwork of fluoridation practices. Some communities add fluoride and others are still vigorously opposed to the idea.

Oscar Adams, Director of Engineering for the Virginia State Health Department, was quoted in the *Baltimore Sun,* 27 December 1977:

> I think most cities are going to be put into a real hard position in that to meet the standards is going to require considerable capital costs . . .
>
> The real kicker is that it would mean a raise in water rates of 27 percent a year . . . or $3.91 per person . . .

Is there justification for setting standards and, as in this case, costing the consumer money, when there is no clear-cut demonstration of risk? Do you feel that the protection you gain from toxic organics is worth a raise in your water bill? Although the cost is not high,

the jump in the average water bill, for a system that installs GAC, will be noticeable. Does the cost make a difference in your attitude? Suppose it were a few cents a year? a few hundred dollars?

Do you feel you should be more directly consulted about matters related to your pocketbook and the protection of your health? Would you rather have the EPA decide them?

What do you suppose the cost of 62.8 excess cancer deaths per year might be to society as a whole? Compare this to the cost to the U.S. population as a whole each year if everyone were paying for a GAC system. Is this comparison valid?

---

* Letters to the Editor, *Science,* **212** (5 June 1981), p. 1086.
† D. J. Baumgartner, "Drinking Water: Sources and Treatment," *Science,* **197** (22 July 1977), p. 324.
‡ *Drinking Water and Health,* National Academy of Sciences, Washington, D.C., 1977.
§ H. Stokinger, "Toxicology and Drinking Water Contaminants," *Journal of the American Water Works Association,* July 1977, p. 399.

Questions

1. Epidemiological studies, such as the study of the occurrence of cancer according to locality, are useful but cannot prove that certain substances or occupations lead to cancer. One of the reasons is that people are exposed to many factors at the same time. Thus, it is not easy to sort out which one or ones actually cause a disease. Give examples of factors or habits that could confuse a study on whether the smoke from a factory is causing lung cancer in a nearby neighborhood.

2. Why is it a problem that a long time goes by between exposure to a carcinogen and clinical diagnoses of cancer?

3. Why do we have to use animal studies to determine whether substances are carcinogenic? What problems does this cause?

4. What is the Ames Assay?

5. What is meant by quantitative risk assessment? How is it related to risk–benefit or cost–benefit analysis?

Further Reading

Fourth Annual Report on Carcinogens, U.S. Department of Health and Human Services, Public Health Service, December 1983.
  All known or believed human carcinogens to which a significant number of people are exposed are listed and described in this publication.
Assessment of Technologies for Determining Cancer Risk from the Environment, Congress of the United States, Office of Technology Assessment, June 1981.

A well-written summary covering the major points in this chapter. The almost 400 references enable the reader to delve further into the literature.
Krause, Carolyn, "Genes and Cancer," Oak Ridge National Laboratory Review, 1981.
  Current theories on cancer causation are reviewed in this clearly presented paper.

References

Weisburger, John H., and Gary M. Williams, "Carcinogen Testing: Current Problems and New Approaches," *Science,* **214** (23 October 1981).

Squire, R. A., "Ranking Animal Carcinogens: A Proposed Regulatory Approach," *Science,* **214** (20 November 1981).

Gori, Gio B., "Regulation of Carcinogenic Hazards," *Science,* **208** (18 April 1980).

Wilkins, J. R., N. A. Reiches, and C. Kruse, "Organic Chemical Contaminants in Drinking Water and Cancer," *American Journal of Epidemiology,* **IID,** No. 4 (1979), 420.

Drinking Water and Cancer: Review of Recent Findings and Assessment of Risks, Council on Environmental Quality, Washington, D.C., 1980.

Drinking Water and Health, Vol. 3, National Academy of Sciences, Washington, D.C., 1980.

CHAPTER THIRTY

Environmental Carcinogens

Asbestos
Diseases Caused by Breathing Asbestos/Uses of Asbestos/Estimating the Risk

Toxic Substances in the Workplace
Types of Toxic-Substance Hazards on the Job/Matching Diseases with Toxic Substances/Laws Protecting Workers/Economics of Controls

Radiation: Microwaves, Radiowaves
The Electromagnetic Spectrum/Biological Effects of Nonionizing Radiation/Microwaves and Public Health Standards

Radiation: x-Rays, Gamma Rays, and Particles
Kinds of Ionizing Radiation/Sources of Human Exposure to Ionizing Radiation/Radiation Exposure and the Nuclear Power Industry/Biological Effects of Ionizing Radiation/Should Radiation Standards Be Made More Strict?/Uranium Mine Tailings: An Unnecessary Hazard

Ultraviolet Light and Chlorofluorocarbons
Ozone: An Essential Gas/Threats to the Ozone Layer/ Effects of Decreasing the Ozone Layer/How the Future Looks

Skin Cancer

CONTROVERSIES:

30.1: ***When Do We Have Enough Information to Act?***

30.2: ***The Cost of Cancer Surveillance***

30.3: ***The Economics of Protecting Workers***

30.4: ***Women's Rights and the Workplace***

30.5: ***Is Unilateral Regulation Similar to Unilateral Disarmament?***

In this chapter we look at some of the areas in which humans have little personal control over their exposure to environmental carcinogens. Exposure to asbestos, ionizing radiation, and a variety of workplace carcinogens is for the most part not affected by individual actions. Several air pollutants that may contribute to the burden of cancer (e.g., particulates) and some water pollutants (e.g., trihalomethanes) also fall into this "involuntary exposure" category. They were discussed in Chapters 10, 11, and 21. In the next chapter we discuss some of the factors over which individuals do have control (e.g., smoking and diet).

Asbestos

Diseases Caused by Breathing Asbestos

A great deal of time and effort were spent before people discovered that the microscopic creatures we call "germs" cause diseases like pneumonia or scarlet fever. Imagine how hard it would be to find the reason for a sickness if the symptoms did not show up until 20 or 40 years after a person was exposed to the causative agent. This is the case, however, with diseases brought on by breathing asbestos.

Asbestosis is a disease in which breathing is made difficult by the presence of asbestos fibers in the lungs. The tissue around the fibers becomes tough; oxygen cannot be transferred to the blood by such tissues. Asbestosis usually shows up 20 years or more after a person starts working with asbestos.

Breathing asbestos dust can also cause cancer, both of the lungs and of the membranes covering the lungs. It seems that even a short exposure to asbestos can cause cancer 20–40 years later. In one case, a woman went to school for a short while near an asbestos field in South Africa. She and her classmates used to slide down piles of asbestos wastes on the way home from school. After her family moved away, she was never exposed to asbestos again. Fifty years later she died from a rare type of lung cancer (mesothelioma) caused by the dust she breathed when she was five years old. Men who have worked in shipyards are at an increased risk of developing mesothelioma. This is true even if they only worked there for periods as short as a few weeks and even if they did not directly handle the asbestos insulation used in ships. Although most asbestos-related disease is seen among people who have worked in industries where asbestos is produced or used, there seems to be a risk also to the families of people who work with asbestos. Asbestos brought home on work clothes appears to have caused mesotheliomas in the children or spouses of asbestos workers. Because such a small amount of asbestos has been shown to cause disease, experts feel that no level of asbestos can be called safe.

Uses of Asbestos

The mineral nature of asbestos ensures that it does not burn. For this reason, asbestos is useful in fireproofing materials. It is used in firemen's clothing, in heatproof gloves and mats, for oven linings, and for furnace ducts. Asbestos is also included in brake linings, ceiling and floor tiles, and in roofing tar and cement. In some products, the asbestos is firmly bound so that it is unlikely to contaminate the environment. Floor tiles containing asbestos, for instance, fall into this category. In other cases, asbestos is used in ways that permit its release into air and water (Figures 30.1 and 30.2).

In some cases, asbestos is found where it is not meant to be. Several prescription drugs, as well as some brands of beer and gin, have been found to contain asbestos fibers. Most likely, the asbestos fibers are washed into these products when the liquid passes through filters made of asbestos. For over 20 years, asbestos fibers contaminated the drinking water in Du-

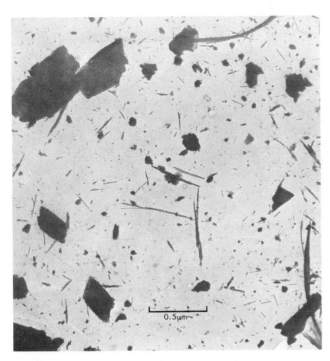

Figure 30.1 Asbestos fibers are found in several products used by the home handyman. The above photograph, taken with an electron microscope, shows asbestos fibers in a consumer spackling product. Asbestos can contaminate the air when the spackling is mixed or when the dried spackling is sanded. Old floor tiles and linoleum should not be sanded off floors because of the risk of releasing asbestos fibers. [Photograph from A. N. Rohl et al., *Science,* **189** (15 August 1975), 551. © 1975 by the American Association for the Advancement of Science]

Figure 30.2 Asbestos-containing coatings have been used on ceilings in a number of schools and public buildings. The coating has, in some cases, begun to flake off, scattering asbestos fibers into the air. In the above photograph, parents at Ramtown School in Hopewell, New Jersey, examine the flaking ceiling that has scattered asbestos onto the hall runner (lower right). A similar situation in a building at Yale University was improved when the entire building interior was scraped and repainted at great cost. In New Jersey, where the problem has been studied most thoroughly, 10% of the schools were found to be painted with the asbestos coating. In 1982, the EPA ruled that all schools must be inspected for asbestos coatings. The agency will provide assistance to schools in dealing with asbestos removal. (Frank Dougherty/NYT Pictures)

luth, Minnesota, and Superior, Wisconsin. Both cities draw their water from Lake Superior. For 22 years, Reserve Mining Company dumped 67,000 tons of mining wastes into Lake Superior every day. A major component in the waste is asbestos. No excess of gastrointestinal cancer has so far been demonstrated in the area. However, asbestos fibers have been found in the urine of people drinking water from Lake Superior.

Estimating the Risk

Although asbestos is a known carcinogen when breathed in, the effects of drinking asbestos are not clear. Twenty years is not long enough to tell whether an excess of cancers will develop among residents who drank contaminated water.

Low levels of asbestos fibers are definitely present in the air in cities. The main cause appears to be construction or demolition of buildings built with asbestos fireproofing and asbestos wallboard. Fibers are also found in the air where asbestos is mined, milled, or

When Do We Have Enough Information to Act?

> If that agency [EPA] has forbidden the use of quarried material on the basis of the evidence described . . . it is a disgrace.
>
> *John T. Hack*

> The agency is to be applauded and its scientists commended for their judgment . . .
>
> *A. N. Rohl*

In 1976, a high school science teacher began to wonder whether stone quarried near his home in Maryland might be a source of asbestos fibers in the air. Don Maxey, a rock collector and a person concerned about the environment, knew the stone was serpentine rock, a type known to contain asbestos. The crushed stone was being used to surface roads, playgrounds, and parking lots all around the Rockville, Maryland, area. When samples of dust and rock that he took appeared to confirm his suspicions, he called in researchers in the field of asbestos pollution.

There was indeed a very high concentration of asbestos in the air around Rockville: almost 1000 times higher than the average of levels in 49 American cities. The use of the asbestos-containing crushed stone from the Rockville quarry appears to be the most likely explanation for the unusually high concentration of airborne asbestos. (Early measurements by the Environmental Protection Agency showed levels to be as high along some roads as they were at Reserve Mining Company's taconite ore processing plant at Silver Bay, Wisconsin.)

When the problem was made public, there was a range of responses. Some parents kept their children home from a school near where a road was being resurfaced with crushed stone. A few real estate agents suggested hushing up the matter lest property values decline. There was scientific controversy, too, about the effects of exposure to asbestos dust at the levels found in Rockville. John T. Hack wrote in a letter to *Science:*

> . . . [L]ittle information is available about any actual danger that results from exposure to airborne dust containing such minerals in the kind of environment described. Even the authors of the report say only that there is a possibility of danger.
>
> The medical aspect of the problem is more to the point than the mineralogy of the quarry. If some medical research group would come up with some hard data that related illnesses or mortalities to degree of exposure to the mineral dust then it might be possible to judge the hazard . . .
>
> I do not know the reasons for the recent Environmental Protection Agency decision relating to this matter, but if that agency has forbidden the use of quarried material on the basis of the evidence described in the *Science* report, it is a disgrace.★

Hack thus appears to be calling for more evidence before action is taken on the problem.

Rohl, Langer, and Selikoff replied:

> . . . Unhappily, there is much information available about asbestos-related deaths—from mesothelioma, lung cancer, cancer of the esophagus, stomach, colon, rectum, and other cancers. But most of these have been elsewhere, in other

environments—mines and mills, factories, construction sites, shipyards, households of asbestos workers (contaminated by dust brought home on clothes and shoes), and neighborhoods around asbestos mines and factories. Do chrysotile ores have deadly potential when excavated and crushed in Quebec and not in Rockville? Should the state of Maryland, Montgomery County, and the Environmental Protection Agency (EPA) allow children in schoolyards to play on surfaces spread with gravel containing asbestos? Should "hard data"—sought after a 20- to 30-year latency period—from Bethesda, Silver Spring, Chevy Chase, Baltimore, Alexandria, College Park, and Washington be added to the experiences of Long Beach, Paterson, London, Dresden, Newark, Rochdale, Johannesburg, Thetford Mine, Barking, and Sverdlovsk before environmental asbestos air pollution is controlled?†

In fact, the EPA did take action by formally advising the county not to use the stone any longer. The agency also recommended that the county repave areas in which the rock had been used.

The problem, however, may not be limited to the Rockville area. Similar rock is found in many places in the United States. Rock quarried in all these areas may be high in asbestos and thus unsafe to use for paving. Decisions must be made in this case, as in many others dealing with environmental pollutants, on information that suggests a hazard to human health but is not sufficient to prove one.

Do you feel the EPA should wait for more information before issuing regulations outlawing the use of this particular kind of rock for paving? (Such regulations would, of course, be a great economic hardship to quarry owners.) Or, do you think the possible effects of the continued use of asbestos-containing rock are so serious that its use should be banned immediately? What should we do when the evidence is not conclusive?

---

* John T. Hack, *Science*, **197** (19 August 1977), 716.
† A. N. Rohl, A. M. Langer, and I. J. Selikoff, *Science,* **197** (19 August 1977), 716.

---

made into various products. Unfortunately, there is no good method for determining low levels of airborne asbestos fibers. In some areas of the country, natural deposits of asbestos-bearing rock can cause unsafe levels of asbestos fibers in the air (Figure 30.3 and Controversy 30.1). Researchers recently recommended closure of dirt-bike trails in the Clear Creek Recreational Area in San Benito County, California. Dust samples along the trail were shown to be 90% asbestos.

We do know that almost all of us—not just people who work with asbestos—have asbestos bodies in our lungs. (Asbestos bodies are fibers of asbestos that the body coats with a special protein after they are breathed into the lungs.) Unfortunately, we do not know whether the presence of asbestos bodies means that cancers or other asbestos-related diseases will some day develop. Remember, cancers caused by asbestos may not become visible for 20–40 years. Because we know so little at the moment about the effects of small amounts of asbestos, it seems wise to prevent as much asbestos as possible from getting into air, water, and food.

Toxic Substances in the Workplace

Types of Toxic-Substance Hazards on the Job

To most people, the term "job safety" brings to mind safeguards against immediate hazards to life and limb: machinery accidents or chemical burns. These are hazards workers rightfully expect their employers to help them guard against. However, other, less visible, but equally crippling job hazards exist, and the extent of these problems is only now becoming known. A study published in 1977 by the National Institute of Occupational Safety and Health (NIOSH) stated that one out of every four U.S. workers may be exposed during their working lifetimes to a toxic substance that can cause disease or death. This means that almost 22 mil-

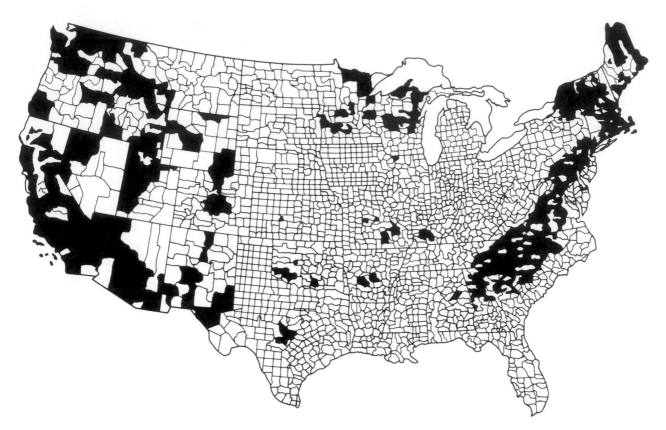

Figure 30.3 The mineralogy of the counties and regions blacked in above is such that quarries found in them may produce asbestos-bearing rock. The map was prepared by the Environmental Defense Fund from information derived by the Mining Enforcement and Safety Administration from reports by Batelle-Columbus and the U.S. Geological Survey. [Source: L. J. Carter, "Asbestos: Trouble in the Air from Maryland Rock Quarry," *Science,* **197** (15 July 1977), 237. © 1977 by the American Association for the Advancement of Science.]

lion workers have been or are being exposed to toxic substances: solvents, mercury, lead, or pesticides, etc. (Figure 30.4). Furthermore, some 880,000 of these workers, or 1% of the entire work force, are currently being exposed to substances known to be carcinogenic, such as asbestos, chromate compounds, arsenic, or chloroform (Table 30.1). Often, the workers or even employers are not aware of the hazards because the toxic substances are in products known by trade name only.

Not only blue-collar workers are in danger. White-collar workers are not exempt. Secretaries and management employees may be exposed to toxic air pollutants at plants, as may the families of asbestos, lead, or pesticide workers exposed to these substances from the workers' clothing. Anesthetists and other operating-room personnel are more likely than the general population to suffer kidney and liver diseases or cancers or to have babies with birth defects (Figure 30.5).

Figure 30.4 At one time, miners were concerned mainly about hazards to life and limb. However, breathing dust is now recognized as a serious health hazard. Among coal miners, it leads to black lung disease (Chapter 16). Workers in the cotton industry may contract brown lung disease from breathing cotton dust, while uranium miners are at an increased risk of lung cancer from breathing uranium dust. (U.S. Department of Energy)

Table 30.1 The Most Hazardous Industries and Some of the Carcinogens Used in Them[a,b]

Industrial and scientific instruments	Solder, asbestos, thallium
Fabricated metal products	Lead, nickel, solvents, chromic acid, asbestos
Electrical equipment and supplies	Lead, mercury, solvents, chlorohydrocarbons, solders
Machinery	Cutting oils, quench oils, lube oils
Transportation equipment	Formaldehyde, phenol, isocyanates, amines
Petroleum and products	Benzene, napthalene, poly-cyclic aromatics
Leather products	Chrome salts, tanning organics
Pipeline transportation	Petroleum derivatives, welding metals

a When industries are ranked according to how likely workers are to be exposed to a carcinogenic material, some surprises are found. Industries commonly thought to be "clean" show up near the top of the list. For instance, manufacturing of industrial and scientific instruments requires very little in the way of carcinogenic materials. However, those hazardous substances that are used are involved in hand assembly of machines; thus worker exposure is great. In contrast, the chemical industry, which may manufacture tons of carcinogenic materials, may employ only a few people to carry out the operation. Thus total worker exposure is relatively small.

b Source: *Science*, **197** (23 September 1977), 1268.

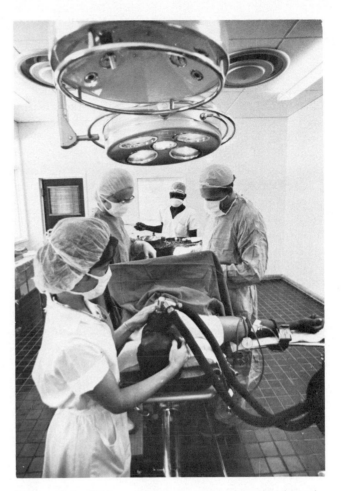

Figure 30.5 White-collar workers are exposed to hazards in industry, research laboratories, and hospitals. People who work as anesthetists in hospitals have an increased risk of developing cancer and other serious diseases, apparently due to exposure to chemicals in their work environment. (© Marc and Evelyne Bernheim/Woodfin Camp & Associates)

Matching Diseases with Toxic Substances

How could a problem of this size have been ignored for so long? One reason is the use of trade names and the absence of labels that list ingredients by common or chemical names. The 1977 National Institute of Occupational Safety and Health (NIOSH) study took two years to complete, partly because 70% of the substances found at workplaces were identified only by trade name. NIOSH had to contact 10,000 manufacturers to find out the ingredients. Still, this problem can be remedied.

A more difficult problem is to determine which workplace materials are toxic. As was explained in Chapter 29, epidemiological studies of workers' health are necessarily indirect investigations, since the information must be taken as it is found. Neat experiments cannot be planned to test interesting theories. As part of the National Toxicology Program (NTP), chemicals are being tested using animal experiments in a priority system. The system is based on both the suspected cancer-causing ability of the chemicals and on how many workers or members of the general public are likely to be exposed to the chemicals.

In the past, workers have often been accidental guinea pigs for toxic materials. Asbestos, arsenic, and vinyl chloride were all recognized as carcinogens after studies showed that workers handling these materials developed more cancers than comparable groups in the population. However, some diseases, such as cancer, do not appear for many years after exposure to toxic material. Even if a company keeps careful records on employees, it is easy to lose track of workers who are no longer employed and who may become sick after they leave. In some cases, companies have been reluctant to release data on employee health, for a variety of reasons. Fear of being sued by employees or confined by additional government regulations are probably among these reasons.

Laws Protecting Workers

The Occupational Safety and Health Administration (OSHA) is by law responsible for protecting the health of workers. OSHA sets limits for exposure to various toxic materials or determines what safety equipment is needed by workers. Furthermore, OSHA inspects workplaces to see that employers are following regulations.

The problems involved in proving that certain materials are hazardous to workers have made it difficult to set standards. A great many regulations were challenged by manufacturers in hearings or in court. This has resulted in long and costly fights among OSHA, manufacturers, and workers rights groups. During its first nine years of existence, OSHA managed to regulate only 20 substances. In the past, in order to justify regulations, OSHA estimated the risk of hazardous substances to workers in a general way best described as qualitative risk assessment. However, a recent court decision directs OSHA to make a

stricter justification for its regulation, rather than assuming that reduction of exposure is always a worthwhile course of action. That is, OSHA must present the actual benefits of measures to reduce worker exposure compared to the cost of those measures.

Economics of Controls

Understandably, manufacturers feel that worker protection devices will increase their production costs. We can expect these costs to be passed on to consumers in one way or another. However, society already pays costs comparable to these in medical care, lost productivity, and welfare payments to miners and millers crippled with lung diseases, to cancer victims, and to those poisoned by lead, mercury or pesticides. (See Controversies 30.2 and 30.3.)

Unions also worry about new restrictions. They fear that increased manufacturing costs will cause plants to close, especially small ones, with the result that many jobs will be lost. In reply to this concern, Douglas Costle, then Administrator of the Environmental Protection Agency, wrote in the December 1977 issue of the *EPA Journal,* p. 4:

> It must also be recognized that environmental regulations create jobs. The facts show that more people have been employed now than would have been without the major pollution control programs. Approximately 19,000 job losses have been attributed to pollution control compared to perhaps half a million jobs that were generated because of cleanup efforts. Such jobs are generated in three ways. First, construction of equipment and plants required by environmental programs create the largest number of jobs . . .
>
> The second way in which jobs are created is in the pollution control equipment manufacturing industry.
>
> Finally, many more indirect jobs are stimulated by these expenditures.
>
> I want to stress that when we talk about the issue of jobs versus the environment we are caught in the old mindset of looking at pollution controls as unproductive, profit-decreasing expenditures. Rather we need to explore calculating productivity in a larger and more meaningful perspective, one that includes protection of workers' health.

The Cost of Cancer Surveillance

> . . . There is now the urgent problem of surveillance and management and treatment of people who [are] at increased risk of developing cancer.
>
> **Dr. Irving Selikoff**

> . . . It could cost as much as $54 billion to provide warnings and health surveillance services . . .
>
> **The New York Times, 3 October 1977**

An area that needs attention is the notification of workers who were exposed to toxic materials in the past but may not know it. Dr. Irving J. Selikoff said in an interview in the November–December 1977 *EPA Journal,* page 8:

> . . . Both in the workplace, and in the environment in general, there is now the urgent problem of surveillance and management and treatment of high risk groups—people who inadvertently were exposed in the past to agents which we now know places them at increased risk of developing cancer in the future. At present, there is little surveillance or care for them. I consider this a social lapse and I strongly urge that attention be devoted to this as rapidly as is possible.

However, this could cost a great deal of money. According to an article in *The New York Times,* page 1, 3 October 1977:

> In discussing the costs involved, according to one Government analysis circulating within the Carter Administration and Congress, it could cost as much as $54 billion to provide warnings and health surveillance services to the 21 million Americans now believed to be exposed to harmful conditions while working.

Yet this cost must be offset against the cost to society of treating occupational disease. Each year, 3–5 billion dollars is spent on treatment for cancer victims and about 12 billion is lost in wages. Early diagnosis can cure or prevent the spread of many cancers.

Do you feel that the government should attempt to ferret out and warn all workers who have been exposed to toxic materials in the past and who may develop a disease from this in the future? (Remember how much this will cost you in increased taxes. Remember, too, that there are savings to be gained from decreased health and welfare costs.) Do we owe these people treatment as well as warnings?

Can you think of benefits to yourself from keeping track of the health of these people?

The Economics of Protecting Workers

> When the pollution-oriented health administrators and the public alike begin to focus clearly on the enormity of the bill that would be required to reduce pollution to meet unnecessarily severe standards . . . then will come the day of reckoning . . .
>
> **H. E. Stokinger, Chief,**
> **Laboratory of Toxicology and Pathology,**
> **National Institute for Occupational Safety and Health**

> . . . liability for asbestos-related deaths of workers will exceed $38 billion . . . more than the combined book value of the major asbestos defendants and 51 insurance companies involved . . .
>
> **The New York Times,** *March 9, 1982*

> But we paid dear and we will pay dear.
>
> *Billie Walker, widow of asbestos worker*

The price of environmental protection is a recurring question. How much are we willing to pay? H. E. Stokinger presents one view:

> When the pollution-oriented health administrators and the public alike begin to focus clearly on the enormity of the bill that would be required to reduce pollution to meet unnecessarily severe standards . . . precipitously prepared from undigested, dubiously related facts . . . on which the public has been ill-advised or misled . . . then will come the day of reckoning and rude awakening to the folly of past antipollution actions. Already industry has felt the bite; shortly, the public will. Hardest hit are the mineral and chemical industries. On top of multimillion-dollar outlays for air pollution control, and sums of similar magnitude for water, are multibillion-dollar legal suits that stagger the imagination, cripple large industry, and eliminate small industries. Two consequences of profound economic importance are the increased price of basic chemicals and the loss of employment. Already a number of small manufacturing plants have been forced to close, unable to bear the burden of meeting pollution standards. Heavy industry, unable to survive on repeated annual financial losses or to continue on less than a 4 to 6 percent profit margin, will ultimately pass the needless charge on to the consumer.
>
> It thus should be evident that such actions, with their unbearable consequences, should only be taken when it is clear beyond a shadow of scientific doubt that human health is in imminent danger.★

Until a few years ago, there was an asbestos plant in Tyler, Texas. Workers there were exposed to asbestos dust from 1954 to 1972. Although public health officials told plant managers that the plant violated health standards for working with asbestos, the management did not install the necessary safety equipment. Neither the public health officials nor the plant management told the workers of the dangers involved. Some 25–40 of the 900 workers at the plant have died from breathing the asbestos dust, and the death toll may reach 200. The remaining workers sued the plant owners and the government. A 20-million-dollar settlement was agreed upon, to be provided by the owners, the government, and the asbestos suppliers.

> William Morris, a 50 year-old former asbestos worker, is afraid that he is dying. And he is angry, because for years neither his employer nor visiting Federal health inspectors warned him that clouds of dust he was sucking into his lungs at work caused cancer.
>
> "I may live for another six months," he said, panting heavily as he sat in his darkened living room, a shoe box full of pain killers and other drugs at his feet.
>
> "It was pretty damned dirty of them not to let us know" . . .
>
> Reports of the Tyler settlement have reached many of the workers and their families. "It sounds like a whole lot of money," said Billie Walker, a 51-year-old widow whose husband worked at the plant for almost 11 years before dying in November of 1973. "But we paid dear and we will pay dear," she added as she sat at her kitchen table in Tyler with her two teen-age sons.†

We might take the viewpoint that industry (and society) will have to pay in one way or another: either for pollution control or for the consequences of not controlling pollution. Compare the ways in which the 20 million dollars involved in this case were spent and could have been spent.

Suits brought by former asbestos workers have been increasing yearly. In 1982, some 10,000–12,000 cases were pending on behalf of over 20,000 workers. Dr. Irving Selikoff, a medical expert on the effects of asbestos exposure, estimates that as many as 9 million workers have been exposed to significant levels of asbestos. The liability as a result of asbestos-caused deaths could total more than $38 billion over the next 20 years. Beleaguered asbestos and insurance companies question whether court suits are the right way to solve such an enormous problem. In addition, because of normal delays in the judicial system, many, if not most, asbestos workers die before their cases are settled.

> [The] tort recovery system is running a little wild, enormously large businesses are threatened by it and a tremendously large number of people must be compensated. That may be a societal problem rather than a legal problem.‡

Do you agree that private companies cannot be expected to handle a problem this large? Should Congress provide a remedy? If so, of what sort?

---

* H. E. Stokinger, *Science,* **174** (12 November 1971), 662.
† *The New York Times,* 20 December 1977.
‡ Dennis Connolly, Senior Counsel, American Insurance Association, in *Business Week,* April 13, 1981, p. 166.

Radiation: Microwaves, Radiowaves

During the years 1953–1976, the U.S. Embassy in Moscow was subjected to low-level microwave irradiation. Why the Soviets chose to irradiate the embassy is still unclear. Possibly the intent was to cause some sort of neurophysiological condition. Certainly there was an unsettling mental effect on embassy personnel. However, as far as current medical methods have been able to determine, the effect was due more to a fear of what microwaves might do than an actual physical effect. We know that radiation can affect human health. However, we also know that the kinds of effects seen depend on what sort of radiation is involved and how high the dose is.

The Electromagnetic Spectrum

The whole electromagnetic spectrum is shown in Figure 30.6. As you can see from this diagram, there are many kinds of radiation. They range from radiation of very long wavelengths, such as that used in power generation, all the way to radiation of very short wavelengths, such as x-rays and cosmic rays. The wavelengths visible to the human eye are also part of the electromagnetic spectrum, but they are only a very small portion of the entire range.

The health effects of radiation depend very much on the wavelength involved. The effects most commonly associated with the term radiation—that is, radiation poisoning and the various x-ray- or atomic-bomb-caused cancers—are caused only by the shorter

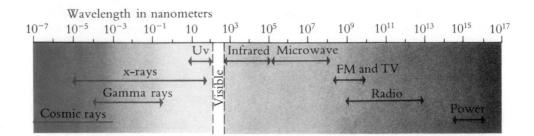

Figure 30.6
The Electromagnetic
Spectrum.

wavelengths. These types of radiation are known as **ionizing radiation.** In contrast, the longer wavelengths, from the ultraviolet region to the power generation region, are called nonionizing radiation and their health effects are quite different.

Microwaves are in this nonionizing range, while x-rays, gamma rays, and cosmic rays are ionizing. The effects of nonionizing radiation will be covered first and the effects of ionizing radiation after that. Finally, the health effects of ultraviolet rays, mostly skin cancers, are detailed in the section on ozone.

Biological Effects of Nonionizing Radiation

Nonionizing radiation can cause thermal motion of the molecules in living tissue. This results in a temperature rise in the tissue and may lead to harmful effects such as burns, cataracts, and birth defects. There is also the possibility that complex biological systems, such as those existing in cell membranes, may be disrupted. Systems such as cell membranes depend on an orderly arrangement of molecules for proper functioning. Thus, nonionizing radiation could cause effects over and above those caused by a simple increase in temperature. The experimental evidence for such effects, however, is incomplete.

Most experimental results on nonionizing radiation concern radiofrequency wavelengths. These results show that doses above 100 milliwatts per square centimeter produce definite thermal damage, as well as cataracts in the eye. Between 10–100 milliwatts/cm^2, changes due to heat stress, including birth defects, are seen. From 1–10 milliwatts/cm^2, changes have been noted in the immune system and the blood–brain barrier. In the range of 100 microwatts per square centimeter and 1 milliwatt/cm^2, almost no confirmed effects have been found.

In general, it appears that the immediate effects of nonionizing radiation, such as the effects of excess heat on a body tissue, are the important ones. No evidence exists for delayed genetic or nongenetic effects, such as cancer. In the Moscow embassy, microwave radiation reached a maximum of 18 microwatts/cm^2 and no effects on the personnel there could be found.

Microwaves and Public Health Standards

Against this lack of apparent effect due to low-level microwave exposure, must be set the fact that the growth in use of microwaves is at least 15% per year. Besides their use in microwave ovens, microwaves are used in radar and as transmission links for TV, telephone, and telegraph. The United States does not have a standard for exposure to nonionizing radiation, although OSHA has recommended that workers be exposed to no more than 10 milliwatts/cm^2. The Soviet Union has set a standard of 1 microwatt/cm^2 for the general public.

Industrial workers involved with heating, drying, and laminating processes may be at some risk, as may technicians working on live broadcasting and relay towers and military service personnel. Worker compensation suits have been filed charging that microwaves have contributed to disabilities, and in at least one case, the board ruled in the worker's favor.

Meanwhile, as the number of sources of microwave exposure increase, concern about the exposure of the general public has been growing. Construction of a microwave TV transmission antenna atop New York's World Trade Center was halted when engineers realized that it would subject some office workers, as well as tourists atop the building, to nonionizing radiation in the range of 360 microwatts/cm^2. The Coast Guard was denied permission to construct a microwave transmission tower as part of a vessel-traffic monitoring system in New York Harbor due to public concern about the safety of microwaves.

Government agencies and public interest groups would like to see standards set for the various forms of nonionizing radiation. Many industrial groups would like to see federal standards both to help them in designing equipment and also to prevent the growth of a patchwork of local ordinances.

Radiation: x-Rays, Gamma Rays, and Particles

Kinds of Ionizing Radiation

Looking at the electromagnetic spectrum again (Figure 30.6), we can see that the shorter-wavelength end of the spectrum is made up of x-rays, gamma rays, and cosmic rays. These rays possess enough energy to free an electron from the atom of which it is a part. This leads to the formation of ions (which is why these types of radiation are called ionizing radiation). The eventual effects of ionizing radiation on living cells are due to the formation of these ions. Besides electromagnetic rays, certain types of particles produce ions.

The nuclear decay of unstable elements gives rise to particles and rays. The unstable elements, which we refer to as **radioactive,** emit alpha particles, beta particles, and gamma rays. Gamma rays have the greatest penetrating ability of any radiation from radioactive decay. They may pass through several centimeters or more of lead without a significant weakening of energy. Those who work near substances that emit gamma radiation must exercise the greatest caution to limit their exposure.

x-Rays are commonly used to examine the internal structure of the human body in order to diagnose and treat disease or injury. Physicians and dentists regard x-rays as one of their most important tools for diagnosis. Like gamma rays, x-rays are highly penetrating; several centimeters or more of lead are necessary to block these beams.

Cosmic rays, which consist of both particles and electromagnetic radiation, are constantly bombarding us from outer space. Although a portion of cosmic rays can be blocked out by several thicknesses of lead, another portion penetrates even into the deepest mines. The intensity of cosmic rays increases at higher altitudes, so much so that jet crews might one day

conceivably be classified as radiation workers. Cosmic-ray intensity also increases as one moves toward the polar latitudes.

Sources of Human Exposure to Ionizing Radiation

Radioactive elements exist naturally in the earth's crust. They are more concentrated in some places than in others. Nonetheless, since the era of the atom bomb and the beginning of nuclear electric generation, the sources of exposure to radioactive elements have multiplied.

Human exposure to ionizing radiation comes via one or more of these routes: decay of radioactive elements, x-rays, and cosmic rays. We measure exposure to radiation most commonly in "rems" and "millirems" (one one-thousandth of a rem), units reflecting both the intensity of the radiation and its effect on human tissues. Radiation standards are also set in rems and millirems. If we exclude radiation from medical x-rays and other man-made sources, we have what is referred to as natural background radiation. This is the radiation we would receive if there were no exposure to man-made radiation.

Background radiation exposure in the United States is most commonly in the range of 100–150 millirems per year. Leadville, Colorado, at two miles (3.3 kilometers) above sea level, has one of the highest annual radiation exposures in the U.S. at 160 millirems, due to high intensity of cosmic rays reaching the city. Radioactive elements in the earth's crust, such as potassium-40 (a form of potassium) and radium, are also contributors to background radiation.

x-Rays ordered by physicians and dentists are a common source of exposure to ionizing radiation. Although such exposure has been estimated at 90 millirems per year, this is an average figure. Many individuals receive far higher doses and some none at all. The annual chest x-ray for tuberculosis is now thought to be not such a good idea. Much older x-ray equipment is still in use and delivers radiation doses far higher than necessary.

During the era when atomic and hydrogen bombs were tested in the atmosphere, radioactive elements were scattered around the globe in clouds of particles. Rain washed these particles from the atmosphere in "fallout." The fallout in areas very near atomic explo-

Women's Rights and the Workplace

> If American Cyanamid . . . can get away with removing women of childbearing age from these jobs, we will have established the principle of altering the worker to the configuration of the workplace instead of altering the configuration of the workplace to protect the worker.
>
> *Anthony Mazzocchi, Vice President,*
> *Oil, Chemical, and Atomic Workers Union★*

A number of companies have started to exclude or remove women workers of childbearing age from jobs where they might be exposed to teratogenic substances (substances that cause birth defects). In one case, four women workers at American Cyanamid claimed they had themselves sterilized because talks with company officials led them to believe they might lose their jobs otherwise. Although some companies are transferring women to less hazardous positions with salaries equal to their former jobs, in other cases women allege they have been discharged or offered only lower-paying positions.

Anthony Mazzocchi, Vice President of the Oil, Chemical, and Atomic Workers Union, notes that the reproductive capacity of men as well as women can be affected by many hazardous substances (e.g., the pesticide DBCP, which can cause sterility in males). Furthermore, lead workers, asbestos workers, and men who worked at a Kepone plant have brought home enough dust in their work clothes to affect other members of their families.

Dr. Karrh, corporate Medical Director at DuPont, agrees that cleaning up workplaces to make them safe for everyone would be best, but:

. . . first of all it is not technically and economically feasible to clean it up to a safe level. And, second, we don't have the data to know what is a safe level.★

Sue Nelson, Director of the Office of Policy Analysis for the Occupational Safety and Health Administration, adds:

What this is really about is that the employers are trying to save themselves from expensive lawsuits . . . It is easier for a woman worker than a man to bring a lawsuit against a company on behalf of a fetus.★

Do you think women able to bear children should be excluded from jobs involving exposure to hazardous chemicals? What about men able to father children? Supposing there is some doubt about whether a chemical is teratogenic, who should make decisions (the company, the government, the worker), or should everyone be protected in any case? Suppose protective measures would cost so much that the company might not produce the product?

★ All quotes are from *The New York Times,* p. A-1, 15 January 1979.

sions, such as in the Pacific Ocean islands, was large enough to leave measurable levels of radiation in the soil, but the minor amount of continuing fallout is negligible when compared to background radiation exposures. These exposures are summarized in Figure 30.7.

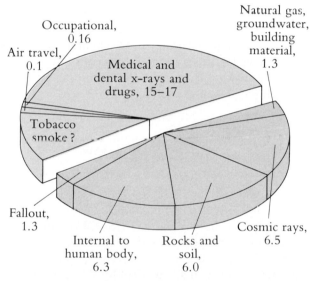

Figure 30.7 Estimates of Exposure to Ionizing Radiation (measured in million person-rems per year, where one rem to 1 million people in a year equals one million person-rems per year). On the average, about one-half of human exposure to radiation is due to natural environmental sources. One-third of this natural background is from cosmic rays, 1/3 from natural radioactive materials in soils and rocks, and 1/3 from radioactive materials (such as potassium-40) which are incorporated into the human body. People using groundwater for drinking or using natural gas are exposed to additional natural radon from these sources. Further, certain building materials (notably stone, brick and plaster board made from phosphogypsum) emit some radiation from naturally occurring radioactive materials. Of the other sources of radiation, medical x-rays and radioactive drugs are the most significant, nearly equalling natural background. Air travel, which increases cosmic ray exposure, is significant for some people, as are the radioactive particles in tobacco smoke. (Data from: *Pilot National Environmental Profile,* US EPA, 1980)

Radiation Exposure and the Nuclear Power Industry

Exposure to radiation from radioactive elements is a concern that has grown with the arrival of the nuclear electric power industry. Routine discharges of low levels of radiation from electric power plants were allowed in the first 15 years of the nuclear industry. The older nuclear plants were to be operated so that no individual in the general population would receive more than 500 millirems of radiation per year, and the average exposure of people away from the site would be kept below 170 millirems per year. Such operation reflected the then-current standards for exposure of people in the general population. In the middle 1970s, the Atomic Energy Commission came under attack for these standards and responded with a tightening of the allowed emissions of radioactive elements from nuclear plants. The largest annual allowable dose was reduced from 500 millirems to 5 millirems and the average dose to less than 1% of natural background, or 1 millirem.

Under these new exposure levels, the normal operations of nuclear power plants are quite clearly not a significant concern; yet radiation remains a problem. One remaining source of radiation exposure involves fuel reprocessing plants (see Nuclear Power, Chapter 18). Another source is the worked and discarded uranium ore (tailings). Of greater concern, however, is the possibility that terrorists or warring nations could hijack spent reactor fuel or fissionable material and build nuclear weapons—an atom bomb from fissionable material such as plutonium, or a dispersal device that spreads the deadly contents of a spent fuel rod. The detonation of these weapons could bring exposure to radiation on a massive scale.

Finally, mention should be made of the fact that a number of radioactive elements can be concentrated in food chains **(biologically magnified).** An example is phosphorus-32, which was found to be concentrated as much as 5000 times above levels in the water in whitefish downstream from the Hanford Atomic Power Plant on the Columbia River. Blue gills and crappies in the same river were found to contain 20,000–30,000 times the concentration in water of phosphorus-32, while filamentous algae were found with up to 100,000 times the water level of phosphorus-32. Zinc-65, iron-59, and iodine-131 are other examples of radioactive elements concentrated by living organisms. These are sources of radiation exposure to the general public when certain foods are eaten. Fish may contain phosphorus-32; oysters and clams, zinc-65; and milk, iodine-131. Unfortunately, not enough is known about the paths such elements take through food chains to estimate the hazards involved.

Biological Effects of Ionizing Radiation

The effects of radiation on the health of people alive today can be divided into two categories. The acute symptoms that result from intense short-term exposures occur within days or weeks. Such exposures are very unlikely except as a result of a nuclear war or a severe, accidental occupational exposure. The effects of long-term, low-level radiation exposures are likely not to be seen for a number of years. These delayed symptoms cannot be distinguished from the common diseases of aging, especially cancer. We know that ionizing radiation can cause breast cancer, thyroid cancer, leukemia, lung cancer, gastrointestinal cancer, and bone cancer. These diseases have been observed in individuals exposed to radiation levels of 100 rems or more in an accident or catastrophe such as the atomic bomb explosions at Nagasaki and Hiroshima.

Although radiation can induce cancer, a person who has cancer cannot ordinarily point to radiation as the cause. Radiation hazards to which the general public or workers are exposed are generally low-level, 0.1–5 rems. The cancers caused by radiation are not caused only by radiation, but are caused by other agents as well. For these reasons, in order to prove by ordinary statistical methods whether or not certain cancers are caused by these low levels of radiation, very large numbers of exposed people would have to be studied. These numbers would be on the order of hundreds of thousands of people. It is usually not possible to find enough exposed people or enough cancers of a specific type to use epidemiological studies as proof of the harm caused by low-level radiation. Some investigators have examined the available information with more complicated statistical methods,[1] but their results are not accepted by most other scientists.

Instead, the results of exposure at doses over 100 rems are used to estimate the probable effect of lower doses. There is a heated argument among radiation

specialists about how to extend the data from high doses to predict low-dose effects, and whether there is a threshold below which no effects occur. This is similar to the arguments about the effect of low doses of chemical carcinogens explained in Chapter 29. Most scientists agree that there is no proof of a threshold level, especially for certain types of radiation such as alpha particles.

In addition to cancers, radiation might be expected to cause genetic damage; that is, mutations that can be passed on to future generations. Evidence from animal and cell culture experiments lead scientists to believe such damage is likely. However, no such effects can yet be demonstrated in humans, even those exposed to atomic bomb blasts. This is probably due to the low average dose to the gonads of parents (50 rems) and also to the small population (78,000 children to parents who survived Hiroshima and Nagasaki). It is estimated that natural background radiation (about 100–150 millirems) causes perhaps 0.1–2% of genetic disease.

Should Radiation Standards Be Made More Strict?

On the basis of population studies of atomic energy workers and others exposed to low-level radiation, some experts argue that the standard for occupational exposure of 5 rems per year should be cut to 0.5 rems, or 500 millirems, per year. A recommendation for such a standard was made in 1978 by a Committee of the National Academy of Sciences. The recommended level is only about twice the level of a chest x-ray exam.

Nearly 5000 of the 70,000 atomic energy workers were exposed to levels greater than 2 rems in 1977. In addition to the employees of nuclear power plants, the nation has about 170,000 x-ray technicians. Although they should be limited to less than the occupational standard of 5 rems per year, their actual exposures are largely unknown.

Uranium Mine Tailings: An Unnecessary Hazard

Uranium miners are exposed to radiation on their jobs and apparently, as a result have experienced a high rate of lung cancer. The radiation to which the miners are

---

1 See the reference by Land for more about this problem.

exposed comes from the natural radioactive decay of uranium. After a number of steps, beginning with the initial uranium decay, a radioactive gas known as radon is produced. Atoms of the element radon are dispersed thoroughly in the air of the mines because radon is a gas.

The decay of radon results in new radioactive elements in particle form. Since the particles are only single atoms, rather than being of any size, they remain suspended in the air for a long while. This contaminated air is inhaled by the miners, and the radioactive "daughters" of radon are deposited in the miners' lungs. Masks or filters are ineffective in capturing particles consisting of a single atom. Hence, the only way to control exposure of the miners is to replace the air in the mine with fresh air on a frequent basis.

Miners are not the only people exposed at this step in the atomic fuel cycle. Wastes, called **tailings,** remain after the uranium oxide has been extracted from uranium ore. This rubble still contains some uranium that could not be removed. Since radon is a product of uranium decay reactions, the air in the vicinity of the rubble also contains radon gas at levels up to 500 times the natural background. The air near the tailings is likewise contaminated with radon "daughters."

Through oversight and a lack of awareness of the hazards, tailings have been used to make concrete for homes and buildings. Rubble from the Old Climax Mill in Colorado is about 0.03% uranium; it was used as building material and as landfill for homes in the town of Grand Junction, Colorado. The tailings were not only used to level land surfaces; they were also used in mortar, concrete, and backfill around basement walls.

Levels of radon decay products in the air of these homes and buildings are up to five times the concentration permitted for uranium miners. Hence, the exposure of people who live in these homes to a potential cancer-causing agent has been substantial. A number of homes have had foundation materials, even walls and chimneys, replaced and backfill removed and replaced; the federal government and the state of Colorado assisted with the costs of these repairs.

Most of the deposits that are rich enough in uranium to be mined are found in the western portion of the United States. The states of New Mexico, Wyoming, Colorado, and Utah are the principal suppliers of ore, and here mounds of tailings may still be found. When milling operations moved on to more radioactive pastures, the piles of tailings were simply abandoned. In this way, piles of tailings fell into private hands. Often, the new owners were ignorant of the dangers of the material. In eight states, such piles still exist, some very near large communities. The smallest such pile is only 2 acres (0.8 hectares) in area; the largest is 107 acres (43 hectares) in extent. One pile, next to an abandoned mill, is within 30 blocks of downtown Salt Lake City, Utah. In all, the Western states are known to have 23 uranium tailings piles over which no control is exercised. The mills have closed; the piles remain (Figure 30.8).

Controlling the problem is difficult. Several feet of soil can be deposited over the piles and vegetation established. This prevents the wind from lifting the dust-

Figure 30.8 Radiation physicist standing on a 250-foot (76-meter) pile of uranium tailings. The city of Durango, Colorado, is seen clearly on the other side of the pile. (Photo courtesy of Oak Ridge National Laboratory Review)

like particles into the air. However, it is thought that 10–20 feet (3–6 meters) of earth will be necessary to prevent the escape of radon gas. This suggests the immediate need to bury the tailings that are currently being produced from continuing mining and milling operations.

Ultraviolet Light and Chlorofluorocarbons

Ozone: An Essential Gas

High above the earth, in that part of the atmosphere called the stratosphere, is a relatively little-known gas essential to life on earth. That gas is **ozone.** Each molecule of ozone is made up of three atoms of oxygen. Ozone in the stratosphere absorbs over 99% of the ultraviolet light that comes from the sun. These ultraviolet rays are often welcomed because they cause skin to tan. They can also cause it to burn and, over long periods of time, ultraviolet rays cause skin cancers.

High concentrations of ultraviolet rays are harmful to many forms of plant and animal life. They are sometimes used to sterilize objects because, in high concentrations, they kill bacteria. Scientists generally believe that life on land did not develop until the earth's protective ozone layer formed. Because ozone in the stratosphere is essential to life, reports that a number of human activities can destroy ozone are understandably alarming.

Threats to the Ozone Layer

Fears about the possible effects of supersonic transport planes (SSTs) on the ozone layer were a strong factor in the U.S. decision to stop development of a fleet. These planes fly in the stratosphere and release two contaminants, water and nitrogen oxides, that can destroy ozone. However, high costs have slowed the growth of supersonic travel to a point where it is not now considered a serious hazard to the ozone layer. On the other hand, newer subsonic planes are designed to fly in the stratosphere, where thinner air can lead to more miles per gallon of fuel. These planes could become a significant problem in the future.

In 1974, a more serious threat was uncovered. Ozone is known to be destroyed by chlorine in the atmosphere, and a source of chlorine had been discovered. Compounds called chlorofluorocarbons (or fluorocarbons), widely used as propellants in spray cans and in refrigerators and air conditioners, were found to be slowly diffusing into the stratosphere. These compounds do not react easily with other materials. For this reason, they are ideal for use in spray cans. However, it has been found that they are broken down by ultraviolet light, with the release of chlorine. This chlorine can then break down ozone. How much ozone are the chlorofluorocarbons destroying? Unfortunately, there are many things we still don't know about ozone and the stratosphere. A 1982 report by the National Academy of Sciences predicts a 5–9% decrease in stratospheric ozone by late in the next century if fluorocarbons are released at about the rate they were in 1977.

To this must be added the effects of other substances that destroy ozone. Nuclear explosions in the stratosphere destroy ozone by the release of nitrogen oxides. Thus, in the event of a nuclear war, the resulting increase in ultraviolet rays could be as serious a problem as nuclear fallout. Automobile exhausts and fertilizers in soil release nitrogen oxide, which can react with ozone. The use of fertilizers is increasing yearly as we try to feed an increasing world population. Thus, fertilizers may come to pose a serious threat to the ozone layer. Bromine is known to destroy ozone in the laboratory. As methyl bromide, it is widely used as a fumigant in agriculture. How much bromine escapes to the stratosphere is not yet known. Similarly, the effect of quantities of industrial chemicals such as carbon tetrachloride and methyl chloroform is not clear, although it is felt that they could release significant amounts of ozone-destroying chlorine.

Effects of Decreasing the Ozone Layer

The average concentration of ozone in the stratosphere is probably about three parts per million. However, ozone concentrations vary geographically. In addition, ozone concentration can vary, normally, by as much as 30% from day to day. Over a period of years, the variation averages out to 10% due to what seems to be a natural balance between the formation and destruction of ozone. There are two major concerns about the additional decreases caused by human activities.

Is Unilateral Regulation Similar to Unilateral Disarmament?

A nation already burdened with rising costs should not take another costly regulatory step for insufficient gain.

Editorial, The New York Times, *October 29, 1980*

If we wait until 1990 to make the decision it could be too late.

Steven Jellinek, Assistant Administrator for Toxic Substances,
U.S. Environmental Protection Agency

In 1973, the U.S. accounted for half of all fluorocarbons released into the atmosphere. As a result of regulations banning almost all uses of fluorocarbons as propellants in spray cans, the U.S. now releases only 1/3 of the world total. There are no good substitutes for fluorocarbons as refrigerants, however, and increasing use of these compounds in refrigerators and air conditioners in the U.S. and worldwide threaten to offset gains from the ban on fluorocarbons as spray-can propellants. U.S. EPA officials are considering, as one option, further limiting U.S. uses of fluorocarbons (in refrigerators and air conditioners, foam insulation, industrial solvents, and fast-food freezing) to 1979 levels. This would probably lead to small price increases for such items as air conditioners, refrigerators, and foam insulation, and might hurt sales of U.S.-made fluorocarbon-containing products abroad.

In addition, the rest of the world does not see the problem as clear-cut. Great Britain's Department of the Environment, for instance, says that "strict regulation is not warranted at present." The European Common Market has asked member countries to achieve a 30% reduction compared to 1976 levels. Combined figures for India, Argentina, and Communist countries show an increase in production of fluorocarbons through 1980.

Should the U.S. set an example environmentally and unilaterally put the lid on fluorocarbon use? Should we seek international cooperation? If we don't get cooperation, should we go ahead anyway? Take into account economic effects as well as the size of U.S. contributions to the problem, scientific uncertainty, and the possible results of nonregulation.

A decrease in the ozone layer would allow more ultraviolet rays to reach the earth, which would almost certainly increase the occurrence of skin cancer. (More about skin cancer is in the supplemental material at the end of this chapter.) It seems quite certain that a proportion of skin cancers are caused by exposure to sunlight over long periods of time. Fair-skinned people are much more apt to develop skin cancers, since the pigments in dark skin help to screen out ultraviolet rays. People living in tropical climates are more subject to skin cancer, apparently because they are exposed to more ultraviolet light. Estimates are that each 1% decrease in the ozone layer will cause a 2–5% increase in basal cell skin cancer, and a 4–10% increase in the less common but more serious squamous cell skin cancer. Ultraviolet-caused skin cancer is almost always curable, although the treatment is not pleasant. Plants and animals also show sensitivity to ultraviolet rays. An increase in ultraviolet light could be expected to cause the extinction of some microscopic life forms and to damage other species or decrease the living space available to them.

The second concern is that changes in the stratosphere might cause changes in climate. Increases in ultraviolet light may possibly increase the melting of polar ice, with resultant flooding. Chlorofluorocarbons, in addition to effects they have on ozone, absorb infrared radiation. This could trap heat in the atmosphere and cause a warming of the earth. To complicate the issue further, carbon dioxide accumulating in the atmosphere (see Chapter 19) may lead to a warming of the atmosphere near the earth's surface but a cooling of the stratosphere. This would, in turn, slow the rate of ozone destruction. At the present time, there is a great deal of uncertainty about the final effect of chlorofluorocarbons and changes in the ozone layer on climate. Predictions actually range from a new ice age to temperature increases of up to 1°C. What should be remembered is that life on earth has evolved under the present set of conditions. Lasting changes in one direction or another will have profound effects on life as we know it.

How the Future Looks

Because chlorofluorocarbons are believed to be a serious threat to the ozone layer, U.S. government agen-

Figure 30.9 Balloons such as this are used to carry aloft equipment for measuring ozone concentrations in the stratosphere. (Photo courtesy of NASA)

cies have decided to prohibit unnecessary uses.[2] The use of chlorofluorocarbons as spray-can propellants was outlawed after 1979. Even with controls, the quantity of chlorofluorocarbons in the stratosphere will continue to increase for some time because those compounds already released are still around and are diffusing very slowly upward. Remember, they are not broken down except by ultraviolet light in the stratosphere. Although the evidence is not conclusive, scientists believe they are beginning to see a small but measurable decrease in the ozone layer (Figure 30.9).

Because the U.S. is not the only nation to use fluorocarbons, however, the problem involves other countries as well (see Controversy 30.5).

---

2 A few drug products, certain pesticides, some aircraft, and some electrical maintenance products are exempt.

Skin Cancer

Skin cancers are of three types. The two most common types, basal cell and squamous cell carcinoma, rarely spread to other parts of the body. The rate of cure is 95% or greater by chemotherapy (using drugs such as fluorouracil), surgery, x-rays, or cauterization. The third type of skin cancer, melanoma, does spread rapidly to other parts of the body. It is a much less common type, however, and is not as strongly linked to exposure to sunlight. For instance, the rate of occurrence of melanoma is 75% greater in southern states than in states on the Canadian border, while the rate of the other two types of cancer is 250% higher.

The rate of skin cancer in the U.S. is reported to be 300,000–400,000 cases of basal cell carcinoma and 100,000 cases of squamous cell carcinoma per year. Each 1% decrease in ozone is believed to cause a 2% increase in ultraviolet radiation, and as much as a 2–5% increase in skin cancer. Proportionately then, each 1% decrease might cause 10,000–30,000 extra cases of skin cancer each year in the United States.

Questions

1. List the ways you, personally, might be exposed to asbestos fibers. If you can only think of one or perhaps none, make up a hypothetical person and list as many sources as possible from which that person might be exposed to asbestos dust.
2. How would you prevent the person in question 1 from breathing so much asbestos dust? (You can list ways to avoid the dust or to avoid breathing it.)
3. What sort of evidence exists that some jobs carry an increased risk of developing cancer or reproductive defects?
4. Should the benefits of certain products be balanced against risks to the workers who produce them? Give an example. Where does the concept of personal freedom fit in with your answer?
5. Explain the differences between ionizing and nonionizing radiation. Note which parts of the electromagnetic spectrum are ionizing and which are nonionizing, and what different biological effects they cause.
6. What are the three most common sources of ionizing radiation?
7. What is meant by natural background radiation?
8. How do average radiation releases from nuclear power plants compare with background radiation levels in the U.S.?
9. Suppose there were a continuing low level of radiation exposure in a particular industry. How would you suggest health authorities determine if the exposure level is safe?
10. Suppose the frequency of leukemia in male workers in a particular industry with radiation exposures were three times that for males of the same age in the general population. Could a worker who has cancer prove that the cancer was the result of his exposure on the job? Why?
11. Why is the ozone layer important?
12. What are the major threats to the ozone layer?
13. Will the human race save itself by accident? It has been predicted that several pollutants will affect the earth's climate in opposite directions. Perhaps the effects will all cancel. Briefly contrast the possible effects of a decrease in the ozone layer, an increase in the carbon dioxide level, and an increase in the level of particulates in the air. How much of a gambler are you?

Further Reading

Radiation

Upton, Arthur C., "The Biological Effects of Low-Level Ionizing Radiation," *Scientific American,* **246,** No. 2 (February 1982), p. 41.

A clear explanation of what is known about ionizing radiation and its biological effects. Very good if you would like to know more about the actual cellular events leading to radiation-caused cancers, etc.

Marshall, Eliot, "New A-Bomb Studies Alter Radiation Estimates," *Science,* **212** (22 May 1981), p. 900.

Marshall, Eliot, "Japanese A-Bomb Data Will Be Revised," *Science,* **214** (2 October 1981), p. 31.

These two articles detail some of the ongoing controversy about radiation standards and their safety.

Ultraviolet Light and Chlorofluorocarbons

Maugh, T. H., "New Link Between Ozone and Cancer," *Science,* **216** (23 April 1982), p. 396.

Occupational Health

"OSHA on the Move," *Environmental Science and Technology,* **11**(13) (December 1977).
> A good history of the problems of regulating toxic substances in the workplace up to this date.

Asbestos

Selikoff, I. J., and D. H. Lee, *Asbestos and Disease.* New York: Academic Press, 1978.
> A comprehensive and readable treatment of the asbestos problem written by a public health authority who is probably the foremost expert in the field.

"Asbestos and Home Improvements," *Environmental News,* EPA Region I, June–July 1981.

"Asbestos Guidance to Schools," *Environmental News,* U.S. EPA, Office of Public Awareness, Washington, D.C. 20460, March 16, 1979.
> These two EPA reports cover asbestos problems most likely to affect the general public.

References

Radiation

Koslov, Samuel, "Radiophobia: The Great American Syndrome," Johns Hopkins Applied Physics Laboratory Technical Digest 2, No. 2 (1981), p. 102.

Land, Charles E., "Estimating Cancer Risks from Low Doses of Ionizing Radiation," *Science,* **209** (12 September 1980), p. 1197.

"Ionizing Radiation and Its Health Impact: 1980," from *Pilot National Environmental Profile,* U.S. Environmental Protection Agency (1980).

Carter, L. J., "Uranium Mill Tailings: Congress Addresses a Long-Neglected Problem," *Science,* **202** (13 October 1978), p. 191.

Genetic and Cellular Effects of Microwave Radiations, U.S. EPA, Health Effects Research Laboratory, May 1980, EPA-600/1-80-027.

Ultraviolet Light and Chlorofluorocarbons

Causes and Effects of Stratospheric Ozone Reduction: An Update, National Academy Press, Washington, D.C., 1982.

Occupational Health

Higginson, John, "Proportion of Cancers Due to Environment," *Preventive Medicine,* **9** (1980), p. 180.

Identification, Classification and Regulation of Potential Occupational Carcinogens, Federal Register, January 22, 1980.

Asbestos

Workplace Exposure to Asbestos, U.S. Department of Health and Human Services Publication #81-103, November 1980.

Peterson, D. L., et al., "Studies of Asbestos Removal by Direct Filtration of Lake Superior Water," *Journal AWWA* (March 1980), p. 155.

Hazards of Asbestos Exposure, Hearing before the Subcommittee on Commerce, Transportation and Tourism, Committee on Energy and Commerce, House of Representatives, 97th Congress, January 19, 1982.

CHAPTER THIRTY-ONE

The Voluntary Factors: Diet, Drugs, and Smoking

Smoking: A Personal Form of Air Pollution

The Problem/What's in a Puff?/The Health of Smokers/The Effects of Smoking on Nonsmokers/Smoking Laws: Should There Be More?/Smoking and Fires

Diet and Disease

Intentional Food Additives/Food, Drug, and Cosmetic Laws/Artificial Sweeteners and the Delaney Clause/Artificial Colors in Foods/Preservatives/The Special Case of Nitrates and Nitrites/Added Vitamins and Minerals/Unintentional Food Additives/Natural Toxins/Pesticide Residues in Food/Drugs in Animal Feeds/PBBs: Story of an Accidental Disaster/A Healthful Diet

Toxic Substances in Drugs and Cosmetics

Hair Dyes/The Case for Testing Before Marketing/Alcoholic Beverages/Weighing Risks and Benefits of Drugs

CONTROVERSIES:

31.1: ***Government Protection for An Illegal Activity? Marijuana Smoking***

31.2: ***What is the Government's Role in Controlling Smoking?***

31.3: ***Does Everything Cause Cancer?***

31.4: ***Protecting People From Themselves: Saccharin***

31.5: ***Life-Style and Disease: Health Foods***

31.6: ***Pitfalls of Regulation: Tris, Pajamas, and Children's Safety***

Cancer has been labeled by some experts an "environmentally caused disease." However, they are also quick to point out that we have a great deal of control over some of the environmental factors that cause cancer. The factor over which we have the greatest control is probably smoking. Fully 30% of U.S. cancer deaths are due to smoking. An enormous improvement in public health would thus result if no one smoked.

The role of diet in cancer is also believed to be large. Thirty-five percent of cancers are probably due to the kinds of food people choose, and a smaller percentage, perhaps 2–4%, are due to food additives. However, the role of diet in cancer is not as well understood as that of smoking. The following sections examine more closely the role of smoking, diet, and drugs in relation to health in general and cancer in particular.

Smoking: A Personal Form of Air Pollution

The Problem

Scientists studying the health effects of air pollution often decide that they must separate the people in the study into two groups: those who smoke and those who do not. The reason is that people who smoke are adding a variety of pollutants to the air they breathe. Some of these pollutants are already present in city air; some are new. It has been said that smoking is a personal form of air pollution. While the effects of breathing in some of the substances in tobacco smoke are known, the effects of others are not. Still, there is a growing acceptance in this country of the fact that smoking is hazardous to health. Over the past 15 years, surveys by the National Clearinghouse for Smoking and Health have noted that a smaller and smaller percentage of adult Americans smoke. In 1965, 51% of American men smoked. The rate has dropped to 36.9%. For women, the rate has declined from 32% in 1965 to 28.2% now.

A 1975 survey also noted that 61% of those who smoke have tried seriously, at least once, to stop smoking. In fact, 9 out of 10 people say they would stop if it were easier to do.

Scientific studies on the effects of smoking fall into two general groups. First, there are studies in which people who smoke are compared with people who do not smoke. People who smoke are found to suffer not only more lung diseases, such as lung cancer and emphysema, but also more heart disease.

Second, there are studies in which tobacco smoke is examined to see what chemicals it contains. These chemicals are then often given to animals to determine whether they are harmful. An examination of the makeup of tobacco smoke shows that it contains a number of chemicals known or suspected to cause cancer. Other harmful chemicals, such as lead and carbon monoxide, are also found in cigarette smoke.

What's in a Puff?

Major toxic components. Table 31.1 lists some of the major toxic components of cigarette smoke. The table is not complete, however. Many other compounds are found in smaller amounts. These may eventually be found to have serious effects.

Table 31.1 Major Toxic and Tumor-Producing Agents in Cigarette Smoke

In vapor phase

Carbon monoxide	toxic (heart disease)
Nitrogen oxides	toxic
Hydrogen cyanide	toxic
Acrolein	toxic
Acetaldehyde	toxic
Formaldehyde	carcinogenic
Hydrazine	carcinogenic
Vinyl chloride	carcinogenic
Urethane	carcinogenic
2-Nitropropane	carcinogenic
Quinoline	carcinogenic
Nitrosoamines	carcinogenic
Nickel carbonyl	carcinogenic

In particulate phase

Benzo(a)pyrene	carcinogenic
5-Methylchrysene	carcinogenic
Polonium-210 (radioactive)	carcinogenic
Cadmium	toxic

Source of data: 1982 Surgeon General's Report on the Health Consequences of Smoking

Carbon monoxide. People who smoke expose themselves to rather high levels of carbon monoxide. Carbon monoxide, which is also a pollutant in city air (Chapter 20), binds to the hemoglobin in blood. Hemoglobin so bound is incapable of carrying oxygen to tissues in the body. Hemoglobin binds 200 times more tightly to a molecule of carbon monoxide than to a molecule of oxygen. Thus, the body tissues in a smoker receive a lower-than-normal supply of oxygen.

The air-quality standard for carbon monoxide is determined by averaging carbon monoxide concentrations over an eight-hour period. This standard, 10 milligrams of carbon monoxide per cubic meter of air, is the level that should not be exceeded more than once a year. When we live in such a concentration year round, about 2% of our hemoglobin is tied up as **carboxyhemoglobin** (the combination of carbon monoxide with hemoglobin). Compare this percentage to that of a smoker (about a pack a day). This person may have further tied up about 6% of his hemoglobin. Thus, we can say smokers have greatly decreased the quality of the air they breathe.

How much of a problem is this? It probably depends on where they live and what they eat. Most cities already have a high concentration of carbon monoxide in the air. Concentrations are highest in traffic jams, underground garages, and tunnels. Levels in these areas can reach 70 milligrams per cubic meter or more. Someone breathing air at the 70-milligram level for eight hours would have a blood concentration of about 10% carboxyhemoglobin. At this level, tests have shown a decreased driving ability. People cannot respond as quickly to such stimuli as brake lights and changes in the speeds of cars. Since smokers have inactivated 6% of their hemoglobin due to the carbon monoxide in cigarette smoke, we can see that in certain situations they may well have decreased further their ability to act quickly when they need to be alert and quick.

Increased blood levels of carbon monoxide may also increase the risk of fatal heart attacks. This is explained later.

What does a smoker's diet have to do with the problem? Nitrites, which are used to cure meats such as hot dogs and corned beef, also tie up hemoglobin.[1] Nitrites react with hemoglobin to form methemoglobin. Methemoglobin, like hemoglobin when it is tied to carbon monoxide, is incapable of carrying oxygen. A large corned-beef sandwich [about one-quarter pound (110 grams) of meat] could inactivate 1.5–5.7% of an adult's hemoglobin by converting it to methemoglobin. Thus, the effects of air pollution, a food additive, and smoking may all combine to inactivate a significant portion of a person's hemoglobin. The combined, or synergistic, effect of smoking and various air, food, and water pollutants are just beginning to be uncovered.

Cigarettes also contribute particles of nickel, arsenic, cadmium, and lead to the lungs of a smoker, in addition to a variety of gaseous and particulate organic materials. Individually and in high quantities, many of these compounds have been known to have harmful effects on humans. Together and in lower quantities, their effects are not clear.

---

1 Nitrites in food are covered on page 599.

Arsenic and lead. In an earlier era, the pesticide lead arsenate was used on tobacco. Because neither lead nor arsenic is broken down in the soil, they are still found wherever lead arsenate was once used. Tobacco plants grown in that soil today absorb lead and arsenic. The lead level in tobacco leaves is variable, depending on where the tobacco is grown. One study, however, estimates the lead in a cigarette at an average of 13 micrograms. Of this quantity, about 1.5 micrograms of lead appear in the smoke. This is the quantity inhaled by an individual while smoking one cigarette. Twenty cigarettes in a day mean 30 micrograms of lead inhaled. About one-third of this quantity is absorbed into the blood.

Lead has many adverse effects on health (Chapter 21). Smoking alone does not lead to levels that cause apathy, sluggishness, and brain damage. However, lead is also found in food and in water. It is also in air, due to lead additives in gasoline. Smoking a pack a day adds about 50% more lead to the daily respiratory intake for an individual who lives in a polluted city.

Arsenic is a cumulative poison; that is, many small doses can accumulate over a period of time until a poisonous level is reached. Tobacco smoke does not contribute enough arsenic to kill anyone in the classic manner. However, it is suspected that at lower levels, arsenic is carcinogenic.

Cadmium and nickel. A pack of cigarettes contains about 30–40 micrograms of cadmium and 85–150 micrograms of nickel. In large enough quantities, cadmium has several effects in the body. It interferes with the body's use of calcium, and may also contribute to high blood pressure and heart disease. A smoker inhales about 2 micrograms of cadmium in a pack of cigarettes, and about $\frac{1}{2}$–$1\frac{1}{2}$ micrograms of nickel. Smokers have, on average, about twice the amount of cadmium in their kidneys and liver as nonsmokers. The effect of cadmium and nickel on human lungs is not clear. Studies in animals, however, have shown that both nickel and cadmium can, under certain circumstances, be carcinogenic.

Flavorings. A number of flavoring materials are added to cigarettes. Experts have voiced concern over cocoa, coumarin, angelica root, and triethylene glycol, which are carcinogens or give rise to carcinogens when burned.

The Health of Smokers

When groups of smokers are compared to groups of nonsmokers, one finds a number of ways in which smokers appear less healthy than their counterparts who do not smoke.

Heart and circulatory problems. According to a 1979 mortality study by State Mutual Life Assurance Company of America, people who smoke suffer death rates double those of nonsmokers at any age. Smokers are at greater risk with respect to both heart disease and lung disease. For instance, people who smoke are approximately twice as likely to suffer an immediately fatal heart attack as those who do not smoke. When a person stops smoking, the risk decreases. In 10 years, the risk to former smokers is about the same as for nonsmokers. Researchers suspect that there is a relationship between the carbon monoxide in cigarette smoke and an increased risk of sudden death. Possibly, carbon monoxide, because it ties up hemoglobin in blood, reduces the amount of oxygen available to the heart muscle itself. This puts a strain on the heart and makes it pump harder. This strain on the heart may explain the higher likelihood of sudden death from heart attacks in groups of smokers.

Smoking also seems to increase the possibility of cerebral hemorrhages. It further increases the occurrence of peptic ulcers and the chance of dying from them.

Effects on unborn babies. Women who smoke tend to have smaller babies. There are more spontaneous abortions, more still births, and up to one-third more deaths of newborns among babies born to mothers who smoke during pregnancy. In general, it can be said that there are more frequent complications of pregnancy and labor among women who smoke. These effects seem to be due to a shortage of oxygen in the blood during pregnancy. Again, carbon monoxide in cigarette smoke, which ties up the oxygen carrier, hemoglobin, is the culprit.

Effects on lungs. It is not surprising that many adverse effects of smoking involve the lungs. Smokers suffer more ordinary respiratory illnesses and have been shown to have more complications, such as pneumonia, after surgery than nonsmokers. Smoking is a major cause of emphysema. This is a disease in which the alveoli, tiny sacs at the end of each bronchiole, do not expel air as they normally should in exhaling (Figure 31.1). People with emphysema have trouble getting the used air out of their lungs so that they can take a fresh breath. Those who stop smoking usually improve.

The disease most commonly associated with cigarette smoking is lung cancer, although smokers also have more of a chance than nonsmokers of contracting cancer of the larynx, esophagus, mouth, bladder, kidney, and pancreas.

Lung cancer accounts for 25% of cancer deaths in the U.S. Cigarette smoking is believed responsible for 85% of lung cancer cases. (Despite attempts at early diagnosis, the survival rate for lung cancer is only about 10%, a rate that has not changed much over the past 15 years.)

The rate of lung cancer for women, which has historically been lower than that for men, has increased significantly in recent years. This probably reflects the fact that women did not begin smoking in large numbers until after World War II. Because of the long latency period before cancers are diagnosed, women's

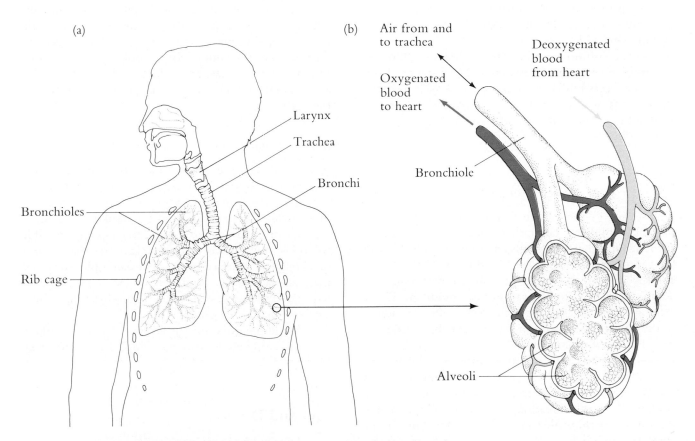

Figure 31.1 The Respiratory Tree. (a) Air is breathed in through the nose and mouth, moves down the trachea into the bronchi and bronchioles, and finally enters the alveoli. (b) The alveoli are tiny sacs at the end of the bronchioles. Oxygen enters the bloodstream at this point. Capillaries carry blood around the alveoli, allowing oxygen to diffuse from the alveoli into the blood. (Adapted from *Life: The Science of Biology,* copyright 1983 by William K. Purves and Gordon H. Orians. Used with permission of Sinauer Associates and Willard Grant Press.)

lung cancer rates are only now increasing. Lung cancer is expected to soon surpass breast cancer as the leading cause of cancer deaths in women.

When people stop smoking, their risk of developing lung cancer drops to that of nonsmokers over a period of 10–15 years. Thus, cigarette smoking appears to be the largest single *preventable* cause of cancer in the U.S. today.

The choice of a filter-tip cigarette or a low-tar brand also seems to reduce the risk of lung cancer, although rates are still much higher for smokers of these cigarettes than for nonsmokers. In addition, these cigarettes do not necessarily reduce the amount of carbon monoxide the smoker inhales. For this reason, low-tar cigarettes do not protect a smoker from smoking-related heart disease.

Tobacco smoke can increase a person's risk of developing cancer from other causes. For instance, a person who both smokes and works with asbestos has a nine times greater chance of developing asbestos-caused lung cancer than an asbestos worker who does not smoke.

The Effects of Smoking on Nonsmokers

Two-thirds of the smoke from a cigarette is not inhaled by smokers, but rather goes into the environment around them. Actually, a larger portion of both the cadmium and nickel are found in the "side-stream smoke," that is, the smoke not inhaled by the smoker. Thus, about 9–13 micrograms of cadmium and 12–21 micrograms of nickel per pack of cigarettes are added to the environment around the smoker. In addition to nickel and cadmium, this side-stream smoke contains twice the tar and nicotine, three times the 3,4-benzo-pyrene, and five times as much carbon monoxide as the smoke the smoker inhales. Studies on "passive smoking" (simply breathing in a room where people are smoking) show that in a very smoky room, non-smokers can inhale as much nicotine and carbon monoxide in an hour as if they smoked one cigarette themselves. Recent studies appear to show that working in smoky offices can have harmful effects on lung function in nonsmokers. Further, some studies seem to show that the nonsmoking wives of men who smoke may be at a higher risk for lung cancer than wives of men who do not smoke.

Children exposed to tobacco smoke at home suffer more respiratory illness, according to several studies. Some doctors are recommending that people with heart disease avoid smoky places because they could inhale enough smoke to injure their health.

Smoking Laws: Should There Be More?

In public vehicles transporting people between states, federal regulation requires separate sections for smokers and nonsmokers. Some 70% of the people queried in a 1975 study by the National Clearing House on Smoking and Health felt that smoking should be prohibited in more places, although only 51% of smokers felt this way. A total of 78% felt that smoking should be prohibited in places of business if the management wishes. And three out of four people believe that doctors, teachers, nurses, and others in the health professions should set us all an example by not smoking themselves.

Smoking and Fires

A dropped cigarette will burn for 20–45 minutes. If it falls on upholstered furniture, it will commonly smolder and cause flames in 15 minutes. Nationwide, almost 10% of fires are caused by smoking and more than 1/4 of fire deaths are related to smoking. Possible solutions would involve either cigarettes that go out quickly when dropped or fire-retardant upholstery and mattress fabrics. Cigarette manufacturers claim to have had little success developing a self-extinguishing cigarette that tastes good.

Should the government take a more active role in trying to stop people from smoking? See Controversy 31.2.

Diet and Disease

Many nutritionists now agree that the foods and drinks Americans choose may be contributing to the development of disease. As much as 40% of cancer deaths may be related to eating or cooking American-style. Broiling, frying, and charcoal grilling meats may cause carcinogenic substances to form. High-fat diets may cause the production of excess bile and lead to intestinal cancers. Excess fat can also stimulate the

Government Protection for an Illegal Activity?
Marijuana Smoking

> The United States Government has a responsibility to insure that its actions do not foreseeably endanger the health and safety of its citizens.
>
> **U.S. Senator Charles Percy of Illinois**

> They [marijuana smokers] don't have to smoke.
>
> **Lee Dogoloff, Office of Drug Abuse Policy**

In 1977, many marijuana smokers were jolted by the news that the pot they were smoking could be contaminated with a poisonous herbicide, paraquat. This came about because the Mexican government, supported and encouraged by the U.S., began a new program to stamp out marijuana farming in Mexico. Scouts located fields of marijuana by plane and then destroyed them by spraying with the herbicide paraquat, a program for which the U.S. government helped pay. Paraquat is absorbed by marijuana plants and kills them by preventing photosynthesis. However, paraquat does not always act immediately. It can take up to 2–3 days to kill plants. Thus, some enterprising farmers managed to harvest their crops and get them on the market even after they were sprayed. Paraquat levels in some samples of marijuana were found to be as high as 2264 parts per million. At this level, one to three cigarettes a day for a few months could cause irreversible lung scarring.

Public reaction to the problem varied. Marijuana is, after all, an illegal substance. An official in the White House Office of Drug Abuse Policy stated:

> The government does feel some responsibility to smokers, but individuals do have some responsibility and choice in the matter—they don't have to smoke.★

What do you think? Does the government have a responsibility to protect the health of its citizens even if they are doing something illegal? The fact that the U.S. government helped pay for the spray program can be viewed as causing harm. Does the government have the responsibility to at least not harm its citizens' health directly? In 1982, a field in the U.S. was sprayed with paraquat. Does the fact that the government was attempting to stamp out trade in an illegal drug have any bearing on this case?

★ *Science,* **200** (28 April 1978), 418.

What Is the Government's Role in Controlling Smoking?

> If we have to make a mistake we should make a mistake in the direction of protecting public health.
>
> *Gus Speth*

> . . . He should first remove the products about which there is no mistake.
>
> *Rodney Adair*

The New York Times quoted the Chairman of the Toxic Substances Strategy Committee as follows:

. . . Today cancer is our second leading killer, and a concerted, Government-wide attack on the substances that cause much of it is essential.

This strategy of prevention is bound to fail, however, if those whom it is designed to protect do not understand the basis for chemical regulation. The recent controversy on the proposed Food and Drug Administration ban on saccharin treated us to a dangerous amount of hilarity about the high dosage levels used in animal tests and demonstrated the prevalence of misunderstanding in this area.

So far, we have no scientific basis for setting a safe threshold dose for a carcinogen. Until such a scientific basis is demonstrated, if we have to make a mistake we should make a mistake in the direction of protecting public health. And that means we assume, for the present, that there is no safe level for a carcinogen.[*]

In a letter to the editor of *The New York Times,* Rodney Adair commented:

Mr. Speth should surely realize that most of the hilarity is cynical, and is extremely well justified, because of the attitude of Government toward the carcinogenic effect of tobacco smoke. We have the Surgeon General's report, the statistical evidence of 100,000 new lung cancer cases a year, of which 80 percent are caused by smoking, and the approximately 90 percent fatality rate associated with lung cancer.

How can Mr. Speth and the F.D.A. expect anything more than cynical laughter toward a Government that subsidizes the tobacco industry and at the same time threatens the removal of other products which are often only marginally suspect and on which there is no statistical evidence of any undue incidence of cancer among its human users?

If he truly believes his own statement in reference to removal of carcinogenic materials from the market, i.e., "if we have to make a mistake, we should make a mistake in the direction of public health," he should first remove the products about which there is no mistake.[†]

What is the responsibility of the government to its citizens when there is a proven health hazard such as smoking? Do you agree with Mr. Adair that the government should remove tobacco products from the market?

---

[*] Gus Speth, member of the Council on Environmental Quality and Chairman of the Carter Administration's Toxic Substances Strategy Committee. *The New York Times,* 9 March 1978, p. A–21.

[†] Rodney Adair, *The New York Times,* 29 March 1978, p. A–26.

production of hormones that could promote breast cancer.

Alcohol is suspected of causing birth defects and also of increasing the risk of respiratory and digestive tract cancers. Excessive amounts of salt can contribute to high blood pressure, while excess sugar leads to tooth decay. Furthermore, we are all concerned about various other additives or contaminants that are found in small amounts in foods, drugs, and cosmetics and which may cause diseases such as cancer.

Later in this chapter, we will examine a diet that is recommended as a possible means of reducing the risk of cancer. First, however, we take a closer look at chemicals which, intentionally or unintentionally, become part of the food supply.

Intentional Food Additives

Why do food manufacturers use additives? A trip to the supermarket should convince anyone that many additives are used to color, preserve, or otherwise "improve" food, drugs, and cosmetics. Over 2000 different chemicals are added to foods alone. The term "additive" covers a wide variety of materials. In foods, additives fall into three main groups. The first includes natural substances such as sugar, salt, or vitamin C. A second group consists of laboratory copies of natural substances, an example of which is vanillin, the chemical that is the main flavoring in natural vanilla bean extract. There are also the substances that are entirely synthetic, or invented in the laboratory: BHA, EDTA, and saccharin, among others (Figure 31.2).

Additives are used for many reasons, all of them understandable. However, some are more justifiable than others. Many are used to increase a product's appeal to consumers. Drugs are flavored to help hide bitterness or other unpleasant tastes. Foods may be colored to help identify flavors (yellow for lemon candies, pink for strawberry ice cream). But colors and flavors are also used as substitutes for expensive ingredients that are left out of cosmetics or foods. For instance, real fruit is often missing from artificially colored and flavored soft drinks.

Modern methods of selling food have made certain additives necessary. Chemicals that kill mold and keep foods moist mean that bakery products and candies can be shipped across the country and still remain

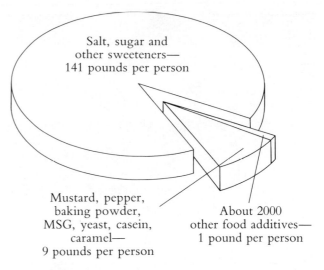

Figure 31.2 Americans and Food Additives. Americans consume about 150 pounds (68 kilograms) of food additives per person each year. Most of this is salt, sugar, and other sweeteners.

fresh-tasting for long periods of time. Antioxidants, which prevent fats from becoming rancid, make convenience foods such as boxed cake mixes possible. In fact, the whole group of convenience foods and special diet foods probably could not exist without the food additives that flavor, color, and stabilize them. In some cases, additives make a wider variety of foods possible. Certain foods could not be canned, frozen, or packaged for shipment or out-of-season selling without additives.

There is not complete agreement, of course, that all or even most additives are necessary. For instance, it has been argued that the consumer does not gain from the use of preservatives:

> Even assuming that these chemicals are harmless, the advantage in selling bread that does not go stale for a week or more, to take an example, seems to lie more with the baker and retailer than with the consumer.[2]

It should be noted, however, that a shorter shelf life would mean higher costs to manufacturers. In a com-

---

2 Dr. W. Lijinsky, testifying at Senate hearings on food additives. *The New York Times,* 22 September 1972, p. 1.

Figure 31.3 In some schools, parents who are concerned about the amount of artificial flavors, colors, and preservatives that children eat and drink have banned the sale of foods containing additives in the school lunch room and vending machines. Soft drinks are replaced by milk and fruit juices. Other snack foods are replaced by fruit or additive-free cookies and potato chips.

petitive market, this would be passed on as higher food prices for the consumer.

In another instance, at Senate hearings in 1972, Dr. Michael Jacobson, a co-leader of a consumer action group, demonstrated several dozen pairs of similar food items, one of which had no chemical additives while the other had preservatives, emulsifiers, dyes, or anti-foaming agents. He stated, "This shows that such chemicals are rarely, if ever, needed in the production of quality foods if manufacturers use good manufacturing practices."[3] (See Figure 31.3.)

However good the reasons for using food, drug, or cosmetic additives may seem, we would like to be sure that the additives are safe.

---

3 *The New York Times,* 22 September 1972, p. 1.

Food, Drug, and Cosmetic Laws

The safety of food, drug, and cosmetic additives is governed by the Federal Food, Drug, and Cosmetic Act of 1938 and its amendments. Adulteration of food is prohibited. This means harmful substances cannot be added and also that unexpected ingredients (such as low-cost fillers) cannot be substituted for materials the consumer would normally expect to find in a product.

In 1958, a food additives amendment was passed. Under this law, the manufacturer is required to prove the safety of an additive before it can be marketed. It is estimated that it costs a manufacturer between $200,000 and $3,000,000 to test a new food additive and prove it safe enough to market.

Also in the 1958 amendment is the Delaney Clause, which forbids the use of an additive, in any amount, if it has been shown to cause cancer in humans or animals. This clause has aroused a great deal of argument because some people feel it prevents reasonable judgments about risks compared to benefits. (See Chapter 29 for a discussion of risk–benefit comparisons.) Some of the arguments for and against the Delaney Clause are detailed in the section on artificial sweeteners.

These are Federal laws, which apply to foods that are sold across state boundaries. Foods sold only within a state are subject to the laws of that state.

The GRAS List. When Congress passed the 1958 Food Additive Amendment, which required manufacturers to prove the safety of food additive chemicals they wished to manufacture, a list of some 600 chemicals was prepared and designated the GRAS List (Generally Regarded As Safe). These chemicals were exempted from the amendment. That is, they were allowed to be added to food, even though they had not undergone a great deal of testing to prove their safety, since they had been in use for a number of years. Chemicals were put on the list if the opinions of experts in nutrition and toxicology were that no unhealthful effects had ever been traced to these additives. Since then, some chemicals have proved to be less safe than was believed. In 1969, the Food and Drug Administration began an extensive testing program to determine whether the items on the GRAS list were indeed safe.

About 450 items on the list have been tested so far. Of these, 80% have been declared safe. Another 14% are probably safe but are undergoing further study.

About 4% were declared uncertain and are also being studied.[4]

The sections that follow cover in detail some of the additives, intentional and unintentional, found in foods, drugs, and cosmetics.

Artificial Sweeteners and the Delaney Clause

Substitutes for sugar have come to seem not only desirable but necessary to a nation beset by weight problems. In the United States, almost five million pounds of the artificial sweetener saccharin were sold in 1976 (Figure 31.4). Thus, it came as unpleasant news in March 1977 that the FDA was considering a total ban on the use of saccharin in foods. New studies had shown that saccharin caused bladder cancer in rats.

Saccharin was not the first artificial sweetener to be declared unsafe. In 1969, the FDA banned cyclamates. Sales of this group of artificial sweeteners had reached around a billion dollars per year when it was found to cause bladder cancer in experimental rats. At the time of the cyclamate ban, manufacturers of diet foods were upset, but not desperate. After all, they had saccharin to fall back on. Diet soda recipes were changed to leave out cyclamates and rely on saccharin.

When the saccharin ban was proposed, reactions were much stronger. Mail flooded congressional offices. One Representative said he had received more mail about saccharin than about any other subject during the three years he was in Congress. Most of the mail opposed the saccharin ban. In addition, soft-drink manufacturers began a vigorous lobbying campaign. Three-quarters of the saccharin used in 1976 went into diet soft drinks. The controversy that followed highlights many of the problems involved in regulating food additives and drugs.

Experimental evidence. In the first place, experiments showing that saccharin causes cancer were challenged. The experiments were done on rats, a common experimental animal. The rat, however, concentrates its urine more than other species before excreting it. Rat bladders were thus exposed to higher

Figure 31.4 Artificially Sweetened Products. Saccharin is not only used in products designed for weight control, but is also used in products where sugar would be too bulky, e.g., in pills, to make them easier to swallow. Saccharin is also used to flavor many toothpastes, because sugar is felt to contribute to the formation of dental caries (cavities). The product on the extreme right does not contain saccharin. It uses the new artificial sweetener aspartame (see page 599).

concentrations of saccharin than might be the case in other animals.

Saccharin does not cause mutations in bacterial cells (the Ames Assay), but it has been shown to cause mutations in test-tube cultures of hamster cells and human cells. Although this does not prove that it is carcinogenic, it is cause for suspicion. (These tests are described on page 555.)

In the rat experiments, very high doses of saccharin were used, much higher than humans would be expected to consume. The rats were fed saccharin as 5% of their diet for two generations. This prompted a number of people to calculate that a human would have to chew 6700 pieces of artificially sweetened bubble gum daily or drink 800 diet drinks each day to receive the same dose the rats did. Are these reasonable comparisons? Not really. Rats live for only $2\frac{1}{2}$ years compared to humans' 70–80-year lifetime. Thus, researchers using high doses are trying to compensate for the longer exposure time humans have (and thus recreate the higher accumulated dose). Furthermore, we are interested in finding out if a chemical is carcinogenic even if it only causes one case in 100,000 people per year. (That would be 2000 cases of cancer in the United States each year caused by saccharin.) In order

---

4 In 1978, the Center for Science in the Public Interest published a table in which common food additives are grouped according to estimates of their safety. The table can be ordered as a poster from C.S.P.I., 1755 S Street, N.W., Washington, D.C. 20009.

to detect a carcinogen of this strength at normal human levels of exposure, scientists would have to run experiments on hundreds of thousands of rats. The expense and work involved are enormous. Thus, higher doses are used in the belief that there is a relationship between dose and rate of cancer causation: the higher the dose, the more likely a carcinogen will cause cancer. There is evidence that this is a reasonable assumption.

Leaving aside questions about the experiments themselves, many people felt that a saccharin ban smacked of the prohibition era. They argued that, if there was a risk, they would rather be told about it and left to make their own decision.

The Delaney Clause. The FDA, however, was on firm legal grounds in proposing a saccharin ban. A 1958 amendment to the Food, Drug, and Cosmetic Act states "no additive shall be deemed to be safe if it is found, after tests which are appropriate for the evaluation of the safety of food additives, to induce cancer in man or animal" [Section 409(c)(3)(A)]. This is known as the Delaney Clause after the Representative who proposed it.

There has been much criticism of the Delaney Clause, mainly because it allows administrators no flexibility. That is, they are not given the opportunity to decide whether a very small amount of a carcinogenic substance might either cause no cancers, or be worth the risk. Second, foods such as meats and dairy products usually contain detectable amounts of pesticides such as DDT. This is a carcinogenic chemical that has become widespread in the environment. It is incorporated into foods without being intentionally used in their production. The Delaney Clause could be used to justify stopping the interstate shipment of such foods. In practice, the clause is not used this way, but it is felt by some to be a law that could not be enforced and, therefore, one that is invalid.

Those in favor of retaining the Delaney Clause point out that we do not know if there are threshold (or "no-effect") levels of carcinogens. The safest policy is to assume that any amount of a carcinogen can cause some cancers. Further, since even weak carcinogens can act together to cause greater hazards, the Delaney Clause provides a safeguard against the additive effects of small amounts of carcinogens in the environment.

Another important feature of the Delaney Clause is that it allows decisions to be made on fairly black-and-white issues. Either a substance causes cancer in experimental animals or it does not. This prevents the kind of delays in removing carcinogens from the marketplace that would occur if regulatory agencies had to fight long drawn-out court battles over risks versus benefits. Nonetheless, the saccharin issue has provided the strongest challenge to the Delaney Clause since it was passed in 1958.

Is a saccharin ban the right solution? In the end, Congress bowed to strong public pressure and directed that the proposed ban be put off while the evidence was sifted. Meanwhile, the FDA required stores selling saccharin-containing products to display a warning sign stating that saccharin has been shown to cause cancer in rats. This warning was also required on the packages of foods containing saccharin. The original House version of the bill did not include a warning label, which prompted Representative Andrew Maguire to jokingly propose that manufacturers should include an "assurance" label stating:

> Assurance: Saccharin does not cause cancer, in the opinion of your Congressman, despite all scientific evidence that it does.[5]

A study committee of the National Academy of Sciences and the Institute of Medicine concluded that saccharin was a "moderate carcinogen."[6] Nonetheless, the committee did not recommend a total ban on the sweetener. Rather, the committee suggested consumers be warned of the possible hazards of saccharin use. Further, since studies in 1973–1978 showed that one-third of children under 10 years of age consumed food products containing saccharin, mainly diet soft drinks, the committee recommended that steps be taken to reduce children's exposure to saccharin.

In contrast, the Canadian government did ban saccharin in foods in 1977. Doctors there say that this has had no visible effect on the health of diabetics in Canada. Further, the diet soda industry has developed two new types of diet sodas for the Canadian market. One is a group of low-sugar sodas that have about 60 calories per can, about half that in the regular product. The second type are no-sugar sodas that can be sweetened

---

5 *The New York Times,* 18 October 1977, p. 16.
6 *Food Safety Policy: Scientific and Policy Considerations,* National Institute of Medicine and National Academy of Sciences, Parts I and II (1979). (Available from National Technical Information Service.)

Does Everything Cause Cancer?

> FDA's absolute zero risk standard—if generally and sincerely applied to all synthetic and natural chemicals would ban most of the food supply, most industrial jobs, going outdoors and staying indoors and much of the rest of the Universe.
>
> **Congressman James G. Martin***

> . . . the public wants to live in an environment free from the risk of cancer, asking, and rightly so, that food, air and water be clean and wholesome . . .
>
> **World Health,** *September–October, 1981*

It is argued that any substance in large enough amounts is carcinogenic. For instance, Representative James Martin,† speaking before the Manufacturing Chemists Association, said:

> The rapidly growing abundance of over 2,000 suspect chemicals seems limited only by the resources of rat breeders and feeders. It seems probable that for every chemical there is a dose above which the metabolic defenses and DNA repair and immune systems of some cancer-prone rat can be overwhelmed, especially if concentrated *in utero* by a transplacental effect. Ironically, the only chemicals unlikely to be overdosed to a carcinogenic level are those that are lethal poisons. If cyanide were a sweetener, it would pass.

This type of argument has been used to protest bans on artificial sweeteners, hair dyes, and food colors, but experiments do not really bear out the claims made. Although researchers have tested 7000 chemicals (by feeding them to animals at the maximum amount that did not kill or poison them directly), less than 7% of the chemicals caused cancers. If this small a proportion of tested chemicals causes cancer, it seems reasonable to assume that the ability of a chemical to cause cancer is a relatively rare phenomenon.

Suppose you were debating Rep. Martin on the floor of the House as he attacked regulations banning a popular food additive suspected of being a weak carcinogen. Could you explain the arguments that everything does not cause cancer, and that dosing animals with high levels of suspected carcinogens is a reasonable way to prove whether or not these chemicals are carcinogenic?

---

* Letters to the Editor, *Science,* **208** (6 June 1980), p. 1086.

† Mr. Martin is also a Ph.D. chemist. The extract above is taken from his speech in October 1978.

Protecting People from Themselves: Saccharin

> All day long we've been taking calls from people, some of them in tears, demanding that we leave saccharin alone.
>
> **FDA Official***

> . . .—put a warning label but don't ban. Let people make their own choice.
>
> **Representative James Martin†**

> Surely the argument that people should be allowed to make their own decisions about using possibly harmful materials is not serious, particularly when those subject to the greatest risk are children.
>
> **Dr. William Lijnsky†**

> Establishing "low-level" residue tolerances for any one carcinogen in food would, by precedent, permit unlimited numbers of other carcinogens also to be added.
>
> **Dr. Samuel S. Epstein§**

Dr. Morris Crammer, head of the National Center for Toxicological Research, has written about saccharin:

> I think most of us are unprepared, in the absence of persuasive evidence, to replace all those items of human convenience, even though an ultraconservative mathematical hypothesis may imply a relatively high level of risk.‡

Do you feel that the FDA should ban saccharin on the basis of the Delaney Clause (no substance that has been shown to cause cancer may be added to food), or should people simply be warned about the hazard and left to make their own decision? This is the way alcoholic beverages (known to cause birth defects) and cigarettes (known to cause lung cancer) are treated. Are there basic differences between alcohol, tobacco, and foods that mean they should be regulated differently?

---

* Quoted in *The New York Times,* 11 March 1977, p. 28.
† *Chemical and Engineering News,* 27 June 1977.
‡ Quoted in *The New York Times,* 20 July 1978, p. A17.
§ Testimony before the Senate Select Committee on Nutrition and Human Needs, 20 September 1972.

with saccharin tablets, which are still available to consumers. The president of the Canadian Diabetes Association noted that most people use only one or two 15-milligram tablets per can. This should be compared to the 120 milligrams per can used in commercially prepared diet soda.

Aspartame. In 1981, a new sweetener, aspartame, was approved by the FDA. Aspartame is composed of two amino acids that are present in the human body: aspartic acid and phenylalanine. There are still some scientists who feel the compound should be tested more thoroughly. There is incomplete evidence from a few studies that it may cause brain tumors in rats. However, other studies seem to show no effect. The FDA commissioner noted that more studies would be helpful, but he felt the weight of evidence was on the side of no serious effects at the levels at which aspartame is expected to be consumed.

Artificial Colors in Foods

A fact known to every food manufacturer is that grape soda tastes better when it has a rich purple color. Ice cream seems creamier if its slight yellow color hints at thick cream and egg yolks. And no one wants to buy oranges that are pale orange or slightly greenish. These and other problems and effects call for the use of food colors: substances added to food to cover up or improve natural colors. Any material added to food to color it is called artificial color, although the material used might be called "natural." For instance, carotene, the substance that makes carrots orange, is often used to color other foods orange. The food must then be labeled artificially colored.

Food colors have a somewhat unpleasant history. Before strict regulations were enforced, it was not uncommon for colors to be added to cover up spoiled or adulterated food (e.g., food to which some unlabeled, cheaper ingredient had been added). Furthermore, the food colors used were sometimes poisonous. Mercuric sulfide and red lead were used on occasion to color cheese. Candies were often colored with lead chromate, red lead, or white lead. Forty-six percent of the candy sampled in Boston in 1880 contained one or more of these poisonous coloring agents.

There are two general groups of food colors in use today. The first group, called uncertified colors, consists of colors that are for the most part natural products. That is, they are usually extracts of various plants. The second group of colors is called Food, Drug and Cosmetic Act Certified (FD&C) colors. These are synthesized or extracted from nonplant materials, such as coal tar, and "certified," or tested for purity by the FDA.

The industry using the largest amount of certified color is the beverage industry. About 30% of all soft drinks sold contain certified color. Sixty percent are colored with caramel coloring (cola drinks), while 10% are uncolored (club soda and lemon–lime drinks).

In 1963, Congress directed the FDA to check the safety of the certified and uncertified dyes in use. Nine certified food colors were taken off the approved list after this was done because it was decided they were unsafe or their safety was not proven. Since then, three more dyes, red #2, red #4, and carbon black, have been removed from the approved list when experiments showed they might be carcinogenic. (Red #2 is still allowed for coloring orange skins, however, on the theory that people do not eat orange skins.)

Preservatives

One of the major problems in the food industry involves keeping foods from becoming stale or spoiling between the time they are made in the factory and the time they are bought and served. The reaction of fats with oxygen leads to the formation of a variety of bad-tasting, smelly compounds. Oxidized fats are said to be rancid. Antioxidants are added to foods to prevent such reactions. BHA (butylated hydroxyanisole) and BHT (butylated hydroxytoluene) are examples of antioxidants used in foods. Other common antioxidants are propyl gallate and ethoxyquin.

The Special Case of Nitrates and Nitrites

Nitrites and methemoglobin. Nitrates (NO_3^-) and nitrites (NO_2^-) are a special case among food preservatives. They were originally added to meats and fish to prevent them from spoiling before refrigeration was available. Over the years, however, people have come to like the cured flavor these additives give to meat. Thus, we now eat cured meats not only because they are a safe way to keep meat but because we like the flavor.

The combination of sodium nitrate and sodium nitrite has three effects on meats. It prevents the growth of bacteria that cause various kinds of food poisoning, such as botulism. It creates the typical pink color of cured meats, such as ham. And nitrates and nitrites give meat that special "cured" flavor. Meat was originally cured by adding only potassium nitrate. It was later found that bacteria were converting some of the nitrate to nitrite. The nitrite is actually the chemical that prevents the growth of bacteria and gives the meat a pink color.

Nitrites (NO_2^-) are not harmless compounds. Hemoglobin in the blood, which carries oxygen to all the body cells, reacts with nitrites to form methemoglobin. Methemoglobin cannot carry oxygen. When 70% of the hemoglobin in the blood is converted to methemoglobin, death results from suffocation. At lower levels of methemoglobin, there may be problems such as dizziness or difficulty in breathing. For this reason, legal limits exist to the amount of nitrite that can be added to meat or fish to cure it. Cases have occurred in which children have been poisoned by bologna and frankfurters containing more nitrite than the legal limit.

Although a large portion (40%) of the nitrite (NO_2^-) we eat is from cured meats, the nitrate (NO_3^-) we eat comes mostly from vegetables. Only about 2% is from cured meats. Spinach, beets, eggplant, radishes, celery, lettuce, and greens all can have high levels of nitrates. Since nitrates (NO_3^-) are not as poisonous as nitrites (NO_2^-) because they do not oxidize hemoglobin, problems do not usually arise unless bacteria change the nitrates to nitrites. This has happened in some cases, for instance, when opened jars of baby-food spinach have been left unrefrigerated. Infants have been poisoned in this manner.

Nitrites and cancer. Besides the fact that nitrites (NO_2^-) lead to the formation of methemoglobin, there is another problem with respect to adding nitrites to foods. Nitrites can react with certain amines, chemicals that are also present in foods, to form nitrosoamines. In one study, 75% of a variety of nitrosoamines tested were carcinogenic; that is, they caused cancer in test animals. Cancers were produced in all

species of animals tested (dogs, monkeys, rats, parakeets, hamsters, guinea pigs, mice, and trout). Nitrosoamines also seem to act with weak carcinogens to make them stronger.

Nitrosoamines are also found as air pollutants, especially in cities. It is believed they are formed from nitrogen oxide pollutants in urban air. Further, nitrosoamines are found in cutting oils (used in industry where metals are cut or ground) and in some pesticides, drugs, and cosmetics. Nitrosoamines are also known to be components of tobacco smoke and certain alcoholic beverages. Beer and whiskey manufacturers were able to reduce significantly the amount of nitrosoamines in their products after this was discovered. The total effect of nitrosoamines in food and the environment is not yet known. Certain scientists feel that the effect, in terms of causing cancer, will eventually be found to be very large.

Should nitrites be banned as food additives?
Before 1978, it was illegal for cured meats such as sausage, ham, or bacon to contain more than 200 parts per million of nitrites. This limit was determined by considering the effects of nitrites on an average person. At this level, 100 grams of meat (1/4 pound of corned beef on rye) would convert 1.5–5.7% of the consumer's hemoglobin to methemoglobin. If the person were also a heavy smoker, living in a city, a further amount of hemoglobin would be inactivated by the carbon monoxide in the cigarette smoke and urban air. The combined effects of the air pollution, smoking, and corned beef sandwich would be just short of causing the first sign of oxygen starvation: headaches.

In 1975 it was shown that when bacon containing legal levels of nitrites was cooked until it was crisp, nitrosoamines formed at the parts per billion level. There is some evidence that vitamin C can prevent this reaction. The FDA considered a ban on the use of nitrites in cured meats such as bacon. However, the question of safety from food poisoning was also involved. That is, would people be in more danger of dying from food poisoning if nitrites were not allowed in cured meats, than they would be in danger of suffering a nitrosoamine-caused cancer if nitrites were allowed to be added?

The main problem appeared to involve cooked bacon because it is usually cooked at high temperatures and has a very high fat content, conditions that encourage nitrosoamines to develop. For this reason, the FDA lowered the amount of nitrite allowed in bacon. Another preservative, potassium sorbate, is added to prevent botulism food poisoning and vitamin C is added to reduce nitrosoamine formation. You can see from the ingredient list on the package that some manufacturers of sausage meat and bacon no longer add nitrites to their product.

Added Vitamins and Minerals

A number of manufactured food products contain added vitamins and minerals. In some cases, this is required by law to replace nutrients lost during processing. Thus, white bread is enriched with thiamine and riboflavin, vitamins lost when wheat is milled to white flour. In other cases, vitamins are added to commonly used foods to ensure good nutrition. For instance, ready-to-eat breakfast cereals, which are of little nutritional value but are eaten for breakfast by millions of children, are enriched with vitamins. Artificial orange breakfast drinks are enriched with vitamin C because they are widely used as a substitute for orange juice, a natural source of the vitamin.

Such supplements are not without hazard, however. Certain vitamins, notably A and D, can accumulate to toxic levels in the body. Children given large doses of vitamin A have developed a deformity in which one leg is shorter than the other. Some pregnant women who received calcium and vitamin D supplements, drank vitamin-D enriched milk, and sunbathed (a natural source of the vitamin, which is formed in the skin by the action of sunlight), have borne babies with excess calcium deposits in the skull. Some scientists have voiced concern that excess iron may cause people to become susceptible to disease. Such an effect has been seen in animal experiments.

The FDA is responsible for ensuring that people do not receive excess doses of vitamins and minerals from foods to which they have been added. In the face of the uncertainty that exists over what is and is not an excess dose, however, the consumer would probably be wise to avoid foods that must be enriched to be nutritious. Instead, it would be safer to rely on less processed foods, traditional safe sources of the vitamins and minerals necessary for good health. (See Controversy 31.5).

Unintentional Food Additives

Four groups of toxic substances in foods are not added to them on purpose. The first is naturally occurring toxins. The second group includes traces of pesticides (residues) left from the spraying of crops. Third, there may be drug residues in meat, left from chemicals added to animal feeds. The fourth category includes additions that occur because of some accident during the manufacture or shipment of food. The PBB disaster, discussed later in this chapter, provides an example of such an accidental contaminant.

Natural Toxins

Some toxic materials are formed in foods by natural processes during storage. Mycotoxins are produced by molds, especially under damp storage conditions. Aflatoxins, produced by the mold *Aspergillus fluvus,* are potent carcinogens that have been found in rice, peanuts, corn, wheat, and food products from Asia, Africa, and North and South America. (See Figure 29.2 for an estimate of the carcinogenicity of aflatoxins.) In Georgia in 1977, a drought was followed by an aflatoxin plague that destroyed the value of the remaining corn crop. Corn cannot be sold in interstate commerce if it contains more than 20 parts per billion of aflatoxin.

Charcoal broiling meats may even be hazardous. Two types of possibly carcinogenic materials are produced. One is related to the tars in cigarette smoke, the other is a breakdown product of proteins in the meat. (See the reference by Ames, in the Further Reading section, for more on this subject.)

Pesticide Residues in Food

Regulation of pesticides in food: a problem for EPA. In 1978, the House Commerce Committee began hearings on cancer-causing chemicals in foods. Many problems in regulating carcinogenic substances, such as pesticides[7] in foods, were brought out in testimony before the committee.

---

7 Pesticides and the problem of growing enough food for the world's people are discussed in Chapter 7. Pesticides as water pollutants are noted in Chapter 11.

Life-Style and Disease: Health Foods

> Don't let any food faddist or organic gardener tell you there is any difference between the vitamin C in an orange and that made in a chemical factory. . . .
>
> *Dr. Frederick J. Stare,*
> *Chairman of Department of Nutrition,*
> *Harvard School of Public Health*

> The guru of food faddism is not Adelle Davis, but Betty Crocker.
>
> *Nutrition Action*

The American diet has become a controversial subject. One group of people, sometimes labeled "health food faddists," argue that we should eat "natural foods," grown without fertilizer or pesticides and packaged without additives or fortifiers of any kind. Others reply that there is no difference between natural foods and those grown with fertilizers and pesticides, or between vitamin C from rosehips and that produced in a laboratory.

> The composition of our bodies can be stated in terms of water, protein, fat, carbohydrates, vitamins, and minerals. And so can the composition of an orange or a loaf of bread. And don't let any food faddist or organic gardener tell you there is any difference between the vitamin C in an orange and that made in a chemical factory and added to grape or apple juice—[or] between the B-vitamin thiamine in pork and that purchased from the chemical factory and added to flour as a part of the enrichment procedures.[*]

According to Dr. Bernard Oser, former member of the National Academy of Sciences Food Protection Committee, not only are some additives harmless, they are essential.

> Were it not for food additives, baked goods would go stale or mold overnight, salad oils and dressings would separate and turn rancid, table salt would turn hard and lumpy, canned fruits and vegetables would become discolored or mushy, vitamin potencies would deteriorate, beverages and frozen desserts would lack flavor and wrappings would stick to the contents.[†]

The other side of this argument can be seen in the following discussion:

> Food faddism is indeed a serious problem. But we have to recognize that the guru of food faddism is not Adelle Davis, but Betty Crocker. The true food faddists are not those who eat raw broccoli, wheat germ, and yogurt, but those who start the day on Breakfast Squares, gulp down bottle after bottle of soda pop, and snack on candy and Twinkies.
>
> Food faddism is promoted from birth. Sugar is a major ingredient in baby food desserts. Then come the artificially flavored and colored breakfast cereals loaded with sugar, followed by soda pop and hotdogs. Meat marbled with fat and alcoholic beverages dominate the diets of many middle-aged people. And, of course, white bread is standard fare throughout life.
>
> This diet–high in fat, sugar, cholesterol, and refined grains—is the prescription for illness; it can contribute to obesity, tooth decay, heart disease, intestinal cancer, and diabetes. And these diseases are, in fact, America's major health problems. So if any diet should be considered faddist, it is the standard one. Our far-out diet—almost 20 percent refined sugar and 45 percent fat—is new to human experience and foreign to

all other animal life. . . .

It is incredible that people who eat a junk food diet constitute the norm while individuals whose diets resemble those of our great-grand-parents are labeled deviants. . . .‡

How do you feel about "health foods"? What does make up a good diet?

* Frederick Stare, *Food Additives, What They Are/How They Are Used,* Manufacturing Chemists Association, Washington, D.C., 1971, page 16.

† B. Oser, "Food Additives—Mountains or Molehills," speech given before the National Newspaper Food Editors Conference, San Francisco, California, 25 September 1970.

‡ Editorial, *Nutrition Action,* Center for Science in the Public Interest, April 1975.

How do pesticides, toxic chemicals intended to kill pests, get into food? In many cases, small amounts of pesticides used to treat crops or animals remain on produce or in meat when they are marketed. These small amounts are called **residues.** Certain government regulations state exactly how high residues can legally be and list procedures that must be followed to keep residues below the legal limit.

In other cases, however, foods become contaminated with pesticides that were not even used on food crops. This is a process of secondary contamination in which pesticides in dust or rain land on food crops. For instance, most food products from animals (milk, meat, cheese) contain measurable amounts of the pesticide DDT. This comes about because of widespread contamination of the environment with DDT, which is picked up and blown around the world on dust particles. Rain washes DDT out of the air and onto pastureland. Grazing cattle eat DDT-contaminated grass and incorporate the DDT into their meat and milk.

In the United States, the Environmental Protection Agency is responsible for protecting society from the risks involved in using pesticides. The most recent major pesticides law, passed in 1972, is the Federal Environmental Pesticide Control Act. Under this law, manufacturers must supply the EPA with data showing that a pesticide they wish to sell is effective against pests and will not harm humans or the environment. If the EPA is satisfied with the data, it will register the pesticide. In some cases a **tolerance** is allowed. That is, a certain amount of pesticide residue may safely remain on food or feed when it is marketed.

Regulation of pesticides has proven to be a major problem for the EPA. By 1976, $2.4 billion in pesticides were sold every year. The industry is expected to grow to $3.3 billion by 1984. Industries on this scale have political muscle. The EPA itself estimates that one-third of the 1500 active ingredients in pesticides are toxic and up to one-fourth may be carcinogenic.

Figure 31.5 Farmer studying a manual on pesticide use during a pesticide applicator training course. A number of pesticides have been classified for restricted use only. This means that only persons who have passed a state test and received a permit may use them. Almost all commercial operators and about one-half of the nation's farmers are certified to use restricted pesticides. Training sessions for applicators stress the hazards to people and to the environment of certain pesticides. Safe use, storage, and disposal of these pesticides are taught. Examples of pesticides in the restricted-use class are: endrin, paraquat, sodium cyanide, and strychnine. (USDA Photo)

Critics point out that although a number of pesticides are restricted to use by certified operators (Figure 31.5), only a few pesticides have so far been banned except for emergency use.[8] Part of the reason for this

8 For instance, DDT, heptachlor/chlordane, Mirex, DBCP, and aldrin/dieldrin.

record may be that industry as well as other special-interest groups are almost certain to challenge regulations through law suits and by stimulating congressional pressure to reverse decisions. For instance, when aldrin/dieldrin was suspended, both Shell Oil, the manufacturer, and the Environmental Defense Fund filed suit, because neither group was happy with the decision (although for exactly opposite reasons). In the same vein, when Mirex was banned for control of fire ants, congressmen from the southern states where the ants are a problem pressured EPA into allowing the use of ferramicide, a mixture of Mirex and two other chemicals. Ferramicide was alleged to break down more quickly in soil than Mirex, but no hard data existed on whether it actually did so. The EDF filed a successful suit to delay the use of ferramicide until further tests were made. In a climate such as this, knowing a long legal battle will result, officials may be reluctant to propose new restrictions.

This sort of pressure has, to some extent, caused a situation in which the EPA seems to make regulations only when it is forced into complying with toxic substances laws by lawsuits brought by public interest groups.

Problems in setting tolerances. The House Commerce Committee has been critical of EPA's tolerance-setting program. In many cases, pesticide tolerances have been set without all the needed information. On pesticides registered more than 10 years ago, we need information on carcinogenicity and data on birth defects and mutations. Furthermore, EPA relies on safety data supplied by the pesticide manufacturer. The committee, noting that the manufacturer is

hardly a disinterested party, suggests that a better program to check the accuracy of such data is needed.

In addition, problems arise when EPA tries to calculate how much of a pesticide people are exposed to. When setting tolerances for pesticide residues in foods, EPA usually divides the total amount of a particular food sold in the U.S. by the total U.S. population. This is the yearly per capita consumption. An obvious problem with this method is that it averages consumers with nonconsumers, making no allowance for the wide variation in dietary habits in the U.S. The procedure leads to underestimates of the amounts of certain foods consumed. For example, a 1965–66 study showed that only 1 in 20 families eats fresh pears. However, the tolerance for pesticides used on pears is set using the assumption that everyone in the U.S. eats pears. What this means is that, since those who do eat pears are eating more of them than the tolerance calculation assumes, they are exposed to higher residue levels than calculated (Table 31.2). A better system for estimating consumption of various foodstuffs is needed.

Problems in monitoring pesticides. The U.S. Department of Agriculture is responsible for checking that meat and poultry do not contain more than the legally allowed residue of hazardous chemicals such as pesticides, drugs (e.g., antibiotics or hormones),[9] and other substances (e.g., lead or mercury).[10] The Food

---

9 More about why these are used and the resulting problems are on p. 605.
10 The human health hazard of mercury is discussed on page 216; that of lead on page 411.

Table 31.2 EPA Estimates of Consumption for Certain Foods: Less than 7.5 Ounces (213 g) of the Following Foods per Year per Person

Artichokes	Eggplant	Musk mellons	Swiss chard
Avocadoes	Figs	Nectarines	Tangelos
Barley	Honeydew melon	Okra	Tangerines
Black-eyed peas	Hops	Plums	Turnips
Blueberries	Horseradish	Radishes	Walnuts
Brussel sprouts	Kale	Raspberries	Winter squash
Coconut	Mangoes	Rye	
Cranberries	Molasses	Safflower	
Dates	Mushrooms	Summer squash	

It is easy to see that, in many of these cases, if someone eats the food at all, he or she will eat more than 7.5 ounces (213 g) per year.

and Drug Administration checks all foods except meat and poultry. The FDA is also the agency that is supposed to actually take contaminated foods off the market if the USDA or FDA monitoring programs find excessive residues. In addition, FDA is the agency that can prosecute growers for sending contaminated foods to market.

However, neither FDA nor USDA monitors all toxic substances known to occur as residues in food. This is partly a matter of economics. Both agencies rely mainly on tests that detect whole groups of chemicals at one time. The tests for some chemicals are difficult and take a great deal of time to carry out. Furthermore, in some cases there are no good tests for a particular chemical. Thus a number of hazardous chemicals are not tested for at all.

Even if USDA finds high levels of a chemical in a meat sample, the meat may not be taken off the market. The reason is that meat leaves the slaughterhouse where samples are taken within 24 hours, while test results are not received for up to a week. Since meat is not tagged after leaving the slaughterhouse, there is no way to retrieve a contaminated carcass by the time the test results come back. In 1976, the USDA sampling program found that 5–15% of meat and poultry samples contained excess chemical residues. Since almost none of the products from which the samples came were removed from the market, it is estimated that 1.9 million tons (1.7 million metric tons) of beef and 1.1 million tons (1 million metric tons) of pork were sold in 1976 with illegal chemical residue levels. The FDA has similar problems in locating and removing contaminated shipments of other foods.

Do we need more regulation? How serious a health problem is this apparent pattern of tolerance violations and contaminated food reaching the marketplace? It is probably impossible to say at the present time. We are dealing with quantities that are almost always too small to make people sick immediately. Rather, we are concerned that some of the chemicals may be carcinogenic; that they may have additive effects with other carcinogens in the environment; and that we probably will not know the final effects until after the long time it takes for cancers to develop.

The Commerce Committee recommended more personnel and better monitoring procedures for sampling and detecting illegal chemical residues. In addition to this, a method of retrieving contaminated meat and produce from the market is needed.

Drugs in Animal Feeds

Toxic residues left in foods. Various chemicals are added to animal feeds to make the animals gain weight more quickly. This is a matter of concern because small amounts (or residues) might be left in the meat and thus be eaten by humans. The hormone diethylstilbesterol, or DES, has been used as a growth promoter in cattle. DES is known to have caused cancer in children born to women who took the drug while they were pregnant. There is also evidence that it has increased the risk of cancer to the women themselves. Whether DES should be allowed as a feed additive depends in a large part on whether any remains in the meat when it is marketed. Arsenic, another known carcinogen, is approved as a feed additive on the basis that it is all excreted by the cattle before they are slaughtered.

Farmers are encouraged to look for growth-promoting additives because of the high cost of cattle feed. Anything that causes the cattle to put on weight faster or shortens the time to market means higher profits.

According to testimony in 1978 before the House Committee on Interstate and Foreign Commerce, the Department of Agriculture is having a "major problem with antibiotic residues in dairy cattle and calves. . . . We find 10 to 15 percent of the dairy cattle and four percent of the calves we test with violative levels of residues. . . . The causes of this are not completely clear, but it appears to be primarily a situation where farmers are treating sick animals with antibiotics. When the animals fail to respond, some farmers may send them to slaughter without first observing FDA drug withdrawal requirements."[11]

Drug resistance risk. A second reason for concern about drugs in animal feed involves the use of antibiotics (Figure 31.6). In the U.S., nearly 100% of the poultry, 90% of the pigs and veal calves, and 60% of cattle are given antibiotics in their feed. Animals raised in the close quarters of feed lots or large chicken ranches respond by producing more meat per pound of feed consumed, although healthy, well-nourished animals raised in exceptionally clean quarters show no such response. It is known, however, that adding anti-

11 C. T. Foreman, Asst. Secretary of Food and Consumer Services, USDA, "Cancer-Causing Chemicals in Food," Report to the Committee on Interstate and Foreign Commerce, 95th Congress, December 1978.

Figure 31.6 Feeding Beef Cattle in Missouri. Farmers found antibiotics sharply cut feed bills. (USDA Photo)

biotics to feed increases the possibility of developing strains of bacteria resistant to the antibiotics used. Whether this resistance could be passed on to bacteria causing human diseases is not clear. There was a case in England in which the injection of large doses of antibiotics into veal calves appeared to start a human epidemic of antibiotic-resistant salmonellosis. Since then, antibiotics for animals in England have been available only by prescription. Many scientists believe the health risk is real. They recommend that, at a minimum, those antibiotics especially useful in treating humans, such as pencillin, not be allowed as feed additives.

PBBs: Story of an Accidental Disaster

During the summer of 1973, 10–20 bags of the flame retardant Firemaster were mixed in with a shipment of the feed additive Nutrimaster. Firemaster is the Michigan Chemical Company's trade name of PBB (polybrominated biphenyl). The contaminated feed was eaten by animals on dozens of Michigan farms in what is now viewed as a classic example of how a toxic chemical can cause an accidental environmental disaster. Fredrick Halbert, a dairy farmer whose farm is near Battle Creek, Michigan, received a shipment of the PBB-contaminated feed in late summer, 1973. His cows ate the feed for about 16 days, each consuming up to one-half pound (about one-quarter kg) of PBBs in the feed. During this time they began to show signs of serious illness: loss of weight, runny eyes and noses, overgrowths of the hoofs, and sharply decreased milk production.

Halbert and the veterinarians he called in were unable to determine what was wrong. They sent all sorts of samples out for analysis, but no one could find anything unusual in the cattle or in their feed. By this time, Halbert had changed his cattle feed because he was still suspicious that there might be something toxic in it. All told, Mr. Halbert estimates that he spent $5000 tracking down experts and sending out samples. Finally, he located George Fries, a scientist at USDA's Agricultural Research Center in Maryland, who, by lucky chance, was familiar with PBBs. Fries was able to determine that the feed was contaminated with large amounts of PBBs. The men also realized that Michigan Chemical Corporation manufactured both a fire retardant containing PBBs and Nutrimaster, a feed additive with a similar appearance. In the investigation that followed, it was found that Firemaster was usually packaged in bags with red lettering. However, due to a shortage of preprinted bags, for a time in 1973 both chemicals were packaged in plain brown bags with the names stenciled on in black. The final link was the finding of partially used bags of Firemaster at the Farm Bureau Service mill where Halbert's feed had been mixed.

Although the authorities were now aware that there was a serious problem, some 11 months had gone by since Halbert had received his contaminated shipment of grain. Meanwhile, other farmers had received some of the contaminated feed and had fed it to their animals. State agricultural investigators visited farms suspected of receiving contaminated feed and quarantined those animals that had high PBB levels.

By this time, several thousand Michigan farm families and their neighbors had eaten contaminated meat, eggs, and milk. An undetermined but presumably lesser amount of PBBs had entered the food supply of the country as a whole.

Although the effects on the most heavily contaminated cattle were obvious, it is still not clear what effects the PBBs have had on people who ate contaminated food. Researchers are having difficulty sorting out symptoms and levels of exposure. In part, this is because the investigations are being carried out 2–4 years after people were exposed. It is also true that individual susceptibility to PBB poisoning may vary. Since the related chemicals, PCBs, are suspected carcinogens, groups of individuals exposed to PBBs are being followed for the next 20 years to determine the long-term effects of the exposure.

The state of Michigan set the limit for PBB in meat at 0.3 parts per million. Thirty thousand cattle, 6000 hogs, 1500 sheep, and 1.5 million chickens died or were condemned. Further, 18,000 pounds (8000 kg) of cheese, 2500 pounds (1100 kg) of butter, 34,000 pounds (1500 kg) of dry milk products, and 5 million eggs had to be destroyed.[12] Many farmers whose animals tested out below the legal limits reported sickness among their animals and their families. They were not able to destroy sick animals and receive compensation from the state because their animals were below the legal PBB limit. Some state officials have blamed these particular problems on poor animal husbandry. Some officials also point out that no one has been able to find a correlation between blood levels of PBBs and symptoms in humans. The farmers claim the state did not want to lower the PBB limit because the government didn't want to pay the additional compensation to farmers whose herds would be condemned under the new limit.

Gerald Woltjer, 39, who had a $1 million investment in his dairy farm, lost everything at auction and still owes $500,000. Unable to work because of a wide range of health problems, including painful joints, dizzy spells, extreme fatigue and blurred vision, Mr. Woltjer, his wife and five children are forced to live on welfare.

A man with a big smile who was once, according to his wife, jolly and even-tempered, Mr. Woltjer says he is now very moody, anxious, irritable and is also extremely impatient with the children.

Mr. Woltjer and a number of others, examined by Dr. Sidney P. Diamond, the neurologist on a team of investigators sent by Mount Sinai School of Medicine, were suffering from a loss of confidence and ambition and diminished sexual activity.

"How much of the problems are organic and how much is emotional overlay may be a valid scientific question, but in terms of the lives of these people, it's really not a relevant issue," Dr. Diamond remarked.

"These people's lives are destroyed, and that is as important as pain in their joints," he continued. "You can't take fluid out of the soul and show PBB levels."[13]

Almost all of the PBBs have probably disappeared from the country's food supply by now, although many farm families still live with the disastrous effects on their lives and health. But Dr. Irving Selikoff has this to add:

. . . This disaster we have seen was caused by at the most only 2,000 pounds of PBB. . . . But Michigan Chemical made and sold 12 million pounds of Firemaster, and they aren't saying where it went.

Most of the chemical is believed bound up with plastics to reduce flammability. But, Dr. Selikoff asks: "How much PBB escapes into the general environment? Will we discover 40 years from now that we have another problem like PCBs on our hands?"[14]

---

12 In October 1974, levels of PBBs were found as high as 595 parts per million in milk; in poultry, as high as 4600 ppm; in eggs, up to 59.7 ppm; and in meat, up to 2700 ppm.

13 *The New York Times*, 8 November 1976, p. 1.
14 *The New York Times*, 12 August 1976, p. 20.

The spread of an accidental contaminant through our complex food supply is illustrated by Figure 31.7, which shows how PCB-contaminated animal feed from a single plant contaminated meat, eggs, bakery goods, and soup in 17 states.

Better coordination of FDA, USDA, and EPA efforts at monitoring for and responding to such incidents, or perhaps the designation of one agency to deal with these events, could improve the response to incidents of accidental contamination.

A Healthful Diet

Up to this point, we have mainly discussed additives and contaminants that may be harmful in foods. The other side of the coin is also important, however. If, as

has been claimed, dietary patterns are responsible for as much as 35–40% of cancer cases, what *should* we eat to prevent disease and maintain health?

In 1982, the Committee on Diet, Nutrition and Cancer of the National Research Council issued a report on dietary patterns and cancer risk. The committee noted that the understanding of how diet is related to cancer is about at the stage that the understanding of how smoking is related to cancer was about 20 years ago. Nonetheless, the committee stated that several general agreements seem to be emerging from the mass of experimental evidence. The evidence is most persuasive that high fat and high alcohol intake increase cancer risk. On the other hand, vitamins A and C, as well as vegetables in the cruciferous family (broccoli, cabbage, cauliflower, and brussels sprouts),

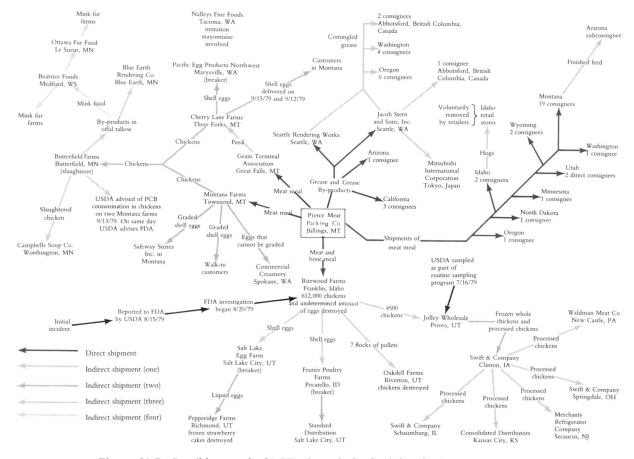

Figure 31.7 Possible spread of PCBs through the food distribution system after accidental contamination of feed at Billings, Montana. This chart illustrates the scope of the federal government investigation to date of the PCB contamination incident in the western states. Not all material identified contains PCB. (Source: Food and Drug Administration)

appear to decrease cancer risk. The committee recommended that:

1. A 1/4 reduction in the amount of fats in the average diet (from 40% to 30%) would be wise.

2. The diet should include vegetables, fruits, and whole-grain cereals, especially: citrus fruits, which are high in vitamin C; green leafy vegetables and deep yellow vegetables, which are high in β-carotene that the body can convert to vitamin A; and vegetables of the broccoli–cabbage group.

3. The consumption of salt-cured, salt-pickled, or smoked foods should be minimized.

4. Alcoholic beverages should be consumed only "in moderation," particularly by smokers, both because of cancer risks and other health problems [cirrhosis of the liver, fetal alcohol syndrome (p. 611), high blood pressure, etc.]

The committee also stated that, although there is no evidence as yet that food additives and contaminants are causing significant numbers of cancers, neither have these groups of substances been found in foods for long enough time to say that they are safe. Thus, efforts should be made to minimize intentional or unintentional contamination of the food supply with possible carcinogens. Care should also be taken to minimize food contamination with substances found to be mutagenic in bacterial tests (which implies that they could be carcinogenic). Evidence was not felt to be strong for dietary fiber as a cancer preventive or for the need for more vitamin E than Americans already consume.

Most of these diet recommendations are not new. They are, in fact, similar to those on the "prudent diet" recommended by the American Heart Association and to the "Dietary Goals for the United States" issued by the Senate Select Committee on Nutrition and Human Needs in 1977. (Also, oddly enough, they are similar to what mothers and grandmothers have been urging children to eat for years.)

Undoubtedly, further recommendations will be made as further research uncovers more associations between diet and disease. Nonetheless, according to Clifford Grobstein, Chairman of the panel that prepared the report, "it is time to further spread the message that cancer is not as inevitable as death and taxes."[15]

---

15 *Science*, **217** (July 1982), p. 37.

Toxic Substances in Drugs and Cosmetics

Hair Dyes

In the past, cosmetics were viewed simply as materials that stay on the outside of the body. Cosmetics were thus felt to be much less able to cause harm than foods and drugs. We now know that this view is too simple. Many substances are, in fact, absorbed through the skin.

A case in point involves hair dyes. When the 1938 Food, Drug, and Cosmetic Law was passed, hair dyes were specifically not included in the regulations. The reason was that they were known to cause allergic reactions in some people and it was felt that this might, under the wording of the law, be enough to ban them from the market. So they were exempted, with the provision that warning labels be put on them to tell consumers about possible hazards. This action set the stage for a later problem. In 1975, certain hair dyes were found to cause mutations (changes in the genetic material) in bacteria (the Ames Assay). This indicated that dyes might be carcinogenic. Studies by the National Cancer Institute showed that at least one dye did, indeed, cause cancer when fed to laboratory animals. The FDA was already aware that the dyes were absorbed through the scalp. In fact, there are many complaints on file that hair dyes have caused a temporary color to appear in the urine of consumers. The industry, however, argued that the feeding studies were irrelevant since hair dyes are not eaten but are applied to the skin. They further pointed out that studies of hairdressers showed no excess occurrence of cancers. Nonetheless, the FDA probably would have had enough evidence to ban certain dyes if they had not been specifically exempted from the law. The agency had to settle for warning labels on products containing 2,4–DAA (2,4-diaminoanisole), 4-MMPD (4-methoxy-*m*-phenylenediamine), or 4-amino-2-nitrophenol.

The Case for Testing Before Marketing

In past years, the FDA has been able to ban several cosmetic ingredients, including hexachlorophene, an antibacterial agent used in soaps, deodorants, and skin creams. Hexachlorophene was found to cause brain damage in infant monkeys. Another material banned was vinyl chloride, now known to cause birth defects

Pitfalls of Regulation: Tris, Pajamas, and Children's Safety

> The use of an untested chemical [tris-BP] as an additive to pajamas is unacceptable.
>
> *Arlene Blum and Bruce Ames*

> Sleepwear treated with tris-BP was being sold in late 1977 (after the ban).
>
> *Marian Gold et al.†*

There are two problems a regulatory agency must face when it considers taking a potentially dangerous product off the market. In the first place, something just as dangerous may be substituted. Second, the product may be sold anyway but not labeled. Both problems occurred when tris-BP was banned as a flame retardant in children's pajamas.

In 1975, the Consumer Products Safety Commission established standards for the flammability of children's pajamas. Arguing that as many as 3000 injuries and 100 deaths each year resulted from flammable children's sleepwear, the commission required all fabrics used in children's sleepwear to pass a standard test proving the fabric would not burn.

Since making a piece of clothing fabric flame-retardant adds 10–30% to its cost, it was felt that many people, especially the poor, would not, by choice, buy flameproof garments. In large part to protect children in families with low incomes, then, the commission did not simply require a label stating whether a fabric was flammable or not.

Most manufacturers met the new standards by treating sleepwear with tris-BP. This was the most effective and least expensive way to meet the standards. All treated garments were to be labeled fire-retardant and the material used was to be specified. Up to 5% of a fabric's weight consisted of tris to make it fire-retardant. However, in 1977, it was discovered that tris-BP was mutagenic in bacterial assays (the Ames Assay). This indicated strongly, although it did not prove, that tris-BP might be carcinogenic. Further, tris painted on rabbit skin at doses of 2.27 grams per kilogram caused male rabbits' testicles to atrophy. As a result, tris-BP was banned for use in children's sleepwear and all treated garments were recalled.

This is not the end of the story, however. Researchers checking in late 1977 found tris-BP-treated garments still being sold—unlabeled. There is evidence that a replacement flame retardant, Fyrol FR2, may also be mutagenic.‡

Thus, we have a dangerous situation that may be partly caused by a regulation. That is to say, sleepwear is required by law to be flameproof, but the safety of the chemicals being used to meet the regulation are in doubt.

Adding flame-retardant chemicals to almost all children's pajamas, as a consequence of the Consumer Product Safety Commission's standards, most probably is reducing the number of burns and deaths due to children's nightwear catching fire, although statistics are unavailable. As we have indicated, there are also other ways of reducing fire injuries.

The risk of the exposure of tens of millions of children to a large amount of a chemical must be balanced against the risk of fire. A calculation . . . suggests that the risk from cancer might be very much higher than the risk from being burned. Flame retardants (and most other large volume industrial chemicals) either have not been tested or have not been adequately tested for carcinogenicity. The use of an untested chemical as an additive to pajamas is unacceptable in view of the enormous possible risks.*

Can you think of other ways to help solve the problem of fire-related injuries? Do you think children's pajamas should be required by law to meet flammability standards? (You may want to read the two articles quoted above for more information.)

* A. Blum and B. Ames, *Science,* **195** (7 January 1977), 21.
† M. Gold, et al., *Science,* **200** (19 May 1978), 785.
‡ The manufacturer has now removed this substance from the market.

and otherwise rare liver cancer. Vinyl chloride was once widely used as a propellant in aerosol cans of hair spray and pesticides.

However, even premarket testing by manufacturers may not assure the safety of a product. Such testing depends in part on good faith by the manufacturer. Cases have been uncovered in which a manufacturer has submitted false data or incorrect conclusions. A case in 1976 involving the Searle Company led to the tightening of FDA regulations on animal testing procedures, as well as a more careful review of data submitted to the FDA.

Alcoholic Beverages

Another toxic material, sometimes classified as a drug, that is exempt from certain provisions of the Food, Drug, and Cosmetic Act, is alcohol. The authority to label alcoholic beverages is given by law not to the FDA, as is the labeling of all other foods, drugs, and cosmetics, but to the Bureau of Alcohol, Tobacco, and Firearms (BATF) in the Treasury Department.

This has caused two recent problems. In the first case, the FDA wanted to require the listing of ingredients on beer, wine, and liquor labels. This would make it easier to recall products containing an ingredient found to be harmful (a food coloring, for instance). In addition, some people are allergic to a few of the ingredients used in alcoholic beverages. Some ingredients that might cause such reactions include yeast, fruit, malt, molasses, spices, preservatives, egg white, and fish glue (the last two are used as clarifying agents in wine). Labels listing ingredients would make it possible for people to avoid those products that might cause an allergic reaction.

Winery owners and distillers opposed the labeling requirements. In part, they opposed listing ingredients because, although they know what goes into making any particular batch of beverage, without expensive tests, they cannot be sure what is left in the final product or what new compounds have formed during fermentation or aging. The FDA lost a federal court suit in 1976 in which it was confirmed that only the BATF was in charge of labeling alcoholic beverages. BATF had earlier refused to require ingredient listing.

The problem again came up when new data was found on the effects of alcohol on unborn children. As little as three ounces (90 ml) of alcohol a day, or one drinking binge, during pregnancy can cause fetal alcohol syndrome. Babies with the syndrome may show reduced growth, mental retardation, smaller head size, and defects in other organs. In addition, it is not known whether lesser amounts of alcohol might cause smaller but still unwanted effects. Again the FDA was unable to require labeling, this time warning of the dangers of drinking alcohol during pregnancy.

In addition to its teratogenic effects, alcohol acts synergistically with tobacco smoke. That is, alcohol and smoking act together to greatly increase the risk of cancers of the mouth, larynx, and esophagus.

Weighing Risks and Benefits of Drugs

As a final caution, we should note that even if all the regulations and procedures work properly, the consumer can never be completely sure that approved drugs and cosmetics are safe. There is uncertainty in the testing procedures themselves, and certain effects, notably the production of cancers, take many years to become apparent.

New data comes to light periodically on the hazards of drugs once thought to be quite without serious side effects. For instance, 15 years after use of "the pill" became common, benign liver tumors (which grow but do not spread to other organs) were found to be a (small) risk to women taking certain oral contraceptives. (Those containing mestranol are suspect. This includes Enovid, Ovulen, Ortho-Novum, Norinyl, and Norquen.) As other examples, birth defects are now believed to result from the use of the common tranquilizers Valium, Miltown, and Librium if they are taken in early pregnancy. Phenacetin, an ingredient in several common analgesics, is suspected of causing renal tumors in humans. The use of diethylstilbesterol (DES) was found to cause vaginal cancer 15–20 years later in a percentage of daughters born to mothers who had taken it to prevent miscarriage during pregnancy.

It seems only reasonable that consumers should know as much as possible about the benefits and possible harmful effects of the drugs or cosmetics they are using. In this way they can take some part in the decision about long-term risks compared to present benefits.

Questions

1. Suppose a friend asked you whether you thought smoking was really dangerous. What would you tell him? Why?
2. How would you answer some of these questions, which were asked in the National Clearing House on Smoking and Health Surveys?
   a. Should cigarette advertising in magazines, newspapers, and on TV be banned? Why?
   b. Should management be able to prohibit smoking in places of business?
   c. Should smoking be prohibited in other public places, i.e., schools, eating establishments, courtrooms, jails, bars, airplanes, and trains? (The American Lung Association can supply a booklet, "Second-Hand Smoke," which has interesting information for discussion.)
3. Do you think laws requiring fire-retardant upholstery and mattress coverings are a good idea? Why or why not?
4. How would you feel about the government's banning all but natural plant extracts as flavoring agents and food colors? Flavorings and colors would be available, but their cost might be so high that many foods (soft drinks, candies, cake icings, fruit drinks, etc.) could not be economically produced. Think about the artificially colored and/or flavored products you normally use, which might not be available. Would this bother you?
5. In some countries, bread is sold almost exclusively without preservatives. Most people buy it fresh each day. Would that work in the U.S.?
6. What is the Delaney Clause? Are you in favor of letting it stand as is or would you rather see agencies allowed to balance the risks from carcinogens against the possible benefits from their use?
7. Do you believe the FDA should forbid the use of nitrites in foods on the basis that they are at least suspected, if not proven, carcinogens? (Remember, you could then have only gray hot dogs or cold cuts and there might be an increased risk of food poisoning.) Or do you feel the consumer should be able to choose whether or not to buy nitrite-containing products?
8. Are there people who would not be able to make a choice about nitrites intelligently?
9. Can you think of ways in which the food industry could obey a nitrite ban and still provide safe products? Consider economic effects of any alternatives you suggest.
10. What do you see as the advantages, if any, to "health foods" as sold in a typical health-food store? What are possible problems, if any?
11. What sort of toxic substances in foods could be called unintentional additives?
12. How do pesticide residues get into foods?
13. Do you believe the Department of Agriculture and the FDA should try to monitor all possible toxic residues in foods? What about the cost of this?
14. Should the EPA set food tolerances for substances suspected of being carcinogens?
15. Why might EPA's method of calculating the average daily consumption of foods be especially hazardous for milk?
16. What rules would you make about toxic chemical manufacturing to protect the public against a disaster similar to that which occurred with PBBs? What kinds of information and employees would you want on hand so that you could determine quickly if a dangerous accident had occurred?
17. Who should determine the levels at which contaminated livestock must be condemned? Do the farmers deserve compensation, and if so, from whom?
18. Do you think alcoholic beverages should carry a label warning about the dangers of fetal alcohol syndrome or synergistic effects with tobacco smoke? Can you think of any problems this might cause?
19. Would you find it useful to have the ingredients listed on cosmetics and alcoholic beverages? Why?

Further Reading

Introduction to Lung Diseases, American Lung Association (1975).

The American Lung Association has a number of booklets, pamphlets, and posters on smoking. Call or write to the association in your state for a listing of what is available.

The Health Consequences of Smoking: Cancer, Report of the Surgeon General, U.S. Department of Health and Human Services, 1982.

Fenstermaker, C., "Can Our Food Supply Ever Be Completely Risk-Free?" *GAO Review,* Spring 1982.

A good summary of the problems involved in regulating possible carcinogens in the food supply.

Brody, J. E., "Food Additives, Do They Hurt?" *The New York Times,* 12 July 1978, p. C-10.

Food Additives, Who Needs Them, Manufacturing Chemists Association, Inc., 1825 Connecticut Avenue, Washington, D.C.

Industry's side of the argument is presented in this well-written booklet.

"Should the Delaney Clause be Changed?" *Chemical and Engineering News,* 27 June 1977.

Several scientists and legislators give their views pro and con on the vexing question of carcinogens, risk benefit, and the food supply.

Marijuana and Health, 1982, Department of Health and Human Services.

Current views on the hazards of marijuana use are summarized in this publication.

Ames, Bruce N., "Dietary Carcinogens and Anticarcinogens," *Science,* 23 September 1983, 1256.

References

Adult Use of Tobacco 1975, Center for Disease Control, National Clearing House for Smoking and Health, Bureau of Health Education, Building 14, Atlanta, Georgia, 30333.

The results of an extensive study of smoking habits of U.S. Citizens is published in this booklet.

"The Changing Cigarette," *Mortality and Morbidity Weekly Review,* U.S. Department of Health and Human Services, February 6, 1981.

Report of the Advisory Committee on Smoking and Health to the Surgeon General, U.S. Department of Health, Education, and Welfare (1979), available from Office on Smoking and Health, Park Building, Room 158, Rockville, Maryland, 20857.

The volumes summarize the huge amount of data that has now been collected on smoking and health.

The Health Effects of Nitrate, Nitrite, and N-Nitroso Compounds 1981, Alternatives to the Current Use of Nitrite in Food 1982, National Research Council, National Academy Press.

Hartman, Philip E., "Nitrate/Nitrite Ingestion and Gastric Cancer Mortality," *Environmental Mutagenesis,* **5** (1), 1983.

Drugs in Livestock Food, Office of Technology Assessment, Washington, D.C., 20510, 1979.

Environmental Contaminants in Food, Office of Technology U.S. citizens is published in this booklet.

Cancer-Causing Chemicals in Food, Report by the Subcommittee on Oversight and Investigations of the Committee on Interstate and Foreign Commerce, 95th Congress, December 1978.

"Diet, Nutrition and Cancer," Committee on Diet, Nutrition and Cancer, National Research Council, National Academy Press, 1982.

Toxic Substances Control and Hazardous Waste Disposal

Toxic Substances Control
Toxic Substances Control Act (TOSCA)/Risk or Risk-versus-Benefit

Disposing of Hazardous Wastes
A Hazardous Waste Dump/What Hazardous Wastes Are/Old Dumpsites: Love Canal/Superfund/Methods of Hazardous Waste Disposal/Siting Landfills: A Political Matter

CONTROVERSY:

32.1: **Toxic Substance Control in the Soviet Union**

Toxic Substances Control

In the United States, we have gradually tightened the rules governing both the use and the disposal of materials that threaten human health or the health of the environment. Congress has passed the Clean Air Act, the Federal Water Pollution Control Act, the Resources, Conservation and Recovery Act, the Safe Drinking Water Act, and the Pesticides Control Act in an attempt to reduce air and water pollutants in the environment. The Occupational Safety and Health Act is designed in part to protect workers from exposure to toxic substances where they work, while the Food, Drug, and Cosmetic Act is meant to assure consumers that harmful quantities of toxic substances do not appear in foods, drugs, and cosmetics.[1]

Toxic Substances Control Act (TOSCA)

Despite the fact that all these laws were already on the books, in 1976 Congress passed another toxic substances law, the Toxic Substances Control Act (TOSCA). TOSCA was designed to fill gaps where other laws did not apply in the regulation of toxic substances. Basically, TOSCA's main provisions fall into two categories. The first category deals with gathering data on hazardous substances so that the Environmental Protection Agency (EPA) and other regulatory agencies can make reasonable regulations. The second concerns premarket review of chemicals. These provisions attempt to ensure that hazardous substances are identified before they reach the environment, rather than afterwards. This is a major change from previous laws, which were generally designed to control toxic substances after they were discovered in the environment. Pesticides, radioactive substances, and food additives are not covered by TOSCA because they are regulated by other laws.

Risk or Risk-versus-Benefit

It is interesting to note the basis for regulation of the various laws dealing with toxic substances. Even when dealing with carcinogens, laws often allow consideration of the possible benefits of a substance or the cost of regulating it (Table 32.1).

At the present time, only part of the Food, Drug, and Cosmetic Act (the Delaney Amendment), part of the Clean Air Act, and the Resources Conservation and Recovery Act are risk based; that is, once a substance is shown to be carcinogenic, agencies are directed to reduce human exposure as close as possible to zero. No consideration of benefits is allowed. Other laws, such as the Clean Water Act and part of the Clean Air Act, direct regulators to use the best available technology or best practicable technology to reduce exposures. There is, of course, cost-balancing involved in deciding what technology is best practicable or even best available.

---

1 For more discussion of these regulations, see Clean Air Act, Chapter 20; Federal Water Pollution Control Act, Chapter 14; RCRA, see later in this chapter; Safe Drinking Water Act, Chapter 11; Pesticides Control Act, Chapter 6; Occupational Safety and Health Act, Chapter 30; Food, Drug, and Cosmetic Act, Chapter 31.

Table 32.1 Regulatory Basis of Laws Governing Toxic Substances

Law[a]	What is regulated	Basis of regulation
Food, Drug and Cosmetic Act (1938, 1958, 1960)	Food additives	Risk to health
	Drugs	Risk versus benefit
	Cosmetics	Risk to health
Occupational Safety and Health Act (1970)	20 substances in workplace	Risk versus benefit, considering technology available
Clean Air Act (1970)	Stationary sources of some 20 substances, including asbestos, mercury, and arsenic	Risk
	Vehicles; diesel particulates and control attempts	Technology available
	Fuel additives	Risk versus benefit
Clean Water Act (1972)	Toxic discharges	Technology available
Pesticides Control Act (1972)	Pesticides	Risk versus benefit
Resource Conservation and Recovery Act (1976)	Hazardous wastes	Risk to human health or environment
Safe Drinking Water Act (1974)	Substances in drinking water: pesticides, trihalomethanes, metals	Risk versus benefit
Consumer Product Safety Act (1972)	Hazards in consumer products, e.g., tris, asbestos, benzene, vinyl chloride	Risk versus benefit
TOSCA (1976)	Anything not covered in above laws, e.g., PCBs in environment	Risk versus benefit

a Dates in parentheses are the date the act was first passed and those dates on which major amendments were added.

All the rest of the laws allow benefits such as increased food production or the cure of a disease to be considered when considering risk. None of these laws, however, specifies that a formal cost–benefit analysis (as described in Chapter 29) be done. In a formal analysis, values might have to be attached to human life or happiness. Congress seems willing to accept more general value judgments from regulatory agencies about the magnitude of risks and benefits. (See Controversy 32.1 for a comparison of U.S. and Soviet approaches to regulating toxic substances.)

Disposing of Hazardous Wastes

A Hazardous Waste Dump

Recently, public concern has focused on the disposal of a specific category of toxic materials, mostly generated by industry, known as hazardous wastes. The disposal problems faced by the State of Maryland when it became necessary to choose a dump site for these wastes illustrate the problems faced all over the nation.

Hawkins Point is a small area in Maryland bordered on three sides by water and bisected by the Balti-more Beltway. On the landward side of the beltway is a small residential community. Some two dozen families own their own homes in a peaceful, almost rural setting on this point, which juts out into the bay. On the other side of the beltway, the State of Maryland is constructing a hazardous wastes dump.

Why did the state choose Hawkins Point as the place where hazardous wastes generated in the whole state of Maryland will be dumped? Can an environmentally safe dump be built there? What effect could this dump have on the people living in the Hawkins Point area? What exactly are hazardous wastes?

What Hazardous Wastes Are

Hazardous waste is a term applied to any waste materials that could be a serious threat to human health or the environment when stored, transported, treated, or disposed of (Figure 32.1).

Toxic substances are a hazard if they leach into groundwater and enter drinking-water supplies. Many such instances have been documented. Toxic chemicals from a nearby waste disposal facility have leaked into individual wells used for drinking water in small towns such as Tome, Tennessee. In other incidents, the drinking-water supplies of whole communities

have been endangered. Accidental discharges of waste carbon tetrachloride into the Kanawha River have more than once threatened the water supply of Huntington, West Virginia.

Another category is *ignitable wastes,* those which present a potential fire hazard. In Elizabeth, New Jersey, a warehouse filled with stored wastes exploded in a thunderous roar in April, 1980. A mushroom-shaped cloud of smoke rose over the city as 55-gallon drums of alcohol, solvents, pesticides, and mercury compounds exploded like bombs and rocketed into the air (Figure 32.2).

Corrosive wastes such as acids present another type of hazard. These wastes can dissolve storage containers and contaminate soils and water supplies.

A fourth category of hazardous wastes includes *reactive wastes* (those materials that can react with each other or with air or water in a dangerous fashion). In 1978, a young truck driver in Tennessee was killed while discharging wastes from his truck into an open pit. He was killed by fumes formed by liquids mixing in the pit.

Officials in Pennsylvania recently began monitoring an abandoned mine which they learned had been used as a dump for cyanide-containing chemicals. They fear that the chemicals could react with acid mine waters and form deadly cyanide gas that could then escape to the surface.

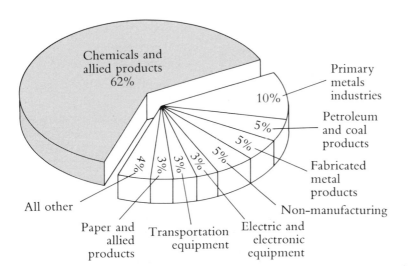

Figure 32.1 Sources of Hazardous Wastes. The major portion of hazardous wastes are generated by the chemical products industries. (Data from U.S. EPA, 1980)

Total: 41,235 thousand metric tons

Figure 32.2 A fireball illuminates the sky over Elizabeth, New Jersey. An abandoned storage warehouse for hazardous wastes blew up in 1980 when chemicals stored there ignited spontaneously. Waste disposal officials had moved some of the chemicals out several weeks before, probably preventing a worse explosion. (Wide World Photos)

Old Dumpsites: Love Canal

This last example illustrates that there are actually two major problems with hazardous wastes today. One is what to do with the more than 34.4 million metric tons of hazardous waste generated yearly. The other problems is what to do about the hundreds of abandoned dumpsites across the country that were used for hazardous wastes in the past.

The most infamous example of an abandoned dumpsite is Love Canal near Niagara Falls, New York. At the turn of the century, William T. Love dreamed of building a model community based on the cheap power he would generate by digging a canal between the upper and lower Niagara Falls. But the project failed, leaving only a partly dug ditch. In the 1920s, the city and various industries began to use the ditch as a waste disposal dump. In 1953, Hooker Chemical Company, which then owned the dumpsite, capped it with clay soil and sold it to the School Board for $1. In the deed was a warning that hazardous chemicals were buried in the old canal ditch.

The events that followed are still not entirely clear, but it appears that construction activities (a school and 100 homes were built encircling the canal), combined with unusually heavy rains, caused the dump to fill up with rainwater and then overflow. It is clear that chemicals from this dump and three others in the area leached into groundwater. Corroding drums of waste collapsed, forming sinkholes from which chemicals evaporated into the air. Backyards, basements, swimming pools, and playing fields built over or adjacent to the old dumpsite were contaminated by a variety of noxious and hazardous chemicals. Among these chemicals were dioxins, some of the most toxic chemicals known. Preliminary studies seemed to show a higher than normal incidence of birth defects and chromosome damage in residents living near the canal. Over 850 families were eventually moved from the area and their houses purchased using federal disaster emergency funds (Figure 32.3).

It is still unclear what kind of health effects residents of the area may have suffered and how severe these effects might be. This tragedy pointed up how poorly equipped the government was to deal with hazardous waste emergencies at former dumpsites. The mechanisms did not really exist to clean up these old sites or to protect the people living near them.

Superfund

Spurred by the Love Canal disaster, in 1981 Congress passed a toxic wastes cleanup bill. Known as the "Superfund," this law provides 1.6 billion dollars to be used to clean up hazardous waste sites when there is immediate danger to public health or the environment. Eighty-five percent of the fund comes from taxes on petroleum and selected chemicals. The government will sue to recover any portion of the fund expended if it can be determined that someone was at fault (i.e., if it can be proven that a firm dumped chemicals illegally). In many cases, however, the source of the chemicals at a dump cannot be determined or the dump operator has gone out of business. In these cases, the superfund can cover cleanup costs. The fund does not provide for the compensation of people injured by hazardous waste dumping. EPA has been identifying 400 sites needing top priority in terms of cleanup, as directed by the bill.

Methods of Hazardous Waste Disposal

There are already four dumps, two of them no longer in use, at Hawkins Point, where the State of Maryland will construct a new hazardous waste dump. Groundwater in the area is contaminated by a variety of chemical wastes and could not be used by industry or even to water lawns. Fortunately, residents in the area have not been using well water but have city water piped into their homes. The state health department has a master plan for cleaning up the groundwater at Hawkins Point, partly in anticipation of the new hazardous waste dump that the state wants to locate there. In fact, some state officials claim that the area could be better off because of the new dump. Is this possible? To answer the question it is necessary to look at the state of the art of hazardous waste disposal.

Cradle-to-grave monitoring. Disposal of hazardous waste is governed by the Safe Drinking Water Act of 1974, the Toxic Substances Control Act of 1976, and the Resource Conservation and Recovery Act of 1976. Regulations devised under this last act require a "cradle-to-grave" monitoring system for hazardous wastes. Any company that produces more than 45 pounds (100 kg) of hazardous waste per month must fill out a manifest for any wastes leaving the plant site. The company that transports the waste must sign the manifest and give a copy to the company handling final disposal. This company must return a copy of the manifest to the company that originally generated the waste. In this way, officials at the company generating the waste can be sure that their hazardous wastes have not simply been abandoned in some remote area or otherwise disposed of improperly. Generators are responsible for notifying EPA if they don't receive notice that their waste has been properly handled.

Figure 32.3 Love Canal home owners protest contamination of their neighborhood. Many families were forced to stay in their homes long past the time they became convinced it was unhealthy because they could not afford to move. State and federal governments eventually moved 850 families out of the Love Canal neighborhood and bought their homes. However, this was only after several years of meetings and protests. Worry about health effects, especially effects on their children, has taken a sad toll in stress-induced mental, physical, and social effects on former residents. (Citizens Clearinghouse for Hazardous Wastes)

Toxic Substance Control in the Soviet Union

Even a brief comparison of the standards for exposure to workplace air pollutants in the U.S. and the Soviet Union shows an astonishing difference in the standards (Table C.1). Of the 100 substances for which both countries set standards, the Soviet standards are more strict 80% of the time, often by a significant amount. In only two cases are American standards more strict than Soviet standards. From the standards alone, it would appear that the Soviets are providing more protection for their workers than we do in the U.S.

Table C.1 A Comparison of Representative American and Soviet Workplace Standards for Air Pollutants[a]

Substance	American standard[b] (mg/cubic meter)	Soviet standard[b] (mg/cubic meter)
Aldrin	0.25	0.01
Aniline	19.0	0.1
Carbon monoxide	55.0	20.0
Dioxane	360.0	10.0
Ethyl alcohol	1900.0	1000.0
Ethyl mercaptan	25.0	1.0
Ethylene oxide	90.0	1.0
Heptachlor	0.5	0.01
Hydrogen cyanide	11.0	0.3
Methylchloroform	1900.0	20.0
Vinyl chloride	1300.0	30.0
Acrolein	0.25	0.7
Anisidine	0.5	1.0

a From Ekel, George J., and Warren H. Teichner, *An Analysis and Critique of Behavioral Toxicology in the USSR,* NIOSH, Dec. 1976. U.S. Government Printing Office, Washington, D.C. 20402

b U.S. standards are 8-hour weighted average concentrations, while Soviet standards are ceiling values never to be exceeded. Thus, Soviet standards are actually more strict than appears from the numbers alone.

This appearance is partly due to the methods of testing used. Russian toxicologists give a great deal of weight to psychological and neurophysiological effects of toxic substances; that is, effects on the sense of smell, on electroencephalogram responses, and on animal motor responses. These effects can be very subtle. They occur at much lower levels than those causing the more visible physical effects that concern U.S. toxicologists, such as production of tumors, decreased respiratory function, etc. Further, not all scientists would agree that such responses prove any harm has been done.

Besides differences of opinion about the kinds of experiments to use, there seems to be a major difference in the philosophy of setting standards in the two countries. One Soviet author★ identifies four possible bases for regulating decisions.

1. A normal unpolluted environment is the natural condition and any change caused by a pollutant should be considered not permissible.

2. The major consideration should be what levels of pollution control are technologically feasible.

3. Economic considerations should be the most important. That is, the allowed pollution level should be set so that the cost of controlling a pollutant is equal to the cost to society of the damage that will result if the pollutant is not controlled.

4. Standards should be set at levels for which there are no direct or indirect harmful or unpleasant effects on humans, in terms of effects on their ability to work as well as their physical and emotional well-being. (It is worthwhile noting that any measurable effect is felt to be potentially harmful; that is, an odor in the workplace, even if generally considered pleasant, would be not permissible, in the same way that adding sugar to the water supply would be undesirable.)

The author then goes on to state that the first option of allowing no change in the environment is unnecessarily strict, because not all changes lead to measurable effects in humans. The third option, which considers only the costs of controlling pollution vs the cost of not controlling it, he feels is unnecessarily commercial, since it involves setting a price on human life, health, and well-being. He feels the second option is not strict enough, since in most cases the level of control technically feasible at the present would still allow harmful effects. Thus, standards set at such levels tend to hinder development of better pollution control. The fourth option he considered the only proper approach.

However, it should be noted that while standards in the Soviet Union are often set much lower than in the U.S., the methods for achieving these standards are not always defined. In fact, it is likely that these levels are often exceeded, at least where the levels are lower than what is technologically feasible or where achieving those levels is extremely costly. The Soviet standards are thus ideals, or "targets" to work toward, rather than limits that must not be exceeded. It is expected that decisions will be made at the local level on whether or not standards will be met.

In the U.S., although we do not seem to have a uniform philosophy about setting standards for toxic pollutants, once standards are set, it is expected that they will be enforced. There are certainly environmental protection groups in the U.S. who feel that option 1, no change in the environment, is the safest, most desirable course. The Delaney Amendment, which prohibits any substance in foods once it has been shown to be a carcinogen, is an example of this most restrictive category of U.S. law. The law allows no consideration of possible costs or benefits involved.

But it is also, in the strictest sense, unenforceable. Some carcinogenic substances such as pesticides do find their way into foods through indirect means. Congress was even forced by public pressure to exempt from control the food additive saccharin, a weak but known carcinogen.

A number of U.S. standards are set according to the second option: what is technologically feasible at the moment. Some water-pollution discharge standards, for instance, are based on the best available technology. So also are OSHA workplace standards, which require that worker exposure to carcinogens be lowered to the lowest possible levels. In other cases, Congress has set future standards below levels that were currently technically possible in order to stimulate industry to develop the necessary technology; standards for emissions from automobiles are an example. In many U.S. standards, economic considerations are given weight.

Which of the four possible bases for regulatory action do you feel are reasonable? Should one of them be the major basis for U.S. decisions, or does the present patchwork serve better? Why?

Do you agree or disagree with the Soviet basis of standard setting and decision making? Why?

Is the third option, which allows the cost of regulation to be compared to the cost of damage to society, the same as risk–benefit analysis? Can it be made to work? Why or why not?

---

* V. A. Riazanov, in Ekel, George J., and Warren H. Teichner, *An Analysis and Critique of Behavioral Toxicology in the USSR,* NIOSH, Dec. 1976. U.S. Government Printing Office, Washington, D.C. 20402

Recycling. Of course, the best disposal solution for hazardous waste would be to *recycle* it into other uses or to recover valuable materials from the waste. In some cases, this can be done. For instance, waste solvents from the electronics industry are still of high enough quality that they can be substituted for virgin materials used in some other industries. EPA has been encouraging the establishment of waste clearinghouses to encourage more matching of useable waste materials with potential users. About three percent of all industrial wastes could probably be reused in this way.

Chemical treatment. Wastes that cannot be reused in some way can sometimes be made much less hazardous by simple *chemical treatment.* For instance, corrosive acids or bases can be neutralized. This sometimes causes toxic heavy metals to be released from solution as solid materials that can be filtered out. Such processes are usually more economical to carry out at large regional waste reprocessing centers. Sometimes such centers can even mix wastes from different companies in order to detoxify both. Acidic and basic wastes can be matched in this way.

Composting and land farming. Organic materials that have little or no heavy metals can be detoxified biologically. *Composting and land farming,* in which materials are spread out over a large land area so that microbes can decompose them, are examples of biological treatment of hazardous waste. Pesticides, phe-

nols, TNT wastes, and paper-mill wastes have all been detoxified by these methods. Monitoring is necessary, however, to insure that the wastes do not percolate down into groundwater before they are detoxified.

Incineration. A more expensive but possibly safer method of disposal is *incineration* in special high-temperature incinerators. Modern incinerators are designed to destroy at least 99.99% of the organic waste material they handle. Any materials that escape up the flue are scrubbed out of the flue gas with special equipment. All of this makes incineration an expensive alternative.

Communities near incinerators have objected to

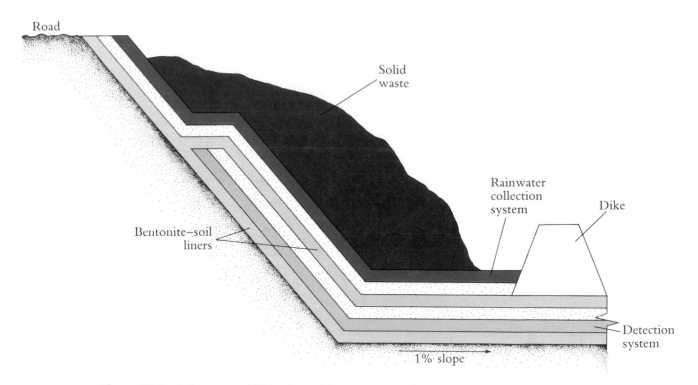

Figure 32.4 A Secure Landfill Design. This cross section shows an example of a secure landfill. It was built by Olin Chemical Corp. to dispose of chlor-alkalai wastes from its Charleston, Tennessee, plant. The cells have a bottom liner of soil and sodium bentonite clay. A detection system is embedded in this layer to monitor any possible chemical leaks. Above this is another bentonite–soil liner, and over this is a rainwater collection system with a pump. Rainwater that seeps in is pumped out and treated. Landfills designed to handle organic liquids require a synthetic membrane liner rather than clay, since clay liners allow organic liquids to leak out eventually.

them because of fears about possible emissions. Some experts feel the answer is to incinerate on ships in mid-ocean, far from any people. It is speculated that these ships would operate without scrubbers, since the trace amounts of hazardous wastes emitted would be a small population burden when mixed into the ocean. At least two such ships already operate: the *Vulcanus,* a Dutch ship, and the *Matthias II,* a German ship. The *Vulcanus* has incinerated large quantities of Agent Orange, a very hazardous herbicide contaminated with even more toxic dioxins, for the U.S. Air Force. EPA monitored the incineration, which took place in mid-Pacific, and found essentially no hazardous emissions.

Deep wells. In some areas, hazardous wastes have been pumped into *deep wells.* Controversy has arisen over the eventual fate of such materials. Proponents argue that most such wells are drilled in geologic formations that have naturally held brine in isolation from fresh water for millions of years. Also, if a use is eventually found for materials in the wastes, they can be pumped up again. Critics express concern about possible leakage and poorly designed wells. They note that explosions and even earthquakes have apparently resulted from waste injection techniques, although additional safeguards might have prevented such incidents.

Landfills. Despite these alternatives, there are still large quantities of materials that are not recycled or reused; that have heavy metal concentrations too great to degrade or incinerate; or that are too thick to inject into wells. These wastes are commonly buried in *landfills.* A variety of factors must be considered in making a landfill "secure"; that is, able to contain hazardous wastes without leakage to the environment.

The best soil for a landfill is clay because clay is less permeable than other types of soil. The clay pits in the landfill must usually be lined with another material because certain organic chemicals react with clay and cause leaks. Some liners are natural materials such as crushed limestone or nut shells; others are synthetic membranes. The liners help contain hazardous wastes by absorbing, or chemically reacting with them, or by providing an impermeable barrier to leaks. A leachate system collects liquid that escapes from the pits. The leachate is monitored to detect any escaping hazardous materials. In some cases, the term landfill is a misnomer, since in order to keep hazardous materials

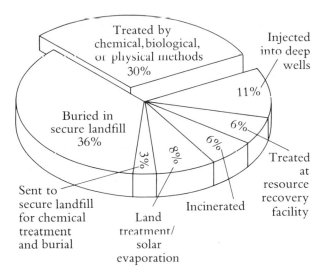

Figure 32.5 Current Disposal Practices. Most hazardous waste generated in the U.S. is disposed of or treated on the grounds of the industry that produces the waste (83%, according to *Civil Engineering—ASCE,* September 1981, p. 82). When hazardous wastes are transported somewhere else for disposal (above), about one-third is treated to make it nonhazardous and the remainder is incinerated, injected into deep wells, applied to land, or recycled. (Data from: U.S. Environmental Protection Agency, 1980)

above the water table, the dump may be entirely above ground, forming a huge mound 30 or more feet high (Figure 32.4).

Materials disposed of in a landfill can be further secured by solidifying them in materials such as cement, fly ash from power plants, asphalt, or organic polymers.

Current disposal practices. Most hazardous waste today is disposed of on the grounds of the company generating it. Figure 32.5 shows how the waste is handled that is transported off a company's grounds.

Siting Landfills: A Political Matter

It is important that monitoring of a landfill go on for a long time, at least 20–30 years, but possibly forever, to be sure the site is secure and that nothing has disturbed the cover. Disturbance of this cover was probably the major cause of leakage from the Love Canal site in upper New York State. Because of the possibility of leaks, and in spite of the technology that exists to build a modern, secure hazardous wastes landfill, the public as a whole remains wary. This is understandable in light of past disposal practice and its legacy of chemical contamination of the environment. To quote Jackson B. Browning, Director of Health, Safety and Environment for Union Carbide:

> It is not terribly difficult to elicit a political consensus in support of the need for siting sound hazardous waste treatment facilities. The majority of voters will readily agree that such sites are necessary and that you should move with all deliberate speed to locate them—at the other end of the state.

Local community opposition often makes it nearly impossible to site hazardous waste landfills. In such a climate, politics rather than technology decides where a disposal site will go. A case in point is the Maryland hazardous waste dump at Hawkins Point. Although the state considered over 40 possible sites, Hawkins Point was chosen in large measure because of a lack of opposition. Most of the land was already owned by the state. The local political jurisdiction, Baltimore City, agreed not to protest such use of the land because the city will get, as part of the deal, a badly needed trash landfill adjacent to the hazardous waste dump. The state health department, which must approve the site, was concerned with cleaning up the already contaminated groundwater and seeing that the three existing chemical landfills at Hawkins Point operate in a secure manner. The one source of possible opposition were the residents of Hawkins Point, who were generally not affluent and had little political influence. Although state officials felt residents would be better off after the Maryland hazardous waste landfill was sited there, due to better fire protection and stricter regulation of all the dumps in the area, residents didn't expect much improvement from the siting of two additional dumps in their area. Eventually, the state offered to buy the homes of residents who wished to move out of the area.

Questions

1. How does the Toxic Substances Control Act differ from the many other laws dealing with toxic substances?
2. What are hazardous wastes?
3. Why are abandoned dumpsites a problem? How can a community deal with the problem if such a dump is discovered?
4. What methods are available for disposal of hazardous wastes?
5. How would you feel if you discovered that your state planned to site a hazardous waste dump on vacant property near your home? What questions would you ask of state authorities? What safeguards would you feel were absolutely necessary to protect yourself and your family?

Further Reading

Walsh, J., "EPA and Toxic Substances Law: Dealing with Uncertainty," *Science,* **202** (10 November 1978), 598.
A good review of TOSCA and also of the difficulties involved in making laws in an area of scientific uncertainty.

The effects of living near a leaking chemical waste site can be at least as damaging psychologically as physically. The following articles give some idea of the stress Love Canal residents lived with.
"Love Canal: A Boyhood is Poisoned," *The New York Times,* 9 June 1980, p. B1.
"A Tangle of Science and Politics Lies Behind Study at Love Canal," *The New York Times,* 27 May 1980, p. A1.

"Love Canal Residents Under Stress," *Science,* **208** (13 June 1980), p. 1242.
Protecting America's Land, *EPA Journal,* July/August 1982. This entire issue is devoted to hazardous waste: history, laws, and possible solutions.
Witherspoon, P. A., N. G. W. Cook, and J. E. Gale, "Geologic Storage of Radioactive Waste," *Science,* **211** (27 February 1981), p. 894.
Hill, D., et al., "Management of High-Level Waste Repository Siting," *Science,* **218** (26 November 1982), p. 859. Radioactive waste disposal is part of hazardous waste disposal, but poses extra problems that must be considered. These papers provide an entry into the literature.

References

Maugh, Thomas H., "Burial is Last Resort for Hazardous Wastes," *Science, 204* (22 June 1979), p. 1295.

Civil Engineering—ASCE, September 1981, pp. 68–81.

Maugh, Thomas H., "Biological Markers for Chemical Exposure," *Science, 215* (5 February 1982), p. 643. Summarizes recent books and conferences on the problems of monitoring toxic substance escape from dumpsites and of determining effects on neighboring populations.

"Research Needs for Evaluation of Health Effects of Toxic Chemical Waste Dumps," Proceedings of a Symposium, Environmental Health Perspectives, December 1982.

Environmental Monitoring at Love Canal, Volume 1, U.S. Environmental Protection Agency, #600/4-82-030a (May 1982).

Ekel, George J., and Warren H. Teichner, *An Analysis and Critique of Behavioral Toxicology in the USSR,* U.S. Department of Health, Education, and Welfare, December 1976.

(U. S. Department of the Interior, National Park Service photo)

# PART EIGHT

Land Resource Issues

I conceived that the land belongs to a vast family. Of this family, many are dead, few are living, and countless members are still unborn. . . .

A Nigerian Chief★

An often heard phrase is "the quality of life." It refers to society's and an individual's well-being. One kind of well-being is economic; another is social. Elements of economic well-being include jobs, pay, and the cost of food, housing, clothing, transportation, medical care, and other basics. An assessment of social well-being asks how civilized and stable a society is; that is, whether there is widespread crime, alienation, or prejudices. Well-being also takes in our concept of freedom to move, change jobs, and travel. Beyond the economic and social components of the quality of life is still another concept, less distinct and less measur-

able, but nonetheless real and important. It has to do with the quality of our environment.

The quality of the environment does not lend itself to neat characterization. Instead, a set of questions might be used to grope toward the meaning of quality of the environment. One might ask, for instance, if there is unspoiled forest nearby that is open to the public to hike or backpack and observe wildlife? Are there parks where families can picnic? Are there coastal beaches within reach that have not been locked up in private ownership? Are historic buildings preserved so that people can see the origins of their settlement? Are farms preserved nearby, both for the food they supply and their contribution to our need for green? Or does surburbia with shopping centers and strip development sprawl to infinity? Is there a place where children can buy pumpkins from the farmer who grew them? Or better yet, where the children can grow pumpkins themselves? Each individual is likely to have a different set of questions to reveal the quality of the environment. What we ask obviously reveals our own personal measures of environmental quality.

★ Quoted in *Land Use and the Environment,* Virginia Curtis, ed., Environmental Protection Agency, Office of Research and Monitoring, Environmental Studies Division, 1973.

One element stands out as a key in all these questions and touches every one of these concerns: how we use the land. If we use the land badly, the questions will be answered in a negative way. If we use the land wisely, the questions can be answered in a positive way.

The subject of land use is an unusual blend of issues. To improve the use of the land, we need to know how we can influence the numerous private decisions that go into creating a pattern of land use. We also need to know where private decisions are likely to fail to bring desired patterns of use; that is, where it is necessary for the public to acquire and preserve the land and its qualities. And when the public comes to own such lands, how can their qualities best be protected?

The following chapters discuss how humans use land areas and how we are attempting to preserve some of these areas and the natural communities living there. Chapter 33 treats private land-use decisions, the numerous microscale decisions that go into the formation of a pattern of land use. We discuss the concept and practice of zoning, a tool to enforce the preservation of established land-use patterns in urban and suburban areas. We also discuss new methods that are evolving to control the loss of farmland to development. In Chapter 34, we take up the issue of public lands and their purposes. There we explore the historical development of public ownership of land and the various forms it currently takes.

CHAPTER THIRTY-THREE

Private Land-Use Decisions

A Changing Tradition

Enforcing Urban/Suburban Land-Use Plans
Zoning by Use Classification/Zoning as a Tool for Environmental Improvement/Zoning to Protect the Public/Growth Management Plans

Preserving Rural Land
Farmlands Are Becoming Suburbs/Taxation Is a Root Cause of Farmland Loss/Ways of Preserving Rural Land/Summary of Ways to Preserve Rural Land

Other Methods to Secure the Preservation of Rural Land

A Changing Tradition

Traditionally, decisions about land use that are made by private individuals and companies have been "let alone"—to a very large degree. This tradition is changing, however, as we discover that our land resource is limited and that the impacts of many private decisions have a way of adding up. In the past, a new factory meant jobs and purchasing power. Stores clustering along a major highway meant more access to goods and services. Now we also recognize that certain kinds of industry and commercial development may bring air and water pollution or high volumes of traffic.

Stores, shopping centers, and the like are beginning to face controls on their freedom of action as citizens express their concern about the noise, the traffic, and the visual aspect caused by new commercial development. We are faced with difficult choices about what is desirable or needed and what is undesirable or can be done without.

Finally, we now realize that the steady development of suburbs is consuming prime farmland at an agonizing rate. The consumption of prime agricultural land for homes is a trend that somehow must be checked.

Enforcing Urban/Suburban Land-Use Plans

Zoning by Use Classification

The most common form of land-use control is that practiced at the local level by community and county governments. Referred to as **zoning,** it is another way of saying that land has been classified as to its appropriate uses. By and large, such zoning is not concerned with air and water pollution, but with the uses of neighboring land. "Is the proposed activity a reasonable 'neighbor' to the activities that already exist in the area?" is the question asked by zoning boards.

The idea of zoning is young. The first comprehensive municipal zoning law in the United States was established in 1916 by New York City. The appeal of the zoning idea was enormous and immediate. In only 10 years, more than 400 American cities had adopted zoning laws. In that year, 1926, the Supreme Court of the United States upheld zoning for the Village of Euclid, Ohio, as a legitimate means of controlling nuisances. The legal justification of zoning is that it protects the health, safety, and welfare of citizens. While zoning may, in fact, serve these purposes, zoning is used most often to protect the values of privately held land.

Zoning laws. Zoning begins when a state government passes legislation that "enables" local and county governments to create land-use plans if they wish to do so. The local government then draws a map that labels each area or zone of the locality as to its allowable uses. Typically, the uses fall into one of four categories, in addition to agriculture. These four categories are industrial, commercial, multi-family residences, and single-family residences. These four categories may be further divided as well.

Figure 33.1 shows the four categories in an arrangement that has been termed "cumulative zoning." Within this planning framework, a given category of use can accept all uses above it in the chart. That is, single-family residences, multi-family residences, and

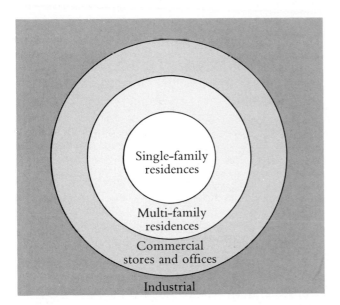

Figure 33.1 The Cumulative Zoning Concept. Single-family and multi-family residences and commercial activities are allowed within an area zoned "industrial." Single-family and multi-family residences are allowed within an area zoned for "commercial" activities. Single-family residences are allowed in an area zoned for "multi-family" units.

commercial activities are all allowed within an area zoned "industrial." In a similar fashion, land zoned in the "commercial" category can be used for its designated use (stores and offices), but single- and multi-family dwellings are also allowed. On the other hand, land earmarked as being in the "multi-family" zone cannot be used for either commercial or industrial purposes. The "single-family residences" category is often referred to as the "highest," or most restrictive, grouping because no other uses are allowed.

Zoning used to protect a land-use pattern. In addition to stating the allowed uses of the land, zoning may be utilized to enforce a uniform pattern of use. For example, in the single-family residence category, the minimum size of the lot may be specified, or the distance of the dwelling from a sidewalk (setback) may be given, or the minimum space between dwellings may be spelled out, as well as allowable construction materials and building heights.

Zoning is typically not cast in concrete. Changes in the ordinances can be made by the body that enacted them in the first place. More often, though, those who

are putting land to "nonconforming" uses may apply for "variances" or "special exceptions." Such exceptions are considered by a planning board or zoning commission on a case-by-case basis.

The idea of having a procedure to consider uses different from those normally allowed is quite reasonable. However, the results of the procedure are often not so reasonable. In the past, and probably in the future as well, commercial interests, apartment developers, and the like often have obtained exceptions for their activities. They do this simply by pleading hardship or by pointing to existing nonconforming uses. A common technique is to begin construction without a variance, but with the excuse that a variance is expected. When the variance is not forthcoming, the developer pleads hardship because of his already "sunk" costs.

When a sufficient number of exceptions have been made, the original labeling of an area loses meaning and further exceptions become easy to obtain. Zoning, then, as a barrier to certain kinds of development, resembles a windbreak. It can help shelter an area from the winds of change, but it cannot always protect it.

Enforcement of zoning laws. Although zoning is interpreted and administered by a planning board or zoning commission, it is enforced by the courts. Those who violate the zoning statutes can be enjoined from doing so; that is, they may be given a court order to prevent their violation. The court's muscle is its ability to fine those who disobey its orders.

We noted, though, that zoning sometimes fails. Failures may occur as individual instances or they may accumulate to produce wholesale changes in a neighborhood. Such failures may be due in large part to the way variances have traditionally been granted.

Requests to rezone a given parcel of land are heard, as we noted, by a zoning board or commission; their meetings are generally open to the public. The commissioners or board members take note at the meeting of community members who come to protest the exception that has been requested. A lack of protest by citizens is often interpreted as meaning that the residents of an area do not object to the proposed exception in use of the land. In fact, citizens may not know the meeting is being held, may have prior commitments, or may be unable to attend. If only a few people appear at the meeting to protest a proposed change in land use, the board may grant the exception to the person or firm requesting it. Thus, making zoning work requires alertness and involvement on the part of citizens.

The publicly proclaimed purpose of zoning is to maintain stable patterns of settlement, but the goal of zoning in practice is quite different. People use zoning primarily to protect the value of the residential property they own. Nearby factories and traffic are seen as threats to property value, and zoning is used to exclude uses that could diminish property value. Zoning, then, has not generally been an instrument to protect the general environment—only the environment of some people.

Shortcomings of zoning. It has been argued that some forms of zoning amount to economic discrimination. Consider, for instance, the common requirement that lot sizes be an acre or more. Such a requirement will cause a builder to erect expensive homes in order to earn enough profit on his investment in the land. Expensive homes are bought by people with upper levels of income; lower-income people are unable to afford these dwellings. While the zoning of acre lots does seem to foster upper-income ownership, the original reason for acre lots may have been the protection of groundwater from septic tank discharges.

It has also been argued that zoning can lead to a monotonous conformity in the landscape. The diversity of having shops, playgrounds, and public buildings near residential sites can create an atmosphere of interest and appeal. Those who have visited Europe will have seen neighborhoods with a wider variety of activities than in United States suburbs. In Europe, each neighborhood, in addition to its homes and apartments, is likely to have a school, a bakery, a vegetable store, a meat market, a drug store, a candy store, a cigar store, and so on.

Suburbs in the United States, in contrast, are often simply miles of homes with services very distant. Residential areas in the United States that exclude commercial shops tend to waste energy. A purchase at the grocery store by a resident of the suburbs often requires a trip by car. Most Europeans can find a sufficient selection of food within a few blocks of their homes or apartments. They not only save on gasoline by walking to market, but get useful exercise as well. Thus, the exclusion by zoning of any kind of com-

merce in a residential area results in an added use of energy.

Zoning as a Tool for Environmental Improvement

However, for all its shortcomings, zoning is firmly in place in the United States. And despite its flaws, zoning can be used in clearly positive and beneficial ways. New zoning concepts are providing these improved possibilities for zoning.

Natural area zoning. One new concept is *natural area zoning*. First adopted in New York City in 1974, natural area zoning can be used to protect wetlands such as swamps, bogs, ponds, and creeks, as well as forested areas and rugged landscapes. The greenbelt area of Staten Island, a borough of New York City, was the first area to be classified by the New York City Planning Commission as a protected natural area, and hence not open to development.

Cluster zoning. We noted that zoning has been criticized for the economic barrier it often sets up. A concept that would not set up such economic barriers is *cluster zoning*. This concept allows homes to be built much closer together than conventional zoning allows, so long as the total ratio of dwellings to total land area of the development remains at the desired level. The homes may be built with shared exterior walls, common walkways, and small yards. Such common wall construction requires fewer materials, shorter lengths of water and sewer lines, less sidewalk area, less pavement, and less labor. As a result, the dwellings can be sold at a relatively moderate price.

As an example, suppose a community insisted on one-half acre per home (or two homes per acre). The community might be willing to accept 40 homes built on a total of 20 acres, with the 40 homes clustered onto just 10 of those acres. To get the savings of clustered construction, the remaining 10 acres surrounding the clustered development would have to be left undeveloped to meet the requirement of no more than two homes per acre. For the entire area of 20 acres, the ratio of homes to acres would be the required two to one. The added open space can enhance the community and provide resources for recreation and leisure to the development itself. In addition, the clustered homes may be less expensive than individual dwellings, thereby providing more access to home ownership for more people.

Zoning to Protect the Public

Zoning has also been used to protect the public from hazards. Although hazard zoning is used much less often than general land-use zoning, its most common example is flood-plain zoning. The flood plain of a particular river is the area outside the river's normal channel over which its flood waters may flow. In the past, it has been common to see dwellings built in a flood plain. They have been built there because memories of floods may have faded and because the homes have access to the water and an attractive natural setting. The arrival of a flood, however, refreshes memories and makes the setting less attractive.

When an area is classified by zoning as a flood plain, certain land uses may be banned. These prohibitions make a great deal of sense because local, state, and federal governments bear much of the burden of flood damages. Such burdens include rescue, cleaning up, and restoration of homes, shops, and factories devastated by flood. Governments may even go to the extent of building flood-control structures such as levees and costly dams in order to protect dwellings that ought never to have been located on the flood plain in the first place.

Some years ago, Professor Gilbert White recognized that the problem of flood control was being approached much like that of the proverbial horse who escaped from the barn. Only after the horse's escape was the barn door being closed. Far better, reasoned White, to avoid the misery, hardships, and enormous expense that were caused by homes and commerce built in a flood plain. The expenses of restoration and of flood-control structures could be avoided by simply barring new development in the path of a potential flood. Hardships and expense could also be avoided by removing flood-threatened dwellings from their current sites.

White's ideas were translated into federal law in 1968. Flood insurance was provided under the National Flood Insurance Program enacted in that year, and this insurance was used to entice local communities to adopt flood-plain zoning. This desirable, low-

cost insurance that the act made available is paid for, in part, by the federal government. However, the insurance can be made available only to residents of those communities that have taken active measures to prevent further development in the flood plain. By the mid-1970s, more than 13,000 communities had indicated a desire to participate in the insurance programs and hence prevent further development in the flood plain.

Growth Management Plans

There are still other ways to influence private decisions on the development of land. Several of these fall under the heading of "growth management," and their history is quite recent. While growth management plans appear legitimate for the moment, they are, nevertheless, controversial. Although general land-use zoning has been approved in principle through decisions of the U.S. Supreme Court, challenges to growth management plans have, to this time, never been heard by the Court. Their present status could, therefore, change.

First in Ramapo, New York, and then in Petaluma, California, communities attempted to control the rate at which they grew. Ramapo was then being developed as a suburb of the New York Metropolitan area. To slow and control its development, Ramapo drew up an 18-year plan that included an orderly expansion of its road network and of its water and sewer utilities. If developers had been allowed to build homes and shopping centers wherever they wanted, the network of roads and water and sewer lines would have had to grow rapidly in many directions to meet new needs. The costs of rapidly expanding this network to go wherever developers chose to build would have been prohibitive to the community. Hence, the Ramapo plan was not only aimed at preserving the rural environment of the community, but also at preventing the explosion of local taxes. Developers who wished to build in areas as yet unimproved by the community had to bear the costs of roads and of water and sewer lines themselves. Challenged in court by the developers, the town's plan was upheld in 1972.

Petaluma, California, under pressure for housing from the San Francisco Metropolitan area, adopted in that same year (1972) a plan that limited the construction of new houses to 500 per year. This figure was far lower than their growth rate of nearly 2000 homes per year in the recent past. Petaluma's plan was at first challenged in Federal District Court, where the plan was ruled unconstitutional for infringing peoples' "right to travel" where they wished. When the case was brought to the Court of Appeals, however, the ruling was reversed, and Petaluma's authority to limit its growth was upheld.

Two features have been common to all methods of influencing private land-use decisions that we have discussed so far. First, they have been aimed at urban and suburban land-use decisions, as opposed to decisions on rural land. Second, the methods have all involved using the police power of government (the power of the courts) to enforce a set of rules. Private individuals are bound by law to follow these rules, unless exceptions are granted in response to specific requests.

Preserving Rural Land

In contrast to the methods that use government enforcement power are techniques that use economic means to convince or require private owners to make better decisions on land use. These economic methods fall short of actual purchase of the land, but in terms of accomplishing their more limited objectives, they are very powerful. Moreover, they are generally more suitable for use on rural land, especially farmland, than the enforcement power of government.

Farmlands Are Becoming Suburbs

The rural land surrounding most of our cities is in genuine danger of extinction as the suburban fringe expands, foundation by foundation, out into the countryside. We value and admire the farmland that surrounds us because it reminds us of an earlier era in which certain aspects of the quality of life were better—when traffic, noise, and congestion were less and when the proportion of our population that worked the land was far larger than today. The suggestion of "a ride in the country" is a very common reference to our feelings for rural land. The tranquil, less hurried atmosphere and the visual pleasure it may impart touch a responsive chord in many of us.

There is more, however, beyond our feelings for rural land. The rural lands on the fringe of the suburban front contain among them the richest agricultural lands near our cities. Some refer to such land as "prime." It must be surprising to be told that the rural farmland being gobbled up by suburban sprawl is among the best and most productive in the country. It seems almost perverse that the rural land most in need of protection for our own economic well-being is also the land most threatened. The explanation is simple.

When our cities and towns were first established, dirt roads connected them one to another, and dirt roads reached out into the surrounding countryside where vegetables, fruit, and animal stock were grown. Naturally, these roads went to the richest farm lands, for these were the roads farmers would use to transport their animals and produce to the town and city markets. This fact of historical geography has brought us to the present, in which those same roads are now widened and paved. These roads, which once carried produce-laden wagons into the cities, now carry commuters in steel-clad vehicles back and forth between homes and commerce and industry. The result is that the rich farmland along these roads is taken for new housing developments. Thus, more and more farm products come from farther away at greater expense. This increased expense is due to the increased distance of shipment and to the fact that the produce may possibly be grown on less productive land. It is as though we were eating not merely the bread but the breadbasket as well.

How severe is this problem? How fast is prime farmland being gobbled up? While the statistics are patchy, the fact of farmland development can be observed by simply taking a car ride to the city fringes.

In 1981, an estimate of the rate of consumption of farmland was prepared by the National Agricultural Lands Study, a cooperative study by twelve U.S. government agencies. The study estimated that during the ten-year period from 1967 to 1977, about 30.8 million acres of farmland were taken; these acres were used for suburbs, for transportation, for shopping centers, for water-resource development, and for production of minerals such as coal. That is a loss of some 3 million acres per year or 12 square miles per day (1.2 million hectares per year or 32 square kilometers per day).

An estimate in the late 1970s by the President's Council on Environmental Quality placed the agricultural land consumed by suburban "sprawl" at 1,000,000 acres (400,000 hectares) per year. This is the equivalent of 1520 square miles (4000 square kilometers). California, at about the same time, estimated its annual rate of agricultural land conversion at 21,000 acres (8400 hectares), or about 3% of the annual loss in the entire country. These figures, however, must be viewed in the perspective of how much land we have and how it is currently being used.

The land area of the United States, excluding Alaska and Hawaii, is 3,522,000 square miles (9,100,000 square kilometers). This is equivalent to 2.27 billion acres (0.91 billion hectares). Of this quantity, about 2.7% of this land area is built over with homes, commerce, and industry. About 18% (413 million acres) of our total land area is actually in use as farmland, according to the U.S. Department of Agriculture. Another 127 million acres is available for farmland, but is presently used as pastureland, rangeland, or forests.

We noted earlier that about 1,000,000 acres of agricultural land were being converted to urban use each year. An annual loss of 1,000,000 acres out of the 413 million acres of producing farmland is about 0.25% of our producing farmland lost each year. This loss is the more significant because of its location. Cities are the obvious focal point for urban expansion. Suburban development spreads in much the same manner as the ripples caused by a stone falling into a pool of water. Cities are also consumers of agricultural products. Continuing loss of farmland in the urbanized East

means higher expenditures for produce because of the increased transportation costs and the higher consumption of energy. It also means greater dependence on other portions of the country for food (Figure 33.2).

Taxation Is a Root Cause of Farmland Loss

We indicated that the roads of agricultural commerce are the paths along which people from the city move to suburbia; this is but one reason for the threat to farmlands. Another and more powerful reason is the way we tax land. While the states and the federal government derive most of their income from taxes on industrial profits, taxes on wages, and taxes on certain luxuries, such as alcohol, the counties and local governments draw much of their income from taxing the land.

To tax the land, the local government must first attach a value to the land. The attachment of value is called assessment of the property; the value is known as the assessed value. Practices differ from locality to locality, but the basis of the assessed value is very frequently the price the land would bring if sold on the market. Some local governments may assess a prop-

Figure 33.2 Farmlands are being converted to commercial buildings, dwellings, and pavement. Prime producing farmlands are being consumed as the suburbs of the cities sprawl across the countryside. In Massachusetts, for instance, if present trends continue, more than half of its prime farmland existing in 1977 will be consumed by the year 2000. (Photo courtesy of USDA)

erty at value equal to the value in the marketplace; others may assess a property at 50% of the market value; and so on. The tax that the owner pays on the property is determined by multiplying the assessed value by the tax rate. The tax rate is stated in dollars per $1000 of assessed value and is a uniform value throughout the taxing locality.

As housing sprawls across the landscape, land once distant from the suburbs comes closer and closer to spreading development, and the value of that land may soar. Land once of value only for farming may become many times more valuable for homesites. The increase in the value of rural land is a natural outcome of suburban expansion.

Since the market value of the land has increased, the assessed value increases and the taxes on the land rise. Farming, therefore, becomes less profitable because of the farmer's obligation to pay higher taxes. At the same time that farming is becoming less profitable because of higher taxes, the opportunity for profits in the real-estate market begin to tempt the farmer who lives on the suburban fringe.

Land in the family for generations or purchased decades earlier can become a source of wealth for a farm family. Land is often now a farmer's money in the bank for retirement. A farmer would have to love the land a great deal not to be tempted. At first, he might sell only a lot or two on the main road and then some less-used acreage. Each sale makes his land more valuable for homesites, makes his taxes higher, and the temptation to retire from farming becomes greater. Some people jestingly call farmland "the farmer's last cash crop."

Still another force pushing farmland into suburbs are federal inheritance taxes. These taxes are levied on the value of the estate if it were sold on the open market. Even if the heirs to the property wish to continue to farm the land, the higher dollar value reflecting the property's use for development purposes, causes a very high tax to be paid by the heirs. To pay such a tax, portions or even all of the farm must be sold (Figure 33.3).

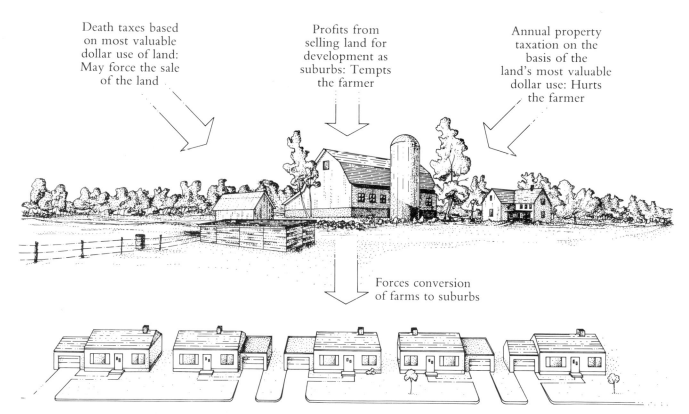

Figure 33.3 Forces that convert farmland to suburbs.

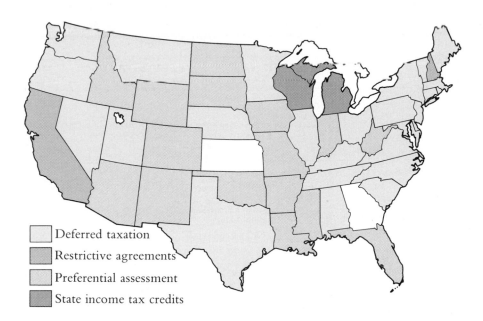

Deferred taxation
Restrictive agreements
Preferential assessment
State income tax credits

Figure 33.4 States where farmland is assessed on the basis of current use. (Note: Michigan and Wisconsin allow farmers to credit property tax against State Income Tax.) (Source: Natural Agricultural Lands Study)

Ways of Preserving Rural Land

Assessing land on the basis of use. The vicious circle of increasing temptation for the farmer has been observed for some time, and efforts have been made to interrupt the economic forces that bear down on farmland. Many states have decided not to value rural land on the basis of its most valuable use; instead, they are assessing it on its actual use. That is, even if a parcel of farmland could be sold for a high price, its assessed value reflects its use as farmland. Property worth $4000 an acre for homesites but only $1000 an acre as a farm will be taxed as a farm so long as that use continues. The tax threat to the farmer's pocketbook from an unexpected rise in value of the land is thereby eliminated.

By the end of 1980, all states except Georgia and Kansas had passed some form of tax legislation to decrease the pressures on farmers. Seventeen states allowed taxation of farmland at its value for agricultural use. This "preferential assessment" clearly reduces a farmer's taxes.

Another 28 states have opted for a program of "deferred taxation." This form of taxation involves a combination of lower taxes for farmland with a penalty if the land is sold. If the land is sold for development purposes, the penalty the owner must pay usually depends on the taxes that were avoided. The taxes that were avoided is the difference between the taxes

that would have been owed if the land had been valued at its market price and the taxes actually paid while the property was valued as farmland. This amount, added up over the years during which the special tax was applied, might have to be paid if such a sale is made (Figure 33.4).

Not all states add up the taxes avoided; seven states levy a "land-use change" tax equal to a portion of the taxes avoided or as a percentage of the fair market value at the time the land is developed. Special requirements exist in New Hampshire and California. In these states, a farmer must sign a long-term agreement not to develop the land, in order to obtain "preferential assessment" of his land. (See Figure 33.4.)

Deferred taxation and long-term contract programs are designed to prevent speculation and slow still further the conversion of farmland. However, these added requirements and penalties are expected to decrease the number of farmers willing to participate in the programs.

Although the tax threat is decreased by such systems of taxation, the temptation to the farmer from the increased value of the land remains. If profits from the sale of the land are banked, they draw interest. When the annual interest on the proceeds of the sale begins to approach the annual profit the farm family earns through their labor, farming may be abandoned, whether there is a lower tax or not. Furthermore, the method of putting lower taxes on land used for farming can have unintended results. The farmer could sell his land to a speculator who might simply hold it or rent it out for farming while the price of the land continues to rise. The speculator may be rewarded with low taxes while he waits to make a killing on the real-estate market. The speculator may, in fact, be attracted to the property because it will cost so little to hold it. Special property tax treatment for the farmer, it is now generally agreed, cannot by itself halt the expansion of suburbia.

Incorporating family farms. Another reason that farms are declining in number on the suburban fringe is also rooted in taxes, but the taxes are not on land; they are on inheritances. When a farmer dies, estate taxes must be paid on the value of the estate. Since the farmlands in the path of advancing suburbs become increasingly valuable, the value of the estate soars. Federal taxes on the estates of deceased farmers can be considerable, amounting to as much as 25% of the value of the estate. The sons and daughters who inherit the farm have to pay such taxes to receive the inheritance. A farm in Cutchogue, New York, valued at only $800 per acre for agriculture, was valued by the Internal Revenue Service (IRS) at $3500 per acre. The tax came to about $1200 an acre. As a neighbor described the situation, ". . . they [the Federal Government] forced its sale, very simply."[1] (See Figure 33.5.) Although the value of an estate that can be inherited tax-free was increased in 1981, the problem of farm inheritance remains because of the high value of farm estates.

To preserve their family farms, farmers have increasingly incorporated. As the owners of their corporations and holding the stock in them, they are able to make stock gifts each year of up to $3000 to individuals, in this case their sons and daughters. These gifts can be made in the form of stocks, and no taxes need be paid as long as the value of the gift is less than $3000 annually. If they were simply to give deeds for the land to their children, the decreased value of their property would affect their credit at the bank. The banks would see the farm decreasing in value and would be less likely to make loans the farmers need for seed, fertilizer, and equipment. Giving stock does not dilute the corporation, which remains a single entity despite having many owners. Incorporation obviously does not

---

1 "Estate Taxes Drive Farmers Off the Land," *The New York Times,* 14 May 1972.

PUBLIC SALE
—OF—
**VALUABLE REAL ESTATE
& FARM MACHINERY**

On Thursday, April 5, 1979 at 12:30 P.M. at R.D. No. 4, Spring Grove, Penna. along Roth's Church Road next to Spring Grove School in Jackson Twp. The undersigned Executrix of the John A. Roth Estate will offer at public sale the following —

**REAL ESTATE
110 ACRE FARM**

Farm consisting of approx. 110 acres of land with the majority being good fertile farming land, some in pasture land, small stream goes over edge of farm. Lots of very good road frontage

It also offers some prime bldg. sites and should be very highly considered for development use. Must be seen to be appreciated. Zoned residential.

Terms on Farm Machinery — Cash or Approved Check

LOUELLA M. DEARDORFF
Executrix

JACOB A. GILBERT, Auctioneer

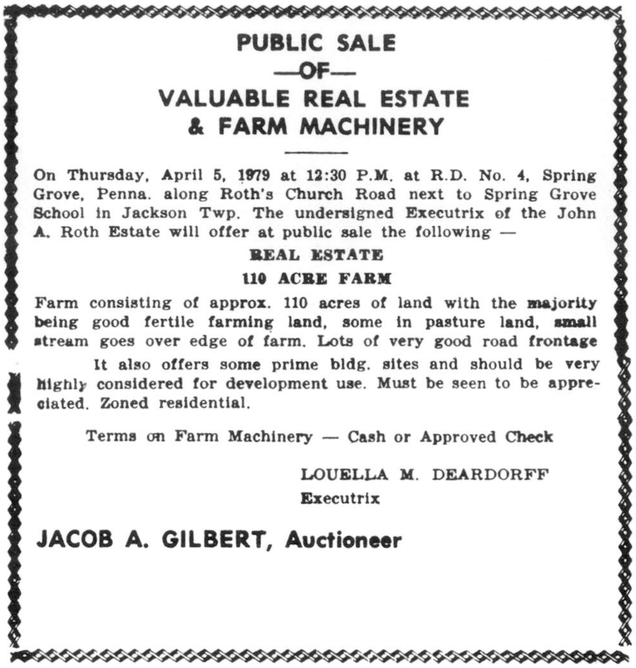

Figure 33.5 A farm falls to suburbia. Note the advice to developers that this fertile farmland is likely to be useful for a subdivision. Advertisements like this are common in many newspapers.

answer all the farmer's estate tax problems. Clearly some change in federal taxation of farm estates is needed.

The federal government has begun to recognize its role in hastening the conversion of farmland by estate taxes. Congress, in 1976, passed a tax reform act that attempted to address the issue. Nevertheless, the eligibility requirements and the penalties for farmland conversion by the heirs to the property are thought to be so strict that very few farms would be saved from development by this law. Further changes in estate tax are needed. Some states have also adopted laws to value farm estates at the farm-use value (Table 33.1).

Purchase and transfer of development rights by the public. The problem of the loss of farmland is occurring nationwide. Around nearly every major city, suburbs lick greedily at the farms on the rural fringe. At a few places in the nation, however, local governments are attempting to preserve their rural character by an interesting and powerful method. The method involves separating the right to develop the land from the land itself.

Table 33.1 States with Laws that Value a Farmer's Estate on the Basis of Its Current Use

Alabama	Mississippi
Alaska	Missouri
Arizona	Montana
Arkansas	New Mexico
California	New York
Colorado	North Dakota
Connecticut	Oregon
Florida	South Carolina
Georgia	Tennessee
Illinois	Utah
Iowa	Vermont
Kansas	Virginia
Kentucky	Washington
Michigan	Wisconsin
Minnesota	

(Source: National Agricultural Lands Study)

The idea of outright public purchase of development rights was first put into practice on Long Island, which feeds commuters by the hundreds of thousands into the nation's largest city, New York. Suffolk County, one of several counties on Long Island, was once noted for fine agriculture, but through the decades of the 60s and 70s, it came under increasing pressure from real-estate developers. Farms were gobbled up by the apparently unstoppable movement of New York City's suburbs.

In 1972, John Klein, county executive, proposed that it would be in the citizens' interest if the county government could somehow halt the conversion of farms to homes. He argued persuasively on economic grounds that the loss of farmland would bring higher tax rates to the community. Because of the increased school-age population that would result, he argued, new schools would be required, and taxes would have to be raised to pay for them. It took the county legislature until 1977 to pass Klein's program.

In September of that year, John Klein sat down at a card table in a potato field with Nathaniel Talmage and signed an agreement guaranteeing that Mr. Talmage's 133-acre farm would remain undeveloped forever (Figure 33.6). Klein's program is aimed at preserving 15,000 acres in Suffolk County and is the model that counties, states, and even nations are watching.

When Klein and Talmage sat down at the card table, Mr. Talmage transferred to Suffolk County the ownership, not of his land, but of the "development rights" to his land. He was paid a substantial amount for giving up those rights to Suffolk County, but not the amount he would have received if he had sold the land itself to real-estate developers. With the signing of the agreement, ownership of the Talmage property was now split into two pieces. Nathaniel Talmage still possessed the land and all dwellings and improvements on the land. He retained the right to farm it or even to let it go wild. He may not subdivide it to build homes on the lots, but he may, if he wishes, sell off parcels of land. The buyer of those parcels, however, will not be able to build new dwellings because the right to develop the land has passed to Suffolk County. As a consequence of the transfer of the development rights, the market value of the land has decreased because of the restriction on how the land may be used.

There are distinct benefits for Mr. Talmage, in addition to the money he was paid, benefits that go far

Figure 33.6 Saving a Farm. John V. N. Klein, right, Suffolk County Executive, and Nathaniel A. Talmage signing land agreement in a potato field on Mr. Talmage's farm. (Louis Manna/NYT Pictures)

to compensate him for the decreased market value of his land. First, as he clearly planned to continue farming, he has lost no privilege that he needs. His choice of crops, whether or not to sow, his choice of equipment—these items remain his decisions. Second, the loss in market value of the land means lower real estate taxes to pay, since taxes are calculated by multiplying a tax rate times some portion of the market value.

Third, Mr. Talmage's children are more likely to be able to continue farming this land after his death because the federal inheritance taxes on the Talmage estate will be much lower than on land with the potential for development. The estate or inheritance taxes are essentially determined by a tax rate multiplied by the market value of the estate. The reduced market value of the land, now shorn of its development rights, will lead to lower estate taxes. A tax burden that could have brought the family to dispose of the farm is now lightened, and it is more likely that the family will be able to continue farming.

It cost Suffolk County $357,000 to obtain the development rights to the Talmage land; the county had to raise the money by selling bonds, but they still have a bargain. Their payment to Mr. Talmage was 80% of the market value of his land. That is, the possibility for developing the land is by far the most important factor in determining its market value. With development rights for the new land in the hands of the county, the value of the land is expected to fall to a bare 20% of the earlier market price. Put another way, the value of the land when sold for homesites is five times its value as farmland only.

As time goes by, that purchase will become more of a bargain. Pressures for new homes in the county will only increase with time as the suburbs continue to expand. Such pressures will drive the market value of the land that can still be developed for homes ever higher; the value of the same land used only for agriculture will rise more slowly.

By 1981, nine local or state governments had established programs for the purchase of development rights. The state of New Jersey passed such a law as well, but terminated the program before any development rights were purchased. The experience of these programs and their funding levels are indicated in Table 33.2. Apparently, little consideration has been given to the compactness of the lands preserved by these programs. The notion here is that if such properties are clustered, the equivalent of an agricultural dis-

Table 33.2 Programs for the Purchase of Development Rights

Jurisdiction	Year first funded	Acreage under easement as of 1981	Total authorized funding ($ million)
Suffolk Co., N.Y.	1976	3214	21.0
Maryland	1977	2400	6.3
Massachusetts	1977	1349	15.0
Connecticut	1978	2585	9.0
Howard Co., Md.	1978	0	1.5
Burlington Co., N.J.	1979	810	3.0
King Co., Wash.	1979	0	50.0
New Hampshire	1979	0	3.0
Southampton, N.Y.	1980	0	6.0

Source: National Agricultural Lands Study, 1981.

trict is created. Farm activities, such as manure spreading, are much more likely to be tolerated by farms than by neighboring homes.

Agricultural districts, right-to-farm laws, and agricultural zoning. Farming is not always a perfectly compatible activity with the suburban pattern of land use. Noise from equipment operation, odors from livestock wastes, pesticide spraying, manure spreading, and the like can put a suburbanite's nose out of joint. A farmer may find himself sued as a "public nuisance" if the suburbanite leans toward that form of coercion. Alternatively, home owners may force the passage of local laws to control the farmer's operations. Such activities as these on the part of home owners make a farmer's profession that much more difficult. It may seem especially unfair to the farmer since his farming activity is likely to have preceded the suburban homes in the area.

State governments have responded to the farmer's plight in two basic ways. One is the creation of agricultural districts within the states. By the end of 1980, California, New York, Virginia, Maryland, Illinois, and Minnesota had each created programs to allow establishment of Agricultural Districts; no two programs were the same.

Three features of these agricultural district programs seem to be most important. The first is the allowance of "differential assessment" of farms within the district. That is, the farms in such a district are taxed on the basis of their current use, not on the basis of their market value as homesites. Four of the six states had such tax laws for agricultural districts. A second important feature of these districts is that local governments in the districts are prevented from establishing laws that interfere with a farmer's activities so long as the farmer's operations do not threaten the public health or safety. Five of the six states had such a feature incorporated in their programs for agricultural districting. A third common feature of these laws is that public investment in an agricultural district is limited by law as a means to deter the attraction of development; four of the six state programs included this concept.

Parallel in intent to the creation of agricultural districts is the passage in a number of states of general Right-to-Farm Laws. By 1980, sixteen states had passed such laws as a means of protecting farmers from nuisance suits and having to plead their cases in court. The laws shield farmers from both the expense and annoyance that come with such proceedings.

The laws differed from state to state, but two basic themes show up. One theme is the prevention of the passage of local laws that unreasonably interfere with farming activities. Of course, laws designed to protect either the public health or public safety would always be in order. Five of the eleven states with laws controlling local government statutes on farming chose to apply the law only in agricultural districts. The other major theme of state right-to-farm laws is the protection of farmers from private law suits by limiting their liability.

Right-to-farm laws are very new; the earliest was passed in 1971 in New York State. Thus, the impact of their enforcement on preserving farming is as yet unknown and possibly unknowable. Further, the boundary between a nuisance and endangering of the public health can be very faint.

Not only have state governments taken up the cause of the beleaguered farmer; in a number of cases local government has been sympathetic to the farmer's exposed situation as well. In these cases, the zoning tool, a local government power, has been used to pro-tect farming. Agriculture zones have been added to the land-use categories shown in Figure 33.1. Such agriculture zones can either permit nonagricultural uses or exclude them.

Summary of Ways to Preserve Rural Land

In the second portion of this chapter on private land-use decisions, we have examined the conflicts between the expansion of suburbs and the maintenance of farming. Until the past decade, the balance favored continued suburban sprawl, but in the 1970s, state and local governments began to respond creatively to the problem.

The first efforts were directed toward removing the unfair tax burden that fell on farmers whose farms lay in the path of suburban expansion. In a first response to the problem, states began to tax farms on the basis of their current use rather than on their value as homesites. A second response has come from the farmers themselves. To avoid crushing estate taxes, farmers have been forming family corporations.

A third response with much potential is the purchase by government of the development rights to a parcel of land. Once the development rights are owned by the government, no future owner can develop the land. It remains farm or wildland forever. Such a purchase not only puts cash in the farmer's pocket; it is proof that the land cannot be used for homesites and qualifies the farmer for taxation on the property only as farmland.

Finally, we have seen that states have created Agriculture Districts as a means to foster farming and have passed Right-to-Farm Laws. In some cases, local governments have established Agricultural Zoning as well. These last efforts are directed toward protecting the farming activity from nuisance suits and other legal action from new suburban neighbors.

Can any of these techniques reverse the trend toward farmland loss? The complete results will not be in for several decades. In the meantime, the programs should be watched carefully for early signs of success or failure.

Other Methods to Secure the Preservation of Rural Land

Purchase and Transfer of Development Rights by Private Corporations

The purchase of development rights is not restricted to governments, although many states are watching the Suffolk County experiment. Such rights may also be purchased by private individuals, by corporations, and by nonprofit institutions. In fact, any legal entity that can hold property can hold development rights as well.

Of what use could these development rights be to corporations? To understand how a home-building corporation could make use of purchased development rights, we need to reemphasize the nature of the transaction on Long Island. Suffolk County purchased the development rights on certain farms; those development rights now in the hands of the county are likely to *never* be exercised. It is possible, however, that the development rights purchased by a corporation on a parcel of farmland in one place can be used by that company *elsewhere* than on that parcel. Once used elsewhere, the land from which they were withdrawn can no longer be developed and is thereafter taxed only as agricultural or undeveloped land.

The first use of this concept to preserve rural lands took place in the small town of Saint George, Vermont. This town, which then had no commercial development whatsoever (no store, gas station, post office, etc.), was in the path of the spreading suburbs of Burlington, and its population had gone from about 100 in 1960 to 500 in 1970. To guide its future growth, the town purchased 48 acres of land to be used as the nucleus for future homes and commercial enterprises. These 48 acres were made available to developers for building, but only if the developers agreed to keep land open elsewhere in the town.

A description of how the town will use this 48-acre project area to guide the expected development reads:

> To achieve the objective of concentrating settlement and preserving the rural character of most of the rest of Saint George, the town may oblige a developer to transfer to the town development rights purchased from owners outside the project area in exchange for the opportunity to develop in the core village area. For example,

a developer wishing to construct twenty units of housing in the village area would have to purchase twenty acres of land zoned at one family to the acre elsewhere in Saint George and transfer his acquired right of twenty units of housing to the project area. The twenty acres from which the rights were transferred will remain open land in perpetuity or until the town releases it to meet future needs. The land will be taxed only at its value as undeveloped land.[2]

It should be noted that those 20 acres purchased elsewhere need not have been land suitable for building. It could be swamp, ravine, mountaintop, or hillside. The point is that a one-for-one trade of preservation area for housing development is taking place.

Conservation Easements

State and local governments have a number of options available to them, short of outright purchase, to prevent the development of agricultural or wildland. One such option is to secure a conservation easement of the land, a statement attached to the deed specifying the uses to which a parcel of land may (or may not) be put. These uses may include whether or not buildings may be erected, the maximum height of buildings, the cutting of trees, the use of billboards, and the like.

How can a government go about attaching such statements to deeds? The simplest way is for the community to buy the property outright when it comes up for sale. The local government attaches the statement on allowable uses and then puts the land back on the market. The land will probably sell for less than the community paid for it because its use is now restricted. The loss in value, however, may be viewed as the amount the community had to pay to put the restriction in the deed.

Although the method of purchase/resale is the easiest way to attach a deed restriction to the land, desired parcels are not likely to come on the market just at the time when a community is ready to act. More than likely, the local government will have to

---

2 L. Wilson, "Precedent-Setting Swap in Vermont," *Journal of the American Institute of Architects,* **61**(3) (March 1974), p. 51.

convince the owners that donating the restrictions will bring them benefits. The benefits may be direct payment for the attachment by the government, or the benefits may be tax advantages in terms of federal estate taxes, federal income-tax deductions, or reduction in assessed valuation and hence in property tax.

Questions

1. What is the purpose of zoning? Contrast the stated purpose and the actual purpose.
2. What social and environmental shortcomings are often associated with zoning? How can zoning be used to protect the environment and people?
3. How do growth management plans differ from traditional zoning?
4. Explain how taxation and the growth of suburbs are "eating up" prime farmlands in the U.S.
5. Describe some ways of keeping privately held land rural. What advantages do such methods have? That is, why not just have the government buy the land outright?

Further Reading

Leopold, A., *Sand County Almanac.* Oxford University Press, 1949. Reprinted, Sierra Club/Ballantine, 1970. Leopold observes nature and humans through 12 months on his Wisconsin farm. This highly readable, classic work has a fine essay, "The Land Ethic," in which the author describes the relationship that ought to exist between people and the land.

Platt, R., *Land Use Control: The Interface of Law and Geography.* Resource Paper 75–1. For sale by Association of American Geographers, 1710 Sixteenth Street, N.W., Washington, D.C. 20009.

Stover, E., Ed., *Protecting Nature's Estate,* Bureau of Outdoor Recreation, U.S. Department of the Interior, 1976. For sale by the Superintendent of Documents, U.S. Government Printing Office, Washington, D.C.

Untaxing Open Space, the President's Council on Environmental Quality, by the Regional Science Research Institute, 1976. Available from the U.S. Government Printing Office.

Whyte, W. H., *The Last Landscape.* Garden City, N.J.: Doubleday and Co., 1968.

National Agricultural Lands Study (a report in four volumes from 12 federal agencies), for sale by the U.S. Superintendent of Documents, Washington, D.C., 1981. Very readable and nontechnical; a comprehensive study on the transition of farmland to other uses.

Miner, D., "Land Banking in Canada: A New Approach to Land Tenure," *Journal of Soil and Water Conservation,* July/August 1977, p. 158.

America's Soil and Water: Conditions and Trends, Soil Conservation Service, U.S. Department of Agriculture, 1980 (booklet).

Pease, J., and P. Jackson, "Farmland Preservation in Oregon," *Journal of Soil and Water Conservation,* November/December 1979, p. 256.

Dunford, R., "Saving Farmland: The King County Program," *Journal of Soil and Water Conservation,* January/February 1981, p. 19.

Ognibene, P., "Vanishing Farmlands," *Saturday Review,* May 1980, p. 29.

Pierce, J., "Conversion of Rural Land to Urban: A Canadian Profile," *Professional Geographer,* **33,** No. 2, 1981, p. 163.

International Regional Science Review, **7,** No. 3, December 1982. Special Issue on Regional Development and the Preservation of Agricultural Land. The issue contains five articles on the problems of urban sprawl and its consumption of farmland.

CHAPTER THIRTY-FOUR

Preserving Public Natural Areas

The Public Preservation Movement

Why Should Lands Be in Public Hands?/A Brief History of the Federal Preservation Effort

Federal Natural Areas

A Fragmented System/National Wildlife Refuges/National Forests/The National Park System/Wilderness/National Wild Rivers and National Trails

Problems of Accessibility

Reviving an Old Tradition of Land Preservation

 Expanding the System of Natural Areas

The Land and Water Conservation Fund/Donations and Tax Benefits/Conservancies or Land Trusts/Coordinated Federal and Private Actions

CONTROVERSIES:

34.1: ***The Baxter Fire—Would You Let It Burn?***

34.2: ***Wilderness***

34.3: ***Should Wilderness Be Accessible?***

34.4: ***The ORV Fight***

The Public Preservation Movement

Why Should Lands Be in Public Hands?

State and National Parks, State and National Forests, National Wildlife Refuges, National Monuments, National Seashores, National Lake Shores, Wild and Scenic Rivers—the list of ways that the public sets aside land for recreation and preservation is longer still.

Wetlands may be rescued from development because they serve migrating birds on their annual journeys up and down the continent. Wildfowl may winter at such places or simply stop at them en route. A forest or desert may be set aside for its unique vegetation. The Redwood National Park and numerous California state parks are devoted to preserving the two species of Redwood tree, the largest living species. The Joshua Tree National Monument and the Saguaro (cactus) National Monument are used to preserve remarkable desert plants. Some parks are used to preserve stunning natural landscapes. Death Valley National Monument, Yellowstone National Park, and Grand Canyon National Park are examples that come immediately to mind. Other parks, such as the Everglades, are chosen to preserve whole ecosystems of plant and animal species. And for every preserved area, it would not be an exaggeration to say that there is another area yet to save.

Save from what? From timbering in the case of the Redwoods; from vacation homes in the case of the wetlands; from mining in the case of Death Valley; from commercial exploitation in the case of Grand Canyon and Yellowstone; from being drained or made into a jetport in the case of the Everglades. The threats are abundant. Although sometimes innocent, often they are from those who are ignorant of the values that other people place on such natural resources. Threats to these landscapes and ecosystems continue to occur.

The road to preservation of our landscape, our species, and our ecosystems has been marked by several disasters, not all in the distant past. Our first spectacle was Niagara Falls, which fell quickly to commercial development around 1820–1840. Mills were built at the foot of the falls; the surrounding forests were cut; and tourist operators set up at the rim of the falls. Visits to the falls were highly commercialized. One pair of English visitors who observed the commercial activities at and around the falls during this period called for protection of the falls as "the property of civilized mankind." In California in the 1850s, one of the largest of the giant sequoia trees in the mountains of Southern California was cut down and transported to New York and to Britain for exhibition. Such acts led to pressure to put the Yosemite region in public hands.

As late as the 1960s, over 100 years later, the awesome and beautiful rock walls of Glen Canyon were sealed finally from view by construction of a massive dam. And even into the 1980s, we find mining of existing claims officially approved in Death Valley by act of Congress. Preservation seems to require not only the initial act of setting aside, but constant watchfulness as well. As if to prove the point, in the early 1980s, the Reagan Administration and, in particular, the Secretary of the Interior, began an evaluation of wilderness and even of National Parks for their mineral potential.

A Brief History of the Federal Preservation Effort

Creation of the federal parks. In 1864, the tradition of federal creation of park lands began. The Yosemite Act gave 44 square miles (113 square kilometers), which had been in the possession of the U.S. simply as public land, to the State of California to preserve for the public. The act was introduced by a senator from California who justified his unusual proposal by noting that the lands were "for all public purposes worthless," by which he meant that no mineral resources or usable waterpower could then be derived from the land. The preserve included the Yosemite Valley, the towering mountains that formed its sides, and one other tract. The giant sequoia was at last protected. In 1890, the park would be taken into federal possession by act of Congress.

Eight years later, in 1872, Congress created a park in the Yellowstone region of Wyoming. The park preserved not only the fabulous falls and gorge of the Yellowstone River, but the unique geysers of the region, of which Old Faithful is the most famous example. This time the park was put directly under federal direction, presumably because Wyoming was still only a territory. Only a year before the creation of the park at Yellowstone, the fabulous geysers had been in danger of being claimed as private property, apparently because of their value as a tourist attraction. Nevertheless, the oratory in Congress that supported the act stressed the "worthlessness" of this park land. Again, the lack of value referred to a lack of mineral resources or of potential for agricultural use.

An 1895 act of Congress prohibited hunting in Yellowstone National Park, but the thrust of wildlife protection by setting aside preserves devoted only to animals was still in the future. Indeed, the protection provided by National Wildlife Refuges evolved in a tradition entirely separate from that of the National Park System. In the 1980s, the geyser region was again threatened, this time by the possible development of geothermal power in the area directly adjacent to the park. (See Controversy 26.1.)

The low commercial value of the lands committed to parks was one theme heard again and again. Another theme was that the areas set aside represented one aspect of America's claim to greatness. The beauty of the parks was compared frequently to the architecture and art of Europe. Though we did not have the wealth of history of the European continent, we had nonetheless our own natural history treasure that was worth preserving. In its day, this viewpoint served us well, for the notion of preserving an ecosystem (then still undefined) could never have occurred to people in that era.[1]

No one was concerned then with preserving a forest, for we had at the time what seemed an abundance of land, species, timber, and mineral resources. Wilderness was not far from anyone's doorstep. Indeed, in those early years, the nation saw its business as taming and subduing the land rather than preserving it intact. The realization that land and forest were limited resources dawned only gradually; and when it did occur to people, it was in the perspective of making profit. Mark Twain poked fun at the madcap commercial atmosphere of land grabbing that occurred after the Civil War. "Buy Land; they ain't making it any more!" one of his characters suggested.

Forest preserves and the U.S. Forest Service. In the period 1880–1910, however, a genuine appreciation of the role of natural resources and the public land was beginning to surface. Individuals like Franklin Hough, Bernhard Fernow, and Gifford Pinchot helped to shape a policy of *forest protection* as opposed to scenery or wildlife protection. Thus began a tradition different from the National Parks Movement and different from the movement to protect wildlife. Their efforts, which came out of the Division of Forestry within the Department of Agriculture, were directed at forest protection and at soil and water conservation.

With the passage of the Forest Reserve Act of 1891, the President was empowered to set aside areas of land in the public domain as public reservations to ensure that adequate timber would be available to the nation in the future. Although President Harrison created, within two years, 15 forest preserves with over 13 million total acres (5.2 million hectares), no means was actually provided to protect these reserves until 1897. In that year, Congress passed an amendment to an unrelated Appropriations Bill that did offer protection for the reserves.

The Pettigrew Amendment, as it is known, directed the Secretary of the Interior to make rules for protection of the reserves and authorized the sale only

---

1 The discussion of the history of the National Parks Movement is drawn from "The National Park Idea" by Alfred Runte, *Journal of Forest History*, **21**(2) (April 1977). The article is insightful, exciting, and beautifully illustrated. It reaches back and recaptures an early era in the environmental movement.

of mature or dead timber within the reserves. The timber was to be marked for removal before cutting could begin. The amendment of 1897 shaped the nation's forest management policies for 63 years, until the Multiple Use Act of 1960 expanded its provisions.

Effective protection, however, did not exist until 1905, when the administration of the Forest Preserves was transferred from the Department of the Interior to the Bureau of Forestry within the Department of Agriculture. The man who was given the responsibility for the 63 million acres (25 million hectares) now in the preserves was Gifford Pinchot. He renamed his agency the U.S. Forest Service, the name still in use today. The preserves were renamed as well; they became the National Forests. By 1907, there were over 150 million acres (61 million hectares) of preserves in the national forests.

Wildlife protection and the refuge system.
Still another tradition of land preservation was growing at about the same time. Whereas the purpose of the national parks was to preserve magnificent scenery and the purpose of the national forests was to preserve timberland, the purpose of wildlife refuges was the preservation of animals by protection of their habitat.

The federal government did not become active in the preservation of wildlife until the turn of the century. In 1900, Congress passed a law instructing the Secretary of Agriculture to take steps to protect and restore game and wild birds. This was the time at which the nation was witnessing with sadness the disappearance of the passenger pigeon, and this loss undoubtedly influenced passage of the law.

In 1903, the United States created its first preserve for animals, a refuge on Pelican Island off the Florida coast. This was soon followed by the establishment of wildlife ranges in the Wichita National Forest (1905) and the Grand Canyon National Forest (1906). In 1906, Congress prohibited hunting on all preserves that were now to be set aside for the protection of wildlife. In 1908, Congress set aside a National Bison Range in Montana for the threatened buffalo.

The fate of the passenger pigeon helped move Congress to pass the Migratory Bird Treaty Act in 1918, which placed migratory game under the protection of the federal government. Two years later, the Supreme Court declared that the federal government's power to regulate migratory wildfowl came before that of the states. Justice Holmes, writing for the majority, noted that without federal regulation, ". . .

Figure 34.1 The fate of wild birds, as seen from the 1930s (courtesy of the *Des Moines Register*). This series of cartoons was the work of J. N. Darling, a national syndicated cartoonist with the *Des Moines Register*. In the middle 1930s, because of his interest in wildlife protection, Darling was appointed Chief of the Bureau of Biological Survey in the federal government. His efforts were instrumental in building and strengthening the National Wildlife Refuge System.

there soon might be no birds for any powers to deal with . . . It is not sufficient to rely upon the states. The reliance is vain."[2]

The 1918 Act, however, had not provided for the acquisition of bird habitat. This defect was corrected by the 1929 Migratory Bird Conservation Act. The act authorized the Secretary of the Interior to create refuges for migratory wildfowl and to operate these refuges as "inviolate sanctuaries" (Figure 34.1). A portion of the funds to purchase wildfowl sanctuaries

---

2 Quoted from *Evolution of National Wildlife Law,* prepared for the Council on Environmental Quality by the Environmental Law Institute, 1977.

from states and private owners have come, as we mentioned in our discussion of endangered species, from the sale of hunting stamps. Revenue for types of refuges other than wildfowl refuges has not been so readily available.

Federal Natural Areas

A Fragmented System

These three traditions—the National Parks, the National Forests, and the National Wildlife Refuges—all evolved over the years. Uses and purposes have changed and blended as new needs and new pressures emerged. Today these three systems are still managed by different agencies, in keeping with their differing purposes at the outset. The National Park Service administers the National Park System within the Department of the Interior; the U.S. Forest Service manages the National Forests in the Department of Agriculture; and the U.S. Fish and Wildlife Service is responsible for the system of National Wildlife Refuges in the Department of the Interior. Table 34.1 lists the land areas these agencies administer, as well as the supply of state parks and forests.

Not all the land the federal government holds for the public is in the National Parks, the National Forests, or the National Wildlife Refuges. A considerable portion of the government's holdings are in the National Resource Lands, some 174 million acres (70 million hectares) of public land in the lower 48 states and another 70 million acres (28 million hectares) in Alaska. These lands, which are supervised by the Bureau of Land Management (BLM), are the remainder of the "public domain." The bulk of the lands we refer to as the public domain are those lands that remain from the great land purchases of the past, such as the Louisiana Purchase. Much of these lands were granted to the railroads, and settlers acquired a portion of them by homesteading. The Forest Preserves and the National Parks and Wildlife Refuges withdrew other lands from the public domain. But the land that has never been sold and which remains in the hands of the BLM is vast and often trackless.

Some of the land held by the BLM is timberland. By a quirk of history, some 2.4 million acres (about one million hectares) of the most productive timberland in the U.S. fell into the hands of the Bureau of Land Management. Two large tracts of land granted to railroads in Oregon reverted back to the Bureau of Land Management when a violation of the grant occurred. These are the 2.4 million acres of highly productive timberland that the BLM manages today. They apparently are administered by the BLM from much the same viewpoint as the Forest Service.

The National Resource Lands, as BLM lands are called, are available for mining, grazing, and timbering. Within the last decade, the recognition of valuable wilderness in the National Resource Lands has prompted an evaluation of these lands for their wilderness and recreation potential. The management of the

Table 34.1 Land Areas Administered by State and Federal Agencies

Agency/Office	Name of land holding	Millions of acres		
		"Lower 48"	Alaska	Total
Bureau of Land Management	National Resource Lands	174	70	244
U.S. Forest Service	National Forests	168	23	191
National Park Service	National Park System	27	52	79
U.S. Fish and Wildlife Service	National Wildlife Refuges	13	76	89
State and Local Parks and Forests	—	about 25	—	—

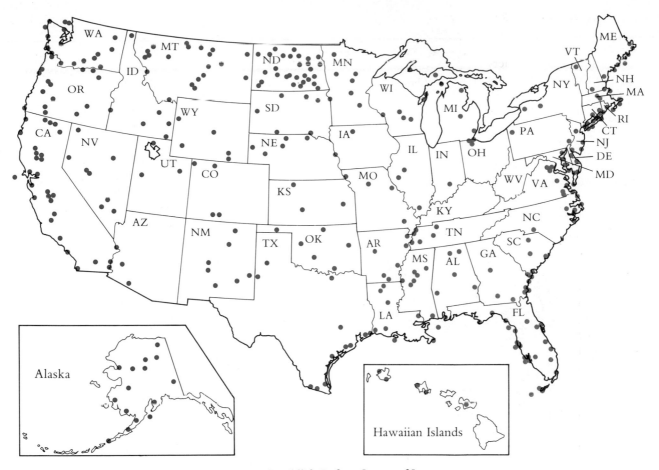

Figure 34.2 Map of the National Wildlife Refuge System, 30 September 1981. Most, but not all, of the Wildlife Refuges managed by the Fish and Wildlife Service are included in this map. The Fish and Wildlife Service publishes a directory of the nearly 400 refuges. The directory lists each refuge by state and county, noting its size, date established, and the primary species protected in the refuge. (Map courtesy of the Fish and Wildlife Service, U.S. Department of the Interior)

National Resource Lands for mineral purposes has been a long-term issue. In the pages to come, we will survey the management of natural areas by the federal agencies involved in their protection.

National Wildlife Refuges

Although originally set aside as "inviolate sanctuaries," the National Wildlife Refuges have evolved over the years so that their use includes other purposes. The Migratory Bird Hunting Stamp Act of 1934 set up the sale of "stamps" or permits to hunters. Revenue from the sale of hunting stamps has been used to fi-

nance in part the purchase of refuges for wildfowl. The use of these revenues from hunters may have influenced decisions by Congress in 1948 and 1959 to permit the Secretary of the Interior to authorize hunting on the refuges.

Although hunting was originally limited to no more than 25% of the refuge area by the 1948 legislation, the limit was raised to 40% in 1959. The Secretary was allowed to use his/her judgment to decide if hunting on a particular refuge was compatible with the purposes of the refuge. No challenge to this use has yet been successful in a court of law. In 1962, the Refuge Recreation Act authorized the Fish and Wildlife Service to open wildlife refuges to public recreation. The authorization gave to the Secretary of the Interior the responsibility to determine whether recreation at a particular site would interfere with the basic purpose of the refuge.

In 1966, the wildlife system, which consisted of game ranges, wildlife management areas, wildlife refuges, and other units, was consolidated into the National Wildlife Refuge System under the administration of the Fish and Wildlife Service (Figure 34.2). The allowable uses of these areas were expanded as well. Now, in addition to bird hunting and public recreation, the uses, when found to be compatible with the central purpose of the refuge, could include hunting of other species, fishing, and public accommodations. Just as the National Forests were expanded to become multiple-use lands in 1960 (see next section), the National Wildlife Refuges were now becoming multiple-use, with the added proviso that their original purposes as species habitat should not be compromised. This notion of accompanying but secondary uses of the refuges has led to the refuges being referred to as "dominant-use" lands, as opposed to "multiple-use."

National Forests

The multiple-use concept. The term **multiple-use** has come to be applied most frequently to the use and management of the National Forests. Indeed, an act of Congress known as the Multiple-Use–Sustained Yield Act was passed in 1960 to define in a formal way the meaning of multiple-use management in the National Forests.

The act stated that the National Forests were to "be administered for outdoor recreation, range, timber, watershed, and wildlife and fish purposes." It further stated that "The establishment and maintenance of areas of wilderness are consistent with the purposes and provisions of this act." The National Forests were also to be managed for a "sustained yield" of forest products. "Sustained yield" was defined as "the maintenance in perpetuity of a high-level annual or regular periodic output . . . without impairment of the productivity of the land."

Surprisingly, the act, which is now the principal policy directive of the Forest Service, was opposed both by conservation organizations *and* by the timber industry. In fact, the Forest Service saw the law as a

means of making legitimate the management practices it had followed since the earliest days of the agency. The National Forests had always been administered for all these purposes, and for mining, in addition. The opposition by both conservationists and by timber people stemmed from their suspicion of the motives of the Forest Service.

Conservationists were concerned that the act would be the tool by which the Forest Service could authorize extensive new cutting in the National Forests. Because of pressure from the Sierra Club, a nationwide organization of conservationists and environmentalists, the statement on wilderness was added to the act. The Sierra Club was deeply distrustful of the motives of the Forest Service in supporting the act and never supported the bill, though they had a hand in shaping it. The distrust of the Forest Service is explained by Steen:[3]

> Several events coincided during the 1950s which cost the Forest Service the support of many conservationists. Since Pinchot's time, the national forests had been held in reserve to supply lumbermen when privately owned timberlands had been cut over and were in the regrowth cycle. Year after year and chief after chief reaffirmed the policy to log a substantial portion of the national forest system—at the appropriate time. The appropriate time followed World War II, when increases in population and affluence concurrently multiplied recreational pressures on all public lands. Couched in terms of timber famine and devastation, decades of propaganda by the Forest Service to justify federal regulation of logging had created the popular image of rangers protecting forests against rampaging, greedy lumbermen. Now the postwar public came to the forests—camping, fishing, hiking—and "discovered" logging in the national forests. To many of the public, national forests and national parks were the same, and logging either was bad. Implementation of long-term timber management plans collided with the results of an extremely effective public relations program against "destructive" logging. As a result, the public felt deceived.

Whereas conservationists were concerned with excessive logging being allowed in the National Forests,

---

3 H. K. Steen, *The U.S. Forest Service: A History*. Seattle: University of Washington Press, 1976, p. 302.

industry, in contrast, saw the bill as giving weight to public uses. Such uses, they felt, could only work against the interests of the timber companies.

Four years later, in 1964, conservation interests were finally successful in passing the Wilderness Act. This act made it possible for areas within the National Forests to be specifically set aside as wilderness areas only. The act was essentially a restatement of distrust of the intentions of the Forest Service. That distrust, which surfaced first in 1960, reflected a concern that the Forest Service would not protect wilderness unless specifically ordered to do so.

The Multiple-Use Act of 1960, while it affirmed wilderness as a legitimate "use," did not order the establishment of such areas. The Wilderness Act gave Congress the right to designate areas of the National Forest as wilderness and hence protect them from logging. By 1982, 19.7 million acres (7.9 million hectares) of the National Forests had been included in the National Wilderness Preservation System from National Forest Land in the "lower 48" states.

Timber management in the National Forests. In addition to the order from Congress to the Forest Service to set aside wilderness areas, court cases arose over the Forest Service's application of the Multiple-Use Act. Specifically, these cases claimed that the Forest Service was not giving due consideration to wilderness in its timber sales.

Controversy on the way in which the Forest Service has administered sales of timber has focused not only on where timber sales have been allowed, but also on the manner in which trees have been cut. One specific method has been attacked again and again; it is the practice known as **clear-cutting.** Clear-cutting means simply the cutting of all trees in an area, regardless of size, quality, or age. The practice of clear-cutting on National Forests has grown through the years. Lumber companies prefer it because it is obviously more economical than selective cutting of mature and dead trees. Since there is a cost of moving lumberjacks and equipment to each site, the companies prefer to take all the timber they can get before moving on to a new area. Clear-cutting, then, is generally preferred over selective cutting by the timber companies for economic reasons.

Nonetheless, there are costs to the public when clear-cutting takes place. In addition to the visual impact of a forest cut to the ground, clear-cutting can affect rainfall retention in the area, and it can affect erosion. More rapid runoff over wide areas can lead to widely fluctuating stream flows, including damaging flood flows. High runoff rates can speed the erosion process as well. Furthermore, species habitat is completely wiped out by clear-cutting. Wilderness values are completely destroyed in the region of the clear-cut (Figure 34.3).

The timber companies argue for clear-cutting for

Figure 34.3 A poorly designed clear-cut can ruin the view of a forested mountain area. This photo was provided by the Bureau of Land Management, but does not necessarily depict activities on BLM lands. Clear-cutting has been a recent practice on Forest Service Lands.

Figure 34.4 Construction of a logging road near a creek. This exposure of loose, unanchored soil causes rapid erosion from the land. The particles will enter the creek, producing soil-choked waters. Sediment in the creek can decrease the ability of the stream to carry high flows, causing flooding downstream. (This photo was provided by the Bureau of Land Management, but does not necessarily depict activities on BLM lands.)

While clear-cutting is displeasing visually, certain areas are definitely harmed by the practice. A clear-cut on a steep slope may lead to erosion of the best soil off the slope. With the best soil eroded away, a very long time is required for the forest to return. An area that includes a stream channel can also be damaged if the trees are cut too close to the banks. Erosion of the stream banks damages the forest, but the stream is damaged as well by the heavy load of soil particles that it carries. Removal of trees also opens the stream to sunlight, which increases the temperature of the flowing water (Figure 34.4).

Other methods of cutting can be and are used by the timber industry. These include selection cutting, seed-tree cutting, and shelterwood cutting. **Selection cutting** removes mature timber, usually the oldest and largest trees. The cutting cycle involves a return to the area for timber cutting every 5–20 years to remove the newly generated mature timber. This method of cutting produces an uneven-aged stand, and has been widely practiced.

Seed-tree cutting is a poor name for this logging method, which actually leaves seed trees uncut (Figure 34.5). The seed trees left to regenerate the stand may be left singly or in groups. Although seed-tree cutting

other reasons than economics. One commercial timber species, the Douglas Fir, is thought to grow best in direct sunlight. Company foresters point to clear-cutting as the way to assure optimum growth of a new forest of Douglas Fir, but the claims that direct sunlight is necessary for the regrowth of the Douglas Fir have been disputed by forest scientists.

One aspect of clear-cutting deserves mention because of its bearing on the ecological system of the harvested site. Recall that the clear-cut virtually wipes out the entire habitat. The species that return to an area regrown from a clear-cut are expected to be somewhat different from the species that existed prior to the clear-cut. The reason is that the regrowing trees will all be about the same age. A natural forest, one which has had time to mature through several growing cycles, has trees of many different ages. The forest regrown from a clear-cut is referred to as an even-aged stand; the forest with trees of different ages is called an uneven-aged stand. The uneven-aged stand is thought to be ecologically superior in terms of the balance of species present.

Figure 34.5 Seed-tree cutting prior to log removal. Some trees are left to seed and regrow the cut-over area. (Bureau of Land Management)

has been used for commercial harvesting in Montana, it is still relatively experimental.

Shelterwood cutting proceeds in three stages. In the first, or "preparatory" stage, dying trees, defective trees, diseased trees, and trees of unwanted species are removed, leaving space for new trees to grow. About 10–15 years later, a stage called "seed cutting" opens the stand further so that seedlings can receive adequate sunlight and heat. In the final phase, called "shelterwood removal cutting," when seedlings have been established, the remainder of the mature trees are cut.

It is important to note that without supervision, any of these logging systems can produce unwanted results. As an example, selection cutting, if it proceeds unsupervised, can be used not to select mature trees as it is supposed to do, but to select the best or most marketable trees.

In 1973, clear-cutting was taken to court, and it was found that the 1897 law that formed the basis for management of the forests still has an impact in the present. In Izaak Walton League vs Butz,[4] a district court examined the timber harvesting that the U.S. Forest Service was allowing in the Monongahela National Forest in West Virginia. The court declared that clear-cutting in the National Forest, the removal of all trees in an area regardless of age, was in violation of the 1897 law. The 4th Circuit Court of Appeals upheld that decision in 1975, citing the rules of the 1897 act, which limited cutting to the trees individually marked and trees of mature growth or dead.

These interpretations of the original law, however, were hollow victories. Industry and the Forest Service wanted other options and Congress responded to the pressure. In 1976, the Forest Management Act finally replaced the original law, putting in the hands of the Forest Service the responsibility to manage the National Forests as they judged best. Although the Forest Service was instructed by the act to consider topography, type of forest, size of cut, climate, and the like, the result has so far been a return to the practice of clear-cutting so clearly prohibited in the 1897 law. (See *The Tragedy of One-Shot Forestry*, p. 654).

How much timber comes from public lands? In the 1900s, virtually all the lumber produced in the

U.S. was cut from private forestlands. After World War II, housing pressures forced the government to open the National Forests to companies in a serious way, although private logging had been going on in the National Forests to a minor extent since the establishment of the preserves. The share of the annual production of lumber from public lands had reached 15% by 1950. By the early 1970s, the lumber from public lands accounted for 40% of annual timber production. In the early 1980s, that percentage dipped substantially to about 20%, as a severe recession and historically unprecedented high interest rates slashed home-building activity.

The Forest Service has obviously "upped the cut" from the period before World War II, and the explanation lies in part in the initial mission of the service. The preserves, now the National Forests, were set aside for the day when timber supplies on private lands had fallen so low that the preserves would be needed to meet demands. Rising demand for lumber in the decades after World War II was the signal to open the forests. Now, with the prices of lumber at near record levels, there is pressure from many directions to keep the national forests producing. Lumber companies, home builders, construction workers, and consumers want lumber at relatively low prices. Management of the forests is a very large problem in the national economy.

If we listen to the debate between conservationists and the timber companies, two terms are heard again and again; these are "allowable cut" and "sustained yield." A **sustained yield** cut is a quantity of timber less than or equal to the natural increase (in cubic feet of lumber) occurring in the area in the past time period. The allowable cut is the quantity of timber that a company is given the right to harvest from an area over a set period. A forest area that has an allowable cut equal to the new growth will, when the cut is complete, still have the same quantity of standing timber it had before the new growth took place. That quantity of new growth can be expected again, and the yield from the forest can be sustained.

Should the rate of cutting in the National Forests be allowed to approach the new growth? Remember that the National Forests are our final reserve of timber, to be cut when the yield from private forests is insufficient to meet demands. Although we are now cutting extensively in the National Forests, we must ask if it is because of high demand or because private

---

4 Earl Butz was then Secretary of the U.S. Department of Agriculture and thus was in charge of the U.S. Forest Service, which is housed within the Department of Agriculture.

forests have been poorly managed. Private forestry efforts must be encouraged so that the use of the National Forests for meeting timber demands can be decreased. There is, in addition, the issue of whether we should allow timber cut from the National Forests to be sold in export to foreign nations, as it now is.

If the cut in the National Forests is equal to sustained yield, the stock of standing timber will not decrease, but it will be standing still in the face of growing demands. Unless private forests are substantially regrown, the pressure for cutting in the National Forests can only increase. Allowing the cut to approach

---

The Tragedy of One-Shot Forestry

The developing pattern of forest management in the 50 eastern national forests is like a personal tragedy to me. I am puzzled and dismayed by the five new management plans I have seen for national forests from Vermont to South Carolina. They call for clearcutting and even-aged management for most of the mixed hardwood types of forest area close to heavily populated areas. The plans show a relative insensitivity to nontimber values and propose an almost exploitative, primitive silviculture, with the objectives of short-term financial returns to the timber operator and ease of administration and future management by the Forest Service. . . . Our bureaucrats will designate a boundary of timber, and the operator will simply harvest all the trees, large and small, regardless of quality or species, and the next clearcut will be 80–120 years hence.

How can this be? The Forest Management Act of 1976 has noble and true words about the importance of environmental values, about the relative unimportance of dollar values alone, and says in so many words that clearcutting shall not be used unless it is the "optimum" method. But there is a catch. The discretion and the judgment are left to the professionals in the Forest Service. . . . The "discretion" has been exercised all in one direction—cheap timber production on the bulk of the forests. How can any professional agency prescribe *one* silvicultural system for the vast diversity of types, topography and climates that constitute eastern United States forests? . . . now the Forest Service seems to believe it has a legal basis for almost universal clearcutting, and it is acting accordingly. . . .

What are the consequences of large-scale block clearcutting of eastern hardwoods? . . . The consequences . . . are greater waste of timber, lower environmental values and greater danger of damage to soil, site and water. Let me explain. Most of the individual trees in eastern hardwood forests are now below mature saw-timber size. But clearcutting harvests these smaller trees, just when they are growing fastest. This is a waste. . . .

This crisis in forestry has been caused by failure to reconcile and balance the values of the timber as a commodity, the forest as an environment, and the integrity of the forest–site–soil–water ecosystem. Such a balance cannot be achieved by the one-shot system of clearcutting and even-aged management now being proposed for the eastern national forests. . . . If the Forest Service will not or cannot reconcile and balance forest values, then Congress or the courts will have to act again.

Leon S. Minckler

Appeared in *Sierra Club Bulletin,* July/August 1978

(Leon S. Minckler is now retired from the U.S. Forest Service; he has taught environmental science and forestry at Virginia Polytechnic Institute and the State University of New York. Mr. Minckler is considered one of the shapers of forestry opinion.)

sustained yield in the National Forests makes sense only if the private forests are being regrown. It may be that decreasing the allowable cut in the National Forests could help encourage adequate private forestry efforts. The problem is that people want houses now and forests take a minimum of 25 years to grow.

Fire—No stranger to the forest. A careless match drops glowing on the forest floor amid fallen and dried needles. The tiny embers slowly ignite nearby needles; at first there are only embers, but dried needles are everywhere thick beneath the match; the embers grow hotter and tiny flames lick at the jumble of dead twigs. The twigs ignite and a flame leaps higher to catch fallen and dead branches. For a moment it looks as though the fire will go nowhere; then the bark on a dry branch crackles and flame encircles it. The branch itself begins to burn.

Now the fire spreads rapidly on many fronts; the small dry materials on the ground spread it quickly to more branches. At first, the living trees and fallen logs do not catch fire, but as the heat grows, the leaves, twigs, and stems of the live trees start to ignite. As the intensity of the fire increases further, fallen logs catch fire and even the trunks of live trees begin to burn. Though the live trees could not catch fire by themselves, in the heat of the forest fire, they will burn.

The fire front moves quickly through the dry underbrush, but fallen logs continue to burn long after the fire front has passed. Thus, though the fire may be stopped at a fire break, the fire within the forest continues. The embers from these burning logs may later feed a new phase of the fire.

Fire is no stranger to the forests. Lightning may set a forest fire as well as a match. When rain from high clouds evaporates before reaching the earth, a lightning bolt that strikes dry ground may ignite a blaze. Forest ecologists have been able to study the frequency of fires by observing fire scars embedded in the annual growth rings of trees. By careful counting of both rings and scars, they have been able to show a remarkable phenomenon. A study in a California forest found that forest fires reoccurred in about eight-year cycles as far back as 1685.

A study in the Boundary Waters Canoe Area of Minnesota, now part of the Wilderness System, found through a study of lake sediments that fires have reoccurred again and again in these forests for thousands of years. In the jack-pine forests of this region, fire is thought to be the agent that regenerates the stand, since the cones on the jack-pine are opened by intense heat. Once the cones are opened, the seeds fall on a ground where competing plants have been burned away and where sunlight finds better entry. The fire has made it possible for new jack-pine seedlings to grow. Fire, then, can be an agent to regenerate a forest.

Fire can also be a means of protecting the forest from a holocaust. How can fire protect a forest? It can do so by consuming the litter and underbrush in the forest, thereby preventing a major buildup of deadwood that could lead to a truly deadly fire. If a light surface fire burns quickly through a small buildup of underbrush, the trees in such a forest may be only scarred, not destroyed. These are the scars ecologists counted in the tree rings in order to establish the frequency of fires. If the buildup of underbrush is reduced by light surface fires, any individual fire may not reach the searing intensity that consumes the living trees themselves. Fire can be an agent of forest survival and/or regeneration; after years of warnings from Smokey the Bear, it may seem hard to accept.

Smokey's warnings are sincere nonetheless; in order for a fire to be of value to a forest, the forest must be in a condition that will not lead to complete destruction. If large amounts of dead fuel have accumulated over decades of careful protection, the hazard to the forest is very great, for the fire is likely to reach an intensity that will consume live trees as well as underbrush. No one but the professional forester or forest ecologist can know the hazard or value of fire to a forest. Smokey is still telling the truth.

How is the tool of fire to be utilized? Prescribed or controlled burns have been used in some Western forests to prevent buildup of dead underbrush, but in other forests, the careful protection of the Forest Service since the turn of the century makes prescribed burning impossible to use. Any attempt at burning in such areas would lead not to surface fires but to major fires, which are difficult to control. Such major fires would occur because of the large quantities of dead underbrush accumulated at ground level. In an area where a major fire has already occurred recently, however, controlled burning might be used in following years because of low levels of deadwood buildup on the forest floor. (See Controversy 34.1.)

The Baxter Fire—Would You Let It Burn?

> Just think if you can, but don't strain yourself, do-gooder, of all the comfortable homes that could have been built with that wasted lumber.
>
> *E. D. Chasse**

> If you could look at the park as something besides a giant woodlot you might not be so concerned about the burn. This beautiful park has burned many times in the past, and it will burn in the future.
>
> *Melvin Ames†*

As we have seen, fire can play two important functions in the forest ecosystem. First, light surface fires clear away dead underbrush, which, if it were to accumulate, could cause massive and intense forest fires. In this century, our very success in preventing forest fires has led to conditions in many forests that could lead to terribly severe fires, fires far more severe than might otherwise have occurred. "Prescribed" burning has been used by professional foresters in some forests to prevent such buildups. Prescribed burning is undertaken with caution; days chosen are moist and windless; all forest uses are curtailed.

The second function of fire is replacement or renewal of forest stock. Fire may act as nature's instrument of forest renewal and has acted in this fashion for hundreds and thousands of years without the aid of the forester. Only recently have we discovered the significance of fire to the forest ecosystem. Such knowledge of the role of fire will increasingly be put to use in "managing" forests, but it is not yet clear precisely how the knowledge will enter into future decisions.

In the summer of 1977, a massive fire occurred in Baxter State Park in Maine. The fire was started by lightning; its intensity was fed by downed timber. Let us, for a moment, enact the scene. Prime recreation land is being destroyed, but the wilderness is renewing itself. Will you let the fire burn?

You will want to know all the circumstances of the fire, how it spread and how it is being fought, what resources may be destroyed. The reasons people are giving to fight it or let it burn will interest you as well.

The story begins many decades ago when the will of Percival Baxter, former Governor of Maine, was opened by his lawyers. To the people of Maine, Baxter bequeathed a 200,000-acre (81,000-hectare) forested tract with 46 mountain peaks and ridges. The jewel of the tract is towering Mount Katahdin. This mountain, at 5240 feet (1600 meters), is the northernmost point of the Appalachian Trail, the wilderness foot path that runs from Georgia to Maine. Governor Baxter's will instructs that the land is

> forever to be held by the State of Maine in trust for public park, public forest, public recreational purposes, and scientific forestry, the same also forever shall be held in its natural wild state and except for a small area forever shall be held as a sanctuary for wild beasts and birds.

Over the years, the park had been developed, so that by 1977 it included an automobile road and seven campgrounds. The camps included space for trailers and tents, as well as bunkhouses and shelters. Because of the popularity of the park, reservations were suggested for those who needed to be sure of space. When the fire struck Baxter Park in 1977, it was a foregone conclusion that the park authority would fight it. The battle, however, was exceedingly difficult because of the rough terrain and the presence of numerous downed trees.

These downed trees were from a fierce winter storm that had struck the southwest slope of Mt. Katahdin in 1974. Wind and ice had combined to snap tall trees in two, leaving a pileup of dead logs on the slopes. The Great Northern Paper Company had been allowed into the park to harvest the blowdown for use as pulpwood in the hopes of lessening the risk of fire, but the harvest was never completed.

Legal action had been taken by the Baxter Park Defense Fund, and a judge's order brought the harvest to a halt in August 1976. While the judge did not forbid the removal of the downed trees, he directed that the heavy equipment being used had to be removed from the park. The heavy equipment was cutting into the forest floor, leaving gouge marks that might be used as roads or that could reroute streams into the depressions. Increased erosion from scarred areas would be likely as well. The equipment was removed by the paper company and the harvest of the blowdown came to an abrupt end. The following summer, when lightning ignited the blaze, much of the tangle of deadwood still lay on the ground.

The primary means chosen to fight the fire was to scrape out a 16-mile (27-kilometer) fire line around the perimeter of the fire. Bulldozers and skidders were borrowed from the Great Northern Paper Company to cut the 12-foot (3.6-meter)-wide fire line; the company, which owns forest adjacent to the park, sent firefighters as well. Beaver seaplanes were also used in an effort to control the fire. The planes loaded water from nearby ponds and dropped the water in 150–gallon (564-liter) loads on fire spots. On the ground, firefighters carried hoses and pumps into the forest. The pumps were set up on streams and ponds; the hoses, connected to the pumps, were dragged to the fire front.

In the heat of the blaze, the Baxter Defense Fund, which had prevented removal of the blowdown, let it be known that it planned once again to go to court. Their aim, as before, was to force out the heavy pieces of equipment that were scarring the slopes of the park and crisscrossing the mountain streams in an effort to create the fire line. Although public reaction to the threatened suit was intense, the Defense Fund would have gone ahead with their legal action had they not seen that the fire was in fact being brought under control faster than court action could be taken. The legal steps to cause removal of the machinery were not undertaken, and hence no injunction against the firefighters was obtained.

The fire line was completed by the eighth day of the blaze, and the fire was brought under control within those lines. Nonetheless, weeks of watching were required to be certain that the fire was truly out, since fire can smolder in the dry litter that makes up the forest floor. The Baxter Park Authority, in an effort to slow erosion, built water bars in the deep gouges left by the bulldozers. These miniature log dams slowed the flow of water in the depressions and allowed earth particles to settle rather than be swept away by flowing water.

Although the fire was out, the controversy raged on. Letters and editorials in the newspapers of Maine insulted the environmentalists. Letters also appeared expressing sympathy for the concerns of the Baxter Defense Fund. The quotations that began this discussion were from such letters as these. There were aftershocks, including court action by the Defense Fund to ensure a natural reclamation of the scars left from the firefighting; but eventually the controversy, too, appeared to go out. Like the remains of a forest fire itself, however, it is only smoldering beneath the surface. If the fire rages again, would you let it burn?

---

* Letter to *Bangor Daily News,* 27 July 1977.
† Letter to *Bangor Daily News,* 19 August 1977.

The National Park System

The National Park System, now grown to more than 79 million acres (33 million hectares), is administered by the National Park Service. The Park Service was created in 1916 to administer a system of 15 national parks and 22 national monuments already in existence. The Park Service was charged with conserving:

> . . . the scenery and the natural historic objects and the wildlife therein and to provide for the enjoyment of the same in such manner and by such means as will leave them unimpaired for the enjoyment of future generations.[5]

By the end of World War I, the Park Service was able to take over protection of the parks from the U.S. Army, whose troops had been used to build roads, fight fires, and generally protect these federally owned areas.

Although the Park System lists 287 units, the Park Service administers other areas as well. The more familiar responsibilities of the Park Service include the 38 National Parks, 61 National Monuments, 52 National Historic Sites, 16 National Historical Parks, 16 National Recreation Areas, and 14 National Seashores and Lakeshores. In 1982, the Park Service recorded 245 million visits to its units, over a 100% increase from the 1965 level of use (Figure 34.6). The less familiar responsibilities of the Park Service include areas that the Service does not even consider part of the Park System: the National Natural Landmarks, National Environmental Education Landmarks, and National Historic Landmarks.

Over the years, the Park Service has accepted other tasks in addition to marshalling people carefully through and past their holdings. One of their important functions is the preservation of historical landmarks, both natural and man-made. They not only preserve but also restore early settlements and interpret early events for visitors. For instance, at DeSoto National Memorial, a relatively out-of-the-way site in Florida, a camp much like that of the early Spanish explorers has been set up. Weapons, such as the crossbow and musket, are displayed and demonstrated by the staff and a movie recreates the DeSoto expedition. Such educational efforts are common in other units of the National Park System.

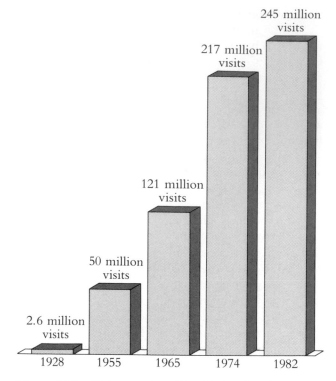

Figure 34.6 Visits to units of the National Park System. Though the National Park System has been growing in size, the growth in visitors is principally at the larger and more famous members of the system. (Source: Statistical Unit of the National Park Service)

The Park Service also conducts surveys of historic buildings and engineering works and maintains a register of sites significant for historical and educational reasons. In addition, the Service provides grants for surveys of sites possibly worthy of preservation. Under agreements with other agencies, the Park Service supervises recreational use of land not in their system.

New lands enter the National Park System by a number of methods, the most basic of which are (a) purchase; (b) condemnation; (c) gift; and (d) exchange or transfer. Lands have had to be purchased frequently in recent years, as the system of National Seashores has grown. Often, where no purchase can be negotiated, the Park Service has had to resort to condemnation. In legal terms, condemnation is made possible by the power of "eminent domain," the privilege of the government to take lands for its use. Though lands may be condemned, they must still be purchased. Just compensation, to use the legal term, or "fair market value," must be paid for lands so taken.

---

5 *Preserving Our Natural Heritage,* prepared for the U.S. Department of the Interior by the Nature Conservancy, 1977, p. 32.

Over the years, the National Park System has benefitted significantly from gifts from states and from private individuals. Two gifts from the Rockefeller family have been especially important. One formed the nucleus of Acadia National Park in Maine, the first National Park in the East. The other gift provided land on the Island of St. John, in the Virgin Islands (a U.S. territory), for the Virgin Islands National Park, a treasure of beach, ocean, and coral reef.

Lands may enter the system by what might be called "barter," in which land in the park system is traded for land outside. More important, however, is the transfer of land from the jurisdiction of another federal agency to the park system. Voyageur's National Park in Minnesota was acquired by gifts, transfer, purchase, and condemnation. Minnesota is donating about 50% of the 220,000 acres (89,000 hectares) of land and water in the park; the Forest Service, which controls 10% of the proposed park, is transferring another component to the Service; and the Park Service will have to purchase the remaining 40% of the area. Those purchases may be willing sales or may require condemnation.

The National Park System still bears the imprint of the early laws with which it began. Justified as the preservation of spectacles and grandeur, the parks were to serve people in a way that would enable them to see the most memorable scenes within the park. Since the parks were initially remote, this required development at the park itself. Such development included accommodations, food, information, and the like. Initially, such development was not intrusive, since relatively few people had the means to visit the parks. Now, however, the Park Service is increasingly called on to usher vast crowds past its most important possessions. Balancing its need to serve and its need to preserve is becoming increasingly difficult for the Park Service.

Wilderness

The Wilderness Act of 1964 cut across agency boundaries and called for the preservation of wilderness in all public land holdings of the government. Units that entered the wilderness system could be areas in the National Parks, National Forests, National Wildlife Refuges, and the National Resource Lands (Figure 34.7). Private holdings could be acquired as well. The language of the Wilderness Act explained its purposes in this way:

> In order to assure that an increasing population accompanied by expanding settlement and growing mechanization does not occupy and modify all areas within the United States and its possessions, leaving no lands designated for preservation and protection in their natural condition, it is hereby declared to be the policy of Congress to secure for the American people of present and future generations the benefits of an enduring resource of wilderness. For this purpose there is hereby established a National Wilderness Preservation System to be composed of Federally owned areas administered for the use and enjoyment of the American people in such manner as will leave them unimpaired for future use and enjoyment as wilderness, and so as to provide for the protection of these areas, the preservation of their wilderness character, and for the gathering and dissemination of information regarding their use and enjoyment as wilderness. . . .[6]

Congress, to guide recommendations on what to include in the Wilderness Preservation System, attempted to define wilderness. They suggested that:

> A wilderness, in contrast with those areas where man and his own works dominate the landscape, is hereby recognized as an area where the earth and its community of life are untrammeled by man, where man himself is a visitor who does not remain. An area of wilderness is further defined to mean in this chapter an area of undeveloped Federal land retaining its primitive character and influence, without permanent improvements or human habitation, which is protected and managed so as to preserve its natural conditions and which (1) generally appears to have been affected primarily by the forces of nature, with the imprint of man's work substantially unnoticeable; (2) has outstanding opportunities for solitude or a primitive and unconfined type of recreation; (3) has at least five thousand acres of land or is of sufficient size as to make practicable its preservation and use in an unimpaired condition; and (4) may also contain ecological, geological, or other features of scientific, educational, scenic, or historical value.[6]

---

6 Wilderness Act of 1964.

(a) Wilderness in lower 48

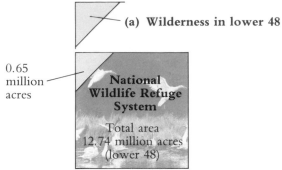

(b) Alaska Wilderness

Agency	Total Alaska Holdings (million acres)	Area of Wilderness (million acres)
National Wildlife Refuge System	76.1	18.6
National Park System	52.1	32.4
National Forests	23.2	5.45

0.65 million acres

National Wildlife Refuge System

Total area 12.74 million acres (lower 48)

2.98 million acres

National Park System

Total area 26.9 million acres (lower 48)

19.70 million acres

National Forests

Total area 167.5 million acres (lower 48)

Figure 34.7 (Opposite page) (a) The boxes represent the Wilderness System as a Component of Federal Parkland, Refuges, and Forests in the lower 48 states. (b) The table shows that in recent years Alaska has added dramatically to our National Parks, Refuges, and Forests. Much area classified as Wilderness is included in these additions. Unfortunately, the Alaska wilderness is not accessible to most people. (Source: Appropriate Federal Agencies, December 1982. Photos: top, National Park Service photo by Richard Frear; middle, American Airlines; bottom, Appalachian Mountain Club photo by Roger Chapman.)

The 1964 Act first instructed both the Department of Interior and the Department of Agriculture to study their holdings for their suitability as units in the National Wilderness Preservation System. The act then quickly designated 54 areas, totalling over nine million acres (3.6 million hectares), as the first components of the Wilderness System. In 1975, the Eastern Wilderness Areas Act supplemented the first act by focusing attention on Eastern wild areas that had recovered from earlier impacts of human activity. A number of Eastern areas scarred by fire or logging and now largely regrown or reforested were placed in the Wilderness System. Other Eastern areas were designated for study for possible inclusion.

The Wilderness Act, in historical perspective, ranks with the setting aside of Yosemite and Yellowstone, the Forest Reserve Act of 1891, and the establishment of wildlife refuges. Each of these earlier acts responded to an emerging awareness. The first of the national parks set aside grandeur; the Forest Preserves were aimed at preserving our timber resources. The Wildlife Refuges were created to protect species from extinction. The Wilderness Act went further; it demonstrated both how far our intelligence of our environment had come and the concern we had to preserve not one item but nature in its entirety. Grandeur was not enough; timber resources were not enough; species alone were not enough. The preservation of nature's intricate pattern as seen through wilderness was required.

National Wild Rivers and National Trails

Four years after the Wilderness Act of 1964, the preservation of wilderness was again before the Congress, this time in two related acts. The wilderness under consideration in 1968 was not simply land and mountains. The remainder of the nation's free-flowing rivers whose banks were still free of development were the subject of the Wild and Scenic Rivers Act. The historic wilderness footpaths—the Appalachian Trail, the Pacific Crest Trail, and others—were the subject of the National Trails Systems Act. The passage of these two acts, while they had defects, strengthened the wall that preservationists were building around the wilderness.

National wild and scenic rivers. The 1968 Act began the process of setting aside the yet unspoiled rivers of the nation. Eight rivers were tagged for immediate entry into the system; these were:

1. Clearwater River, Middle Fork, Idaho
2. Eleven Point River, Missouri
3. Feather River, California
4. Rio Grande River, New Mexico
5. Rogue River, Oregon
6. Saint Croix River, Minnesota and Wisconsin
7. Salmon River, Middle Fork, Idaho
8. Wolf River, Wisconsin

By 1978, the system had been expanded by congressional, state, and federal agency action to include 28 rivers. Taken together, these 28 rivers cover some 2300 miles (3800 kilometers). The Park Service has sole administration of 5 of the 28 and shares two others with the Forest Service. The Forest Service administers 10 of these as the single guardian, and shares other rivers with the Bureau of Land Management. Five of the rivers are administered solely by the states and one by the Bureau of Land Management. A river may be recommended for inclusion by individual state legislatures. Such a river, if approved by the Secretary of the Interior, becomes part of the system without further Congressional action. The Little Beaver River in Ohio entered the system in this way and is administered by the state of Ohio. In all, five state-administered wild rivers have joined the system: the Allagash in Maine; the Little Miami and Little Beaver Rivers in Ohio; the New River in North Carolina; and the Lower St. Croix in Minnesota and Wisconsin.

Wilderness

> All this land is being set aside for non-use at a time when
> public interest in the maintenance of single-use wilderness . . .
> is evaporating.
>
> *R. L. Brown*

> The U.S. Forest Service should think in terms of generations
> if not of centuries.
>
> *Editorial,* **Business Week,** *29 January 1979*

Additions to the National Wilderness Preservation System are surrounded by controversy. A study by the U.S. Forest Service made public in 1979, considered the potential uses of 62 million acres of roadless land in the National Forests. Known as RARE II, for the second Roadless Area Review and Evaluation, the study concluded that the areas in question should be largely given over to other uses than wilderness. Of the 62 million acres, only 15 million acres were proposed for wilderness, while 36 million acres were to be "open" for multiple-use activities. These activities include "mineral entry" (prospecting and mining), grazing of stock, and timber cutting. Eleven of the 62 million acres were not designated either "open" or for wilderness but were recommended for further study. The next step after the Forest Service proposal was for Congress to consider the study and then allocate areas to wilderness as they saw fit.

Some people thought the area of land that the Forest Service had recommended for wilderness to be quite ample and perhaps even too generous. In a letter to *The New York Times*, R. L. Brown asserted:

> All this land is being set aside for non-use at a time when public interest in the maintenance of a single-use wilderness, which appears to have been a fad of some three or four years' duration, is evaporating. Entry into wilderness areas last year totaled only seven million man-days, a fraction of the entry by Americans into national parks and onto other types of Federal land. Sporting-equipment manufacturers report that sales of tents, sleeping bags and other equipment used by hikers for treks into the wilderness are decreasing. . . .
>
> It is not true that all use of land results in permanent loss of wilderness characteristics. Most exploration for mineral deposits, for an example, fails. All signs of preliminary exploration are usually obliterated by new growth in a very short time. Even clearcut log areas reforestrate, and wilderness groups now suggest that many reforested areas are in fact wilderness and should be so designated.★

(To put some of Brown's figures in perspective, the National Parks include about 27 million acres in the lower 48 states. Visitor days to the parks in 1982 reached a total of about 245 million.)

Other people felt the Forest Service had recommended too little wilderness.

> The U.S. Forest Service should think in terms of generations if not of centuries. But continuing pressure from lumbermen, mining companies, and enterprising recreational developers makes it hard for Forest Service officials to think beyond the day's schedule of appointments. The result is a built-in bias in favor of early utilization, which shows in the recommendations the Forest Service has just drawn up for classifying some 62 million acres of undeveloped land in the national forests.†
>
> They [the recommendations] would designate only 15 million acres as wilderness and 11 mil-

lion for future study, while opening 36 million for various kinds of development. There are strong arguments for at least reversing these proportions, designating sufficient wilderness areas besides barren ice and rock, and being sure possibilities are not overlooked for both preserving wilderness and attaining necessary development.‡

Is Brown right that wilderness use is a passing fad? Brown implies that wilderness is only set aside for human uses such as hiking and camping. Is that a correct interpretation? What criteria should be used in de-

ciding how much public land should be set aside as wilderness compared to the amount of land on which multiple uses are allowed? Are visitor-use days a good measure of the value of parks or of wilderness?

* R. L. Brown, Letter to the Editor, *The New York Times,* 9 May 1979, p. A24.
† Editorial, *Business Week,* 29 January 1979, p. 134.
‡ Editorial, *Christian Science Monitor,* 8 January 1979, p. 28. Reprinted by permission from the *Christian Science Monitor,* © 1979 The Christian Science Publishing Society. All rights reserved.

The rivers or portions of rivers included in the system are to be kept free of dams or other structures; that is, their free-flowing character is to be preserved. Further, a corridor on both sides of the river is to be kept free of all development, thereby maintaining the wilderness setting of the river. The rivers placed in the system are classified into one of three categories, according to the legislation that created the system:

(1) Wild river areas—Those rivers or sections of rivers that are free of impoundments and generally inaccessible except by trail, with watersheds or shorelines essentially primitive and waters unpolluted. These represent vestiges of primitive America.

(2) Scenic river areas—Those rivers or sections of rivers that are free of impoundments, with shorelines or watersheds still largely primitive and shorelines largely undeveloped, but accessible in places by roads.

(3) Recreational river areas—Those rivers or sections of rivers accessible by road or railroad, that may have some development along their shorelines, and that may have undergone some impoundment or diversion in the past.[7]

The protection provided a river in the national system is of several kinds. First, the Federal Power Commission is prohibited from licensing new power dams on the protected portion of the river. Rivers under study for inclusion are also protected in this way while they are being studied. Power developments above and below the protected portions of the rivers

are not prohibited, but if they conflict with the flow in the protected stretch they may still be prevented.

In addition to banning dams and diversions, the U.S. government can no longer sell or exchange public lands within the boundaries specified for the rivers. Although mineral claims in these areas that were established before the act were not voided, new mineral-related activities are under the regulation of the Secretary of the Interior to prevent either pollution or other degradation of the area.

National scenic and national historic trails. Although the Appalachian Trail is today under the primary administration of the Secretary of the Interior and the Park Service, and the Pacific Crest Trail is under the primary administration of the Secretary of Agriculture and the Forest Service, it would be a mistake to assume that the federal presence is the reason these trails exist today. In particular, the bulk of the Appalachian Trail is not a result of federal effort. The federal presence does now provide some added protection to the trails and helps to ensure their preservation, but at the time the trails were assembled, ordinary citizens and private organizations furnished the creative force.

The original proposal for the Appalachian Trail came in 1921 in an article by Benton MacKaye, a forester and regional planner, in the *Journal of the American Institute of Architects.* MacKaye suggested linking a number of trails, as well as building new trails. Hiking and outdoor clubs expressed great interest in MacKaye's idea, which initially was for a trail without end, and they undertook to build MacKaye's dream.

7 Title 16, Section 1273 of the U.S. code.

Figure 34.8   Part of the Appalachian Trail. Pinnacles Overlook, Shenendoah National Park. (National Park Service photo)

Within a year of the proposal, the first section of the trail was opened in the Palisades Interstate Park in New York and New Jersey. Of the more than 2000 miles ultimately assembled, only 350 miles of trails existed at the outset. All of the trails that already existed were in New England and New York. The formation of the Appalachian Trail Conference, a federation of clubs, government agencies, and individuals, in 1925 gave hope that the trail could become a swift reality. The complexity of assembling the trail, however, slowed its progress, especially south of Pennsylvania where hiking clubs were still to be organized.

Completed in 1937, the Trail runs along the mountain backbone of the East, the Appalachian Mountains, from Mount Katahdin, Maine, to Springer Mountain, Georgia (Figure 34.8). Public land in the National Forests and in the National Parks from Virginia south helped to make the final assembly possible (Table 34.2). To protect the portion of the trail through public land, the conference signed an agree-

ment with the Park Service and Forest Service, as well as most of the state governments involved in 1938. Under its terms, the federal government promised not to construct incompatible projects within a one-mile corridor on either side of the trail; the states agreed to a quarter-mile strip on each side.

Nevertheless, establishing the connections between such sections proved to be a large task, as the trail often wound its way across private lands. Permissions had to be requested of owners; at the time, simply an oral agreement was usually needed. As the pressure from use grew, however, individual owners became more reluctant to extend their hospitality. The trail was threatened by its very popularity.

At about the same time the Appalachian Trail was beginning to be assembled, Clinton Clark of California proposed (1932) a similar footpath extending from Canada to Mexico on the West Coast (Figure 34.9). The Pacific Crest Trail was also planned as a mountain range path. Its 2600 miles (4300 kilometers) includes

Table 34.2 Approximate Division of
National Scenic Trail Jurisdictions

Agencies with jurisdiction	Appalachian Trail		Pacific Crest Trail	
	miles	kilometers	miles	kilometers
U.S. Forest Service	719	1158	1856	2989
National Park Service	215	346	249	401
Bureau of Land Management	—	—	204	329
States	289	465	43	69
Private	805[a]	1296	106	170

a No cooperative agreements for use exist on 75% of the 805 miles (1296
kilometers) in private ownership.

the Cascade Crest Trail and the Oregon Skyline Trail in Washington and Oregon. The Pacific Crest Trail follows the Lava Crest Trail, the Tahoe–Yosemite Trail, and the John Muir Trail, among others, in California. Fully 85% of the Pacific Crest Trail is on federal lands, such as national park or national forest, so that the trail's existence is less threatened by development than the Appalachian Trail.

After several years of effort, Congress passed and President Johnson signed the National Trail Systems Act in 1968. The act created a system of scenic and recreation trails across the country and made the Ap-

Figure 34.9 Along the Pacific Crest Trail. (U.S. Forest Service)

palachian Trail and its sister trail, the Pacific Crest Trail, the first members of the system. Federal purchase of the needed right-of-ways for the trails was authorized by the 1968 Act, but funds were not appropriated for another 10 years. If negotiations with private land owners proved unsuccessful, the government was given the right to acquire the land by "condemnation."

The Land and Water Conservation Fund administered by the National Park Service was also supposed to assist states in purchasing portions of the corridor needed for the trail, but this federal effort was not a part of the National Trail Systems Act. Individual trail clubs and the Appalachian Trail Conference still maintain funds to buy trail right-of-way, and they continue to make such purchases to the present time. The struggle to create the trail and to preserve it still falls largely to individual citizens and the trail clubs.

The National Trail Systems Act did more than promise to stabilize the historic mountain backbone trails of the east and west coasts. While the Appalachian and Pacific Crest Trails were designated the first National Scenic Trails, the Act directed that other trails be studied to see if they would qualify for inclusion in the Scenic Trail System. The Continental Divide Trail has since been designated as a National Sce-

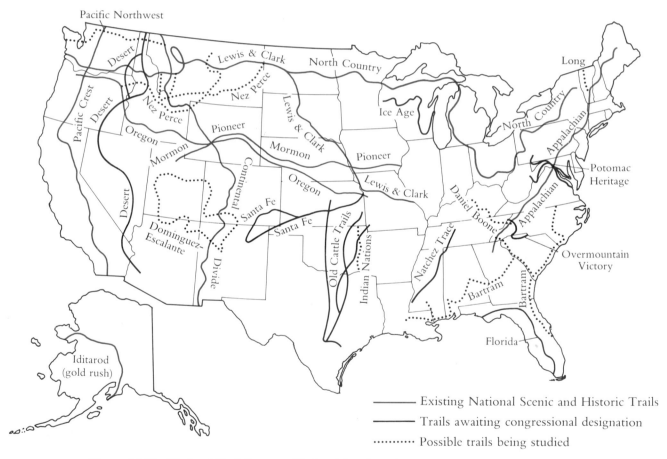

Figure 34.10 National Trail Systems: National Scenic and National Historic Trails, 1980 (Reprinted by permission from *National Parks & Conservation Magazine,* October, 1980. © 1980 by National Parks & Conservation Association.)

nic Trail (1978). From Glacier National Park in Montana, the trail follows the continental divide down to New Mexico and the Mexican border (Figure 34.10). In 1980, the 3200-mile-long North Country Trail was added to the Scenic Trail System.

Recognizing the historical significance of a number of the proposed trails, in 1978 Congress created a new category of trail, the National Historic Trail. Four trails have been taken into the system as National Historic Trails. The Oregon Trail (2000 miles from Missouri to Washington); the Lewis and Clark Trail (3700 miles from Illinois to Oregon); the Mormon Pioneer Trail (1300 miles from Illinois to Utah); and the Iditarod (an Alaskan Gold Rush trail, 2000 miles from Seward to Nome) thus entered the National Trail System. Other routes are still under study.

The trails will not necessarily follow the original routes; development has in some cases wiped out entire sections of the original trail. Thus, some rerouting will be necessary to achieve unbroken trails. In addition, federal protection is provided only to those portions of the trails that are on federal lands.

For a number of years, the promise of the National Trail Systems Act to preserve the nation's trails has been a hollow commitment. The hope that the Appalachian Trail could be finally secured by federal purchase of threatened areas has not yet been fulfilled. Congress finally appropriated some money under the act for such purchases in 1978, 10 years after the initial commitment to acquire the needed land. During those 10 years, development marched closer to the Trail in many places, and the cost of acquiring right-of-way rose considerably. Now, as before, the Trail's continued existence as an unbroken footpath is the result of the pooled private efforts of individuals and hiking organizations. There is a lesson in the history of the Appalachian Trail.

Problems of Accessibility

A stone-and-mortar walkway with wooden handrails guides visitors away from the picnic area down along a steeply sided stream. The visitors witness a quickly changing scene of rock cliffs and evergreen-covered earth banks. Then the valley opens with a thunderous roar; a magnificent waterfall confronts the visitors. It is a spectacle worth remembering—a vivid demonstration of the remarkable power of flowing water, cutting through the rock over countless centu-

ries. The visitors, an elderly couple, would have been unable to see it had not the walkway and handrails been present to guide and steady their steps. The walkway was the legacy of a work crew from the Civilian Conservation Corps who, almost 50 years ago, helped to make parks more accessible to less athletic people. It was an effort of noble proportions and one of benefit to millions of people.

Yet such development at park after park, year after year, leaves its mark on the wilderness setting. Construction of roads, walkways, picnic grounds, swimming areas, restaurants, campgrounds, shops, service stations, and even hotels begins to take on a pattern. Nature is being enclosed in an amphitheater. A supporting staff of ticket takers, ushers, waiters, and waitresses are needed for an orderly display. Each "improvement" makes the park more accessible and accommodates more and more people. And the crush of people justifies the improvements; the park authorities are simply attempting to meet "demand" for public recreation.

Many argue that such developments, which are designed to meet the need for public recreation, degrade wilderness values to an unacceptable extent. Indeed, the word recreation, when broken into its parts, means "to be created again." Only by usage has its meaning evolved to include amusement, exercise, or educational activities.

The measurement of success in wilderness preservation is very difficult. It is far easier to count visitors and use the annual visitor count as the measure of success of a park operation than somehow assess how well the land has been preserved. At a famous Midwestern site managed by the National Park Service, one of the goals of management is to keep the average time of a visit to less than 50 minutes. If the average time a party takes there is much more than 50 minutes, backups on the access road to the site will occur and motor homes will boil over. The Park Service needs to move people in and move them out.

Although an increased visitor count is a possible measure of success in a park, in a wilderness area an increased visitor count may be a danger signal. Congested trails in the John Muir Wilderness of California led the U.S. Forest Service to go to a permit system for trail camping there. The number of hikers on the trail to Mount Whitney was limited to no more than 75 per day. Since two days are usually needed to hike the 8-mile trail, the number of overnight campers is generally kept to less than 150 on any given evening. In the early 1970s, more than 1200 hikers had been on the trail during one day of the Labor Day weekend before the permit system was begun.

Often the development of tourist facilities at the site of the spectacle itself diminishes the experience of visiting. Old Faithful in Yellowstone, with its parking lot and visitor center, is often mentioned as a natural area that has been overdeveloped. The development in the Yosemite Valley is also seen as overdone. Such developments began in a era when visitors to the parks were few and the modest facilities needed for the visitors did not intrude on their appreciation of the park's values, even if the facilities were built at the site itself. But as the number of visitors grew, the development required to serve them had to expand as well. It was easiest, apparently, simply to "add on."

The alternative to development at the site, worthy of careful consideration, is a visitor center at or near the park gate. Such a visitor center, remote from the scenes of the park itself, was wisely built in the early 1970s at Acadia National Park in Maine. At other parks, however, existing facilities are unlikely to be torn down to be rebuilt at less scenic locations; the expense is seen as too great. Nonetheless, the concept is attractive because not only development, but also litter that accompanies tourism, can, to a degree, be kept from the park interior.

Wilderness and recreation come in conflict in more ways than number of people and development. Certain human activities allowed in a natural area can also diminish wilderness values. Prime examples are motor boating, off-road vehicle operation, and snowmobile use. The activities are supported, however, not only by a large public, but by business interests who organize lobbying efforts so that their products can continue to be used.

Reviving an Old Tradition of Land Preservation

When the Forest Preserves were first created in concept at the federal level in 1891, Congress was responding to growing conservationist sentiment throughout the nation. This sentiment for preservation of our forests was expressed in conservation efforts by the states as well as by the federal government. The most ambitious of these efforts, the Adirondack Mountain Preserve, predates the nation's Forest Preserves by six years.

Created in 1885, the Adirondack Mountain Preserve was meant to protect the exquisite wilderness of mountains, forests, lakes, rivers, and wild species that the region possesses. Threatened by lumbering and mining interests, the land was snatched from exploitation by the preserve. A portion of the Constitution of the State of New York calls for the region to "forever be kept as wild forest lands."

In 1894, the New York legislature penciled a blue line around a six-million-acre region of the Adirondacks. The area enclosed by this line became known as a greenline park, an area within which wilderness would be preserved. The movement to designate such areas for needed protection became known as the "greenline movement." (Color blindness has been suggested for the shift in colors in naming the movement.)

Yet the Adirondack Forest Preserve is not a pristine and totally undeveloped wilderness. About 120,000 people are permanent inhabitants of the region, scattered in 106 towns and villages, and many more people come to spend the summer in the Adirondacks in cottages and homes that dot its lakes. Even more people come to vacation at the numerous public campgrounds that the state runs and at the many private resorts in the region. Although the state holds 2.3 million acres of the area within the greenline, private interests hold 3.7 million acres within the preserves. Some 62% of the land area of the preserves, then, is in private hands.

Commerce of all sorts proceeds in the preserves. Businesses open and grow. Developers speculate and subdivide throughout the region in search of profits. All these activities co-exist with the state-owned wilderness. One must ask how this can constitute "protection."

The answer is that protection comes from an elaborate set of controls on the use and development of privately held land. An example is found in the rules for using privately held land that fronts on lakes. Shoreline modifications and the building of structures require permits. The cutting of live trees within 50 feet of the shoreline is strictly limited. Septic tanks must be kept suitably distant from shoreline. Such controls as these are placed on land in private hands when the condition of the lands directly influences the visual aspect of the preserve. Development within the business

Should Wilderness Be Accessible?

To be precious, the heritage of wilderness must be open to those who can earn it again for themselves. The rest, since they cannot gain the genuine treasure by their own efforts, must relinquish the shadow of it. We must earn again for ourselves what we have inherited.*

Garrett Hardin

As a young forestry student in 1939 . . . I roamed the high country west of the Three Sisters, mingling with deer, elk, and bear. . . . Thirty-five years later I returned. It was only three months after open-heart surgery so I could not climb very much. . . . It was a logging road that brought a middle-aged college professor within walking distance of his goal.†

Stephen H. Spurr

The views of Garrett Hardin and of Stephen Spurr are in conflict over the issue of whether wilderness ought or ought not be easily accessible to people. The sentiments expressed by Hardin suggest that the definition of wilderness as offered by Congress in the Wilderness Act (p. 659) is somehow inadequate.

Hardin is suggesting that wilderness should not be easily accessible. There are, of course, practical reasons to limit accessibility. You should be able to provide such practical reasons. Hardin's point of view, however, is one of principle. The practical implications are not of concern to him, at least not in this passage.

The sentiments of Spurr, on the other hand, suggest that wilderness should be used by people, and that accessibility provides the wilderness values to many, rather than to the few.

You may have thought about these points of view before when you were on the trail and looking to escape crowds of people. You may consciously have chosen less-traveled paths. Your opinion on the accessibility of wilderness, however, may have something to do with your age and condition. Age and condition may have influenced Stephen Spurr's opinion. You should examine your views of this issue because the shape of future wilderness will be influenced by them. Is there a middle ground or compromise between inaccessibility and accessibility?

---

* Quoted from the Foreword by David Brower of *Earth and the Great Weather: The Brooks Range*. San Francisco: Friends of the Earth, 1971. (Introduction by John Milton; Photos by Peter Martin.)

† Stephen H. Spurr, *American Forest Policy in Development*. Seattle: University of Washington Press, 1976, 86 pp.

The ORV Fight

Just put your gang on Suzuki's DS trail bikes. And head for the boonies. Doesn't matter where you go. Peaks or valleys, it's all the same to those rugged off-road machines.

Suzuki advertizing copy, **Cycle World** *17(3) (1978) 110–111*

St. Francis of Assissi himself while driving an off-road vehicle on wild land could not avoid diminishing the recreational experience of many non-ORVers in the same area.

CEQ report, 1979

One of the most bitter controversies over the use of public lands involves off-road vehicles (ORVs): trail bikes, snowmobiles, dune buggies, and four-wheel-drive vehicles. Some 25% of the American public enjoys the use of these vehicles. ORVs allow their users the physical thrill of conquering rugged terrain and also allow them to penetrate far into wilderness areas for hunting, fishing, and camping. But ORVs can destroy that very wilderness, as well as ruin the wilderness experience for hikers, backpackers, and other nonmotorized users.

ORVs destroy the vegetation they ride over, leaving soil exposed to erosion from wind and rain. They compact soil, killing plant roots and decreasing the soil's ability to absorb water, which further increases erosion. In the arctic, ORV use leads to the melting of permafrost and the formation of water-filled tracks. The noise from ORVs can drive away wildlife or even damage vital hearing ability in certain species.

Non-ORV users are not only worried about these environmental effects, but also extremely angry at the intrusive noise ORVs produce.

. . . [O]ne ORV operator can effectively restrict a large public area to his own use through the emission of loud engine noise, obnoxious smoke, gas and oil odors and dangerously high speeds. Whereas previously many persons of all ages and wealth could observe the beauty of unspoiled land, now a single ORV can reign supreme.★

Consider the snowmobile . . . and the often-heard argument that this machine makes it possible to "get way back in there, away from it all." There are, of course, several other ways to "get back in there," including snowshoes and skis. Maybe if you need an engine to get there, you don't belong there in the first place! To my mind, "getting away from it all" means, foremost, getting away from our society's overdependence on the combustion engine.†

[We need to] provide some place on God's green earth for man to spend some time without hearing a damned motor.‡

I hope there is some way we could outlaw all off-road vehicles, including snowmobiles, motorcycles, etc., which are doing more damage to our forests and deserts than anything man has ever created. I don't think the Forest Service should encourage the use of these vehicles by even suggesting areas they can travel in. . . . I have often felt that these vehicles have been Japan's way of getting even with us [for World War II].§

In some areas, attempts have been made to restrict ORVs to particular areas or trails. TVA, for instance, has tried this on an experimental basis. TVA ranger Scott Seber says,

I used to hate ORVs. Now I feel they can be worked with. We have demonstrated that they don't have to be running amuck everywhere.‖

This has not satisfied everyone, however.

> [We] object to the continual enhancement of non-ORV recreation at the expense of the off-road vehicle enthusiast. We do not feel that all compromises should be made at the expense of off-road motorcyclists.★★

How do you feel about ORV use? Do you think all public lands should be open to ORV users? Or do you feel their use should be forbidden on public lands; perhaps restricted to privately run ORV parks?

★ Gary A. Rosenberg, *Environmental Affairs,* 1976.
† Jerry Buerer, Professor of Sociology, Marquette University, and organizer of a group to protect the rights of non-ORVers, 1975.
‡ Ben Huffman, Vermont Department of Forests and Parks, 1974.
§ Senator Barry Goldwater, letter to William D. Hurst, Regional Forester, Region 3, U.S. Forest Service, Albuquerque, New Mexico, 23 March 1973.
‖ *Off Road Vehicles on Public Land,* Council on Environmental Quality, 1979, p. 14.
★★Robert Rasor, AMA, 1976, Environmental Impact Statement, supra note 19, at 260.

districts of incorporated towns is governed by the zoning within those towns, but elsewhere development is controlled by the Adirondack Park Agency, the preserver/enforcer of the preserve. In areas still wild, though privately owned, the agency allows no more than one building for every 42.7 acres.

To many who make their home in the region, the regulations are "big government," unnecessary interference with people's natural rights. To millions of others, though, in New York State, the rules are the means by which the Adirondacks were preserved for their use and the use of their children. The preservation of the Adirondacks without actual purchase of the land creates inevitable conflicts between people and the administering agency. Without these conflicts and tensions, however, the land could only be preserved at enormous expense—by the actual purchase of the land outright. The Adirondacks are being preserved, even in some cases against the will of its residents. The land-use controls are working, and working, in fact, to the long-term benefit of the residents of the region, whose income derives largely from tourism.

The success in preserving the Adirondacks is now leading to parallel efforts elsewhere in the United States. As in the Adirondacks, we will find that preservation by controlling land uses meets with opposition from commercial interests.

In 1978, the nation's second greenline park was created, this time in the unique pitch-pine forests of southern New Jersey. The action did not come a moment too soon. Under pressure for rapid vacation-home development from the nearby metropolitan areas in New York, New Jersey, and Pennsylvania, the pine barrens, as the forests are called, had already shrunk from 2000 to 1500 square miles. The sandy soil of the region had largely prevented farming and subsequent development in the area, so that much virgin forest and many unspoiled rivers still remained. The region of the pine barrens harbors rare plant and animal life, such as Pickering's morning glory, the Karner blue butterfly, and the carpenter frog.

The region of the pine barrens was too large to purchase outright, so Congress created the Pinelands National Reserve, again an area defined by a "greenline" in which federal and state land would coexist with private land. Development on the privately held land would be carefully controlled. Just as in New York, where the Adirondack Park Agency administered the land-use controls, an intergovernmental commission has been established to oversee the regulation of development in the Pinelands. Just as in New York, the plan is to acquire the most ecologically sensitive lands and limit further development in the remainder of the million-acre area. At the outset, some 20% of the area was in state hands and another 5% of the area was to be purchased by the federal government.

As proof that the pinelands were under intense pressure and needed rescuing, would-be developers have filed suit to have the land-management plan set aside.

Greenlining may well be used to protect other natural areas of the U.S. These areas would be ones where some development is already in place and cannot be removed, and where unique features still remain to be preserved. These are areas also where the cost of outright government purchase of land would be too great. The areas surrounding a number of National Parks fall in this category; these areas have developed to accommodate vacationers and often possess landscape and wildlife qualities worth preserving. Further, the development has often increased land values and made purchase at a fair price very difficult.

The National Parks and Conservation Association

has identified a number of areas where greenlining, a combination of land purchases and land-use controls, could make an impact on the preservation effort. One such area is the Big Sur coastal region of Northern California. Rich in redwood forests, this area boasts a rugged seascape of towering cliffs and misty rock-encrusted shores. Its exquisite beauty is as haunting as its ocean waters are cold.

The Jackson Hole region of Wyoming is another area where greenlining could make a difference. The rugged Teton Mountains look over this area where exciting western history was made. Tourist developments have pushed up property values and taxes so that ranchers are looking for relief from high taxes. Taxation based on current use rather than on value for development is one possibility (see Chapter 33), but care must be taken that such taxation does not simply reward speculators with higher profits. The purchase of scenic easements is another possibility. Here payment

is made for a promise not to develop. A question arises here, however. Should the ranchers be compensated by purchase of easements in this way? Land owners in the Pinelands and the Adirondacks do not receive compensation for leaving the land undeveloped; they simply *cannot* develop it.

A third area identified by the National Parks and Conservation Association is the Gorge region of the Columbia River. The Columbia's majesty has been reduced by numerous hydropower developments, but many waterfalls remain, and a number of unusual plant species are found in the Gorge. Although development pressures come from nearby urban areas, only a patchwork of state parks and national forests now protect the region. Because the Gorge region takes in portions of both Washington and Oregon, effective protection has been elusive. In 1982, legislation lay before the U.S. Congress to establish the nation's third greenline park around the Gorge of the Columbia.

Expanding the System of Natural Areas

The Land and Water Conservation Fund

The Land and Water Conservation Fund (LWCF) is a recent and powerful means to expand public natural areas from the local level to the national level. Created by Congress in 1965, the Land and Water Conservation Fund has in the past furnished federal money to match local money for the purchase of natural areas. The Fund is presently administered by the Division of State, Local, and Urban Programs of the National Park Service, but was formerly administered by the Heritage Conservation and Recreation Service (HCRS), housed in the Department of the Interior. The original administering office was begun as the Bureau of Outdoor Recreation and was created in 1962 by the Secretary of the Interior at the request of President Kennedy.

The Land and Water Conservation Fund (LWCF) Act of 1965 gave clout to the mission of the office. The act established a fund of money to be allocated to state and federal agencies for the purchase of land. The fund is fed by revenues from the sale of surplus properties, from the leasing of the outer continental shelf for oil and gas, and from other sources, including direct appropriations by Congress. Of the fund's revenues, 60% have been for the use of the states. Three activities were supported by this portion of the fund: the planning related to a specific project; the acquisition of the area; and the development of recreation facilities and resources at the site. The fund has supplied up to one-half of the total cost of a specific project. For a state to be eligible to apply for money, it must have already developed a comprehensive outdoor recreation plan that includes an assessment of outdoor recreation needs and recreation resources in the state. The remaining 40% of the fund is for the use of federal agencies involved in outdoor recreation.

Contributions of the fund to purchases are often made in cooperation with the efforts of state governments and conservancies. Federal laws giving tax benefits for donations of land are often involved in the transactions. After describing the functions of conservancies, we will illustrate the acquisition process with several examples. These examples will show how conservancies and state governments, using the Land and Water Conservation Fund, have been combining their talents and using the benefits of the federal tax law to enrich the public with new natural areas.

Donations and Tax Benefits

We mentioned that the National Park System has received a number of gifts, but we did not describe in any detail how gifts are made. To someone interested in giving a valuable property to the public, the monetary advantages that can be obtained from the gift are important.

The simplest type of gift, called an outright donation, transfers the land directly from owner to recipient at the time of agreement. Depending on whether the recipient is a conservancy (a nonprofit land trust) or a government agency, different but nonetheless attractive tax benefits to the donor are available. Land may also be given in a will and thus be transferred after the death of the owner. In still another type of gift, the transfer can be made in such a way that the donor retains for his lifetime the privilege of living on the land, even though the land has already changed hands. Still another popular form of gift is the bargain sale, in which the owner sells the land to a conservancy or government agency for less than its fair market value. The difference between what would have been obtained by sale on the open market and what actually was obtained in the bargain sale represents a gift or donation, and the value of the difference determines the tax benefits to the donor.

It is this bargain sale procedure that we will mention further in describing how the LWCF, conservancies, and state governments have been functioning together to expand the system of natural areas. First, however, we need to dwell for a moment on the tax benefits of donation, because these benefits are a part of the argument that convinces would-be donors to become the generous individuals they wish to be. The value of the land is first established through an appraisal by a professional appraiser. The appraiser normally considers what similar tracts of land in the geographical vicinity have been sold for on a per-acre basis and applies that to the land in question, with adjustments for the specific factors of its location, such as road frontage, distance from water and sewer, etc. The appraised value reflects the price the property would bring if placed on the open market.

Specific federal tax benefits differ, depending on whether the property is given to a conservancy, a local government, or an agency of the federal government. If the donation is made to a conservancy or land trust, the donor may deduct from his taxable income either the value of the gift or 30% of his adjusted gross income for that year, if the gift is larger than this amount. If the value of the gift exceeds 30% of his adjusted gross income, the donor may carry over the remainder and deduct that remainder the following year, subject to the same percent limitation. Any ex-

cess may be carried over to the following year for a total of six years.

Here is an example.

Mr. Sandune donates a beach property worth $75,000 to the Nature Conservancy. His adjusted gross income is $130,000 per year. He can deduct from his taxable income up to $0.3 \times (130,000) = \$39,000$ of the $75,000 gift. The income on which he must pay taxes is now reduced to $130,000 - \$39,000$ or $91,000. His tax savings are considerable. The following year, if he has the same level of income, he can deduct up to $75,000 - 39,000 = 36,000$, the remainder of his contribution. Again, he obtains large tax savings.

Tax benefits are even greater if the gift is to a local government or government agency. The full value of the gift may be deducted from the adjusted gross income without any limitation calculated as above. In the simplified example above, Mr. Sandune can deduct $75,000 from his adjusted gross income; that is, he will not need to pay any tax on $75,000 of his income because of the gift. Corporations may also donate land, but special rules apply for the tax treatment of their donations.

Conservancies or Land Trusts

A number of private nonprofit organizations exist that assist in the process of obtaining land for the public and which themselves obtain lands for future transfer. Most prominent among these is the Nature Conservancy, a private, nonprofit organization that buys ecologically and geologically valuable land. The Conservancy has purchased nearly 1800 parcels of land, totalling 1.1 million acres, since it was founded in 1950. About two-thirds of these purchases have been transferred to government agencies, mostly to state governments, for their permanent protection as natural areas. The remainder are managed by the Conservancy itself.

The Conservancy, which has many state branches, does more than buy and transfer land. It is active in identifying and inventorying natural areas across the country. Because of the Conservancy's stature in preservation activities and because of the tax advantages that come to donors of land, the Nature Conservancy is often successful at obtaining land by simply describing to potential donors their purposes and the tax advantages from gifts.

A 138-acre parcel at Stevens Creek, South Carolina, was obtained by the Conservancy in this way from the Continental Group. The land, which harbors several endangered species, was then transferred to the protection of the state of South Carolina. In a similar fashion, the Conservancy was able to obtain a portion of Jupiter Island off the East Coast of Florida. The manatee, a remarkable sea mammal resembling nothing so much as a large cigar in shape, lives at Jupiter Island. This distant relative of the elephant, known more commonly as the "sea cow," is one of America's endangered species. The 500-acre tract on Jupiter Island will aid in protecting the manatee.

Another example of the Conservancy's work is the preservation of the 200-acre farm of Dr. and Mrs. Wright near Albany, New York. Given to the Conservancy outright by Dr. and Mrs. Wright, the area is now maintained as a nature preserve. One of the most significant of the Conservancy's purchases is the Dewey Ranch in Kansas. The 7200-acre tract preserves a portion of the tall-grass prairie. The prairie that drew so many pioneers west for settlement is now virtually gone. Only small pockets of this unique ecological landscape still exist, so the Conservancy's purchase is an especially important one. The area is to be leased at present to Kansas State University for ecological research. (See Chapter 5 for a discussion of the tall-grass prairie.)

Coordinated Federal and Private Actions

Federal tax laws, as we pointed out, often make it economically attractive for people to donate land for natural areas. The Nature Conservancy helps explain these advantages to donors and may assist in the transactions themselves.

Take the case of a 3100-acre parcel in Rutland County, Vermont, which was prime land for a vacation-home development. This parcel, which adjoins a state forest, was put up for sale in 1971. Its appraised value was $610,000. Vermont's Agency of Environmental Conservation had no money to acquire the property; not even enough to go halves with the Land and Water Conservation Fund in purchasing the land. The agency proposed to the owners that they sell the property to the state for one-half of its market value, or $305,000, money they felt they could obtain from the Land and Water Conservation Fund, whose guidelines allow them to contribute up to half of the market value for a tract of land. The difference between the full market value of $610,000 and the bargain sale price of $305,000 they were suggesting could be treated as a donation for federal tax purposes. Although interested, time was important to the owners. They wished to move faster than the grant from the Land and Water Conservation Fund could be received. The Nature Conservancy stepped in to protect the potential sale.

They offered the owners $305,000 in cash and pointed out the tax advantages of donating the remainder of the property's market value. The owners could deduct up to 30% of their individual shares of the donation from their taxable incomes. Furthermore, they would save realtor fees and save taxes on a portion of their investment profits as well. All of these advantages were theirs, said the Conservancy, in addition to public recognition of their gift.

The owners sold the property to the Conservancy and the Conservancy held the land in its name until the state agency obtained its grant from the Land and Water Conservation Fund. With the grant in hand, the Agency purchased the land from the Conservancy. The price was the original $305,000 plus miscellaneous interest and legal fees. Here the Conservancy both acted as an influential spokesman for land preservation and actually took possession of the land on a temporary basis to ensure that it would not be sold to developers.

The linkage and the coordination of the Nature Conservancy, the states, the Land and Water Conservation Fund, and federal tax laws has been in the past a very powerful tool in saving land. Although the LWCF fell on lean times under the Reagan Presidency, the Division of State, Local, and Urban Programs continued to allocate small amounts of LWCF money in the early 1980s. This was money that had been obligated for state projects in prior years but for which actual purchases had fallen through.

The Reagan Administration was recommending zero funding for the office during this time, but Congress was trying to respond to local sentiment and attempted to provide funds to be channeled to the states. Each new session of Congress brought new attempts to restore the money and functions of the office. Although these changes took place in the early 1980s, the successes of the LWCF in the 70s make it clear that the LWCF was an effective tool to bring new lands into public ownership at low cost to the states. As such, the anticipation is that its functions will one day be completely restored so that the state–federal–private partnership can be renewed.

Questions

1. Why are so many different agencies involved in administering public lands? Briefly outline the agencies and their responsibilities.
2. Is fire always a bad thing in a forest? Explain.
3. Are the National Forests a subsidy to timber companies, even though the timber companies must bid against one another to log a particular parcel of forest? Should the National Forests be logged at all? Should we allow timber cut from National Forests to be sold in export? How could we prevent this if logging continues in the National Forests?
4. What is a greenline park? How does it preserve the natural characteristics of the land?
5. Are trail shelters compatible with a wilderness designation? That is, should trail shelters be built or allowed to continue to exist on land that has been given wilderness status? (In Olympic National Park, Washington, such a controversy arose when the National Park Service began to remove trail shelters. Users of the wilderness complained and were conducting a campaign to have the shelters restored.)
6. Which do you think is to be preferred: federal or state ownership of a public natural area?
7. Suppose you wanted to see a particular parcel of land, now privately held, become part of the public lands system. Describe how you might go about it if (1) you are the owner (What tax advantages are available to you?); (2) you do not own the land.

Further Reading

Dodge, M., "Forest Fuel Accumulation: A Growing Problem," *Science,* **177** (14 July 1972), 139.
This well-written article can lead you into the literature of this controversial area.

Fitzsimmons, A., "National Parks: The Dilemma of Development," *Science,* **191** (6 February 1976), 440.
A nontechnical article with details of park development problems and methods to deal with intrusions on the setting.

Krieger, M., "What's Wrong with Plastic Trees?" *Science,* **179** (2 February 1973), 446. See also Letters and Reply in *Science,* **179** (25 May 1973), 813.
We don't all think alike and this article proves it. But can you argue with the point of view? The letters show you how to begin.

Marsh, G. P., *Man and Nature.* New York: Scribner's, 1864. (Reprinted by Harvard University Press, 1965.)
From an historical standpoint, this is the first warning of how badly we were using the land. Marsh points to overgrazing of land and overcutting of forests as having the potential to "destroy the balance which nature has established."

Moore, W., "Fire!", *National Wildlife Magazine* (August 1976), p. 4.
A readable account of how fire is used in forest management.

Preserving Our Natural Heritage: Volume I, Federal Activities, prepared for U.S. Department of the Interior by The Nature Conservancy, 1975, for sale by the Superintendent of Documents, U.S. Government Printing Office, Washington, D.C. 20402.
No other book we have found brings together so much information on the role, holdings, and management practices of the federal agencies concerned with land preservation. Because it is so well organized and devoid of pictures and maps (these are referred to but never reproduced), there may be a problem with maintaining an adequate attention span. If you can remember your question on federal activities in conservation long enough, the answer is quite likely to be in this book, or a reference on the subject may be cited.

Steen, H. K., *The U.S. Forest Service: A History.* Seattle: University of Washington Press, 1976.
Wonderfully detailed historical account of the beginnings and evolution of forest management from the post-Civil War Era to the early 1970s. Recommended for those keenly interested in the past of the forestry movement and its influence on the shape of the present.

Waver, R., and W. Supernaugh, "Wildlife Management in the National Parks," *National Parks,* July–August 1983, p. 12.

Journals in which you may wish to browse on land preservation are the *Journal of Forest History; American Forests; Journal of Forest Ecology; National Parks and Conservation Magazine;* and *Living Wilderness.*

Conclusion

Two Voices

Our Narrow Niche in Time/Our Narrow Niche in Space/A Thank-You Across the Generations/Concern with the "Here and Now"

Restoring the Environment—Personal Choices

A Choice of Methods/Environmental Decisions

Two Voices

In Part VIII, we noted the words of a Nigerian Chieftain who, sensing the briefness of our own span of years compared to the infinity of years belonging to our ancestors and descendants, said

I conceived that the land belongs to a vast family. Of this family, many are dead, few are living and countless members are still to be born.[1]

In Chapter 3, we quoted Henry Beston, a biologist, who provided us a unique view of nature and its relationship to humankind.

For the animal shall not be measured by man. In a world older and more complete than ours they move finished and complete, gifted with extensions of the senses we have lost or never attained, living by voices we shall never hear. They are not brethren, they are not underlings; they are other nations, caught with ourselves in the net of life and time, fellow prisoners of the splendour and travail of the earth.[2]

We quoted these two people not only because of their eloquence but also for the wisdom in their messages.

Our Narrow Niche in Time

The concept of the Nigerian Chieftain places our years on earth in some perspective. Being aware of the finiteness of our time span makes us sensitive to the fact that depletable resources, such as the fossil fuels that we use and consume, are lost to future generations. The land we consume—by building houses, by paving, by surface mining, by allowing unchecked erosion—can only be restored over centuries. The species destroyed by human activities are lost forever. It is true that technology is constantly finding new resources, but this is not a process upon which we can depend. Nor should we depend on it, since each new application of technology seems to carry new environmental burdens. Shale oil may one day be produced in large quantities to power the engines of society, but the upheaval of the land and pollution of the water from producing shale oil make it a potentially damaging technology. Commercial actions that maximize our well-being today may lead to degraded conditions for future generations. The sacrifices that parents make for their children are an example of how the human species instinctively looks to the protection of future generations.

Our Narrow Niche in Space

The second concept is that we share our space. We are not alone, as Beston points out, nor can we survive alone. We are part of a wider ecological system. We cannot chart all the interrelationships in this system and so we cannot see its limits, just as we cannot see to the limits of the universe itself, no matter the power of our telescopes. Sadly, because we cannot see the limits of our ecosystem and find all the species of plants and

---

1 *Land Use and the Environment,* V. Curtis, ed., EPA, Office of Research and Monitoring Environmental Studies Division, 1973.
2 Henry Beston, *The Outermost House.* New York: Viking Press, 1962, p. 25.

animals upon which we depend, and upon which these species in turn depend, some of us are tempted to believe that what we can see is the whole of the system. Europeans of the 15th century felt that way about the extent of the earth, until Columbus sailed beyond the limits of contemporary knowledge. In the same way, new Columbuses show us, every day, that our knowledge of the environmental relationships on which we depend is incomplete. We are not alone, nor are we free to be independent of the environment that surrounds us.

A Thank-You Across the Generations

These ideas on sharing time and space lead us to consider actions that have long-term benefits. That is, the benefits of the actions we may take today are distant in time and are to be enjoyed principally by future generations, rather than by us. These ideas also lead us to consider the prevention of actions that will have long-run negative effects, even though the effects may fall outside of our own time span. We sense the necessity of protecting the welfare of future generations, but we have no way to measure the resulting satisfaction and comfort of those future generations. Few of us will be present 100 years from now to hear a "thank you," though it be a thunder that echoes from every succeeding generation. Yet, if you have ever seen the Redwoods in California, or hiked the Pacific Crest Trail or the Appalachian Trail, or viewed the waterfall and geysers of Yellowstone, you may yourself have said a thank you to generations past. The wisdom that inspired the setting aside of these lands is both an inspiration and an instruction.

Long-run benefits stem from other actions than setting land aside. If population growth is slowed and finally halted, our future impact on the earth can be slowed and controlled. When we turn away from the fossil fuels and nuclear power to limitless sources of energy such as the sun and wind, we preserve petroleum and coal for future uses. These steps also limit the degradation of the land from surface mining and help check oil pollution of the oceans. Turning away from nuclear power can lessen for future generations the burden of pollution by radioactive elements, and it can slow the spread of nuclear weapons, perhaps the most awesome threat to the future of life on earth.

Concern with the "Here and Now"

Concern about the narrowness of our niche in time and in space is complemented by a third concern, which also motivates us to improve the environment. The third concern is with the "here and now." In some cases, human health and lives are threatened by pollution and toxic substances in the environment. In other cases, the welfare of plants and animals is threatened by pollution and human activity. To remove these threats requires action in the present. Such actions as controlling automobile emissions or scrubbing sulfur oxides from the stack gases of power plants have present benefits in decreasing the frequency of respiratory illnesses such as bronchitis. Removing such hazardous substances as asbestos and vinyl chloride from the workplace will lower the number of cancer deaths among the work force. Preventing oil spills in the oceans will protect waterfowl and aquatic species from being destroyed in local environments. Using closed-cycle cooling on power plants will decrease the quantity of water required by power plants and prevent organisms from being sucked through a condenser at enormous speeds or from experiencing thermal shock. Treating wastewater on a particular river to remove organic wastes will protect the oxygen content of the river and thereby ensure the survival of fish and other aquatic species.

This concern for the present is one point of focus for improving the environment. Indeed, a large portion of this book was devoted to the subject of cleaning up pollution in water, in the air, and in the workplace. Fortunately, at the same time that we are making present-day improvements in the environment, we can also make long-term changes for the better. Reducing the pollution burden today is plainly a commitment to maintain the quality of the air and water and workplace for future generations. Furthermore, since many toxic chemicals can also cause birth defects and changes in the genetic information, the removal of toxic substances from the workplace has long-term as well as short-term benefits.

Because of this implied promise to maintain the quality of the environment once it has been improved, our short-term interest in improving our surroundings merges with our long-term concern for the quality of the environment of future generations.

Restoring the Environment— Personal Choices

A Choice of Methods

We began this book by observing some of the thoughts people have when faced with the challenge of restoring and preserving the environment. Now we examine some of the ways that people act on behalf of the environment. It would be far too simplistic for us to offer you a formula for environmental progress. What works in one time and place may be terribly inappropriate in another. And it would be foolish and arrogant for us to advise you in what manner to act. As you can see in the statements that follow, people respond to the challenge in ways that differ not only in their impact, but also in their intent.

> My contribution to the Sierra Club is helping to influence government action on environmental/conservation problems.[3]
>
> **Sierra Club member**

Fix it up
Wear it out
Make it do
Do without

> **Old Yankee saying**

Save the Whales, Don't Buy Japanese Products

> **Bumper Sticker**

We want to build a self-sufficient community, complete with organic gardens, solar heat, wind power, and—especially—the chance to provide an environment God intended for raising healthy families. . . .

> From the advertising section of **Mother Earth News**, No. 36, November 1975.

We should applaud and look up to those who adopt life styles that are modest in terms of the amount of space they monopolize or the amount of materials and energy they consume. . . .[4]

> **Maurice Strong**

---

3 Sierra Club Environmental Survey, *Sierra Club Bulletin* (March/April 1979).
4 Maurice Strong, first Executive Director of the United Nations Environment Program, now Chairman of Petro-Canada. From *Mazingira*, No. 3/4, 1977.

In 1976, we tried to save [the harp seal] pups by spraying their coats with a harmless green dye. . . . In 1977, we were back again, placing our bodies over the pups to save their lives.[5]

> **Greenpeace**

Political action groups work within the system and, if they are successful, become a part of the system. Their implied purpose is to influence the government's position on environment and conservation issues and possibly, therefore, influence the lives of many other people. The choice of an alternate life-style, on the other hand, is an act that usually influences only one or a few lives. Although it does not touch the system, it has its special satisfactions. Civil disobedience, in contrast, threatens or blocks the system itself. It draws attention, but requires large sacrifices. Still another way to influence the environment is to choose a profession that is concerned with environmental improvement. These are the ways people have chosen to influence the environment:

political action groups
consumer boycotts
alternate life-styles
civil disobedience
a conserving ethic
an environmental profession

Environmental Decisions

How you decide to influence environmental progress is up to you. What is important to remember, however, is that decisions about the environment are being made right now. Decisions are being made by politicians, administrators, company executives, and individual citizens. Often, these decisions require inputs from experts on matters such as the biology and technology of a situation; the effects of pollutants at various levels; the cost of cleaning up; and the people who gain or lose benefits when a substance or technology is changed or set aside.

Nevertheless, we must all decide the level of costs we are willing to bear, and who should bear them; what benefits we can give up, and what risks we are willing to endure. Experts can only lay out the choices; you must help make the decisions. No one is more

---

5 Greenpeace, letter requesting support for their activities, January 1980.

qualified to decide social, moral, and economic is-sues—which come up again and again—than you, the individual citizen. This privilege and this burden are yours in a democratic society. We have exposed you to the clash of opinions over values, over risk, and over science to show you where and how your own opin-ions and views are needed and are valuable in the envi-ronmental debate. And we have tried to give you as much information as possible about present-day envi-ronmental problems. In the future, however, you will have to face new problems, problems now only dimly seen or, perhaps, problems not yet dreamed of, even by the most far-seeing environmental expert. We can only leave you with a plea to enter the decision process because of the importance of your views. We ask that you look carefully at actions and solutions affecting the environment.

A coal-fired power plant interrupts the landscape in the rural area near Shakertown, Kentucky. Although the picture suggests the presence of air pollution from the plant, Ken-tucky has chosen to protect its environment another way—by excluding nuclear power plants. Kentucky is the only state east of the Mississippi with no nuclear power plants operating or under construction. (Richard H. Gross, Motlow State Community College)

Glossary

abiotic Nonliving; for instance, water, light, etc., the nonliving components of ecosystems.

acclimation The biochemical changes that enable an organism to withstand changed temperatures (either higher or lower).

acid mine drainage Water that has dissolved iron pyrites (ferrous sulfide), which were left behind from coal-mining operations. The water becomes acidic and deposits ferric hydroxide (yellow boy) on stream bottoms. The acid makes the water undrinkable and is harmful to aquatic life. The acid waters are also low in dissolved oxygen, a substance needed by aquatic life for survival.

acid rain The rainfall downwind of major fuel-burning areas has been observed to be acidic. Sulfur oxides from the burning of fossil fuels are the culprit. Acid rain may stunt the growth of plants and turn lakes and streams acidic, driving out the normal aquatic species.

activated sludge plant (a secondary treatment process) A device for removing dissolved organics from wastewater. The plant is a well-aerated tank in which microbes using oxygen convert the dissolved organics to simpler substances.

adaptation A characteristic that helps an organism survive in a particular environment.

aerobic An environment in which oxygen is present.

algae Simple, often microscopic, plants that live in water or very moist land environments.

amino acids Chemical compounds from which proteins are made.

anaerobic In the context of water pollution, a condition of water in which all the dissolved oxygen has been used up or removed. Only a few specially adapted species can survive in anaerobic (oxygen-depleted) waters.

arable Able to be farmed.

area mining of coal In this surface-mining method, the overburden is removed and laid up in successive parallel rows; the technique is used on relatively flat lands.

bacteria Single-celled organisms visible with a microscope. Some bacteria are harmful and cause plant and animal diseases. Many more are harmless or even useful. Bacteria aid in the decay and recycling of organic matter and are used industrially (e.g., to produce drugs or ferment dairy products).

ballast Weight used in ships in certain areas of their cargo space to make them stable and easy to steer. Oil tankers use as ballast either the oil they carry or seawater.

barrel of oil One barrel of oil equals about 43 gallons. A ton of oil varies in volume according to what kind of oil it is (e.g., gasolines are lighter than diesel oils), but as a general statement, one ton of oil is about seven barrels or 300 gallons.

biochemical oxygen demand (BOD) The amount of oxygen that would be consumed if all the organics in one liter of polluted water were oxidized by bacteria and protozoa; it is reported in milligrams per liter. The number is useful in predicting how low the levels of oxygen in a stream or river may be forced to go when organic wastes are oxidized by species in the stream.

biodegradable Able to be broken down by living organisms.

biological controls Control methods that use natural predators, parasites, or diseases to control pests or that rely on the use of naturally produced chemicals such as insect pheromones.

biological magnification The process by which certain, often toxic, materials become more concentrated as they move up food chains. That is, organisms at the top of the food chain contain more of the substance than do organisms on the bottom of the food chain, or than does the environment itself.

biomass The weight of the living creatures in a given area.

biomass energy Energy derived from crops (trees, sugar cane, corn, etc.) by either direct burning or by conversion to an intermediate fuel such as alcohol.

biomes Climax communities characteristic of given regions of the world.

biota The living organisms, plant and animal, in a region.

birth rate The number of babies born each year per thousand people in a population.

bituminous coal The most plentiful form of coal in the U.S. Its high heating value and abundance also make it the most widely used coal in the U.S. Its principal use (about 2/3) is in steam electric power plants.

black lung disease The inhalation of coal dust over a relatively long period and the implanting of the particles in the lung leads to a condition in which the elasticity of the lung is destroyed. Breathing becomes severely labored for the coal miner who falls victim to black lung disease, and many coal miners are permanently disabled by the disease.

BLM *See* Bureau of Land Management.

bloom A rapid overgrowth of the water plants called algae.

blowdown The water removed from that circulating in a cooling tower to prevent solids from building up in the tower water. This water is replaced by fresh water called "make-up."

blowout Explosive release of gas and/or oil from an oil well.

BOD *See* Biochemical Oxygen Demand.

boiling-water reactor (BWR) The water passing through the core of this nuclear reactor is heated by the fission process and is converted directly to steam, which turns a turbine. The BWR is one of the two major types of nuclear reactors used for electric production. *See* Pressurized Water Reactor for contrast.

breeder reactor A nuclear reactor in which the coolant and heat-transfer medium is molten sodium. The fuel is plutonium-239 which is continuously bred from uranium-238.

Btu The British thermal unit, the quantity of heat needed to raise the temperature of one pound of water one degree Fahrenheit. One Btu is equivalent to 1.054 kilojoules (a common metric energy unit).

bulb turbine A French invention, this is a turbine whose blade face is oriented perpendicular to the direction of water flow. In contrast to the Peyton, Francis, and Kaplan wheels, which require water elevations of 100 feet or more, the bulb turbine can generate electricity from only rapidly moving water. It thereby makes possible both tidal power and hydro development at many sites not previously feasible.

Bureau of Land Management The managing agency of the National Resource Lands. The Bureau is within the Department of Interior.

BWR *See* Boiling Water Reactor.

Calorie A measure of the heat or energy content in food. Human food requirements are usually measured in kilocalories (the amount of heat needed to raise the temperature of 1000 grams of water by one degree centigrade) and the term is written with a capital c, "Calories."

carbon cycle Carbon dioxide is produced when plants and animals respire and is consumed by green plants during photosynthesis. Carbon dioxide also dissolves in the oceans and precipitates as limestone or dolomite.

carbon dioxide A colorless, odorless gas at normal temperatures, composed of one atom of carbon and two of oxygen, comprising about 0.03% of the atmosphere by weight. Carbon dioxide is consumed in photosynthesis by green plants and produced by the respiration of plants and animals and the burning of fossil fuel.

carbon monoxide The principal source of this major air pollutant, formed from the incomplete combustion of carbon, is the automobile. The carbon monoxide molecule is one atom of carbon bonded to one of oxygen. Its action on human health is a result of its "tying up" of hemoglobin, the protein that carries oxygen to the cells.

carboxyhemoglobin Carbon monoxide combines with the oxygen-carrying blood protein hemoglobin to form carboxyhemoglobin. This new combination cannot transport oxygen in the bloodstream.

carcinogen Something that causes cancer.

carnivores Meat-eaters.

carrying capacity The largest population a particular environment can support indefinitely.

catalytic converter The device used on automobile exhaust gases to combust carbon monoxide (to carbon dioxide) and hydrocarbons (to carbon dioxide and water). An additional converter may be used to convert oxides of nitrogen back to oxygen and nitrogen.

chlorinated hydrocarbons Chemicals composed mainly of carbon and hydrogen plus one or more atoms of chlorine. Examples are the pesticides DDT, aldrin, dieldrin, chlordane, and heptachlor.

chlorinated organics Chemical compounds composed of carbon, hydrogen, oxygen, and one or more atoms of chlorine. These chemicals can be formed in drinking water by the action of the disinfectant chlorine on organic chemicals found in some water supplies.

chlorination The controlled addition of chlorine to water destined for drinking and to wastewaters being discharged into receiving bodies. Chlorine kills bacteria and viruses and so renders the water safe for human consumption and use. The process is also known as disinfection.

cholera An intestinal disease caused by specific bacteria. The disease can be spread by water polluted by human wastes.

clear-cutting The practice of cutting all trees in an area, regardless of size, quality, or age. The practice hastens erosion, is visually displeasing, and leads to a loss of species habitat. Better practices are *selection cutting, seed-tree cutting,* or *shelterwood cutting.*

climate A complex of factors affecting the environment. Climate includes: temperature, humidity, amount of precipitation, rate of evaporation, amount of sunlight, and winds.

climax community The characteristic and relatively stable community for a particular area.

coagulation A process for the removal of suspended material from drinking water. A "floc" of insoluble material is created by the addition of alum or ferrous sulfate. When the floc settles in a detention basin, suspended material is captured in the floc

and is settled out as well.

coal cleaning The removal of sulfur from coal by washing and chemical steps. Cleaned coal is less likely to require "scrubbers" to remove sulfur oxides from the stack gases.

coal gasification The conversion of the carbon in coal to a gas that can burn. Coal is burned in the presence of oxygen and steam to produce carbon monoxide and hydrogen gases. This low-Btu gas can be burned directly or can be converted to a high-quality methane by the addition of hydrogen in further reactions. The latter product can be substituted for natural gas.

coal liquefaction The conversion of coal to a hydrocarbon liquid. Pyrolysis is one method for this conversion. A modification of coal gasification will also produce a hydrocarbon liquid. Solvent refining of coal is still a third possible process.

coal-slurry pipeline A technology with wide potential in which coal is transported via pipeline as a mixture (about 50/50) of coal particles and water.

coliform bacteria A category of bacteria largely derived from fecal wastes. The presence of these bacteria in a river or stream is taken as evidence of fecal pollution and indicates the possibility that pathogenic (disease-causing) bacteria may be present.

combined cycle power plant In this concept, oil or gas is first burned in a turbine (jet) engine, generating turning power in the engine for electric generation. The hot gases are then used to boil water to steam to turn a conventional turbine for power generation. The efficiency of conversion of heat to electricity is increased by this two-stage process.

combined sewers Most American cities have one sewer system that carries both domestic wastewater and the water from rainstorms. During storms, the total of the flows is too large to be treated and hence enters the water body virtually without treatment.

community All of the living creatures, plant and animal, interacting in a particular environment.

competition In ecological terms, competition refers to the struggle between individuals or populations for a limited resource.

condense To change from a gas to a liquid, as when steam is condensed to water in a power plant condenser.

consumers Organisms who eat other organisms. Primary consumers eat producers, secondary consumers eat primary consumers, and so on.

contour mining of coal In this surface-mining method, L-shaped cuts are made into the hillside in long curving arcs that follow the contour of the hill.

contraceptive A device, chemical, or action that prevents pregnancy.

cooling tower A structure designed to cool the water that was used to condense steam at a power plant.

core (of a nuclear reactor) The concrete and metal shielded structure that houses the nuclear fuel in a reactor; the place where the fission process and heat production is occurring.

crude oil Oil as it comes from the ground, in its natural state. Crude oil is a mixture of many chemical compounds.

DDT A member of the chlorinated hydrocarbon family of pesticides, DDT was banned in the U.S. in 1972 because it was interfering with reproduction in certain bird species.

death rate The number of people who die each year per thousand people in a population.

deciduous forest biome Most of the eastern U.S. is part of this biome, characterized by trees that lose their leaves each fall.

decomposers Organisms that take part in the decay of organic materials to simple compounds, e.g., bacteria and fungi.

demographic transition Pattern of change in which birth rates fall as, or after, death rates fall. After the transition, a country's birth rate is closer to its death rate and the population does not grow rapidly.

demography The study of populations.

desertification The severe degradation of a land environment to the point where it resembles a desert.

detritus Dead organic matter, composed of plant and animal remains.

developing nation A term applied to countries that have little or no technological development.

development rights The rights, most often accompanying ownership of the land, that enable the owner to construct buildings, build roads and sewers, and otherwise alter the land. Such rights can be sold or transferred to other parties, separating the rights from the land itself.

digester A water pollution control device used to further degrade and stabilize the organic solids that arise from primary and secondary wastewater treatment.

dissolved oxygen (DO) The amount of oxygen dissolved in water, reported in milligrams per liter. Levels of 5 milligrams per liter or above indicate a relatively healthy stream. The maximum level of dissolved oxygen is ordinarily 8–9 milligrams per liter, depending on the water temperature.

district heating The spent steam from a steam-electric power plant can be carried in underground pipes to homes, offices, factories, etc. for heating during the winter months.

diversity A measure of the number of species in a given area. The more species there are per square meter, the higher the diversity.

DNA (Deoxyribonucleic acid) The hereditary material contained within cells that determines the characteristics of an organism and determines the characteristics of its offspring.

DO *See* dissolved oxygen.

drift Particles of liquid water that escape from cooling towers and cause corrosion or other environmental problems.

ecology The study of where organisms live, or their environment.

ecosystem All of the living organisms in a particular environment plus the nonliving factors in that environment. The nonliving factors include such things as soil type, rainfall, and the amount of sunlight.

effluent A liquid or gaseous waste material produced from a physical or chemical process.

electric power plant (thermal) The heat (hence the word "thermal") from a burning fossil fuel or from the fissioning of uranium is used to boil water to steam and the steam is used to turn a turbine, generating electricity.

electrostatic precipitator A device that removes particulate matter from stack gases; the particles are given an electric charge and then attracted to a collecting electrode. The precipitator is very efficient and is widely used on electric power plants.

emergency core cooling system A spray system designed to inject water rapidly into a core of a nuclear reactor that has experienced a loss-of-coolant accident.

emigration The movement of organisms out of a population.

endangered species A species with so few living members that it will soon become extinct unless measures are begun to slow its loss.

enrichment The process of converting uranium from 0.7% uranium-235 to 3% uranium-235, the higher concentration being needed for the fuel rods of the reactor of a nuclear electric power plant.

entrainment Organisms that are caught in the condenser water pipes and drawn into a power plant are said to be entrained.

EPA Environmental Protection Agency.

epidemiology The study of disease in groups of people or populations.

erosion The loss of soil due to wind or as a result of washing away by water.

estuary A coastal body of water, partly surrounded by land, but having a free connection with the ocean.

eutrophic Water that has a high concentration of plant nutrients.

evolution Change in the frequency of occurrence of various genes in a population over a period of time.

fauna The animal life of a particular region.

FDA Food and Drug Administration.

fertility rate The total number of live births a woman in a particular country is expected to have during her reproductive years. (Also called total fertility rate.)

fertilizer A material that promotes the growth of plants. It can be natural or artificial.

filtration In the context of water supply, this is the practice of forcing water through beds of sand to trap the bacteria that cause disease and thereby preventing their presence in drinking water.

first law of thermodynamics Energy can be changed from one form to another, but it cannot be created and it cannot be destroyed.

fission When the nucleus of certain heavy atoms is struck by neutrons, it breaks apart into two or more fragments (fission products) with the production of heat and more neutrons; fission means "breaking apart."

food chain A picture of the relationship between the predators in an area and their prey, i.e., who is eating whom. When the relationships are simple and few creatures are involved it is a food chain.

food web In most environments, food chains are connected into complicated food webs made up of many organisms, with many interrelationships.

fossil fuels Fuels like coal, oil, and natural gas, derived from the remains of organic matter deposited long ago.

fuel cell A device in which oxygen combines with hydrogen or carbon monoxide, producing direct current electricity. Fuel cells are expected to be relatively clean producers of electricity, unless coal is burned to provide the carbon monoxide.

fuel reprocessing Spent nuclear fuel rods are broken apart. Then the highly radioactive fission products are separated and uranium and plutonium are recovered to be re-used. No reprocessing plant operated in the United States through most of the 1970s, although reprocessing is taking place in Europe.

fusion The combination of the deuterium atom (a form of hydrogen) with either another deuterium or a tritium atom (still another hydrogen form). The combination releases enormous heat, which scientists hope some day to capture to produce electrical energy. Fusion is the basis also for the hydrogen bomb.

gamma ray A highly penetrating form of ionizing electromagnetic radiation that is produced by one type of radioactive decay.

geopressured methane Natural gas known to be dissolved in salt water in deep caverns at extremely high pressures and temperatures. It is not known if geopressured methane can be recovered profitably.

geothermal heat Heat from the earth's interior carried to the surface as hot water or steam. The steam and hot water can be used to heat homes, offices, and factories or can be used to generate electricity.

geothermal power plant An electric generating station that uses hot water or steam from the earth's interior as the energy source. Releases of steam from the earth can be used directly to turn a turbine. Alternatively, hot water releases can be "flashed" to steam for such uses. Or hot water can be used to vaporize a fluid such as isobutane, which will then be used to turn a turbine.

greenhouse effect The effect of carbon dioxide in the atmosphere is compared to that of the panes of glass in a greenhouse. Sunlight passes the glass into the greenhouse without effect, but heat radiated outward from inside the greenhouse is blocked. Sunlight reaches the earth without effect, but outward heat radiation is blocked by carbon dioxide molecules that absorb the energy. The greenhouse effect suggests the earth will undergo a warming trend as carbon dioxide from fossil fuel combustion accumulates in the atmosphere.

green revolution The term given to the new developments in farming, including the use of high-yielding grains, which promise to enable farmers to grow much more food on the same number of acres than with conventional techniques and older crop varieties.

groundwater Water beneath the earth's surface.

growth promoter A substance that makes an animal grow better or more quickly.

habitat The physical surroundings in which an organism lives.

half-life The time required for one-half of a given quantity of a chemical to disappear from the environment (or to be excreted by the body, if it is a chemical absorbed by a living organism).

hazardous wastes Any waste materials that could be a serious threat to human health or the environment when disposed of, transported, treated, or stored.

heat pump The reverse of a refrigerator or air conditioner; the heat pump in the heating mode draws in cold air and exhausts this air at an even colder temperature. The heat captured is transferred via a refrigerant liquid to the indoor air. The heat pump has the potential to cut in half electric requirements for home or hot-water heating.

hectare A metric unit of area. One hectare is 100 meters by 100 meters and equals 2.47 acres.

hepatitis A disease caused by a virus, in which the liver becomes inflamed. Epidemics of hepatitis have been traced to contaminated water supplies.

herbicides Chemicals used to kill weeds.

herbivores Plant-eaters.

highwall The wall of overburden and coal left behind after contour mining of hillsides.

high-yielding grains New varieties of corn, wheat, and rice developed by agricultural research. These varieties produce much more grain per acre than older varieties. They also require more water, pesticides, and fertilizers.

humus Large, stable organic molecules formed in the soil from the breakdown of organic waste materials. Humus contributes to soil fertility by helping to retain water and keeping the soil loose or friable.

hydrocarbons A class of air pollutant, principally from the operation of internal combustion engines of motor vehicles; the pollutants contribute to photochemical pollution by the reactions they undergo.

hydroelectric energy Electric energy derived from falling or moving water. The water is commonly stored behind a dam and released through penstocks to turn a turbine and generate electricity.

hydrogen economy An energy system that uses hydrogen to store and produce energy. Hydrogen from chemical processing of fossil fuels or from the electrolysis of seawater would be used in fuel cells. The cells would generate electricity by combining the hydrogen with oxygen or carbon monoxide.

hydrologic cycle The cycling of water in the environment, from rainfall to runoff to evaporation and back again.

immigration The movement of individuals into a population.

indicated and inferred reserves (of oil or gas) Oil or gas known to exist and likely to be recoverable with the application of additional technology (indicated) and oil or gas for which we already have some limited evidence of existence (inferred). (*See also* proved reserves.)

inversion A weather phenomenon in which cold air lies close to the earth's surface, trapped by a warm air mass above it. This is the inverse of the normal situation in which temperature decreases with increasing distance from earth out into space.

ionizing radiation Rays or particles that possess enough energy to separate an electron from its atom. Some ionizing radiation is electromagnetic, for instance, x-rays and gamma rays. Other forms are particles such as electrons (beta radiation), neutrons, and the nuclei of helium atoms (alpha radiation).

IPM Integrated pest management, a combination of

techniques designed to control pests using a very minimum of chemical sprays.

irrigation scheme To supply water, other than natural rainfall, to farmland.

kilowatt 1000 watts, which is a rate of providing electrical energy.

kwashiorkor A children's disease caused by too little protein.

laeterization The process by which certain tropical soils containing iron harden into a material called laeterite when vegetative cover is removed. The resulting soil is hard enough to cut into blocks for building and is no longer suitable for plant growth.

land application of wastewater A set of three different processes in which wastewater is applied to the land as a means of treatment. The wastewater should previously have undergone primary treatment. The processes are overland flow, spray irrigation, and infiltration–percolation.

leaching The movement of a chemical through the soil.

leaching field A system of underground tiles designed to channel the flow from a septic tank into porous soil.

lead Lead is a cumulative poison that can cause brain damage and death; it is present in air (from the lead in gasoline), food, and water. It was once in paint as well, and the eating of paint chips by children is a frequent cause of lead poisoning.

lignite A lower form of coal with a little over half the heating value of bituminous or anthracite coal; it constitutes about four percent of the U.S. coal energy resource.

limiting factor Whatever nutrient is in shortest supply compared to the amount needed for growth. This factor then limits the growth of plants in a particular environment.

liquefied natural gas (LNG) Natural gas made liquid at very low temperatures ($-162°C$) in order to transport it economically via special tankers.

liquid metal fast breeder reactor (LMFBR) In this advanced concept, a liquid metal (sodium) circulates through the core of a reactor, removing the heat from the fission of plutonium. The heat in the liquid metal is eventually used to boil water to steam to turn a turbine. The LMFBR fuel is plutonium-239 created by neutrons striking the nucleus of uranium-238. The scheme is designed to use uranium-238 rather than uranium-235 because uranium-238 is so much more abundant.

LMFBR *See* liquid metal fast breeder reactor.

longwall mining Imported from Europe, this is a relatively new method of mining coal underground. In this method, the mine roof is allowed to collapse in a controlled way as the mine "room" moves across the coal seam. Greater coal removal and prevention of acid mine drainage are claimed advantages.

loss-of-coolant accident (LOCA) If the water in the core of a nuclear reactor is lost because of a pipe break, the temperature in the core would rise rapidly as the heat from fission builds up. The fuel rods could melt and radioactive substances would be released through the pipe break if the core is not cooled quickly by the emergency core cooling system.

low-head hydropower In contrast to large dams and their hydropower installations, low-head hydroelectric energy can be generated using water elevations of fifty feet or less. The bulb turbine, a new technology, makes possible relatively efficient capture of energy from low-head dam sites not currently in use for electric generation.

magnetohydrodynamics (MHD) In this concept, hot gases from the burning of a fossil fuel are seeded with potassium, which then ionizes. Electric current is extracted from the hot gases, which are then used to boil water to steam for conventional electric generation.

malnourished Someone who does not get enough of the various nutrients needed for good health is malnourished.

mariculture The "farming" of marine organisms. This may be done in pens or rafts in coastal waters or in artificially maintained salt-water environments.

marine Having to do with the oceans or salt waters of the earth.

megawatt A term describing the rate at which electricity can be generated by a power plant. One megawatt is 1000 kilowatts.

metabolism The chemical reactions that take place within a living organism or cell. These reactions include those yielding energy for life processes and those synthesizing new biological materials.

methemoglobin A form of hemoglobin in which the iron is oxidized. Methemoglobin is not able to carry oxygen in the bloodstream as hemoglobin can.

metric ton (or long ton) 1000 kilograms or about 2200 pounds, 10% larger than the (short) ton of English measure.

MHD *See* magnetohydrodynamics.

middle distillates That portion of crude oil refined to diesel fuel, kerosene (jet fuel), and home heating oil.

migration The periodic movement of organisms into or out of an area.

monoculture The cultivation of a single species of plant as opposed to mixtures of species, as is usually found in nature.

multiple-use The concept applied to the management of lands in the National Forest. The concept allows timbering, mining, recreation, grazing, watershed protection, fishing, wildlife protection, and wilderness as legitimate uses of the same land. The official designation of wilderness, however, most often excludes timbering, mining, and grazing.

mutagen Substance that causes an inheritable change in a cell's genetic material.

mutation An inheritable change in the genetic material.

National Park Service The managing agency of the National Park System. The National Park Service is within the Department of Interior.

natural gas Derived from chemical and physical processes operating on buried ocean plankton, natural gas consists mainly of methane, a simple hydrocarbon gas. Nitrogen may be present in the gas as well.

natural selection A difference in reproduction whereby organisms having more advantageous genetic characteristics reproduce more successfully than other organisms. This leads to an increased frequency of those favorable genes or gene combinations in the population.

NEPA National Environmental Policy Act. One of the most important parts of this act is that it requires environmental impact statements. These statements are reports based on studies of how a proposed government project will affect the environment.

niche Where an organism lives and how it functions in this environment (i.e., what it eats, who its predators are, what activities it carries out).

NIOSH National Institute for Occupational Safety and Health.

nitrate (NO_3^-) A salt of nitric acid. Nitrate is a major nutrient for higher plants and also a food additive and water pollutant.

nitrite (NO_2^-) A salt of nitrous acid. Nitrite is a food additive but is very toxic above certain concentrations.

nitrogen oxides (NO_x) A contributor to photochemical air pollution, nitrogen oxides are produced during high-temperature combustion of fossil fuels. Oxygen and nitrogen *from the air* produce the pollutant gas. The oxides of nitrogen (nitric oxide and nitrogen dioxide) are both air pollutants, and nitrogen dioxide has been linked to an increase in respiratory illnesses.

non-point-source water pollution Polluted water arising typically from rural areas and which enters a receiving body from many small widely scattered sources.

nucleic acids Chemical compounds from which important biological materials (such as the hereditary materials DNA and RNA) are made.

ocean thermal energy conversion (OTEC) A concept now being tested on a pilot scale to use the temperature difference between the warm surface waters of the ocean and the cold deeper waters to produce electrical energy. The warmer waters would boil a working fluid to a "steam," which would turn a turbine. The colder water would condense the vapor for reuse.

oil Derived from chemical and physical processes operating on buried ocean plankton, oil consists mainly of liquid hydrocarbons (compounds of hydrogen and carbon). Nitrogen and sulfur are other elements that may be present.

oil shale Shale rock containing oil; *see* shale oil.

oligotrophic Water that has low concentrations of plant nutrients.

OPEC Organization of Petroleum Exporting Countries.

organic chemical One composed mainly of carbon, hydrogen, and oxygen.

organic wastes A class of water pollutants composed of organic substances. When these substances are oxidized by bacteria and other species, oxygen is removed from the water. When high concentrations of organic wastes are present, aquatic species may thus be deprived of the oxygen necessary for survival.

organophosphates A major group of synthetic pesticides. They are organic molecules containing the element phosphorus. A number of them are extremely toxic to humans; however, they do not persist in the environment for long periods of time.

OSHA Occupational Safety and Health Administration.

oxidant A chemical compound that can oxidize substances that oxygen in the air cannot. Ozone, a prominent photochemical air pollutant, is an oxidant, as are nitrogen dioxide, PAN compounds, and aldehydes. Levels of photochemical pollution are often reported as oxidant levels.

ozonation The process of treating water intended for drinking with ozone gas in order to kill microorganisms. Although practiced widely in Europe, most water engineers in the U.S. prefer chlorination because of the simple test for free chlorine.

ozone A compound made up of three atoms of oxygen (O_3). The oxygen we need to breathe is O_2.

particulate matter A class of air pollutants consisting of solid particles and liquid droplets of many different chemical types. Examples are fly ash, compounds from photochemical reactions, metal sulfates, sulfuric acid droplets, lead oxides. Particles will soil and corrode materials; they may also soil and react chemically with the leaves of plants. Particles are linked firmly to increases in human respiratory illnesses, and some particle types are suspected of causing human cancer.

pathogens Disease-producing microorganisms. These include bacteria, viruses, and protozoa. In the context of drinking water, cholera and typhoid are two diseases spread by specific bacterial pathogens.

PCBs Polychlorinated biphenyls; a family of chemicals similar in structure to the pesticide DDT and having a variable number of chlorine atoms attached to a double ring structure.

peak-load pricing The practice of raising the price of electricity during the hours (or season) of peak demand and lowering the price during times of slack demand. The goal is to level the rate of electric usage through time and decrease the need for bringing into service new electrical-generating capacity to meet peak demands.

peat A low-heating-value fossil fuel derived from wood which decayed while immersed in water.

permafrost Frozen layer of soil underlying the arctic tundra biome.

persistence The length of time a pesticide remains in the soil or on crops after it is applied.

pesticide A substance that kills pests such as insects or rats.

petroleum A mixture of organic compounds formed from the bodies of organisms that died in prehistoric times. This complex mixture is separated into less complex fractions which we know as gasoline, heating oil, asphalt, etc.

pH A measure of the acidity or alkalinity of solutions. Solutions with a pH of 8 or above are alkaline or basic; solutions with a pH of 6 or below are acidic.

photochemical air pollution Air pollutants such as nitrogen dioxide, ozone, aldehydes, PAN compounds, etc., produced as a consequence of sunlight-stimulated reactions involving nitrogen oxides and hydrocarbons. The various compounds have differing effects, but plants are damaged by ozone and PAN compounds. Eye and throat irritation and respiratory illness are common effects of these substances on people.

photosynthesis The process by which green plants use carbon dioxide, water, and the energy in sunlight to synthesize various organic materials, e.g., sugars.

photovoltaic conversion A process in which silicon cells are used to convert sunlight directly to electricity. Although costs of cells are high, attempts to make photovoltaic conversion economical are underway.

phthalates Chemical compounds that are esters of phthalic acid and various alcohols; for example, diethyl phthalate.

physical–chemical wastewater treatment A relatively new set of methods aimed at replacing conventional primary and secondary wastewater treatment processes. Physical–chemical treatment is designed to remove phosphorus by precipitation and settling and to remove dissolved organics by adsorption on carbon particles.

phytoplankton Microscopic, drifting plant species.

pica The habit of eating nonfood items; it occurs among about 50% of all children, independent of social or economic class, beginning at about one year of age. Such items as paper, string, dirt, paint chips, etc. may be eaten. Lead poisoning may occur if flakes or chips of leaded paint are eaten by such children.

plankton Microscopic plants and animals that drift in water, mostly at the mercy of currents and tides.

plutonium A radioactive element of high atomic weight, produced in nuclear reactors. Plutonium can be used to make atomic weapons or as fuel for nuclear reactors.

point-source water pollution Those water pollutants arising in urban areas or from industries and which enter a receiving body from a single pipe.

population (natural) The members of a species living together in a particular locality.

population profile A bar graph showing the number of people in each age group in a population.

power tower A method (in the testing stage) to generate central station electricity using the sun as the energy source. The sun's rays are focused by banks of mirrors on a tower to provide the heat to boil water to steam. The steam turns a turbine for conventional electric generation.

precipitation In ecological terms, precipitation refers to the amount of water that falls as rain or snow on a given area.

predator A creature that eats another.

pressurized water reactor (PWR) The water passing through the core of this nuclear reactor is heated by the fission process but does not boil because it is under great pressure. This extremely hot water is used to boil to steam a parallel but separate stream of water in a steam generator. The steam in this second loop is used to turn a turbine for electric generation. *See* boiling water reactor for contrast.

prey A creature that is eaten by another.

primary production The energy captured by plants in photosynthesis. Gross primary production measures the amount of energy stored as organic materials, as well as that used in respiration by the plant. Net production includes only the amount stored.

primary recovery The quantity of oil that flows out of a well by natural pressure alone; it averages about 20% of the oil in place.

primary treatment of wastewater The first (in sequence) major wastewater treatment process in the typical set of processes at American cities. Primary treatment consists of allowing the organic particles that were in suspension to settle out of the flowing water.

producers Organisms who produce organic materials by photosynthesis.

productivity Amount of living tissue (plant or animal) produced by a population in a given period of time.

promoter A substance that does not cause cancer itself but which can act to cause another material to be carcinogenic.

proved reserves (of oil or gas) Oil or gas *known* to be contained in the portions of fields that have already been drilled and which is profitable to recover. Said another way, it is oil or gas we know we have.

pumped storage At times when spare electric capacity is available, water may be pumped to a high elevation. The water can then be released from its high pool through the turbine pump that raised it in order to generate electricity. Such releases are made when the existing basic generating capacity is insufficient to meet electrical demand.

PWR *See* pressurized water reactor.

quad One quadrillion (a million billion) Btu. The annual energy use of an entire nation is often given in quads. The U.S. consumed 77 quads of energy in 1980.

radioactivity Particles or rays emitted by the decay of unstable elements. Examples are alpha, beta, and gamma radiation (*see also* ionizing radiation).

rate of growth A country's growth rate is the rate of increase (birth rate minus death rate) plus the rates of change due to immigration (people arriving) and emigration (people leaving).

rate of increase The birth rate minus the death rate of a population is its rate of increase.

recharge Applied to groundwater, recharge is the replacement of water that has been withdrawn from the ground. The replacement can occur from surface or other underground sources.

reclamation of surface-mined land In the mining of coal, this step is the replacement of overburden and topsoil and revegetation of the land.

refining (oil) Crude oil before it is used is refined, or separated into less complex mixtures of substances, i.e., gasolines, kerosenes, heating oils, waxes, tars, asphalts.

refuse banks (or "gob" piles) The wastes from a coal preparation plant, which washes and grades coal. The wastes include low-grade coal, shale, slate, and coal dust.

residue The amount of a chemical left in a food by the time it reaches the consumer.

resistance When a particular germ or pest is no longer killed by a drug or pesticide, that organism is said to have developed resistance.

room-and-pillar coal mining In this method of underground coal mining, pillars are left in the mine in a regular pattern in order to support the mine roof while the coal mining is going on.

runoff Water that comes to the earth as rain and runs off the land into lakes, rivers and the oceans.

salinity Salt content.

salinization The accumulation of salts in soil. Eventually, the salt buildup prevents plant growth.

sand filtration *See* filtration.

satellite solar power station A concept in which photovoltaic cells are installed in a satellite orbiting earth. Electricity generated by the cells would be beamed to earth via microwave. The system would be enormously costly if ever undertaken.

scrubbers A class of devices (differing widely in their chemical process steps) that remove sulfur oxides from the stack gases of coal-burning power plants.

secondary recovery Techniques such as gas injection or water injection to place increased pressure on oil in a deep reservoir and thus force more of it to the surface.

secondary treatment of wastewater The second (in sequence) major wastewater treatment process in the typical set of processes at American cities. Two processes, the trickling filter and the activated sludge plant, are used to remove dissolved organic material from wastewater.

second law of thermodynamics When energy is changed from one form into another, some energy is always unavailable or lost as heat. Another way of stating this explains the inefficiency of electric power generation, namely, that a natural limit exists on the extent to which heat can be converted to work and hence to electricity. The limit in the

conversion of heat to work is about 45–50% conversion.

sediment The fine particles of soil washed off land by the erosion of water; these are the particles that become suspended in flowing water and ultimately settle to stream or lake bottoms or form river deltas.

seed-tree cutting A logging practice that leaves behind trees to seed the area for regrowth.

selection cutting Cutting only mature timber from an area.

septic tank A device used in rural areas for partial treatment of wastewater. The device is a concrete or metal tank that detains wastes from a home for up to 3 days, providing settling of solids and partial treatment of dissolved organics.

shale oil A hydrocarbon liquid derived by retorting (cooking) crushed oil-bearing rock. The hydrocarbon liquid is known as kerogen. The U.S. may have the equivalent of 600 billion barrels of oil in the shale rock of Colorado and neighboring states. The economics of shale oil recovery have prevented previous development of this resource. Potential environmental impacts of shale mining and oil production are thought to be very large.

shelterwood cutting A three-stage cutting plan in which (1) defective trees are first removed; (2) the stand is "opened" by further cutting 10–15 years later; and (3) the mature trees are cut after seedlings are well-established.

siltation The dropping or settling of sediments to the bottom of a body of water. Water brought to a halt behind a dam drops much of its load of sediment into the reservoir; the effectiveness of the reservoir for water storage is thus decreased.

smog The term originally coined to describe the "fog" of photochemical air pollution.

solvent-refined coal *See* coal liquefaction.

species A species is made up of all those organisms who are able to breed successfully (if they are given the opportunity to do so), who share ties of common parentage, and who share a common pool of hereditary material.

stratosphere The layer of the earth's atmosphere directly above the troposphere, which is the lowest layer. Scientists are concerned because ozone levels in the stratosphere are declining and carbon dioxide levels in the stratosphere are increasing.

subsidence As a result of the empty spaces left underground by coal mining, the surface of the land may collapse. Roads may buckle and sewer lines and gas mains may crack as a result of severe subsidence if it occurs in developed areas.

succession A natural process in which the species found in a given area change conditions to make the area less suitable for themselves and more suitable for other species. This continues until the climax vegetation for the area grows up.

sulfur oxides A class of air pollutants from the burning of fossil fuels (mainly coal and oil) containing sulfur. The sulfur is oxidized to sulfur dioxide and sulfur trioxide when the fuel is burned. The oxides of sulfur and the acids they form with water vapor will damage building materials like marble, mortar, and metals and they damage plants and the health of people. Respiratory illnesses are increased during times when the levels of sulfur oxides are elevated.

surface mining (or strip mining) The practice of removing coal by excavation of the surface *without* an underground mine. Area strip mining is practiced on flat lands. Contour strip mining is utilized on hillsides.

sustained yield The sustained timber yield in an area is a quantity of timber less than or equal to the natural increase in that area occurring over the past time period. Increase is measured in volume units, such as cubic feet.

synthetic pesticide One that was invented in a laboratory and not produced by any natural system.

taiga Northern coniferous forest biome, characterized by spruce and other coniferous trees.

tailings When a substance, for instance, uranium or iron, is mined, it is usually found mixed with various other materials. After the desired substance is extracted from this ore, the remaining waste material, along with any other unwanted mining waste, is called tailings.

tar sands A sand coated with bitumen, an oily black hydrocarbon liquid. Extensive tar sand deposits exist in Alberta, Canada, and smaller deposits are in Utah. In the Alberta sands, about two barrels of hydrocarbon liquid can be recovered from three tons of sand; the liquid can be refined to all grades of petroleum.

teratogen A substance that causes a defect during development before birth.

tertiary treatment of wastewater The third major component of wastewater treatment plants. In tertiary treatment, which consists of a number of processes, nitrogen in its various chemical forms is removed; phosphorus is removed by precipitation and settling; and resistant organics are removed by passage through towers containing activated carbon.

tetraethyl lead An organic lead compound, which has been added to gasoline to increase octane and decrease "engine knocking."

thermal pollution Pollution of the environment with heat.

threatened species One that is not yet endangered but whose populations are heading in that direction.

threshold The level or concentration at which an effect can be detected.

tidal power The capturing of the energy of the tides as electrical energy by the use of dams and a turbine/generator unit. Both the receding and arriving tides can be channeled through the bulb turbine to generate electricity.

tolerance The amount or residue of a pesticide or drug legally allowed in food.

topsoil The top few inches of soil, which are rich in organic matter and plant nutrients.

TOSCA Toxic Substances Control Act.

toxic substance A material harmful to life.

toxin A naturally produced poisonous material secreted by certain organisms.

trickling filter (a secondary treatment process) A device for removing dissolved organics from wastewater. In the device, the wastewater is distributed across a bed of stones on which a microbial slime is growing. The microbes remove the dissolved organics by converting them to simpler substances.

trophic structure How the various organisms in a community obtain their nourishment.

tundra Arctic biome characterized by a permanently frozen subsoil and low growing plants such as mosses and lichens.

turbidity A measure of how clear water is, turbidity depends on the amount of suspended solid materials or organisms in the water.

typhoid An intestinal disease caused by specific bacteria. The disease can be spread by water polluted by human wastes.

ultraviolet light A part of the electromagnetic spectrum, ultraviolet or UV light waves are shorter than visible violet light rays but longer than x-rays. These light rays can contribute to the development of skin cancer but are normally absorbed in the upper atmosphere by ozone.

undernourished A person who does not get enough Calories to maintain body weight with normal activity is undernourished.

undeveloped nation A country with little technological development is said to be undeveloped.

undiscovered resources (of oil and gas) Oil or gas not yet discovered by drilling, but which, on the basis of geological and statistical evidence, is expected to be found eventually.

upwelling Upwelling is a result of offshore winds that "push" surface waters away from shore and allow nutrient-rich bottom waters to rise from the deeper oceans.

uranium-235 An isotope of uranium normally found in a mixture with uranium-238. Uranium-235 serves as the fuel for light-water nuclear reactors. It can also be used for making atomic bombs.

U.S. Fish and Wildlife Service The managing agency of the National Wildlife Refuge System. The service is within the Department of Interior.

U.S. Forest Service The managing agency of the National Forests. The Forest Service is in the Department of Agriculture.

waste stabilization lagoon A large shallow pond into which wastewater is discharged for biological treatment. Solids settle out and dissolved organics are removed by microbes in the pond. The device is used primarily where waste loads are small and land with no conflicting uses is available inexpensively.

watershed The land area that drains into a particular river, lake, or reservoir.

weather Weather is the day-to-day variation in temperature, humidity, air pressure, cloudiness, and the amount of precipitation. Thus, weather describes the state of the atmosphere at a certain moment.

windmill or wind turbine A machine whose blades are rotated by the wind and which converts the wind energy into electricity or work (for example, pumping water).

zero population growth (ZPG) The situation in which birth rates equal death rates. Assuming immigration and emigration are not significant, then the population does not grow.

zoning The practice in which local governments designate the allowable uses of a tract or tracts of land as a means of preventing incompatible land uses. Typical zoning classifications are single-family residential, multi-family residential, commercial, industrial.

zooplankton Microscopic forms of animal life that drift about in water, moving mainly in whatever currents there are.

Index

A

Abalone, depredation of, 75(f)
Abiotic components of ecosystems, 14(f), 15, 16(f), 17–21
Abortion, 60–61
ABS, 255
Abyssal plain, 164, 165(f)
Accelerated development, 518, 519
Acclimation, 449–450
Acid mine drainage, 216, 300, 303, 307–311
Acid rain, 368, 402–406
  causes, 378, 402
  environmental impact of, 95, 403–405
  formation of, 17–18, 402
Acre-foot (AF), 181(n), 330
Activated sludge treatment of wastewater, 260–261, 265
Acts
  Black Lung Benefits Reform Act, 313
  Clean Air Act, 388, 398, 424–426, 615, 616(t)
  Clean Water Act, 615, 616(t)
  Coastal Zone Management Act, 444
  Consumer Product Safety Act, 616(t)
  Eastern Wilderness Areas Act 661
  Endangered Species Act, 73, 80–85
  Energy Policy and Conservation Act, 537
  Environmental Pesticide Control Act, 603
  Food, Drug and Cosmetic Act, 594, 596, 599, 609, 611, 615, 616(t), see also Delaney Clause
  Forest Management Act, 653
  Forest Reserve Act, 647, 661
  Land and Water Conservation Fund (LWCF) Act, 672–673
  Marine Mammal Protection Act, 75(f), 88
  Migratory Bird Conservation Act, 647
  Migratory Bird Hunting Stamp Act, 649
  Migratory Bird Treaty Act, 647
  Multiple Use Act, 647
  Multiple-Use-Sustained Yield Act, 650, 651
  National Energy Act, 520
  National Environmental Policy Act (NEPA), 86
  National Trails Systems Acts, 661, 665–667
  Noise Control Act, 460
  Occupational Safety and Health Act (OSHA), 615, 616(t), 621
  Paperwork Reduction Act, 82
  Pesticides Control Act, 615(t)
  Refuge Recreation Act, 649
  Regulatory Flexibility Act, 82
  Resource Conservation and Recovery Act, 615, 616(t), 619
  Safe Drinking Water Act, 230–232, 560, 616(t), 619
  Surface Mining Control and Reclamation Act, 301, 304, 306
  tax reform, 638
  Toxic Substances Control Act (TOSCA), 4, 162, 225, 615, 616(t), 619
  Water Pollution Control Act, 206, 229, 267
  Wild and Scenic Rivers Act, 661–663
  Wilderness Act, 651, 659–661
  Yosemite Act, 646
Adair, Rodney, 592
Adams, Oscar, 561
Adaptive zone, 14
"Adopt-a-plant," 84
Advanced waste treatment, 271
Aeration, 203, 263, 266
Aflatoxin, 554, 556(f), 601
Africa, 95–96, 108, 131, 141, 143–144
  demographics, 51–53, 63, 65, 67
  wildlife protection, 76, 77, 86
Agency for International Development, U.S., 86, 120
Agricultural Districts, 640
Agriculture
  energy input and productivity, 169(t)
  modernization problems, 147
Agriculture, U.S. Department of (USDA), 116, 226, 230, 604–605, 635
  land administration by, 646–648
Air, components of, 371
Air pollutants
  accumulation of, 378, 383, 386, 411
  combustion by-products, 419–420, 431
  potentiation of, 414
  See also Fossil fuels; Gasoline
Air pollution, 367–368, 381–393, 395–432
  alerts, 397, 401, 407, 426–427, 430
  asbestos, 407
  from burning fossil fuels, see Fossil fuels; Gasoline
  from burning wastes, 221, 226, 286, 313
  detecting, 381
  environmental impact of, 367–368, 396, 407, 421, 430
  episodes, 381–382, 396, 397, 414
  and health, see Health, and increase in, 422
  and inversion, 378
  natural sources of, 567
  by nitrogen oxides, 419–426, 600
  patterns of, 420, 431–432
  radioactive, 579
  range of, 396, 408
  smoking, diet, and, 587, 600
  sources of, 287(t), 308, 415–416, 534, 564–565
  See also Acid rain; Photochemical air pollution
Air-pollution control
  automobiles and, see Emissions control
  costs of, 398–399
  of emissions, see Emissions control
  need for, 382
  policy beyond capability, 424
  progress in, 422, 425
  standard violations, 401, 414, 430
  standards for, see Air-quality standards
Air quality, and health precautions, 427–428
  trends in, 412–413
Air Quality, National Commission on, 388, 401, 410, 422
Air-quality standards
  auto emissions, 425–426
  carbon monoxide, 387, 389, 426(t), 427(t), 428(t), 587
  difficulty in meeting, 411, 422, 425–426
  hydrocarbons, 426(t), 428(t)
  inadequacy of, 410–411
  lead, 411
  nitrogen dioxide, 427(t), 428(t)
  nitrogen oxides, 426(t)
  oxidants, 430
  ozone, 421, 427(t), 430
  particles, 407, 410, 414, 427(t), 428(t)
  photochemical air pollutants, 421
  setting the levels for, 388–389, 420, 424, 429–430
  sulfur dioxide, 414, 427(t)
  sulfur oxides, 397, 428(t)
  workplace, 620(t)
Airports, 458
  noise control for, 462
Alaska, 90, 101, 648(t), 661(t)
  environmental threats to, 444
  native culture, 87–92
  oil in, 320
  natural gas in, 326
Albatross, in food chain, 168
Alcohol (fuel), sources of, 515, 516
Alcoholic beverages, 600, 609
  harmful effects of, 593, 611
Aldehydes, 420–421
Alderfly, mercury in, 217(t)
Aldrin, 120, 126(t), 230, 603(n), 604
Alfalfa, 35, 38, 123
Algae, 16(f), 20, 99, 135, 156, 160, 163, 166(f), 170, 174, 479
  blooms, 156, 248, 251, 255
  control of, 203
  effect of temperatures on, 450
  nitrogen fixation by, 35, 249
  nutrients limiting, 249–253
  in polluted water, 249–253, 255, 341(f), 342(f), 343, 451, 577
  salt concentration in, 18
Algeria, U.S. gas imports from, 326
Ali, Dr. Salim, 10
Alkyl benzene sulfonate (ABS), 255
All American Canal System, 186
Allied Chemical Corp., 229–230, 552
Allied General Nuclear Services, 350, 351
Allowable cut, 653, 654
Alsands oil project, 330, 332
Alum, water treatment with, 204
American Dental Association, 233
American Electric Power Co., 400
American Heart Association, 222, 609
American Medical Association, 150

American Water Works Association, 198
Ames, Bruce, 610
Ames Assay of carcinogenicity, 555, 595, 610
2-Aminofluorene, 556(f)
4-Amino-2-nitrophenol, 609
Ammonia, 34(f)
  reacts with chlorine, 205, 211, 273
Amoco Cadiz, 435
Amoeba, pathogenic, 197
Anaerobic condition, 239–240
Anaerobic digester, gas from, 327
Anchovy, overfishing, 133
Anemones, 437
Animal Welfare Institute, 90
Antarctic, temperatures in, 386
Antelope, managed, 77
Anthracite coal, 295–297
Antibiotics, 49, 74
  residues of, in foods, 605
Anticyclone, 378
Antioxidants, in food, 593, 599
Aphids, control of, 119
Appalachian Trail, 663–664
Appalachian Trail Conference, 664, 666
Apple worms, control of, 122
Apples, pest control for, 124
Appropriate technology, 148
Aquaculture, 135–136, 170, 452, 453(t)
Aquifers, 173, 174, 179–182, 186
  persistance of contamination, 215
  See also Groundwater
Archimedes, 503, 511
Arco Solar, 510
Arctic regions, 22
  biomes in, 101–102
  DDT in, 119
  food chain in, 28
  protection of, 90
Area mining, 302, 303
Argonne National Laboratory, 341
Arms Control and Disarmament Agency, 356
Arsenic, 119, 587–588
  and cancer, 552, 570
  drinking water standard for, 235(t)
  persistance of, 118, 588
  sources of pollution by, 118, 216, 295, 313, 588
d'Arsonval, 516
Asbestos in the environment, risks of, 407, 552, 564–567, 570, 572–573
Asbestosis, 564
Ash, from burning, 407, *see also* Fly ash
Aspartame, 595(f), 599
Atmosphere, 372
  cooling of, 375–376, 384–385
Atomic Energy of Canada, Ltd., 362
Atomic Energy Commission, 327, 341, 345, 359, 519, 577
Atoms for Peace program, 341
Austria, growth rate, 50
Automobiles, 442, 530
  efficiency of, 536–537
  electric, 6, 540
  emissions control, 3, 389–391, 406, 416, 421–426
  fuel economy standards for, 537

fuel-efficient, 537–539
fuel-efficient speeds for, 539
hybrid, 540
hydrogen as fuel for, 289
inspection, 426
pollution by, 3, 368, 378, 402, 406, 411, 419–421, 425, 580, *see also* Gasoline

B
Backfilling, 303
Background radiation, 575–578
Backswimmers, 160(f)
Bacteria, 38
  in aquifer, 174
  for carcinogenicity screening, 555
  chemical transformation by, 216, 239, 600
  coliform, 201, 202, 207
  elimination of, from water supplies, 197–199, 204–207, 265
  in food chains, 30, 160, 167(f)
  in septic tank, 266
  and thermal pollution, 451
  in wastewater treatment, 229, 255, 260–261, 270, 272
Badgers, 99, 104(f)
Baghouse to capture fly ash, 409
Bailey, William, 504, 505
Bangladesh, 52–53, 131
  demographics, 55(f), 66(t)
  population control, 61, 63, 64
Barnacles, 479
Barro Colorado Island, conservation plans for, 86
Bass, 160(f), 217, 224, 241, 255
BAT, *see* Best available technology
Batelle Memorial Institute, 399
Bathyal zone, 165(f)
Baumgartner, D. J., 560
Baxter, Percival, 656
Bay of Fundy, tidal power potential, 475, 477
Bears, 103(f)
  grizzly, threatened, 14(f), 498
Beatti, Bruce R., 184
Beavers, 27(f)
  giardiasis infected, 200
Bebbington, W. P., 353
Beech trees, 100, 103(f)
Bees, in arctic regions, 19
Behavioral regulation of temperatures, 19, 449
Benefits vs. costs (risks)
  carcinogens, 556–559
  consumer products, 252, 610
  development, 441, 498
  drugs, 611–612
  environmental controls, 6, 208, 268, 455, 570, 572
  food additives, 596–598
  hunting, 84
  nuclear technology, 347
  wildlife protection, 75(n), 81, 83–85, 498
Benthos, 167(f)
Benzene, 569(t)
Benzidine, 554(n)
Benzo(a)pyrene, 555, 559
Best Available Technology (BAT), 267, 615, 621
Beston, Henry, 76, 677
Bethlehem Steel Works, 457

BHA (butylated hydroxyanisole), 599
BHT (butylated hydroxytoluene), 599
Bills
  Outer Continental Shelf Bill, 444
  Ocean Mammal Protection Bill, 88
  Toxic Wastes Cleanup Bill, 618
Bioaccumulation, 221, 223, 228
  measuring pesticide, 126(t)
Biochemical oxidation, 239
Biochemical oxygen demand (BOD), 244–245, 267, 451
Biodegradability, of detergents, 251, 256
Biological controls, 74, 79, 122(t), 124–125
Biological extrapolation, 556–557
Biological magnification, 37, 117–118, 163, 216, 226, 577
Biomass, 38, 101
Biomass fuels, 513–516
Biomes, 99–105
  world, 99–105
Biosphere reserves, 80
Biotechnology, 142, 147
Birch trees, 103(f)
Birds, 16(f), 168
  effects of pesticides on, 22, 79, 118
  and magnetic fields, 464
  migratory, 647
  and oil pollution, 436
  potential hazards to, 488
Birth control, 60–61, 62, 63
Birth defects and environmental hazards, 224, 568, 574, 575, 593, 609, 611, 612, 618
Birth rate(s), 48–50, 63
Bishop, Mayor William, 215
Bison, *see* Buffalo
Bitumen, 330
Bituminous coal, 295–297, 307
Black Mesa Pipeline Co., 316
Black Lung Benefits Reform Act, 313
Black lung disease (CWP), 300, 312–313
Blenheim-Gilboa Pumped-Storage Project, 480(f)
Bloodworms, 21, 160, 242
Blooms (algal), 156, 248, 251, 255
Blowdown, 454
Blowouts (oil-well), 434–436
Blue gills, contaminated, 577
Bluefish, pesticide in, 229
Blum, Arlene, 610
Bobcats, 99
BOD, 244–245, 267, 451
Boeing Engineering, 490
Boiling-water reactor (BWR), 342–344
Boll weevil, control of, 119, 122, 124
Bonneville Power Administration, 490
Boulding, Kenneth, 536
Bradley, William, 332
Borates, in detergents, 251
Botulism, preventing, 601
Brazil, 77, 352, 358, 515
Britain, nuclear power and, 352, 360
Brittle stars, 168(f)

Bromeliads, 78
Bromine, 580
Brown, R. L., 662–663
Browning, Jackson B., 624
Btu, 285(n), 528
Buck, Alan, 498
Buffalo (bison), 104(f)
  endangered, 76, 498
  population restored, 84
Buffalo Mining Company, 308
Builders, phosphate, 250–251
Bulb turbine, 473, 476–477
Bureau of Alcohol, Tobacco, and Firearms (BATF), 611
Bureau of the Census, U.S., 65
Bureau of Land Management (BLM), 648, 661
Butterfly (Karner blue), 671
Butz, Earl, 121, 653
BWR, 342, 343, 344

C
Cacti, 105(f)
  threats to, 80
Caddisfly, 29(f), 30, 217(t)
Cade, Dr. Tom, 13–15, 21, 22
Cadmium, 221, 235(t), 587–589
Cadmium sulfide, 507
Calcium hypochlorite, 197, 199
Caliche, 107
California, 474
  water problems of, 188–190
California Aqueduct, 189, 190
Calories, 131, 132
Canada goose, threats to, 498
Canadian Diabetic Association, 599
Cancer, 547, 552–554
  causes of, 548(t), 578, 580, 582, 583
  decreasing risk of, 608–609
  diet and, 590, 593, 608, 609
  drug for, 74
  latency of, 552, 559, 564
  preventable cause of, 590
  promoters of, 559
  radiation, 345, 578
  and smoking, 589–590
  synergism, increasing risk of, 611
  ultraviolet rays and, 580, 582, 583
Car pool, fuel efficiency of, 541
Carbamates, 121(t), 126(t)
Carbon adsorption, 265, 270, 272
Carbon cycle, 33, 382–383
Carbon dioxide, 23, 371, 382–386
  atmospheric, 33–34, 376, 382, 383, 386
  global effects of atmospheric build-up, 384–386
  removal from atmosphere, 383
  sources of, 33, 313
Carbon filtration, 204, 208, 560
Carbon monoxide, 387–393
  emissions control of, 389–391
  physiological and health effects of, 391–392, 587, 588
  sources, 387, 425, 588–589
  standards, 426(t), 427(t)
Carbon polishing of wastewater, 272
Carbonates, 33, 251
Carboxyhemoglobin, 392, 587
Carcinogenesis, 559
Carcinogenic substance(s), 414
  experimental tests, 554–556, 595–596

Carcinogens, 203, 224, 226, 229, 230, 313, 355, 411, 414, 430, 438, 552, 554, 555, 559, 560, 563–564, 609
in cigarette smoke, 587(t), 588
in food, 590, 595, 599, 600–604
basis for evaluating, 551, 557
Delaney Clause for, 551, 594–598
justifiability of regulation based on laboratory experiments, 554–559, 595–597
risk assessment of, 556–558, 570
in water supplies, 206–208, 265, 269
Cardiovascular disease
drugs for, 74
environmental factors in, 222, 389, 392–393, 426–427, 458, 460
and smoking, 587–588, 590
Carey, Governor (N.Y.), 228
Caribou, 28, 102(f)
Carnivores, in food chain, 28
Carotene, 599
Carp, 255
Carriers, disease, 194
Carruthers, H. M., 505
Carrying capacity of ecosystems, 43–45
Cartel, 335, *see also* OPEC
Carter, Luther J., 217
Catalytic converter, 330, 389, 406, 416, 422
Catfish, 255
Cattails, 160(f)
Center for Science in the Public Interest, 595(n)
Central Arizona Project, 186, 188, 190
Cesspool, 265
Chemical oxygen demand (COD), 245–246
Cherokee, land loss by, 81
Chevron Resources Co., 496
Chicago Drainage Canal, 197
Child labor, 52(f), 53, 61, 63
Children, 49, 51–53, 63, 66
decline in death rate, 49
health, 56, 131
in noisy environments, 458, 460
poisoning in, 411, 415–416, 600
pollution hazards to, 414
as a resource, 9
of smoking mothers, 588
and vitamin excess, 601
Chimpanzees, threats to, 85
China, People's Republic of, 55, 60, 64, 66(t), 358
Chironomids, 21, 160
Chloramines, 272
Chlordane, 120, 121(t)
Chlorinated hydrocarbons, 117–119, 207, 208, 223, 273
drinking water standards for, 235(t)
listed, 126(t)
toxicity of, 126(t)
Chlorination, 49, 197, 199–201, 203, 205, 207, 265, 272
alternatives to, 451
hazards of, 207, 208, 265, 273, 451
inadequacy of, 201
Chlorine
in the atmosphere, 580

removal from wastewater, 273
toxicity for fish, 273
water treatment with, 199, 205, 207, 208, 211, 451
Chlorine-caustic soda plants, 217
Chlornaploazin, 554(n)
Chlorofluorocarbons, 376
ban on, 582
effects of atmospheric pollution with, 580, 582
Chloroform in drinking water, 553(n), 560
Chlorohydrocarbons, 569(t)
Cholera, 49, 194–197
Chriswell, Colin, 128
Chromosomes, and air pollutants, 430
Cigarette smoke, 554(n), 588, 600
Cities, 53–54
air pollution in, 217, 387, 390, 395, 397, 401, 408(t), 420, 431, 565
air pollution patterns in, 420, 431
nitrogen dioxide levels in, 430
noise levels in, 460
solar heating in, 520
Citrus fruits, pest control for, 125
Citrus trees, and air pollution, 421, 430
CIVEX process, 361
Civilian Conservation Corps, 667
Cladocera, 160(f)
Clams, 141(t), 160, 167(f), 168, 241, 479
contaminated, 200, 438, 578
farming, 170, 452
Clarifier, 260
Clark, Clinton, 664
Clark, Wilson, 514
Claude, Georges, 516
Clayton, Cubia L., 398
Clean Air Act, 388, 398, 424–426
Clean Lakes Program (EPA), 253
Clean Water Act, 615, 616(t)
Cleanup, superfund for, 618–619
Clear-cutting, 19, 77, 651–654
Clebenger, Richard, 123
Climate, 372, 373–378
factors influencing, 582
global shifts in, 375–376, 385–386, 408
human influence on, 376, 378, 408
Climax (solar water heater), 504
Climax communities, 99–101, 107
Clover, 35, 116(t)
Cluster zoning, 632
Coagulation of colloids in water, 204
Coal, 295–317
cleaning, 397, 399, 400
consumption of, 298, 383–384
future options, 314
gasification, 314, 315, 329–330, 332, 333
impact of burning, 300, 313–314, *see also* Fossil fuel burning
increasing reserves of, 330
kinds of, 295
liquefied, 330
mining, 300–314
natural gas from, 327
resources, 185, 291
synthetic fuels from, 329–330
tax on, 304, 313
transport of, 297, 315–317, 542

uses of, 286, 291, 298
Coal-slurry pipelines, 297, 315–316, 542
Coastal zone ecosystem, 165, 169(t)
Coastal Zone Management Act, 444
Cocarcinogens, 397, 414, 559
Cocoa, 588
COD, 245–246
Cogeneration of electricity, 531
Cold-blooded organisms, 19
Coliforms, 201, 202, 207
permissible counts of, 202, 235(t)
Collectors (solar), 505, 510, 519, 521–522
Colloids, 203, 204
Colony shale oil project, 332
Colorado River, as a water resource, 182–190
Colorado River Compact, 183, 186
Columbia, 60–61, 65, 66(t), 68
Combustion by-products, 419–420, 431
Commerce, U.S. Department of, 444
Commons, the, 64
Communities, 159
effect of heat changes on, 450
field, 30, 99
intertidal zone, 479
lake, 160, 450
marine, 165–168
stability of, 30, *see also* Climax communities
stream, 18(f), 241–242
succession in, 99–101
Community, 15
climax, 99–101, 107
trophic structure of, 28–30
Composting hazardous wastes, 622
Coniferous forests, 101, 103
Conservancies, 673–674
Conservation
cost-sharing for, 116
energy, 528–544
fuel, 320
gasoline, 537–541
oil, reasons for, 334, 336–337
of water, 178, 182
worldwide attempts at, 80
Conservation groups, 10, 89–90, 99, 657, 664, 666, 671–674
Conservation programs, soil, 116
Conservation restrictions, 642
Conservation tillage, 269
Construction, pollutants from, 407, 565, 579
Consumer products, safety standards for, 609–611
Consumer Products Safety Commission, 610
Consumers (food chain), 28–31, 39
Consumption of
certain foods, 604(t)
coal (U.S.), 298, 383–384
energy, 469(f), 528
gasoline, 383
natural gas (U.S.), 326
oil (U.S.), 323, 325(f)
Continental shelf, 164, 165
Contour mining, 301, 303–304
Contour plowing, 115, 269
Contraceptives, 61, 611
Cooling ponds, 453
Cooling systems, 448–457
emergency, 345

entrainment in, 451–452
pollution by, *see* Effluents
Cooling towers, 453–456
geothermal, 500
disadvantages of, 456–457
Cooling water, impact of, 368
Copepods, 160(f), 166(f), 450
Copper, 451
Copper sulfate, 126(t), 203
Coprocessing, 361–362
Coral reefs, 101
Core (reactor), 342–344
emergency cooling system, 345
Corn, 114, 123, 421
aflatoxin standard for, 601
as fuel source, 516
new species, 76(f)
production (U.S.), 114
yields, 113, 122, 180–181
Corona, 462
Corps of Engineers, U.S., 473, 477
Corrosion, causes, 396, 405, 407
Cosmetics, 593, 600, 609
Cosmic rays, 575, 577(f)
Cost-benefit analysis, *see* Benefits vs. costs
Costle, Douglas M., 455, 570
Cotton, 119, 181(t)
pest control for, 119, 124, 125
Cotton, F. A., 146
Coumarin, 588
Courts, environmental role of, 4, 199, 458, 462, 631, *see also* Supreme Court
Cowbirds, 27
Coyotes, 27, 79, 99(f), 104(f), 200
Crabs, 75(f), 128, 167(f), 230, 437, 439, 449, 479
Crammer, Dr. Morris, 598
Crandall, Robert W., 268, 269
Cranes (whooping), in captivity, 84
Crappies, contaminated, 577
Crawford, W. Donham, 288
Crayfish, threats to, 310
Crop(s), 75, 180–181, 421
cross-breeding, 74
fallow, 115(n)
mixed-field, 122
new varieties of, 76(f), 142–143
pest resistant, 122
rotation of, 115, 122
Crop insurance, 149
Cropland
acres convertible to, 186
for energy crops, 516
erosion losses, 110, 112
percent irrigated, 181, 185
sewage sludge for, 262
Cross-breeding, danger of, 74
Crowding, effects of, 45, 59–60
Crude oil refining, 321
Crumett, Dr. Warren, 127–128
Crustaceans, 168
world catch, 141(t)
Curlews, 102(f)
Current, 162, 164–165, 375(f), 378
Cyclamates, FDA ban on, 595

D
Dams, 186–188, 307–308, 663
environmental impact of, 81, 147, 175–176
need for, 186, 471, 473
Dang, Hari, 10
Darling, J. N., 647(f)

Day-and-Night water heater, 504–505

Daylight Savings Time, 535

DDT, 13, 22(f), 75, 79, 117–121, 126, 596
   banned, 120
   effect on birds, 22(f), 79, 118
   in food, 603
   half-life of, 118
   human levels of, 121(t)
   toxicity of, 126(t)

Death rate, 9–10, 48–50, 539, 547
   defined, 9, 48
   highest, 63
   air pollution and, 396, 407, 414

Decibel (dB) levels, common, 459, 460

Deciduous forest, 26(f), 101, 103, 106

Decisions, 2–7, 60, 81, 679
   private land-use, 633
   water allocation, 179–192

Decomposers in food chain, 28, 30, 39, *see also* Detritus feeders

Decomposition, 33, 105, 160

Deep sea bottom, 34, 37, 168

Deep-water ports, 439–441

Deer, 75, 77, 99, 103(f)

Deer mice, 21

Deerflies, 27(f)

Defoliants, pollution by, 216

Delaney Clause, 551, 594–598, 615, 621

Delta-Mendota Canal, 189

Deltas, 112, 189–190

Demographic transition, 48, 53, 65

Demography, 68–70

Denison, Edward, 268, 269

Denmark, growth rate, 50

DES, *see* Diethylstilbesterol

Desai, Prime Minister, 62

Desalinization, sea water, 173, 453(t)

Desert(s), 20, 22, 105, 107, 169(t)
   reclamation of, 305, 306

Desert basins, aquifers in, 178

Desertification, 95–96, 108(f)
   causes, 96, 141, 145

Detergents, 249–252, 255–256
   for oil cleanup, 437

Detoxification of hazardous wastes, 622

Detritus, 30, 37

Detritus feeders, 160, 167(f), *see also* Decomposers

Deuterium, 287, 343

Developing nations, 49, 54–57, 131–132, 147, 151, 200, 219(t)
   aid to, 144(f), 149
   demographics, 49–53, 66(t), 68–70
   food production in, 54–55, 114, 139, 142–144
   population control, 61–63
   priorities of, 57
   technology in, 143, 148

Development
   controlled, 630–634, 642, 668, 671
   at National Parks, 659, 667–668

Development rights, 638–640, 642

Devils claw, 516

Diamond, Dr. Sidney P., 607

Diatoms, 166(f), 479

Dieldrin, 119, 121(t), 126(t)
   banned, 120, 603(n), 604

Diet
   the chemicals in, and disease, 587, 590–611
   and health foods, 602
   meat eating, 138–139, 150
   recommendations, 609

Diethylstibesterol, 554(n), 605, 612

Digester tank, 261

Diker (Texas), 182

Dinoflagellates, 160(f), 166(f), 479

Dioxins, pollution by, 618

Diptheria, prevention of, 49

Disasters, environmental, 4, 58, 229, 308, 311, 313, 370, 381, 397, 449, 606, 618, 645

Disease(s), 85, 119, 139, 141
   associated with air pollution, 313, 381, 396–397, 407, 414–415, 426
   associated with toxic substances, 564
   from chemical pollutants, *see* Poisoning
   diet and, 131–132, 602
   environmental noise and, 458
   epidemics of, *see* Epidemics
   among miners, 30, 312–313, 564, 578
   prevention of, 49, 119, 197, 199
   protozoan, 200–201
   for which smoking increases risk, 588
   spread by water supplies, 194–197, 200–201, 267
   viral, 200
   See also Cancer; Carcinogens; Cardiovascular diseases; Respiratory disease

District heating, 531

Diversity (species), 21–22, 30, 101
   vs. food production, 141

DNA, 559

Dogoloff, Lee, 591

Dollar, oil imports and value of, 337

Dolomite, production of, 383

Dow Chemical, *Silent Autumn*, 123

Down-dip/up-dip mining, 310

Dragonflies, 160(f)

Drift, from cooling towers, 456–457

Drinking water, 202–208, 215–234
   acidic, 404–405
   asbestos in, 564–565
   carcinogenicity of, 553–554, 560, 564, 565
   chemical pollution of, 215–232, 404–405, 617
   chloroform in, 553(n), 560
   EPA regulations for, 202, 204, 206, 208, 560
   fluoridation of, 560–561
   impurities in, 155–156, 174
   laws, 230–232
   removing organic compounds from, 208
   safety of, 202, 207, 211
   spread of disease by, 194–197
   standards for, 221, 222, 234
   taste and odor in, 203, 208

Drone fly, in polluted water, 241

Drucker, C. B., 136

Drug Abuse Policy, White House Office of, 591

Drugs, 49, 74
   marijuana, 591

resistance to, 119, 605
   safety of, 611–612

Drum (fresh water), 255

Dryfoos, George, 347

DuBay, W. H., 90

DuBoff, Richard B., 268

Dumpsites, 4, 221
   for hazardous wastes, 616–618

DuPont, 217

Durham, Jimmie, 81

Dust, 376, 385
   coal, 307, 313
   volcanic, 375–377, 385, 408

Dust Bowl, 114

Dust storms, 113–114

E

Eagles, 27, 75(f), 79, 118, 404

Ear, human, 458(f)

Earthquake, 444, 623

Earworm (corn), 122

Easements, scenic, 672

ECHO viruses, 200

Ecological reserves, 80

Ecology, 12

Economics
   and cancer protection, 560, 571
   of environmental controls, 228, 268, 398, 455, 570, 572, 620
   and fuel-efficient cars, 538
   of High Plains water transfers, 191
   of nuclear power, 360
   of organic gardening, 123
   of public power ownership, 288
   of strip-mining expansion, 309
   of synthetic fuels, 332
   of western growth and water supply, 184

Ecosystem(s), 15–17, 30–42
   abiotic components of, 14–21
   climax, 100
   cycles in, 30–36
   oil pollution and, 437–438
   temperature changes and, 450–451
   forest, 26(f), 27(f), 656
   freshwater, 160–162
   incomplete, 37
   island, 79
   marine, 169(t)
   mixed-field, 30
   reserves for major types of, 80, 645
   soil in, 105
   stable, 99
   transfer of energy in, 37–42
   water in, 17–18
   watershed, 159

Ectotherms, 19

Education, 56, 63

Efficiency
   of automobiles, 536–537
   of car pools, 541
   of electric cars, 540
   of electric generation, 283, 285–286, 448, 471, 500, 507
   of home heating, 532
   of public transportation, 540–541

Effluents (power plant)
   pollution by, 447–452
   uses for, 452–453, 457

Electric fields, shock hazards and effects of, 462–463

Electric generation, 283–293, 313–314, 422, 448, 473

coal-fired, 283, 285, 286, 298, 313
   efficiency, 283, 285–286, 471, 500, 507
   future alternatives, 287
   geothermal, 492–496
   meeting peak loads, 542–543
   new concepts, 292–293
   nuclear, *see* Nuclear power
   OTEC, 517–518
   photovoltaic, 507, 509–510
   potential resources for, 495, 497
   projected capacity for (U.S.), 542
   solar, 507–513
   ways of improving efficiency of, 531
   wind-powered, 486–488, 490–491
   See also Hydropower; Nuclear power; Nuclear power plants; Power plants

Electric inverter, 488

Electric power
   potential sources, 474

Electricité de France, 475, 479

Electricity
   consumption of, 283, 488
   home heating with, 532
   photovoltaic, 507–510
   process of generating, 285, 342–343
   See also High-voltage power lines

Electromagnetic spectrum, 573

Electrostatic precipitator, 401, 409

Elephants, 42, 77(t)

Elk, 77, 498

Emigration, 50

Eminent domain, power of, 658

Emissions, 419–426
   auto, 419–426
   hydrocarbons in, 420–421
   power plants, 286, 313–314, 344, 419, 422
   nitrogen oxide in, 419–420, 422, 425

Emissions control
   auto, 389–391, 406, 416, 421–422, 425–426
   power plants, 313–314, 399–402, 409, 422
   standards, 389–390, 410, 420, 425–426, 577
   success of, 390–391, 425

Endangered cultures, 87–92

Endangered species, 10, 73
   birds, 79, 80
   control of trade in, 84
   vs. endangered cultures, 87–92
   as environmental barometers, 81, 86
   fish, 73, 86(f)
   listing of, 81–82
   plants, 74(f), 77, 80, 81
   protection of, 80–92, 674
   recovering populations of, 75(f), 85, 86, 89
   sea mammals, 87(f), 88
   threats to, 10, 22, 74(f), 75–80, 443

Endangered Species Act, 73, 80–84, 85(f)

Endangered Species of Wild Fauna and Flora, Convention on International Trade in (CITES), 84

Endemics, 82(t)

Endotherms, 19

Endrin, 79(f), 126(t), 235, 603(f)
Energy, 17, 23, 101
  consumption of, 289–291, 469(f), 528
  conversion of, 281–283
  electric, 281–283, 462–464
  geothermal, 492–500
  from hot rocks, 495, 497
  meat as a source of, 138
  pollution by, 368, 446–464
  sources of, 281, 295–297, 467–468
  transfer of, in ecosystems, 37–42
Energy, U.S. Department of, 351, 490, 491, 495, 496, 509, 512, 520, 540
Energy budgets, of ecosystems, 39
Energy conservation, 528–544
  vs. business as usual, 529–530
  means of, 528–541
  reasons for, 528
Energy crisis, 10, 145, 278
Energy Information Administration, 283, 491, 519, 542
Energy Policy and Conservation Act, 537, 520
Energy Regulatory Commission, 474
England, coal burning in, 58
Engman, Lewis A., 120
Enrichment of uranium-235, 349
Entrainment, 452
Environment(s), 17, 159–162
  carrying capacity of, 43(f), 45
  fleeting, 44
  and population growth, 55–56
  quality of, 627, 632
  sound levels of, 460
  zero-risk, 207
Environmental control
  costs vs. benefits, *see* Benefits vs. costs
  crash programs for, 206–207
  regulatory problems, 601–605
  strip mining and, 301, 304–305
  worldwide, 84–87
  See also Air pollution control; Emission controls; Pollution control; Thermal pollution control; Water pollution control
Environmental Defense Fund (EDF), 4, 206, 553
Environmental engineering, 155
Environmental groups, action by, 4, 73, 90, 120, 206, 230, 253, 267, 443
Environmental Impact Statement, 334
Environmental Pesticide Control Act, 603
Environmental Protection Agency (EPA), 227, 229, 230, 232, 310, 457
  actions on asbestos, 565(f), 566–567
  and air quality, 3–4, 388–389, 400, 408, 412–413, 420, 422, 581
  Clean Lakes Program, 253
  control standards from, 202, 204, 221, 222, 234, 388, 411, 460–461
  and detergents, 437
  and drinking water, 208, 224–225, 553, 560
  enabling laws, 615

hazardous waste monitoring, 621
  and pesticides, 230, 601–604
  suits against, 4, 206, 230(n), 267
  waste disposal settlement agreement, 267
Environmental Quality, President's Council on, 198, 444, 488, 518, 529, 530, 634
EPA, *see* Environmental Protection Agency
Environmentalists, 6, 314, 444
Epidemics
  antibiotic resistant, 606
  chemical, 216, 217, 223
  diarrheal, 174
  typhoid, 162
  water supplies and, 194–197, 199, 200–201
Epidemiology, 548
Epilimnion, 170
Epiphytes, 78(f)
Erosion, 95, 479
  causes, 77, 110–112, 115, 141, 142, 175, 300, 302(f), 651–652
  control, 114, 115, 253(n), 269, 303
  results, 35, 108, 112–114, 116, 269
Eskimos, 88–92
Estuaries, 18, 159(f), 162–163, 169(t), 175, 190
Ethics
  of animal research, 85
  of foreign aid, 5, 146, 150–151
  of hunting whales, 91
  of ORVs, 670–671
  of overseas pesticide sales, 12
  of population control policies, 62, 64, 146, 150
Ethoxyquin, 599
Euphorbias, 105(f)
European Atomic Energy Community (EURATOM), 358
European Common Market, 581
Eutrophication, 35, 36, 100, 248–250, 255, 271
  control of, 250–253
Evaporative cooling, 453
Evapotranspiration, 31(n)
Evolution, 22
Ewald, Charles, 505
Exponential growth pattern, 42–43, 46
Externalities, 86
Extinction of species, 78
Exxon, 330, 332, 333

F
Failures
  of environmental control policies, 3, 61, 116, 251, 255, 391, 401, 410, 425
  of pesticides, 119
  of regulatory agencies, 224
Falcons, 79 *see also* Peregrine falcons
Fall overturn of lake water, 171
Fallout, 575–577
Fallow period, 141
Family planning, 62, 66, 67
Family size, 9, 66, 150
  factors influencing, 51–53, 61–64
Famine, 131, 133
  relief, 144(f), 149
FAO, *see* Food and Agricultural Organization
Farben, 333
Farming, 74, 121–122, 135–138

conservation techniques, 114–115, 269
  conventional vs. organic, 122–123
  "slash-and-burn," 77, 107
  return to dry-land, 181, 182
  subsistence, 77
  traditional, 135–137
  in undeveloped countries, 138
  in water, 135–136, 170, 452, 453(t)
Farmland
  control of pollution from, 269
  erosion of, 112–113
  loss of, 634–638
  percent irrigated, 138
  prime, 95, 304, 306, 634–635
  ways of preserving, 637–643
  See also Cropland
Farms, 137, 267
  corporate ownership of, 137(n)
  incorporation of, 638
  inheritance taxes on, 635, 638, 640
Fatigue, metal, 487
FDA (Food and Drug Administration), 594–596, 599–601, 605, 609, 611
Federal Energy Administration, 442, 518
Feed additives, 605–606
Feedlots, 249, 253, 267
  usable gas from, 327
Fernow, Bernhard, 646
Ferriamicide, 230(n), 604
Ferrous sulfate, 204
Ferrous sulfide, 310
Fertility rates, 60, 65, 66–68
Fertilizer(s), 39, 138
  chemicals added to environment by, 34, 35, 36, 580
  natural, 35
  need for, 57, 107, 142
  pollution by, 113, 222, 249, 253, 269
Fetal alcohol syndrome, 611
Filtration, 197, 201, 204–205, 208, 248, 272
Fir trees, 103(f), 652
Fire, 108, 655
  advantages of forest, 20, 655, 656
  caused by smoking, 590
  in coal refuse banks, 308
Fire ants, eradication of, 230, 604
Fire retardants, toxicity, 606–607, 610
Firemaster flame retardant, 606–608
First law of thermodynamics, 39
Fish, 16(f), 133, 141, 160
  acclimation of, 449–450
  chemical contamination of, 119, 162, 216–218, 221, 223, 224, 226, 229, 230, 255
  chlorine poisoning, 273
  loss of, 18, 254–255, 403, 438
  and oil pollution, 437–438
  oxygen levels required by, 241
  pesticide toxicities in, 126(t)
  radioactive contamination of, 577–578
Fish kills, 244, 437, 450
Fish ladders, in reservoirs, 176
Fish and Wildlife Service, U.S., 82–84, 648–650
  definition of species, 13, 14
Fisheries, 165
  changing, 254–255
  collapse of, 133, 141(n)

loss of, through pollution, 224–225, 229, 403, 435, 449
  threats to, 444
Fishing
  commercial, *see* Fisheries
  industry, demands of, 75(n)
  restricted, 218
  sport, 80, 650
  See also Whaling
Fission, nuclear, 283, 287, 342–345, 349–352
Flamingos, 16(f)
Flavorings, 588
Floc, 204
Flocculation tank, 205(f)
Flood control, 175, 188, 633
Flooding, causes, 32, 77, 113, 302(f), 651
Flood insurance, 633
Flood-plain zoning, 633
Florida, 438
Flounder, entrainment of, 452
Flow-by, 175
Fluidized bed combustion, 400–401
Fluoridation, 233, 234, 560–561
Fluorides, drinking-water standard, 235
Fluorocarbons, regulation of, 581
  See also Chlorofluorocarbons
Fly ash, 253, 313, 400, 409
Food, 117, 221, 596, 601, 602
  carcinogens in, 590, 595–604
  contamination of, 118, 120, 162, 194, 200, 216, 217, 223–224, 226, 230, 601–608
  effects of cooking, 590, 600, 601
  influences on cost of, 133, 134, 181, 182, 594, 634, 635
  new sources of, 141, 168(f), 170
  nitrites in, 587, 599–601
  world programs for, 144(f), 149, 150–151
Food additives, 551, 556(n), 588, 593–602, 609
  laws for safety of, 594–596
  problems of control, 595–599
Food and Agricultural Organization, (FAO), 133, 152
Food chains, 28, 30, 160, 165–168
  biological magnification in, 37
  effect of thermal pollution on, 450
  loss of energy in, 41
  oil contamination of, 438
  radioactive elements in, 577
Food colors, FD&C certified, 599
Food crisis, 10, 133
Food and Drug Administration, *see* FDA
Food, Drug, and Cosmetics Act, 594, 596, 599, 609, 611, *see also* Delaney Clause
Food production, 52, 54–55, 133–144, 150, 377
  limits on, 39–40
  losses in, 117
  successional stage for, 107
  use of pesticides in, 117–121
  ways to increase, 138–140, 142
Food supply
  estimating, 132
  grain reserves, 149–152
  problem categories, 148
  world, 133
Food webs, 28–30, 74–75
Ford Foundation, 345

Forest(s), 101, 169(t)
  biomass in, 38
  cloud, 78(f)
  deciduous, 26(f), 101, 103, 106
  fire in, 20, 656
  global impact of destroying, 33, 384
  northern coniferous, 101, 103
  pine, 20, 26(f), 27(f)
  protection of, *see* National Forests
  rain, *see* Rain forests
  rainfall in, 101
  soil in, 20, 107
  species diversity in new, 22
  threats to, 18, 19, 77, 107
Forest management, 650–657
  fire as, 20, 655–657
Forest Service, U.S., 647–653, 661–663, 667
Forestry, private, 655
Formaldehyde, 569(t)
Fossil fuels
  alternatives to, 513–516
  burning of, pollutants released by, 17, 33, 36, 58, 216, 217, 286, 313, 314, 368, 395–396, 400–401, 407, 419–420, 422
  categories of consumption, 291
  conservation of, *see* Energy conservation
  dependence of agriculture on, 138
  efficiency of burning, 283, 285, 448
  future reliance on, 529–530
  need for, 58, 282
  See also Gasoline
Foster, J. S., 356
Foxes, 30, 79, 103(f), 104(f)
France, 352, 358, 476
Franklin, Benjamin, 535
Freshwater habitats, 159–162
Freshwater resources, *see* Water resources
Fries, George, 606
Frigate birds, 168
Frogs, 27(f), 160, 162, 404, 671
Fruit flies, control of, 124
Fruit trees, 396
Fruit worms, control of, 125
Fuel cells, 287, 289, 488, 510
Fuel rods (reactor), 342–345, 349–351, 362–364
  storage of, 350–351
Fuels
  alternative, 513–516
  hydrogen, 510
  for hydropower, 471–472
  synthetic, 327–335
  wood, 515
  See also Coal; Fossil fuels; Methane; Natural gas; Nuclear fuel
Fuller, George, 199
Fungi, 30, 38, 160
Furylfuramide (AF-2), 556
Fusion, nuclear, 286–287
Fyrol FR2, 610

G

Gallium arsenide, 507
Game management, 76, 84
Gamma rays, 343, 575
Gandhi, Indira, 62
Garzel, Andrew, 224
Gas stripping, 272
Gasoline, 323, 383, 437, 537

air pollution from, 216, 407, 411, 415–416, 419–420, 431, *see also* Emissions
  from coal, 330
  conservation of, 537–541
  unleaded, 422
Gasohol, 516
Gaston, John M., 231
Geese (Canada), management of, 84
Gene pool, and loss of species, 75
General Electric Co., 224–225, 228, 350, 490, 540
General Motors, 537
Genes, 75
Genetic engineering, 75, 142–143
Gentry, Bobby, 424
Geological Survey, U.S., 327, 497
Geopressured systems, 497
Geothermal energy, 492–500
Germany, nuclear power and, 352, 360
Geysers, 646
  geothermal plant, 492–493, 496, 498–500
Giardiasis, 200–201
Gillberg, Bjorn, 252
Glaubers salt, 524
Gob piles, 307
Gold, Marian, 610
Goldfish, 255
Goodall, Jane, 85
Gophers, 104(f)
Gori, Gio Batta, 558
Government
  acquisition of land by, 658–659, 666, 672
  aid for cleanup, 565(f), 579, 618
  cost-sharing, 116
  disaster aid, 618
  investment in oil, 333
  market intervention by, 518–519
  price regulation by, 326, 328–329, 544
  and public transportation, 541
  role in control of smoking, 592
Government policy
  on air quality control, 424, 426
  determinants of, 55
  on food production, 55
  on marijuana, 591
  on oil, 435
  population control, 60–64
  on sea mammals, 87–92
  on wilderness preservation, 645–650, 654, 674
Grain, high-yield, 142, 144
  reserves of, 149–152
GRAS (Generally Regarded As Safe) List, 594
Grasses, 98–99, 101, 108, 115, 116(f), 163
Grasslands, 20, 98–101, 104, 105, 107, 169(f)
Grazing, impact of, 107
Great Northern Paper Co., 657
Great Plains Coal Gasification, 329, 333
Green Revolution, 133, 142, 143, 145, 147
Greenhouse, solar, 507
Greenhouse effect, 384
Greenlining, 668, 671–672
Greenpeace, 679
Griffin, Mr., 424
Grimm, Wayne, 83

Grobstein, Clifford, 609
Groundwater, 95, 173, 174, 177–180, 184, 266, 267
  chemical pollution of, 215–216, 221, 222, 226
  radon in, 577(f)
  recharged, 221, 271
  See also Aquifers
Growth management, 633, 642
Guano, 35
Guayule, rubber from, 515–516
Guckert, William, 304(n)
Gulhati, Kaval, 62
Gully erosion, 110, 113
Guppies, 226
Gypsy moth, biological control of, 124

H

Habitat(s), 21, 159–163
  critical, 81–82
  loss of, 77, 113, 301(f), 472, 652
  protection of, 73, 650
  stable, 44, 168
  threats to, 141, 479
Hack, John T., 566
Halbert, Fredrick, 606–607
Hamilton Standard Corp., 490, 491
Hardin, Garrett, 5(n), 64, 146, 669
Hardpan, 107
Hares, 28(f), 45
Hasui, Shigeru, 91
Hawaii, species competition in, 79
Hawkins Point, 616–617, 619, 624
Hawks, threats to, 79
Hay, for ground cover, 116(f)
Hazardous substances
  justifying regulation of, 570
  regulating and monitoring, 601–605
  worker exposure to, 567–570, 576
Hazardous wastes, 616–617
  disposal of, 616–624
Health, 56, 84, 85, 564–583
  and air pollution, 381–382, 389, 391–393, 396–397, 401, 408, 426–427, 430
  and coal mining, 300, 309, 311–313
  effects of noise on, 458, 459–460
  effects of population growth, 55–56
  effects of smoking on, 587–590
  environmental hazards to, 404, 564–583
  See also Cardiovascular disease; Respiratory disease
Health, Education and Welfare, U.S. Department of, 232
Hearing loss, and noise, 459–460
Heat
  from coal, 295–297
  disposal of waste, 447–448, 453–457, 499
  energy equivalents, 285
  geothermal, 492–500
  storage of, *see under* Solar energy
  use of waste, 453, 531
Heat exchangers, 505, 506, 516–517, 523, 524, 534
Heat pump, 533
Heating
  energy used for, 290–291
  geothermal, 492–493, 497
  home options for, 532–533

new ideas for, 292
  solar, 503–508, 519–524
  waste heat for, 452, 531–532
  wood for, 513
Heavy-water reactors, 343, 362
Heber Project, 494–495, 496
Heckler, Honorable Kenneth, 306
Hedgerows, pest control with, 122
Helgrammite, 240
Heliostats, 511–512
Hemoglobin, effect of pollutants on, 387, 391, 411, 429, 587, 600
Hepatitis, 85, 119, 200
Heptachlor, banned, 120
Herbicides, 114, 269
  levels of, in marijuana, 591
  toxicity of, 126(t), 127, 128
  in water treatment, 203
Herbivores, in food chain, 28, 39
Heronemus, William, 488
Herring, 42, 255
  fisheries, 141(n)
Hevea plant, hydrocarbons from, 515
Hexachlorophene, banned, 609
Hickory trees, 103
High Plains, water problems of, 179–182
High-voltage power lines, 473, 478
  pollution from, 462–464
Highwall, 301–304
Highways, impact of, 269
Hijacking, 352, 355, 356, 361–362
Historic landmarks preserved, 658
Holmes, Justice, 647
Homeotherms, 19
Honey-creeper, threatened, 79
Hooker Chemical and Plastics Corp., 4, 230, 618
Hormones
  in animal feeds, 605
  pest control with, 122(t)
Hosmer, Craig, 191
Hough, Franklin, 646
Houseflies, mutant, 119
Hubbert, M. King, 323, 324
Hudson Canyon, 164
Hudson River, 158, 159, 162–163
  epidemic from, 197
  PCBs pollution of, 224
  reclamation program, 225
Hughes, Thomas, 332
Human actions, spill-over effects of, 86
Human rights
  freedom of choice, 596, 598
  right to farm, 641
  right to food, 150
  right to hunt, 88–92
  right to know, 2, 231, 356, 612
  women's, 576
Human Rights, Universal Declaration of, 64
Humans
  effect of climate change on, 377
  decisions and actions *now*, 677–680
  in food chain, 28
  levels of chemical pollutants found in, 121(t), 218, 223
  population growth, 9–10, 46–53
Humus, 20, 105, 107
Hunting, 80, 84, 646, 647, 649–650
  environmental effects of, 75, 76
  subsistence, 88, 89–92

Hydro–Quebec, 473, 478
Hydrocarbons
    air pollution by, 420–422, 425, 430–431
    controlling emissions of, 421–422, 425
    liquid, 321, 322, 330–331, 407
    plant species supplying, 515–516
Hydroelec power plant, 474
Hydroelectric power, *see* Hydropower
Hydrofracturing, 495
Hydrogen
    as fuel, 510
    future use of, 289
    in gas and oil, 321
Hydrogen sulfide, sources of, 174, 178, 203, 211, 240, 321, 492, 500
Hydrologic cycle, 31–32, 159, 343
Hydropower, 282, 286, 467, 471–475, 543
    environmental impact, 472, 479, 480
    pumped storage for, 479–481
Hypolimnion, 170

I
Idso, S. B., 386
Igorots, rice growing by, 135–136
Immigration, 50–51, 59
Impact (environmental) of
    acid rain, 95, 402–405
    air pollution, 367–368, 396, 407, 421, 430
    clear-cutting, 651–652, 654
    dams, 81, 147, 175–176
    deep-water ports, 441
    electric generation, 283, 286, 313–314, 368, 646
    fossil fuel burning, 300, 313–314
    hydropower, 472, 479, 480
    mining, 108, 300–312, 331
    offshore drilling, 439, 441–444
    oil pollution, 435–444
    ozone, 421, 430
    paper industry, 107, 217, 657
    peak-load pricing, 544
    population growth, 9–10, 41–43, 48–68
    producing biomass fuels, 516
    recreation, 198, 667–668
    reservoirs, 175–177
    shale oil production, 334
    thermal pollution, 449–452
    using geothermal energy, 498
Impact (social)
    of environmental controls, 570, 581
    offshore drilling, 439, 441–444
Imperial Oil, 330
Incinerators, 622–623
    air pollution by, 226, 407
Income, per capita, decreasing, 56
India, 52, 54, 117, 358
    demographics, 63, 66(t), 69(t), 69(f)
    food production, 133, 140, 142
    population control programs, 62, 64, 68
    wildlife conservation, 86
Indicator organisms
    coliforms, 202, 209
    for Minimata Disease, 218
Indonesia, 66(t)

Industrial processes
    cancers and carcinogens associated with, 552, 568
    pollution from, 287(t), 402(f), 407
    worker hazards in, *see* Workers
Industrialization, effects of, 32, 33, 35
Industry
    environmental poisoning by, 216–218, 221, 224, 227–230, 378
    hazardous wastes from, 616–624
    liability of, 573
Infants, and nitrates, 221–222
Infiltration-percolation of wastewater, 270–271
Inflation, oil imports and, 336–337
Insecticides, 49, 128, *see also* Pesticides
Insects, 19, 29(f), 119, 217(t), 450
Institute of Ecology, 334
Institute of Medicine, 596
Insulation (home), 533–534
Integrated pest management, 122, 123–125
Interior, U.S. Department of, 490
International Atomic Energy Agency (IAEA), 357–358
International Nickel Company, 396
International Whaling Commission (IWC), 89–92
Interstate and Foreign Commerce, House Committee on, 601, 604, 605
Intertidal zone, 165
    threats to, 479
Inversion, 370, 378, 382
Iodine, 350
    -131, contamination by, 577, 578
Ionizing radiation, 574–578
IPM, *see* Integrated pest management
Iron
    excess of, in diet, 601
    -59, contamination by, 577
    pyrites, 308, 310, 397, 399
Irrigation, 139, 144, 179–190
    canals, hydropower from, 474
    disadvantages of, 95, 108, 271
    environmental effects of, 139–140
    impact on crop yields, 180–181
    methods, 181(f), 182
    with wastewater, 270, 271
    water for, 175, 177, 179–182, 185–190, 453(t), 497
Irruptive growth pattern, 43
Island Park Geothermal Area (IPGA), 498–499
Israel, 139(f), 222, 271, 358
Italy, birth rate, 50
Ixtoc, 435

J
Jacobson, Dr. Michael, 594
Japan, 89, 91, 139(f), 222
    chemical disease in, 216, 221, 223
    population density, 55(f)
    zero growth rate, 68(n)
Japanese beetles, 124
Jellyfish, 166(f), 479
Johnson, George, 199
Jojoba, 515

K
K-selected organism, 44
Karrh, Dr., 576
Keepin, G. R., 356, 357

Kelp, in food web, 75(f)
Kemp, Clarence, 504
Kennedy, David, 191
Kentucky, 680
Kenya, *see* Lake Nakuru
Kepone, 227, 229–230
Kerogen, 331
Kilbey, B. J., 252
Killdeer, 27
Kinetic energy, 471
Kissinger, Henry A., 150
Klein, John, 639
Korea, growth-rate decline, 65
Krill, 166(f), 168
    as human food source, 141–142, 168(f)
Kuck, David L., 309
Kwashiorkor, 131

L
LaRance Power Station, 473, 475–478, 479, 481
Labeling, 596, 609
    alcoholic beverages, 611
Ladybugs, 75, 119, 123
Lagoons, waste stabilization, 266, 267
Lake Erie, death of, 254–255
Lake Nakuru (Kenya), 16(f), 28
Lakes, 100, 160, 170, 472
    acidic, 403–404, 406
    aging, 248
    communities in, 16(f), 21, 160, 450
    eutrophic, 248–255
    oligotrophic, 248
    oxygen-poor, 20, 170–171
    polluted, 254–255
    restoration of, 253, 255
    stratification of, 160–162, 170–171
Lamprey (sea), 255
Land
    acquiring public, 658, 666, 672–674
    arable, 55(f), 140–141, 635
    area cultivated, 169(t)
    degradation of, 95–96, 300–312, 670
    federal funds for purchase of, 672
    ownership of, 52, 642
    preservation of natural areas, 645–674
    public, for preservation, 645–674
    reclamation of mined, 301–304
    results of clearing, 141
    rural, 634–643
    value of, 635–638, 646
Land farming hazardous wastes, 622
Land trusts, 673
Land use, 628, 635
    control, 630–633, 638–640, 642
    results of changing, 78, 637
    trends, 635
Land-use programs, need for, 54
Land and Water Conservation Fund, 672–674
Landfills, 262, 327, 623–624
    pollution by, 174, 215, 226, 579
    secure, 623
Landslides, 303
Langer, A. M., 566
Larderello (Italy) geothermal power plant, 492–493
LAS, 256
Laser beams, 287

Laterite, 107
Law of the Sea Treaty, 444
Laws (environmental), 182, 217, 218, 251, 570, 615, *see also* Acts; Bills
    exemptions to, 88, 92, 267, 457, 582(n), 609, 611
    flexibility of, 81–83, 596
    preservation, 646–649
    Public Law, 83–84, 149
    regulatory basis of, 615–616, 620–621
    Right-to-Farm, 641
    smoking, 590
    tax reform, 630
    unenforceable, 621
    zoning, 630
Leaching, 35, 95
Leaching field, 266
Lead, 118, 128, 216, 235(t), 569(t)
    in cigarette smoke, 587–588
    in food, 599
    health effects of, 588
    poisoning, 411, 415–416
    pollution by, 216, 404–405, 407, 588
Leaf beetle, 45
Lebbinl, Keith, 184
Legal action, *see* Suits
Legumes, 35, 115, 249
Lemmings, 28(f), 45, 60
Leopold, Aldo, 87
Leroy-Somer Co., 474
Lichens, 21(f), 101
Life Science Products Corp., 229, 230
Life-style
    changing, 469
    and disease, 602
    need for changes in, 2–3
    trends, 54
Lifeboat ethic, 5, 146
Light-water reactors, 343, 344–348, 362
Lighting, interior, 510–511, 530–531
Lightning, 34(f), 249
Lignite, 295–297
Lijnsky, Dr. William, 598
Lilienthal, David, 514
Limestone, production of, 383
Limiting factor concept, 248–249
Lindane, 126(t), 203, 235
Linear alkyl sulfonates (LAS), 256
Liquid-metal fast breeder reactors (LMFBR), 362–364
Liquefied natural gas (LNG), 326
Livermore, K. C., 123
Lizards, temperature regulation by, 19
LMFBR, 362–364
Lobsters, 75(f), 170, 449, 452
Locust tree, 35
Logging methods, 651–654
    impact of, 100
Logistic growth pattern, 43(f), 44, 46
London Fog, 58, 313, 381, 397, 414
Longwall mining, 306
Loon, 404
Loss-of-coolant accidents, 345–348
Loss of species, 74–80, 101, 142, 176, 437–438, 449, 450, 452
    importance of, 74–80
Love, William T., 618
Love Canal, 4, 618, 623

Lovins, Amory, 514
Low-flow augmentation, 175, 263
Low-head hydropower, 473–474
Luce, Charles, 347
Lurgi process, 329–330
Lynx, 45, 103(f)

M
MacAvoy, Paul, 388–389
McDonnel-Douglas, 512
MacKaye, Benton, 663
Maggots, 44, 242
Magma, potential of, 495
Magnetic fields, power-line, 464
Magnetohydrodynamics, 292, 293, 531
Maguire, Andrew, 596
Malaria, 49, 119
Malcolm, Andrew, 441
Malnutrition, 56, 131–133
Malthus, R. I., 43
Malthusian growth pattern, 43, 150
Mamane tree, threatened, 79
Manatee, endangered, 674
Maple trees, 100, 103
Marasmus, 131
Mariculture, 170, *see also* Aquaculture
Marijuana smoking, 591
Marine ecosystems, 169(t)
    impact of oil pollution on, 437–438
Marine habitats, 159(f), 163–168
Marine Mammal Protection Act, 75(f)
Martin, J. D., 146
Martin, James G., 597, 598
Martin Marietta Corp., 512
Maughan, Ralph, 498
Maxey, Don, 566
Mayflies, 29(f), 241, 255
Mazzocchi, Anthony, 576
Meadowlarks, 27
Measles, 49
Medicine, 49
    value of species to, 74, 85
Melanoma, 583
Meltdown, 345, 346, 364
Mencken, H. L., 6
Mercury, 37, 118, 216–221, 255, 404, 451, 569(t), 599
    poisoning by, 128, 216–218, 221
    safety standards for, 221, 235(t)
Mesosphere, 372
Mesquite, 105(f)
Methane, sources of, 326–327, 329, 420, 497
Methemoglobin, 429, 587, 600
Methemoglobinemia, 271, 272
4-Methoxy-*m*-phenylenediamine, 609
Mexico, 66–67, 89
    water treaty with, 183, 188
Michigan Chemical Co., 606–607
Microflagellates, 166(f), 479
Mircoorganisms, 19, 36, 265
Microplankton, 168
Mircowaves, health effects of, 574
Milk, radioactive contamination of, 578
Miller, Arnold, 309
Millipore filter, 209
Millirem, 235(t)
Milorganite, 262
Minckler, Leon, 654

Miners, hazards to, 300, 311 313, 564, 578
Mines, abandoned, 310–311, 617
Minimata Disease, 216, 218
Mining, 298, 309
    environmental impact of, 300–312, 331, *see also* Acid mine drainage
    methods, 301–302, 306, 310
    pollution by, 221, 242, 407, 565, 579
    See also Strip–mining; Surface mining; Underground mining
Mink, 404
Minnows, 241
Mirex, 227, 230, 255, 603(n), 604
Mist of sulfuric acid, 396, 407, 414
Mitchell, J. M., 386
Mixed-field ecosystems, 30
Mixed oxide fuel, 354, 364
Modified block cut mining, 303–304
Moisture, environmental, 17
Molluscs, world catch of, 141(t)
Mongooses, 79
Monitoring
    hazardous wastes, 619, 622, 623
    pesticides, 604–605
Monkeys (experimental), 59
Monopoly, 334
Monsanto Chemical Co., 224
Moose, 103(f)
Moratorium, whaling, 89–90, 91
Morning glory (Pickering's), 671
Mosquitoes, 19, 27(f)
    control of, 49, 119
Mosses, 101
Motor vehicles, pollutants from, 387, 425, *see also* Automobiles
Mouchot, 503
Mud, 21, 168
Multiple-use of public lands, 650
Mungai, Professor, 67
Muskellunge, 255
Muskgrass, 160(f)
Muskie, Senator Edmund, 388, 424
Mussels, 437, 479
Mutagens, 555, 609, 610
Mutants, insect, 119, 122
Mutation, 119, 252, 555, 578, 595
Mycotoxins, 601

N
Nannoplankton, in food chain, 169
Napthalene, 569(t)
2-Napthylamine, 554
NASA, 512
National Academy of Sciences, 222, 304, 314, 386, 426, 560, 578, 580, 596
National Agricultural Lands Study, 634
National Cancer Institute, 609
National Clean Lakes Group, 253
National Clearing House for Smoking and Health, 586, 590
National Coal Association, 298
National Flood Insurance Program, 633
National Forests, 498, 645–657
    management of, 650–657
    multiple-use, 650–651, 662
National Institute of Occupational Safety and Health (NIOSH), 567, 569
National Monuments, 645

National Park Service, 648, 658, 661, 663, 666, 667, 672
National Park System, 658–666
National parks, 99, 493, 498, 645–648, 662
    Katmai (Alaska), 14(f)
    Borivli (India), 10
National Reactor Testing Station, 358
National Research Council, 191, 232, 608
National Resource Lands, 648–649
National Science Foundation, 520
National Solar Heating and Cooling Institute, 519
National Toxicology Program (NTP), 570
National Trail System, 666, 667
National Wilderness Preservation System, 651, 655, 659–662
National Wildlife Refuge System, 80
Natural area zoning, 632
Natural areas, preservation of, 645–674
Natural gas, 319, 321, 396, 577
    home heating with, 532
    liquified (LNG), 326
    price control, 326–329
    proved reserves/consumption of, 326
    sources, 320, 322, 326–327
    synthetic (SNG), 313, 315, 329
    uses, 286, 291, 292, 326
Nature Conservancy, 673–674
Nature Protection and Wildlife Preservation in the Western Hemisphere, Convention on, 86
NAWAPA concept, 190, 192
Nelson, Sue, 576
Neritic zone, 165(f)
Netherlands, 55(f), 68(n), 271, 490
Neusiedler Lake (Austria/Hungary), 100
Neyrpic, 473
Niches, 21, 22, 77
Nickel, 569(t), 587–589
Nicotine sulfate, 119
Nigeria, 50
Nitrates, 271, 272
    drinking water standard, 235(t)
    effect on infants, 221
    in food, 599–601
    physiological effects, 587, 600
    removal from wastewater, 272
    sources of, 222, 249–250
Nitrogen cycle, 35
Nitrogen dioxide, 419–421, 428–432
    air standards for, 426(t), 427(t)
Nitrogen oxides, 34(f), 368, 402, 419–432, 462, 580
    air pollution by, 287(t), 419–428, 430, 600
Nitrosoamines, 600
Noise
    defined, 259
    desirable levels of, 460
    pollution by, 447, 458–462
Noise Control, Office of (EPA), 458
Noise Control Act, 460
Nonionizing radiation, 574
North American Water and Power Alliance, 190
North Sea, oil from, 335
Northeast Utilities, 474
NPT, 357–358
Nuclear accidents, 345, 346–348

Nuclear explosions, effect of, 580
Nuclear fuel reprocessing, 349–353
    alternative (CIVEX) process, 361–362
Nuclear Fuel Services, 351
Nuclear Non-Proliferation Treaty (NPT), 357–358
Nuclear power, 341–360, 678
    fuel for, *see* Fuel rods; Nuclear fuel reprocessing
    fuel cycle, 348, 355(f)
    hazards of, 347
    risk assessment, 6, 345–346
    uncertain future of, 351–352, 360, 529
Nuclear power plants, 341, 344, 346, 449, 451(f), 454(ff)
    decommissioning, 344, 348
    efficiency of, 286, 448
    efficiency of controls on, 344, 351
    environmental controls on, 457
    locating, 355, 457
    pollution from, 344, 345, 577
    radiation standards, 577, 578
Nuclear Regulatory Commission, 345, 351, 353
Nuclear safeguards, international, 355–358
Nuclear theft, risks of, 352, 355, 356
Nuclear weapons, 353–358, 577
    and radiation exposure, 576–577
Nutrients, 162, 163, 165
    limiting factor concept, 248–249
Nutrimaster, 606
Nutrition, basic requirements, 132–133
Nutrition and Human Needs, Committee on, 609

O
Oak trees, 103
Oats, 123
Occupational Safety and Health Administration (OSHA), 460, 570, 574
Ocean(s), 18, 31, 33, 35, 168–170
    carbon dioxide in, 33, 382–383
    nutrients in, 163, 165
    pollution of, 226, 434–436
Ocean Mammal Protection Bill, 88
Ocean Thermal Energy Conversion, 516–518
Oceanic zone, 165(f)
Ochero, Louis, 67
Office of Technology Assessment (OTA), 334, 421
Offshore drilling, 439, 441–444
Offshore oil and gas, 320, 326, 335, 434–435
Ogallala Aquifer, 179–184
Oil, 320–326, 333–336, 434–444
    estimates of undiscovered, 322–324, 435
    imports (U.S.), 323, 325(f), 336, 518
    monopoly of, 334–335
    plant substitutes, 515
    prices, role of, 57, 277–278, 323, 332–335, 468
    re-refining, 440
    residual, 395–396
    resources, 322–323
    from shale, 186, 325, 331–334
    from tar sands, 330–331
    toxicity of, to marine life, 437

Oil (*continued*)
  uses, 283, 286, 291
Oil pollution, 368, 434–445
  impact of, 435–444
  persistance of, 438–439
  sources, 434–436
Oil spills, 434–440, 444
Oil tanker accidents, 434–435, 439, 441
Oils, toxic, 569(t)
Oilseed crops, 516
Ojibway, mercury levels in, 217–218
Olin Chemical Corp., 622(f)
OPEC, 323, 325, 333
  oil embargo, 336
  power of, 334–335
Optical brighteners, 252
Orchids, threats to, 80
Ore processing, pollution from, 217
Organic chemical pollution, 208, 222–230, *see also* PCBs; Pesticides; Chlorinated hydrocarbons
  drinking water standards, 235(t)
Organic farming, 121–122, 123
Organization of Petroleum Exporting Countries, *see* OPEC
Organophosphates, 121(t), 126(t), 127
Orlans, F. B., 85
Oser, Dr. Bernard, 602
OSHA, 460, 570, 574
Ospreys, 27(f), 404
OTEC, 516–518
Otters, 27(f), 75(f), 404, 443
Ouch-ouch disease, 221
Outer Continental Shelf Bill, 444
Overburden, 301, 303–309
Overland flow, 270
Owen, Mr., 441
Owls (great horned), 27
Oxidants, 420–422, 430
Oxidation pond for wastewater, 266
Oxygen, 23, 321, 371
  in the body, 391–392
  oceans and, 165, 168, 170
  in polluted water, 239–246, 251
  removal of, from water, 160, 170–171, 177, 239–246
  in water, 19, 162, 239, 451
Oxyhemoglobin, 392
Oysters, 141(t), 163, 435, 438, 449, 451, 479
  contaminated, 200, 229, 438, 578
Ozonation, 207, 211, 273
Ozone, 313, 371, 376, 408, 420–425, 426(t), 427(t), 430–432
  air pollution by, 408, 420–422, 430–432
  air standards for, 421–425, 426(t), 427(t)
  sources, 420, 462
Ozone layer, 580–582

P
Pacemakers, and power lines, 463
Pacific Crest Trail, 664–665
Pacific Gas and Electric, 491–493, 496, 498, 499
Packard, A. S., Jr., 85
Paints, poisons in, 221, 411, 415–416
Pakistan, 61, 66(t), 67, 352, 358
Panama, 86
Paper industry, 107, 217, 622, 657

Paraquat, 591, 603(f)
Parasites, 14(n), 45
  pest control with, 124
Parathion, 126(t)
Parker, Frank L., 455
Parks
  greenline, 668, 671
  national, *see* National Parks
Parsons, Ralph M., 190
Particulates (airborne), 313, 371, 376, 385, 395, 406–411, 425
  atmospheric levels, 417(f)
  concentration standard, 407, 410, 414
  control of, 409–410, 414
Passenger pigeons, 100, 647
Pasteurization, 49
Pathogens, 194
Payne, Dr. Roger, 90, 92
PBBs (polybrominated biphenyls), 606–607
PCBs, 223–226
  human levels of, 121(t), 223
  pollution by, 162, 224–225, 255
  safe disposal of, 226
  toxicity of, 162
Peabody Coal Co., 302(f), 316
Peaches, 452(f)
Peak-load pricing, 543–544
Peanuts, 421, 601
Peat, 295
Pelicans, 16(f), 79(f)
Penguins, 119
Penstock, 471
Perch, 255
Percy, Senator Charles, 591
Peregrine falcons, 12–15, 21, 22, 27, 118
Peripheral Canal, 190
Periwinkle, drug extracted from, 74
Permafrost, 101
Peroxyacyl nitrate (PAN), 420–421, 430
Persistance of chemical pollutants, 118–119, 126(t), 217, 230, 251, 588
Persistent trillium, endangered, 74(f)
Pest(s), 117
  creation of new, 119
Pest control, 122, 123–127
  hazards of, 230
  need for, 117
  See also Biological control; Pesticides
Pesticides, 117–127, 204, 216, 227–230, 600, 622
  banned/restricted, 120, 603–604
  bioaccumulation of, 37, 126(t)
  biological effects of, 451, 576
  biological magnification of, 117–118
  disadvantages of, 121
  drinking water standards, 235(t)
  environmental pollution by, 79, 112, 113, 120, 142, 216, 227–230, 588
  in food, 596, 601–604
  human levels of, 121(t)
  laws, 603
  need for, 58, 107, 115, 117, 123
  persistence of, 118–119, 126(t), 230, 588
  rating for safety, 127
  regulatory problems, 603–605
  resistance to, 119–121

risks vs. benefits, 6
  selected list of, 126(t)
  selective use of, 122, 123–127
  toxicity of, 126(t)
Pesticides, Commission on, 113
Pesticides Control Act, 616(t)
Petrels, in food chain, 168
Petro-Canada, 330
Petroleum, 320–323, 395
  primary/secondary recovery of, 321
  U.S. strategic reserve of, 333
  See also Oil; Tar sands
Pettigrew Amendment, 646
pH, 403
Phenacetin, safety of, 612
Phenols, 569(t), 622
Pheromones, 122(t), 124
Phillips Petroleum Co., 496
Philippines, 65, 135, 147
Phosphates, 249–253, 264
Phosphorus
  cycle, 35
  removal, from wastewater, 271–272
  role in lake eutrophication, 249–250
  -32, biological magnification of, 577
Photochemical air pollution, 419–432
Photolytic cycle, 431
Photosynthesis, 17, 19, 37, 372, 382–383, 513
  chemistry of, 23
Photovoltaic conversion, 507, 510
Phthalates, pollution by, 222, 226
Physical-chemical wastewater treatment, 270
Phytoplankton, 160, 166(f), 450, 451, 479
Pike, 254, 255
Pilchard fisheries, 141
Pimentel, Professor David, 122
Pinchot, Gifford, 646, 647
Pine barrens, greenline for, 671
Pine trees, 20
  threats to, 79, 99, 396, 404
Pipelines
  coal-slurry, 297, 315–316, 542
  natural gas, 326
  oil, 320, 443, 444
Pittston Coal Co., 308
Planes, environmental threat of, 580
Plankton, 160, 169, 479
  as human food source, 141–142, 170
  petroleum from, 321
Plants
  classification, 14
  drugs from, 74
  effects of pollution on, 396, 404, 407, 421, 430, 450
  as fuel sources, 513–516
  limiting growth factors, 248–253
  pioneer, 44
  temperature of, 19
Plastics, pollution by, 226
Plowshare Program, 327
Plutonium, 341, 349, 352–355, 361–364
  transport of, 352, 354–239, 363
Pneumonia, 49
Poaching, 77, 85

Poison ivy, 27(f)
Poisoning by pollutants, 120, 216–218, 221–223, 229, 411, 415–416, 600
Polar ice, melting of, 386, 582
Polio, 49
Politics, in environmental decisions, 624
Pollutant Standards Index, 426
Pollutants
  atmospheric accumulation of, 381, 383, 384, 386
  banned, 120, 225, 603–604
  biological magnification of, 37, 117–118, 163, 216, 226
  carcinogenic, 564–583
  in coal, 295
  in consumer products, 219–221, 223–224, 226, 252, 581
  in detergents, 249–256
  dispersion and dilution of, 41
  environmental impact of, 156
  kinds of, 267, 368
  non-point-source, 267–269
  particulate, 313, 371, 376, 385
  permanent, 219
  persistance of, 118–119, 126(t), 217
  radioactive, 344, 579
  in water, 17, 215–232
  See also Air pollutants; Emissions
Pollution, 10, 17, 170
  and disease, *see* Disease
  by fuel combustion, *see* Fossil fuel burning
  increasing, 58–59
  by mining, *see* Mining
  non-point-source, 253–254
  pesticide, *see* Pesticides
  radioactive, 344, 358, 577, 579
  range of impact, 224, 229, 396, 408, 435, 439
  reducing, 518
  and species diversity, 21, 22
  types of, 368
  See also Acid rain; Air pollution; Noise pollution; Oil pollution; Thermal pollution; Water pollution; Water resources, pollution of
Pollution control
  affordability of, 334
  air, *see* Air pollution control
  efficiency of nuclear plant, 344, 351
  emissions, *see* Emissions control
  mining, 310–311, 313
  oil, 440, 442, 444
  of ozone, 581
  standards for, 267, 461, *see also* Air quality standards; Drinking water, standards; Wastewater treatment standards
  wastewater, 258–274
  water supplies, 197–211
Polychlorinated biphenyls, *see* PCBs
Ponds, communities in, 160–162
Poplar trees, 115, 515
Popular Science, 520, 534
Population (human)
  density, 55(f)
  profiles, 68–70
  urban, 53
  world, 50–51, 66
  See also Population growth

Population control, methods of, 60–
64, 68, 131
Population growth, 9–10, 41–43,
48–68
environmental impact of rapid,
10, 41–42, 54–60, 108
food production and, 134, 144
rate of increase, 49–51
trends, 65–68
world, 49–51
zero, 50, 68
Populations, 14, 15(f)
restoration of endangered, 75(f),
84–86, 89
factors in regulation of, 44–46
r- and K-selected, 44, 46
Potentiation, of air pollutants, 414
Power
centralized vs. decentralized
sources of, 418, 512, 514
generation of, 447, 471
geothermal, 492–496
See also Electric generation;
Hydropower; Nuclear power;
Power plants; Tidal power;
Wind power
Power parks, 457
Power plants, 283, 285–287, 510–512
combined cycle, 292
ecological effects of cooling for,
449–452
environmental impact of, 283,
286, 313–314, 368
future, 360
pollution by, see Emissions; Fossil
fuel, burning; Thermal pollu-
tion repowering, 512
where to build, 314, 355, 451,
457, 663
See also Electric generation
Power tower, 511–512
Prairie dogs, 104(f)
Prairies, 98–99, 674
Precipitators (stack), 313, 314, 401,
409
Predator(s), 14(n), 21, 27, 30, 45,
75, 118, 119
Preneoplasia, 559
Preservatives, food, 593, 599
Preserves (wildlife), 84, 86
Pressure, atmospheric, 372
Pressurized-water reactor (PWR),
342–344, 362
Pretreatment standards for toxic
wastes, 268–269
Prey, 21, 27, 30, 45
Price regulation, 326–329, 544
Primary production, 38
Primary recovery of oil, 321
Primary treatment of wastewater,
260, 262, 265, 267, 270
Primrose (evening), 80
Producers (food chain), 28, 33, 160,
168
Production
coal, 298
food, see Food production
oil, 323–324
projected, of wind-powered
electricity, 491
Productivity
agricultural, 112, 138, 147
ecological, 38–39, 107, 163, 169–
170
political restraints on, 143–144

Progress, pollution control, 390–
391, 401–402, 409, 425
Project Independence report, 518
Project Sanguine (USN), 464
Project Tiger, 86
Prometone, toxicity of, 126(t)
Promoters (cancer), 559
Propoxur, use of, 119
Propyl gallate, 599
Protozoa, 200, 204
Proved reserves
natural gas, 326
oil, 322, 325
Pteropods, 166(f)
Public, informing the, 2, 231, 356,
426
Public Health Service, U.S., 199,
226, 232, 552
Public interest vs. the environment,
443
Public land, agencies for, 648–667
Public Law, 83–84, 149
Public opinion, and nuclear power,
360
Public Service Company of New
Mexico, 493, 496
Public transportation, 3, 6, 391, 534,
540–541
Pumped storage, 479–481, 490, 543
Pupfish (Devil's Hole), endangered,
86(f)
PUREX process, 352, 353, 361
Putnam, John, 10
PWR, see Pressurized-water reactor
Pyramids, ecological, 38–39, 169
Pyrolysis, coal liquefication by, 330

Q
Quad, 285
Quality of life, 627, 634
maintaining, 59
OPEC influence on, 334
Quarles, John, 538
Quotas, whaling, 89, 92

R
r-selected organisms, 44
Rabbits, 30
Raccoons, 27
Races, 14
Radiation
from coal-fired plants, 315
exposure to, 351, 574, 575–579
exposure standards, 577, 578
health effects of, 573–574, 578
in nuclear power cycle, 342(f),
344–345, 350
solar, 373
Radin, Alex, 288
Radioactive elements, 295, 575
biological magnification of, 577
in power plant emissions, 313,
344
Radioactive wastes, 344, 354, 358–
359
Radioactivity, 287, 327
drinking water standards, 235(t)
Radon, 577(f)
air pollution by, 579–580
Rain
acidity of, see Acid rain
purity of, 31
Rain forests, 77, 101
diversity in, 30, 101

impact of logging on, 100, 107
primary production in, 39
rainfall in, 101
soil in, 15, 107
succession in, 100
Rainfall, 31–32
contamination of, 119, 226
in deserts, 105
effect of cities on, 378
effect on sulfate levels, 409
influence of air pollution on, 408
required, 101
Randomness, 41
Rasmussen, Professor Norman, 345
"Ratchet effect" of food prices, 134
Rats, 59, 121
pesticide toxicities in, 126(t)
Reactors
breeder (LMFBR), 362–364
BWR/PWR, 342
heavy water (CANDU), 343, 362
light water, 343–348, 362
safety of, 344–348
Reclamation
of deserts, 305, 306
Hudson River, 225
after strip-mining, 301–314
Recreation
impact of, 198, 667–668
ORVs vs. preservation, 95, 670–
671
provisions for, 175, 198, 248, 632,
649–650, 663, 672
Recreational areas, threats to, 59
Red blood cells, 430
Red spider mites, 75
Red tides, 479
Reddy, Amulya, 148
Reduced tillage farming, 114–115
Reed communities, 30, 100
Refining, 321
coal, 399
crude oil, 321, 395
Refrigeration, waste heat for, 453
Refuges, wildlife, 80, 647–650
Refuse banks, 307–308
Reid, Ogden, 228
Reindeer, 43
Reindeer lichen, 28(f)
Remmert, H., 13, 41
Rems, 575
Repowering, with solar energy, 512
Reprocessing hazardous wastes, 622
Re-refining oil, 440, 442
Research, use of animals in, 85
Research and development in solar
energy, 490–491, 496, 498–499,
507, 510–512, 516–517
Reserpine, 74
Reserve Mining Company, 565, 566
Reserves
ecological (biosphere), 80
forest, 646
Reservoirs, 173–177
hydropower from, 474, 479–481
recreational use of, 198
sedimentation of, 113, 175, 186
tidal, impacts of, 479
Residual oil, 395–396
Residues, pesticide, 603
Resistance (biological), 119
to drugs, 119, 605–606
to pesticides, 119–121
Resistance (social)
to environmental controls, 67, 81,

88, 182, 208, 228, 268, 398,
421, 426, 537, 570, 603–604,
650, 671
to food, drug, and cosmetic
controls, 594–599, 611
to grain reserve programs, 151–
152
to landfill siting, 624
to new water projects, 178, 191
to offshore drilling, 443
to nuclear controls, 358
to nuclear power, 286, 341, 347,
360
to population control programs,
67
to radioactive disposal sites, 344
to water sharing, 179, 190
Resource(s), 35, 58
air quality as, 398
children as, 9
coal, 185, 291
conversion of, 281–282
energy, 277–278, 468(f)
estimating undiscovered, 322–324
finite sea, 141
geothermal, 497–499
hydropower, 473
natural areas, 645–674
new gas, 330
offshore, 320, 326, 335
oil, 322–323, 443
political role of uranium, 352
potential energy, 495
potential tidal, 477–478
renewable and nonrenewable, 324,
334, 467, 497, 510
solar, 509(f)
substitution, 58, 278
water, 177–178
wildlife as, 72–92
wind (U.S.), 488
Resources Conservation and Recovery Act, 615
Respiration, 23, 33
Respiratory disease
and air pollution, 313, 381, 389,
396–397, 407, 414–425, 430
and smoking, 588–589
Retrievable surface storage facility
(RSSF), 359
ReVelle, C., 538
Revelle, Roger, 5, 146
Reykjavik (Iceland), 493–494, 497
Rice, 221, 601
growing of, 135–136, 151
high–yield, 142
Rice–oil disease, 223
Rill erosion, 110, 113
Risk assessment, 556–557, 570
Risk vs. benefit
and regulatory law, 615–616
Risk-based laws, 615
Risk-benefit analysis, 551, 558, see
also Benefits vs. costs (risks)
Rivers, 140, 162, 177, 263
chemical contamination of, 217,
218, 221, 223, 228–230
Colorado, 182–190
communities in, 162
Hudson, 158, 159, 162–163, 197,
224, 225
hydropower from, 471, 474
preservation of, 661, 663
sewage pollution of, 158, 162,
194–197, 200

Rivers (*continued*)
  wild, 186, 217, 661, 663
Roadless Area Review and Evaluation (RARE), 662
Rohl, A. N., 566
Room and pillar mining, 306
Roosevelt, Franklin D., 475, 514
Rotifers, in food chain, 160(f)
Rubber, sources of, 515
Ruckelshaus, William D., 2(n)
Runoff, 31
  in hydrologic cycle, 31, 32
  effect of erosion on, 112
  oil in, 434
  pollution by, 208, 249, 253, 259, 269
  reducing, 115, 269
  in sewer systems, 260, 273–274
Rustin, Bayard, 398
Russia, 89, 91, 133, *see also* Soviet Union
Rye, planting of, 114

S

Saccharin, 555(f), 559, 595–599
  ban on, 595–596
Safe Drinking Water Act, 230–232, 560
Safety
  of drugs, 611–612
  of nuclear power, 287, 344–348
  at reprocessing plants, 351
Sage brush, 105(f)
Sail power, 486
St. John's wort, control of, 45
Salamanders, 404
Salinity, 18, 451
  Colorado River, 186
  well water, 182
Salinization, 95, 139, 142, 174
Salmon, 163, 224, 241
Salt beds, waste storage in, 359
Samuelson, Paul, 269
Sanitary Engineering, 155
Sardine fisheries, 141(n)
Satellite, solar power, 512–513
Sauger, 254
Saury, tar lumps in, 438
Sawhill, John C., 332
Scarlet fever, 49
Scenic resources, 645, 663
Schachle Company, 491
Schistosomiasis, 139
Schneider generator, 474
Scott, Sir Peter, 91
Scott, Senator William, 83
Screw worms, control of, 125
Scrubbers, 313, 314, 399–400, 406
Sea bottom, 34, 37, 163–164, 168
Sea cucumber, 168(f)
Sea mammals, 167(f), 168
  government policy on, 87–92
Sea urchins, 168(f), 479
Sea water, hydrogen from, 289
Seals, 75(f), 119
  (fur) protection of, 88
Searle Company, 611
Seaweeds, 166(f), 479
  world catch of, 141(t)
Seber, Scott, 670
Second law of thermodynamics, 41, 138, 283, 285, 378, 448
Secondary recovery methods, 321
Secondary treatment of wastewater, 260–262, 265, 267, 269, 270

Sedges, 28(f)
Sedimentation, effects of, 113, 175, 186
Sediments, 35, 113, 209
Seed-tree cutting, 652
Seeger, Pete, 87(n)
Selection cutting, 652–653
Selenium, 313
Selikoff, Irving J., 566, 571, 573, 607
Semiconductors, 507, 510
Septic systems, 216, 253, 265, 269
Septic tanks, 216, 265
Sequoia, protected, 646
Settling chamber (stack), 409
Settling tank, 204, 260, 261
Sewage
  diversion of, 253
  ecological impact, 240–243, 248–249, 255
  pollution by, 158, 162, 194–197, 200
  gas from treatment of, 327
  removal of phosphates from, 251, 264
  toxic chemicals in, 249–255
  treatment of, 251
Sewage sludge, disposal of, 262
Sewer systems, 259–260, 274
Shad (Hudson River), 163
Shale mining, 331
  natural gas from, 327
Shale oil, 186, 325, 331–334, 677
Shannon, Mike, 123
Sheep (mouplan), threatened, 79
Sheet erosion, 110
Shell Oil, 604
Shellfish, 75(f)
  contaminated, 37, 216, 221, 229, 230, 438
Shelterbelts, 115
Shelterwood cutting, 653
Shipping, 434–435, 439, 441
  flags of convenience, 444
Shipworms, 45(f)
Shore-zone communities, 160
Shrimp, 162, 163, 229, 230, 449, 479
Shuman and Boys, 503
Sierra Club, 650, 679
Sierra Pacific Power Co., 496
Silicates, in detergents, 251
Silicon, 507
Singapore, 68(n)
  population control measures, 61, 64
Sleep, effect of noise on, 460
Sleeping sickness, 141
Sleepwear, flammability standards for, 610
Sludgeworm, 160(f), 242
Smelt, 255, 452
Smelters, air pollution by, 396, 401
Smith, Morgan, 184
Smith, R. Jeffrey, 388
Smith-Putnam Generator, 486
Smog, 3, 35, 407, 430
  biological effects of, 430
  cause and effect, 59, 79, 368, 370
Smoking
  and government responsibility, 592
  hazards of, 221, 387, 392, 547, 548(t), 577(f), 586–590, 600, 611
  prohibition of, 590

Snail darter controversy, 73, 81, 83
Snails, 44, 139, 160(f), 167(f), 479
Snakes, 19, 160
Snow, Dr. John, 195–196
Social effects
  of natural gas pricing, 328
  of oil development, 441
Social responsibility, 252, 388
Sodium
  in drinking water, 222
  liquid, in breeder reactor, 363
Sodium chloride in oceans, 18
Sodium cyanide, 603(f)
Sodium cyclamate, 559
Sodium tripolyphosphate (STPP), 250
Soil, 95–96, 123, 266
  erosion of, 110–115
  formation of, 106–107, 112
  management of, 115
  pesticides in, 112, 118
  profiles, 105
  salinization of, 95, 139, 142
  sensitivity of, to acid rain, 405
  types, 20, 21, 105, 374
Soil Bank, 149
Soil conservation, 114–116
Soil Conservation Service (USDA), 110, 116
Soil and Water Conservation Districts, 116(f)
Solar breeder, 510(f)
Solar cells, 507, 510
Solar Development Bank, 519
Solar energy, 503–525
  advantages of, 518–521
  government incentives, 519–520
  storage of, 504, 505, 507, 510, 512, 519, 523–525
Solar heating
  active, 506, 519, 522–524
  building design for, 507, 508
  cost of, 519
  cost effectiveness of, 520–521
  hot water, 503–505, 523
  passive, 506, 507, 524–525
  space, 505, 506, 516, 523–524
  systems designs, 522–525
Solar One, 511–512
Solar Research Institute, 491
Solutions
  environmental, types of, 7
  to water depletion, 182
  to foaming water, 256
  to phosphate pollution, 251
Solvent refining of coal (SRC), 330, 333, 399
Sonic booms, 460
Sorghum, 115, 181(t)
Sound
  evaluation of, as noise, 460, 461
  measurement of, 459
South Africa, 358
Southern California Edison, 410, 493, 494, 496, 512
Soviet Union
  toxic substance control, 620–621
  See also Russia
Sowbug, 217(t)
Soybeans, 35, 122, 123, 421
Species, 13–14, 74–80
  benefits of variety in, 84, 142–143
  change in, 254–255, 438, 652
  competition, 21, 22, 79, 101
  diversity, 21–22, 30, 101, 141

endangered, *see* Endangered species
  endemic, 79
  exotic, 13
  and heat pollution, 450
  intrinsic value of, 75–76
  loss of, *see* Loss of species
  new, 22, 76(f), 142
  pollution-sensitive, -tolerant, and intermediately tolerant, 241(f), 242–243
  threatened, *see* Threatened species
  threats to, 80, 85, 163, 165, 404
  value of, 74–80
Speth, Gus, 592
Spruce trees, 100, 103(f)
Spurr, Stephen H., 669
Squid, 167(f)
  world catch, 141(t)
Squirrels, 103(f)
Sri Lanka, growth rate, 65
Stack gases
  cleaning methods for, 313, 314, 344, 399–400, 406, 409
  flamed, 395
Standards
  consumer safety, 610
  drinking water, 221, 222, 234, 235(t)
  fuel economy, 537
  radiation exposure, 574, 577, 578
  wastewater treatment, 267
  See also Air quality standards; Emissions control, standards
Stare, Dr. Frederick, 602
Starfish, 479
Starvation, 130, 132
State Mutual Life Assurance Company of America, 588
Steam (geothermal), 492–493, 495
  dry, 493
Steen, H. K., 650
Sterilization
  (human) compulsory, 61, 62
  (insect), pest control by, 122(t), 124–125
Stewart, Charles A., 560
Stipples (on plants), cause, 430
Stokinger, H. E., 560, 572
Stoneflies, 29(f), 241
Storro-Patterson, R., 90
Stotts, Dale B., 90
STPP, 250
Strategic Petroleum Reserve, 333
Stratification (thermal), of water, 160, 170–171
Stratified-charge, dual-carburetor engine, 389
Stratosphere, 372
  ozone in, 580
Stream communities, 18(f), 30, 162, 241
Stream ecosystem, 37
Streams, 162, 177
  carrying capacity of, 113
  polluted, 310
Strip-cropping, 116(f), 253
Strip-mining, 108, 295, 308, *see also* Surface mining
Strong, Maurice, 679
Strychnine, 603(f)
Sturgeon, 255
Sub-bituminous coal, 295–297
Subsidence, 308, 310

Subsoil, 113
Subspecies, 14
Suburbs, 632
  sprawl of, 634–636, 638
Succession, 99–101
Sugar cane, 515
Sugar-cane borer, control of, 74
Suits
  by asbestos workers, 573
  environmental, 4, 206, 230(n),
    267, 653
  by FDA on labelling, 611
  for microwave disability, 574
  pesticide, 604
Sulfates, 407
  air pollution by, 395–396, 407,
    408
  hazardous levels of, 408
Sulfur, 295, 297, 313, 321
  reactions of, in air, 397(f)
  as scrubber product, 400
  sources of pollution with, 395–
    396
Sulfur cycle, 36
Sulfur dioxide, 36, 313, 382, 395–
    399, 559
  air standards for, 426(t), 427(t)
  reducing levels of, 286
  See also Sulfur oxides
Sulfur oxides, 368, 395–402, 414
  air pollution by, 287(t), 313, 395–
    397, 425
  concentration of, 412(f)
  control of, 414
  impact on health, 396–397
Sulfuric acid, air pollution with,
    396, 402, 407, 408
Sun Oil, Suncor oil project of, 330
Sunfish, 160(f)
Sunlight, 17, 23, 373
  conversion to electricity, 507
  decreasing, 385, 386
  interior lighting by, 510–511
Superfund, 618
Superport, 441
Supersonic transport planes (SSTs),
    580
Supreme Court rulings, 73, 86(f),
    190, 630, 633, 647
Surfactants, 255–256
Surface mining, 296, 298, 300–306,
    407
  expansion of, 298, 309
  an unresolved issue, 304–306
  See also Strip mining
Surface Mining Control and Recla-
    mation Act, 301, 304, 306
Surface water
  allocation of, 182–190
  pollution of, 174, 215–216
  quality of, 175
  See also Lakes; Reservoirs; Rivers;
    Streams
Surgeon General's report, 592
Sustained yield, 650, 653–654
Swamps, methane gas from, 420
Sweeteners, artificial, 595–599
Swordfish, 167(f)
Sycamore trees, 515
Syncrude project, 330
Synthetic fuels, 327–335
Synthetic Fuels Corp., 33
Synthetic natural gas (SNG), 329
Systems analysis, water resources,
    178–179

T

Taiga, 101
Tailings, radioactive, 579
Takeuchi, T., 218
Talbot, Lee M., 78
Talmage, Nathaniel, 639–640
Tamarack trees, 103(f), 396
Tambora eruption, 376
Tar lumps, 438
Tar sands, oil from, 330–333
Tax benefits, to land donors, 673
Tax incentives, to buy solar, 519–
    520
Taxation
  land loss to, 672
  threat to farmland, 635–638
Taylor, Dr. L. H., 196
Taylor, T., 356
Technical fix, 259
Technology, 5–6, 57, 59, 141, 142–
    143, 147–148
  appropriate, 148
  best available (BAT)/best conven-
    tional, 267
  export of nuclear, 352–355, 358
  inevitable result of, 347
  mining, 300
  public knowledge of solar, 520
Technology forcing, 424–425
Television reception
  corona effects on, 462
  windmill interference with, 488
Temperature, 17
  adjustment to environmental, 19,
    449
  atmospheric, 372
  differences in tropical oceans,
    516
  in earth's core, 492
  ecological effects of changes in,
    449–451
  global trends of, 375–376, 384–
    386
  importance to life, 449–451
Tennessee Valley Authority (TVA),
    73
Teratogens, 576, 611, *see also* Birth
    defects
Tern (sooty), 460
Terracing, 115, 135, 253, 269, 303
Tertiary treatment of wastewater,
    264, 270–272
Thailand, 65, 66(t)
Thalidomide, 554
Thallium, 569(t)
Thermal energy (ocean), conversion
    of (OTEC), 516–518
Thermal pollution, 447–457, 532
  avoiding, 453
  source, 448
Thermal pollution control, 455–457
Thermocline, 170
Thermodynamics
  first law of, 39
  second law of, 41, 138, 283, 285,
    378, 448
Thermosiphon, 504, 524
Thermosphere, 372
Third world, investment in, 150
Threatened species, 14(f), 75, 79,
    81, 84, 85, 498
  trade controls for, 84
Three Mile Island, 346–348
Threshold dose
  for carcinogens, 592

for harmful substances, 556
  for radiation, 578
Thrushes, 103(f)
Tidal power, 165, 473, 475–479, 481
Tidal Power Corporation of Can-
    ada, 477
Tidal Power Review Board, 477
Tides, 163, 165
  in Bay of Fundy, 475, 477
  cause of, 475
  power generation from, *see* Tidal
    power
Tiger (endangered), 10, 86
Timber, forest management for,
    646–648, 650–651
Timberland, protection of, *see*
    National Forests
Titmice, 103(f)
Tmetuchl, Roman, 441
Toads, threats to, 404
Tokamak concept, 287
Tolerance, pesticide, 603, 604
Tomatoes, 125, 404
Ton, metric, 296(n)
Topsoil, 105
  loss of, 112–113
  in rain forests, 107
Torrey Canyon, 6, 437
Townsley, John, 498
Toxaphene, 119, 128(t)
  drinking-water standard, 235
Toxic Effluent Guidelines Settlement
    Agreement, 206
Toxic substances, control of, 592,
    615–617, *see also* Hazardous
    substances; Hazardous wastes
Toxic Substances Control Act
    (TOSCA), 4, 162, 225, 615,
    616(t), 619
Toxic Substances Strategy Commit-
    tee, 592
Toxic wastes
  pretreatment of, 268–269
  See also Hazardous wastes
Toxicity, of pesticides, 126(t)
Trace concentration units, 127–128
Traffic control, failure of, 3
Trails, scenic and historic, 663–667
Train, Russell, 78, 224, 248
Tranquilizers, safety of, 612
Trans-Alaska pipeline, 320, 443, 444
Transpiration, 31(n)
Transportation
  of coal, 315–317, 542
  effect on nitrogen cycle, 34(f)
  energy used by, 290
  importance of automobile to, 534,
    536
  of oil, 368, 434–435
  pollution by, 287(t), 419–420, 462
  public, 3, 6, 391, 534, 540–541
  of radioactive materials, 352, 354
  use of waste heat for, 453
Transportation, U.S. Department
    of, 116, 537(f)
Treaties
  CITES (Convention on Interna-
    tional Trade in Endangered
    Species), 84
  Law of the Sea Treaty, 444
  Nuclear Non-Proliferation Treaty,
    357–358
  water, with Mexico, 183, 188
Tree shrews, population control, 45
Triazines, 126(t)

Trichlorethylene (TCE), 215, 216
Trickling filter, 261, 265
Triethylene glycol, 588
Trihalomethanes, 208, 235
Tris-BP, banned, 610
Triticale, 143(f)
Trombe Wall, 525
Tropical rain forest, 78, 101, 104
Troposphere, 372
Trout, 128, 224, 241, 255, 450
Trumpeter swan, threatened, 80,
    498
Trust, issue of, 5–6, 329, 651
Tsetse flies, 141
Tuberculosis, treatment of, 49
Tuna, 167(f), 438
Tundra, 101
Turbidity (of water), 159–160
  removal of, 204
  standard for, 235(t)
Turbine engine, 292
Turbines
  bulb, 473, 476–477
  low-head, 473–474
  types of, 373
Turkey, desertification in, 96
Turkey (wild), management of, 84
Turtles, 27(f), 84
TVA, 73
Twain, Mark, 646
2,4-D and 2,4,5-D
  toxicity of, 126(t)
  drinking-water standard, 235
Typhoid, 49, 164, 194, 196–197

U

Ultraviolet rays, effects of, 580, 582
Uludong, Moses, 441
Underground coal gasification, 330
Underground mining, 296, 300,
    306–311
Underground water storage, 481
Undeveloped nations, 50, 53, 120,
    133, 139, 145
  agriculture, 145(f)
  conservation efforts, 86
  environmental damage, 56
  population problems, 49–51, 53,
    56, 67
Undiscovered resources, 322
  estimating oil, 322–324
  natural gas, 326
Union Oil Co., 332, 333, 436(f),
    493, 496, 498
Unionid clam, 240(f)
Unions, 570
United Nations, 64, 86
  FAO, 144(f), 149
  IMCO, 434
  UNESCO, 80
United States
  agriculture, 95, 114, 137, 138,
    147, 149
  coal reserves, 296–298
  demographics, 50–51, 59, 63, 68–
    70, 547, 552
  electric generation, 471, 473, 485,
    491, 495–496
  energy consumption, 528
  erosion problems, 112(t)
  foreign aid policies, 149
  giardiasis in, 201
  inflation, 337
  land area, 635
  life-style trends, 54

United States (*continued*)
  meat consumption, 139(f)
  moral obligation of, 150
  natural gas, 326
  nuclear policies, 350, 352, 354, 355
  oil, 322–326, 331, 333, 336
  solar energy for, 509(f)
  water problems, 202, 232–233
Upton, Arthur, 557, 558
Upwelling, 165
  productivity in areas of, 169(t)
Uranium, 282, 285, 348–349, 352, 354, 361
  hazards of mining, 577, 579
  political importance of, 352
Uranium-235, 342, 343–344, 349, 354, 362
  reuse of, 352–353
Uranium-238, 343, 349, 352, 361–363
Uranium dioxide, 342
Urban heat island, 378
Urban sprawl, underlying cause of, 10
Urbanization, environmental impact of, 378, *see also* Suburbs
Urea-formaldehyde insulation, 533
USDA, *see* Agriculture, U.S. Department of
Utah Power and Light, 493, 496
Utility companies
  controls on, 231, 475
  ownership of, 288
  peak-load pricing, 543–544
  rate of return, 544
  repowering by, 512
  stand-by charge of, 475

V
Vaccines, 49
  ethics of producing, 85
Vardamis, Alex A., 347
Vegetables
  decreasing cancer risk, 608–609
  high in nitrates, 600
  susceptible to air pollutants, 407, 421, 430
Venezuela, tar sands in, 331
Vincristine, 74
Vinyl chloride, 3, 554(n), 570, 609, 611
Violations of environmental controls, 73, 81, 202, 224, 229, 258, 306, 390, 401, 414, 430, 572, 605, 610, 653
Virgin Islands, 79
Viruses, 200
  removal from water supplies, 204, 265
Vitamins, 559, 600–609
Volcanic eruptions, 492
  global impact of, 375–377, 385, 408
  role, in nitrogen cycle, 35, 249
  succession communities following, 99

W
Wade, N., 85
Walker, Billie, 572
Walker, Frank, 504
Wallace, Henry, 149
Walleye, 160(f)
Walruses, protection of, 88

Ward, Barbara, 147, 150–151
Warm-blooded organisms, 19, 449
  as a "reservoir" of infection, 200
Warning labels, 596, 609
Wasps, pest control with, 74, 124
Waste, disposal of
  geothermal, 499
  hazardous, 616
  heat, 447–448, 453–457
  oil, 434
  radioactive, 344, 350–354, 358–359, 579–580
  See also Dump sites; Incinerators; Landfills; Sewage
Wastes
  agricultural, 267
  heat, uses for, 453, 531
  industrial, 216–218, 221, 224, 228–230, 268
  methane gas from, 327
  mining, 221, 307–308, 331
  organic, 239–246
  pretreatment of industrial, 268
  radioactive, 287
  recycling oil, 440, 442
  scrubber, 400
  sulfur, 395
  See also Waste, disposal of
Wastewaters, 156, 260–274
  reusing, 222, 270–272, 457
  geothermal, 499
Wastewater treatment
  methods, 260–264, 270–273
  standards and pretreatment standards for, 267
  tertiary, 251, 262, 264–265
Water
  density of, 17, 170(n)
  evaporation and transpiration of, 31, 32(f)
  foaming, 255–256
  hard, 156, 250, 251, 406
  hazards of softening, 222
  and life, 17, 159
  light/heavy, 343, 362
  odors in, 203, 208, 244
  quality, 113, 170, 174, 177, 178, 472
  thermal layering of, 18(f), 160, 162, 170–171
  use and abuse in power-plant cooling systems, 448–449, 454
Water conservation, measures for, 182
Water fleas, 226
Water heaters
  heat-pump, 533
  solar, 503–505, 520
Water lilies, 160(f)
Water pollution, 243–259
  eutrophication and, 248–255
  measurement of, 243–246
  non-point-source, 259, 267–269
  oil, *see* Oil pollution
  organic, 239–243
  point-source, 258–259
  radioactive, 577
  testing for, 201, 202, 209
  See also Water resources
Water pollution control, 156, 258–274
Water Pollution Control Act, 206, 229, 267
Water Pollution Control Federation, 271

Water Quality, National Commission on, 191, 269
Water resources, 173–192
  allocation and conflict, 182–192
  conjuctive use of, 178
  depletion of, 179–182
  geothermal, 493–494, 497–500
  meeting increased demands for, 178, 179, 190–192
  pollution of, 162, 174, 215–232, 269, 307
  systems analysis, 178–179
  transfers of surplus, 182, 184, 186, 190, 192
  transport of, 186, 189, 190
  uses, 314, 316–317, 330, 453(t)
  See also Tidal power; Tides
Water Resources Council, 185
Water supplies
  chemical contaminants in, 203, 216–230
  disease and, 196–206
  disease-causing organisms in, 202
  measuring contamination of, 202
  treatment of, 49, 197, 199–211
Water table(s), 180
  falling, 95, 182, 186
  and septic systems, 266
Water treatment, 258, *see also* Water supplies, treatment of
Waterfalls, energy from, 282, 471
Waterfowl, 249, 498
Watersheds, 159, 162, 406–407
Waterstriders, 160(f)
Watt, James, 443
Watt, K. E., 377
Waxman, Henry, 388
Weather, 372
  and air pollution, 381–382
  influences on, 385, 407–408
  potential influences on, 457
  prediction, 385
Weber, Jerome, 127
Weeds, 123, 124
Weevil (brown), 19
Weinberg, Alvin, 514
Weir, D., 120
Wellman-Lord System, 400
Wells, 174, 177, 180, 266
  contaminated, 194–196, 215, 221–222, 617
  geothermal, 494
  waste disposal in, 623
Wetlands, protection of, 632, 645
WFS (World Fertility Studies), 66, 67
Whales, 42, 87(f), 88–92, 167(f), 168
Whaling, 89–92
Wheat, 115, 122, 221, 421
  high-yield, 142
  pest-resistant, 122
  toxins in, 601
  U.S. production, 181(t)
White, Professor Gilbert, 633
White, L. F., 426
Whitefish, 255, 577
Whitehead, Rev. Mr., 196
Whooping cough, 49
Wild Rivers, 186, 217, 661, 663
Wilderness
  accessibility of, 667–669
  defined, 659
  development vs. preservation of, 186, 191

  preservation of, 647–649, 651, 659–662, 667–672
  threats to, 10, 59, 645, 651, 667–668, 670
Wildlife
  effect of climate change on, 376
  effects of noise on, 460
  management of, 22, 76, 84
  protection of, 80–87, 645–649
  threats to, 10, 301, 443, 498, 671
  warm-water ponds for, 453(t)
Wildlife Refuges (National), 647–650
Williams, Robert, 538
Willow trees, in shelterbelts, 115
Willrich, M., 356
Wind, 112–114, 375(f), 378, 468, 483–491
  resources, 488
  speeds, 486–488
Wind farm, 490
Wind power, 484–491
  environmental impact of, 489–490
  projected productivity of, 491
Wind turbines, *see* Windmills
Windbreaks, 534
Windfarms Limited, 491
Windmills, 484–488, 491
Winikoff, Beverly, 147
Wolman, Dr. Abel, 179
Woltjer, Gerald, 607
Wolves, 28, 75, 79, 103(f)
Women in the work force, 63, 576
Wood, producing, for fuel, 515
Woodlots, regeneration of, 513, 514
Woodpeckers, 103(f)
Woods Hole Oceanographic Institute, 438
Workers
  health hazards, 216, 229–230, 357, 462, 567–570, 574, 578, *see also* Miners
  laws protecting, 615
  rights of, 576
Workplace standards, 620(t), 621
World Bank, 152
World Fertility Studies (WFS), 66, 67
World Food Programme (WFP), 144(f), 149
World Health Organization, 120
World oil situation, 334–335
Worms, 167(f), 168, 479
Wright, Dr. and Mrs., 674
Wy Coal Gas Company, 332

X
X-rays, 575

Y
Yellow boy, 308, 310
Yellow-cake, 348
Yellow fever, 49, 119

Z
Zebra fish, 226
Zero population growth (ZPG), 50, 68
Zero risk, 207, 597, 615
Zetterberg, G., 252
Zinc-65, 577, 578
Zoning, 628, 630–633, 643
Zooplankton, 160, 166(f), 168, 479
ZPG, *see* Zero population growth